OXFORD CLASSIC TEXTS IN
THE PHYSICAL SCIENCES

D0161903

PRINCIPLES AND APPLICATIONS OF FERROELECTRICS AND RELATED MATERIALS

BY

M. E. LINES

AND

A. M. GLASS

CLARENDON PRESS · OXFORD

OXFORD

UNIVERSITY PRESS

Great Clarendon Street, Oxford OX2 6DP

Oxford University Press is a department of the University of Oxford.
It furthers the University's objective of excellence in research, scholarship,
and education by publishing worldwide in

Oxford New York

Athens Auckland Bangkok Bogotá Buenos Aires Calcutta
Cape Town Chennai Dar es Salaam Delhi Florence Hong Kong Istanbul
Karachi Kuala Lumpur Madrid Melbourne Mexico City Mumbai
Nairobi Paris São Paulo Shanghai Singapore Taipei Tokyo Toronto Warsaw
with associated companies in Berlin Ibadan

Oxford is a registered trade mark of Oxford University Press
in the UK and in certain other countries

Published in the United States
by Oxford University Press Inc., New York

First published 1977
First published in paperback 1979
Published in the Oxford Classics Series 2001

A catalogue record for this book is available from the British Library

Library of Congress Cataloging in Publication Data
Data available
ISBN 0 19 850778 X

Printed in Great Britain
on acid-free paper by
Redwood Books, Trowbridge, Wiltshire

PREFACE

IT IS NOW some 55 years since the phenomenon now known as ferroelectricity was first recognized in Rochelle salt. For well over half this period the phenomenon was thought to be a great rarity in nature and attempts to understand it, at least at the microscopic level, were formulated in terms of the very specific characteristics of each of the very few crystal structures which supported a ferroelectric instability. Since some, if not most, of these structures were relatively complex, the subject appeared to be a difficult and somewhat unattractive one from the theoretical standpoint at least. Even as recently as 1960 Jona and Shirane found it possible to write a book which gave a fairly comprehensive survey of nearly all materials which were then known to support ferroelectricity including a detailed discussion of most of the basic crystal structures. Significantly, theoretical discussions were included separately for each structure, with the unifying concepts appearing only at the macroscopic (free-energy) level.

The microscopic breakthough came in 1960 with the recognition of the fundamental relationship between lattice dynamics and ferroelectricity and, most importantly, of the existence of a soft-mode instability at a ferroelectric transition. This led very quickly to an intense study of soft modes in general by infrared-optic, X-ray, and neutron-scattering techniques and to a classification of the ferroelectric instability as just one of several possible classes of lattice instability (i.e. structural transition), each definable in terms of a diffuse or propagating soft mode. From the ferroelectric standpoint the great value of the soft-mode concept is that it enables us to form a reasonably unified microscopic picture of ferroelectricity (and antiferroelectricity), the basic simplicity of which is not affected by the possibly immense complexity of any particular crystal structure. It then becomes possible to focus on those microscopic characteristics which are common to all ferroelectrics and to grasp the relationship of ferroelectricity to the more general field of structural transitions and even to critical phenomena in general.

The enormous increase in the number of known ferroelectrics over the last two decades no longer makes it even remotely possible to attempt a comprehensive survey of the field. Fortunately, in the light of what has been said above, such is no longer necessary or even desirable in a discussion of the fundamentals of ferroelectricity. The present book lays emphasis on the definition of those basic microscopic properties which allow for a categorization of ferroelectrics into different classes and on the modern techniques of measurement which enable such a differentiation to be made. Thus fundamental theoretical principles are introduced without reference to specific materials and only the practically more important or best understood specific materials are discussed in detail. Most sections on experimental techniques (in particular those on X-ray and neutron scattering, infrared and Raman spectroscopy, and magnetic resonance) have a tutorial introduction, and the role of each in a ferroelectric context is discussed by means of carefully chosen examples for which the relevant method has been of particular value. Readers requiring detailed numerical data in encyclopaedic form can find it elsewhere—in recent years there have been several compilations of experimental data for ferroelectrics and related materials, the most comprehensive being the two volumes (3 and 9) of the *Landolt–Bornstein New Series on Crystal and Solid State Physics (group III)*. These volumes provide an essentially complete set of data up to the end of the year 1972.

The present book covers a wide range of topics right through from basic theory at the fundamental level all the way to devices. We have tried to maintain a balance throughout, not only with respect to theory and experiment, principle and application, but also in terms of the tutorial and review aspects of the text. We assume that the reader has some background in elementary thermodynamics, statistical mechanics, and quantum theory at the harmonic-oscillator level, but other than that the mathematical level of the text rarely requires anything more advanced than the solution of a simple differential equation. Thus the book is suitable for all levels of interest in this field from student to research worker.

In many parts of the book the important feature of ferroelectricity is the effect which a spontaneous polarization can have on the macroscopic and microscopic behaviour of a crystal and not so much the reversibility of the polarization. For this reason we include the wider field of pyroelectric phenomena within the scope of the book. On the other hand, to limit the scope of the text, we have chosen not to include piezoelectric and dielectric materials and applications except where these are directly related to ferroelectric behaviour. These fields have been the subject of entire texts of their own and we can obviously not do justice to them in the wider context. In the same vein we have only discussed non-linear optics in the limited context of its relevance to the field of ferroelectrics. We have, however, included short sections on electrets and liquid crystals, both because of the increasing overlap of these fields with that of ferroelectricity and because we believe that it is important to distinguish clearly between the metastable polar configurations frequently found in these materials and the stable polar configurations of pyroelectrics and ferroelectrics.

Because of our emphasis on principles and applications in a general sense rather than a detailed coverage of specific ferroelectric materials, our reference list, though extensive, is far from complete, and many workers who have made significant contributions to the field will unfortunately have gone completely unmentioned. To them we apologize, but a glance at the

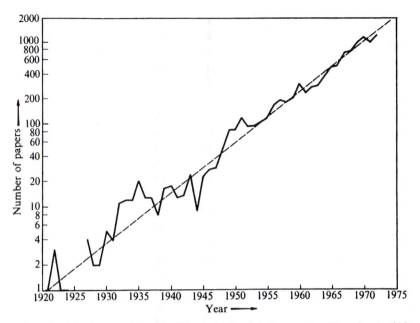

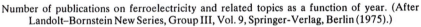

Number of publications on ferroelectricity and related topics as a function of year. (After Landolt–Bornstein New Series, Group III, Vol. 9, Springer-Verlag, Berlin (1975).)

preface figure shows how publications in the field of ferroelectricity have expanded exponentially and explains why many readers who turn to the bibliography for their own references may well be disappointed. The latter will fall particularly into two groups—those who contributed significantly to the early development of the field and whose work has been well documented in earlier texts (e.g. the books by Känzig 1957, Jona and Shirane 1962, and Fatuzzo and Merz 1967), and those who have worked on the more recently discovered but less completely documented ferroelectric crystals. Our sense of guilt is lessened, however, by the knowledge that comprehensive bibliographies covering both the fields of ferroelectricity (by Connolly, Turner, and Hawkins 1970, 1974) and pyroelectricity (by Lang 1974) are available.

We should like to acknowledge the helpful discussions and interactions that we have had with our colleagues at Bell Laboratories both before and throughout the preparation of this text, and we are deeply appreciative of the permission given by the Bell Laboratories administration and Book Committee to undertake such a time-consuming venture as the preparation of a book of this size. Last, but by no means least, we wish to thank Mrs. Nancie Geiger for her diligence and patience in typing the final manuscript and Mrs. Kathleen Lines for undertaking the unenviable task of assembling and checking the reference lists.

CONTENTS

CONTENTS xiii

1

BASIC CONCEPTS

1.1. Introduction

1.1.1. Historical

THE PHENOMENON of pyroelectricity, or the possession by some materials of a temperature-dependent spontaneous electric dipole moment, has been known since ancient times because of the ability of such materials to attract objects when they are heated. In the eighteenth and nineteenth centuries many experiments were carried out in an attempt to characterize the pyroelectric effect in a quantitative manner, particular attention being paid to tourmaline (for instance, Gaugain 1856). These studies eventually led to the discovery of piezoelectricity, which is the production of electrical polarity by application of stress, by J. Curie and P. Curie in 1880. The Curies realized that the difference between the charge developed upon uniform and non-uniform heating was due to the thermal stress created in the pyroelectric.

None of the early known pyroelectric materials were ferroelectric in the sense of possessing a reorientable electric moment. The principal reason that ferroelectrics were discovered so much later was because the formation of domains of differently oriented polarization within virgin single crystals led to a lack of any net polarization and very small pyroelectric and piezoelectric response. It was not until 1920 that Valasek discovered that the polarization of sodium potassium tartrate tetrahydrate ($NaKC_4H_4O_6 \cdot 4H_2O$), better known as Rochelle salt, could be reversed by application of an external electric field. Valasek (1920, 1921) recognized ferroelectricity by experiments which showed that the dielectric properties of this crystal were in many respects similar in nature to the ferromagnetic properties of iron in that there was a hysteresis effect in the field-polarization curve, a Curie temperature T_c (in fact two—since Rochelle salt exhibits a spontaneously polarized phase only between −18°C and +24°C), and an extremely large dielectric and piezoelectric response in and near the ferroelectric region. Although the term 'Curie point' was used by Valasek to describe the onset of polar ordering in Rochelle salt, the dielectric phenomena associated with the anomaly were for a considerable period known as Seignette-electricity (the salt was first prepared by Seignette around 1655 in La Rochelle, France) and the term ferroelectricity was not commonly in use much before the early 1940's. This was at least partly due to the fact that Rochelle salt remained the only known example of the

phenomenon for over a decade after the initial discovery, but also because
the full significance of the phenomenon in terms of a theoretical interpretation was not given until 1933 (see for example Kurchatov 1933). This
theory, now obsolete, focused primarily on the dipolar interactions between
the constituent waters of crystallization, ascribing the ferroelectricity (along
the lines of the Weiss theory of magnetism) to a spontaneous orientation of
the assumed rotatable water dipoles.

Another reason for the slow acceptance of ferroelectricity as a subject
worthy of more general study was the fact that any small deviation from the
correct chemical composition of Rochelle salt seemed to destroy the
phenomenon completely, leading to experimental problems of reproducibility and a growing conviction that this was one of nature's great accidents. In
addition, since the detailed structure of the crystal was unknown and in all
probability immensely complicated, any simple microscopic models and
attempts at a theoretical explanation could at best be little more than
speculative. In fact Rochelle salt is now known to contain four formula units
per cell (112 atoms) and is indeed one of the more complicated ferroelectric
materials known today. As a result interest in it has waned over the years
and, with more and more research quite naturally becoming focused on
simpler examples of the basic ferroelectric phenomenon as they were
discovered, the Rochelle salt transitions remain less than quantitatively
understood at the time of writing, and we shall not focus other than historic
interest on them in this book.

In 1935 to 1938 the first *series* of ferroelectric crystals was produced in
Zurich (Busch and Scherrer 1935, Busch 1938). Perhaps the greatest
significance of this event was that the discovery was of a whole series of
isomorphous crystals and not just another isolated example, welcome
though even the latter would have been after so many years. The crystals in
question were phosphates and arsenates, of which the principal example was
potassium dihydrogen phosphate, KH_2PO_4 (commonly abbreviated as
KDP), with a single transition temperature at about 122 K. However, all
other crystals isomorphous to KDP seemed to show ferroelectricity or
something very closely related to it. For example, they all exhibited very
marked dielectric anomalies indicating a quasi-divergence of dielectric
response to a uniform static electric field. However, the ammonium salts,
typified by $(NH_4)H_2PO_4$ (abbreviated as ADP), unlike the others, did not
seem to acquire a spontaneous polarization below the Curie point. Only 20
years later would it become apparent that the ammonium salts were
antiferroelectric instead.

Like Rochelle salt, KDP and ADP are piezoelectric even above T_c, and
indeed most of the technical applications using these materials focused on
their piezoelectricity rather than their ferroelectricity. In particular, ADP,
with a 30 per cent electromechanical coupling efficiency at room tempera-

ture (T_c for ADP being 148 K), became the principal underwater sound transducer and submarine detector in World War II, replacing the very temperature-sensitive Rochelle salt. However, aside from technical usefulness the importance of the new materials was the fact that their structure (with two formula units or 16 atoms per prototype primitive cell, see Fig. 9.1) is much simpler than Rochelle salt and hence more amenable to theoretical understanding. Although there is no water of crystallization in KDP there are hydrogen bonds for which different possible arrangements of the hydrogens can result in different orientations of the resulting $(H_2PO_4)^-$ dipolar units. From this concept Slater (1941) presented the first basic microscopic model of a ferroelectric which has, in its essentials, withstood the test of time.

Allowing for only one hydrogen per (double-well) bond and two hydrogens per tetrahedral phosphate group, Slater considered the six possible ways in which two hydrogens could become associated with the four tetrahedral corners, according to two of them (corresponding to 'up' and 'down' dipolar units) a lower energy than the other four (corresponding to transversely oriented dipolar units). This model assumed the very ordering of the hydrogen atoms on passage to the polar phase which was later beautifully confirmed by neutron analysis. Slater's theory was qualitatively sound, but not surprisingly represented an oversimplification of the problem and led to certain unphysical features which could, however, be lessened by relaxing some of the model's constraints (e.g. allowing for non-zero probabilities for the formation of H_3PO_4 and $(HPO_4)^{2-}$ configurations and for the inclusion of long-range interactions). One seemingly insurmountable stumbling block for the Slater-type theories did remain after all these embellishments, namely the observation by Bantle (1942) of a very large isotope effect on T_c upon deuteration. The explanation of this effect would require the concept of quantum tunnelling between the hydrogen-bond double wells and would come much later (Blinc 1960).

After the discovery of the KDP series another decade was to pass without further experimental breakthrough, and there was a growing conviction that ferroelectrics were indeed comparative rarities in nature. It was thought quite likely that the existence of a hydrogen bond was a necessary, if not sufficient, condition for the polar instability to occur, and there was therefore little motivation for looking for ferroelectricity in materials such as oxides which did not contain hydrogen. The next discoveries of ferroelectrics arose out of a search for new dielectrics to replace mica in the early 1940's. Titania had been established as a ceramic with a high dielectric constant in 1925 and the logical approach was to study modifications of titania to provide even higher permittivities. In 1945 barium titanate ceramic was found to have a dielectric constant of 1000 to 3000 at room temperature and even higher values as the temperature was raised. Shortly thereafter ferroelectricity in this material was reported (Wul and Goldman 1945, 1946). With this the hydrogen hypothesis of ferroelectricity had to be abandoned. The discovery was important in a number of other ways, only some of which were immediately recognized. It represented, for example, the establishment of quite a number of 'firsts': first ferroelectric without hydrogen bonds; first ferroelectric with more than one ferroelectric phase; first ferroelectric with a non-piezoelectric prototype or paraelectric phase. In addition the crystal structure of the prototype was cubic centrosymmetric perovskite with very high symmetry and only five atoms per unit cell. As a result of this simplicity and also because of its practical utility (it is chemically and mechanically very stable, is ferroelectric at room temperature, and could easily be prepared and used in ceramic form although a technique for growing large single crystals was not to be perfected until 1954), barium titanate rapidly became by far the most extensively investigated ferroelectric material. What was not realized, however, was that this material was to be the forerunner of what is now perhaps the largest single class of all ferroelectrics—the oxygen octahedral ferroelectrics made up from basic BO_6 building blocks. Nevertheless, owing to the extraordinarily structure-sensitive character of the phenomenon of ferroelectricity it was only logical that after the discovery of $BaTiO_3$ considerable effort would be expended on a search for additional ferroelectrics with the same perovskite structure. Although success came slowly, come it did with the discovery of ferroelectric activity in $KNbO_3$ and $KTaO_3$ (Matthias 1949), $LiNbO_3$ and $LiTaO_3$ (Matthias and Remeika 1949), and in $PbTiO_3$ (Shirane, Hoshino, and Suzuki 1950).

With the simplicity of the perovskite lattice structure it was natural to expect some theoretical progress at the microscopic level. Slater (1950) assumed that the ferroelectric behaviour of $BaTiO_3$ was caused by the long-range dipolar forces which (via the Lorentz local effective field) tended to destabilize the high-symmetry configuration favoured by the local forces. This explanation became the basic model for displacive (as opposed to order–disorder) transitions and met with considerable success, although it suffered in general from an embarrassingly large number of variables unless some very restrictive assumptions were made concerning which ion was primarily responsible for the ionic instability. For example, the concept of the 'rattling' titanium ion became popular when acceptable results were obtained by assuming that one could focus on the motion of the titanium ion in a rigid framework of the rest of the lattice. Problems of this nature were never successfully overcome until Anderson (1960) and Cochran (1960) realized that the theory should properly be cast within the framework of lattice dynamics and that one should focus on one of the lattice *modes* (the 'soft' mode) involving the ionic motions of all constituent atoms as the basic variable in terms of which to describe the displacive lattice instability.

At the macroscopic level theory had progressed more rapidly. The great advantage of proceeding macroscopically is that one can overlook the microscopic nuances concerning displacive or order–disorder character, ionic or electronic displacements, long- or short-range interactions, etc., and focus solely on thermodynamic concepts. Limitations in terms of a basic understanding are obvious, but the ability to relate macroscopic measurables and to include with ease changes of external constraints such as stress prove to be extremely valuable. There is, for example, a distinct difference between ferromagnetic and ferroelectric theories in that one can nearly always neglect mechanical coupling in the former (where magnetostrictive effects are usually measured in parts per million), whereas strains associated with the onset of ferroelectricity are frequently of essential importance at 1 per cent or larger. Mueller (1940 a, b) appears to have been the first to apply thermodynamics in a ferroelectric context—in this case to ferroelectric Rochelle salt. The idea was to write a free energy as an expansion in powers of polarization and strain, and to determine the parameters involved by associating them with measurables. Often only one of these parameters (usually the dielectric stiffness, i.e. reciprocal permittivity) is grossly temperature dependent and the temperature variation of all the other thermodynamic measurables can then be predicted in terms of it. The success of the theory therefore lies in the fact that it is capable of explaining dielectric, piezoelectric, and elastic behaviour at any temperature from a free-energy polynomial involving a limited number of terms. The technique, which assumes that the same energy function is capable of describing both polar and non-polar phases, was perfected by Ginzburg (1945, 1949) and Devonshire (1949, 1951, 1954) with specific reference to $BaTiO_3$, for which the accumulation of an immense amount of data, coupled with the existence of three different ferroelectric phases with different polar axes, enabled a thorough test of the method. The extension of the method to antiferroelectrics (Kittel 1951) was straightforward.

In the mid-fifties, as a result of a not very systematic search for new ferroelectrics, guanidine aluminium sulphate hexahydrate $C(NH_2)_3$-$Al(SO_4)_2 \cdot 6H_2O$, abbreviated as GASH, was discovered (Holden *et al.* 1955). GASH and its isomorphs, though all ferroelectric, no longer exhibit a Curie temperature since they decompose before their ferroelectricity is lost. This group of crystals was very reminiscent of the alums, which might have led to a search for ferroelectricity in the latter except that the alums had been investigated for over a century already. Fortunately Pepinsky, Jona, and Shirane (1956) ignored the existing evidence and found ferroelectricity in methylammonium aluminium alum $CH_3NH_3Al(SO_4)_2 \cdot 12H_2O$. In a similar vein Matthias and Remeika (1956) ignored 40 years of data on ammonium sulphate, $(NH_4)_2SO_4$, to find ferroelectricity in this system. Finally the lesson was learned that most reports earlier than the forties were unreliable. The dam was finally broken and crystals for which dielectric anomalies had endlessly been reported in the past all suddenly turned out to be ferroelectric. In fact virtually all transitions involving a dielectric anomaly seemed to be either ferroelectric or antiferroelectric, and the list of known ferroelectrics quickly ran into the hundreds—a long way indeed from the early concept of ferroelectricity as one of nature's great accidents. By contrast, the problem today is of the reverse nature. In addition to the now enormous number of known genuine ferroelectric and antiferroelectric materials there is a growing tendency to classify every antidistortive crystal as 'antiferroelectric' and to include materials with merely metastable polar properties as 'ferroelectric'.

A 'coming of age' of the ferroelectric phenomenon in the 1960's and 1970's is essentially the story which we attempt to portray in this book. The modern era of theoretical understanding really began with the articles of Anderson and Cochran referred to earlier. Since 1960 the emphasis has been very dominantly on the lattice-dynamical or soft-mode description of ferroelectricity, at first solely for displacement systems but later in terms of a fundamental model Hamiltonian which is general enough to include essentially all types of ferroelectric instability. In a word, unification has been the theoretical theme of recent years—an effort to get away from the idea that each ferroelectric structure poses a specific problem only marginally related to others and to focus on those general aspects of the phenomenon which are common to all. An important part of this has been the realization that both ferroelectric and antiferroelectric transitions are but particular cases of the more general concept—the structural phase transition. Accompanying the recognition of the significance of the soft mode (whether diffuse, tunnelling, or phonon like in character) and of its periodic nature has been an explosion of experimental activity using techniques capable of measuring its frequency and/or wavevector-dependent properties. These are primarily scattering or resonance techniques involving X-rays, neutrons, light, and ultrasound, in support of the more traditional dielectric measurements. As regards basic concepts, a recognition of the importance of the coupling between ferroelectric modes and non-polar modes proved to be very important leading to concepts like extrinsic ferroelectricity (i.e. ferroelectrics driven by a coupling to non-polar instabilities) and ferroelasticity. The recognition of the phenomenon of ferroelasticity by Aizu (1969) was particularly significant, since ferroelasticity is essentially the complementary phenomenon to ferroelectricity; all even-rank polar property-tensors changing with a ferroelastic state shift and all odd-rank polar property-tensors changing with a ferroelectric state shift. Careful definition of all these properties will be given in the following section.

While progress in understanding the fundamental nature of ferroelectrics has been fairly monotonic, this is not the case with the application of ferroelectrics—in fact the effort devoted to ferroelectric devices has seen many ups and downs. The large dielectric and piezoelectric constants of ferroelectrics immediately made these materials attractive candidates for a variety of applications. For many years ferroelectrics dominated the field of sonar detectors, phonograph pickups, and so on. Although in more recent years quartz has become the leading candidate for most piezoelectric applications, there are still many areas where ferroelectrics can play an important role in both piezoelectric and dielectric applications. However, none of these devices directly utilizes the ferroelectric nature of the material, namely the large reversible spontaneous polarization.

In the 1950's came the demand for high-capacity computer memories, and ferroelectrics seemed prime candidates since a bistable polarization offered the potential for binary memories. For instance a reversible polarization of $10 \, \mu C \, cm^{-2}$ corresponds to an available charge for destructive read-out of the memory of about 10^{14} electrons per cm^2. This allows enormous storage density with high signal/noise for read-out if a material with fast and reliable switching characteristics were available. Enthusiasm for this kind of device was diminished by practical rather than fundamental reasons after a few years. Although the ferroelectric polarization could be reversed rapidly enough for some application, the switching characteristics were not reliable and generally fatigued with time. Eventually more reliable magnetic and semiconductor memories became favoured.

Devices using dielectric, piezoelectric, and pyroelectric properties of ferroelectrics received continued attention, however. In particular the use of the pyroelectric effect for infrared detection, and more recently infrared imaging, is particularly promising both since the pyroelectrics offer particularly simple schemes for imaging and because there is no viable alternative scheme for directly imaging a thermal scene with a room-temperature target.

With the advent of the laser came the need for materials with large non-linear polarizability at optical frequencies for second-harmonic generation, electro-optic modulation, parametric oscillators, and so on. It was soon realized that a large non-linear polarizability was generally to be found in materials having a large linear polarizability and once again ferroelectrics became prime candidates. In addition, the crystalline anisotropy of ferroelectrics generally gives rise to a large optical birefringence which is frequently necessary for phase matching the interactions. A large number of new ferroelectrics were discovered during this materials search during the second half of the 1960's. Application of these materials for optical devices imposed severe requirements on crystalline perfection but, as these problems were overcome, a host of new applications of ferroelectrics for optical memories and display were evolved. The production of optically transparent ceramics greatly increased the likelihood of widespread commercial use of ferroelectric optical devices. More recently there has been a concerted effort toward the production of ferroelectric films for optical waveguides which allow the efficient utilization of the non-linear optical properties of ferroelectrics with relatively low optical and electrical power dissipation.

In retrospect it appears that ferroelectrics have offered the potential for many applications only a few of which have yet become commercial devices. Part of the problem is that, because of the complexity of the microscopic processes involved and the difficulty in preparing (and discovering!) the materials, the delay between the conception of the device and the production of a suitable material has been too great; the demands made on ferroelectrics were premature. Nevertheless the potential still remains, and for this reason we shall include in this book a discussion of several of the practical goals which the experimentalist has attempted to achieve in addition to the more academic treatise leading to a basic understanding of the ferroelectric phenomenon itself.

1.1.2. Definitions

Depending on their geometry crystals are commonly classified into seven systems: triclinic (the least symmetrical), monoclinic, orthorhombic, tetragonal, trigonal, hexagonal, and cubic. These systems can again be subdivided into point groups (crystal classes) according to their symmetry with respect to a point. There are 32 such crystal classes (they are detailed in Appendix A of this book) and 11 of them possess a centre of symmetry. The latter, of course, possess no polar properties. If, for example, a uniform stress is applied to such a *centrosymmetric* crystal, the resulting small movement of charge is symmetrically distributed about the centre of symmetry in a manner which brings about a full compensation of relative displacements. The application of an electric field does produce a strain, but the strain is unchanged on reversal of the field, i.e. the effect is quadratic. This property is termed *electrostriction* and occurs naturally in all substances, crystalline or otherwise.

Of the remaining 21 non-centric crystal classes, all except one exhibit electrical polarity when subject to stress. The effect (and also its converse, the production of strain by application of an electric field) is linear, with reversal of the stimulus resulting in a reversal of the response, and is termed the *piezoelectric* effect. Of the 20 piezoelectric crystal classes 10 are characterized by the fact that they have a unique polar axis. Crystals belonging to these classes are called *polar* because they possess a *spontaneous polarization* or electric moment per unit volume. Frequently this spontaneous polarization cannot be detected by charges on the surface of the crystal because, unlike the situation in the magnetic equivalent, the depolarizing field which results from such a charge distribution can be compensated by the flow of free charge within the crystal and in the surrounding medium. However, the spontaneous polarization is in general temperature dependent and its existence can be detected by observing the flow of charge to and from the surfaces on change of temperature. This is the *pyroelectric* effect and the 10 polar classes are often referred to as the *pyroelectric classes*.

A crystal is said to be *ferroelectric* when it has two or more orientational states in the absence of an electric field and can be shifted from one to another of these states by an electric field. Any two of the orientation states are identical (or enantiomorphous) in crystal structure and differ only in electric polarization vector at null electric field. Understood in this definition is the fact that the polar character of the orientation states should represent an absolutely stable configuration in null field. Although we shall later include discussion of materials in which a persistent metastable polarization may be produced and switched by application of an electric field, we shall not consider them to be properly classified as ferroelectric in spite of a growing tendency in the literature to refer to them as such. It is recognized that an empirical definition of this kind cannot be completely rigorous and exclusive. If ferroelectrics are to be considered that subgroup of the pyroelectric class which includes only those crystals capable of being switched in some manner, then whether or not a material is ferroelectric depends on experimental limitations. Crystal perfection, electrical conductivity, and the temperature and pressure are all factors which affect the reversibility of polarization. It follows that ferroelectric character cannot be determined solely by a crystallographic determination, although a possibility of switching usually implies a polar structure characterized by only small distortions from a higher-symmetry non-ferroelectric phase. Even when polarization reversal has been effected it is not always straightforward to determine whether or not the polarization represents a stable configuration due to a co-operative effect or a metastable configuration due to the temporary application of the switching field. The latter difficulty is to a large extent removed if at some temperature there is a well-defined phase transition to a non-polar phase, but difficulties still remain if the transition is diffuse owing to structural disorder.

The highest symmetry phase compatible with the ferroelectric structure (i.e. in terms of which the ferroelectric phase can be described by small perturbational structural changes) is termed the *prototype* phase. Although it is not necessarily of non-polar character it proves to be so for the great majority of known ferroelectrics, and we shall in general make this assumption in what follows. For most ferroelectrics the prototype phase actually exists as the highest temperature phase of the crystal, although in some instances the structure may melt before the prototype phase would otherwise become stable. As the result of its small structural displacements from an (assumed non-polar) prototype, a typical ferroelectric frequently possesses a spontaneous polarization $\mathbf{P}_s$ which decreases with increasing temperature T to disappear continuously, or more often discontinuously, at a *Curie* point T_c. However, since polar crystal classes do not always belong to systems of lower crystal symmetry than non-polar classes, transitions from non-polar to polar phases as a function of *increasing* temperature can and do occasionally occur, and $d\mathbf{P}_s/dT$ is accordingly not always necessarily negative. Nevertheless for most known ferroelectrics the onset of ferroelectricity occurs as a function of decreasing temperature.

A ferroelectric phase change represents a special class of structural phase transition denoted by the appearance of a spontaneous polarization. Above the Curie point the approaching transition is often, though not always, signalled by a diverging differential dielectric response or permittivity ε, which close to T_c varies with temperature in an approximate Curie–Weiss manner $\varepsilon = C/(T - T_0)$, where T_0 is the Curie–Weiss temperature which is equal to the Curie temperature T_c only for the case of a continuous transition. The phase which transforms to the ferroelectric form at T_c is often termed *paraelectric*, although some authors confine this term to situations with diverging response. It is not, of course, necessary that the paraelectric phase be the prototype, although such is quite often the case. Below T_c, in the absence of applied field, there are at least two directions along which a spontaneous polarization can develop. To minimize the depolarizing fields different regions of the crystal polarize in each of these directions, each volume of uniform polarization being called a *domain*. The resulting *domain structure* usually results in a near complete compensation of polarization and the crystals consequently exhibit very small, if any, pyroelectric effects until they are poled by application of a field.

As will be shown later a ferroelectric transition can usually be associated with the condensation of a *soft* (or low-frequency) mode of lattice motion at the Brillouin-zone centre. Structural transitions triggered by zone-centre soft modes are generally termed *ferrodistortive*, and in this sense we can say that ferroelectrics constitute a subgroup of the class of ferrodistortive transitions—specifically that subgroup which involves the condensation of a polar or optically active mode and whose condensation therefore causes the appearance of a long-range polar order. If the transition is very strongly of first order then mode softening may not occur to a significant degree, and, in this situation, there is also the possibility that the large polarization which sets in discontinuously at T_c may not be reversible, i.e. the low-temperature phase may be pyroelectric only. Ferroelectrics are also often categorized as being of either *displacive* or *order–disorder* character. In earlier literature the latter distinction was generally made in terms of whether the paraelectric phase was microscopically non-polar (displacive) or only non-polar in a macroscopic or thermally averaged sense (order–disorder). Thus, for example, for a centrosymmetric non-polar phase one might ask whether the thermal distribution of ionic motion in a primitive cell above T_c is peaked at the centrosymmetric sites themselves or at some double- or multi-well configuration about these sites. More recently there has been a tendency to define displacive or order–disorder character in terms of the dynamics of the phase transition, specifically as to whether the soft mode is of a propagating or diffusive character respectively. Although the two definitions coincide in the extremes of well-localized single wells on the one hand and deep double wells with negligible quantum tunnelling on the other, it is now apparent that many ferroelectrics fall into a middle ground of shallow double wells, grossly anharmonic single wells, or tunnelling (e.g. H-bonded) wells, for which the two definitions do not necessarily coincide and for which the distinction is perhaps better not made at all. Where the distinction can be made, the displacive or propagating soft mode is a damped optic phonon, representing small quasi-harmonic motion about the mean position, while the diffusive soft mode is not a phonon at all but represents large-amplitude thermal hopping motion between the wells.

When a mode condensation occurs at a place in the Brillouin zone other than at the centre we refer to an *antidistortive* or *antiferrodistortive* structural transition. Most frequently this occurs as a cell-doubling transition with a soft-mode condensation at the Brillouin-zone boundary of the high-temperature phase. It is obviously now tempting to define an antiferroelectric transition as that subgroup of antidistortive transitions which involves a polar mode. In this picture the ordered phase would then necessarily involve an ordered arrangement of dipoles with a zero resultant polarization. This classification, however, would tell us little about the dielectric properties associated with the ordered phase or the transition since these involve dominantly the zone-centre modes of the structure. Traditionally the term *antiferroelectric* has been reserved for antipolar phases in which the free energy of the antipolar dipole arrangement is comparable to that of the polar crystal. In soft-mode terms this often implies the existence of a low-frequency polar zone-centre mode in addition to the condensing antipolar mode. For this reason we shall call a general transition involving a zone-boundary polar mode an *antipolar* transition and reserve the term *antiferroelectric* for that fraction of antipolar systems which exhibits large dielectric anomalies near the Curie temperature (which we shall still denote as T_c) and which can be transformed to an induced ferroelectric phase by application of an electric field. This definition of antiferroelectricity is then exactly analogous to that of ferroelectricity, and has the same drawback in terms of rigour since it leaves the boundary between antiferroelectric and antipolar phases a purely empirical one. It will be obvious, though, that in contrast to pyroelectricity, which is characterized by a non-zero dipole moment per unit cell, the antipolar phase is only a useful concept if at some temperature or pressure an antidistortive phase transition can be observed since the unit cell has oppositely directed dipoles with no net dipole moment which is characteristic of a large group of centrosymmetric crystal structures. A schematic representation of the various basic structural transitions and definitions is shown in Fig. 1.1. One should also bear in mind the fact that materials can be

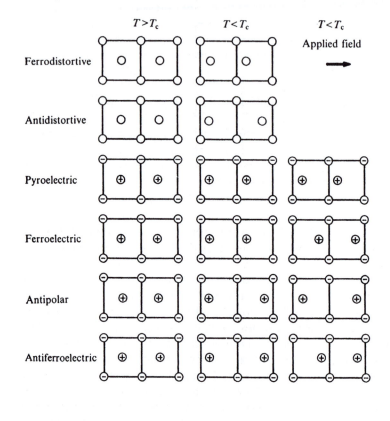

Fig. 1.1. A schematic representation of some fundamental types of structural phase transition from a centrosymmetric prototype.

ferroelectric in one direction and pyroelectric or antipolar (or even antiferroelectric) in another. A few examples of transitions to phases of this more complicated kind are sketched schematically in Fig. 1.2.

Having emphasized the close relationship between ferroelectrics and ferrodistortive structural transitions on the one hand and antiferroelectrics

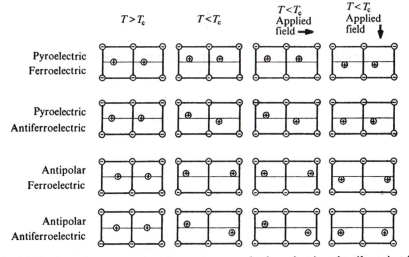

Fig. 1.2. A schematic representation of some more complex ferroelectric and antiferroelectric phase transitions.

and antidistortive transitions on the other, we now have to make it clear that although most ferroelectrics are indeed ferrodistortive (common examples being barium titanate, sodium nitrite, and triglycine sulphate) some are not. To understand this it is necessary to recognize that, because of the existence of coupling between modes, it is not a necessary condition for ferroelectricity that a zone-centre polar mode should be the driving instability. Sometimes a driving antidistortive mode can couple directly or indirectly to a zone-centre polar mode and upon condensation thereby induce a small spontaneous polarization in an indirect fashion. We shall refer to such a transition as being *intrinsically* antidistortive but *extrinsically* ferroelectric. In this case the *primary* order parameter is of antidistortive character while the spontaneous polarization is said to be a *secondary* order parameter of the transition. There can of course be only one primary order parameter (at least for a continuous or near-continuous transition), but there may be many induced or secondary order parameters resulting from couplings to the primary order parameter. Examples of extrinsic ferroelectrics are gadolinium molybdate and the boracites, and such crystals are often termed 'improper' ferroelectrics in the literature although we shall generally avoid this term. All the known antiferroelectrics (examples being lead zirconate and ammonium dihydrogen phosphate) are intrinsically antidistortive, although one can conceive of a ferrodistortive antiferroelectric as one having an antiparallel arrangement of electric dipoles occurring within a primitive cell of the higher-symmetry phase. Such a phase would be precipitated by the condensation of an antipolar zone-centre soft mode.

Once the importance of coupling between polar modes and other modes has been recognized it is clear that, via the piezoelectric interaction (or coupling to acoustic modes), a spontaneous strain will be virtually a universal characteristic of ferroelectrics since all ferroelectrics are piezoelectric. If this strain can be switched by application of stress then an obvious parallel in elastic terms exists with the concept of ferroelectricity. This property is termed *ferroelasticity*, and a crystal is said to be *ferroelastic* when it has two or more orientation states in the absence of mechanical stress (and electric field) and can be shifted from one to another of these states by mechanical stress—here any two of the orientational states are identical or enantiomorphous in crystal structure and different in mechanical strain tensor at null mechanical stress (and null electric field). Concepts such as *ferroelastic domains, paraelastics*, and possibly even *antiferroelastics* can be defined in analogy with their dielectric counterparts. Intrinsic ferroelastic transitions are associated with the condensation of long-wavelength acoustic phonons and many are known. In a book devoted to electrically polar materials we are obviously concerned primarily with the phenomenon of ferroelasticity only in conjunction with ferroelectricity, and in this context the ferroelastics are most often extrinsic—being driven by a coupling between strain and the primary ferroelectric order parameter, the spontaneous polarization. As a result the ferroelastic and ferroelectric properties of ferroelastic ferroelectrics are often strongly coupled. Finally, for crystals containing magnetic constituent ions the possibility of a coupling of dielectric and/or elastic properties with magnetic properties must also be recognized. However, in this book we have included only minor reference to concentrated magnetic systems since relatively few which undergo a ferroelectric transition are yet well studied. In particular, coupling phenomena such as piezomagnetism and the linear magnetoelectric effect (the inducement of an electric moment by application of a magnetic field) have not been discussed.

Although most theoretical work and experimental studies of fundamental ferroelectric properties are carried out on large single crystals, the use of polycrystalline materials or *ceramics* is of particular importance in the ferroelectric context because of the possibilities of preparing a wide range of compositions (and the corresponding ability to adjust their properties for different applications) and because the presence of grain boundaries gives rise to effects which have important practical implications. A *ferroelectric ceramic* is an aggregate of ferroelectric single-crystal grains (or crystallites) with dimensions typically between 0·5 and 50 μm. Each ceramic grain unless it is *very* small has properties not unlike a single crystal but, because of the imperfect alignment of the crystallographic axes of the grains within the aggregate, the macroscopic properties of a ceramic in general differ significantly from those of a single crystal. For example when poled the grains can never be perfectly aligned, and also the resulting macroscopic symmetry of the elastic, piezoelectric, and dielectric constants is axial with an axis of rotation in the polar direction of infinite order.

1.2. A simple-model Hamiltonian

The value of simple models is well established in solid-state physics. From a theoretical standpoint it is of great value to deduce the simplest Hamiltonian which appears to embody the essential features of a co-operative phenomenon and to study its statistical behaviour. In such a way the basic properties of an idealistic system can be probed in considerable detail using relatively sophisticated techniques, and subsequent experimental deviations from this fundamental behaviour can then be interpreted in terms of the (hopefully small) deviations of a real-system Hamiltonian from the model. Excellent examples from other fields are the BCS model of superconductivity and the Ising and Heisenberg models of ferromagnetism.

The basic-model Hamiltonian which describes structural phase transitions (and ferroelectrics in particular) takes advantage of the fact that such transitions are often associated with the rearrangement of only a few atoms in the unit cell, whereas the positions of others are relatively undisturbed. Thus for a simple model it seems essential to take account only of these particular co-ordinates while treating the rest of the crystal lattice as a heat bath. In this section we indicate the manner in which such a model is set up, noting the approximations made in order to appreciate those properties which we choose, in this simple model, to neglect.

The basic Hamiltonian describing a solid is conventionally written

$$\mathcal{H} = \mathcal{H}(\text{ion}) + \mathcal{H}(\text{electron}) + \mathcal{H}(\text{electron–ion}) \qquad (1.2.1)$$

where $\mathcal{H}(\text{ion})$ describes a collection of ions interacting via a potential which depends only on the positions $\mathbf{R}_i, \mathbf{R}_j, \ldots$, of the ion centres, $\mathcal{H}(\text{electron})$ describes the valence electron motion, and $\mathcal{H}(\text{electron–ion})$ is a suitably chosen potential which represents interactions between valence electrons and ion cores. It is well known that the electronic and ionic motions can be separated to an excellent approximation by use of the adiabatic principle (see, for example, Ziman 1960). In this approximation the electrons are treated as if they respond so quickly to the motion of the ions that their state is always just a function of the ionic co-ordinates. It allows us to write $\mathcal{H}(\text{electron–ion})$ as a potential energy contribution $E(\mathbf{R}_i, \mathbf{R}_j, \ldots)$ to an effective Hamiltonian for ionic motion alone in the form

$$\mathcal{H}_{\text{eff}}(\text{ion}) = \sum_i (p_i^2/2m_i) + U(\mathbf{R}_i, \mathbf{R}_j, \ldots) + E(\mathbf{R}_i, \mathbf{R}_j, \ldots) \qquad (1.2.2)$$

where, in an obvious notation, the first and second terms are respectively the kinetic and potential energy of the lattice of ion cores themselves.

The next approximation is usually to assume $E(\mathbf{R}_i, \mathbf{R}_j, \ldots)$ to be independent of electron configuration or, in the subsequent statistical calculation, of temperature. In this way the potentials $U \cdot$ and E can be combined into a resultant effective ion–ion potential $V(\mathbf{R}_i, \mathbf{R}_j, \ldots)$. This is in fact quite a good approximation even for narrow-band-gap materials since, by the exclusion principle, most valence electrons are always in the same state even when thermal electronic excitations occur. On the other hand the effects of valence to conduction band thermal excitations are by no means negligible when band gaps are small and they will be considered explicitly in Chapter 14. For the present we neglect any temperature dependence of E and arrive accordingly at an effective ionic Hamiltonian

$$\mathcal{H}_{\text{eff}}(\text{ion}) = \sum_i (p_i^2/2m_i) + V(\mathbf{R}_i, \mathbf{R}_j, \ldots). \tag{1.2.3}$$

A little care in terminology must now be exercised. The potential V in eqn (1.2.3) is an effective ionic potential (depending only on ion-centre co-ordinates) and so we shall term it. It is nevertheless the sum of two distinct parts, one of which is physically of electronic origin. Thus Hamiltonian eqn (1.2.3) does not describe the motion of rigid ions but, to use the shell-model picture (for example Cochran 1960, 1961), includes a relative motion of rigid core and valance-shell electrons for each ion. This distortion has traditionally been described as resulting from the electronic polarizability of ions in the local field of their neighbours and discussed apart from the 'rigid-ion' motion of the lattice. In the present context this separation is not necessary (nor even desirable) and we shall refer to eqn (1.2.3) as a description of ionic, as opposed to rigid-ion, motion and as defining ionic energies.

The adiabatic principle allows us to think of electronic and ionic energies as essentially independent. This enables us to concentrate on eqn (1.2.3) to find any lattice-driven instabilities responsible for structural transitions. The companion effective electronic Hamiltonian involving $\mathcal{H}(\text{electron})$, and dependent on instantaneous ion configuration via $\mathcal{H}(\text{electron–ion})$, is not to be forgotten. It is, for example, responsible for the electronic contribution to polarizability which remains at optic frequencies (when ion motion is completely relaxed) and will feature prominently in subsequent discussions of electro-optic and non-linear optic properties. Nevertheless it is not known to play any active role in the driving of ferroelectric instabilities and can be set aside for the moment.

Phase transitions in crystal lattices usually involve some particular type of co-ordinates. This is perhaps best illustrated by examples: the rotations of BO_6 octahedra in antidistortive perovskite (ABO_3) transitions, the displacement of B with respect to the O_6 oxygen cage in ferroelectric perovskite

transitions, the coupled proton–lattice motion in hydrogen-bonded ferroelectrics, etc. An enormous simplification results if a theory can be
constructed from the effective ionic Hamiltonian which takes into account
only the motion of these particular co-ordinates, with the rest of the crystal
lattice as a bath (Lines 1969 a, Thomas 1971). Very often a single 'local
mode' of appropriate symmetry is sufficient for an adequate description, and
accordingly we define local canonically conjugate momentum and displacement co-ordinates for each unit cell.

The non-kinetic part of the effective ionic Hamiltonian is most con-
veniently expressed in terms of the displacements $\mathbf{q}_{lb}$ of the bth ion in the lth
unit cell from a fixed reference configuration representing the averaged
positions of the ions in the high-temperature phase. This reference configuration (the prototype phase) is the structure of highest symmetry which can be
related by 'small' displacements to the actual dynamic configurations
involved. Accordingly we now define 'local-mode' generalized momentum
π_l and displacement ξ_l variables in terms of the 'local-ion' momentum $\mathbf{p}_{lb}$
and displacement $\mathbf{q}_{lb}$ co-ordinates by the equations

$$\mathbf{p}_{lb} = m_b \mathbf{u}_{lb} \pi_l, \qquad \mathbf{q}_{lb} = \mathbf{u}_{lb} \xi_l \qquad (1.2.4)$$

$$\pi_l = \sum_b \mathbf{u}_{lb} \cdot \mathbf{p}_{lb}, \qquad \xi_l = \sum_b m_b \mathbf{u}_{lb} \cdot \mathbf{q}_{lb} \qquad (1.2.5)$$

where $\sum_b$ runs over all ions in the lth cell, m_b is the mass of the bth ion, and
$(m_b)^{1/2}\mathbf{u}_{lb}$ is the bth component of the normalized local eigenvector and
satisfies

$$\sum_b m_b \mathbf{u}_{lb} \cdot \mathbf{u}_{lb} = 1. \qquad (1.2.6)$$

Thus ξ_l is the scalar amplitude of that local motion which broadly characterizes the phase transition. If this mode should be degenerate then it
becomes necessary to define ξ_l as a vector of appropriate dimension. The
dynamics of the ionic system (eqn 1.2.3) is now described in the local-mode
approximation by the model Hamiltonian

$$\mathcal{H} = \sum_l \tfrac{1}{2}\pi_l^2 + V(\xi_1, \xi_2, \dots, \xi_N) \qquad (1.2.7)$$

in which N labels the number of unit lattice cells in the macroscopic crystal.

The potential V of eqn (1.2.7) can now be decomposed into a sum of
single-cell contributions $V(\xi_l)$ and an intercell interaction part. We shall
write the latter as a sum of bilinear two-body interactions $v_{ll'}\xi_l\xi_{l'}$. Although a
generalization to more complex forms of interaction is perhaps not without
physical interest (allowing, for example, for a wavevector-dependent anharmonicity), a very informative discussion of structural phase transitions in

general and ferroelectric transitions in particular can be achieved with the simple bilinear formalism.

The final form of our model Hamiltonian is therefore

$$\mathcal{H} = \sum_l \{\tfrac{1}{2}\pi_l^2 + V(\xi_l)\} - \tfrac{1}{2}\sum_l \sum_{l'} v_{ll'}\xi_l\xi_{l'}. \tag{1.2.8}$$

In this model the local potential function $V(\xi)$ can be anything from quasi-harmonic to deep double-well form, and the interaction potential $v_{ll'}$ can be short range and/or long range in character. With this flexibility the Hamiltonian eqn (1.2.8) can be used to describe in a fairly realistic way a whole range of polar and non-polar displacive, tunnel-mode, and order–disorder transitions. Careful analysis of its static and dynamic statistical behaviour in both classical and quantum theories is therefore highly desirable. Statistical analysis of eqn (1.2.8) is as yet by no means as advanced as for its magnetic counterparts but considerable progress has been made and will be discussed in Chapter 2.

The whole philosophy surrounding Hamiltonian eqn (1.2.8) is analogous to the well-known treatment of magnetic problems in terms of a spin Hamiltonian. The various constants needed to specify $V(\xi_l)$ and $v_{ll'}$ are treated as adjustable parameters. Their derivation in terms of the more fundamental concepts of atomic masses and force constants presents a separate problem akin to that of calculating the exchange and anisotropy constants of a magnetic system from first principles.

The great value of eqn (1.2.8) is its generality and simplicity. On the other hand, by the rather drastic approximation of excluding all local modes except that one deemed to characterize the transition, use of eqn (1.2.8) in its basic form precludes any discussion of coupling to other modes. In particular the coupling to acoustic modes and to elastic strains is absent. In addition the description of ferroelectrics for which the dielectric anomaly is driven by a direct coupling to a non-polar instability (such as occurs for example in gadolinium molybdate) is also outside the realm of Hamiltonian eqn (1.2.8). Nevertheless (see for example § 10.1) the model Hamiltonian eqn (1.2.8) is readily embellished to allow for such couplings and the basic single-mode theory can be expanded to describe these phenomena.

From a purely formal point of view some mathematical difficulties are contained in the local-mode description and, although we shall not dwell on such, it is perhaps necessary to make some passing reference to them. For example the precise mathematical construction of a complete set of independent local-mode co-ordinates appears to be very difficult. When there are ions on the faces or edges of the unit cell the same atomic displacement is involved in the construction of neighbour-cell local co-ordinates, which implies that the latter are no longer independent but satisfy certain relations expressing consistency conditions for the displacements of face and edge atoms. The difficulty is really evaded if the local co-ordinates are restricted, as in eqns (1.2.5) and (1.2.8), to a single representation per cell, since the consistency conditions then involve other representations and do not have to be taken into account explicitly. For this reason these difficulties will not concern us in the study of the Hamiltonian eqn (1.2.8). On the other hand it is well to be aware of the problems concerning a more complete local-mode picture and Elliott (1971) has recently given some attention to it in the limit of quasi-harmonic vibrational motion. In the latter case it appears that the concept of independent local-mode co-ordinates is most clearly established when the motion is dominated by a small region of reciprocal lattice space such as found close to the condensation of a soft lattice mode.

1.3. An outline of the statistical problem

As is so often the case in many-body physics even a basic-model Hamiltonian stripped of all but its most essential features still defies exact solution. Eqn (1.2.8) representing a simple-model structural transition is no exception. The potential function is made up of both local terms $V(\xi_l)$ and intercell interaction terms $v_{ll'}\xi_l\xi_{l'}$. The absence of either would render the Hamiltonian exactly soluble at least in principle. Thus for $v_{ll'} = 0$ the problem reduces to one of non-interacting localized motion, while for $V(\xi_l) = 0$ or $V(\xi_l)$ of harmonic form the Hamiltonian can be diagonalized in terms of running waves (phonons).

The competition between local-mode and running-wave motion is a not uncommon feature of many-body dynamics, and there are two obvious methods of attacking the problem. One is to use the running waves, or non-interacting phonons, as zeroth-order basis states and to represent local anharmonicity in terms of phonon–phonon interactions which can be statistically approximated in some way to allow diagonalization in terms of 'renormalized' phonons. The other is to approximate statistically the interaction potential $v_{ll'}\xi_l\xi_{l'}$ in some manner which reduces the many-body Hamiltonian eqn (1.2.8) to non-interacting cell form. Looked at from the former standpoint structural transitions are caused by lattice anharmonicities; from the latter viewpoint they are precipitated by intercell forces. Both pictures are informative.

Historically, as discussed in § 1.1, early efforts to describe ferroelectric transitions were cast in terms of an individual-ion Lorentz-field picture. The importance of a lattice-mode description (at least for displacive systems) was first stressed in the pioneering work of Anderson (1960) and Cochran (1960) which recognized the existence of a soft transverse optic phonon mode of long wavelength and suggested a connection between the strong temperature dependence of this mode near the phase transition and the dielectric anomaly observed in many ferroelectrics. The idea was that a group of modes near the Brillouin zone centre would have negative energies in the basic (bare-phonon) zeroth-order harmonic approximation, but that the anharmonic phonon interaction terms would provide a renormalization sufficient to stabilize these modes at higher temperatures (paraelectric phase). As the temperature is lowered renormalization decreases with a resulting lowering (or softening) of these 'soft-mode' frequencies until the paraelectric phase becomes unstable. The qualitative correctness of this idea was established by subsequent calculations (Silverman and Joseph 1963, 1964, Silverman 1964, Cowley 1965), and with these papers the efforts to solve the associated statistical problem in the interacting-phonon framework began.

The traditional theoretical approach to the interacting-phonon problem in lattice dynamics goes right back to Born and von Karman (1912) and employs a perturbational approximation about the harmonic (bare-phonon) basis making use of the assumed smallness of the correction terms (see for example Born and Huang 1954, Maradudin and Fein 1962). However, since the bare-phonon soft-mode frequencies are imaginary in the displacement-phase transition problem, the contribution of these terms in the (renormalization) perturbation calculation leads to divergencies indicating that the perturbation series does not in principle exist. Faced with this difficulty the initial approach was to ignore it by merely neglecting the unstable-mode contribution to renormalization. This amounts to neglecting the interactions between the soft modes themselves and to assuming that all significant stabilization of the soft modes is accomplished by the harmonically stable excitations (Kwok and Miller 1966, Vaks 1968).

The notion that a really satisfactory soft-mode theory should treat the entire frequency spectrum self-consistently was first voiced by Silverman (1964) and by Boccara and Sarma (1965 a). Such a theory had been constructed earlier in a series of papers by Hooton (1955 a, b, 1958) in connection with the lattice dynamics problem in solid ^{4}He. It involves employing a renormalized phonon basis at the outset with adjustable parameters which are optimized such that the low-lying excitation spectrum of the harmonic-model system approximates as closely as possible the corresponding spectrum of the actual many-body system. The formal treatment of what is now referred to as the self-consistent phonon theory has since been given in a variety of forms (see for example Werthamer 1969, 1970) and was first set out for the soft-mode problem by Boccara and Sarma (1965 b). Only now, however, is it beginning to reach the stage where explicit descriptions of measured quantities for specific systems are being attempted (Pietrass 1972).

In spite of the promise of these running-wave-based approximations, they are obviously physically valid in low order only to the extent that the motion about thermal equilibrium is of damped harmonic form. For grossly anharmonic local potential $V(\xi_l)$ it seems more promising to concentrate on the picture of the local motion of an individual cell mode in the potential due to the self-consistent motion of all the others. The simplest approximation in this scheme is the static mean field one (Aizu 1966 a, Lines 1969 a) in which all cells except the one of immediate interest are replaced by their thermally averaged configuration. From this mean-field state a lattice vibration (of arbitrarily anharmonic form) can then be found as a collective mode of response of the mean-field array of cells to a disturbance of arbitrary wavevector (Thomas 1969). The philosophy is that of the random-phase approximation and the method, which now goes by the name of self-consistent field theory (or time-dependent Hartree approximation), was first introduced into lattice dynamics by Fredkin and Werthamer (1965) following an earlier discussion by Brenig (1963). Specific application to the ferroelectric problem has been given by Miller and Kwok (1968), Vaks, Galitskii, and Larkin (1966), and Vaks, Larkin, and Pikin (1966).

The time-dependent Hartree approximation contains one very glaring weakness. For equilibrium calculations it reduces to the mean-field picture and therefore neglects all static correlations between neighbour cells. The extension to allow for these has been made by Lines (1972 a) who, extending the Onsager (1936) reaction-field concept, retains the local-field picture but includes in the static part both a random-phase (mean-field) and an in-phase (fully correlated) contribution. The relative magnitude of the two components is temperature dependent and is determined by using a constraint imposed by the fluctuation theorem (Kubo 1966).

The two methods of attacking the many-body problem defined by eqn (1.2.8) have recently been analysed classically by determining which can produce the lower free energy or the better approximation to exact high- and low-temperature series expansions (Eisenriegler 1974). The general conclusion seems to be that the local-mode (independent-cell) scheme is to be preferred except in the quasi-harmonic limit at low temperatures where the self-consistent phonon method is the more accurate.

For local potentials $V(\xi_l)$ which deviate markedly from quasi-harmonic form quantum solutions are enormously difficult to obtain (even in the simplest mean-field approximation) since they involve quantum motion in a grossly anharmonic well. Classical solutions can sometimes be obtained analytically (Onodera 1970) but quantum solutions have to be approached numerically (Koehler and Gillis 1973, Gillis and Koehler 1974). The exception is for $V(\xi_l)$ of deep double-well form when, to the extent that only the two lowest states (one near the bottom of each well) in each cell are populated at temperatures of interest, the quantum problem can be transformed into a 'two-level' pseudo-spin half-formalism. In this form $S^z = +\frac{1}{2}$, $-\frac{1}{2}$ correspond to 'up' and 'down' wells, and an allowance for intracell quantum tunnelling between wells can be made by use of spin-shift operators. The tunnel terms then appear as transverse field energies (de Gennes 1963).

The restriction of the number of 'spin' states in this so-called transverse Ising model simplifies the associated many-body problem to the point where one can conceive of going beyond the context of a simple self-consistent field description (such as that of Brout, Müller, and Thomas 1966 and Miller and Kwok 1968) to make use of the sophisticated many-body techniques which have been developed over the years in connection with the extensive theoretical investigation of the magnetic Ising and Heisenberg models. Thus both the static and dynamic theories for this deep double-well limit have progressed beyond their single or shallow double-well counterparts to include Green's function, series expansion, and diagrammatic techniques (Wang and Cooper 1968, Elliott and Wood 1971, Pfeuty and Elliott 1971, Moore and Williams 1972).

In the transverse Ising model the soft-phonon picture of phase-transition dynamics is replaced by the concept of the tunnel mode. For cases where the tunnelling probability is small the peak in frequency response which corresponds to the tunnel mode may be quite distinct from that corresponding to the associated phonon mode (oscillation within the wells). In this situation the latter peak may remain well defined and relatively temperature independent through the phase transition, whereas the former moves to lower energies, defining a 'soft' tunnel mode, as the instability is approached. In the limit of zero tunnelling probability the 'spin' dynamics are described only by interactions with the thermal bath of other modes and the response takes a Debye relaxation form with 'soft-mode frequency' imaginary (no propagating component).

One dynamic feature, which has recently been found experimentally to be a fairly common characteristic of structural phase transitions in general, is conspicuously absent from all the above theoretical schemes. This is the existence of an extremely narrow 'central peak' in the frequency response in addition to the propagating or diffuse soft mode response itself. Although its existence may well be due to couplings (e.g. to impurities or to entropy fluctuations) which are not contained in eqn (1.2.8), recent computer simulations (Schneider and Stoll 1973, 1976) and analytic solutions (Krumhansl and Schrieffer 1975, Varma 1976) for double-well models of restricted dimensionality have revealed the existence of excitations of a type completely overlooked by the conventional mean-field and phonon based approximations. These are excitations which may be thought of as moving domain walls or dislocations, for which the order parameter changes from one potential minimum to the other through a wall region of well defined character. Near a phase transition this leads to the appearance of microscopically substantial clusters (or microdomains) of the new structural phase appearing before the transition temperature is reached. The associated dynamics do indeed take the form of a strong narrow central peak of frequency response but there is some doubt as to whether the experimentally observed features of this nature are really manifestations of this effect or, indeed, to what extent the stability of this microdomain picture is an artifact of the low dimensionality (i.e. one and two dimensions) of the lattices for which relevant calculations have as yet been performed. Nevertheless, experimental evidence is beginning to accumulate favouring the existence of just such long-lived polarized clusters near T_c in some KDP ferroelectrics (Adriaenssens 1975, Müller et al. 1976).

One additional complication which makes structural transition theory more difficult than that of conventional magnetic models is that the interaction energy $v_{ll'}\xi_l\xi_{l'}$ (even if we assume that the bilinear formalism is applicable, and in general this potential can be quite complicated and perhaps not even expressible as a sum of pairwise interactions) is generally of unknown range and directional dependence. Only in ferroelectric transitions can we feel any confidence in making detailed statements. Here the long-range forces, to the extent that they are not screened by free carriers, are thought to be dominantly of electric dipolar origin. However, dipolar interactions are both long range and directionally dependent and even the simplest Ising model has not been discussed beyond the simpler statistical approximations with forces of this complexity. Nevertheless theory has developed beyond the mean-field and random-phase level in the discussion of dipole–dipole critical phenomena (e.g. Larkin and Khmel'nitskii 1969) and suggests logarithmic corrections to simple phenomenological, or mean-field, behaviour.

2

STATISTICAL THEORY, SOFT MODES, AND PHASE TRANSITIONS

2.1. Mean-field theory and the soft-mode concept

IN ORDER to acquire a simple physical picture of the dynamic mechanism of a phase transition it is sufficient at the outset to use the simplest of many-body approximations. It is instructive, in particular, to study the mean-field response of the model system (eqn (1.2.8)) to a time-dependent applied field. In this way one can obtain considerable insight into the nature of the collective excitations and into the relationship between the static aspects of a phase transition and the occurrence of temperature-dependent (i.e. 'soft') modes and of critical fluctuations. In this we shall follow the work of Thomas (1969, 1971).

As a preface it is necessary to discuss the static aspects of mean-field theory and the nature of the static singularities which accompany a second-order phase transition. Mean-field dynamics can then be described in terms of deviations from the equilibrium mean-field state.

2.1.1. Statics

In the mean-field approximation we consider a representative lattice cell l and, in particular, those local-mode co-ordinates π_l, ξ_l which describe the phase transition of interest, replacing all other cells by their thermally averaged states. Thus in eqn (1.2.8) we replace operators $\xi_{l'}$, for all $l' \neq l$, by their thermal averages $\langle \xi_{l'} \rangle$ which are to be determined self-consistently. In this way Hamiltonian eqn (1.2.8) is immediately reduced to non-interacting form. We can now define a mean-field Hamiltonian for the lth cell in the form

$$\mathcal{H}_l(\mathrm{mf}) = \tfrac{1}{2}\pi_l^2 + V(\xi_l) - h\xi_l - \sum_{l'} v_{ll'}\xi_l\langle\xi_{l'}\rangle \qquad (2.1.1)$$

where we have included a term describing the introduction of a uniform static field h. This field is in general an internal field and may differ from a corresponding externally applied field if long-range (e.g. dipolar) forces are included in $v_{ll'}$. However, if we assume a needle-shaped macroscopic specimen with field parallel to the long axis, the difference vanishes and we can refer to h as an applied field without ambiguity.

In eqn (2.1.1) interactions have been replaced by an effective (or 'Lorentz') field of magnitude $\sum_{l'} v_{ll'}\langle\xi_{l'}\rangle$. In this way the many-body problem has been reduced to one of an ensemble of independent single-ion oscillators. We can therefore use the familiar statistical result for thermal averages

$$\langle(\ldots)_l\rangle = \sum_i \frac{\langle i|(\ldots)_l|i\rangle \exp(-E_i/kT)}{\sum_i \exp(-E_i/kT)} \quad (2.1.2)$$

where E_i is the ith eigenvalue of local Hamiltonian (2.1.1) and where $(\ldots)_l$ refers to any operator involving only local co-ordinates π_l, ξ_l, with $\langle i|(\ldots)_l|i\rangle$ being its diagonal matrix element in the ith eigenstate $|i\rangle$.

Of particular interest is the equation for the static displacement average $\langle \xi_l \rangle$ itself. Since $\langle \xi_{l'} \rangle = \langle \xi_l \rangle = \langle \xi \rangle_h$, say, for all l' in a uniform field h (assuming a ferrodistortive system, an antiferrodistortive situation in the ordered phase requires the definition of cell sublattices), it follows that the eigenvalues E_i are functions of this order parameter $\langle \xi \rangle_h$ and field h and that eqn (2.1.2) is an implicit equation for displacement as a function of field. In particular one can calculate static susceptibility as the derivative of $\langle \xi \rangle_h$ with respect to h in the zero-field limit.

In quantum theory even the static mean-field approximation is very difficult to pursue quantitatively for general $V(\xi_l)$ since it requires the quantum solution for motion in a well of arbitrary complexity. As a result mean-field calculations are often performed classically (Aizu 1966a, Lines 1969a, b) using the classical ensemble average

$$\langle \xi \rangle_h = \frac{\int_{-\infty}^{\infty} \xi_l \exp(-W_l/kT)\, d\xi_l}{\int_{-\infty}^{\infty} \exp(-W_l/kT)\, d\xi_l} \quad (2.1.3)$$

where

$$W_l = V(\xi_l) - h\xi_l - v(0)\xi_l\langle \xi \rangle_h \quad (2.1.4)$$

and we define

$$v(0) = \sum_{l'} v_{ll'}. \quad (2.1.5)$$

The integrals in eqn (2.1.3) can be computed numerically for arbitrary local potential.

Expanding the exponentials in eqn (2.1.3) to first order in the order parameter $\langle \xi \rangle_h$ when $h = 0$ allows us to study the approach $\langle \xi \rangle_0 \to 0$ to a phase transition at $T = T_c$. One readily establishes that

$$kT_c = v(0)\langle \xi^2 \rangle_{h=0} \quad (2.1.6)$$

where temperature T_c represents a second-order Curie point if the zero-field order parameter $\langle \xi \rangle_0$ has arbitrarily small values as $T \to T_c^-$ (i.e. $T \to T_c$, $T < T_c$). If the order parameter goes to zero as $T \to T_c^+$ (i.e. $T \to T_c$, $T > T_c$), then the solution (now usually symbolized as T_0) represents the limit of metastability of the non-polar phase below a first-order transition temperature (see § 3.3.2).

Some numerical results for a local potential with simple quartic stabilizing anharmonicity

$$V(\xi_l) = \tfrac{1}{2}\omega_0^2\xi_l^2 + A\xi_l^4 \qquad (2.1.7)$$

have been given by Lines (1969 a), where ω_0^2 and A are both positive constants. A transition to an ordered phase is found at finite temperatures if $v(0) > \omega_0^2$. It is always of second order for potential eqn (2.1.7) and with Curie temperature given by

$$kT_c = A^{-1}\{0\cdot338v(0)f\}^2 \qquad (2.1.8)$$

where f as a function of $\omega_0^2/v(0)$ is shown in Fig. 2.1. The temperature dependence of the normalized order parameter for several values of $\omega_0^2/v(0)$ is shown in Fig. 2.2.

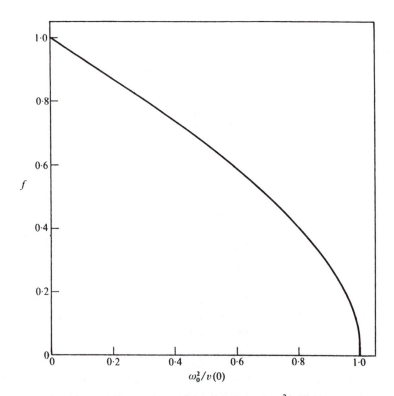

Fig. 2.1. The parameter f of eqn (2.1.8) as a function of the ratio $\omega_0^2/v(0)$ of harmonic intracell to bilinear intercell energy parameters. (After Lines 1969 a.)

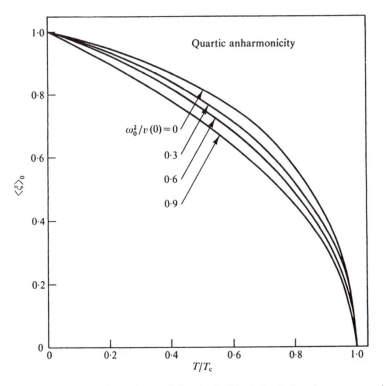

Fig. 2.2. The temperature dependence of the classical 'polarization' order parameter $\langle\xi\rangle_0$ (normalized to its value at $T = 0°$ K) as a function of reduced temperature for several values of $\omega_0^2/v(0)$. (After Lines 1969 a.)

From an experimental point of view a more interesting relationship can be derived between T_c and the zero-temperature saturation value of displacement $\langle\xi\rangle_{T=0}$ (which within the simple model is proportional to the spontaneous polarization at $T = 0$). It is that to a good approximation $kT_c = K\langle\xi\rangle_{T=0}^2$, where K is largely independent of local potential and primarily dependent only on the intercell potential v. It should therefore follow that, if v is dominated by long-range dipolar forces, K should be essentially constant over classes of structurally similar materials. A relationship of this type was first postulated by Abrahams, Kurtz, and Jamieson (1968) and found to be in reasonable accord with experimental data for a range of displacement systems. However, the theoretical basis is quite classical and one might have reservations as to whether it would hold for $T_c \to 0$. Indeed, from (2.1.6) one might query whether the presence of zero-point motion might not prevent the existence of structrual phase transitions arbitrarily close to $T_c = 0$ in this context. Such a proposition has been advanced recently by Kurtz (1976) on more general grounds.

Static uniform susceptibility $\chi(0)$ is found to be of limiting Curie–Weiss form as $T \rightarrow T_c$, with

$$\{\chi(0)\}^{-1} = \beta(T - T_c), \qquad T > T_c \qquad (2.1.9)$$

and

$$\{\chi(0)\}^{-1} = 2\beta(T_c - T), \qquad T < T_c. \qquad (2.1.10)$$

The factor of 2 difference between the paraelectric-phase and ferroelectric-phase constants (the so-called Curie–Weiss constants) is familiar from the thermodynamic theory of ferroelectricity (§ 3.3.1). Finally, the specific-heat anomaly at T_c is found to take the form of a finite discontinuity.

The physical picture is one of competition between a local constraint (ω_0^2) favouring a high-symmetry phase with $\langle \xi \rangle_0 = 0$ and an interaction field ($v(0)$) preferring an ordered phase. For values $\omega_0^2 / v(0) > 1$ the interaction is too weak to overcome the local constraint at any temperature and the high-symmetry phase remains stable right down to absolute zero. When $\omega_0^2 / v(0) < 1$ the ordered phase is stable at low temperatures but is destroyed by thermally induced disorder as the temperature is raised.

Computations are also available (Lines 1969 a, b, Zook and Liu 1976) for a potential

$$V(\xi_l) = \tfrac{1}{2}\omega_0^2 \xi_l^2 + A\xi_l^4 + B\xi_l^6 \qquad (2.1.11)$$

with ω_0^2 and B positive but A negative, corresponding to developing double side wells as A increases in magnitude. For this potential a range of parameter values is found for which the transition is first order. There is, however, no direct connection between the sign of quartic anharmonicity and the order of the transition, and care must be taken not to confuse eqn (2.1.11) with the corresponding free-energy expression

$$G_1 = \tfrac{1}{2}\omega_0'^2 \langle \xi \rangle_0^2 + A'\langle \xi \rangle_0^4 + B'\langle \xi \rangle_0^6 \qquad (2.1.12)$$

for which an oft-quoted statement is made (§ 3.3) that a first-order transition results when A' is negative and a second-order transition when A' is positive. In fact the latter statement is true only if A' and B' are both taken to be temperature independent (B' positive), whereas the free energy G_1 when calculated from eqn (2.1.11) using conventional statistical mechanics does indeed assume the form shown in eqn (2.1.12) but with A' temperature dependent (Lines 1969 a).

2.1.2. Dynamics

To discuss phase-transition dynamics within mean-field theory we study the response of the mean-field equilibrium state to time-dependent external influences (Thomas 1969, 1971). We start by considering the behaviour of the single-cell displacement co-ordinate ξ_l in the absence of intercell forces (i.e. $v_{ll'} = 0$). We now allow the uniform applied field h in eqn (2.1.1) to undergo a time-dependent perturbation

$$h_l(t) = h + \delta h_l \exp(-i\omega t). \tag{2.1.13}$$

Treating the rest of the crystal as a heat bath the co-ordinate ξ_l can be written

$$\xi_l(t) = \langle \xi_l(t) \rangle + \Delta \xi_l(t) \tag{2.1.14}$$

where the ensemble average

$$\langle \xi_l(t) \rangle = \langle \xi \rangle_h + \delta \langle \xi_l \rangle \exp(-i\omega t) \tag{2.1.15}$$

is made up of a uniform static part $\langle \xi \rangle_h$ and a linear response part $\delta \langle \xi_l \rangle$ (higher-order response being neglected), and where $\Delta \xi_l(t)$ describes spontaneous fluctuations about the equilibrium state.

The static part is obtained from the equilibrium calculations set out above and will generally be a nonlinear function of h. In the limit $h \to 0$ it defines the zero-field order parameter $\langle \xi \rangle_0$, and its derivative with respect to h gives the static susceptibility $\chi_s(0)$. The amplitude $\delta \langle \xi_l \rangle$ of linear response is related to δh_l by the generalized susceptibility, which we symbolize as $\chi_s(\omega)$ and define by the equation

$$\delta \langle \xi_l \rangle = \chi_s(\omega) \, \delta h_l \tag{2.1.16}$$

the subscript s denoting 'single-cell' response. This single-cell dynamic susceptibility can be calculated by conventional linear response theory (Kubo 1957, Zubarev 1960). An excellent classical calculation for the anharmonic oscillator, including both single-well and double-well $V(\xi_l)$, has been given by Onodera (1970). We shall not pursue the details of linear response theory here but merely assume, for illustrative purposes, a single-cell response of damped-harmonic-oscillator form

$$\chi_s(\omega) = \chi_s(0) \frac{\Omega_s^2}{\Omega_s^2 + i\Gamma\omega - \omega^2} \tag{2.1.17}$$

where Ω_s is a resonance frequency in the absence of damping and Γ is a damping constant. Such a form is often an excellent approximation for single-well local potentials.

We now introduce the intercell interactions in a mean-field manner. We also allow the perturbing part of the applied field to be non-uniform. From eqn (2.1.1) the mean-field Hamiltonian can now be written

$$\mathcal{H}_l(\text{mf}) = \tfrac{1}{2}\pi_l^2 + V(\xi_l) - h_l^{\text{eff}}(t)\xi_l \tag{2.1.18}$$

where the effective field is given by

$$h_l^{\text{eff}}(t) = h_l(t) + \sum_{l'} v_{ll'} \langle \xi_{l'}(t) \rangle. \tag{2.1.19}$$

The mean-field Hamiltonian describes a motion in a local potential $V(\xi_l)$ perturbed by an effective field $h_l^{\text{eff}}(t)$. The latter can be decomposed as

$$h_l^{\text{eff}}(t) = h^{\text{eff}} + \delta h_l^{\text{eff}} \exp(-i\omega t) + \Delta h_l^{\text{eff}}(t) \qquad (2.1.20)$$

into a static part

$$h^{\text{eff}} = v(0)\langle\xi\rangle_h + h \qquad (2.1.21)$$

a linear response part

$$\delta h_l^{\text{eff}} = \sum_{l'} v_{ll'} \, \delta\langle\xi_{l'}\rangle + \delta h_l \qquad (2.1.22)$$

and the statistical fluctuations

$$\Delta h_l^{\text{eff}}(t) = \sum_{l'} v_{ll'} \, \Delta\xi_{l'}(t). \qquad (2.1.23)$$

In mean-field theory it is assumed that the response to the total effective field can be treated in a linear response approximation, thereby decoupling statistical fluctuations from thermal averages. One therefore writes

$$\delta\langle\xi_l\rangle = \chi_s(\omega)\left(\delta h_l + \sum_{l'} v_{ll'} \, \delta\langle\xi_{l'}\rangle\right) \qquad (2.1.24)$$

where $\chi_s(\omega)$ is a function of h^{eff} and in the ordered phase depends, via eqn (2.1.21), on the static order parameter. Introducing a wavevector $\mathbf{q}$ by the Fourier transforms

$$\delta\langle\xi_l\rangle = N^{-1/2} \sum_{\mathbf{q}} \delta\langle\xi(\mathbf{q})\rangle \exp(-i\mathbf{q}\cdot\mathbf{l}) \qquad (2.1.25)$$

$$\delta h_l = N^{-1/2} \sum_{\mathbf{q}} \delta h(\mathbf{q}) \exp(-i\mathbf{q}\cdot\mathbf{l}) \qquad (2.1.26)$$

where there are N cells in the macroscopic lattice, eqn (2.1.24) is decoupled in the form

$$\delta\langle\xi(\mathbf{q})\rangle = \chi_s(\omega)\{\delta h(\mathbf{q}) + v(\mathbf{q}) \, \delta\langle\xi(\mathbf{q})\rangle\} \qquad (2.1.27)$$

where

$$v(\mathbf{q}) = \sum_{l'} v_{ll'} \exp\{i\mathbf{q}\cdot(\mathbf{l'}-\mathbf{l})\}. \qquad (2.1.28)$$

Defining a collective susceptibility

$$\chi(\mathbf{q}, \omega) = \frac{\delta\langle\xi(\mathbf{q})\rangle}{\delta h(\mathbf{q})} \qquad (2.1.29)$$

we calculate it from eqn (2.1.27) in the form

$$\chi(\mathbf{q}, \omega) = \frac{\chi_s(\omega)}{1 - v(\mathbf{q})\chi_s(\omega)}. \qquad (2.1.30)$$

We recognize immediately the possibility of an instability. For systems in thermodynamic equilibrium this instability occurs, if it occurs at all, at $\omega = 0$ and we find a divergence of collective response (eqn (2.1.30)) when

$$v(\mathbf{q})\chi_s(0) = 1. \tag{2.1.31}$$

The instability occurs (as $T \to T_c^+$) with respect to that mode $\mathbf{q} = \mathbf{q}_0$ for which $v(\mathbf{q})$ takes on its maximum value. If $\mathbf{q}_0 = 0$ we refer to a ferrodistortive ordering; if $\mathbf{q}_0 = \frac{1}{2}\mathbf{K}$, where $\mathbf{K}$ is a reciprocal lattice vector, we have simple 'up-down' antiferrodistortive ordering. For a general $\mathbf{q}_0$ the ordered structure can be described in terms of a more complicated mode.

The normal-mode frequencies $\omega(\mathbf{q})$ are found as the poles of the dynamic response $\chi(\mathbf{q}, \omega)$. From eqn (2.1.30) it follows that they are the solutions of

$$v(\mathbf{q})\chi_s(\omega) = 1. \tag{2.1.32}$$

Using the damped-oscillator form $\chi_s(\omega)$ of eqn (2.1.17) and substituting in eqn (2.1.30) yields a collective response also of damped harmonic form but now with renormalized frequencies $\Omega(\mathbf{q})$. Specifically we obtain

$$\chi(\mathbf{q}, \omega) = \frac{\chi(\mathbf{q}, 0)\Omega^2(\mathbf{q})}{\Omega^2(\mathbf{q}) + i\Gamma\omega - \omega^2} \tag{2.1.33}$$

where

$$\frac{\Omega^2(\mathbf{q})}{\Omega_s^2} = \frac{\chi_s(0)}{\chi(\mathbf{q}, 0)} = 1 - v(\mathbf{q})\chi_s(0). \tag{2.1.34}$$

The last equation is very significant indeed and relates directly the static and dynamic properties of the system. It shows in particular that a divergence $\chi(\mathbf{q}, 0)$ of static susceptibility is associated with a soft mode $\Omega(\mathbf{q}) \to 0$. More specifically, since from eqn (2.1.9) the static susceptibility diverges as $(T - T_c)^{-1}$ at a second-order phase transition in the mean-field approximation and since neither $\chi_s(0)$ nor Ω_s is anomalous at $T = T_c$, eqn (2.1.34) leads directly to the relationship

$$\Omega^2(\mathbf{q}_0) \sim (T - T_c), \qquad T \to T_c^+ \tag{2.1.35}$$

relating soft-mode frequency to temperature (Fig. 2.3). A similar relationship holds an approaching T_c from the ordered phase with the one important modification that the soft mode is *always* a zone-centre ($\mathbf{q}_0 = 0$) mode in the reduced Brillouin zone of the reciprocal sublattice.

Solving eqn (2.1.33) for the complex poles $\omega^{\pm}(\mathbf{q})$ of the collective response function we find

$$\omega^{\pm}(\mathbf{q}) = \pm\{\Omega^2(\mathbf{q}) - \tfrac{1}{4}\Gamma^2\}^{1/2} + \tfrac{1}{2}i\Gamma. \tag{2.1.36}$$

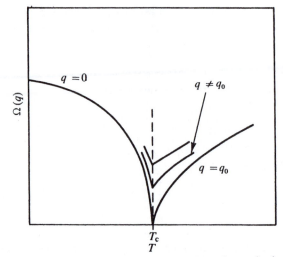

Fig. 2.3. A qualitative representation of the temperature dependence of soft-mode frequency near a second-order displacive phase transition.

At a second-order instability one of these modes ($\mathbf{q} = \mathbf{q}_0$) becomes unstable and, as is readily discerned from eqn (2.1.36), the real and the imaginary parts go to zero at the stability limit in the manner indicated in Fig. 2.4.

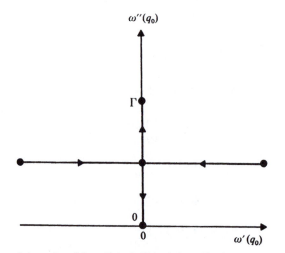

Fig. 2.4. Motion of the pole $\omega(\mathbf{q}) = \omega'(\mathbf{q}) + i\omega''(\mathbf{q})$ of the collective response function in the complex plane on approach to condensation as $T \to T_c^+$, $\mathbf{q} = \mathbf{q}_0$.

2.2. Correlated effective-field theory

Although the mean-field theory of the previous section is able to give a sound qualitative picture of a structural transition, one naturally has some reservations about the quantitative accuracy of a statistical approximation which neglects completely the existence of all spontaneous fluctuations and correlations between the motion of neighbouring cells. Nevertheless until quite recently theorists have tended to be much more satisfied with the mean-field theory in a ferroelectric context than, say, in magnetism. The prime reason has been that the mean-field form for the temperature dependence of many bulk thermodynamic properties is often actually observed, to good accuracy, on approach to a ferroelectric transition but not to a magnetic one. A simple example is the variation of spontaneous polarization as $(T_c - T)^{1/2}$ or of dielectric constant as $|T - T_c|^{-1}$ on approaching a second-order ferroelectric Curie point T_c.

The conventional argument is that for most magnetic transitions the dominant ordering forces are of short range (e.g. exchange interactions), while in ferroelectricity the ordering forces include strong electric dipolar contributions which are of long range. Since it can be established that the mean-field solution becomes exact in the long-range limit of equal interactions between all constituents, there has been a general belief that mean-field theory is consequently less crude in the ferroelectric context. To what extent is this really true? We shall probe this question in the present section by introducing intercell static correlations into the effective-field framework and determining them in a self-consistent fashion (Lines 1972 a, 1974).

Once again we consider the basic model of § 1.2, but now allow for the presence of a position-dependent static field energy $-\Sigma_l h_l \xi_l$ so that our defining Hamiltonian is

$$\mathcal{H} = \sum_l \{\tfrac{1}{2}\pi_l^2 + V(\xi_l) - h_l \xi_l\} - \tfrac{1}{2}\sum_l \sum_{l'} v_{ll'} \xi_l \xi_{l'}. \qquad (2.2.1)$$

The equation of motion for the lth cell co-ordinate ξ_l follows in both classical and quantum theories as

$$\ddot{\xi}_l = h_l - \frac{dV(\xi_l)}{d\xi_l} + \sum_{l'} v_{ll'} \xi_{l'} \qquad (2.2.2)$$

if $V(\xi_l)$ has a well-defined Maclaurin expansion. This can be rewritten as

$$\ddot{\xi}_l = h_l(\text{eff}) - \frac{dV(\xi_l)}{d\xi_l} \qquad (2.2.3)$$

defining a local effective field at the lth cell

$$h_l(\text{eff}) = h_l + \sum_{l'} v_{ll'} \xi_{l'}. \qquad (2.2.4)$$

If correlations are neglected completely we obtain the mean-field approximation for this effective field as

$$h_l(\text{eff}) = h_l + \sum_{l'} v_{ll'} \langle \xi_{l'} \rangle, \quad \text{mean field.} \qquad (2.2.5)$$

In an opposite extreme of fully correlated motion between cells the effective field can be written as

$$h_l(\text{eff}) = h_l + \sum_{l'} v_{ll'}\langle \xi_{l'} \rangle + \sum_{l'} v_{ll'}(\xi_l - \langle \xi_l \rangle), \quad \text{fully correlated} \quad (2.2.6)$$

for which the displacement of the l'th-cell co-ordinate from its average position is put equal to the displacement of the lth-cell co-ordinate from its (possibly different) average position throughout the motion.

It follows that one simple measure of intercell correlations for an actual (partly correlated) thermal-equilibrium situation can be written in a form intermediate between eqns (2.2.5) and (2.2.6) as

$$h_l(\text{eff}) = h_l + \sum_{l'} v_{ll'}\{\langle \xi_{l'} \rangle + A_{ll'}(\xi_l - \langle \xi_l \rangle)\} \quad (2.2.7)$$

where $A_{ll'}$ is a dimensionless temperature-dependent parameter of as yet unspecified form which describes the amplitude of the correlations between cells l and l'. In the language of eqn (2.1.23) we have written $\Delta \xi_{l'}(t) = A_{ll'}\xi_l(t)$. Defining a single temperature-dependent parameter α by writing

$$\alpha \sum_{l'} v_{ll'} = \sum_{l'} A_{ll'}v_{ll'} \quad (2.2.8)$$

and neglecting any site (i.e. field) dependence of the correlation α, the equation of motion (eqn (2.2.2)) now becomes

$$\ddot{\xi}_l = h_l - \frac{\mathrm{d}V(\xi_l)}{\mathrm{d}\xi_l} + \sum_{l'} v_{ll'}\{\langle \xi_{l'} \rangle + \alpha(\xi_l - \langle \xi_l \rangle)\} \quad (2.2.9)$$

and is directly derivable, via the Hamiltonian equations of motion, from an effective Hamiltonian for the lth cell of the form

$$\mathcal{H}_l(\text{eff}) = \tfrac{1}{2}\pi_l^2 + V(\xi_l) - h_l\xi_l - \sum_{l'} v_{ll'}\xi_l\{\langle \xi_{l'} \rangle + \tfrac{1}{2}\alpha\xi_l - \alpha\langle \xi_l \rangle\}. \quad (2.2.10)$$

This Hamiltonian is a function only of the lth-cell variables and we have accomplished the reduction of the many-body Hamiltonian eqn (2.2.1) to a sum of independent one-body terms. It follows that statistical ensemble averages for an operator $X(\pi_l, \xi_l)$ can be calculated in the form

$$\langle X(\pi_l, \xi_l) \rangle = \text{tr}(\rho_l X) \quad (2.2.11)$$

where $\rho_l = A/\text{tr}(A)$ with $A = \exp\{-\mathcal{H}_l(\text{eff})/kT\}$ is the density matrix (or classically the distribution function) for the lth cell.

Defining a long-range order parameter $\langle \xi \rangle_0$ in the absence of an applied field and defining

$$m_l = \langle \xi_l \rangle - \langle \xi \rangle_0 \quad (2.2.12)$$

as the field-dependent contribution to long-range order, it follows that

$$\mathscr{H}_l(\text{eff}) = \mathscr{H}_0 - \xi_l T_l \tag{2.2.13}$$

where

$$\mathscr{H}_0 = \tfrac{1}{2}\pi_l^2 + V(\xi_l) - \tfrac{1}{2}\alpha v(0)\xi_l^2 - \xi_l\langle\xi\rangle_0(1-\alpha)v(0) \tag{2.2.14}$$

is the zero-field-effective Hamiltonian and

$$T_l = h_l + \sum_{l'} v_{ll'}(m_{l'} - \alpha m_l) \tag{2.2.15}$$

is the effective perturbing local field. In eqn (2.2.14) we have written $v(0)$ as the zero-wavevector $(\mathbf{q} \to 0)$ limit of the Fourier transform

$$v(\mathbf{q}) = \sum_{l-l'} v_{ll'} \exp\{i\mathbf{q}\cdot(\mathbf{l}'-\mathbf{l})\} \tag{2.2.16}$$

which conforms with eqn (2.1.5). We can now calculate the ensemble average $\langle\xi_l\rangle$ to first order in field h_l and response m_l by expanding the formal statistical relationship

$$Z\langle\xi_l\rangle = \text{tr}[\xi_l \exp\{-\beta(\mathscr{H}_0 - \xi_l T_l)\}] \tag{2.2.17}$$

in which Z is the partition function (given by the trace of unity over the exponential function in eqn (2.2.17)) and where $\beta = 1/kT$. In such an expansion, however, care is necessary to recognize the fact that $\mathscr{H}_0$ and ξ_l do not commute.

Verifying by direct differentiation the equation

$$\frac{\partial}{\partial\beta}[\exp(\beta\mathscr{H}_0)\exp\{-\beta\mathscr{H}_l(\text{eff})\}] = \exp(\beta\mathscr{H}_0)\xi_l T_l \exp\{-\beta\mathscr{H}_l(\text{eff})\} \tag{2.2.18}$$

we establish the expansion

$$\exp\{-\beta\mathscr{H}_l(\text{eff})\} = \exp(-\beta\mathscr{H}_0)\left[1 + \int_0^\beta \exp(\lambda\mathscr{H}_0)\xi_l T_l \exp(-\lambda\mathscr{H}_0)\,d\lambda\right] \tag{2.2.19}$$

correct to first order in T_l. Substituting in eqn (2.2.17), using eqns (2.2.13) and (2.2.15), and again neglecting any possible field dependence of correlation parameter α, we find

$$\langle\xi_l\rangle = \langle\xi\rangle_0 + \beta(\langle\xi:\xi\rangle_0 - \langle\xi\rangle_0^2)\left\{h_l + \sum_{l'} v_{ll'}(m_{l'} - \alpha m_l)\right\} \tag{2.2.20}$$

where $\langle\ldots\rangle_0$ here and henceforth refers to a static ensemble average in the absence of applied field (e.g. $\langle \xi \rangle_0$ is the zero-field static order parameter) and we define the colon product

$$\langle \xi : \xi \rangle_0 = \frac{\text{tr}\{\exp(-\beta \mathcal{H}_0) \int_0^\beta \exp(\lambda \mathcal{H}_0)\xi_l \exp(-\lambda \mathcal{H}_0)\xi_l \, d\lambda\}}{\beta \, \text{tr}\{\exp(-\beta \mathcal{H}_0)\}} \qquad (2.2.21)$$

Classically it is just the mean-square deviation $\langle \xi^2 \rangle_0$.

Recognizing that zero field averages are independent of l and recalling the definition of m_l as $\langle \xi_l \rangle - \langle \xi \rangle_0$, eqn (2.2.20) can be Fourier transformed with respect to the lattice and solved explicitly for $\langle m(\mathbf{q}) \rangle$. Writing a wavevector-dependent static susceptibility $\chi(\mathbf{q}) = m(\mathbf{q})/h(\mathbf{q})$, the resulting equation is

$$\{\chi(\mathbf{q})\}^{-1} = \tau + \alpha v(0) - v(\mathbf{q}) \qquad (2.2.22)$$

in which

$$\tau = \frac{kT}{\langle \xi : \xi \rangle_0 - \langle \xi \rangle_0^2}. \qquad (2.2.23)$$

We note in particular that uniform static susceptibility is given by

$$\{\chi(0)\}^{-1} = \tau + (\alpha - 1)v(0) \qquad (2.2.24)$$

and also that

$$\{\chi(\mathbf{q})\}^{-1} = \{\chi(0)\}^{-1} + v(0) - v(\mathbf{q}). \qquad (2.2.25)$$

To this point the correlation α has remained unspecified. In particular if we put $\alpha = 0$ we regenerate static mean-field theory. What has been gained? If this were the whole story the answer would be simply one parameter, to be determined perhaps by fitting theory with experiment for some actual physical system. It turns out, however, that an exact relationship (a form of the so-called fluctuation theorem) is readily obtainable which expresses static susceptibility directly in terms of spontaneous fluctuations $\langle (\xi - \langle \xi \rangle_0)^2 \rangle_0$ and therefore determines α as a function of temperature in an unambiguous fashion. It follows that mean-field theory with $\alpha = 0$ is in violation of the fluctuation theorem.

To determine the required fluctuation result we note that in the original many-body Hamiltonian $\mathcal{H}$ of eqn (2.2.1) the applied field enters only in the very simple explicit form $-\Sigma_l h_l \xi_l$ or, in terms of wavevector $\mathbf{q}$, as $\Sigma_{\mathbf{q}} h(\mathbf{q})\xi(-\mathbf{q})$. It follows that an exact expression for $\chi(\mathbf{q})$ can be obtained from the formal definition

$$\chi(\mathbf{q}) = \frac{\partial \langle \xi(\mathbf{q}) \rangle}{\partial h(\mathbf{q})} = \frac{\partial}{\partial h(\mathbf{q})} \text{tr}\{\rho \xi(\mathbf{q})\} \qquad (2.2.26)$$

where ρ is the exact density matrix

$$\rho = \frac{\exp(-\mathcal{H}/kT)}{\text{tr}\{\exp(-\mathcal{H}/kT)\}} \qquad (2.2.27)$$

by direct differentiation. The result to lowest order is

$$kT\chi(\mathbf{q}) = \langle \xi(-\mathbf{q}) : \xi(+\mathbf{q})\rangle_0 - \langle \xi(-\mathbf{q})\rangle_0 \langle \xi(+\mathbf{q})\rangle_0 \qquad (2.2.28)$$

in which

$$\beta \xi(-\mathbf{q}) : \xi(\mathbf{q}) = \int_0^\beta e^{\lambda \mathcal{H}} \xi(-\mathbf{q}) \, e^{-\lambda \mathcal{H}} \xi(\mathbf{q}) \, d\lambda. \qquad (2.2.29)$$

Using this exact relationship in the zero-field limit and summing the wavevector $\mathbf{q}$ over all allowed values in the first Brillouin zone of the cell reciprocal lattice, we find

$$kT \sum_{\mathbf{q}} \chi(\mathbf{q}) = N(\langle \xi : \xi \rangle_0 - \langle \xi \rangle_0^2) = NkT/\tau \qquad (2.2.30)$$

where N is the number of cells in the macroscopic lattice, and we have recognized the equivalence of the Hamiltonian $\mathcal{H}$ and the effective Hamiltonian $\mathcal{H}_0$ of eqn (2.2.14) within the correlated effective-field approximation for operations and averages involving single-cell variables alone. Combining eqns (2.2.22) and (2.2.30) now gives

$$N/\tau = \sum_{\mathbf{q}} \{\tau + \alpha v(0) - v(\mathbf{q})\}^{-1} \qquad (2.2.31)$$

which defines the required temperature dependence of α. The equations are now closed and define the static correlated effective-field theory of structural phase transitions.

Phase-transition dynamics within the correlated field theory are readily described by the linear response approximation. The effective Hamiltonian eqn (2.2.10) can be pictured as describing an isolated local mode moving in a local potential $V(\xi_l) - \frac{1}{2}\alpha v(0)\xi_l^2$ and subject to an effective field of form (2.1.20) but where now

$$h_l^{\text{eff}} = \sum_{l'} v_{ll'}(\langle \xi_{l'}\rangle - \alpha\langle \xi_l\rangle) + h_l \qquad (2.2.32)$$

$$\delta h_l^{\text{eff}} = \sum_{l'} v_{ll'}(\delta\langle \xi_{l'}\rangle - \alpha\delta\langle \xi_l\rangle) + \delta h_l \qquad (2.2.33)$$

$$\Delta h_l^{\text{eff}}(t) = 0 \qquad (2.2.34)$$

the statistical fluctuations having now been incorporated into the definition of the local potential. Assuming that the total effective field can be treated in linear response theory we write

$$\delta\langle\xi_l\rangle = \chi_s(\omega)\,\delta h_l^{\text{eff}} \tag{2.2.35}$$

after which, using eqn (2.2.33) and Fourier transforming, we find a dynamic response (compare eqn (2.1.30))

$$\chi(\mathbf{q}, \omega) = \frac{\chi_s(\omega)}{1 - \chi_s(\omega)\{v(\mathbf{q}) - \alpha v(0)\}}. \tag{2.2.36}$$

In analogy with the mean-field result (eqn (2.1.30)), and using the damped-harmonic-oscillator model (eqn (2.1.17)), it follows that the mean-field eqns (2.1.31) to (2.1.34) remain valid if $v(\mathbf{q}) - \alpha v(0)$ is substituted everywhere for $v(\mathbf{q})$. In particular the square of the soft-mode frequency still varies as the reciprocal static susceptibility when $T \to T_c$. It does not follow, however, that the temperature dependences of susceptibility, specific heat, soft-mode frequency, etc., are necessarily the same in the two theories and in general they are not. For short-range interactions $v_{ll'}$ the form of the thermodynamic critical singularities within the correlated effective-field theory, are essentially those of the spherical model (Stanley 1971 a) with, for example, $\chi(\mathbf{q}, 0) \sim (T - T_c)^{-2}$ when $T \to T_c^+$, $\mathbf{q} = \mathbf{q}_0$. For $v_{ll'}$ of dipolar form (which may well be relevant for many ferroelectric transitions $\mathbf{q}_0 = 0$) the critical predictions are closer to their mean-field equivalents, differing in general only by logarithmic factors (Lines 1972 a).

The important question which we set out to probe was 'how large is α?', particularly in that region close to T_c where the mean-field ($\alpha = 0$) findings are so often quoted. From eqn (2.2.31) we find

$$\alpha = \frac{\Sigma[v(\mathbf{q})/\{\tau + v(0) - v(\mathbf{q})\}]}{\Sigma[v(0)/\{\tau + v(0) - v(\mathbf{q})\}]} \tag{2.2.37}$$

where the sums are over the first Brillouin zone of the cell reciprocal lattice. Using eqn (2.2.24) as $T \to T_c$ (i.e. $\chi(0) \to \infty$), this simplifies to

$$\alpha_c = \frac{\Sigma[v(\mathbf{q})/\{v(0) - v(\mathbf{q})\}]}{\Sigma[v(0)/\{v(0) - v(\mathbf{q})\}]} \tag{2.2.38}$$

where α_c denotes the correlation parameter at the Curie point. Eqn (2.2.38) is rather remarkable in that it is completely independent of any details concerning the form of the local potential. It tells us that the $T = T_c$ correlation is simply a function of the character of the intercell forces and the symmetry of the lattice. For the case of equal nearest-neighbour cell forces $v_{ll'}$ only, α_c can easily be computed from eqn (2.2.38) for different lattice structures. It is, for example, equal to $0\cdot341$, $0\cdot282$, and $0\cdot256$ for the simple cubic, b.c.c., and f.c.c. lattices respectively. For lattices which favour stronger interactions in one or two dimensions than in others the value of α_c

increases markedly and in fact (for finite-range forces) approaches unity (fully correlated motion) in the limit of complete lower dimensionality. Less work has been performed with dipolar forces but somewhat surprisingly (Lines 1972 a) the value of α_c in the cubic lattices (at least) seems to be as large, if not larger, for dipolar $v(\mathbf{q})$ than for the equivalent nearest-neighbour force field. For example, Lines calculates $\alpha_c \approx 0 \cdot 29$ for the f.c.c. lattice with Ising dipolar intercell forces.

The primary inference is that the observation of a mean-field-like temperature dependence of dielectric properties near a ferroelectric Curie point does not necessarily imply a smallness of near-neighbour statistical fluctuations. On the contrary the quantitative errors incurred by making the mean-field approximation $\alpha = 0$ may well be larger for ferroelectrics than for an analogous situation with short-range forces. Indeed there is some experimental evidence (Lines 1970 a, b) that α may be quite close to unity in the region of the phase transition for ferroelectric $LiNbO_3$ and $LiTaO_3$ for example. The reason appears to be associated with the unusual orientational dependence of dipolar interactions for which even the sign of the interaction depends upon the direction of the vector joining the two dipoles. A force field with the dipolar radial dependence but no orientational dependence would approach mean-field static behaviour at all temperatures. A recent study by Friedman and Felsteiner (1974) using high-temperature-expansion statistical techniques has confirmed the importance of fluctuations in dipolar lattices and also supports the belief that these fluctuations may often be larger than those present in the equivalent nearest-neighbour-forces-only situation.

In summary the strength of the correlated effective-field approximation is that it is able to include at least the major effects of these fluctuations while maintaining the prime advantage of mean-field theory, viz. a tractable form for arbitrarily complex local potential $V(\xi_l)$ and bilinear interaction coefficients $v_{ll'}$. Thus in one and the same approximation one can probe both the quasi-harmonic displacive limit and also grossly anharmonic situations which, in the order–disorder limit of a deep double well $V(\xi_l)$, approach the Ising model itself. Its prime weakness is in the microscopic description of the equilibrium state. Here, as in mean-field theory, it reduces to a lattice of effectively non-interacting local modes with an Einstein-like (i.e. dispersionless) excitation spectrum, dispersion only arising as the response to a perturbing field. In this respect it is to be contrasted with the self-consistent phonon theory to be developed in the following section.

2.3. The quasi-harmonic limit and self-consistent phonons

In the first two sections of the present chapter we have approached the many-body problem by approximating the interaction terms in the model

Hamiltonian and making use of the local-field concept. In this picture the ferroelectric transition occurs upon lowering the temperature when the intercell forces, which are assumed to favour a lower-symmetry structure, overcome the combined effects of intracell forces and the entropy of disorder, both of which prefer the higher-symmetry phase.

An alternative approach to the statistical problem, as described in § 1.3, is to take advantage of the fact that the bilinear and quadratic terms in the translationally invariant Hamiltonian eqn (1.2.8) can be directly diagonalized in terms of running waves, i.e. phonons. Higher-order terms can then be included by a suitable approximation in the form of phonon–phonon interactions. In eqn (1.2.8) these higher-order terms arise solely from the anharmonicity of the local potential function $V(\xi_l)$. In a more general model, of course, one should also allow for intercell anharmonicity (i.e. for the presence of intercell terms of higher order than bilinear in the displacement variables), and indeed such terms are more readily accommodated in the interacting-phonon picture than in terms of local-field concepts. However, a practical restriction of the running-wave method is to motion which is not grossly anharmonic in character. Only for systems with a well-defined damped-harmonic response can a simple low-order interacting-phonon picture accurately describe the motion.

The most successful of the simple lowest-order renormalized phonon theories of structural phase transitions is referred to as the self-consistent phonon approximation (Boccara and Sarma 1965 a, b, Pytte and Feder 1969, Nettleton 1969, 1971 a, Gillis and Koehler 1971, 1972 a, b, 1974, Koehler and Gillis 1973, Pietrass 1972, Cohen and Einstein 1973). It has formally been derived by many techniques including variational methods and operator renormalization. We shall set out here that derivation which requires least formal knowledge of many-body methods and in which the physics of the approximation is quite transparent.

Consider the model Hamiltonian eqn (1.2.8) and let the local potential function $V(\xi_l)$ be of the form

$$V(\xi_l) = a_1 \xi_l^2 + a_2 \xi_l^4 \qquad (2.3.1)$$

with a_1 positive. The equation of motion for this Hamiltonian is

$$\ddot{\xi_l} = \sum_{l'} v_{ll'} \xi_{l'} - 2a_1 \xi_l - 4a_2 \xi_l^3 \qquad (2.3.2)$$

and is valid both classically and in quantum theory. We now write the displacement variable ξ_l as the sum of its ensemble (or thermal) average $\langle \xi \rangle_0$ and a contribution η_l representing dynamic deviations from $\langle \xi \rangle_0$, i.e.

$$\xi_l = \langle \xi \rangle_0 + \eta_l. \qquad (2.3.3)$$

The important point is that we are expressing the motion as a deviation from a temperature-dependent base which is to be determined self-consistently. Using eqns (2.3.2) and (2.3.3) the equation of motion for η_l is derived in the form

$$\ddot{\eta}_l = \sum_{l'} v_{ll'}(\langle \xi \rangle_0 + \eta_{l'}) - 2a_1(\langle \xi \rangle_0 + \eta_l)$$

$$- 4a_2(\langle \xi \rangle_0^3 + 3\langle \xi \rangle_0^2 \eta_l + 3\langle \xi \rangle_0 \eta_l^2 + \eta_l^3). \qquad (2.3.4)$$

Taking thermal averages on both sides and assuming quasi-harmonic motion about $\langle \xi \rangle_0$, such that $\langle \ddot{\eta}_l \rangle_0 = \langle \eta_l \rangle_0 = 0$, leads to an equation

$$4a_2\langle \xi \rangle_0^3 + \{2a_1 - v(0) + 12a_2\langle \eta^2 \rangle_0\}\langle \xi \rangle_0 = 0 \qquad (2.3.5)$$

for the static order parameter $\langle \xi \rangle_0$ in terms of mean-square fluctuations from equilibrium.

Subtracting eqn (2.3.5) from (2.3.4) leaves the motion

$$\ddot{\eta}_l = \sum_{l'} v_{ll'}\eta_{l'} - (2a_1 + 12a_2\langle \xi \rangle_0^2)\eta_l - 4a_2\eta_l^3 \qquad (2.3.6)$$

if we neglect the fluctuations in η_l^2, i.e. writing $\eta_l^2 - \langle \eta^2 \rangle_0 = 0$. Fourier transforming this equation of motion by introducing

$$\eta(\mathbf{q}) = (1/N)^{1/2} \sum_l \eta_l \exp(i\mathbf{q} \cdot \mathbf{l}) \qquad (2.3.7)$$

we find

$$\ddot{\eta}(\mathbf{q}) = \eta(\mathbf{q})\{v(\mathbf{q}) - 2a_1 - 12a_2\langle \xi \rangle_0^2 - 12a_2 N^{-1} \sum_{q'} \eta(\mathbf{q}')\eta(-\mathbf{q}')\} \qquad (2.3.8)$$

in which we have retained only those terms involving $\eta(\mathbf{q})$. We now make the basic statistical approximation, which is in the spirit of the random-phase approximation. We assume that in all phonon-interaction terms involving $\eta(\mathbf{q})$ all other pertinent phonon operators can be replaced by their thermal averages. In particular in eqn (2.3.8) we approximate the final term on the right-hand side as

$$-12a_2 N^{-1} \eta(\mathbf{q}) \sum_{q'} \langle \eta(\mathbf{q}')\eta(-\mathbf{q}') \rangle_0$$

or equivalently

$$-12a_2\langle \eta^2 \rangle_0 \eta(\mathbf{q}).$$

The equation of motion (eqn 2.3.8) is now in simple harmonic form

$$\ddot{\eta}(\mathbf{q}) = -\Omega^2(\mathbf{q})\eta(\mathbf{q}) \qquad (2.3.9)$$

with renormalized frequency

$$\Omega^2(\mathbf{q}) = 2a_1 + 12a_2(\langle\xi\rangle_0^2 + \langle\eta^2\rangle_0) - v(\mathbf{q}). \qquad (2.3.10)$$

Using the textbook statistical result for quantum harmonic oscillators

$$\langle\eta(\mathbf{q})\eta(-\mathbf{q})\rangle_0 = \frac{\hbar}{2\Omega(\mathbf{q})} \coth\left\{\frac{\hbar\Omega(\mathbf{q})}{2kT}\right\} \qquad (2.3.11)$$

the self-consistent formalism is finally closed by the equations

$$N\langle\eta^2\rangle_0 = \sum_{\mathbf{q}} \langle\eta(\mathbf{q})\eta(-\mathbf{q})\rangle_0$$

$$= \sum_{\mathbf{q}} \frac{\hbar}{2\Omega(\mathbf{q})} \coth\left\{\frac{\hbar\Omega(\mathbf{q})}{2kT}\right\} \qquad (2.3.12)$$

when eqns (2.3.5), (2.3.10), and (2.3.12) together determine the order parameter, mean-square displacement, and renormalized phonon frequencies self-consistently as functions of temperature.

The static susceptibility $\chi(\mathbf{q})$ is readily calculated using the fluctuation result (eqn (2.2.28)) in the form

$$kT\chi(\mathbf{q}) = \langle\eta(-\mathbf{q}):\eta(\mathbf{q})\rangle_0. \qquad (2.3.13)$$

For an effective Hamiltonian of renormalized harmonic form the colon product, as defined in eqn (2.2.21), is readily evaluated directly in the co-ordinate basis for which the energy is diagonal. The final result is simply

$$\chi(\mathbf{q}) = 1/\Omega^2(\mathbf{q}) \qquad (2.3.14)$$

and, in particular, the uniform static susceptibility is given by

$$\{\chi(0)\}^{-1} = 2a_1 - v(0) + 12a_2(\langle\xi\rangle_0^2 + \langle\eta^2\rangle_0) \qquad (2.3.15)$$

where we have used eqn (2.3.10).

Detailed numerical work using the self-consistent phonon approximation has been performed by Gillis and Koehler (1971, 1972 a) who consider an NaCl structure with a zone-centre soft mode. They include anharmonicity in quartic form and allow for both long- and short-range forces. They confirm, in particular, that, contrary to the widely held earlier view that soft-mode branches do not make an important contribution to their own renormalization, (see § 1.3), they do in fact contribute significantly both directly and possibly indirectly through a coupling to an acoustic branch of the same symmetry. This implies that a self-consistent description of the soft-mode renormalization is essential.

Perhaps the most surprising feature of Gillis and Koehler's results is the finding that the self-consistent phonon equations for the paraelectric phase never admit to a zero-frequency solution. This implies that within the theory the transition to the ordered phase always occurs discontinuously (i.e. a first-order transition). The general situation in terms of free energy is sketched in Fig. 2.5. Coupling to strain was not considered explicitly, but this would normally be expected to increase the sharpness of the transition. Pytte (1972 a) has attacked the finding as spurious and suggests that it results from a breakdown of the random-phase approximation for long-wavelength fluctuations when the latter become large. There is no question that in mean-field theory (which effectively cuts off the long-wavelength fluctuations) the quartic-anharmonicity model exhibits a second-order transition (Koehler and Gillis 1973). For this case the equivalent free energy is also sketched in Fig. 2.5 and can be contrasted there with the analogous curves from self-consistent phonon theory.

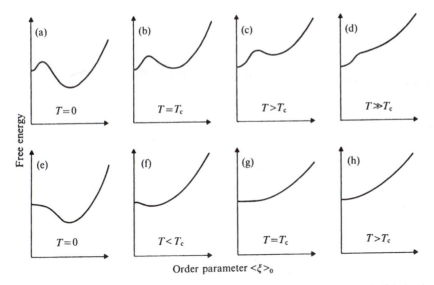

Fig. 2.5. Qualitative behaviour of the free energy as a function of order parameter $\langle \xi \rangle_0$ for the model Hamiltonian eqn (1.2.8) with quartic anharmonicity. Figs. (a) to (d) result from a self-consistent phonon approximation and Figs. (e) to (h) from a mean-field calculation. (After Gillis and Koehler 1972 a.)

The most recent studies, using higher-order statistical approximations and diagrammatic expansion techniques (Sokolov 1974, Sokolov and Vendik 1974), reveal that the order of the transition is a sensitive function of the order of the statistical approximation. They conclude that the true order of a phase transition cannot be reliably determined by *any* simple first-order approximate theory, all such theories breaking down very close to T_c, in which region more powerful techniques (see Chapter 11) are necessary to deal with fluctuation-dominated motion. Thus mean-field theories tend to underestimate the sharpness of phase transitions (e.g. to overestimate the exponent in $\langle \xi \rangle_0 \propto (T_c - T)^\beta$, $T \to T_c^-$) while self-consistent phonon schemes, albeit rather more dramatically, overestimate this sharpness.

Recent work by Benguigui (1975 a) indicates that both the self-consistent phonon and the correlated effective-field approximations are members of a common group of simple closed-form statistical theories which attempt to go beyond the mean field by approximating correlations in a 'self-consistent' fashion, and that as such they should have qualitatively similar properties at a phase transition, with spherical critical exponents on the disordered side and a first-order instability on the ordered side. Numerical work near T_c for the correlated effective-field theory has not yet been reported below T_c but an exact numerical computation for a large but finite lattice (Aviram *et al.* 1975) shows that the correlated effective-field estimate for T_c^+ is very close to the numerical instability (which is indicative of a transition region for the finite sample) for a quartic-anharmonic model with bilinear interactions. This would seem to indicate that the spurious first-order characteristics of the closed-form theories are much less severe in the correlated effective-field scheme than in the equivalent self-consistent phonon calculation.

There are no simple general rules governing the temperature dependences of susceptibility, soft mode, etc., in self-consistent phonon theory as there are (e.g. Curie–Weiss law, etc.) in mean-field theory (§ 2.1). Such details are model dependent, particularly as regards the range of the forces. Nevertheless one can establish some general relationships relating order parameter $\langle \xi \rangle_0$, static susceptibility $\chi(\mathbf{q})$, and soft-mode frequency $\Omega(0)$. Thus, for a zone-centre soft mode we have (Lines 1974)

$$\langle \xi \rangle_0^2 \sim [\chi(0)]^{-1}, \quad \text{ordered phase} \tag{2.3.16}$$

$$\Omega^2(\mathbf{q}) \sim [\chi(\mathbf{q})]^{-1}, \quad \text{both phases} \tag{2.3.17}$$

and hence

$$\langle \xi \rangle_0 \sim \Omega(0), \quad \text{ordered phase.} \tag{2.3.18}$$

There is a particularly close relationship between self-consistent phonon theory and a restricted form of the correlated effective-field approximation in the case of quasi-harmonic local potential. In fact the self-consistent phonon theory and a linearized form of the correlated effective-field method yield an identical classical response to external disturbances (Lines 1974). Even here, however, there is an essential difference between the theories which is most easily visualized in the absence of any disturbance. In equilibrium the correlated effective-field theory regards the crystal as an array of non-dynamically interacting cells with a discrete single-particle excitation spectrum. In the self-consistent-phonon picture phonons are already present as fluctuations in the equilibrium state.

Dynamic response in the simple random-phase phonon-decoupling scheme set out above is just that of an undamped simple harmonic oscillator with frequency $\Omega(\mathbf{q})$, viz.

$$\chi(\mathbf{q}, \omega) = \frac{1}{\Omega^2(\mathbf{q}) - \omega^2}. \tag{2.3.19}$$

Damping can be added phenomenologically, as in eqn (2.1.33), or can in principle be derived explicitly by including phonon–phonon interactions beyond the random phase, i.e. calculating in some approximate fashion those interactions between the *renormalized* phonons which provide a relaxation mechanism for these 'quasi-particles'.

2.4. The deep double-well limit and the Ising model

Our interest so far has centred on a description of lattice dynamics which can be cast in damped-harmonic form. It seems physically unlikely that such a description can be adequate for the grossly anharmonic motion possible in a double-well potential when the thermal energy is of the order of the well depth. Fortunately neither the mean-field theory nor the correlated effective-field approximation is restricted to a response of damped-harmonic form, and it is instructive to deduce the form of the dynamic susceptibility $\chi(\mathbf{q}, \omega)$ for a simple double-well local potential $V(\xi_l)$, e.g. eqn (2.3.1) with a_1 negative and a_2 positive.

In both the above-mentioned schemes the dynamic susceptibility is completely determined once the isolated cell response $\chi_s(\omega)$ is known (see eqns (2.1.30) and (2.2.36)). An extremely interesting *exact* classical calculation of $\chi_s(\omega)$ for the double-well potential (eqn (2.3.1)) with a_1 negative has been carried out by Onodera (1970). Measuring temperature kT in units of well depth $a_1^2/4a_2$ he finds curves for the real $\chi_s'(\omega)$ and imaginary $\chi_s''(\omega)$ parts of the dynamic susceptibility as shown in Fig. 2.6 (a). We note that the absorption spectrum develops a double maximum quite unlike any damped-harmonic behaviour when $kT \sim 0.5$.

Substituting these $\chi_s(\omega)$ into the mean-field (eqn 2.1.30), Onodera has also calculated the real $\chi'(\mathbf{q}, \omega)$ and imaginary $\chi''(\mathbf{q}, \omega)$ parts of uniform $(\mathbf{q} \to 0)$ dynamic susceptibility for mean-field ferroelectrically interacting oscillators at temperatures $T > T_c$. For cases where kT_c is less than but of order of the well depth the absorption spectrum again acquires a double-peaked form as $T \to T_c^+$ (Fig. 2.6(b)). For such a situation the mean-field or

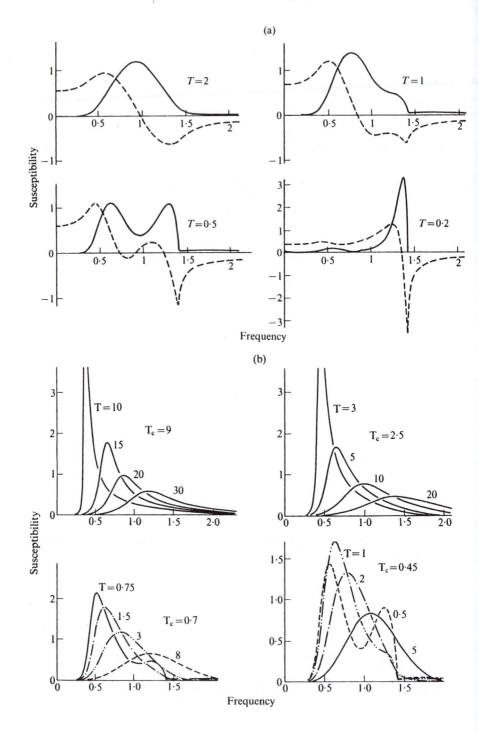

correlated effective-field methods must evidently be preferred to self-consistent phonon theory which, in its simplest form, approximates the absorption spectrum by a single delta function. If, however, kT_c is large compared with well depth, the system behaves very much as if it were of a conventional displacive type with a response of damped-harmonic-oscillator form and a soft-mode frequency proportional to $(T - T'_c)^{1/2}$, where T'_c is a little smaller than T_c. At the opposite extreme, where thermal energy kT_c is much smaller than the well depth (i.e. deep double wells), no soft-phonon modes exist in this theory since the peak in absorption spectrum now corresponds to ($\mathbf{q} = 0$) phonon oscillations localized within either well and these wells remain essentially unchanged (as far as motion about their minima is concerned) right through the phase transition. Although in real systems there are always thermal-hopping and quantum-tunnelling effects present, allowance for these (which will be discussed below and in § 2.5) does not alter this general conclusion so far as the hopping or tunnelling frequency is much smaller than the localized well frequency, and a high-frequency response peak will remain approximately unchanged right through the transition.

In the limit of a deep double well with negligible tunnelling field only two degenerate eigenstates per cell, one near the bottom of each well, enter the statistical problem and we approach an Ising situation with model Hamiltonian.

$$\mathcal{H} = -2 \sum_l \sum_{l'} v_{ll'} S_l^z S_{l'}^z - \sum_l S_l^z h_l(t) \tag{2.4.1}$$

where $h_l(t)$ is a time-dependent perturbing field and we have introduced a z-component spin-$\frac{1}{2}$ Pauli matrix

$$S^z = \begin{pmatrix} \frac{1}{2} & 0 \\ 0 & -\frac{1}{2} \end{pmatrix}. \tag{2.4.2}$$

The static theory of the Ising many-body Hamiltonian has been vigorously pursued for nearly half a century and has now reached a highly developed state (Domb and Green 1972). Indeed, for the case where $v_{ll'}$ takes a simple near-neighbour-only form, one- and two-dimensional models are exactly soluble, the latter being one of the rare many-body problems which is both exactly soluble and exhibits a phase transition. Even in three dimensions the

Fig. 2.6. (a) Real part (broken) and imaginary part (solid) of the dynamic susceptibility $\chi_s(\omega)$ of classical non-interacting oscillators with a double-well potential $V(\xi) = a_1 \xi^2 + a_2 \xi^4$, $a_1 < 0 < a_2$. Thermal energies are measured in units of well depth $a_1^2/4a_2$, susceptibilities in units of $1/|a_1|$, and frequency in units of $|2a_1|^{1/2}$, the natural harmonic frequency when $a_1 > 0$, $a_2 = 0$. (b) Imaginary part of the long-wavelength dynamic susceptibility $\chi(\mathbf{q} \to 0, \omega)$ for interacting classical double-well oscillators $V(\xi) = a_1 \xi^2 + a_2 \xi^4$, $a_1 < 0 < a_2$, in the mean-field approximation. Units as in (a). (After Onodera 1970.)

use of high- and low-temperature series expansion techniques has enabled the thermodynamic singularities near T_c to be quantitatively explored with great precision (see Chapter 11).

Valuable though this work is, particularly within the context of critical phenomena, theoretical complications increase rapidly for more complicated forms of interaction $v_{ll'}$, and Ising theory for general $v_{ll'}$ and general temperature is much less well developed. In particular, effective coupling terms of dipole–dipole form (which are thought to be dominant in many ferroelectric systems) are both long range and directionally dependent. Although a simple range dependence has been discussed in the Ising context (Dalton and Domb 1966) the full dipolar $v_{ll'}$ has not been explored to date except in relatively simple approximations and, more recently, in the critical limit $T \to T_c$. The former, as exemplified by the mean-field and correlated effective-field theories, are often tractable formalisms for intercell bilinear potential $v_{ll'}$ of arbitrary form. The mean-field Ising approximation is now standard textbook material (e.g. Pathria 1972) and the correlated effective-field equations of § 2.2 reduce in the Ising limit (Lines 1974) to another textbook statistical form, viz. the spherical model (Brout 1965, Stanley 1971 a).

Work on the dynamic Ising model has until recently lagged behind the achievements of static theory. The Hamiltonian eqn (2.4.1) does not contain any terms which produce transitions between the two orientations $S^z = \pm\frac{1}{2}$, and all dynamic effects are due to interactions with the surrounding thermal bath which produce thermal hopping and ensure equilibrium in the steady state. These interactions function only in giving rise to spontaneous spin flips by exchanging energy. The interactions $v_{ll'}$ between spins then affect the probability of transition for a given spin (Glauber 1963, Suzuki and Kubo 1968, Blinc and Zeks 1972). We shall discuss briefly the mean-field approach to this problem.

The flip probability of the lth spin in the instantaneous (time-dependent) molecular field

$$h_l^{\text{eff}}(t) = 4 \sum_{l'} v_{ll'} \langle S_{l'}^z \rangle + h_l(t) \tag{2.4.3}$$

will be denoted by $W_l(S_l^z)$. We assume that relaxation towards thermal equilibrium occurs according to an equation (the so-called master equation) of form

$$\frac{\mathrm{d}}{\mathrm{d}t} P(S_1^z, \ldots, S_N^z, t) = -\sum_l W_l(S_l^z) P(S_1^z, \ldots, S_N^z, t)$$

$$+ \sum_l W_l(-S_l^z) P(S_1^z, \ldots, -S_l^z, \ldots, S_N^z, t) \tag{2.4.4}$$

where $P(S_1^z, \ldots, S_N^z, t)$ is the probability of finding the spins in the configuration $(S_1^z, \ldots, S_N^z)$ at time t. The first term on the right-hand side is the loss of probability by flipping out and the second term a gain by flipping in the opposite direction.

Assuming that detailed balance exists in equilibrium we write

$$W_l(S_l^z)P_0(S_1^z, \ldots, S_N^z) = W_l(-S_l^z)P_0(S_1^z, \ldots, -S_l^z, \ldots, S_N^z)$$

$$(2.4.5)$$

where subscript zero signifies equilibrium. It follows that

$$\frac{W_l(S_l^z)}{W_l(-S_l^z)} = \frac{\exp(-S_l^z h_l^{\text{eff}}/kT)}{\exp(S_l^z h_l^{\text{eff}}/kT)} \qquad (2.4.6)$$

which can be written as

$$\frac{W_l(S_l^z)}{W_l(-S_l^z)} = \frac{1 - 2S_l^z \tanh(h_l^{\text{eff}}/2kT)}{1 + 2S_l^z \tanh(h_l^{\text{eff}}/2kT)} \qquad (2.4.7)$$

since S_l^z can take only the two values $\pm\frac{1}{2}$. The mean-field transition probability can therefore be written in the form

$$W_l(S_l^z) = (2\tau_0)^{-1}\left\{1 - 2S_l^z \tanh\left(\frac{h_l^{\text{eff}}}{2kT}\right)\right\} \qquad (2.4.8)$$

where τ_0 is an unknown (relaxation-time) parameter.

Defining the expectation value of the lth spin by

$$\langle S_l^z \rangle = \sum' S_l^z P(S_1^z, \ldots, S_N^z, t) \qquad (2.4.9)$$

where the primed sum is taken over all spin configurations, it follows (using successively eqns (2.4.4) and (2.4.8)) that

$$\frac{\mathrm{d}}{\mathrm{d}t}\langle S_l^z \rangle = -2\sum' S_l^z W_l(S_l^z) P(S_1^z, \ldots, S_N^z, t)$$

$$= -2\langle S_l^z W_l(S_l^z) \rangle$$

$$= \frac{-1}{\tau_0}\left\{\langle S_l^z \rangle - \frac{1}{2}\left\langle \tanh\left(\frac{h_l^{\text{eff}}}{2kT}\right)\right\rangle\right\}. \qquad (2.4.10)$$

In mean-field approximation this reduces to

$$\tau_0 \frac{\mathrm{d}}{\mathrm{d}t}\langle S_l^z \rangle = -\langle S^z \rangle_0 + \frac{1}{2}\tanh\left\{\frac{2v(0)\langle S^z \rangle_0}{kT}\right\} \qquad (2.4.11)$$

in the absence of perturbing applied field.

Let us now consider the response to a perturbing field $h_l(t)$ of form

$$h_l(t) = \delta h_l \exp(-i\omega t). \qquad (2.4.12)$$

Looking for a solution of form

$$\langle S_l^z \rangle = \langle S^z \rangle_0 + \delta\langle S_l^z \rangle \exp(-i\omega t) \qquad (2.4.13)$$

where $\langle S^z \rangle_0$ is the static order parameter, we substitute into eqn (2.4.10), using the molecular-field approximation (eqn (2.4.3)). Collecting in turn the terms of zeroth- and first-order of smallness, we find

$$\langle S^z \rangle_0 = \frac{1}{2} \tanh\left\{\frac{2v(0)\langle S^z \rangle_0}{kT}\right\} \qquad (2.4.14)$$

$$\delta\langle S_i^z \rangle (1 - i\omega\tau_0) = \tfrac{1}{2}(1 - 4\langle S^z \rangle_0^2)\left\{\frac{\delta h_l + \sum_{l'} 4v_{ll'}\delta\langle S_{l'}^z \rangle}{2kT}\right\}. \qquad (2.4.15)$$

Fourier transforming with respect to the lattice we find a frequency response $\chi(\mathbf{q}, \omega) = \delta\langle S^z(\mathbf{q}) \rangle / \delta h(\mathbf{q})$ which can be written as

$$\chi(\mathbf{q}, \omega) = \frac{\chi(\mathbf{q}, 0)}{1 - i\omega\tau(\mathbf{q})} \qquad (2.4.16)$$

where

$$\tau(\mathbf{q}) = \frac{\tau_0 kT}{kT - v(\mathbf{q})(1 - 4\langle S^z \rangle_0^2)}. \qquad (2.4.17)$$

The resulting dielectric constant $\varepsilon(\mathbf{q}, \omega)\varepsilon_0$ is then related to the dynamic susceptibility by the equation

$$\varepsilon(\mathbf{q}, \omega) - 1 = 4\mu^2 N\chi(\mathbf{q}, \omega) \qquad (2.4.18)$$

where ε_0 is the dielectric constant of free space, μ is the magnitude of the (two-position) Ising electric dipole moment, and N is the number of dipoles per unit volume.

Thus, the dielectric constant and resulting spectrum of Ising fluctuations are of a simple Debye form. The relaxation time $\tau(\mathbf{q})$, in contrast with the single-spin relaxation time τ_0 (which is expected to be but weakly dependent on temperature), exhibits a critical slowing down on approaching the transition temperature. If wavevector $\mathbf{q}$ has its critical value $\mathbf{q}_0$ (i.e. $\mathbf{q}_0 = 0$ for ferroelectrics) we find the mean-field relationships

$$\tau(\mathbf{q}_0) = \frac{\tau_0 T}{T - T_c}, \qquad T \to T_c^+ \qquad (2.4.19)$$

$$\tau(\mathbf{q}_0) = \frac{\tau_0 T}{T - T_c(1 - 4\langle S^z \rangle_0^2)}, \qquad T \to T_c^- \qquad (2.4.20)$$

where $T \to T_c^{\pm}$ signifies an approach to the critical temperature T_c from the disordered $(+)$ or ordered $(-)$ phase, and where $kT_c = v(0)$ from (2.4.14).

We note in particular that the pole of the mean-field response (eqn (2.4.16)) is on the imaginary axis and that therefore the soft-mode eigenfrequency $\omega(\mathbf{q}_0) = -i/\tau(\mathbf{q}_0)$ is pure imaginary in this case and approaches zero, as $T \to T_c$, along the imaginary axis. Equivalent results can be obtained from the general mean-field discussion of § 2.1 if a single-cell Debye response

$$\chi_s(\omega) = \frac{\chi_s(0)}{1 - i\omega\tau_0} \qquad (2.4.21)$$

is chosen in place of the damped-harmonic formalism (eqn (2.1.17)).

The parameter τ_0, which is an unspecified 'isolated-spin' relaxation time in the above Ising theory, takes on a more specific character in a physical system with finite well depth E. If we ignore the possibility of quantum tunnelling the inverse relaxation time is evidently related to the thermal-hopping probability for crossing a barrier of height E at temperature T. According to Mason (1947), who first attacked the ferroelectric order–disorder relaxation problem in an attempt to understand the dielectric behaviour of Rochelle salt, it čan be approximately expressed as

$$\tau_0 = 2\pi\frac{\hbar}{kT}\exp\left(\frac{E}{kT}\right). \qquad (2.4.22)$$

For a real order–disorder situation with deep double-well local potential the Ising approximation contains another basic shortcoming. It ignores the fact that cell oscillations within the wells are of finite frequency (corresponding to a lattice phonon) and produce a characteristic phonon response peak at higher frequencies which is not contained in eqn (2.4.21). This phonon peak in the delta-function limit of the Ising approximation has been formally pushed to infinite frequency. The physical system can more properly be considered to have not one, but two, characteristic frequencies belonging to the soft mode: one is given by the quasi-harmonic frequency within either side of the double well, and the other is given by the inverse relaxation (or tunnelling) time characterizing interwell motion. Correspondingly two peaks (with the same symmetry and eigenvector) may be expected in the dielectric response and both, of course, contribute to the dielectric function. In a sense therefore we have an 'extra' mode, i.e. one not predicted by group-theoretical calculations. There is nothing mysterious about this; the reason only one peak in the response function is predicted by group theory for each normal mode is that *infinitesimal* motion about a static configuration of specified lattice symmetry is assumed. In order–disorder systems near T_c very large *finite* displacements (i.e. interwell motion) must be considered as well. Although for the case where the phonon and relaxation frequencies are well separated the usual phonon and Debye responses can be added in a simple fashion (Andrade *et al.* 1974), the situation becomes less transparent as the well depth decreases and the well width increases. The two separate responses then merge and a simple algebraic formalism representing the more complex situation has yet to be developed, although Onodera's approach becomes qualitatively sound outside the low-frequency region.

2.5. Pseudo-spin formalism and the tunnel mode

To this point we have neglected any possibility of quantum tunnelling between double-well minima in our theoretical discussion of order–disorder ferroelectrics. In some instances particularly when the order–disorder motion involves primarily heavy atoms, this neglect is valid and the Ising-like theory of the previous section is appropriate when the double well is deep enough. On the other hand, for light ions, and particularly for protons in hydrogen-bonded materials, tunnelling probability can be quite large and its neglect very serious. In this section we shall discuss the manner in which quantum-tunnelling probability can be incorporated into a theoretical description of double-well ferroelectricity.

For a very deep double-well there are only two degenerate states, one near the bottom of each well, which need be considered to a good approximation. However, in real systems this degeneracy is in fact always lifted, at least in principle, by the quantum-mechanical tunnelling between the minima. The tunnelling creates an energy splitting, Δ say, which goes to zero as the tunnelling probability goes to zero. Associated with this splitting is a tunnelling frequency

$$\Omega = \Delta/\hbar \tag{2.5.1}$$

which (other things being equal) tends to increase as the wells become shallower (Blinc 1960, de Gennes 1963). In addition to the ground states there are, of course, higher-energy vibrational states, but for moderately deep wells the energy separation between ground and higher levels is large compared with the splitting of the ground doublet and it is a good approximation to neglect all except the lowest states (Fig. 2.7).

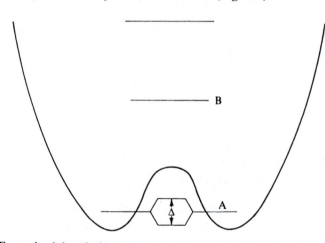

Fig. 2.7. Energy levels in a double well showing the case of relatively small tunnelling in the lowest level A and a large separation from the excited vibration levels B.

Since the spin-$\frac{1}{2}$ Pauli matrices

$$S^x = \tfrac{1}{2}\begin{pmatrix} 0 & 1 \\ 1 & 0 \end{pmatrix}, \qquad S^y = \tfrac{1}{2}\begin{pmatrix} 0 & -i \\ i & 0 \end{pmatrix}, \qquad S^z = \tfrac{1}{2}\begin{pmatrix} 1 & 0 \\ 0 & -1 \end{pmatrix} \qquad (2.5.2)$$

and the unit matrix together form a complete set in the space of Hermitian 2×2 matrices, it follows that all two-level systems can be described in terms of a fictitious spin-$\frac{1}{2}$ operator. Thus in the doublet-only approximation for the double-well situation we can associate a fictitious spin-$\frac{1}{2}$ $\mathbf{S}_l$ with each cell l.

Consider specifically the model Hamiltonian eqn (1.2.8) with a symmetric double-well potential $V(\xi_l)$. There are two 'stable' configurations $\xi_l = \pm\xi_0$, and above T_c the system is as likely to be found in one as the other. For proton tunnelling ξ_0 may refer simply to the co-ordinate of the proton along the bond direction. However, the definition of ξ_0 is more general than this and these 'stable' configurations in general involve *all* ions in the unit cell (via eqn (1.2.5)) although, of course, in any specific example some atoms may be displaced considerably farther than others.

Let us associate the orthogonalized wavefunctions $|l, L\rangle$ and $|l, R\rangle$ with configurations localized in the 'left' and 'right' equilibrium sites respectively of the lth cell. Neglecting the possibility of any *inter*cell tunnelling we write the product wavefunction

$$\Psi = |1, \alpha_1\rangle|2, \alpha_2\rangle \ldots |N, \alpha_N\rangle \qquad (2.5.3)$$

where α may take the values L or R, to describe the macroscopic system of N cells. Taking matrix elements of the model Hamiltonian eqn (1.2.8) between these states and dropping constant terms we obtain a representation

$$\mathcal{H} = -\Delta \sum_l S_l^x - \tfrac{1}{2}\sum_l \sum_{l'} J_{ll'} S_l^z S_{l'}^z \qquad (2.5.4)$$

in which

$$\Delta = -2\langle l, L|\tfrac{1}{2}\pi_l^2 + V(\xi_l)|l, R\rangle \qquad (2.5.5)$$

and

$$J_{ll'} = 4v_{ll'}\langle l, R|\xi_l|l, R\rangle\langle l', R|\xi_{l'}|l', R\rangle \qquad (2.5.6)$$

and where we have used the symmetry restrictions

$$\langle l, R|\xi_l|l, R\rangle = -\langle l, L|\xi_l|l, L\rangle \qquad (2.5.7)$$

$$\langle l, L|\tfrac{1}{2}\pi_l^2 + V(\xi_l)|l, R\rangle = \langle l, R|\tfrac{1}{2}\pi_l^2 + V(\xi_l)|l, L\rangle \qquad (2.5.8)$$

$$\langle l, L|\tfrac{1}{2}\pi_l^2 + V(\xi_l)|l, L\rangle = \langle l, R|\tfrac{1}{2}\pi_l^2 + V(\xi_l)|l, R\rangle \qquad (2.5.9)$$

and neglected all overlap terms except (2.5.5).

For wells of other shapes and for cases where additional overlap terms are deemed to be important it is possible to devise other pseudo-spin formulations. Indeed, even a harmonic $V(\xi_l)$ can be forced into spin form (Elliott 1971) at the expense of going to an infinite spin quantum number. Other generalizations are possible for the case of a degenerate soft-mode situation. For example, in a cubic phase a threefold degenerate mode might correspond to displacements along the three cubic axes. It would involve six wells and a six-fold degenerate ground state of a single cell potential which would again be split by tunnelling. Such a situation could be described by three independent spin-$\frac{1}{2}$ operators at each cell site. Other embellishments of the simple form (eqn (2.5.4)) are required if a coupling to other phonon modes is judged to be significant. These might result, for example, from a lattice motion modulating the distance between the two equilibrium sites and hence the tunnel term. Kobayashi (1968) has suggested that such a pseudo-spin phonon coupling is indeed very important in KH_2PO_4.

In spite of all these possible additional features it does appear that the Ising model with transverse field, which has a Hamiltonian of form (eqn (2.5.4)), may be used to represent a simplified model of an order–disorder structural transition with tunnelling. Indeed, even a coupled spin–phonon Hamiltonian can often be canonically transformed back to a transverse Ising form with modified parameters (Moore and Williams 1972). Let us therefore examine its statistical properties, particularly in the context of a ferroelectric transition ($\mathbf{q}_0 = 0$). Formal extension to cover more general structural phase transitions with $\mathbf{q}_0 \neq 0$ is quite straightforward.

2.5.1. Static mean field theory

In the mean-field approximation (see § 2.1) we concentrate attention on one particular cell, replacing all others by their thermally averaged states. Thus for the lth cell we write the effective Hamiltonian

$$\mathcal{H}_l^{\text{eff}} = -\Delta S_l^x - J(0)S_l^z \langle S^z \rangle_0 \qquad (2.5.10)$$

where $J(0) = \sum_{l'} J_{ll'}$. The molecular field therefore forms a vector $\mathbf{h}^{\text{eff}}$ with components

$$h_x^{\text{eff}} = \Delta, \qquad h_y^{\text{eff}} = 0, \qquad h_z^{\text{eff}} = J(0)\langle S^z \rangle_0. \qquad (2.5.11)$$

Eigenvalues and eigenfunctions of eqn (2.5.10) are readily obtained in the two-dimensional spin-space of $\mathbf{S}_l$ and using simple Boltzmann statistical mechanics we can calculate the ensemble averages $\langle \dots \rangle_0$ self-consistently to obtain the main features of the transition (Fig. 2.8).

Above the transition temperature T_c we find $\langle S^z \rangle_0 = 0$, $\langle S^y \rangle_0 = 0$, and

$$\langle S^x \rangle_0 = \tfrac{1}{2} \tanh(\Delta/2kT) \qquad (2.5.12)$$

representing a paraelectric phase with no spontaneous polarization (since

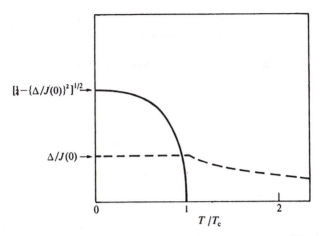

Fig. 2.8. Qualitative temperature dependence of pseudo-spin components $\langle S^z \rangle_0$ (solid curve) and $\langle S^x \rangle_0$ (broken curve) in the mean-field theory of the transverse Ising model. For ferroelectrics $\langle S^z \rangle_0$ is proportional to the spontaneous electric dipole moment. (After Brout *et al.* 1966.)

the z-component of pseudo-spin measures the co-ordinate dipole-moment operator).

Below T_c the ordering is given by the self-consistency condition

$$h^{\text{eff}}/J(0) = \tfrac{1}{2} \tanh(h^{\text{eff}}/2kT) \tag{2.5.13}$$

in which

$$h^{\text{eff}} = [\Delta^2 + \{J(0)\langle S^z \rangle_0\}^2]^{1/2} \tag{2.5.14}$$

is the magnitude of the molecular field. The spontaneous polarization $(\propto \langle S^z \rangle_0)$ has a typical Brillouin form falling from a value

$$\langle S^z \rangle_0 = \left[\frac{1}{4} - \left\{ \frac{\Delta}{J(0)} \right\}^2 \right]^{1/2} \tag{2.5.15}$$

at $T = 0$ to zero at $T = T_c$. In the ordered phase $\langle S^x \rangle_0$ is independent of temperature with value

$$\langle S^x \rangle_0 = \Delta/J(0) \tag{2.5.16}$$

and T_c obtained from (2.5.13) as $\langle S^z \rangle_0 \to 0$, is given by

$$\Delta/J(0) = \tfrac{1}{2} \tanh(\Delta/2kT_c). \tag{2.5.17}$$

This equation gives a curve for T_c *versus* Δ which falls from the Ising mean-field value $kT_c = J(0)/4$ when $\Delta = 0$ to $T_c = 0$ when $\Delta = \tfrac{1}{2}J(0)$. If $\Delta > \tfrac{1}{2}J(0)$ no ordering occurs at any temperature.

In the disordered phase the static susceptibility per spin in the direction z is

$$\{\chi(\mathbf{q})\}^{-1} = 2\Delta \coth(\Delta/2kT) - J(\mathbf{q}) \qquad (2.5.18)$$

leading to a divergence of uniform response of Curie–Weiss form

$$\chi(0) = \text{const}/(T - T_c), \qquad T \to T_c^+. \qquad (2.5.19)$$

It should be noted that eqn (2.5.17) predicts a significant isotope effect in T_c. As the effective mass of the reorientable dipole increases the tunnelling frequency decreases and T_c is shifted to higher temperatures. Such effects are indeed observed in hydrogen-bonded ferroelectrics on deuteration.

2.5.2. Dynamic mean-field theory

To investigate the motion of the pseudo-spins around the mean-field equilibrium we can study the response to a small time- and space-dependent perturbing field (in the direction z) in the spirit of the linear response theory set out in § 2.1.2. The mean-field Hamiltonian in the presence of the applied field is

$$\mathcal{H}_l^{\text{eff}} = -\Delta S_l^x - \sum_{l'} J_{ll'} S_l^z \langle S_{l'}^z \rangle - h_l(t) S_l^z \qquad (2.5.20)$$

where, neglecting spontaneous fluctuations, the ensemble average takes the form

$$\langle S_{l'}^z \rangle = \langle S^z \rangle_0 + \delta \langle S_{l'}^z \rangle \exp(-i\omega t) \qquad (2.5.21)$$

made up of a static part $\langle S^z \rangle_0$ and a linear response part $\delta \langle S_{l'}^z \rangle$ when the field $h_l(t)$ is

$$h_l(t) = \delta h_l \exp(-i\omega t). \qquad (2.5.22)$$

In analogy with eqn (2.5.21) we define the x and y response

$$\langle S_l^x \rangle = \langle S^x \rangle_0 + \delta \langle S_l^x \rangle \exp(-i\omega t) \qquad (2.5.23)$$

$$\langle S_l^y \rangle = \delta \langle S_l^y \rangle \exp(-i\omega t) \qquad (2.5.24)$$

where we have noted from the static findings that $\langle S^y \rangle_0 = 0$ at all temperatures.

The mean-field equations of motion for these expectation values are

$$\frac{d}{dt} \langle \mathbf{S}_l \rangle = \frac{-i}{\hbar} \langle [\mathbf{S}_l, \mathcal{H}_l^{\text{eff}}] \rangle \qquad (2.5.25)$$

and, upon use of eqns (2.5.20) to (2.5.24) and the retention of only terms linear in the deviations $\delta \langle \mathbf{S}_l \rangle$ and δh_l, they can be cast in the explicit form

$$i\omega \hbar \delta \langle S_l^x \rangle + \sum_{l'} J_{ll'} \langle S^z \rangle_0 \, \delta \langle S_l^y \rangle = 0 \qquad (2.5.26)$$

$$i\omega \hbar \delta \langle S_l^y \rangle + \Delta \delta \langle S_l^z \rangle - \delta h_l (\langle S^x \rangle_0$$

$$-\sum_{l'} J_{ll'} (\langle S^z \rangle_0 \, \delta \langle S_{l'}^x \rangle + \langle S^x \rangle_0 \, \delta \langle S_{l'}^z \rangle) = 0 \qquad (2.5.27)$$

$$i\omega \hbar \, \delta \langle S_l^z \rangle - \Delta \delta \langle S_l^y \rangle = 0. \qquad (2.5.28)$$

Fourier transforming with respect to the lattice, this homogeneous system of three linear equations has a non-trivial solution for $\delta h_l = 0$ only if the secular determinant

$$(2.5.29) \quad \begin{vmatrix} i\omega \hbar & J(0) \langle S^z \rangle_0 & 0 \\ -J(0) \langle S^z \rangle_0 & i\omega \hbar & \Delta - J(\mathbf{q}) \langle S^x \rangle_0 \\ 0 & -\Delta & i\omega \hbar \end{vmatrix}$$

is identically equal to zero. The three solutions are (Brout *et al.* 1966, Tokunaga and Matsubara 1966, Novakovic 1966)

$$\hbar \Omega_1(\mathbf{q}) = 0 \qquad (2.5.30)$$

$$\hbar^2 \Omega_2^2(\mathbf{q}) = \hbar^2 \Omega_3^2(\mathbf{q})$$

$$= \Delta^2 - \Delta J(\mathbf{q}) \langle S^x \rangle_0 + \{J(0) \langle S^z \rangle_0\}^2. \qquad (2.5.31)$$

The excitations can be considered as spin waves. The solution $\Omega_1(\mathbf{q})$ corresponds to the longitudinal mode with motion in the direction of the molecular field h^{eff}. The other two modes represent transverse excitations or precession of the quasi-spins around the molecular field. For $T > T_c$ the field h^{eff} is perpendicular to the direction $\langle S^z \rangle_0$ of polarization and only the transverse modes contribute to the polarization fluctuations. In the ordered phase this is no longer the case and all three modes contribute. Indeed, in the limit $\Delta \to 0$ of zero tunnelling probability the eigenvectors of the transverse modes have no components parallel to $\langle S^z \rangle_0$ so that the polarization fluctuations in the ordered phase for this limit are given in the random-phase approximation by a delta function at $\Omega_1 = 0$.

In the paraelectric phase at very high temperatures we have $\langle S^z \rangle_0 = 0$, $\langle S^x \rangle_0 \to 0$, and the transverse speudo-spins behave as free particles with $\hbar \Omega_{2,3}(\mathbf{q}) \to \Delta$. At lower temperatures the second term in eqn (2.5.31) produces some dispersion and the zone-centre excitations eventually approach zero (for a ferroelectric) when (using eqn (2.5.12))

$$2\Delta = J(0) \tanh(\Delta/2kT) \qquad (2.5.32)$$

i.e. at the stability limit $T = T_c$ of eqn (2.5.17). Again one readily shows that the square of the frequency of the soft pseudo-spin wave is, as in the mean-field soft-phonon case of § (2.1), a linear function of $T - T_c$ as $T \to T_c^+$.

Below the transition temperature, using eqns (2.5.16) and (2.5.31), we find

$$\hbar^2 \Omega_{2,3}^2(\mathbf{q}) = \{J(0)\langle S^z \rangle_0\}^2 + \Delta^2 \left\{1 - \frac{J(\mathbf{q})}{J(0)}\right\} \qquad (2.5.33)$$

and note that the soft-mode frequency $\Omega_{2,3}(0) = J(0)\langle S^z \rangle_0$ is directly proportional to the spontaneous polarization and therefore its square goes to zero linearly as $T_c - T$ when $T \to T_c^-$.

Finally we note the results for dynamic susceptibility

$$\chi(\mathbf{q}, \omega) = \delta \langle S^z(\mathbf{q}) \rangle / \delta h(\mathbf{q})$$

which are

$$\chi(\mathbf{q}, \omega) = \frac{\frac{1}{2}\Delta \tanh(\Delta/2kT)}{\Omega_{2,3}^2(\mathbf{q}) - \omega^2} \qquad (2.5.34)$$

and

$$\chi(\mathbf{q}, \omega) = \frac{\Delta^2}{J(0)\{\Omega_{2,3}^2(\mathbf{q}) - \omega^2\}} \qquad (2.5.35)$$

for temperatures above and below T_c respectively. For $\omega \to 0$ eqn (2.5.34) goes over smoothly to the static mean-field susceptibility as calculated from the static statistical theory. This is not so, however, for eqn (2.5.35) because the contribution of the Ω_1 longitudinal mode is not properly included in the dynamic mean-field scheme and is important below T_c.

In general the mean-field approximation is seen to give an excellent qualitative picture of a pseudo-spin phase transition. Nevertheless the approximation is deficient in that it neglects damping and static spin correlations as well as pseudo-spin motion in the direction of the molecular field. In an improved theory damping and longitudinal motion can be added phenomenologically or can be calculated directly by carrying the statistical calculation beyond the stage where such terms appear. Some of these developments of the basic pseudo-spin theory will be discussed further when we consider some specific examples of order–disorder ferroelectrics in Chapter 9.

3

MACROSCOPICS AND PHENOMENOLOGY

3.1. The elastic dielectric

PHENOMENOLOGICAL or macroscopic theories of dielectrics (and ferroelectrics in particular) treat the material in question as a continuum without regard to any underlying atomic structure. In general, fields are postulated in sufficient number to describe the thermal, elastic, and dielectric properties of the macroscopic system, and the laws of thermodynamics and classical mechanics are used to obtain the relationships between them.

It is usually assumed possible to describe a dielectric system by three independent variables, chosen from the pairs (temperature, entropy), (stress, strain), and (Maxwell field, polarization). In the basic microscopic model discussed in Chapter 2 terms representing coupling between the order parameter (e.g. polarization) and strain have been neglected altogether. This is mathematically convenient, of course, but restricts the findings to situations for which such coupling is small. Unfortunately in dielectrics the opposite is more often the case.

Although it is by no means impossible to represent this coupling microscopically, the formalism becomes increasingly complicated and there are few detailed statistical calculations of this type in the ferroelectric literature to date. To add to the problem a statistical calculation of the relationship between a response (say, static susceptibility) for one set of thermodynamic constraints and the same response for a different set requires, in general, a calculation of the relevant thermodynamic potentials (i.e. free energies). This statistically is an integral calculation which can often be performed only numerically. As a result phenomenology and thermodynamics still have a particularly important part to play in the theory of ferroelectricity, much more, for example, than in magnetism, where interactions with strain are generally much smaller and the model Hamiltonians simpler.

Phenomenological theory for ferroelectricity goes back to Mueller (1940 a, b), Ginzburg (1945, 1949), and Devonshire (1949, 1951, 1954). An excellent and quite detailed account of the macroscopic theory of homogeneous dielectrics (e.g. the interaction of elastic, dielectric, and thermal properties of such materials), with specific reference to the field of ferroelectricity, has been given in the monograph by Grindlay (1970) upon which much of this chapter is based.

3.1.1. Strain

Suppose a point in a continuum body is referred to a rectangular Cartesian co-ordinate system R_i ($i = 1, 2, 3$). If the body is deformed such that $R_i \to r_i$, the resulting transformation is described by three functions $r_i = r_i(R_j)$ with $i, j = 1, 2, 3$. An elastic displacement vector can now be defined as $u_i = r_i - R_i$. Consider the line element dR_i at position R_i. After deformation it becomes dr_i at r_i. From the definition of elastic displacement u_i we have

$$dr_i = dR_i + \sum_{j=1}^{3} \frac{\partial u_i}{\partial R_j} dR_j. \tag{3.1.1}$$

The nine derivatives $\partial u_i / \partial R_j$ are assumed to be continuous and a deformation is termed *uniform* if they remain constant throughout the body.

On squaring both sides of eqn (3.1.1) and summing over i we obtain

$$|d\mathbf{r}|^2 - |d\mathbf{R}|^2 = 2 \sum_i \sum_j x_{ij} \, dR_i \, dR_j \tag{3.1.2}$$

where we have assumed an infinitesimal deformation $\partial u_i / \partial R_j$ and where

$$x_{ij} = \frac{1}{2}\left(\frac{\partial u_i}{\partial R_j} + \frac{\partial u_j}{\partial R_i} \right) \tag{3.1.3}$$

is a symmetric tensor of the second rank called the *elastic strain* tensor. The diagonal components are referred to as *normal* and the off-diagonal ones as *shear*. By virtue of its symmetric property the strain tensor is completely defined in general by only six variables, often written as

$$x_1 = x_{11}, \qquad x_2 = x_{22}, \qquad x_3 = x_{33}$$
$$x_4 = x_{23} + x_{32}, \qquad x_5 = x_{31} + x_{13}, \qquad x_6 = x_{12} + x_{21} \tag{3.1.4}$$

a notation due to Voigt. In general these tensor elements are functions of position R_i, but for uniform strains they are constant. For the latter case, in physical terms, x_{ii} is the fractional elongation along the R_i axis and $x_{ij} + x_{ji}$ is the decrease in angle between the R_i and R_j axes upon distortion.

3.1.2. Stress

Consider an arbitrary volume of continuum dielectric V (surface A) subject to a net force $\mathbf{F}$. This force in general can be expressed as the sum of a body force part of density $\mathbf{f}$ and a surface or stress force part of density $\mathbf{s}$, such that

$$\mathbf{F} = \int_V \mathbf{f} \, dV + \oint_V \mathbf{s} \, dA \tag{3.1.5}$$

where $\oint$ denotes a surface integral. The infinitesimal force $\mathbf{s}dA$ on the surface element dA is in general a function of both the position $\mathbf{R}$ and the orientation of dA. Its component normal to dA is called the normal stress and its component perpendicular to this normal the shear stress. We write

$$s_i = \sum_j X_{ji}(\mathbf{R})n_j \qquad (i, j = 1, 2, 3) \qquad (3.1.6)$$

where $\mathbf{n}$ is the unit vector along the outward normal to the surface at $\mathbf{R}$.

Since $\mathbf{s}$ and $\mathbf{n}$ are vectors, X_{ji} is a second-rank tensor (nine components) and is called the *stress tensor*. If the representation (eqn (3.1.5)) is complete, then the stress tensor can be proved to be symmetric ($X_{ij} = X_{ji}$) as the result of a constraint imposed by the conservation of linear and angular momentum (Grindlay 1970). Physically X_{ii} is the normal stress acting on a plane perpendicular to the R_i axis and is, with the sign convention adopted above, positive for tension and negative for compression; X_{ji} is a shear stress along the R_j axis in the plane normal to the R_i axis.

The surface integral in eqn (3.1.5) can be transformed by Gauss's theorem to an equivalent volume integral when we find

$$F_i = \int_V \left(f_i + \sum_j \frac{\partial x_{ji}}{\partial R_j} \right) dV. \qquad (3.1.7)$$

Thus the effect of a stress tensor can be replaced in the force equation by an equivalent body force of density $\sum_j (\partial X_{ji}/\partial R_j)$ and the use of two fields seems to be a redundant notation. Although this is not quite so, and the inequivalence of body force and stress can be demonstrated in their contribution to work, it is indeed customary to put $f = 0$ in the standard treatments of the linear dielectric.

3.1.3. Mechanical work and electrostatic work

Neglecting body forces the contribution of stress force to mechanical work comprises the total and can be written

$$dW_M = \oint_V \mathbf{s} \cdot d\mathbf{u} \, dA = \sum_i \sum_j \oint_V X_{ji} n_j \, du_i dA \qquad (3.1.8)$$

where $d\mathbf{u}$ is an infinitesimal displacement of elementary volume dV at position $\mathbf{R}$. Using Gauss's theorem this becomes

$$dW_M = \int_V \sum_i \sum_j \left\{ \frac{\partial X_{ji}}{\partial R_j} du_i + X_{ji} \frac{\partial}{\partial R_j} du_i \right\} dV. \qquad (3.1.9)$$

For cases where X_{ji} is symmetric and independent of $\mathbf{R}$ we find, using eqn (3.1.3), the simple result

$$dW_M = \int_V \sum_i \sum_j X_{ji} \, dx_{ij} \, dV \qquad (3.1.10)$$

where, under the influence of stress, the strain has been changed from x_{ij} to $x_{ij} + dx_{ij}$. Since $X_{ji} = X_{ij}$ and $x_{ji} = x_{ij}$ this reduces in general to the sum of six terms. Defining

$$X_1 = X_{11}, \qquad X_2 = X_{22}, \qquad X_3 = X_{33}$$
$$X_4 = X_{23} = X_{32}, \qquad X_5 = X_{31} = X_{13}, \qquad X_6 = X_{12} = X_{21}$$

$$(3.1.11)$$

(compare eqn (3.1.4)) enables us to write eqn (3.1.10) in the simpler notation

$$dW_M = \int_V \sum_i X_i \, dx_i \, dV \qquad (i = 1, \ldots, 6). \qquad (3.1.12)$$

From here on, unless otherwise specified, we shall use the Voigt notation (eqn (3.1.4)), eqn (3.1.11) for components of stress and strain. Also as a final simplification we shall adopt the implied summation convention for repeated indices throughout the following thermodynamic calculations. With this convention eqn (3.1.12) becomes

$$dW_M = \int_V X_i \, dx_i \, dV. \qquad (3.1.13)$$

Electrostatic work is defined in all standard texts on electricity and magnetism and can be expressed in terms of the Maxwell electric field $\mathbf{E}$ and electric displacement $\mathbf{D}$ as

$$dW_E = \int_V \mathbf{E} \cdot d\mathbf{D} \, dV. \qquad (3.1.14)$$

It follows that the total work done when the dielectric V is subjected to an infinitesimal change of strain and electric displacement in the presence of uniform stress and electric field is

$$dW = dW_M + dW_E = \int_V (X_i \, dx_i + E_i \, dD_i) \, dV. \qquad (3.1.15)$$

In much of the ferroelectric literature (e.g. Jona and Shirane 1962, Fatuzzo and Merz 1967) it is customary to study the polarization $\mathbf{P}$ explicitly rather than the electric displacement $\mathbf{D}$. Since $\mathbf{D} = \varepsilon_0 \mathbf{E} + \mathbf{P}$, where ε_0 is the dielectric constant of free space, one can naturally define a thermodynamic formalism in terms of the conjugate pair of electric variables $\mathbf{P}$, $\mathbf{E}$ rather than $\mathbf{D}$, $\mathbf{E}$, but we feel that the retention of $\mathbf{D}$ is important for the following reasons. Firstly it is impossible to impose the experimental constraint of constant polarization for changes of state involving ferroelectric systems. It therefore makes little sense to define material compliances (see § 3.2)

requiring this constraint. Secondly, although a set of thermodynamic potentials analogous to those in eqn (3.2.4) are readily defined which generate
equations of state in terms of electric variables **P** and **E**, they do *not* possess
the external properties normally associated with free energies in connection
with questions of stability although the contrary is often assumed. The
reason is that the variable D_i enters the work function (eqn (3.1.15)) and, as
a result, enters the laws of thermodynamics in a simpler and more natural
fashion that does P_i.

3.2. Thermodynamics

Let us assume that the thermal, elastic, and dielectric behaviour of a
homogeneous dielectric is fully described by the six fields temperature T,
entropy S, stress **X**, strain **x**, electric field **E**, and displacement **D**. Since stress
and strain are determined by six variables each and **E** and **D** by three
variables each, this gives a total of 20 thermodynamic co-ordinates to
describe the system.

According to the first law of thermodynamics the change in internal
energy U (per unit volume) when an infinitesimal quantity of heat dQ is
received by a unit volume of dielectric is given by

$$dU = dQ + dW \qquad (3.2.1)$$

where dW is the work done on this same volume (by electric and mechanical
forces) during the resulting quasi-static transformation. Assuming reversibility the second law of thermodynamics relates dQ to the absolute temperature and entropy in the form $dQ = T dS$. Using this and the expression (eqn
(3.1.15)) for dW we can rewrite the first law (3.2.1) as

$$dU = T \, dS + X_i \, dx_i + E_i \, dD_i \qquad (3.2.2)$$

where we have assumed uniform strain and electric displacement increments
over the unit volume. Although U can be expressed in terms of any three
independent variables it is evident from eqn (3.2.2) that the exact differential takes its simplest form when we choose S, **x**, and **D**, and we shall refer to
these as the principal variables for internal energy. With this set it is possible
to calculate T, **X**, and **E** directly as

$$T = \left(\frac{\partial U}{\partial S}\right)_{\mathbf{x},\mathbf{D}}, \qquad X_i = \left(\frac{\partial U}{\partial x_i}\right)_{S,\mathbf{D}}, \qquad E_i = \left(\frac{\partial U}{\partial D_i}\right)_{S,\mathbf{x}} \qquad (3.2.3)$$

These three equations describe respectively the calorimetric, elastic, and
dielectric equations of state.

When the conditions of a specific problem or measurement suggest a
different set of independent variables it is convenient, though not essential,
to define additional thermodynamic functions (or potentials) in terms of
which the equations of state can be stated in an equally concise manner.

Since three independent variables can be chosen in eight different ways from among the conjugate pairs (T, S), (X_i, x_i), and (E_i, D_i), this implies eight thermodynamic potentials. They are, in addition to U,

Helmholtz free energy	$A = U - TS$
enthalpy	$H = U - X_i x_i - E_i D_i$
elastic enthalpy	$H_1 = U - X_i x_i$
electric enthalpy	$H_2 = U - E_i D_i$
Gibbs free energy	$G = U - TS - X_i x_i - E_i D_i$
elastic Gibbs energy	$G_1 = U - TS - X_i x_i$
electric Gibbs energy	$G_2 = U - TS - E_i D_i$ (3.2.4)

and each can be used to generate the equations of state, no additional information being contained in one representation rather than another. The differential forms describing infinitesimal changes in these thermodynamic potentials are, using eqn (3.2.2),

$$dA = -S\,dT + X_i\,dx_i + E_i\,dD_i$$
$$dH = T\,dS - x_i\,dX_i - D_i\,dE_i$$
$$dH_1 = T\,dS - x_i\,dX_i + E_i\,dD_i$$
$$dH_2 = T\,dS + X_i\,dx_i - D_i\,dE_i$$
$$dG = -S\,dT - x_i\,dX_i - D_i\,dE_i$$
$$dG_1 = -S\,dT - x_i\,dX_i + E_i\,dD_i$$
$$dG_2 = -S\,dT + X_i\,dx_i - D_i\,dE_i \qquad (3.2.5)$$

and explicitly indicate the three principal variables for each potential.

Since each thermodynamic potential contains a complete description of the dielectric of interest we may deduce a large number of identities relating derivatives. In fact for each potential (eqn (3.2.4)) we can deduce 27 so-called *Maxwell relations*. For example, from the internal-energy equations of state (eqn 3.2.3) we have

$$\left(\frac{\partial T}{\partial x_i}\right)_{D,S} = \left(\frac{\partial X_i}{\partial S}\right)_{D,x} \qquad i = 1, \ldots, 6 \qquad (3.2.6)$$

$$\left(\frac{\partial T}{\partial D_i}\right)_{x,S} = \left(\frac{\partial E_i}{\partial S}\right)_{x,D} \qquad i = 1, 2, 3 \qquad (3.2.7)$$

$$\left(\frac{\partial X_i}{\partial D_j}\right)_{S,x} = \left(\frac{\partial E_j}{\partial x_i}\right)_{S,D} \qquad i = 1, \ldots, 6; \quad j = 1, 2, 3. \qquad (3.2.8)$$

Thermodynamic identities of this type are a direct consequence of the fact that the variables T, S, X_i, x_i, E_i, D_i are not independent. They can be extremely useful in practice either as a means of checking the internal consistency of two sets of measurements, or in deducing the properties of a system subject to a set of constraints which may be difficult or inconvenient to produce experimentally.

From eqn (3.2.5) we see that the various free energies are, in equilibrium, stationary functions with respect to virtual displacements of the unconstrained variables. From the second law of thermodynamics one can in fact establish that these free-energy functions possess an even more important property for *non*-equilibrium configurations, this time concerning the sign of their change during the various irreversible processes which eventually bring the system to thermodynamic equilibrium. It is that when a particular set of independent variables is held constant the approach to equilibrium minimizes that free energy for which the constrained variables are the principal ones. Thus, for example, in the common experimental situation for static measurements with T, X_i, and E_i fixed, non-equilibrium processes bring about an approach to equilibrium which minimizes the Gibbs free energy G with respect to all its homogeneous and possibly inhomogeneous unconstrained variables. The values of these components which minimize G then define the equilibrium values of the unconstrained variables in terms of T, X_i, and E_i. Finally, using the latter relationships, the minimized free energy (i.e. equilibrium free energy) is readily expressed in terms of its principal variables alone. It can of course happen that two or more states are locally stable for the same set of constraints. When this occurs the state with the least value of G is called *absolutely stable* and the others *metastable*.

The potential G (or whatever function is appropriate for the constraints) is in equilibrium expressible as a function of those constraints. If for a given set of constraints more than one locally stable state exists (say two, $G = G_A$ and $G = G_B$ for simplicity), with G_A lowest and hence absolutely stable, then it may happen that as the constraints are varied $G_B - G_A$ changes sign, whereupon G_B becomes absolutely stable and the system undergoes a phase transition. This transition occurs when $G_A(T, X_i, E_i) = G_B(T, X_i, E_i)$ and this equation therefore determines the transition temperature, i.e. $T = T_c(X_i, E_i)$. The function $T = T_c(X_i, E_i)$ defines a surface in T, X_i, E_i space separating the two phases G_1 and G_2. This surface is called the *phase boundary* and its intersections with different planes are called *phase diagrams*. If the phase boundary in a phase diagram terminates at a point, we refer to a *critical point*. It is conventional to define an *order* of the transition as equal to that order of the partial derivatives of $G_A - G_B$ with respect to its arguments for which non-zero values first occur.

3.2.1. Linear equations of state

In macroscopic theories of ferroelectricity it is often convenient to take the displacement D_i, strain x_i, and temperature T as independent variables. The appropriate thermodynamic potential is then the Helmholtz free energy A, and the equations of state (see eqn (3.2.5))

$$-S = \left(\frac{\partial A}{\partial T}\right)_{x,D}, \qquad X_i = \left(\frac{\partial A}{\partial x_i}\right)_{T,D}, \qquad E_i = \left(\frac{\partial A}{\partial D_i}\right)_{T,x}. \qquad (3.2.9)$$

Consider the linear differential form of these equations, viz.

$$dS = \left(\frac{\partial S}{\partial T}\right)_{D,x} dT + \left(\frac{\partial S}{\partial x_j}\right)_{D,T} dx_j + \left(\frac{\partial S}{\partial D_j}\right)_{x,T} dD_j \qquad (3.2.10)$$

$$dX_i = \left(\frac{\partial X_i}{\partial T}\right)_{D,x} dT + \left(\frac{\partial X_i}{\partial x_j}\right)_{D,T} dx_j + \left(\frac{\partial X_i}{\partial D_j}\right)_{x,T} dD_j \qquad (3.2.11)$$

$$dE_i = \left(\frac{\partial E_i}{\partial T}\right)_{D,x} dT + \left(\frac{\partial E_i}{\partial x_j}\right)_{D,T} dx_j + \left(\frac{\partial E_i}{\partial D_j}\right)_{x,T} dD_j. \qquad (3.2.12)$$

The coefficients in these eqns (3.2.10) to (3.2.12) are called *compliances* and are functions of the unperturbed independent variables. They provide a measure of the coupling between fields. Together they form a square array of 100 compliance matrix elements as generated by the potential A. Because of the Maxwell relations only 55 are independent. Other compliance matrices can be formed using the other thermodynamic potentials, but they are not in general independent and a great number of thermodynamic identities can be obtained relating the compliances of one compliance matrix with those of another. Not all compliances have been specifically named, but among the more important are those relating stress to strain (elastic), displacement to electric field (dielectric constant or permittivity), stress or strain to displacement or electric field (piezoelectric), displacement to temperature (pyroelectric), and strain to temperature (thermal expansion).

   With this large number of compliances it is not surprising, though perhaps unfortunate, that no generally agreed symbolism has developed. However, in the case of isothermal (or adiabatic) elastic and dielectric equations there is some consensus, and we present these equations explicitly below with conventional symbols. Consider for example the isothermal form of eqn (3.2.11) and (3.2.12). It can be written

$$dX_i = c_{ij}^{D,T} dx_j - h_{ij}^{T\dagger} dD_j \qquad (3.2.13)$$

$$dE_i = -h_{ij}^{T} dx_j + \kappa_{ij}^{x,T} dD_j \qquad (3.2.14)$$

in which $h_{ij}^\dagger \equiv h_{ji}$

$$c_{ij}^{D,T} = \left(\frac{\partial X_i}{\partial x_j}\right)_{D,T} = \left(\frac{\partial^2 A}{\partial x_i \partial x_j}\right)_{D,T} \qquad (3.2.15)$$

is the isothermal elastic stiffness at constant displacement,

$$\kappa_{ij}^{x,T} = \left(\frac{\partial E_i}{\partial D_j}\right)_{x,T} = \left(\frac{\partial^2 A}{\partial D_i \partial D_j}\right)_{x,T} \qquad (3.2.16)$$

is the inverse isothermal permittivity at constant strain, and

$$h_{ij}^{T\dagger} = -\left(\frac{\partial X_i}{\partial D_j}\right)_{x,T} = -\left(\frac{\partial E_j}{\partial x_i}\right)_{D,T} = -\left(\frac{\partial^2 A}{\partial x_i \,\partial D_j}\right)_{T} \qquad (3.2.17)$$

is an isothermal linear piezoelectric compliance. For a corresponding adiabatic process the same pair of equations can be derived from the internal-energy potential U, but where the entropy is now held fixed in the various derivatives and the symbols $c_{ij}^{D,S}$, $\kappa_{ij}^{x,S}$, and h_{ij}^S are now used to indicate adiabatic compliances.

In addition to eqns (3.2.13) and (3.2.14), six other linear equations can be derived (either isothermally or adiabatically) by use of the other thermodynamic potentials. Omitting the S or T superscripts they are

$$dx_i = s_{ij}^D \, dX_j + g_{ij}^\dagger \, dD_j \qquad (3.2.18)$$

$$dE_i = -g_{ij} \, dX_j + \kappa_{ij}^X \, dD_j \qquad (3.2.19)$$

$$dx_i = s_{ij}^E \, dX_j + d_{ij}^\dagger \, dE_j \qquad (3.2.20)$$

$$dD_i = d_{ij} \, dX_j + \varepsilon_{ij}^X \, dE_j \qquad (3.2.21)$$

$$dX_i = c_{ij}^E \, dx_j - e_{ij}^\dagger \, dE_j \qquad (3.2.22)$$

$$dD_i = e_{ij} \, dx_j + \varepsilon_{ij}^x \, dE_j \qquad (3.2.23)$$

and together with eqn (3.2.13) and (3.2.14) they make up the elastic and dielectric linear equations of state. The dielectric permittivity ε_{ij} and its reciprocal κ_{ij} are second-rank tensors (we shall often treat them as dimensionless in which case they should more properly be written as $\varepsilon_{ij}\varepsilon_o$ and $\kappa_{ij}\varepsilon_0^{-1}$ in the equations of state), the piezoelectric compliances $d_{ij}, e_{ij}, h_{ij}, g_{ij}$ are of third rank, and the elastic stiffness c_{ij} and elastic compliance s_{ij} of fourth rank. Crystal symmetry imposes restrictions on the form of the tensors and in many cases the full pattern of symmetries imposed by each of the 32 point groups has been worked out (Nye 1964) (see Appendix E).

In dielectric theory, for cases where displacement D_i is small (e.g. non-polar phases or ferroelectrics close to a Curie temperature), it is common practice to define a linear or low-order non-linear theory in which $dD_i = D_i$ and $dx_i = x_i$ are defined as small deviations from a prototype equilibrium state in which $D_i = x_i = 0$. It follows that the compliances, being functions of the unperturbed independent variables, now depend solely on temperature. Expressing the free energy A as a Taylor series in D_i and x_i up to quadratic terms gives

$$A = A_0 + \frac{\partial A}{\partial x_i} x_i + \frac{\partial A}{\partial D_i} D_i + \frac{1}{2} \frac{\partial^2 A}{\partial x_i \partial x_j} x_i x_j$$

$$+ \frac{1}{2} \frac{\partial^2 A}{\partial D_i \partial D_j} D_i D_j + \frac{\partial^2 A}{\partial x_i \partial D_j} x_i D_j \qquad (3.2.24)$$

where A_0 is a function of temperature alone and the derivatives are taken in the state $D_i = 0$, $x_i = 0$. If the stress and electric field are zero in this prototype state then we find from the eqn (3.2.5) dA equation that the linear terms in eqn (3.2.24) are zero. If not, the linear coefficients are referred to as the thermal stress $X_i^0 = \partial A / \partial x_i$ and the thermal electric field $E_i^0 = \partial A / \partial D_i$. They are, respectively, the values of stress and electric field necessary to achieve the condition $D_i = x_i = 0$ at the temperature of interest. The linear elastic and dielectric equations of state (eqns (3.2.13) and (3.2.14)) become, in this scheme,

$$X_i - X_i^0 = c_{ij}^{D,T} x_j - h_{ij}^{T\dagger} D_j \qquad (3.2.25)$$

$$E_i - E_i^0 = -h_{ij}^T x_j + \kappa_{ij}^{x,T} D_j \qquad (3.2.26)$$

with the compliances now functions of temperature alone. In particular, putting applied stress and electric field equal to zero allows a determination of the temperature dependence of spontaneous *polarization* (since $\mathbf{D} = \mathbf{P}$ when $\mathbf{E} = 0$) and strain in the linear approximation. Their temperature derivatives are respectively the pyroelectric coefficients and thermal-expansion coefficients at zero stress and electric field. Finally, noting that entropy $S = -(\partial A / \partial T)_{x,D}$ from eqn (3.2.5), the corresponding calorimetric equation of state follows from a direct differentiation of Helmholtz energy (eqn (3.2.24)) with respect to temperature (A_0 and the Taylor expansion coefficients being temperature dependent). In a consistently linear theory the differential form of this equation should be terminated at terms linear in the temperature deviation.

As mentioned above the demands of material symmetry in general impose conditions on the elements of the compliance tensors. In particular some elements may be required to be zero and others to be interrelated. Three subclassifications of point groups are of special interest. Firstly there are the *centrosymmetric* groups: these are characterized by a centre of symmetry and can therefore possess no polar properties. Eleven of the 32, point groups are centrosymmetric. In such crystals a uniform stress causes displacement of charges which are relatively compensating about the centre of symmetry, and an applied electric field produces a strain (electrostriction) which is not changed when the field is reversed.

Of the non-centrosymmetric groups all except one (the cubic group 432) allow for at least one non-vanishing component of a piezoelectric compliance. These are the 20 *piezoelectric* groups which allow a crystal to exhibit electric polarity when subject to stress. The effect is linear and a reversal of the sign of the stress reverses the induced polarity. In similar fashion a strain is produced by the application of an electric field which is linearly proportional to that field in lowest order.

Finally of the piezoelectric groups 10 are characterized by the fact that they have a unique polar axis. These are termed pyroelectric or *polar* and are spontaneously polarized, i.e. have a non-zero thermal electric field E_i^0. The 32 crystallographic point groups with their international and Schoenflies symbols and their centrosymmetric, piezoelectric, or polar classifications are shown in Appendix A. For strongly polar crystals it is usually experimentally more convenient to express compliances with respect to a spontaneously polarized 'equilibrium' unperturbed state. However, care is required since the spontaneous polarization itself can now induce indirect piezoelectric effects (e.g. since the spontaneous polarization is a dipole moment per unit volume, a volume change from a simple compression gives a piezoelectric-like effect). The literature on these effects has been confused and conflicting, but it now appears (Nelson and Lax 1976) that a *unique* piezoelectric tensor can be defined in the polar base system (in the sense that the *same* piezoelectric tensor enters all the differential equations and boundary conditions of electrostatic and electrodynamic theory) provided only that the spontaneous electric field has been cancelled by extrinsic surface charge. In the absence of charge compensation Nelson and Lax find that new elastic and stress tensor components appear which alter the symmetries of the (charge-compensated) conventional tensors.

3.2.2. Non-linear relations

To this point we have discussed only linear differential equations of state. Some of the most important characteristics of ferroelectrics (e.g. hysteresis loops) and even conventional dielectrics (e.g. electrostriction) are fundamentally non-linear and hence require an extension of the theory to higher orders. In principle there is no difficulty in extending differential forms such as eqns (3.2.10) to (3.2.12) to arbitrarily high orders in order to define non-linear compliances. The practical difficulty concerns the increasingly high order of the tensorial nature of this non-linear characterization. In fact it is generally prohibitively difficult to proceed in a fully consistent manner to higher orders, and in practice physical insight is nearly always used to single out the most significant non-linearities in a particular context.

Phenomenological theories tend to lean heavily on free-energy expansions of the form eqn (3.2.24) about the prototype state $D_i = x_i = 0$. Possible terms of third order are

$$D_iD_jD_k(3), \qquad D_iD_jx_k(4), \qquad D_ix_jx_k(5), \qquad x_ix_jx_k(6) \qquad (3.2.27)$$

the figures in parentheses indicating that the coefficients of such terms in the Taylor expansion are tensors of rank 3, 4, 5 and 6, respectively. In fourth order we find terms

$$D_iD_jD_kD_l(4), \qquad D_iD_jD_kx_l(5), \qquad D_iD_jx_kx_l(6)$$
$$D_ix_jx_kx_l(7), \qquad x_ix_jx_kx_l(8). \qquad (3.2.28)$$

For a given material all the associated coefficients are tensor invariants under the operations of the appropriate symmetry group. It is generally quite difficult to investigate the symmetry restrictions imposed on tensors of high rank, and simplification is often accomplished in non-linear theory by making use of a physical approximation. A common assumption is that strains are small to the extent that only terms $x_i \Pi_j D_j$ and $x_i x_j$ involving them need be retained (where Π_j indicates a product). In addition it is normal to assume that the same free-energy representation describes both the polar and non-polar phases. For crystals with centrosymmetric non-polar phases this implies that all the odd-rank tensor coefficients are zero.

In a situation where all these conditions apply the free-energy expansion is reduced to

$$A = A_0 + \tfrac{1}{2}\kappa_{ij}^{x,T}D_iD_j + \tfrac{1}{2}c_{ij}^{D,T}x_ix_j + \tfrac{1}{2}q_{ijk}^T D_iD_jx_k$$
$$+ \tfrac{1}{4}\xi_{ijkl}^{x,T}D_iD_jD_kD_l + X_i^0 x_i \qquad (3.2.29)$$

where the coefficients q_{ijk}^T are termed electrostrictive constants. The elastic non-linear equation of state resulting from (3.2.29) is

$$X_i - X_i^0 = c_{ij}^{D,T}x_j + \tfrac{1}{2}q_{ijk}^T D_jD_k. \qquad (3.2.30)$$

If strain is measured from the paraelectric equilibrium configuration in the absence of stress, then we can put $X_i^0 = 0$. It follows that spontaneous strain x_i is of the same order of smallness as D_i^2. This justifies the expansion (3.2.29) as written, the terms retained being consistently of lower order of smallness than those neglected.

For a crystal with a piezoelectric non-polar phase the problem can be a great deal more complicated. Again one usually appeals to physical insight. Thus, for example, high-order terms in the free-energy expansion are required only for those variables which exhibit a markedly non-linear response. In particular, if the polar phase has axial symmetry with spontaneous polarization along the axis, then only low-order terms need be retained for displacement normal to the axis. In such ways the number of higher-order terms required for an adequate phenomenological description can be considerably decreased.

3.3. Phenomenological theory for ferroelectrics

The inclusion of non-linear terms in the thermodynamic equations of state allows for a description of possible divergencies in response. With the independent variables used to describe the simple elastic dielectric this divergence can be with respect to elastic response (ferroelasticity) or to dielectric response (ferroelectricity). In this section we shall probe specifically the latter. Linear coupling to other properties (e.g. elastic, thermal, other dielectric, or even magnetic) will, of course, produce associated anomalies in other responses but, at least in principle, one can usually examine the theoretical consequences of removing such couplings and separate the 'driving' from the 'driven' characteristics.

Thus in this section we shall concern ourselves with ferroelectrically driven response anomalies. Most simply we shall examine the divergence of dielectric susceptibility itself (i.e. the ferroelectric phase transition). In thermodynamic theory one stops short of inquiring as to the origin of the driving mechanism, but merely identifies the primary pair of conjugate variables (D_i and E_i for ferroelectrics) and defines lowest-order non-linearities in the equation of state involving them. Results are conventionally obtained using the assumption that the relevant equation of state can be represented by a polynomial and that the same polynomial can be used above and below the phase transition anomaly. The limitations imposed by these basic assumptions, particularly as regards critical phenomena, are discussed in Chapter 11.

An equivalent and more usual starting point is a polynomial thermodynamic potential expressed in terms of deviations from the prototype state $D_i = x_i = 0$. Normally the most convenient potential is the elastic Gibbs function G_1 expressed as a function of temperature, stress, and displacement. The simplest conceivable theoretical situation results with the assumption that $D_i = D$ is directed along one of the crystallographic axes only (i.e. that spontaneous polarization occurs along this direction and the applied field is restricted to this direction also), that all stresses are zero, and that the non-polar phase is centrosymmetric. None of these restrictions is essential, but with them the free energy G_1 can be expressed in the very simple polynomial form

$$G_1 = (\alpha/2)D^2 + (\gamma/4)D^4 + (\delta/6)D^6 \qquad (3.3.1)$$

where energy is measured from the non-polar phase and we terminate the polynomial rather arbitrarily at D^6 for mathematical simplicity. The coefficients α, γ, δ are in general temperature dependent, but simple examples of first- and second-order ferroelectric transitions can be described assuming γ and δ independent of temperature. With this assumption the order of the transition depends on the sign of γ (δ being necessarily positive in eqn (3.3.1) for stability reasons as $D \to \infty$).

3.3.1. Second order ferroelectric phase transitions

Differentiating the elastic Gibbs function (eqn (3.3.1)) with respect to D at constant temperature generates the dielectric equation of state

$$E = \alpha D + \gamma D^3 + \delta D^5 \qquad (3.3.2)$$

where E is the Maxwell field (parallel to D). With parameters γ and δ. positive the curves (eqn (3.3.1)) for G_1 versus D are qualitatively sketched in Fig. 3.1 (a). When parameter α is also positive the free-energy curve has a

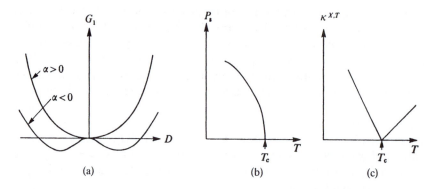

Fig. 3.1. Qualitative temperature dependence of the free energy versus displacement curves, and of the spontaneous polarization (P_s) and reciprocal isothermal permittivity ($\kappa^{X,T}$) near a second-order ferroelectric transition. The free-energy parameter α is proportional to $T - T_c$.

single minimum at $D = 0$, but when α is negative the curve acquires a double-minimum form with minima at non-zero values of displacement. Since $E = (\partial G_1/\partial D)_T$ the extrema (or more accurately the minima) describe the equilibrium values of zero-field displacement (i.e. spontaneous polarization P_s) as a function of temperature.

It is clear that P_s undergoes a continuous (second-order) phase transition as α passes through zero, but from eqn (3.3.2) we see that α is in fact the reciprocal permittivity at zero field in the non-polar phase. Conventional phenomenological theory (often termed Devonshire theory, after Devonshire (1949, 1951, 1954)) assumes that near the Curie point T_c the parameter α depends on temperature in the linear fashion

$$\alpha = \beta(T - T_c) = \kappa^{X, T > T_c} \qquad (3.3.3)$$

with β a positive constant. This is the form most often observed experimentally for reciprocal permittivity and is also that which results from simple mean-field statistical models as $T \to T_c$ (see eqn (2.1.9) where the zero-field susceptibility χ of Chapter 2 is related to the dimensionless dielectric constant ε by $\varepsilon = 1 + \chi$). It is now a simple exercise to calculate spontaneous polarization and isothermal dielectric constant *below* T_c. We find respectively

$$P_s^2 = \beta(T_c - T)/\gamma, \qquad P_s \to 0 \qquad (3.3.4)$$

and

$$\kappa^{X,T} = \beta(T - T_c) + 3\gamma P_s^2, \qquad P_s \to 0. \qquad (3.3.5)$$

Substituting eqn (3.3.4) into eqn (3.3.5) gives

$$\kappa^{X,T} = 2\beta(T_c - T), \qquad T < T_c. \qquad (3.3.6)$$

These temperature dependences are qualitatively sketched in Fig. 3.1 (b) and (c). We note in particular that the slope of the reciprocal permittivity $\kappa^{X,T}$ *versus* temperature curve below T_c is negative and twice as large as above T_c.

In like manner the dielectric equation of state (eqn (3.3.2)) can be sketched through the phase-transition region (Fig. 3.2). Below T_c the curve

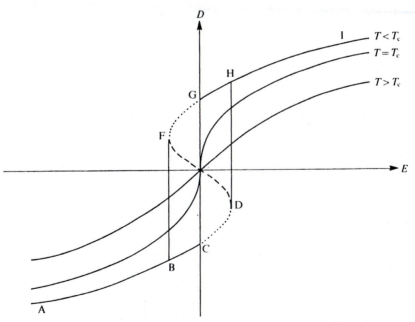

Fig. 3.2. Qualitative sketch of the dielectric equation of state (3.3.2) near T_c. Broken curves represent regions of instability $(\partial E/\partial D)_T < 0$ and dotted curves regions of metastability.

contains regions of negative permittivity (broken in the figure) which correspond to unstable states, and regions (dotted in the figure) which correspond to metastable states. The question of metastability and absolute stability for those states with positive permittivity is settled by evaluating the relevant thermodynamic potential in the states in question. Since we are here concerned with a change of displacement at constant field, temperature, and (zero) stress, the relevant potential is the Gibbs function $G = G_1 - ED$. Using eqns (3.3.1) and (3.3.2) we can express G explicitly as a function of Maxwell field E and temperature. The qualitative form can be obtained by inspection and for $T < T_c$ is shown in Fig. 3.3, where the lettered

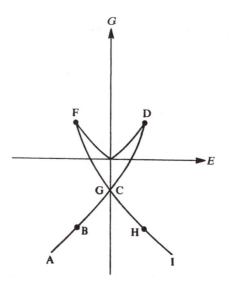

Fig. 3.3. Qualitative sketch of the Gibbs free energy G as a function of field E for temperatures below T_c. The lettered points correspond to their equivalents in Fig. 3.2.

points correspond to their equivalents in Fig. 3.2. We see that the absolutely stable branches as a function of E are ABC and GHI. If the system is restricted to equilibrium states we expect jumps D to H and F to B as field E is raised or lowered (Fig. 3.2), defining a hysteresis loop BCDHGF and a coercive field $\frac{1}{2}(E_D - E_B)$. However, the present treatment has ignored the possibility of non-uniform states (e.g. domains, space-charge effects, etc.), and in real ferroelectrics polarization reversal takes place by nucleation and growth of new domains (see Chapter 4) in a complex and time-dependent manner to the extent that it is often not possible to define a coercive field in any general sense at all. It is even possible that some of the absolutely stable states of the uniformly polarized system are in fact metastable relative to non-uniform states.

Entropy and specific heat can be discussed within the phenomenological framework by noting from eqn (3.2.5) that $S = -(\partial G_1/\partial T)_{\mathbf{D},\mathbf{X}}$. If coefficients γ and δ are temperature independent and $\alpha = \beta(T - T_c)$ according to eqn (3.3.3), then G_1 of (3.3.1) differentiates directly to give a zero-field entropy

$$S = -\tfrac{1}{2}\beta P_s^2. \tag{3.3.7}$$

However, the spontaneous polarization P_s is given by eqn (3.3.4) below T_c and is zero above T_c. It follows that close to T_c the dielectric contributions to entropy are

$$S = \tfrac{1}{2}\beta^2(T - T_c)/\gamma, \qquad T < T_c \tag{3.3.8}$$

$$S = 0, \qquad T > T_c \tag{3.3.9}$$

The corresponding contributions $T\,dS/dT$ to zero-field specific heat are

$$c^{X,P_s} = \tfrac{1}{2}\beta^2 T/\gamma, \qquad T < T_c \tag{3.3.10}$$

$$c^X = 0. \qquad T > T_c. \tag{3.3.11}$$

We note that S is continuous at T_c but that c is discontinuous with a jump $\Delta c = \tfrac{1}{2}\beta^2 T_c/\gamma$.

The continuous nature of the first derivatives of the free energy (e.g. P_s and S) and the discontinuous nature of second derivatives (e.g. c) conforms with the normal definition of a second-order phase transition. Clearly the phenomenological picture is at least qualitatively able to describe transitions of this nature. It is also evident that the formalism can be extended to cover a wider range of temperatures and there is little difficulty in adapting it for use with different external constraints (see Fatuzzo and Merz 1967, Huang and Grindlay 1970). General indications are that the theory is fairly self-consistent close to T_c and parameters β, γ, δ have been deduced by a comparison of theory with experiment for a number of actual ferroelectrics. However, the breakdown of the simple theory in the form of a need for temperature-dependent parameters and higher-order powers of D in eqn (3.3.1) becomes increasingly evident as $|T - T_c|$ increases. In addition, evidence is beginning to accumulate which points to a breakdown when T approaches *extremely* close to T_c. We shall explore this evidence in Chapter 11, where we shall also identify the Devonshire theory of ferroelectricity as a particular example of a wider class of phenomenological theories (known as Landau theories) which necessarily become invalid in real systems as a critical point (e.g. a second-point phase transition) is approached arbitrarily closely. The basic problem is that an expansion of the thermodynamic potential about its value at the critical point is usually not valid because a second-order transition point must be some singular point of the thermodynamic potential. In fact this critical region of Landau breakdown is extremely small in feroelectrics (though not necessarily in antiferroelectrics) and is quite difficult to observe reproducibly. In Chapter 11 we shall see that this is an indication of the long-range nature of the forces responsible for the ferroelectric transition (probably electric dipolar forces) and is perhaps the best evidence yet available for the dominance of long-range interactions close to a ferroelectric singularily.

On the other hand most ferroelectric phase transitions are not of second order but first, with a discontinuity in the first derivatives of the thermo-dynamic potential. Now a first-order transition point is not a singularity but merely a point at which the thermodynamic potentials of the two phases are equal. On either side of the transition each branch of the potential corres-ponds to an ordinary equilibrium state, and at the transition the stable state jumps from one to the other. For this situation we do not anticipate Landau breakdown unless the first-order discontinuities are small (i.e. the phase transition is 'nearly second order').

3.3.2. First-order ferroelectric phase transitions

Consider the free energy (eqn (3.3.1)) for γ negative and δ positive (and both temperature independent). In this case it is possible for the potential G_1 to develop equal minima at $D = 0$ and at non-zero values $D = \pm D_c$ ($P = \pm P_c$) and to define a first-order transition. From eqn (3.3.2) the parameter α is still the reciprocal isothermal permittivity at constant stress in the non-polar phase. The basic Devonshire assumption is again that it can be written in Curie–Weiss form

$$\alpha = \beta(T - T_0) \qquad (3.3.12)$$

but where now, as we shall see below, T_0 is called the Curie–Weiss temperature and is not equal to the transition temperature T_c. Writing $\gamma = -\gamma'$ we have

$$G_1 = (\beta/2)(T - T_0)D^2 - (\gamma'/4)D^4 + (\delta/6)D^6. \qquad (3.3.13)$$

Assuming β, γ', and δ to be positive constants the dielectric equation of state follows as

$$E = \beta(T - T_0)D - \gamma'D^3 + \delta D^5. \qquad (3.3.14)$$

As a function of temperature the curves G_1 take the qualitative form shown in Fig. 3.4 (a). As can be seen a zero-field first-order transition takes place when G_1 and its first derivative with respect to D are both zero for a non-zero value of D; that is when

$$(\beta/2)(T - T_0) - (\gamma'/4)P_s^2 + (\delta/6)P_s^4 = 0 \qquad (3.3.15)$$

$$\beta(T - T_0) - \gamma'P_s^2 + \delta P_s^4 = 0 \qquad (3.3.16)$$

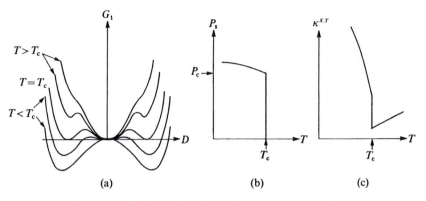

Fig. 3.4. As Fig. 3.1 but for a first-order ferroelectric transition.

are simultaneously satisfied. The solution is

$$T = T_c = T_0 + (3/16)(\gamma')^2(\beta\delta)^{-1} \qquad (3.3.17)$$

Substituting back into eqn (3.3.15) gives

$$P_s^2 = P_c^2 = 3\gamma'/4\delta, \qquad T = T_c. \qquad (3.3.18)$$

for the value of spontaneous polarization at T_c. Zero-field permittivity can be deduced from eqn (3.3.14) as

$$\kappa^{X,T} = \beta(T - T_0) - 3\gamma'P_s^2 + 5\delta P_s^4. \qquad (3.3.19)$$

Above T_c we find simply $\kappa^{X,T} = \beta(T - T_0)$ which was used above to define the Devonshire approximation. Below T_c the spontaneous polarization contributes, and solving eqn (3.3.14) for spontaneous polarization ($D = P_s$ when $E = 0$) and substituting in eqn (3.3.19) gives

$$\kappa^{X,T} = \frac{3\gamma'^2}{4\delta} + 8\beta(T_c - T), \qquad T \to T_c^- \qquad (3.3.20)$$

in the limit of T approaching the transition point T_c from below. Since the paraelectric reciprocal permittivity can be written

$$\kappa^{X,T} = \frac{3\gamma'^2}{16\delta} + \beta(T - T_c), \qquad T \to T_c^+ \qquad (3.3.21)$$

it follows that the permittivity is finite but discontinuous at T_c and that the ratio of the slope $d\kappa/dT$ immediately below T_c to that immediately above T_c is -8. The schematic representations for spontaneous polarization and reciprocal permittivity are shown in Figs. 3.4(b) and (c).

Because a first-order transition does not reflect a singularity in thermodynamic potential the low-temperature phase can exist at temperatures above T_c as a metastable phase. In like fashion the high-temperature phase can persist below T_c as a metastable phase. These are represented in Fig. 3.4 (a) by local minima higher in energy than the stable minima. Theoretically it is obviously a simple exercise (Fatuzzo and Merz 1967) to calculate the lowest temperature for which the nonpolar phase can exist as a metastable state $(T = T_0)$ and the highest temperature for which the zero-field ferroelectric phase can exist metastably $(T = T_1 = T_0 + (\gamma'^2/4\beta\delta))$. In practice such metastable states do tend to persist for first-order ferroelectrics and the actual transition temperature is often found to occur at a somewhat higher value if reached from the low- rather than the high-temperature side. This effect is called thermal hysteresis and $T_1 - T_0$ represents its maximum possible extent. In practice the hysteresis is usually smaller and the dynamic mechanism whereby the system tunnels from a metastable minimum to a stable one is usually complex, involving non-uniform states, and is not amenable to simple thermodynamic theory.

The discontinuous change in polarization (Fig. 3.4(b)) causes a discontinuous change in entropy and hence a latent heat at T_c. Noting that entropy $S = -(\partial G_1/\partial T)_{D,X}$ from eqn (3.2.5) we find a value $S = -\frac{1}{2}\beta P_c^2$ in the ordered phase at T_c. Since the corresponding value in the non-polar phase is zero, the entropy discontinuity at T_c is

$$\Delta S = \tfrac{1}{2}\beta P_c^2 \qquad (3.3.22)$$

with P_c given by eqn (3.3.18). The resulting latent heat is

$$T_c\,\Delta S = \tfrac{1}{2}T_c\beta P_c^2. \qquad (3.3.23)$$

All these simple thermodynamic results for first-order transitions are unfortunately complicated in real transitions by thermal hysteresis, not to mention the experimental difficulties such as sample non-uniformities, heat-rate variations, etc. We are forced by the present thermodynamic formalism to concentrate on two limiting cases: one in which the transition is at T_c and the system passes through only absolutely stable states, and the other for which thermal hysteresis is maximum and the lifetime of metastable states is infinite. With this in mind we now turn to the dielectric equation of state (eqn (3.3.14)). It is convenient to introduce the reduced variables

$$d = \left(\frac{2\delta}{\gamma'}\right)^{1/2} D, \qquad e = \left(\frac{2}{\gamma'}\right)^{5/2}\delta^{3/2}E, \qquad t = 4\beta\frac{\delta}{\gamma'^2}(T - T_0),$$
$$(3.3.24)$$

when the equation reduces to

$$e = td - 2d^3 + d^5 \qquad (3.3.25)$$

with t as the single variable. The family of curves defined by eqn (3.3.25) is

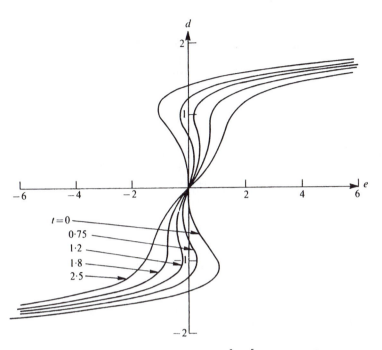

Fig. 3.5. The family of curves $e = td - 2d^3 + d^5$ of eqn (3.3.25).

shown in Fig. 3.5. The d *versus* e isotherms of absolute stability follow from an examination of the Gibbs free energy analogous to that described for second-order transitions; they are shown in Fig. 3.6. For $T < T_c$ ($t < \frac{3}{4}$) the isotherms suffer a discontinuity at $E = 0$ ($e = 0$). For $T_c < T < T_2$, $\frac{3}{4} < t < \frac{9}{5}$ where $T_2 = T_0 + (9\gamma'^2/20\beta\delta)$ is the temperature above which G_1 (Fig. 3.4 (a)) ceases to have inflection points, the discontinuity is for non-zero E. Within this range the ferroelectric phase can exist only in the presence of an electric field. Finally, for $T > T_2$, the ferroelectric phase cannot be induced even by applied field. On the other hand if we admit metastable states we obtain dielectric hysteresis in a variety of possible forms. From Fig. 3.5 it is evident that single loops, double loops, and broken loops (see Fig. 3.7) can all occur in the right temperature range and examples of such loops have indeed been observed experimentally.

Summarizing the findings for a first-order transition resulting from free energy (eqn (3.3.13)); the system undergoes a transition between *absolutely* stable states at $T_c = T_0 + (3\gamma'^2/16\beta\delta)$ with a discontinuity in spontaneous polarization and susceptibility. Reciprocal susceptibility near T_c is a linear function in temperature (i.e. Curie–Weiss like) in both phases with a slope ratio from polar to non-polar of $-8 : 1$. If we allow for metastable states the

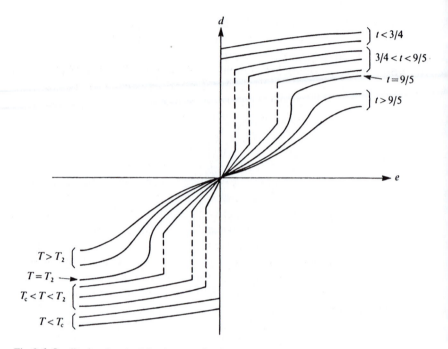

Fig. 3.6. Qualitative sketch of the *d versus e* isotherms of absolutely stable states for a first-order phase transition.

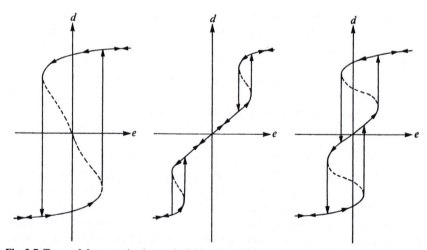

Fig. 3.7. Types of *d versus e* isotherms (solid lines) possible when metastable states are included in the first order analysis and the criterion for switchover from one branch to another is the onset of local instability. The broken lines indicate those states $(\partial e/\partial d)_t < 0$ which are not accessible to this system. The vertical arrows connect the states at which local instability occurs to the new stable configuration.

non-polar phase can remain metastable down to temperature T_0 (supercooling) and the polar phase metastable up to temperature $T_1 = T_0 + (\gamma'^2/4\beta\delta)$ (superheating) giving a maximum possible thermal hysteresis of magnitude $\gamma'^2/4\beta\delta$. Above T_1 but below a higher temperature $T_2 = T_0 + (9\gamma'^2/20\beta\delta)$ the ferroelectric phase can be induced by an applied field of sufficient magnitude. Above temperature T_2 only the non-polar phase can exist.

In practice the theory proves to be very useful for characterizing materials. The parameters β, γ', and δ are not difficult to estimate from a few dielectric measurements close to T_c and self-consistency is often quite good. On the other hand with increasing $|T - T_c|$ the need for a longer series expansion and for temperature-dependent parameters becomes increasingly apparent although self-consistency may still be good (Benepe and Reese 1971). The theory can be extended to cover multi-axial ferroelectrics (with the elastic Gibbs function expanded in terms of the three orthonormal components of displacement), an early application being the one by Devonshire (1949) to the perovskite $BaTiO_3$ which has a cubic nonpolar phase.

The basic strength of the thermodynamic theory of ferroelectricity is its mathematical simplicity and its very wide range of application and corresponding use in correlating the different macroscopic parameters of ferroelectrics. Its limitations are the purely macroscopic picture (which precludes any discussion of the atomic processes responsible for ferroelectricity or of the microscopic character of the transition), phenomenology *per se*, and the restriction to equilibrium phenomena. There must naturally be a statistical bridge betweeen the macroscopics of thermodynamic theory and any particular microscopic model. The manner in which the elastic Gibbs function can actually be deduced from a microscopic ferroelectric picture has been explored with varying degrees of statistical sophistication by many authors (e.g. Slater 1950, Kwok and Miller 1966, Lines 1969 a, b) and the Devonshire coefficients expressed in terms of microscopic parameters. In this manner the origin and extent of the temperature dependence of the various Devonshire coefficients can be investigated.

3.4. Macroscopic phenomenology for antiferroelectrics

The conventional definition of an antiferroelectric is an antipolar crystal whose free energy is closely comparable to that of a ferroelectric modification of the same crystal and which may therefore be switched from antipolar to ferroelectric by application of an external electric field E. This close approach in energy of the polar and antipolar phases leads to high permittivities and to other distinctive and interesting dielectric properties in the antiferroelectric. As was the case for ferroelectrics in the previous section a phenomenological description can be developed in terms of a polynomial expansion of free energy. Obviously, however, an antipolar configuration requires the definition of at least two sublattice polarizations. For mathematical convenience we shall consider the simplest conceivable situation, that of a stress-free centrosymmetric paraelectric prototype phase and a single allowed direction of sublattice polarization and applied electric field.

The possibility of the existence of antiferroelectrics and some of their special properties were predicted by Kittel (1951) before such compounds were discovered experimentally. Kittel's basic two-sublattice model has been discussed and embellished by many subsequent authors (e.g. Takagi 1952, Smolenskii and Kozlovskii 1954, Cross 1967, Okada 1969, 1970, 1974), and we shall outline the basic formalism in this section. Associating displacement components D_a and D_b with the two sublattices a and b of the antipolar configuration it is immediately evident that an additional pair of dielectric variables representing 'staggered' displacement $D_a - D_b$ and its conjugate staggered field are required for a description of antipolar stimulus and response. To make the nomenclature more transparent we shall, for this section alone, write the symbols for normal displacement and Maxwell field and D_F and E_F respectively (i.e. a measure of *ferro*electric response) with their staggered antiferroelectric counterparts designated D_A and E_A. In terms of sublattice displacements we have

$$D_F = D_a + D_b, \qquad D_A = D_a - D_b. \tag{3.4.1}$$

With the new dielectric variables the work function (eqn (3.1.15)) becomes

$$dW = \int_V (X_i \, dx_i + E_F \, dD_F + E_A \, dD_A) \, dV \tag{3.4.2}$$

and, via the laws of thermodynamics, leads to an elastic Gibbs function derivative

$$dG_1 = -S \, dT - x_i \, dX_i + E_F \, dD_F + E_A \, dD_A. \tag{3.4.3}$$

Defining a Gibbs free energy

$$G_A = U - TS - X_i x_i - E_F D_F - E_A D_A \tag{3.4.4}$$

we also obtain

$$dG_A = -S \, dT - x_i \, dX_i - D_F \, dE_F - D_A \, dE_A. \tag{3.4.5}$$

A simple polynomial in D_a and D_b which we shall show to be capable of describing a simple antiferroelectric phase transition for zero stress can be written

$$G_1 = f(D_a^2 + D_b^2) + gD_aD_b + 2h(D_a^4 + D_b^4) \tag{3.4.6}$$

where f, g, and h are in general temperature dependent. It is evident from eqn (3.4.6) that a positive value of g favours an antipolar situation. Terms in

$D_a^2D_b^2$ and $(D_a^3D_b + D_aD_b^3)$ have been omitted from eqn (3.4.6) only for mathematical simplicity. Transforming to co-ordinates D_F and D_A we find

$$G_1 = \tfrac{1}{2}(f + \tfrac{1}{2}g)D_F^2 + \tfrac{1}{2}(f - \tfrac{1}{2}g)D_A^2 + \tfrac{1}{4}h(D_F^4 + 6D_F^2D_A^2 + D_A^4). \quad (3.4.7)$$

From eqn (3.4.3) the dielectric equation of state for staggered field E_A results from the derivative $(\partial G_1/\partial D_A)$ at constant T, X, and D_F. We find

$$E_A = (f - \tfrac{1}{2}g)D_A + h(D_A^3 + 3D_F^2D_A). \quad (3.4.8)$$

The corresponding reciprocal staggered permittivity follows as

$$\kappa_A^{T,X,D_F} = f - \tfrac{1}{2}g + 3h(D_A^2 + D_F^2). \quad (3.4.9)$$

A simple second-order antiferroelectric transition is then described if this staggered permittivity in the paraelectric phase ($D_F = D_A = 0$) is written in Curie–Weiss form, namely

$$\kappa_A^{T,X,D_F} = f - \tfrac{1}{2}g = \beta(T - T_c), \qquad T > T_c. \quad (3.4.10)$$

Potential G_1 now becomes

$$G_1 = \tfrac{1}{2}\{g + \beta(T - T_c)\}D_F^2 + \tfrac{1}{2}\beta(T - T_c)D_A^2 + \tfrac{1}{4}h(D_F^4 + 6D_F^2D_A^2 + D_A^4). \quad (3.4.11)$$

If the external constraints are those of fixed temperature, stress, Maxwell field, and staggered field (the last is necessarily zero experimentally), then the stable configurations can be sought by minimizing the Gibbs free energy $G_A = G_1 - E_FD_F - E_AD_A$ of (3.4.4). Putting $E_A = 0$ and introducing the reduced variables

$$g_A = \frac{h}{g^2}G_A, \qquad t = \frac{\beta}{g}(T - T_c), \qquad e = \left(\frac{2h}{g^3}\right)^{1/2} E_F$$

$$d_F^2 = \frac{h}{2g}D_F^2, \qquad d_A^2 = \frac{h}{2g}D_A^2 \quad (3.4.12)$$

reduces the Gibbs free energy to

$$g_A = (1 + t)d_F^2 + td_A^2 + (d_F^4 + 6d_F^2d_A^2 + d_A^4) - ed_F \quad (3.4.13)$$

with a single parameter t. The locally stable configurations result from minimizing eqn (3.4.13) with respect to d_F and d_A. We first consider the situation in the absence of applied field ($e = 0$) following Okada (1969). When $t > 0$ (i.e. $T > T_c$) the paraelectric phase $d_F = 0$, $d_A = 0$ is stable. When $t < 0$ the antiferroelectric phase $d_F = 0$, $d_A^2 = -t/2$ is stable. In this antipolar phase the order parameter D_A is given by

$$D_A^2 = \frac{\beta}{h}(T_c - T) \quad (3.4.14)$$

and is directly comparable in form to the corresponding ferroelectric result (eqn (3.3.4)). Using eqn (3.4.9) the inverse staggered permittivity below T_c is

$$\kappa_A^{T,X,D_F} = 2\beta(T_c - T), \qquad T < T_c \tag{3.4.15}$$

and we note again an equivalence to the ferroelectric result (eqn (3.3.6)). The dielectric equation of state for Maxwell field E_F is given by

$$E_F = \left(\frac{\partial G_1}{\partial D_F}\right)_{T,X,D_A} = (f + \tfrac{1}{2}g)D_F + h(D_F^3 + 3D_F D_A^2). \tag{3.4.16}$$

The inverse (ferroelectric) permittivity follows as

$$\kappa_F^{T,X,D_A} = f + \tfrac{1}{2}g + 3h(D_A^2 + D_F^2) \tag{3.4.17}$$

and, comparing with eqn (3.4.9), we find that the inverse permittivity differs from its staggered counterpart by the parameter g which, in simple theory, is usually assumed (with h) to be a positive constant. Thus the zero-field ferroelectric permittivity is Curie–Weiss like both above and below the second-order antiferroelectric transition but remains finite *at* the transition.

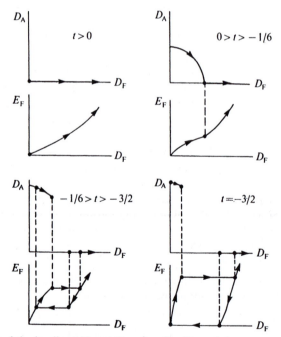

Fig. 3.8. Locus of the locally stable configuration D_F, D_A and the corresponding dielectric curves D_F, E_F for a simple phenomenological model antiferroelectric at various reduced temperatures t of eqn (3.4.12) below and above the zero-field Curie temperature $t = 0$. The sequence of curves is described in the text. (After Okada 1969.)

The most interesting properties of the antiferroelectric phenomenology occur in the presence of an applied field E_F. In Fig. 3.8 we show schematically the locus of the locally stable configuration on a D_F, D_A plot and the corresponding dielectric curves as a function of applied field E_F. When $t > 0$ the zero-field stable configuration point is $D_F = 0$, $D_A = 0$, the origin of the D_F, D_A plot. As the field E_F increases, the stable configuration moves continuously along the D_F axis with no singular behaviour. For $t < 0$ the zero-field stable configuration is antiferroelectric (i.e. on the D_A axis). As a function of applied field the stable configuration moves round the quadrant of an ellipse in the D_F, D_A plane to the D_F axis. If $0 > t > -\frac{1}{6}$ the path is continuous and a second-order transition occurs when the configuration point reaches the D_F axis, after which it continues along the axis. The entire process is reversible and no dielectric hysteresis occurs. When $t < -\frac{1}{6}$ the configuration point becomes metastable and finally unstable as it moves along the D_F, D_A ellipse. It then jumps *discontinuously* to a point on the D_F axis. A first-order phase transition results and hysteresis is observed if the metastable states are considered to be long lived. As a function of decreasing field the configuration point moves back along the D_F axis becoming metastable and finally unstable at which point it jumps back discontinuously to the ellipse. Finally, for $t < -\frac{3}{2}$, the polar phase becomes metastable even in zero applied field. The point $t = -\frac{1}{6}$, at which the phase transition changes from second order to first order, is of some significance in the modern theory of critical phenomena and is referred to as a tricritical point.

The reciprocal permittivity as a function of temperature for various bias fields is shown schematically in Fig. 3.9. At zero bias, as discussed above, the permittivity is continuous at T_c with a Curie–Weiss temperature dependence on both sides. Under a bias field the transition point shifts to lower temperatures and a discontinuity in the permittivity develops at the shifted transition temperature. At the tricritical point this discontinuity reaches a maximum as the permittivity diverges on the low-temperature side (Okada 1969).

The above analysis of the simple polynomial (eqn (3.4.6)) can be extended in a number of ways. Firstly, allowing the parameter f of eqn (3.4.6) to be negative makes possible the description of zero-field antiferroelectric-to-ferroelectric transitions. Also, adding a higher-order stabilizing anharmonicity while letting the fourth-order term be negative enables a discussion of first-order zero-bias antiferroelectric transitions. Finally, lifting the restriction of collinear sublattice polarizations leads to a full three-dimensional analysis at the expense of rather significant mathematical complexity.

As was the case for ferroelectrics one expects the simple phenomenology to break down very close to a second-order transition (within the fluctuation-dominated regime) and also when $|T - T_c|$ becomes large. In

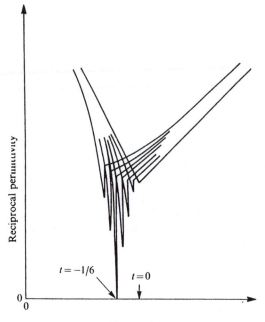

Fig. 3.9. Reciprocal permittivity as a function of temperature for various bias fields. The Curie–Weiss form with continuous behaviour at the transition temperature $t = 0$ corresponds to zero bias. The maximum discontinuity as a function of bias occurs at the tricritical point $t = -\frac{1}{6}$ when the bias field e of eqn (3.4.12) has the value 0·296. (After Okada 1969.)

practice most antiferroelectrics are of first order which makes the theory less simple. Accordingly, less parameterization of materials has been carried out for antiferroelectrics than for ferroelectrics although, in general, the simple theory seems to be qualitatively valid in many cases. In particular, the double hysteresis loop of Fig. 3.8 (the figure only shows the positive-field half of the full cycle) is considered to be somewhat typical of antiferroelectric behaviour although, as we have seen, its existence is not actually necessary nor sufficient as a criterion for the existence of antiferroelectricity.

4

DOMAINS, IMPERFECTIONS, AND POLARIZATION REVERSAL

4.1. Domains

THE THEORETICAL foundations of the preceding chapters have dealt with uniform infinite ferroelectrics. In real crystals important additional considerations arise owing to the presence of crystal surfaces and imperfections, and these are the subjects of this chapter.

The most fundamental consideration is the change of spontaneous polarization at inhomogeneities. In the linear relation between electric displacement, field, and polarization

$$\mathbf{D} = \varepsilon_0 \mathbf{E} + \mathbf{P} \qquad (4.1.1)$$

the polarization arises from both the polarizability of the material in the presence of a field $\mathbf{P}_E = \chi \mathbf{E}$ and from the spontaneous alignment of dipoles in the ferroelectric $\mathbf{P}_s$. If there is a spatial variation of $\mathbf{D}$ the free charge density ρ must satisfy Poisson's equation

$$\text{div } \mathbf{D} = \rho \qquad (4.1.2)$$

so that

$$\text{div } \mathbf{E} = \frac{1}{\varepsilon \varepsilon_0} (\rho - \text{div } \mathbf{P}_s). \qquad (4.1.3)$$

In an ideal infinite ferroelectric the spontaneous polarization is uniform so that div $\mathbf{E} = \rho/\varepsilon\varepsilon_0$ as in ordinary dielectrics. However, at the surface where $\mathbf{P}_s$ decreases to zero, or in the neighbourhood of defects where $\mathbf{P}_s$ may differ from the perfect crystal, div $\mathbf{P}_s$ acts as the source for a depolarizing field. This field can be compensated by the flow of free charge within the crystal (unlike the magnetic equivalent)

$$\rho = \int_0^t \boldsymbol{\sigma} \cdot \mathbf{E} \, dt \qquad (4.1.4)$$

where $\boldsymbol{\sigma}$ is the electrical conductivity of the crystal. Alternatively, free charge in the surrounding medium can compensate for the depolarization field at the crystal surface, but in air this process is usually slow compared with crystal conductivity. Accumulation of surface charge by conduction satisfies the requirement that the electric field vanish both outside the crystal and inside the bulk, but not necessarily in the region just below the crystal surface (see § 4.4).

The energy associated with the depolarization field

$$W_E = \tfrac{1}{2} \int_V \mathbf{D} \cdot \mathbf{E} \, dV \qquad (4.1.5)$$

is zero for a totally compensated crystal in equilibrium. In an insulating crystal (in an insulating environment) this equilibrium is reached very slowly and at short times the energy W_E can reach very high values, the magnitude depending on the crystal geometry and the distribution of $\mathbf{P}_s$ within the crystal.

When such a crystal is cooled from a paraelectric phase to a ferroelectric phase in the absence of applied fields there are at least two equivalent directions along which the spontaneous polarization may occur. In order to minimize W_E different regions of the crystal polarize in each of these directions, each volume of uniform polarization being referred to as a domain. The depolarizing fields which appear on cooling are usually sufficient to prevent any net polarization in a virgin crystal. These crystals consequently show very small pyroelectric and piezoelectric effects, which probably accounts for their relatively recent discovery compared with non-ferroelectric pyroelectrics.

The boundaries separating domains are referred to as domain walls. Since these walls differ from the perfect crystal there is a certain amount of energy W_w associated with them in addition to the electrostatic energy W_E. The final domain configuration is determined by minimizing an appropriate free energy including both these terms. In equilibrium, when all depolarizing fields are compensated, the minimum energy in the absence of defects would correspond to a single domain configuration from considerations of W_E and W_w alone. Indeed in highly conducting ferroelectrics single-domain formation seems to be preferred (DiDomenico and Wemple 1967). However, such an equilibrium state is rarely achieved in a virgin crystal in the absence of applied fields and the observed domain pattern depends on a great many factors including the crystal symmetry, the electrical conductivity, the defect structure, the magnitudes of the spontaneous polarization and elastic and dielectric compliances, as well as the history of the crystal preparation and sample geometry. The influence of these factors on domain geometry will be discussed in § 4.1.2 following a brief outline of the various experimental techniques used to reveal the domain structure in ferroelectrics.

4.1.1. Observation of domains

A number of techniques have now been developed for revealing domain structures. The usefulness of each of these techniques varies from one material to another, with the crystal geometry, the necessary speed and resolution, and whether the electrical sense of the domains is required. Of the techniques described below (a) to (e) can be used for the observation of domains throughout the crystal volume, while (f) to (j) delineate the regions of intersection of domains with the crystal surfaces.

(a) *Optical birefringence.* Just as the low-frequency dielectric suscepti-bility of pyroelectric crystals is anisotropic, so is the optical frequency susceptibility. Thus all pyroelectric crystals are optically anisotropic unless the refractive indices n_i $(= \varepsilon_i^{1/2})$ are 'accidentally' equal at some particular temperature. In optically uniaxial crystals which belong to tetragonal, trigonal, or hexagonal crystal classes the refractive index for light polarized along the polar axis is different from those for light polarized perpendicularly to the polar axis, i.e.

$$n_1 = n_2 \neq n_3. \tag{4.1.6a}$$

Between crossed polarizers domains polarized in the viewing direction (c-domains) appear dark for all rotations about the polar axis, while domains polarized in any other direction are birefringent and appear bright provided the polar axis of the crystal and the polarizer axes are not coplanar.

In the case of optically biaxial crystals which belong to orthorhombic, monoclinic, and triclinic crystal classes all three refractive indices are generally different,

$$n_1 \neq n_2 \neq n_3 \tag{4.1.6b}$$

so that all domains are birefringent. These domains are also revealed between crossed polarizers since the birefringence differs for domains of different orientation.

An example of domains revealed in this way in the tetragonal phase of PbTiO$_3$ is shown in Fig. 4.1. In this case the bright, so-called a-domains are almost perpendicular to the c-domains, the actual angle being $2 \tan^{-1}(a/c) = 86°26'$ at 25°C where a/c is the ratio of tetragonal axes.

This technique has been used for a large number of ferroelectrics of different crystal symmetries and is particularly useful because of its simplicity. However, it cannot generally be used for the observation of antiparallel domains (since the optical indicatrix is invariant under domain reversal) unless there is appreciable scattering from the domain walls or when the walls may be distorted by applied stresses or fields (Merz 1954).

In BaTiO$_3$ antiparallel domains *are* directly visible between crossed polarizers even in the absence of any externally applied perturbation (Miller and Savage 1959). The origin of this visibility has not been conclusively resolved but Kobayashi, Yamada, and Nakamura (1963) suggest that the effect is due to birefringence induced by shear strain in the neighbourhood of the domain walls. This effect does not appear to be a general property for

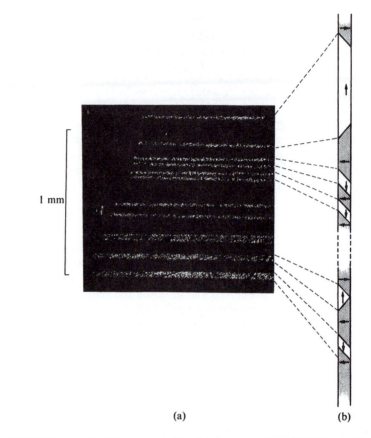

(a)   (b)

Fig. 4.1. (a) Photograph of the transmission of a flux-grown $PbTiO_3$ single-crystal platelet, 0·006 cm thick, between crossed polarizers, showing bright a-domains and dark c-domains. The domain walls at 45° to the platelet are seen as bright regions bordering the larger a-domains. (b) Schematic showing possible wall orientations through the platelet thickness and possible orientations of polarization within the domains.

other materials, and in crystals such as $LiNbO_3$ and triglycine sulphate, where only 180° domains exist, other techniques must be used.

This optical technique for observing domains, when applicable, is probably the most straightforward way of directly observing both the static configurations and the kinetics of domain-wall motion.

(b) *Optical rotation.* An example where optical rotation can be used to reveal 180° domain structures is $Pb_5Ge_3O_{11}$ (Dougherty, Sawaguchi, and Cross 1972). This crystal has an optical rotary power of 5·6° mm^{-1} for light propagating down the polar three-fold axis, and upon inversion of this axis the sense of the rotation is reversed. Consequently, when a c-cut plate (polar axis normal to plate) is viewed in polarized light one set of domains can be set

Fig. 4.2. Domain structure of a c-cut plate of $Pb_5Ge_3O_{11}$, 0·05 cm thick, observed in plane-polarized light with the analyser rotated by 2·5° from the crossed position. Rotation of the analyser by 2·5° in the opposite sense produces reversed contrast. (After Dougherty *et al.* 1972.)

for extinction with an analyser while another set appears bright, as shown in Fig. 4.2.

(*c*) *Second-harmonic generation.* In triglycine sulphate 180° domains have been observed by phase-matched second-harmonic generation (Dolino 1973) (see Chapter 13). This technique can in principle be used for any crystal which can be phase matched for second-harmonic generation with light propagating close to the polar axis. The intensity of second-harmonic light depends on the optical interaction length with a single domain of either sign. On crossing a boundary the second-order non-linear coefficient changes sign and phase cancellation of the second harmonic occurs. Thus regions containing domain boundaries appear darker than the surrounding single-domain regions. Apart from revealing domain patterns second-harmonic generation can be used to measure the width of extremely small domains with periodic geometry (Miller 1964 a).

(*d*) *Electron microscopy.* Optical techniques are limited to domain sizes down to micron dimensions. By transmission electron microscopy of thin $BaTiO_3$ crystals domain structures of width down to 5 nm have been observed (Tanaka and Honjo 1964, Blank and Amelinckx 1963, Yakunin *et al.* 1972). Recently Shakmanov, Spivak, and Yakunin (1970) examined the details of domain motion with transmission microscopy with electric field applied to the crystal during observation. Extremely small domain sizes have also been observed by using replicas of $BaTiO_3$ and Rochelle salt surfaces (Spivak *et al.* 1959). With electron microscopy only thin regions of crystals are examined, so the observed effects may only be characteristic of surface regions (see § 4.4).

(e) *Pyroelectric techniques.* When a ferroelectric is heated P_s changes giving rise to free charge on the crystal surface, the sign of the charge depending on the sense of the domain orientation. An insulating crystal can be uniformly heated and the charge distribution examined with an electron beam, as in the pyroelectric vidicon (see § 16.2) or with charged particles (Bergman *et al.* 1972). Alternatively the crystal can be locally heated with a scanned laser beam. By electroding the crystal normal to the polar axis the voltage response can be recorded with a display device giving an image of the domain pattern (Hadni 1971). A domain pattern obtained in this way for TGS is shown in Fig. 4.3.

Fig. 4.3. Domain structure of TGS obtained by mapping a crystal plate (of 3 mm diameter and 0·1 mm thickness). The pyroelectric response of the plate to a focused laser beam is displayed on an XY recorder. (After Hadni 1971.)

(f) *Chemical etching.* Chemical etchants have been found for many ferroelectrics which etch the positive and negative ends of domains at a different rate thereby revealing the domain structure. Early studies with $BaTiO_3$ using HCl etchant have been extensively documented (see for instance Forsbergh 1956, Jona and Shirane 1962, Fatuzzo and Merz 1967). The main disadvantage of etching techniques is that the process is destructive and slow.

(g) *Powder techniques.* Using colloidal suspensions of charged particles it was shown (Pearson and Feldmann 1958) that certain particles deposit preferentially on either positive or negative ends of domains. For instance sulphur and lead oxide (Pb_3O_4) in hexane deposit on negative and positive ends of domains respectively.

(h) *Dew method.* Isobutyl alcohol condensed onto the surface of tri-glycine sulphate is attracted towards domain walls and can be seen in reflected light (Fousek, Safrankova, and Kaczer 1966). Another variation of this kind of technique is the selective crystallization of anthraquinone on the surface of TGS. The domain morphology and sense can be determined from the crystallization patterns.

(*i*) *Liquid crystal method.* A thin layer of the nematic liquid crystal p-N-(p-methoxy benzilidene) amino n-butylbenzene on ferroelectric crystal surfaces has recently been shown to reveal 180° domain structure (Furuhata and Toriyama 1973) when viewed between crossed polarizers. The proposed alignment of the liquid-crystal molecules relative to the ferroelectric domains is shown in Fig. 4.4. If this technique is generally applicable it would

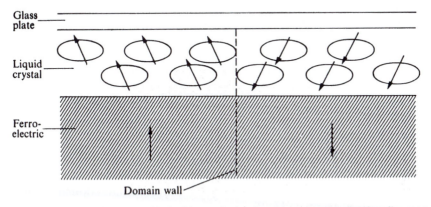

Glass plate

Liquid crystal

Ferro-electric

Domain wall

Fig. 4.4. Domain delineation with liquid crystals showing the proposed molecular alignment where domains meet the crystal surface. (After Furuhata and Toriyama 1973.)

have the advantage over (*f*), (*g*), and (*h*) in that it is fast and the liquid crystal can respond rapidly to changes of the domain configuration, while being simple and having quite high resolution.

(*j*) *X-ray topography.* Anomalous dispersion of X-rays causes a difference between the X-ray intensity reflected from the positive and negative ends of domains. By using X-ray wavelengths close to an absorption edge of a constituent element these differences can be maximized. Domains have been successfully observed by this method in $BaTiO_3$ (Niizeki and Hasegawa 1964), $NaNO_2$ (Takagi Suzuki, and Watanabe 1970), TGS (Autier and Petroff 1964), $LiNbO_3$ (Wallace 1970), and $Pb_5Ge_3O_{11}$ (Sugii *et al.* 1972). With modern topographic techniques examination of crystals can be made very rapidly.

(*k*) *U.V. photoemission.* The photoemission spectrum of a crystal depends on the crystallographic orientation of the illuminated surface. Thus, for instance, **a** and **c** domains at the surface of a perovskite ferroelectric may be directly observed from the difference in photoelectric yield at the two domains (Morlon *et al.* 1970). In principle this technique should also be able to observe 180° domains, as a result of the differences between the positive and negative polar surfaces of a ferroelectric (see § 4.4).

4.1.2. Free energy of domains

The static domain configuration of a ferroelectric is obtained by minimizing the total free energy of the crystal including the energy associated with the crystal surfaces and domain walls. Using the model of § 3.3 we write

$$G_1 = G_1^0 + \int \left(\frac{\alpha}{2} D^2 + \frac{\gamma}{4} D^4 \right) dV + W_w + W_E \qquad (4.1.7)$$

where powers in the expansion higher than the fourth have been neglected. It is readily seen that the properties of a ferroelectric such as the dielectric constant $(\partial^2 G_1 / \partial D^2)^{-1}$, specific heat $-T(\partial^2 G_1 / \partial T^2)$, and of course the electric displacement itself are affected by the presence of domains. For any measurement it is therefore necessary either to understand the domain contribution to G_1 or else to obtain a single-domain crystal. We shall first discuss these extra contributions to G_1 and minimize the total free energy for a specific example.

Depolarization energy W_E. To minimize the depolarization energy at the crystal surfaces before compensation by free charge takes place domains form throughout the volume of the crystal. Indeed without domain formation the energy associated with the crystal surfaces can be greater than the energy associated with ferroelectric ordering, so that ferroelectricity could not exist.

The magnitude of W_E depends on the crystal geometry and on the geometry of the domain configuration at the crystal surfaces. In general the depolarizing field inside a ferroelectric is proportional to the polarization

$$\mathbf{E} = -\frac{L}{\varepsilon_0} \mathbf{P} \qquad (4.1.8)$$

where L is a depolarization factor. Thus the energy of the depolarization field is

$$W_E = \frac{1}{2} \int \mathbf{D} \cdot \mathbf{E} \, dV = \frac{1}{2} \int \frac{\varepsilon}{\varepsilon_0} L^2 P^2 \, dV. \qquad (4.1.9)$$

Only for a thin crystal wafer with uniform polarization normal to the wafer is L equal to unity. For the case of a general ellipsoid in the absence of domains the depolarizing field is uniform and L can be written as

$$L = \frac{abc}{2} A \qquad (4:1.10)$$

where A is an elliptical integral which can be evaluated for specific ellipsoid axes a, b, c (Stratton 1941).

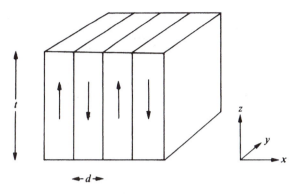

Fig. 4.5. Sketch of a model periodic domain structure.

For a multi-domain crystal evaluation of W_E becomes more complicated. For a simple periodic domain structure, such as that shown in Fig. 4.5, W_E has been evaluated (Kittel 1946, Mitsui and Furuichi 1953) to be

$$W_E = \frac{\varepsilon^* d P_0^2 V}{t} \qquad (4.1.11)$$

where d is the domain width, t is the crystal thickness, P_0 is the polarization at the *centre* of a domain, V is crystal volume, and ε^* is a constant depending on the dielectric constants of the ferroelectric.

For some domain geometries it is convenient (Miller and Weinreich 1960) to calculate W_E from the equation

$$W_E = \int \int \frac{\rho_1 \rho_2}{4\pi\varepsilon_0 r_{12}} \, d\tau_1 \, d\tau_2 \qquad (4.1.12)$$

where ρ_1 and ρ_2 are the total charge densities in volumes $d\tau_1$ and $d\tau_2$ separated by r_{12}.

Domain wall energy W_w. If the energy per unit area of a domain wall is σ then for the domain geometry of Fig. 4.5 the energy W_w is

$$W_w = (\sigma/d) V. \qquad (4.1.13)$$

Minimizing the energy $W_E + W_w$ yields the equilibrium value of the domain width from eqns (4.1.11) and (4.1.13)

$$d = \left(\frac{\sigma t}{\varepsilon^* P_0^2}\right)^{1/2} \qquad (4.1.14)$$

where P_0 must still be obtained from a self-consistent solution of eqn (4.1.7). Domains of width greater than this are suppressed by the depolarization field while domains smaller than this are suppressed by the wall energy. The domain width is seen from eqn (4.1.14) to vary quadratically with crystal thickness in agreement with experiment (Mitsui and Furuichi 1953). When the crystal thickness decreases until the domain width approaches the domain wall thickness the depolarization energy can no longer be minimized by domain formation and ferroelectricity may cease to exist in the crystal (Jaccard, Känzig, and Peter 1953).

Contributions to the wall energy are the depolarization energy due to div **P** at domain boundaries, the dipolar energy (which is due to the fact that ferroelectric dipoles on each side of the wall are not in alignment as in the perfect ferroelectric), and elastic energy.

To minimize the electrostatic energy the orientation of neighbouring domains is usually such that div **P** ~ 0 at the domain boundary. This condition is satisfied for the 90° and 180° walls in perovskites, for example, but not for wedge-shaped domains which are also often found in these materials. However, the wedge angles are so small that the total depolarization energy is small. Another notable exception where div **P** ≠ 0 are head-to-head domains observed in boracites (Schmid 1970), but these are only observed at higher temperatures when crystals become conducting and depolarization fields can be neutralized by free charge. Recently Yakunin *et al.* (1972) observed 90° walls in BaTiO$_3$ separating head-to-head domains. Electron microscopic examination revealed that the actual wall shape was zig-zagged as shown in Fig. 4.6, so that on a microscopic scale all walls were head to tail. The zig-zag arrangement increased the wall length by a factor of 5.

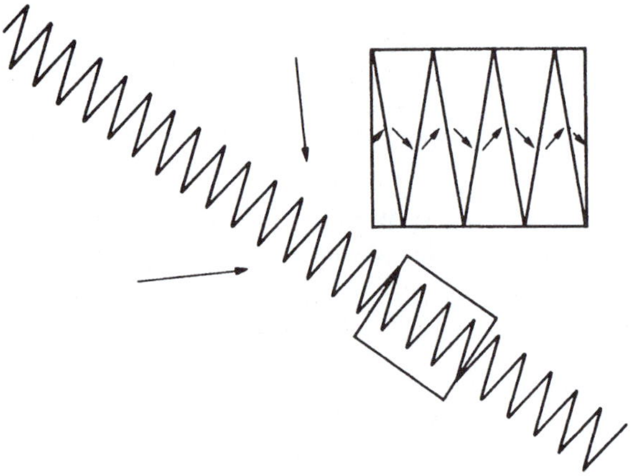

Fig. 4.6. A schematic illustration of a domain wall in BaTiO$_3$ in the boundary region between 90° *a*-domains polarized 'head to head'. (After Yakunin *et al.* 1972.)

The contribution to wall energy from dipolar interaction—the so-called correlation energy—can be calculated by a point-charge model for specific structures (Kinase and Takahasi 1957) or phenomenologically as follows.

The energy of a dipole μ_i at site i in the field due to all other dipoles μ_j is (Mitsui and Furuichi 1953)

$$-\mu_i \cdot \sum_j \alpha_{ij} \cdot \mu_j = -\sum_j \beta_i \beta_j \mathbf{P}_i \cdot \alpha_{ij} \cdot \mathbf{P}_j \qquad (4.1.15)$$

where α_{ij} is determined by the relative positions and orientations of the dipoles, and the β's are constants of proportionality between the polarization and the dipole moments. If μ_i and μ_j are parallel then α_{ij} takes the form $(3\cos^2\theta_{ij} - 1)/r_{ij}^3$, where r_{ij} is the dipole separation and θ_{ij} the angle between them. If $\mathbf{P}_j$ is expanded as a Taylor series about $\mathbf{P}_i$ and the sum in eqn (4.1.15) is replaced by an integration, then the dipolar energy can be expressed in the form

$$W_{\text{dip}} = \tfrac{1}{2} \int_V -\mathbf{P} \cdot (\mathbf{f}_0 \cdot \mathbf{P} + \mathbf{f}_1 \cdot \mathbf{P}' + \mathbf{f}_2 \cdot \mathbf{P}'' + \ldots)\,\mathrm{d}V \qquad (4.1.16)$$

where the $\mathbf{f}_i$ are tensor constants depending on the crystal structure and the primes denote derivatives with respect to lattice displacements. The integrals over the odd derivatives in general vanish and the lowest-order term involving $\mathbf{P} \cdot \mathbf{f}_0 \cdot \mathbf{P}$ corresponds simply to the interaction energy for the uniformly polarized crystal. Thus the major contribution to the increase in energy at the domain wall comes from the term involving $\mathbf{P} \cdot \mathbf{f}_2 \cdot \mathbf{P}''$ and, integrating by parts, this can be expressed in a form

$$W_{\text{dip}} = \int_V \tfrac{1}{2}\nabla\mathbf{P} \cdot \mathbf{f}_2 \cdot \nabla\mathbf{P}\,\mathrm{d}V. \qquad (4.1.17)$$

The form is only valid, of course, if the integration and expansion in a continuum notation is physically reasonable, i.e. if the variations of polarization from one unit cell to the next are small.

The elastic energy of a ferroelectric is

$$W_x = \tfrac{1}{2} \int_V \mathbf{X} \cdot \mathbf{x}\,\mathrm{d}V = \tfrac{1}{2} \int_V c_{ij} x_i x_j\,\mathrm{d}V \qquad (4.1.18)$$

where i and j now refer to a Voigt representation and the repeated-index summation convention is assumed. The elastic strain is related to the electric displacement vector $\mathbf{D}$ by

$$x_i = g_{ij}^\dagger D_j + Q_{ijk} D_j D_k \qquad (4.1.19)$$

where $g_{ij}^\dagger$ and Q_{ijk} are piezoelectric and electrostrictive coefficients. A domain wall can exist only if the two domains may meet without infinite stresses or cracks in the bulk of an infinite crystal. This condition means that the spontaneous strain of neighbouring domains, given by eqn (4.1.19), must be equal in the plane of the domain wall up to a rigid-body motion. The domain-wall orientations satisfying this condition of 'mechanical compatibility' have been tabulated for all crystal symmetries by Fousek and Janovek (1969). The wall orientations may be determined by crystal symmetry alone, by the direction of spontaneous displacement, or by the magnitudes of $g_{ij}^\dagger$, Q_{ijk}, and D_i. It is interesting to note in the latter case that the magnitudes of the parameters are temperature dependent so that such a wall would change its orientation with temperature. This would result in thermal switching since the domain volumes would change. While this effect has not yet been observed walls of this kind may occur in boracites.

These mechanical-compatibility conditions seem to be obeyed for all known ferroelectrics with the possible exception of $Pb_5Ge_3O_{11}$ (Dougherty et al. 1972). Of course slight misorientations of the walls can occur near defects or crystal surfaces or in regions locally stressed by thermal shock (Hadni and Thomas 1972, Chynoweth and Feldmann 1960). The best-known example is a wedge-shaped domain observed in barium titanate.

Mechanical compatibility does not imply that the elastic energy is zero. Examination of Fig. 4.7, which shows structural models of the 90° and 180°

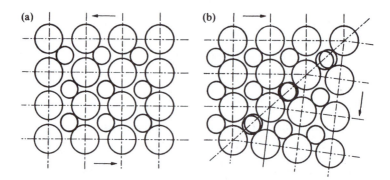

Fig. 4.7. Two-dimensional structural models of a 180° domain wall and a 90° domain wall in a tetragonal perovskite structure. (After Kittel 1972.)

domain walls in a perovskite structure, is helpful to demonstrate this. For both of these models mechanical compatibility is obvious since the spontaneous distortion parallel to the wall, but a few unit cells away, is the same for both domains. The centre of the domain wall has been drawn with a centrosymmetric unit cell characteristic of the paraelectric phase. The spontaneous strain given by eqn (4.1.19) varies with **D** through the domain wall. To prevent macroscopic deformation at the interface there must be additional strains in the wall region given by

$$x_i - x_i^0 = g_{ij}^\dagger (D_j - D_j^0) + Q_{ijk}(D_j D_k - D_j^0 D_k^0) \qquad (4.1.20)$$

where $\mathbf{x}^0$ and $\mathbf{D}^0$ are the spontaneous strain and displacement at the centre of a domain.

4.1.3. Minimizing the free energy

We shall now evaluate the polarization variation across a domain and some equilibrium characteristics of the periodic domain structure of Fig. 4.5 by minimizing the total free energy (Zhirnov 1958, Chenskii 1972). Using eqn (4.1.7) we have

$$G_1 - G_1^0 = W_E + W_{\text{dip}} + W_x + \int_V (\tfrac{1}{2}\alpha D^2 + \tfrac{1}{4}\gamma D^4) \, dV. \qquad (4.1.21)$$

The effect of strain energy W_x can be included by renormalizing the coefficients α and γ. Details of the renormalization have been outlined for domains in $BaTiO_3$ and Rochelle salt by Zhirnov (1958). Dipole energy W_{dip} can also be renormalized to replace P by D. For the geometry of Fig. 4.5 the electric displacement has components only along the z-direction ($D \equiv D_z$) and varies along the x-direction. Thus to determine the variation of D across a domain wall for which $W_E = 0$ we minimize G_1 with respect to D. Using eqns (4.1.17) and (4.1.21) we derive the equation

$$\tfrac{1}{2} f \frac{d^2 D}{dx^2} = aD + bD^3 \qquad (4.1.22)$$

where f, a, and b are the renormalized values of f_2, α, and γ respectively, and obtain the solution

$$D = D_0 \tanh \left(\frac{x}{\delta} \right) \qquad (4.1.23)$$

where $\delta = (-f/a)^{1/2}$ is the wall thickness, and $D_0 = (-a/b)^{1/2}$ is the displacement at the domain centre. The wall energy σ can now be evaluated using this solution in eqn (4.1.21) and integrating over x. Zhirnov finds the value

$$\sigma = -(4\sqrt{2}/3)\, \delta a D_0^2. \qquad (4.1.24)$$

Using this result eqn (4.1.21) can now be rewritten in the form

$$G_1 - G_1^0 = W_E + \sigma/d + \tfrac{1}{2} a D_0^2 + \tfrac{1}{4} b D_0^4 \qquad (4.1.25)$$

per unit volume. The value of D_0 for the cases where W_E is important (i.e. from eqn (4.1.11), for crystals which are not 'thick' in the sense $d/t \to 0$) can now be established by minimizing eqn (4.1.25) with respect to D_0 using eqn (4.1.11) with $P_0 \to D_0$. Although ε^* in eqn (4.1.11) and the numerical coefficient in eqn (4.1.24) may vary slightly with crystal thickness (Chenskii 1972), we can write more generally $\sigma = -K\,\delta a D_0^2$ ($K \approx 1 \cdot 5$–$2 \cdot 0$) and obtain

$$D_0^2 = \frac{-1}{b}\left(a - 2K\frac{\delta}{d}a + 2\varepsilon^*\frac{d}{t}\right). \qquad (4.1.26)$$

Substituting back into eqn (4.1.25) and minimizing with respect to d gives finally the equilibrium domain width

$$d = \left(\frac{-K\,\delta a t}{\varepsilon^*}\right)^{1/2}. \qquad (4.1.27)$$

For the 180° domain walls in BaTiO$_3$ the above analysis (Zhirnov 1958) leads to an estimate for wall thickness $\delta \sim 10^{-7}$ cm in reasonable agreement with experimental results which indicate a width of less than 5 nm (Tanaka and Honjo 1964). An early measurement of 400 nm by Little (1955) using optical microscopy was presumably an upper limit set by the resolution of the microscope. The wall energy estimated from eqn (4.1.24) is $\sigma \sim 10$ erg cm^{-2}.

Estimates for other domain structures obtained by Zhirnov are listed in Table 4.1. It is noticeable that the domain-wall thickness is only a few unit

TABLE 4.1

Domain wall thickness and wall energy for BaTiO$_3$ and Rochelle salt

Crystal and wall type	Wall thickness (10^{-8} cm)	Wall energy (erg cm^{-2})
BaTiO$_3$ 180° walls	5–20	10
BaTiO$_3$ 90° walls	50–100	2–4
Rochelle salt 0°C 180° walls	12	0·06
Rochelle salt 20°C 180° walls	220	0·012

After Zhirnov 1958.

cells wide in sharp contrast with ferromagnetic domain walls. This difference is due to the fact that magnetic exchange energy is much larger than the elastic energy and slow rotation of the magnetization vector occurs over hundreds of unit cells. While the vector magnetization of a ferromagnet has a constant magnitude, the ferroelectric polarization decreases to zero at the centre of a domain wall. This has been demonstrated for 90° walls by electron microscopy (Yakunin *et al.* 1972). It has also been shown by Raman scattering from 180° domain walls in gadolinium molybdate that the walls have at least some component characteristic of the paraelectric phase (Shepherd and Barkley 1972). However, the wall thickness in this case was estimated to be between 0·8 and 3·0 μm—considerably greater than observed from other experiments. This experiment was performed in a dynamic manner with moving domain walls so that this thickness may not be characteristic of the static domain configuration (see § 4.2).

4.1.4. The effect of free carriers

In the expression (eqn (4.1.21)) for the free energy of a ferroelectric the effect of free carriers was neglected. This is a good approximation for a large group of ferroelectrics, since the domain structure is established during time short compared with the relaxation time of the material and the free-carrier concentration can be neglected. Free carriers do, however, play an important role in ferroelectrics. Eventually free carriers compensate for the depolarization fields by conduction, and removal of these fields would then favour single-domain formation. In most cases the domain walls are sufficiently immobile to prevent noticeable changes of the domain structure, and defects or internal strains can stabilize a domain pattern. An increase of the domain size in TGS over a period of several days after the crystal was cooled through the ferroelectric phase transition (Moravets and Konstantinova 1968) may be evidence for the crystal establishing a new free energy state. When a multi-domain crystal is heated into its paraelectric phase the non-uniform distribution of free charge which compensated for $\mathbf{P}_s$ below the Curie temperature gives rise to high electric fields above the Curie temperature during the dielectric relaxation time. For instance if $\mathbf{P}_s \sim 10 \ \mu\mathrm{C \ cm}^{-2}$ and the dielectric constant $\varepsilon \sim 100$ then $E \sim 10^6 \ \mathrm{V \ cm}^{-1}$! These fields can induce ferroelectric behaviour above T_c (Chynoweth 1960). If the crystal is cooled below T_c again the space charge encourages the formation of the original domain structure.

The effect of space charge and surface charge on the static domain structure of insulating ferroelectrics has been considered by Selyuk (1971). Additional terms in the free energy due to the free charges are the electrostatic energy

$$W_\rho = \int_V \rho\varphi \ \mathrm{d}V \qquad (4.1.28)$$

where ρ is the free-charge density and φ is the electrical potential, and a term due to the change in binding energy of the charge carriers with the lattice in their redistributed sites

$$W_\xi = \int_V \xi \ \mathrm{d}n \qquad (4.1.29)$$

where n is the change of carrier density and ξ is the chemical potential of the carriers in their redistributed sites. Of course the space charge also affects the distribution of electric field and displacement in the ferroelectric (see § 4.4). For a complete description eqn (4.1.21) must be solved with these additional terms in a self-consistent manner. If the space charge is located in a surface region of depth L_ρ which is much smaller than the crystal dimensions or the domain dimensions then it has been shown (Selyuk 1971) that the free energy associated with the space charge is independent of the domain structure. Since the depolarizing energy in the bulk of the crystal is then zero a single-domain structure is energetically favourable as mentioned earlier. However, when L_ρ is greater than the domain dimensions or even the crystal dimensions the space-charge energy is influenced by the domain structure. Selyuk's analysis indicates that at temperatures well below the Curie temperature a single-domain structure is still favoured, but very close to the transition temperature when $\mathbf{P}_s$ becomes small (particularly for a second-order transition) a transition to a multi-domain state is predicted. No observations of this effect below the transition temperature have been made.

4.2. Polarization reversal

The primary feature distinguishing ferroelectrics from other pyroelectrics is that the spontaneous polarization can be reversed, at least partially, with an applied electric field. The first demonstration of polarization reversal was by dielectric hysteresis (Valasek, 1920). The observation of hysteresis loops

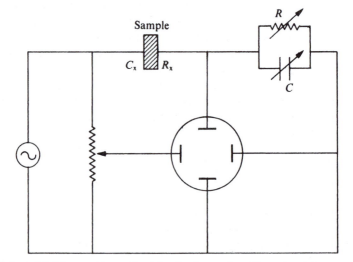

Fig. 4.8. A modified Sawyer–Tower circuit for the observation of ferroelectric hysteresis loops. The parallel RC circuit allows compensation for any phase shift due to conductivity or dielectric loss in the sample. (After Sinha 1965.)

using basically the Sawyer–Tower (Sawyer and Tower 1930) circuit of Fig. 4.8 is still frequently used for the identification of ferroelectrics. A typical loop is shown in Fig. 4.9. At low fields and at very high fields a ferroelectric behaves like an ordinary dielectric (usually with a high dielectric constant),

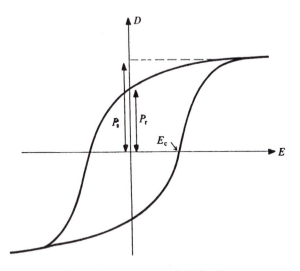

Fig. 4.9. A hysteresis loop illustrating the coercive field E_c, the spontaneous polarization P_s, and the remanent polarization P_r.

but at the so-called coercive field E_c polarization reversal occurs giving a large dielectric non-linearity. The area within the loop is a measure of the energy required to twice reverse the polarization. At zero field the electric displacement within a single domain (saturated value of the displacement) has two values corresponding to the opposite orientations of the spontaneous polarization. In a multi-domain crystal the average zero-field displacement can have any value between these two extremes. In principle the spontaneous polarization is equal to the saturation value of the electric displacement extrapolated to zero field as shown in Fig. 4.9 since $\varepsilon_0 E$ usually gives a negligible contribution to D. The remanent polarization P_r, shown in Fig. 4.9 to be the displacement at zero field, may be different from the spontaneous polarization P_s if reverse nucleation occurs before the applied field reverses. This can happen either in the presence of internal (or external) stresses or if the free charges below the crustal surfaces cannot reach their new equilibrium distribution during each half-cycle of the loop. This effect can be minimized by cycling the loop at very low frequencies.

Some caution must be exercised with this kind of measurement. Firstly, the loop gives only a measure of the switchable portion of the spontaneous polarization. If part of P_s is clamped then a measure of P_s from the loop will

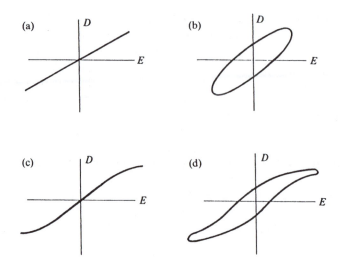

Fig. 4.10. Sketch of D–E loops for (a) a linear lossless dielectric, (b) a linear lossy dielectric, (c) a non-linear lossless dielectric, and (d) a nonlinear lossy dielectric.

be erroneous. Secondly, ferroelectric hysteresis can easily be confused with non-linear dielectric loss. To clarify this in Fig. 4.10 D–E loops have been sketched for four non-ferroelectric dielectrics: (a) a linear lossless dielectric; (b) a linear lossy dielectric; (c) a non-linear lossless dielectric; (d) a non-linear lossy dielectric. While the linear dielectric loss can be compensated for by modification of the Sawyer–Tower circuit so that (b) looks like (a), this cannot be done for the non-linear contribution. Errors in measurement of P_s and even incorrect identification of ferroelectrics can arise from these non-linear contributions to the displacement.

In materials which are simultaneously ferroelectric and ferroelastic and in which these effects are coupled (see § 10.2) the spontaneous polarization may be reversed or partially reversed with an applied stress instead of an electric field, so that D–X hysteresis loops can be observed in an analogous manner to D–E loops. The switchable component of P_s and the 'coercive stress' can be obtained in this way. Of course non-linear piezoelectric effects will give the same errors in this case as the non-linear dielectric effects already discussed.

Hysteresis loops can be observed under adiabatic or isothermal conditions or with clamped or unclamped crystals, each measurement yielding a different value of the coercive field. In coupled ferroelectric–ferroelastic crystals there can be an enormous difference of the coercive stress depending on whether the crystal is switched with the polar faces of the crystal open or short-circuited, since depolarization fields set up during polarization reversal oppose further reversal. In the cases of $Tb_2(MoO_4)_3$ and

$Gd_2(MoO_4)_3$ the open-circuit coercive stress is more than an order of magnitude greater than under short-circuit conditions (Keve, Abrahams, Nassau and Glass 1970). Presumably equivalent differences between clamped and unclamped coercive field could be observed.

The coercive field, defined by Fig. 4.9 as the field at which half the polarization has reversed, is also a function of the frequency of the alternating field since there is some switching time associated with polarization reversal. The shape of the loop consequently depends on the dependence of the switching time on the applied field.

The first quantitative experiments to determine the time and field dependence of polarization reversal were carried out by Merz (1954) on $BaTiO_3$. The experimental procedure involved applying a step-function field to the crystal and measuring the displacement current density $J = dP/dt$ as a function of time using the circuit of Fig. 4.11. Typical results for field pulses

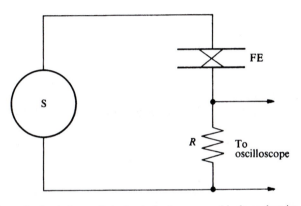

Fig. 4.11. Schematic circuit for studies of polarization reversal in ferroelectrics, where S is a source of applied voltage and FE is the ferroelectric crystal. (After Fatuzzo and Merz 1967.)

applied parallel and antiparallel to P_s of a c-domain crystal are shown in Fig. 4.12, showing the switching time t_s. The electrical time constants of the measuring circuit are of course short compared with t_s. For complete polarization reversal the area under curve A of Fig. 4.12 is $2AP_s + \varepsilon_0\varepsilon AE$ where A is the electrode area normal to P_s and E is the applied field. The area under the curve B of Fig. 4.12 is $\varepsilon_0\varepsilon AE$, so from these two measurements P_s can be obtained.

This procedure is still probably the most direct method for studying the switching behaviour for crystals of very low conductivity. For conducting crystals the conduction current obscures the displacement current, and other techniques based on the measurement of the remanent polarization during or following the application of fields are more appropriate, piezoelectric (Husimi and Kataoka 1960) and pyroelectric (Chynoweth 1956a, Ballman

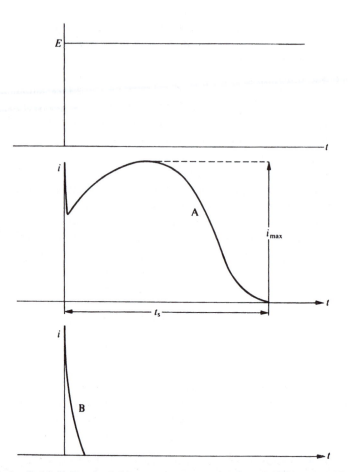

Fig. 4.12. Applied field E and switching current i versus time for a BaTiO$_3$ crystal. The curve marked A is the switching pulse when the applied field is antiparallel to the polarization, and the curve B is obtained when the field is parallel to the polarization and no switching occurs. (After Merz 1956.)

and Brown 1972) techniques probably being the most useful since they give a quantitative, non-destructive measure of the polarization.

For BaTiO$_3$ the switching time followed an exponential law

$$t_s \propto \exp(\alpha/E) \qquad (4.2.1)$$

for fields from 1 to 15 kV cm^{-1}, while at higher fields up to 100 kV cm^{-1} (Stadler 1958)

$$t_s \propto E^{-n} \qquad (4.2.2)$$

where the index n was about 1·5 for BaTiO$_3$.

Several other materials which have been studied also show either an exponential or power-law dependence of t_s on E. The index n has been found to vary from 1 in sodium nitrite (Hatta and Sawada 1965) to 7 in tetramethyl ammonium trichloromercurate (Fatuzzo 1960). The constants α and n and the constants of proportionality are usually temperature dependent, the switching time usually decreasing as the Curie temperature is approached. For instance for TGS and colemanite (Hayashi 1972 a) the activation field α has a temperature dependence

$$\alpha \propto P_s^3 / T. \tag{4.2.3}$$

There has been a great deal of experimental and theoretical work on the mechanism of polarization reversal and domain dynamics. The experimental data have been collected from both direct microscopic observation of domains during switching and electrical measurements on bulk crystals. Most of the effort has been directed towards TGS and BaTiO$_3$. From the complexity of the problem it is not surprising that there is no universal rule governing all ferroelectrics. The switching behaviour can be greatly affected by the nature of the electrodes, (liquid-electrolyte electrodes often giving quite different results from metal electrodes (Fatuzzo and Merz 1967, Camlibel 1969)), the nature of the crystal surfaces, electrical conductivity (Fridkin *et al.* 1972), the domain geometry (Janta 1971), and the presence of defects (see § 4.3).

Polarization reversal can be accomplished either by the growth of existing domains antiparallel to the applied field, by domain-wall motion, or by the nucleation and growth of new antiparallel domains. The domains can grow either along the polar direction or by sideways motion of 180° domain walls. The relative importance of these processes depends on the material and the applied field, although it appears that for most ferroelectrics with the polar axis normal to a plate-shaped crystal, sideways motion of 180° domain walls is preferred to forward motion. Two examples where complete polarization reversal was accomplished by the nucleation of a single domain followed by sideways motion of the 180° domain wall across the entire crystal are afforded by BaTiO$_3$ (Miller 1958) and Gd$_2$(MoO$_4$)$_3$ (Kumada 1969). Miller used liquid electrodes and predetermined the nucleation site by dimpling the crystal surface. Direct measurement of the domain-wall motion and switching current showed that the wall velocity in BaTiO$_3$ varied as

$$v = v_\infty \exp(-\delta/E) \tag{4.2.4}$$

where δ is a constant up to 300 V cm^{-1} which appears to depend on the defect concentration of the crystal (Miller and Savage 1958). This field dependence of v is in agreement with the bulk switching rate of eqn (4.2.1). In Gd$_2$(MoO$_4$)$_3$ the domain-wall velocity varied linearly with field giving rise to a switching rate

$$t_s = L/\mu E \qquad (4.2.5)$$

where μ is the domain-wall mobility and L is the distance moved by the domain wall.

It was recognized very early (Landauer 1957) that, since a domain wall is generally only a few lattice spacings thick, then the energy required to move the wall through one lattice spacing is comparable to the wall energy itself. Since the energy gained in moving the wall by one lattice constant is much smaller than the wall energy, this kind of wall motion is unlikely. The only physical model which seems to explain the experimental data is that the observed sideways wall motion is caused by the nucleation and two-dimensional growth of reversed step-like domains on existing 180° walls. Detailed analysis of this model by Miller and Weinreich (1960) showed that the field and temperature dependence of the domain-wall velocity agreed fairly well with experiment in the low-field regime (eqn 4.2.1). Miller found that the nuclei steps had to be only one unit cell thick and that the rate-controlling process was the rate of nucleation. Stadler and Zachmanidis (1963) extended this theory to account for the high-field behaviour of $BaTiO_3$ (eqn (4.2.2)) by including the effects of nucleation steps two or more unit cells thick.

Recently Hayashi (1972 b) has developed a general formulation for the kinetics of domain-wall motion from which it is possible to determine absolutely the rate of nucleation and domain-wall motion in the low- and high-field regimes. The approach of these theories is briefly as follows.

The total energy change upon nucleation of an antiparallel domain is

$$\Delta W = \sigma A + W_E - \mathbf{D}.\mathbf{E}V \qquad (4.2.6)$$

where the first two terms are the wall energy and depolarization energy of the nucleus discussed in § 4.1.2 and the third term is the electrostatic energy of the nucleus of volume V in the applied field. The polarization and local field E within the nucleus are assumed to be uniform.

The rate of domain nucleation is thus proportional to $\exp(-\Delta W/kT)$. The probability of nucleation of domains by thermal fluctuations in the bulk of a ferroelectric is extremely small since an energy of about $10^8\, kT$ is needed to nucleate such a domain (Landauer 1957). However, at crystal surfaces, inhomogeneities, or at existing domain walls, ΔW can be considerably smaller than this. From energy considerations the most favourable shape of a nucleus forming on an existing wall where it intersects the crystal surface is the triangular shape shown in Fig. 4.13. The requirement that the nucleus shall grow gives rise to a critical size of the nucleus determined by the conditions

$$\frac{\partial W}{\partial a} \leq 0 \quad \text{and} \quad \frac{\partial W}{\partial l} \leq 0 \qquad (4.2.7)$$

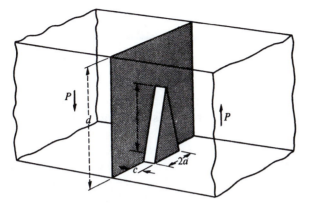

Fig. 4.13. Schematic drawing of a triangular step on a 180° domain wall. The applied electric field is parallel to the spontaneous polarization on the left-hand side of the figure. (After Miller and Weinreich 1960.)

where a and l are the dimensions shown in Fig. 4.13 and the thickness c is taken as one lattice constant in the low-field regime. The critical dimensions decrease with increasing local field E. Using an approach equivalent to that leading to eqns (4.1.11) and (4.1.24) to calculate the depolarization energy and wall energy of the triangular-shaped nucleus, the energy ΔW of the nucleus follows from eqn (4.2.6), and hence the nucleation rate $1/\tau$ can be calculated giving

$$\frac{1}{\tau} \propto \exp\left(-\frac{\delta}{E}\right) \qquad (4.2.8)$$

where δ is only slightly dependent on E. The temperature dependence of δ is

$$\delta \propto \frac{P_s^3}{T\varepsilon_a^{1/2}} \qquad (4.2.9)$$

where ε_a is the a-axis dielectric constant for the geometry of Fig. 4.13. If the nucleation rate is the rate-controlling process of polarization reversal then eqns (4.2.8) and (4.2.9) also describe the switching rate $1/t_s$ and its temperature dependence, in agreement with experiment (eqns (4.2.1) and (4.2.3)). However, the situation is more complicated than this for the general case. For instance nuclei can grow on the growing steps of earlier nucleations as shown in Fig. 4.14 (Hayashi 1972 b).

In general it is not obvious whether the wall velocity depends on the nucleation rate or the growth velocity of the steps. Nucleation of new steps adds to the wall thickness of a moving wall while sideways growth decreases the thickness. If nucleation occurs at the electrodes, as seems energetically the most favourable, then the domain wall could be either concave (nucleation at both electrodes) or oblique (nucleation at one electrode), giving an

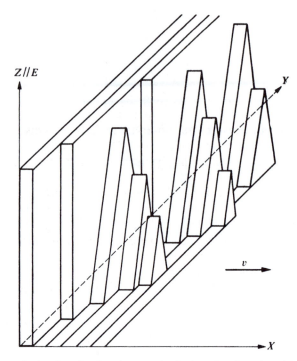

Fig. 4.14. Schematic drawing showing the growth of triangular-shaped nuclei on the steps formed by earlier nucleations. (After Hayashi 1972 b.)

apparent broadening of the wall. The complete description of thick and thin moving walls in high and low fields becomes too complicated to reproduce here, so Hayashi's results will just be summarized. At low fields an exponential law of the form (eqn (4.2.8)) describes the field dependence of the domain-wall velocity and is replaced by a power law at high fields. The value of δ is not strictly constant but changes with increasing field so the transition to the high-field regime is smooth. This is due to the change of the relative importance of nucleation and sideways growth of nuclei on the wall. The transition to a power law at high fields is due to the fact that two-dimensional triangular nuclei which are important at low fields cannot nucleate at high fields since the critical dimension a^* in the a-direction becomes smaller than a lattice constant. Instead, one-dimensional nuclei are formed with their length along the polar axis and these grow in a two-dimensional manner. This explanation of the power-law dependence of domain-wall velocity appears to be more favourable than that of Stadler and Zachmanidis, since for $BaTiO_3$, according to Hayashi, triangular domains cannot nucleate at fields above $25\,\mathrm{kV\,cm^{-1}}$. Experimentally the power law is valid up to

450 kV cm^{-1}. The exponent of E at high fields depends on several material parameters so it is not surprising that different materials show widely different exponents.

4.2.1. Barkhausen pulses

The displacement current observed during the reversal of spontaneous polarization of a ferroelectric is often accompanied by transient current pulses. These pulses were first observed in ferromagnets by Barkhausen (1919) and the ferroelectric analog still bears his name. The main interest in these pulses in ferroelectrics has been for the study of the mechanisms of polarization reversal.

Barkhausen pulses are resolved by monitoring the switching current as the polarization is slowly reversed by an electric field slowly increasing in either a smooth or step-like manner. In an extensive study of $BaTiO_3$ using evaporated metal electrodes Chynoweth (1958) counted typically 10^5 to 10^6 Barkhausen pulses during complete polarization reversal. The total charge in the pulses corresponded to only a small fraction (10^{-3} to 10^{-2}) of the total charge switched. The switched volume associated with each pulse increases with crystal thickness and the pulse height increases with increasing field. The pulse counting rate was found to vary approximately as the switching current so that the total count was field independent.

From these observations Chynoweth concluded that the pulses were associated with the nucleation and forward growth of wedge-shaped domains through the crystal, while the remainder, and majority, of the switching current was due to sideways motion of the domains. Complementary optical and electrical studies (Miller 1958) suggested that Barkhausen pulses occurred when two growing domains came together giving rapid switching of the decreasing volume between them.

Careful measurements by Rudyak, Kudzin, and Panchenko (1972) of the shapes of Barkhausen pulses in $BaTiO_3$ have shown that in general both nucleation and the fusion of domains give pulses of different characteristics. The pulses associated with nucleation and forward motion of domains include a greater volume of repolarizing region and have a much sharper leading edge than the pulses associated with domain fusion.

The large amount of data which has been collected for a number of materials has been reviewed by Rudyak (1970). The general features are similar to $BaTiO_3$. In general the counting rate $N(t)$ varies proportionally with the volume of reversed polarization $P(t)$. This is true as a function of temperature for TGS (Kamaev 1967) and Rochelle salt (Abe 1956), and $N(t)$ even shows the same asymmetry as $P(t)$ in crystals which have biased hysteresis loops (Kamaev 1967). The number and shape of the pulses vary considerably from one material to another and with the defect content of the crystals.

Barkhausen pulses have been observed with applied stress in Rochelle salt owing to ferroelectric–ferroelastic coupling (Shuvalov, Rudyak, and Kamaev 1965), and in the absence of any applied fields near ferroelectric phase transitions in $BaTiO_3$, SbSI, TGS, KDP and Rochelle salt (Bogomolov, Ivanov, and Rudyak 1969) and in SbSI under illumination (Rudyak and Bogolmolov 1967). Thus Barkhausen pulses provide a technique for studying polarization reversal in small volumes under conditions which might be difficult by other techniques.

4.3. The influence of defects

The purpose of this section is to outline the principal effects of inhomogeneities such as impurities and radiation damage on the dielectric properties and switching behaviour of ferroelectrics. The effects of defects on the optical and paramagnetic resonance spectra will be considered in Chapter 12. These discussions refer primarily to dilute defect concentrations (less than 1 per cent), since gross departures from stoichiometry or changes of crystal composition are considered in Chapter 8.

Defects in any crystalline lattice generally cause deformation of the surrounding volume and modification of the local fields. The magnitudes of these effects are more difficult to estimate than the effects of domain walls since there are no 'mechanical-compatibility' conditions and the extent of the crystal deformation depends on the nature of the defect, its site in the crystal, and the host-defect interaction.

In an acentric site a defect has a dipole moment associated with it

$$\overline{\Delta\mu} = \Delta\mu_d + \sum_i q_i \, \Delta x_i \qquad (4.3.1)$$

where $\Delta\mu_d$ is the change of dipole moment at the defect site and Δx_i is the displacement of charge q_i in the surrounding lattice owing to the presence of the defect. In pyroelectrics and ferroelectrics $\overline{\Delta\mu}$ reflects the polar nature of the host so that the sense of $\overline{\Delta\mu}$ is the same for all similar defects within a single domain. If the defect concentration N is sufficiently dilute that the interaction between them can be neglected, the macroscopic polarization change is

$$\Delta P = N\overline{\Delta\mu} \qquad (4.3.2)$$

which must now be included in the expression for the free energy of the ferroelectric. For example we write an internal energy

$$U = f(D) = f(D_p + \Delta P) \qquad (4.3.3)$$

where D_p is the electric displacement of the perfect crystal. To a first approximation this can be expanded

$$U = f(D_\text{p}) + \left(\frac{\partial U}{\partial D}\right)_{D_\text{p}} \Delta P. \tag{4.3.4}$$

Differentiating we have

$$\frac{\partial f(D_\text{p})}{\partial D_\text{p}} = E - \frac{\Delta P}{\varepsilon \varepsilon_0} \tag{4.3.5}$$

where ε is the dielectric constant of the perfect crystal. Phenomenologically the crystals should behave as the perfect crystal biased by the defects, with an equivalent field $\Delta P / \varepsilon \varepsilon_0$.

The macroscopic average value of the polarization ΔP due to the defects is adequate to account for several of the experimentally observed effects. However, this description cannot be used for any effects which depend on the local variation of polarization around the defects.

When the polarization of a crystal is reversed by an applied field, the polarization ΔP due to the defects may or may not reverse. If it does reverse then the coercive field will depend on both the field required to switch the defects and the sign and magnitude of ΔP. In general the presence of defects tends to increase the coercive field. If ΔP does not reverse in an external field the defects can have a marked effect on the switching properties depending on the distribution of ΔP throughout the crystal volume. If all the dipoles have the same sense the hysteresis loops will appear biased as shown in Fig. 4.15 (a). If the dipole orientations are ordered over large regions, but different regions are antiparallel in the same way as ferroelectric domains, then the hysteresis loops could appear as in Fig. 4.15 (b). If the dipoles are

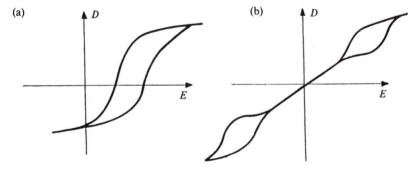

Fig. 4.15. Biased hysteresis loops which may arise owing to the presence of defects in a ferroelectric crystal.

completely random, as could happen if they were introduced into the crystal in a nonpolar phase, then the loop would appear normal with an increased coercive field.

There has been a considerable amount of work on distorted hysteresis loops and the stabilization of the spontaneous polarization in one sense in several ferroelectrics. Some as-grown crystals such as GASH and colemanite show large internal bias fields which have been associated with naturally occurring defects. Many crystals show distorted hysteresis loops such as Fig. 4.15 (a) and (b) (or combinations of the two) following irradiation with X-rays, γ-rays, or neutrons. Similar results are often obtained with various impurities.

The best example of the stabilization of the spontaneous polarization by impurities for which the microscopic mechanism is understood is TGS doped with alanine impurities during growth (Keve, Bye, Whipps, and Annis 1971). A crystal grown below the Curie temperature with about 0·1 per cent (in the crystal) of either L-alanine or D-alanine grows as a single domain and its hysteresis loop appears biased as shown in Fig. 4.15 (a). The internal bias field allows only one stable polarization state in the absence of an applied field. Crystals grown with both L- and D-alanine are multi-domain and hysteresis loops appears as in Fig. 4.15 (b).

This can be understood in terms of the crystal structure of TGS (see § 9.4) and the molelcular structures of glycine and alanine shown in Fig. 4.16. The three glycine groups I, II, and III all participate in the polarization reversal of TGS, but the main reversible dipole is that associated with glycine I. During polarization reversal the glycine molecule changes into its mirror image by rotation about the a axis and its two α-hydrogens interchange roles. The alanine molecule is sufficiently similar, structurally and chemically, to substitute (at least partially) for the glycine molecule, but reversal of this dipole does not take place since the CH_3 group and the hydrogen on the alanine cannot physically interchange by simple rotation. It is clear from Fig. 4.16 that all alanine molecules of the same form substituting for glycine will have their dipoles pointing in the same direction. Thus there is a macroscopic irreversible polarization associated with the alanines which polarizes the host in one sense. The internal bias field obtained from hysteresis loops is directly proportional to the alanine concentration in the crystal.

The internal bias field due to the alanine in TGS also has pronounced effects near the ferroelectric Curie temperature. The effects of the internal bias field E_b are equivalent to an external field of the same magnitude in undoped TGS. The peak value of the dielectric constant decreases and shifts to higher temperature, and the polarization persists at temperatures much higher than in pure crystals as shown in Fig. 4.17. Some broadening of the phase transition also occurs owing to the inhomogeneous nature of the internal fields. Presumably the spontaneous polarization due to the aligned defects persists at all temperatures above T_c, which implies that a true paraelectric state does not exist in alanine-doped TGS.

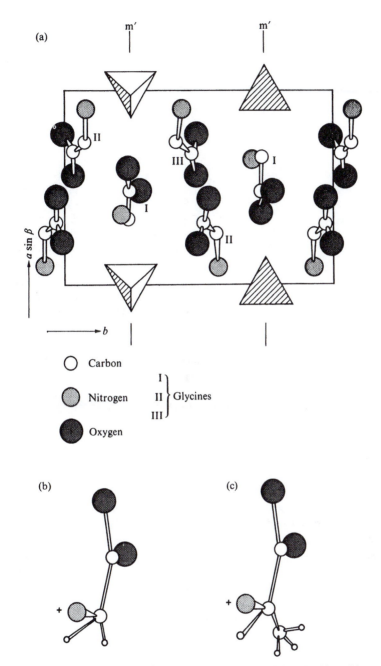

Fig. 4.16. (a) View along the c-axis of the structure of TGS (after Hoshino, Okaya, and Pepinsky 1959, Itoh and Mitsu 1973) showing the three glycine groups I, II, III. The sulphate ions are represented by tetrahedra and the oxygens have been omitted; m' are sets of pseudo-mirror planes in which the glycine I molecules are inverted upon switching. (b) A glycine molecule compared with (c) an L-alanine molecule (after Keve *et al.* (1971). The smallest circles are hydrogen.

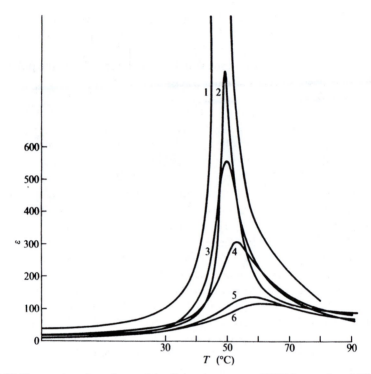

Fig. 4.17. Temperature dependence of the dielectric constant of TGS for a series of different internal bias fields E_b corresponding to various alanine impurity concentrations. Curves 1–6 correspond respectively to values $E_b = 0$, 5·6, 8·0, 19·2, >34, >80 kV cm^{-1} where the magnitude of E_b is obtained from hysteresis-loop measurements. (After Bye, Whipps, and Keve 1972.)

Generally speaking the effects of ionizing radiation are similar to the effects of doping already discussed and can be accounted for in a similar manner. However, in these cases the defects have not been unambiguously identified so that no detailed mechanism can be proposed. Early studies of neutron irradiation of $BaTiO_3$ (Wittels and Sherrill 1957) and $BaTiO_3$ ceramics (Lefkowitz 1959) showed that as neutron doses were increased distortion of hysteresis loops occurred first, and then loops completely disappeared as the domains were clamped by defects. At the same time the dielectric-constant peaks shifted in temperature and broadened. At extremely high doses (10^{20} neutrons/cm^2) anisotropic expansion of the $BaTiO_3$ lattice was observed followed by a transformation to a cubic phase.

Detailed studies of TGS (Chynoweth 1959) showed that even very small doses of X-irradiation can give large changes of the ferroelectric properties—biased loops, double loops, increased coercive field, and decreased dielectric constant—and direct evidence of domain clamping was obtained from optical studies. It is interesting that X-irradiation of single-domain TGS gives biased loops as in Fig. 4.15 (a) while X-irradiation of multi-domain TGS gives double loops as in Fig. 4.15 (b), demonstrating that the polarity of the radiation-induced defects is determined by the polarity of the host during irradiation. With increasing dosage the dielectric-constant peak and polarization curve broaden and move to *lower* temperatures (Bye and Keve 1972, Alemany *et al.* 1973). The effects of X-irradiation on the structure of TGS may offer some clues as to the detailed mechanisms but no definite identification has yet been made (Bye and Keve 1972).

Complex annealing processes occur with both d.c. and a.c. fields for which no explanations have been obtained. It seems that there are slow relaxation processes associated with defect motion which affect irradiated crystals for some time after electric fields have been removed, or after heat treatment above T_c (Chynoweth 1959, Bye and Keve 1972).

Qualitatively similar results to those described for TGS have been observed following irradiation of Rochelle salt (Zheludev *et al.* 1955, Boutin, Frazer, and Jona 1963, Okada 1961).

A final word should also be said about crystal fatigue, which may also be related to the formation or migration of defects during polarization reversal. Upon repeated switching of crystals, notably $BaTiO_3$, the spontaneous polarization becomes clamped and the coercive field increases. The decay of the switchable portion of P_s depends on several factors including (a) the ambient atmosphere (Anderson *et al.* 1955), the decay being slower in oxidizing atmospheres than in reducing atmospheres, (b) the electrode material, the decay being slower with liquid electrodes than with metal electrodes (Kudzin and Panchenko 1972, Williams 1965), and (c) the form of the applied fields, the decay being more rapid with pulsed fields than with low frequencies (Fatuzzo and Merz 1967). No satisfactory explanation of fatigue has been proposed. Although it appears that the effect is associated with the build-up of space charge near the crystal surfaces and the interaction of this charge with domain nuclei, it is not clear whether ionic or electronic processes play a dominant role.

4.4. Surfaces

There have been a great many experimental observations which indicate that ferroelectric behaviour at the surfaces of crystals is different from the bulk behaviour. Some of these observations have been made on as-grown crystals where the chemical nature of the surfaces may not be well characterized; for instance the stoichiometry may be different from the bulk, or chemical adsorption may have occurred. These effects would be particularly important for crystals grown at high temperatures. However, other observations have been made on freshly etched crystals where the effects are presumably intrinsic to interfaces between the perfect crystal and the external medium.

The experiments may be briefly summarized as follows.

(1) Early X-ray and electron microscopic examination of fine $BaTiO_3$ powders (Anliker, Brugger, and Känzig 1954) showed that the structure of the $BaTiO_3$ surfaces is different from the bulk. Indeed, above the Curie temperature the surfaces remained tetragonal even though the bulk of the crystal became cubic. It was proposed that this behaviour was due to high fields in the surface region which induced the tetragonal phase above T_c. However, these anomalous surface regions were not observed with etched single crystals using transmission electron microscopy (Tanaka and Honjo 1964). The structure of the thin (100 nm) crystals was found to be characteristic of the bulk material and at the Curie point all domains disappeared. At temperatures above 500°C new structures appeared to grow on the $BaTiO_3$ surfaces which may be due to surface decomposition of the crystals and which may account for some of the observations of Anliker. Studies on as-grown $BaTiO_3$ by electron mirror microscopy (English 1968) showed the presence of a surface layer of different domain structure but the ferroelectric nature of this layer was the same as the bulk crystal. It was not determined whether this behaviour is also characteristic of etched crystals, nor does this observation necessarily imply the presence of a space-charge field at the surface.

(2) Optical absorption measurements have shown (Coufova and Arend 1962) that the optical absorption of as-grown crystals of $BaTiO_3$ differs from the bulk. Since this region can be etched away it is not intrinsic to the crystal surface.

(3) Evidence for polarized surface layers in $BaTiO_3$ was obtained with pyroelectric measurements (Chynoweth 1956 b). The surface of a single c-domain plate is heated with optical pulses as shown in Fig. 4.18 (a) and (b), the absorption occurring at the electrode. The pyroelectric response below the Curie temperature and just above the Curie temperature is shown in Figs. 4.18 (c) and (d). Fig. 4.18 (c) is the characteristic response of a uniformly polarized crystal (see § 5.4), while the differentiated waveform of Fig. 4.18 (d) is characteristic of a crystal having a polarized surface region and a non-polar bulk. The differentiation is due either to rapid thermal relaxation of the surface region with the bulk of the crystal (see § 5.4) or to the unpolarized bulk acting as a series capacitor as suggested by Chynoweth.

(4) The capacitance and dielectric loss of crystal wafers depends on their thickness (Schlosser and Drougard 1961) and electrode material (Glass 1969 a). These experiments suggest that the surface region has a much higher impedance (lower dielectric constant and dielectric loss) than the bulk and also does not vary as rapidly with temperature near T_c as the bulk of the crystal. The surface layer consequently influences the overall im-

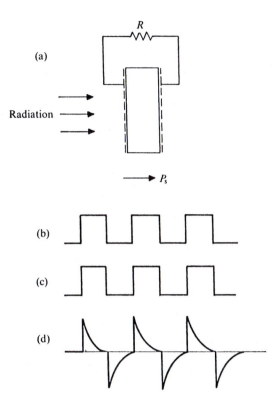

Fig. 4.18. Pyroelectric studies of surface effects showing (a) a sketch of the experimental configuration, (b) the input optical signal waveform, (c) the pyroelectric response from a uniformly polarized crystal, and (d) the pyroelectric response of a crystal polarized only in the surface region. Curves (c) and (d) are typical of those seen in $BaTiO_3$ below and just above the ferroelectric–paraelectric Curie temperature.

pedance and the effect is particularly marked near T_c where the bulk impedance is greatly reduced. These observations can be explained by the presence of intense space-charge fields near the ferroelectric–electrode interface which modify the ferroelectric behaviour (see § 5.3). Application of d.c. fields (Schlosser and Drougard 1961, Triebwasser 1960) leads to a further build-up of space charge with a corresponding drop of the field in the bulk of the crystal. The measured relaxation times in $BaTiO_3$ were greater than 10^{-4} s at 120°C giving rise to dispersion of the dielectric behaviour at low frequencies.

(5) The switching time t_s and coercive field E_c of $BaTiO_3$ vary with the crystal thickness d (Merz 1956) as

$$\alpha = \alpha_0\left(1 + \frac{d_0}{d}\right) \tag{4.4.1}$$

and

$$E_c = E_0\left(1+\frac{d_0}{d}\right)$$
(4.4.2)

where d_0 is the surface-layer thickness and α is the activation field which determines the switching time according to eqn (4.2.1). Likewise the activation field of eqn (4.2.4) which determines the domain-wall velocity varies as

$$\delta = \delta_0\left(1+\frac{d_0}{d}\right).$$
(4.4.3)

These relationships were found to be approximately valid (Callaby 1966) until the crystal thickness d became smaller than the surface-layer thickness d_0 when the switching behaviour became independent of thickness; d_0 was measured to be typically 20 μm. These data can all be adequately accounted for by a surface layer in which the domain-wall mobility is less than the bulk of the crystal (Callaby 1966). This causes the wall to move sideways with a convex shape and with a corresponding increase of depolarization energy of the wall.

Except for the structural investigations all the measurements of surface-layer effects were carried out on electroded crystals. In this case Schottky barriers would be expected at the ferroelectric–metal junction. Above the ferroelectric Curie temperature the Schottky barrier can be described in the usual way using Poisson's equation and the requirement that in equilibrium, for no current flow ($J = 0$), any electrical potential must be balanced by a carrier-diffusion potential. For a crystal plate normal to the z-direction the equations are

$$\frac{dD}{dz} = Ne$$
(4.4.4)

and

$$J = 0 = Ne\mu E + eD'\frac{dN}{dz}$$
(4.4.5)

where μ is the carrier mobility, N the carrier concentration, e is the electronic charge, and D' is the carrier-diffusion coefficient. Using Einstein's relation $eD' = \mu kT$ and integrating

$$N = N_0 \exp\left(\frac{-eV}{kT}\right).$$
(4.4.6)

This is a Boltzmann distribution where the potential $V = \int -E\,dz$ is referred to the crystal surface and N_0 is the surface concentration of carriers determined by the boundary conditions. From (4.4.4) we have

$$\frac{d(\varepsilon E)}{dz} = \frac{N_0 e}{\varepsilon_0} \exp\left(\frac{-eV}{kT}\right). \tag{4.4.7}$$

For a linear dielectric, when ε is independent of the electric field, this equation leads to the well-known result for a Schottky barrier junction

$$V = \frac{2kT}{e} \ln\left(\frac{z}{z_0} + 1\right) \tag{4.4.8}$$

where $z_0 = (2\varepsilon\varepsilon_0 kT/N_0 e^2)^{1/2}$ is the width of the surface layer where the electric field falls to half its value at the surface. For insulating materials N_0 is small (Mott and Gurney 1940) so that the effects of space charge can extend well below the crystal surface.

It was pointed out by Bloomfield, Lefkowitz, and Aronoff (1971) that above the Curie temperature where some of the surface effects were studied the dielectric behaviour is extremely non-linear at the high fields encountered in the surface region. Then eqn (4.4.7) becomes

$$\frac{dE}{dz} = \frac{N_0 e}{\varepsilon\varepsilon_0} \exp\left(\frac{-eV}{kT}\right) - \frac{E}{\varepsilon}\frac{d\varepsilon}{dz}$$

$$= \left\{1 + \frac{d(\ln \varepsilon)}{d(\ln E)}\right\}^{-1} \frac{N_0 e \exp(-eV/kT)}{\varepsilon\varepsilon_0}. \tag{4.4.9}$$

The extra term $d(\ln \varepsilon)/d(\ln E)$ due to the non-linearity is usually negative, so this term acts to enhance the effect of space charge on the internal field.

The space-charge fields evaluated above arise from the boundary conditions which require the carrier concentration at the surface to be different from the bulk value. For instance for metal electrodes N_0 is determined by the Fermi distribution function of electrons in the metal. Even in the absence of electrodes in the paraelectric phase the boundary conditions are determined by the surface states of the crystal and also give rise to large electric fields below the crystal surface. The magnitudes of the fields and the charge distribution depend on the detailed nature of the surface states. These considerations are discussed in detail in several texts (i.e. Many, Goldstein, and Grover 1965).

In the ferroelectric phase the situation is more complicated because of the additional boundary condition which requires that the charge density at the surfaces must neutralize the depolarization fields in the bulk of the crystal owing to the spontaneous polarization. The traditional view is that the compensating charges are located in the electrodes at the crystal surface or in surface states following the poling of the cyrstal. For instance the surface charge density necessary to screen a polarization of $20 \ \mu\text{C cm}^{-2}$ is $1 \cdot 2 \times 10^{14}$

carriers/cm^2. Because of the high density of free carriers in a metal the additional charge has little effect on the energy states of the metal and all the charge can be located very close to the crystal surface.

This view is probably adequate for crystals which are highly insulating both at the poling temperature and the ambient temperature and which are contacted with metal electrodes. Ormancey and Godefroy (1974) have calculated that even if the Fermi level of the metallic electrode is separated from the conduction band of the insulator by 0·5 eV, the injected free carrier density in the crystal is small compared with the 10^{14} carriers per cm^2 required to compensate for the polarization. However, when a crystal is not a perfect insulator, or the electrodes are not metallic conductors, this picture is no longer valid. The free carriers diffuse into (or from) the ferroelectric from (or to) the electrodes and charge depletion or accumulation layers are created at the interface. The complete charge distribution is determined by the electronic states in the bulk and at the surface, the work functions of the electrode and ferroelectric, and the spontaneous polarization. Since the spontaneous polarization of the ferroelectric depends on the free- and bound-carrier density (since a unit cell containing an extra electron has a different dipole moment from the perfect unit cell), the charge distribution must be solved in a self-consistent manner. Several detailed theoretical papers discussing effects of this kind may be found in the Soviet literature (Guro, Ivanchik, and Kovtonyuk 1968, 1970, Vul, Guro, and Ivanchik 1973, Selyuk 1973).

To demonstrate the importance of the spontaneous polarization on the surface behaviour we shall greatly simplify the problem by first considering a semiconducting ferroelectric in which the effect of free carriers on the spontaneous polarization is neglected so that P_s is uniform throughout the entire crystal. Eqn (4.4.8) is still the correct solution for the potential inside the crystal but the value of N_0, the carrier concentration at the crystal surface, is now affected by P_s. For the depolarization field to vanish in the bulk of the crystal well below the surface we require the total charge density

$$\int eN \, \mathrm{d}z = P_s \qquad (4.4.10)$$

where the integral is taken over the surface region. From eqns (4.4.6) and (4.4.8) we have

$$N = N_0 \left(\frac{z_0}{z_0 + z} \right)^2 \qquad (4.4.11)$$

so that if we neglect the other boundary conditions such as deep surface states we find

$$N_0 = \frac{P_s^2}{2\varepsilon\varepsilon_0 kT} \qquad (4.4.12)$$

and

$$z_0 = \frac{P_s}{eN_0} \qquad (4.4.13)$$

Approximate values of the carrier density N_0, the thickness of the surface layer, and the surface field E_0 evaluated from these equations are listed in Table 4.2. For distances of the order of a lattice constant the above analysis

TABLE 4.2

Order-of-magnitude estimates of the surface carrier density N_0, the Schottky barrier width z_0, and the surface field E_0 due to the free charge which compensates for the spontaneous polarization P_s of a semiconducting ferroelectric†

P_s (C cm^{-2})	N_0 (cm^{-3})	z_0 (cm)	E_0 (V cm^{-1})
10^{-5}	10^{21}	10^{-7}	3×10^5
10^{-6}	10^{19}	10^{-6}	3×10^4
10^{-7}	10^{17}	10^{-5}	3×10^3

† A dielectric constant $\varepsilon \sim 100$ and a temperature $T \sim 300$ K has been assumed for the calculation.

is not valid, but nevertheless it is seen that the effect of the polarization on the free-carrier density and the surface fields can be enormous. The electronic energy bands become degenerate for electron densities of this magnitude. It is also important that both E_0 and the charge carriers have the opposite sign for positive and negative polar faces of the crystal. The bending of the energy bands is shown schematically in Fig. 4.19. However, this picture is oversimplified. The neglected surface states are likely to play an important role also, but there is no experimental information on such states in ferroelectrics. If, for instance, the density of surface states was exactly equal to the charge density needed to compensate the polarization, then there would be no surface fields at equilibrium in unelectroded crystals. Undoubtedly the real free ferroelectric surface is somewhere between these extremes.

Since the effect of the electrodes was neglected in obtaining the result eqn (4.4.12) it is evident that large surface fields and carrier densities are present in non-electroded crystals. Experimental evidence for the polarization dependence of the surface charge densities is provided by optical-breakdown measurements on LiNbO$_3$ (Zverev *et al.* 1972). The threshold for optical damage was found to increase as the temperature of the crystal

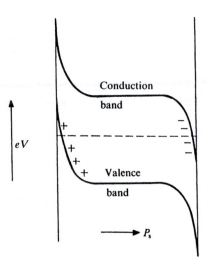

Fig. 4.19. Schematic diagram showing the bending of the conduction and valence bands of a semiconducting ferroelectric due only to the polarization change at the surface. The effect of surface states is to cause further band bending.

was increased, and this was attributed to the decreasing concentration of free carriers available for avalanche breakdown at the crystal surface. At the present time there do not seem to be any reports of experimental work aimed at detecting the different surface behaviour at the positive and negative polar faces of the crystal due to this band bending.

It is also important to notice that similar surface effects should be present at grain boundaries in semiconducting ferroelectric ceramics where div $\mathbf{P}_s \neq 0$. Effects of this kind undoubtedly contribute to the boundary-layer effects discussed in § 15.2.

A second case to consider before concluding this discussion of surface effects is the other limiting situation of a perfectly insulating ferroelectric in contact with semiconducting electrodes. Since the free carriers necessary to screen the polarization cannot be provided by conduction in the ferroelectric they must originate in the semiconductor. The large surface concentration of free carriers may significantly distort the electronic energy bands of the semiconductor and create large depletion regions, or else complete compensation of the polarization may not even be possible.

A situation which has been investigated experimentally (Wurfel and Batra 1973), shown in Fig. 4.20, is the contact of thin, insulating ferroelectric films on silicon substrates. The other electrode is metallic and it is assumed that band bending in this electrode is negligible. With p-type silicon, when the polarization vector is directed toward the silicon electrode (positive polarization), charge compensation must be provided by the accumulation of minority carriers (electrons) and the depletion of majority carriers. When

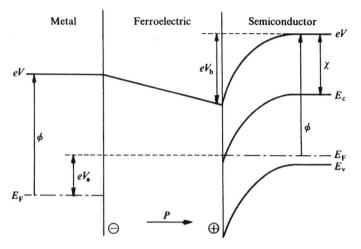

Fig. 4.20. Distribution of the potential V in a ferroelectric thin film sandwiched between a metal and a semiconducting electrode of equal work functions ϕ for an applied bias V_a. The valence and conduction band energies E_v and E_c undergo a band bending eV_b, and V is drawn parallel to the bands but displaced from them by the electron affinity χ. E_F is the Fermi level. (After Wurfel and Batra 1973.)

the polarization is directed away from the silicon electrode (negative polarization), positive charge required for compensation is provided by majority-carrier accumulation.

The behaviour of the polarization was studied experimentally in a triglycine sulphate film 1 μm thick, in contact with a p-type silicon electrode with 10^{15} acceptors/cm^3, by observing hysteresis loops with a Sawyer-Tower circuit. On the time scale of the experiments it can be assumed that there was no carrier transport in the TGS crystal. In the dark (no photo-carriers) only negative polarization values were stable—the ferroelectric could not be switched at a frequency of 100 Hz. Positive polarization values were unstable because of the large band bending in the semiconductor owing to majority carrier depletion which causes large depolarization fields. The thermal generation rate of minority carriers was too slow for compensation at 100 Hz. At lower frequencies ($\sim$0·1 Hz) the time available for building up minority carrier charge increases and the compensation improves. In any case, for negative polarization values the band bending is small because of the accumulation of majority carriers in surface states which compensates for the polarization. When the silicon substrate was illuminated with high-intensity light completely symmetrical hysteresis loops were observed since a high density of photocarriers was available to compensate for the polarization. With n-type silicon substrates exactly the reverse behaviour was noted—only positive polarization was stable at high frequencies.

It must be emphasized that these experiments were performed on thin ferroelectric films where external voltages corresponding to the band gap of the semiconductor contact can give rise to large electric fields in the ferroelectric. For thicker, insulating crystals effects of this kind become correspondingly less important. In ferroelectrics of finite conductivity semiconducting contacts could greatly affect the surface regions. For high enough surface fields only one polarization direction becomes stable. Indeed if both contacts are similar semiconductors then, as the thickness of the ferroelectric sandwich is decreased, the polarization cannot be compensated for with either polarity and the polar state becomes unstable (Batra, Wurfel, and Silverman 1973).

5

EXPERIMENTAL STUDY OF
THERMODYNAMIC PROPERTIES

THIS CHAPTER is devoted to the experimental study of the various low-frequency compliances relating the electric displacement **D**, entropy S, and strain **x** to the electric field **E**, temperature T, and stress **X** which characterize a ferroelectric, and measurement of the spontaneous polarization. In view of the enormous growth of experimental data on a large number of ferroelectrics in recent years it is no longer possible to present an exhaustive review of the results. Instead, particular reference is made only to a few materials which adequately demonstrate the techniques and principles.

5.1. Poling

From the considerations of the last chapter it is clear that the domain-wall contribution to the free energy will affect most of the macroscopic properties of ferroelectrics. Consequently the starting point for the experimental study of a ferroelectric is the preparation of single crystals which are both single domain electrically and untwinned crystallographically. In most materials electrical poling may be achieved by cooling a crystal from the paraelectric phase into the ferroelectric phase in an applied electric field parallel to the polar crystallographic axis under constant **E** conditions. The single-domain state is then thermodynamically stable in an unstressed crystal since depolarization fields are neutralized by charge flow in the external circuit. Short-circuiting the polar faces of a crystal during cooling could, in principle, be sufficient to produce the single-domain state, but in practice this is not usually sufficient because of temperature gradients, internal strains, and surface effects. To maintain constant **E** conditions in the crystal intimate contact of the electrodes with the polar faces, either by metallic films or by liquid electrolytes, is necessary. Placing a crystal between capacitor plates without intimate contact is not likely to produce single-domain material because of depolarization fields.

An alternative procedure for poling ferroelectrics is to switch reverse domains below the Curie temperature with electric fields higher than the coercive field. For several materials high coercivity allows poling in this way only near to T_c.

If the paraelectric crystal system is higher in symmetry than the ferroelectric phase the probability of crystallographic twinning is large. These twins can be prevented or removed by applying a uniaxial stress. In some ferroelectric–ferroelastic crystals the ferroelectric domains and the mechanical twins are fundamentally identical and may be removed either by electrical or mechanical means alone. In some perovskite ferroelectrics such as $BaTiO_3$ it is often straightforward to obtain a single-domain plate by application of electric fields alone, while for others such as $PbTiO_3$ it is easier to remove a-axis-type domains with an applied stress in the plane of the plate while orienting c-axis or $180°$ domains with a field. The crystal $Ba_2NaNb_5O_{15}$ represents a case where crystallographic twinning occurs at a tetragonal–orthorhombic phase transition at $300°C$—considerably below the paraelectric to ferroelectric cubic–tetragonal transition at $585°C$. In this case best results are achieved by simultaneous application of stress and field (Singh, Draegert, and Geusic 1970).

The determination of whether a crystal is fully poled may be made either by direct observation of domains (§ 4.1.1), by measurement of the pyroelectric coefficient, or by second harmonic generation. For the pyroelectric measurement it is necessary to know dP/dT for the poled material or else to assume that the maximum attainable dP/dT corresponds to the fully poled state. With the second-harmonic measurement (SHG) this assumption is not made. The intensity of SHG is measured as the crystal is rotated about the polar axis. For a single-domain crystal Maker fringes (see Chapter 13) of high contrast are observed, while for an unpoled crystal the fringe contrast is poor.

Direct measurement of the spontaneous polarization of a ferroelectric can only be made by methods which destroy the polarization state of the crystal. Switching techniques have already been discussed in § 4.2, where the spontaneous polarization is reversed by an external field and the total charge displaced ($2AP_s$ where A is the electroded area of the crystal normal to the polarization) is measured in an external circuit. Alternatively if the crystal conductivity or the coercive field is too high one of the pyroelectric techniques discussed in § 5.4 can be used. While pyroelectric techniques give an accurate measure of the relative change of the polarization as a function of temperature, for an absolute determination of P_s the polarization at some temperature must be known. For instance, at a second-order Curie temperature P_s becomes zero so that both the sense and magnitude of P_s can be determined absolutely by pyroelectric techniques for this case. Of course none of the procedures can be used to determine P_s in non-ferroelectric pyroelectrics. Neither can the sense of the polarization be determined since the pyroelectric coefficient could have either sign. In such cases one has to resort to more indirect techniques such as detailed structure determination, with some assumptions concerning the charge distribution, or perhaps a non-linear optical technique (see Chapter 13) where it may be possible to infer P_s from the non-linear coefficients.

5.2. Experimental constraints

The importance of the experimental constraints during measurements of ferroelectric materials has already been noted in Chapter 3. For ordinary dielectrics it is well known that measurements at constant stress may differ significantly from measurements at constant strain because of elastic contributions to the various compliances. In ferroelectrics important differences also exist between measurements at constant **D** and constant **E** or between constant T and constant S. In practice clamped measurements are normally made with a dynamic method at frequencies well above the fundamental mechanical resonance frequency ($\omega_m = \pi v/L$, where v is the velocity of sound in the material and L is a crystal dimension), while unclamped measurements are made at low frequencies or in a quasi-static manner. Adequate clamping of a crystal by external means is normally difficult. Similarly adiabatic and isothermal measurements must be made at freqencies above and below the thermal relaxation frequency ($\omega_t = G/\mathscr{C}$ where G and $\mathscr{C}$ are the thermal conductivity and capacity), while open-circuit (const **D**) and short-circuit (const **E**) measurements are performed above and below the dielectric relaxation frequency. External short-circuiting of crystals is adequate for constant-**E** measurements provided that any stresses or temperature changes of the crystal are uniform.

Relationships between the various compliances may be obtained from the thermodynamic potentials of eqn (3.2.4). For instance the relationship between the adiabatic and isothermal dielectric permittivity is

$$\varepsilon_{ij}^{S,X} = \left(\frac{\partial D_i}{\partial E_j}\right)_{S,X} = \left(\frac{\partial D_i}{\partial E_j}\right)_{T,X} + \left(\frac{\partial D_i}{\partial T}\right)_{E,X}\left(\frac{\partial T}{\partial E_j}\right)_{S,X}. \qquad (5.2.1)$$

Using the Maxwell relation from enthalpy H

$$\left(\frac{\partial T}{\partial E_j}\right)_{S,X} = -\left(\frac{\partial D_i}{\partial S_j}\right)_{E,X} = -\frac{(\partial D_i/\partial T)_{E,X}}{(\partial S/\partial T)_{E,X}}$$

so that

$$\varepsilon_{ij}^{S,X} = \varepsilon_{ij}^{T,X} - \frac{(p_i^{E,X})^2 T}{c^{X,E}} \qquad (5.2.2)$$

where $p_i = \partial D_i/\partial T$ is a pyroelectric compliance (or coefficient) and c is specific heat. The difference between the adiabatic and isothermal dielectric permittivity may be appreciable near a phase transition where the pyroelectric coefficient may be large.

Likewise the difference between clamped and unclamped pyroelectric coefficients is

$$p_i^{x,E} = p_i^{X,E} + \frac{d_{ij}^{T,E}\alpha_i^{x,E}}{s_{ij}^{x,E}} \qquad (5.2.3)$$

where α_i is the thermal-expansion coefficient dx_i/dT and compliances d_{ij} and s_{ij} have been defined in § 3.2. The piezoelectric contribution to the pyroelectric coefficient via thermal expansion is usually referred to as secondary pyroelectricity and may be comparable to the primary effect $p_i^{x,E}$ (Lang 1971).

In the intermediate region between high- and low-frequency measurements the frequency dependence of the compliances is determined by the equations of motion describing the relaxation of the system. Dielectric and thermal relaxation are normally heavily damped processes which can be described by a simple relaxation. For instance at a frequency $\omega \sim \omega_t = 1/\tau_t$ eqn (5.2.2) becomes

$$\varepsilon_{ij}^{S,X} = \varepsilon_{ij}^{T,X} - \frac{(p_i^{E,X})^2 T}{c^{X,E}} \frac{\omega\tau_t}{(1+\omega^2\tau_t^2)^{1/2}}. \tag{5.2.4}$$

The situation is usually more complicated for mechanical relaxation since damping may not be small and mechanical resonance may greatly affect the compliances. The piezoelectric resonator is based on the excitation of mechanical resonance by electric fields via the piezoelectric effect, and the solutions for any crystal geometry may be found in standard texts (Mason 1950). Mechanical resonance may be excited thermally, via thermal expansion, so there is a similar effect on the pyroelectric coefficient in polar crystals.

For simplicity consider a long, thin crystal of length L in the 1 or x-direction with the polar axis in the 3- or z-direction. The crystal is heated uniformly in such a manner as to produce a periodic temperature change at frequency ω. The polar crystal faces are electroded and the ends of the crystal at $x = \pm L/2$ are unclamped. In addition to the linear equations appropriate for this problem, viz.

$$dD_3 = d_{31} \, dX_1 + \varepsilon_{33} \, dE_3 + p_3 \, dT \tag{5.2.5}$$

and

$$dx_1 = s_{11} \, dX_1 + d_{31} \, dE_3 + \alpha_1 \, dT \tag{5.2.6}$$

we have Newton's equation of motion

$$\rho\frac{\partial^2\xi}{\partial t^2} = \frac{\partial X_1}{\partial x} \tag{5.2.7}$$

where ρ is the macroscopic material density and ξ is a displacement variable $(dx_1 = \partial\xi/\partial x)$. For simplicity the superscripts on the compliances have been omitted and Maxwell's relation $(\partial D_3/\partial X_1)_{E,T} = (\partial x_1/\partial E_3)_{X,T}$ has been used. From the boundary conditions we have $\partial E_3/\partial x = \partial T/\partial x = 0$ and it follows from eqns (5.2.6) and (5.2.7) that

$$\rho\frac{\partial^2\xi}{\partial t^2} = -\omega^2\rho\xi = \frac{1}{s_{11}}\frac{\partial^2\xi}{\partial x^2}.$$ (5.2.8)

This equation has been solved for the piezoelectric resonator (Mason 1950). Writing $\gamma = (s_{11}\rho)^{1/2}$, which is the inverse velocity of sound, we have

$$\mathrm{d}x_1 = (d_{31}\,\mathrm{d}E_3 + \alpha_1\,\mathrm{d}T)\left\{\frac{\cos(\gamma\omega x)}{\cos(\frac{1}{2}\gamma\omega x)}\right\}.$$ (5.2.9)

Eliminating $\mathrm{d}x_1$ and $\mathrm{d}X_1$ from eqns (5.2.5), (5.2.6), and (5.2.9), the displacement $\mathrm{d}D_3$ may be expressed in terms of $\mathrm{d}E_3$ and $\mathrm{d}T$ and the compliances as a function of x. Integrating over x, the total displacement current density is

$$J = \frac{1}{L}\int_{-L/2}^{+L/2}\frac{\partial D_3}{\partial t}\,\mathrm{d}x$$

and we obtain

$$J = i\omega\,\mathrm{d}E_3\left[\varepsilon_{33} - \frac{d_{31}^2}{s_{11}}\left\{1 - \left(\frac{2}{\gamma\omega L}\right)\tan(\tfrac{1}{2}\gamma\omega L)\right\}\right]$$

$$+ i\omega\,\mathrm{d}T\left[p_3 - \frac{d_{31}\alpha_1}{s_{11}}\left\{1 - \left(\frac{2}{\gamma\omega L}\right)\tan(\tfrac{1}{2}\gamma\omega L)\right\}\right].$$ (5.2.10)

The first term in parentheses in (5.2.10) describes the well-known piezoelectric resonance contribution to the crystal impedance. The second term describes the effect of thermally excited resonance on the pyroelectric coefficient. In practice acoustic and dielectric loss prevent the terms in parentheses from diverging when $\omega = \pi/\gamma L$.

Experimental observation (Glass and Abrams 1970 a) of the response of a long thin $LiTaO_3$ crystal to a modulated infrared beam from a CO_2 laser which produces the sinusoidal temperature variation is shown in Fig. 5.1 as the modulation frequency is varied. A sharp peak in the response is observed as predicted by eqn (5.2.10) at the fundamental longitudinal resonance frequency of the crystal. The smaller peak was attributed to coupling of the longitudinal mode with a low-frequency flexural mode.

The Fourier transform of Fig. 5.1, obtained experimentally by thermally exciting the crystal with a single short optical pulse, is shown in Fig. 5.2. The initial pyroelectric response is followed by a slowly decaying oscillatory piezoelectric signal.

In eqns (5.2.5) and (5.2.6) linear relationships between the variables were assumed. Non-linear effects are usually more important in ferroelectrics than in ordinary dielectrics, especially close to a ferroelectric–paraelectric phase transition, so it is important to define the compliances in terms of

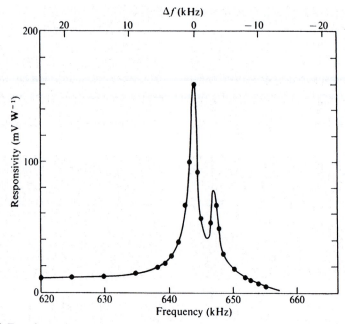

Fig. 5.1. Experimental measurement of the pyroelectric response of a freely suspended LiTaO$_3$ crystal as the modulation frequency of the absorbed radiation is varied. The crystal dimensions are 0·4 cm along the pyroelectric axis and 0·09 × 0·002 cm perpendicular to this axis. (After Glass and Abrams 1970 a.)

infinitesimal changes of the variables. Non-linearities are most simply included by allowing the compliances to be functions of the variables. For instance at constant stress and temperature the dielectric permittivity is defined by (again omitting constraint superscripts)

$$dD_i = \varepsilon_{ij}\, dE_j \qquad (5.2.11)$$

where

$$\varepsilon_{ij} = \varepsilon^0_{ij} + \left(\frac{\partial \varepsilon_{ij}}{\partial E_k}\right)_{X,T} E_k + \left(\frac{\partial \varepsilon_{ij}}{\partial X_k}\right)_{E,T} X_k + \left(\frac{\partial \varepsilon_{ij}}{\partial T}\right)_{E,X} T$$

$$+ \text{higher-order non-linearities.} \qquad (5.2.12)$$

Strictly speaking the dielectric constant is the linear part ε^0_{ij} and the other coefficients are constant non-linear coefficients. In this case ε^0_{ij} corresponds to a measurement at $T = 0$ as well as $X = E = 0$, so that the temperature dependence of the dielectric permittivity, Curie–Weiss behaviour and so on should for consistency be described by the non-linear coefficients. Common practice allows the dielectric permittivity to be temperature dependent but with $E = 0$ and $X = 0$ (or atmospheric pressure), and describes only the field

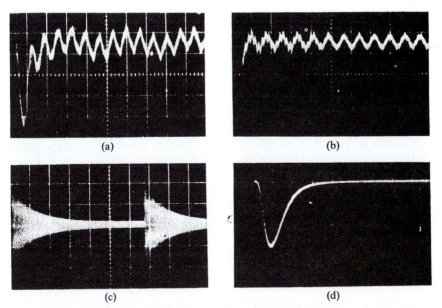

Fig. 5.2. The pyroelectric response of a freely suspended $LiTaO_3$ crystal (a, b, c) and a crystal clamped by cementing it to a substrate (d) to a heat pulse of 200 ns duration from a CO_2 laser. The horizontal time scales are (a) $0.5 \mu s$/div, (b) $2 \mu s$/div, (c) $500 \mu s$/div, and (d) $0.2 \mu s$/div. The vertical scales are arbitrary units. (After Glass and Abrams 1970 a.)

and stress dependence by non-linear coefficients. In what follows we shall prefer to treat all the independent variables in the same way and to consider ε_{ij} to be the total dielectric permittivity, which may be explicitly separated into linear and non-linear parts when it is convenient to do so.

In §§ 5.3 to 5.7 experimental measurements of the compliances are described in the absence of static fields and stresses ($E = X = 0$) and in § 5.8 and § 5.9 the field and stress dependences are discussed.

5.3. Small-signal dielectric measurements

The dielectric constant and dielectric loss may be obtained from a measurement of the real and complex admittance of a crystal in conventional ways, i.e.

$$J/E = \sigma + i\omega\varepsilon = (\sigma - \omega\varepsilon'') + i(\omega\varepsilon') \qquad (5.3.1)$$

where $\varepsilon = \varepsilon' + i\varepsilon''$ has been separated into real and imaginary parts. These measurements are particularly straightforward with impedance bridges which separate the conductance and capacitance of the crystals. For this reason, as well as the fact that qualitative information may be obtained from unpoled crystals, ceramics, and powders, dielectric measurements are most commonly used for the identification of phase transitions and the recording of transition temperatures. At frequencies between 100 Mhz and optical frequencies the crystal must be part of a transmission line, waveguide, or resonant cavity, and the impedance is measured by conventional RF or microwave techniques (Von Hippel 1954). Studies at optical frequencies are discussed in Chapter 7.

Measurements below 100 MHz usually involve electroding crystals, in which case the results may be affected by the space-charge fields just below the contacts. Problems of this kind can become serious close to a phase transition where the bulk impedance of the crystal becomes very low (see § 4.4). These effects can sometimes be identified by using different contact electrodes or crystal geometries. To avoid electrical contact completely dielectric measurements can be made by suspending a rod-shaped crystal in an electric field and measuring the torque on the rod due to the dielectric anisotropy.

An interesting method of measuring the dielectric properties in the limit of zero field is by measurement of the power spectrum of the polarization fluctuations, or electric 'noise', of a crystal (Brophy and Webb 1962). Using Nyquist's theorem to relate current (i) and polarization fluctuations, we have

$$\langle \Delta i^2 \rangle = \omega^2 \langle A^2 \, \Delta D^2 \rangle = \frac{4kT}{R} \tag{5.3.2}$$

where A is the electrode area and $\langle \Delta D^2 \rangle = \langle \Delta P^2 \rangle$ in zero field. The averages $\langle \Delta D^2 \rangle$ and $\langle \Delta P^2 \rangle$ are spectral densities of the fluctuations and R is the real part of the crystal impedance. The short-circuit noise current is directly related to the polarization fluctuations and can be used as the 'source' for the measurement of the dielectric properties of the crystal. For example the measured voltage fluctuations $\langle \Delta V^2 \rangle$ across a load resistor R_L at temperature T_L connected as shown in Fig. 5.3 is

$$\langle \Delta V^2 \rangle = \left(\frac{4kT}{R} + \frac{4kT_L}{R_L} \right) \left\{ \left(\frac{1}{R} + \frac{1}{R_L} \right)^2 + \omega^2 C^2 \right\}^{-1} \tag{5.3.3}$$

where the term in curly brackets is $1/Z^2$, with Z the real part of the parallel crystal–load impedance. By varying the load resistance and measuring the voltage fluctuations the crystal resistance and capacitance can be measured at different frequencies.

The usefulness of dielectric measurements for the characterization of ferroelectrics is clear from an examination of the low-frequency measurements of the c-axis dielectric constant ε_c' in high-purity $BaTiO_3$ and $LiTaO_3$ shown in Figs. 5.4 and 5.5. Extremely large increases of the dielectric constants of these materials are observed at their paraelectric–ferroelectric

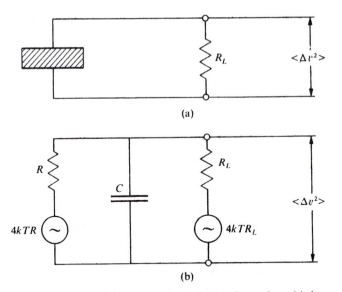

(a)

(b)

Fig. 5.3. Measurement of a noise spectrum of polarization fluctuations; (a) the experimental arrangement, and (b) the equivalent circuit. (After Brophy 1965.)

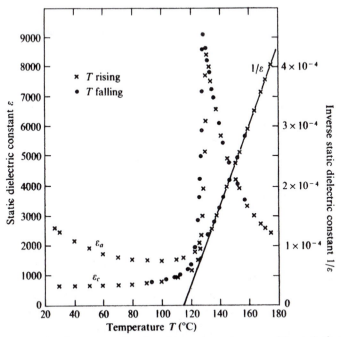

Fig. 5.4. Static dielectric constant $\varepsilon\,(=\varepsilon')$ and its reciprocal measured for a single crystal of $BaTiO_3$ as a function of temperature. Below T_c spontaneous polarization occurs parallel to the c-axis. (After Johnson 1965.)

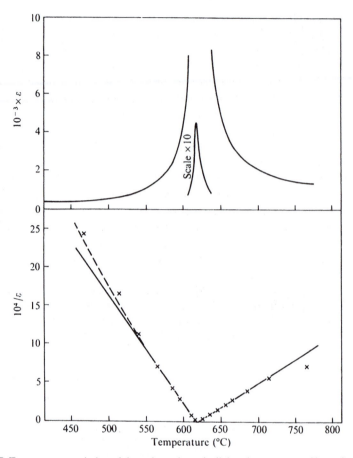

Fig. 5.5. Temperature variation of the polar-axis static dielectric constant and its reciprocal for LiTaO₃. (After Glass 1968.)

phase transitions at 135°C and 618°C characteristic of first- and second-order phase transitions respectively. Both materials show Curie–Weiss behaviour, i.e.

and

$$\text{for BaTIO}_3 \quad \varepsilon'_c = C/(T - T_0)$$

$$\text{for LiTaO}_3 \quad \varepsilon'_c = C/(T - T_c) \tag{5.3.4}$$

for several degrees above the phase transitions with $C = 1 \cdot 8 \times 10^{5} °C$ for BaTIO₃ and $1 \cdot 6 \times 10^{5} °C$ for LiTaO₃. These Curie constants vary somewhat from one sample to another but they are typical values for a wide variety of displacive ferroelectrics. Order–disorder ferroelectrics tend to have Curie constants $C \sim 10^{3} °C$ (Jona and Shirane 1962). The ratio of the slopes $\partial(1/\varepsilon)/\partial T$ below and above T_c is about $-5 \cdot 3$ for BaTIO₃ and $-2 \cdot 1$ for

LiTaO₃ (corrected to isothermal values) to be compared with the values −8
and −2 expected for first- and second-order phase transitions within the
Devonshire theory of Chapter 3. For first-order transitions theoretical
values of magnitude less than 8 can occur if metastability (hysteresis) at the
transition is considered (Draegert and Singh 1971).

Both LiTaO₃ and BaTiO₃ are good examples of ferroelectric transitions
triggered by an instability of a soft optic mode at the Brillouin zone centre,
resulting in large values of the low-frequency dielectric constant. These
large anomalies are observed both in clamped and unclamped measurements. In contrast to these materials the temperature dependence of both
the clamped and unclamped dielectric constant of $Gd_2(MoO_4)_3$ is shown in
Fig. 5.6. Only a weak dielectric anomaly is observed at low frequencies and
no anomaly at all appears in the clamped measurements. The ferroelectric
phase transition at 159°C is clearly not associated with the softening of a
zone-centre optic mode. It is in fact produced via an indirect coupling
through strain to a soft mode elsewhere in the Brillouin zone. Coupling
effects of this kind are discussed in detail in Chapter 10.

Most materials do not provide such clear-cut evidence for a first- or
second-order phase transition as the LiTaO₃ and BaTiO₃ examples cited
above. Any strain or inhomogeneities in the crystal tend to broaden the
phase transition so that there is no apparent discontinuity in ε'. In some
cases for small broadening the order of the transition may be implied by the

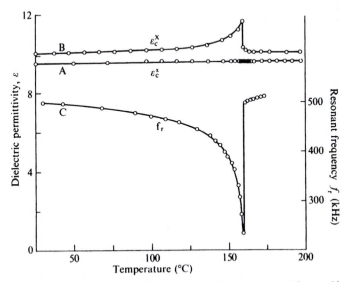

Fig. 5.6. Temperature dependence of the clamped dielectric constant (curve A), the free
dielectric constant (curve B), and the piezoelectric resonance frequency of a single
$Gd_2(MoO_4)_3$ crystal (curve C). (After Cross, Fouskova, and Cummins 1968.)

slopes $\partial(1/\varepsilon)/\partial T$ above and below the phase transition (Draegert and Singh 1971), but in many cases even this is ambiguous.

The dielectric loss ε'' generally follows similar temperature behaviour to the dielectric constant ε', although measurements of ε'' are often masked by the d.c. conductivity σ of the crystals at low frequencies. If the phase transition is sharp, ε' and ε'' peak at the same temperature and both follow Curie–Weiss behaviour as expected from the Kramers–Krönig relations. If the phase transition is not well defined then ε' and ε'' peak at different temperatures, the separation of the maxima depending on the degree of broadening of the transition and the temperature dependence of the dielectric relaxation.

The rate at which the dielectric polarization responds to a change of applied field is determined by the dielectric relaxation times τ of the various contributions to the polarization. Relaxation of the piezoelectric contribution to ε has already been described in eqn (5.2.10). It has been appreciated for many years that dipolar defects in dielectrics also lead to relaxation

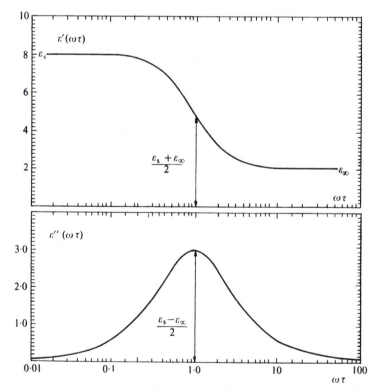

Fig. 5.7. The real part ε' and the imaginary part ε'' of the dielectric constant as a function of $\ln(\omega\tau)$ according to the Debye equation (5.3.5) with $\varepsilon_s = 8$ and $\varepsilon_\infty = 2$.

effects which can be described by the Debye equation

$$\varepsilon = \varepsilon_\infty + \frac{\varepsilon_s}{1 - i\omega\tau} \qquad (5.3.5)$$

where ε_∞ is the high-frequency dielectric constant. The real and imaginary parts of ε are shown in Fig. 5.7. For a symmetrical bistable system of non-interacting dipoles with barrier energy E (see § 2.4) the relaxation time is just given by the Arrhenius equation

$$\tau = \tau_0 = \frac{2\pi\hbar}{kT} \exp\left(\frac{E}{kT}\right). \qquad (5.3.6)$$

In ferroelectrics additional relaxation effects due to domain walls are observed in multi-domain crystals, but of greatest interest in these materials is the relaxation of the ferroelectric dipoles themselves in single-domain material. In an order–disorder context the electrostatic interaction of the dipoles, which leads to the local-field enhancement, results in a dielectric spectrum still of the Debye form but now with a temperature dependence of the relaxation time given by eqns (2.4.19) and (2.4.20) of the Ising model.

Studies of the dielectric constant ε' and dielectric loss ε'' as a function of frequency near the ferroelectric phase transition have revealed this 'critical slowing down' in several materials. The early measurements of ε'' by Hill and Ichiki (1962, 1963) for 'order–disorder' TGS are shown in Fig. 5.8 as

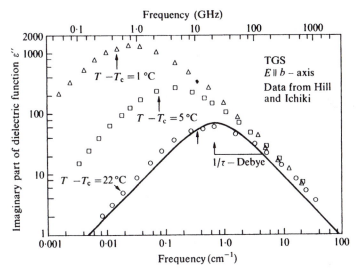

Fig. 5.8. The imaginary part ε'' of the dielectric function of TGS at three temperatures. Arrows mark the relaxation frequency parameter. The solid curve is a fit of a Debye dispersion equation to the data of Hill and Ichiki (1962, 1963). (After Barker 1967.)

they were replotted by Barker (1967). At each temperature the data closely resemble a Debye dispersion curve with a relaxation time $\tau \propto 1/(T - T_c)$ in agreement with mean-field theory. Actually Hill and Ichiki used a Gaussian distribution of Debye functions to account for their data. The frequency at which the dielectric loss peaks is much too low to be a normal propagating phonon mode. However, Barker (1967) has shown that the solid curve in Fig. 5.8 can be equally well described by a Debye relaxation with $1/\tau \approx 0.7$ cm^{-1} or by a highly overdamped oscillator (i.e. eqn (2.1.33)) with a frequency $\omega \approx 200$ cm^{-1} and a damping factor $\Gamma \approx 57\,000$ cm^{-1}. In the limit of high damping the relaxational and oscillator descriptions become identical and it is therefore not always a simple matter to determine whether a system is basically diffusive or oscillatory.

Oscillatory dispersion is often seen at infrared frequencies in soft-mode displacement ferroelectrics (although the motion is often quite highly damped), but even here, as the Curie temperature is approached and the dispersion moves to lower frequencies, it tends to become more and more overdamped and to approach the diffusive limit. In fact most of the ferroelectric systems which have been studied in a region where soft-mode dispersion occurs well below infrared frequencies exhibit a dielectric response which approximates a Debye relaxation. In some crystals such as $AgNa(NO_2)_2$ the relaxation occurs at extremely low frequencies as shown in Fig. 5.9. Practically no dielectric anomaly is observed at frequencies above 100 kHz even though a very sharp transition is indicated at low frequencies.

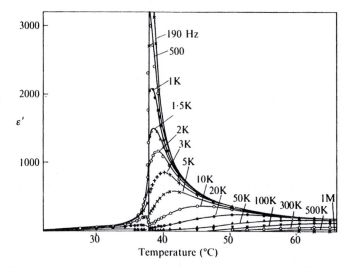

Fig. 5.9. Temperature dependence of the real part of the dielectric function ε' for $AgNa(NO_2)_2$ along the [010] direction at various frequencies. (After Gesi 1970.)

A more general discussion of soft-mode condensation and dispersion near ferroelectric and other structural transitions is given in Chapter 11.

5.4. Pyroelectric measurements

The ancient discovery of pyroelectricity, which is reviewed by Lang (1974), can probably be attributed to the high electric fields which develop across insulating pyroelectrics subjected to relatively small temperature changes. For instance, a crystal with a typical pyroelectric coefficient of 10^{-8} C cm^{-2} K^{-1} and a dielectric constant of 50 develops a field of 50 000 V cm^{-1}—sufficient to break down air—with just a 25°C temperature change.

At equilibrium the depolarization fields due to the polarization discontinuity at the surfaces of a pyroelectric crystal are neutralized by free charge as discussed in Chapter 4. When the crystal temperature is changed the spontaneous polarization changes so that an excess of free charge appears on one of the polar faces of the crystal which gives rise to current flow in the crystal and external circuit, the sense of the current depending on the direction of the polarization change. For most pyroelectric measurements crystals are supported in a manner which allows them to expand freely so that, for a slow uniform temperature change, $\Delta X_i = 0$ and

$$\Delta D_i = \varepsilon_{ij}^{X,T} \Delta E_j + p_i^{E,X} \Delta T. \qquad (5.4.1)$$

For rapid temperature changes $\Delta x_i = 0$ and the ε_{ij} and p_i compliances are replaced by their clamped values. The current density in the crystal is (using a simple scalar formalism)

$$J = \sigma E + \frac{\partial D}{\partial t} \qquad (5.4.2)$$

where σ is the crystal conductivity. If the polar faces of the crystal are connected to an external circuit as shown in Fig. 5.10 (a) then for continuity

$$AJ + aC_L \frac{\partial E}{\partial t} + a \frac{E}{R_L} = 0 \qquad (5.4.3)$$

where C_L and R_L are load capacitance and resistance respectively and where A and a are the electroded area and electrode separation. Combining eqns (5.4.1) to (5.4.3) we have

$$(C_X + C_L) \frac{\partial V}{\partial t} + \left(\frac{1}{R_X} + \frac{1}{R_L} \right) V = -Ap \frac{\partial T}{\partial t} \qquad (5.4.4)$$

where the external voltage $V = Ea$, the crystal capacitance $C_X = \varepsilon A/a$, and resistance $R_X = a/\sigma A$. It is clear that this expression is represented by the

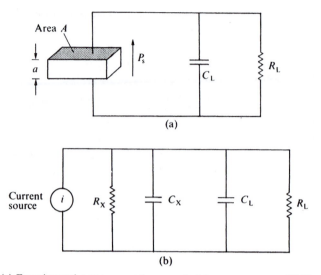

Fig. 5.10. (a) Experimental arrangement for pyroelectric measurements. (b) The equivalent electrical circuit.

equivalent circuit of Fig. 5.10 (b) in which the crystal acts as a current source driving the parallel crystal–load impedance.

A variety of techniques have been described for the measurement of the pyroelectric coefficient by measurement of the voltage, charge, or current developed during a temperature change of the crystal, each technique involving specific solutions of eqn (5.4.4). Solving this equation for the pyroelectric voltage we have

$$\Delta V = -\frac{Ap}{C}\exp\left(-\frac{t}{RC}\right)\int_0^t \exp\left(\frac{t'}{RC}\right)\left(\frac{dT}{dt'}\right)dt' \qquad (5.4.5)$$

where R and C are the parallel crystal–load parameter, $C = C_L + C_X$ and $R = (1/R_L + 1/R_X)^{-1}$. If the temperature change of the crystal is fast compared with the RC time constant

$$\Delta V = -\frac{Ap\,\Delta T}{C}\exp\left(-\frac{t}{RC}\right). \qquad (5.4.6)$$

Early measurements (Ackermann 1915) of high-resistivity pyroelectrics were made in this way using an electrometer voltmeter, $R_L = 0$, and $C_L \gg C_X$ in order to keep the RC time constant long.

If the rate of change of temperature is slow compared with the RC time constant then

$$\Delta V = -ApR\frac{dT}{dt}\left\{1 - \exp\left(-\frac{t}{RC}\right)\right\}. \qquad (5.4.7)$$

The temperature may be changed in a sinusoidal manner (Chynoweth 1956 a) or in a continuous manner (Lang and Steckel 1965). In any measurement it is necessary to keep ΔV sufficiently small that non-linearities in p do not become significant. It is clear that the crystal impedance may be measured by varying R_L and C_L at each temperature, but it is most convenient to have pyroelectric measurements independent of the crystal impedance, especially near a phase transition or at high temperatures where the impedance may vary rapidly with temperature. This is best achieved by measuring the pyroelectric current or charge under short-circuit (constant-E) conditions. An easy way (Glass 1969 a) to maintain short-circuit conditions is with an operational amplifier as shown in Fig. 5.11. The

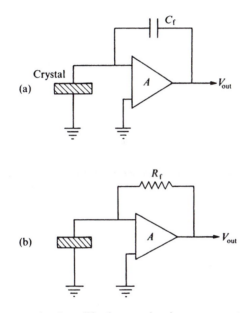

Fig. 5.11. Use of an operational amplifier for pyroelectric measurements. (a) Measurement of pyroelectric charge. (b) Measurement of pyroelectric current under short-circuit conditions.

effective input impedance is the feedback impedance divided by the open-loop gain which is typically 10^5. For the measurement of charge a calibrated capacitor C_f is connected in the feedback loop and the pyroelectric charge $Q = \int_0^t AJ\, dt$ developed across the crystal is instantaneously transferred to the feedback capacitor in order to maintain zero-field conditions across the amplifier input. The output voltage

$$V = Q/C_f$$

gives a direct measure of the charge Q. Since the field across the crystal is

zero the conduction current σE through the crystal is zero and the measurement is not affected by the crystal resistance and hence, from eqns (5.4.1)
and (5.4.2), we have

$$V = \frac{Ap\,\Delta T}{C_{\mathrm{f}}} = \frac{A\,\Delta P_{\mathrm{s}}}{C_{\mathrm{f}}}. \tag{5.4.8}$$

In practice, however, small-input offset voltages of the amplifier and other
small e.m.f.'s E' in the input circuit (i.e. thermo-e.m.f.'s) give rise to a
conduction current which limits measurements to crystal resistances such
that $\sigma E' < \partial D/\partial t$.

Since the integrating time constant of the amplifier is extremely long,
measurement of Q may be made by varying Q either continuously over a
long period of time or discretely. A ballistic galvanometer can only integrate
Q over a much shorter mechanical time constant, so is limited to discrete
temperature changes.

For measurement of pyroelectric current i a calibrated resistor R_{f} is
connected in the feedback loop; then

$$V = R_{\mathrm{f}}i = R_{\mathrm{f}}Ap\frac{\partial T}{\partial t}. \tag{5.4.9}$$

The change of spontaneous polarization as a function of temperature is
easily obtained by integration of the pyroelectric charge. A curve of ΔP_{s} as a
function of temperature is shown in Fig. 5.12(a) for various $Sr_{1-x}Ba_xNb_2O_6$
compositions. When the crystal temperature is well above the Curie temperature T_{c} the polarization vanishes and no further charge integration
takes place, so that the change ΔP_{s} from T_{c} to a lower temperature T gives a

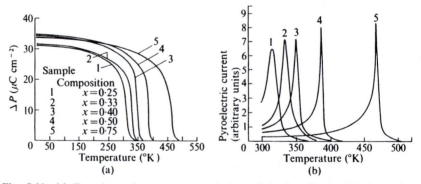

Fig. 5.12. (a) Experimental measurement of the polarization change ΔP for various
$Sr_{1-x}Ba_xNb_2O_6$ compositions by integration of the pyroelectric charge from above the
ferroelectric phase transition. (b) Measurement of the pyroelectric current of the same crystals
as in (a) by the dynamic technique of Chynoweth at 1 kHz as a function of temperature. (After
Glass 1969 a.)

direct measure of the spontaneous polarization. These experiments are always performed on heating, since on cooling through T_c multi-domain crystals may be formed. Differentiation of Fig. 5.12(a) gives the unclamped pyroelectric coefficient. Charge integration is useful for the determination of P_s at low temperatures where polarization reversal may not be possible, provided that calibration of the polarization at one temperature is possible.

One alternative to differentiation of the P_s-T curve for the determination of the pyroelectric coefficient p is by controlling the rate of change of temperature with a carefully programmed oven (Byer and Roundy 1972) and measuring the pyroelectric current (eqn (5.4.2)). Another is by using the dynamic method of Chynoweth (1956 a) as follows. A crystal is heated in a sinusoidal manner using a modulated light beam which is absorbed either in the crystal or in an absorbing layer in intimate contact with the crystal. The wavelength of light is chosen so that free carriers are not excited which may cause photovoltaic or photoconductive effects. If the incident waveform is a square wave the temperature change and short-circuit current response is as shown in Fig. 5.13 (for the case that the modulation frequency is much

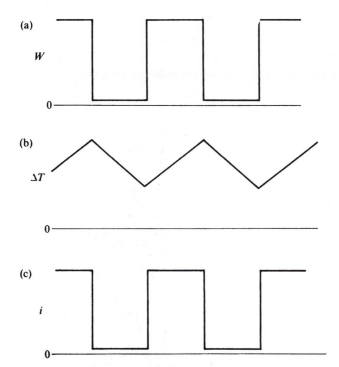

Fig. 5.13. Diagram showing (b) the temperature change ΔT of the crystal and (c) the pyroelectric current response i to incident radiation W modulated as in (a). It is assumed that the modulation frequency is much greater than the thermal relaxation frequency.

greater than the reciprocal thermal relaxation time of the crystal). For incident power of form $W = W_0 \exp(i\omega t)$ the current response is

$$i = \frac{AepW}{\mathscr{C}} \qquad (5.4.10)$$

where e is the fraction of incident light absorbed, $\mathscr{C}$ is the thermal capacity of the crystal, and A is the electroded area normal to the pyroelectric axis. The current reproduces the incident waveform. Extremely small pyroelectric currents can be measured with phase-sensitive detection and for many pyroelectrics temperature changes of $10^{-6}°C$ give easily measurable pyroelectric response. An experimental measurement of the a.c. current response of $Sr_{1-x}Ba_xNb_2O_6$ as the ambient temperature is slowly changed is shown in Fig. 5.12(b). The data gives *directly* the derivative of the charge-integration data if the temperature variation of $\mathscr{C}$ can be neglected. Any discontinuous change of P_s would give a pyroelectric current transient so that integration of the $p–T$ curve gives ΔP only for continuous polarization changes. The discontinuous change of P_s at a first-order ferroelectric phase transition for instance must be measured by a voltage or charge measurement.

While the crystal absorption e in (5.4.10) is usually, to a good approximation, independent of temperature, it is often difficult to determine the absorbed energy with sufficient accuracy for a good determination of the absolute magnitude of p. Consequently the dynamic technique is usually used for the study of the temperature or time variation of P_s and other techniques are used for absolute calibration. The thermal capacity may vary somewhat over a wide temperature range so some correction for the variation of $c_{X,E}$ may be necessary—especially close to a phase transition (see § 5.6).

Both clamped and unclamped pyroelectric coefficients may be measured by the dynamic technique using appropriate modulation frequencies, but a necessary precaution in low-frequency measurements is to ensure that the tertiary pyroelectric effect due to non-uniform heating of the crystal does not contribute significantly to the measurements. Non-uniform heating sets up stress gradients which give additional piezoelectric contributions to the pyroelectric effect (piezoelectric terms must be included in eqn (5.4.1)). Fortunately piezoelectric terms are usually small compared with the primary and secondary effects and can be neglected, but a useful check is to observe the change of P following a short heating pulse. Tertiary effects contribute to P only for times shorter than the thermal diffusion time across the crystal, while primary and secondary effects persist up to the thermal relaxation time of the crystal with its surroundings.

The dynamic technique is particularly well suited for the study of small crystals or weak pyroelectric effects because of its high sensitivity. In addition, crystals of high electrical conductivity can be studied because any temperature modulation of the conduction current is out of phase with the pyroelectric current and can be separated by phase-sensitive detection. Finally, the dynamic technique is well suited for dynamic studies of polarization reversal provided the change of polarization is slow compared with the modulation frequency.

It is clear that from pyroelectric measurements the absolute sense of the polarization change can be determined with respect to any other measured parameter of the crystal. This allows one to relate the atomic displacements obtained from a structural determination to the macroscopic polarization change. In the case of ferroelectrics where P_s goes to zero at the Curie temperature, the absolute sense and magnitude of P_s can then be compared with the atomic displacements below T_c from their positions in the paraelectric phase. In the absence of a non-polar phase it is not usually possible from a structural determination alone to define an absolute sense of the polarization unambiguously since a knowledge of the entire charge distribution within the crystal (including surfaces) would be necessary.

By convention the sign of the pyroelectric coefficient is defined with respect to the sign of the piezoelectric axis of the crystal. The IRE standard (1949) defines the positive end of a crystal axis as that which develops a positive charge when the crystal is in tension along that axis. The pyroelectric coefficient of a crystal is defined as positive if the positive face (determined piezoelectrically) develops a positive charge upon heating. In most ferroelectrics the pyroelectric coefficient is negative since the polarization generally decreases with increasing temperature, but this is by no means the case in all pyroelectrics at all temperatures.

Within the Devonshire theory the pyroelectric coefficient is simply related to the spontaneous polarization (Liu and Zook 1974). If we write eqn (3.3.2) as

$$E = \beta(T - T_c)D + \varphi(D) \tag{5.4.11}$$

where $\varphi(D)$ includes all higher-power terms, then

$$\frac{\partial E}{\partial T} = \beta D + \beta(T - T_c)\frac{\partial D}{\partial T} + \frac{\partial \varphi}{\partial D} \cdot \frac{\partial D}{\partial T}. \tag{5.4.12}$$

For short-circuit conditions with $E = 0$ we have

$$-p = \frac{-\partial D}{\partial T} = \frac{\beta P_s}{\{\beta(T - T_c) + \partial\varphi/\partial D\}} = \varepsilon\beta P_s. \tag{5.4.13}$$

This relationship has been found to hold reasonably well for several ferroelectrics at temperatures well below T_c, as shown in Table 5.1.

TABLE 5.1

Comparison of the pyroelectric coefficient calculated from the relation p =
$\varepsilon\beta P_s$ *of eqn (5.4.13) with the experimental values*

Material	ε	P_s	β	p $(10^{-8}\,\mathrm{C\,cm^{-2}\,K^{-1}})$	
		$(\mu\mathrm{C\,cm^{-2}})$	$(10^{-6}\,\mathrm{K^{-1}})$	calc.	exp.
TGS	43	2·8	308	3·7	2·7
BaTiO$_3$	160	26	5·8	2·4	2·0
LiTaO$_3$	46	50	6·2	1·4	1·9
Sr$_{1/2}$Ba$_{1/2}$Nb$_2$O$_6$	400	27	2·9	3·1	6·0

After Liu and Zook 1974.

5.5. Electrocaloric effect

The electrocaloric effect is the temperature change of a crystal which results
from an adiabatic application of an electric field. The electrocaloric compliances are simply related to the pyroelectric effect by Maxwell's relation
(for constant stress or strain)

$$-\left(\frac{\partial T}{\partial E}\right)_S = \left(\frac{\partial D}{\partial S}\right)_E = \frac{Tp^E}{c^E}. \tag{5.5.1}$$

Alternatively,

$$\left(\frac{\partial T}{\partial D}\right)_S = \left(\frac{\partial E}{\partial S}\right)_D = \frac{T}{c^D}\left(\frac{\partial E}{\partial T}\right)_D. \tag{5.5.2}$$

Using eqn (3.3.14) this can be written in terms of the Devonshire coefficient
β as

$$\left(\frac{\partial T}{\partial D}\right)_S = \frac{\beta DT}{c^D}. \tag{5.5.3}$$

Electrocaloric measurements are usually difficult because of the small
temperature changes produced by the largest electric fields which can be
applied without introducing large non-linearities. The coefficients are
usually calculated from measurements of the pyroelectric effect and the
specific heat, or from the differences between the isothermal and adiabatic
dielectric constants (e.g. eqn (5.2.2)). The small amount of experimental
work on direct measurement of the electrocaloric compliances has been
restricted to temperatures very close to ferroelectric phase transitions where
the effect is largest, and only on highly insulating materials where Joule
heating is negligible.

Over a finite temperature range eqn (5.5.3) gives

$$\Delta T = T_1 - T_2 = \frac{1}{2}\frac{T\beta}{c^D}(D_1^2 - D_2^2) \tag{5.5.4}$$

if the D dependence of c^D can be neglected, as is usually the case. Extrapolation of measurements of ΔT *versus* ΔD^2 to the limit $\Delta T = 0$ gives an indirect measure of the spontaneous polarization ($D^2 = P_s^2$ at $\Delta T = 0$) which may be of use in materials where P_s cannot be measured by more direct techniques.

Measurements are performed on freely suspended crystals, to avoid clamping, in carefully constructed adiabatic calorimeters. Simultaneous measurement of the change of electric displacement and temperature (in the millidegree range) are made as a function of the applied field.

Typical results are shown for triglycine sulphate in Fig. 5.14 (Strukov 1966 a). Fig. 14(b) was obtained by differentiation of Fig. 14(a). The maximum of $(\partial T/\partial E)_S$ observed at T_c is consistent with the anomalous behaviour, via (5.5.1), of the pyroelectric coefficient and specific heat c^E

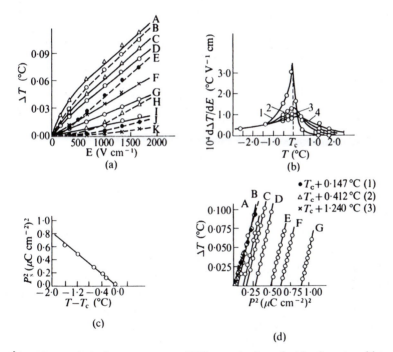

Fig. 5.14. Electrocaloric measurements on TGS near the ferroelectric phase transition. (a) Calorimetric measurement of the crystal temperature change as a function of applied field: A, $T_c-0\cdot041$; B, $T_c+0\cdot147$; C, $T_c-0\cdot289$; D, $T_c-0\cdot535$; E, $T_c+0\cdot412$; F, $T_c-1\cdot375$; G, $T_c-2\cdot637$; H, $T_c+1\cdot240$; I, $T_c-8\cdot004$; J, $T_c+1\cdot784$; K, $T_c+3.176$ (all in °C). (b) Temperature variation of the electrocaloric compliance for various values of applied field (curves 1–4 corresponding respectively to fields of 0, 500, 1000, and 1500 V cm^{-1}) obtained by differentiation of (a). (c) Variation of the square of the spontaneous polarization with temperature. (d) Electrocaloric temperature change as a function of the square of the total polarization: A, $T_c-0\cdot005$; B, $T_c-0\cdot290$; C, $T_c-0\cdot387$; D, $T_c-0\cdot663$; E, $T_c-1\cdot190$; F, $T_c-1\cdot570$; G, $T_c-2\cdot186$ (all in °C). (After Strukov 1966 a.)

near T_c. The polarization is seen to be a continuous function of temperature with

$$P_s^2 \propto (T_c - T) \qquad (5.5.5)$$

in good agreement with the Devonshire theory for a second-order transition. A temperature dependence of the form shown in eqn (5.5.5) has been confirmed near T_c for a number of other materials both by electrocaloric and pyroelectric measurements.

From eqn (5.5.1) it is seen that the unclamped electrocaloric coefficient varies inversely with $c^{E,X}$. It is well known that at low temperatures $c^{E,X}$ approaches zero as T^3 so this may seem encouraging for electrocaloric refrigeration like the magnetic analogue. However, at low temperatures $p^{E,X}$ is dominantly due to the excitation of acoustic modes (secondary pyroelectric effect) and also decreases rapidly (as T^3) as the temperature $T \to 0$, so that the electrocaloric coefficient never becomes large in this limit and actually decreases linearly at low temperatures. Born (1945) has pointed out that linear temperature contributions to low-temperature pyroelectric compliance are also theoretically possible from second-order acoustic contributions to the *primary* pyroelectric compliance. These would give rise to a diverging electrocaloric compliance as $T \to 0$, but evidence for their importance (in terms of magnitude) from direct pyroelectric measurements is essentially lacking, although Lang (1971) did include such a term in his analysis of the pyroelectric effect in $LiSO_4H_2O$.

5.6. Specific heat

The entropy $S = (-\partial G_1/\partial T)_{D,X}$ associated with the long-range ordering of dipoles in a pyroelectric is related in a simple way to the spontaneous polarization within the Devonshire approximation as shown in § 3.3, i.e.

$$S = -\tfrac{1}{2}\beta D^2.$$

Consequently the difference between the specific heat $c = T(\partial S/\partial T)$ measured under short-circuit and open-circuit conditions is

$$c^{X,E} - c^{X,D} = -\tfrac{1}{2}\beta T\left(\frac{\partial D^2}{\partial T}\right)_{X,E}. \qquad (5.6.1)$$

The abrupt changes of the spontaneous polarization at a first-order phase transition have an associated latent heat (eqn (3.3.23)) and anomalous behaviour of $c^{X,E}$. Even at a second-order phase transition there is a discontinuous change of $c^{X,E}$ owing to the discontinuity in $p^{X,E}$.

If we can neglect powers higher than P^4 in the free-energy expansion, then for $E = 0$ $(D = P_s)$ and a second-order transition we have from (3.3.4)

$$P_s^2 = -\frac{\beta}{\gamma}(T - T_c), \qquad P_s \to 0 \qquad (5.6.2)$$

and hence

$$c^{X,E} - c^{X,D} = \frac{1}{2}\frac{\beta^2 T}{\gamma}, \qquad T \to T_c^- \qquad (5.6.3)$$

is only slowly varying with temperature. For studies on insulating ferroelectrics single-domain single crystals with the polar faces short-circuited are essential. Conducting powders or ceramics can be used provided measurements are made at times longer than the electrical relaxation time. Conventional adiabatic calorimeters or differential calorimeters are generally used for accurate absolute measurement of $c^{X,E}$. Temperature control of less than a millidegree can be achieved, which allows detailed study of the heat capacity and latent heat in the vicinity of a phase transition. An a.c. technique has been developed which lends itself particularly well to the study of relative variations of specific heat near a phase transition (Handler, Mapother, and Rayl 1967). This technique uses the same experimental arrangement as the dynamic pyroelectric measurements so that simultaneous measurement of both $c^{X,E}$ and $p^{X,E}$ is possible. The sample is heated with a light beam in a sinusoidal manner and the periodic temperature change of the crystal is detected with a thermocouple (of very low thermal mass) in intimate contact with the rear face of the crystal. The small temperature changes created by the light beam can be detected with phase-sensitive electronics allowing changes of specific heat to be measured with high-temperature resolution. Detailed experimental considerations have been discussed by Salamon (1970). The use of this technique for the study of LiTaO$_3$ (Glass 1968) is shown in Fig. 5.15, demonstrating a specific-heat variation characteristic of a second-order phase transition near T_c. The relative changes of $c^{X,E}$ are more accurate than the absolute calibration. The heat $\Delta Q = \int c^{X,E} \, dT$ and the entropy $\Delta S = \int (c^{X,E}/T) \, dT$ associated with the ferroelectric order are often difficult to obtain from measurements of specific heat, particularly for second-order transitions, owing to the uncertainty in the background contribution $c^{X,D}$ to the specific heat. A reasonably good fit to the data of Fig. 5.15 is obtained using eqn (5.6.1) and the measurements of P_s and β (i.e. from $1/\varepsilon$) for LiTaO$_3$ if the broken line is chosen as the baseline. Values of $\Delta Q = 210$ cal mol^{-1} and $\Delta S = 0.23$ cal mol^{-1} K^{-1} are evaluated from the data.

In most ferroelectrics studied the simple phenomenological model which predicts the specific-heat variation of eqn (5.6.3) is not in good agreement with experiment very close to the ferroelectric transition. For instance in triglycine sulphate (TGS) (Strukov 1964) and triglycine fluorberyllate (TGFB) (Strukov, Taraskin, and Koptsik 1966) the specific heat shows a prominent peak at T_c in contrast to the slow variation predicted by eqn (5.6.3), and the heat of transformation ΔQ_{obs} is much larger than the calculated value ΔQ_{calc}. For the case of TGS shown in Fig. 5.16 $\Delta Q_{obs} = 1.07$ cal g^{-1} while $\Delta Q_{calc} = 0.77$ cal g^{-1}. In these same materials,

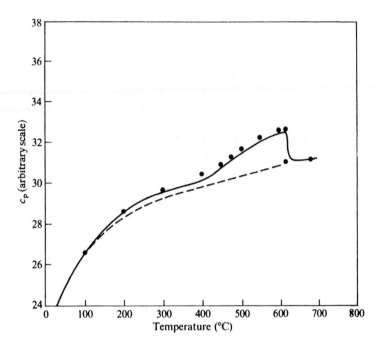

Fig. 5.15. Temperature variation of the specific heat of $LiTaO_3$ measured by the dynamic technique. The solid curve is the experimental measurement corrected for the thermocouple response. The broken curve is the estimated background specific heat not directly associated with the ferroelectric transition, and the points are theoretical values calculated from polarization measurements. (After Glass 1968.)

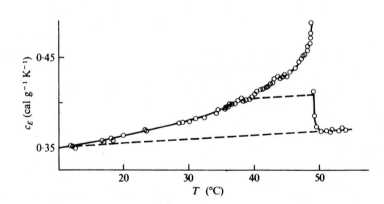

Fig. 5.16. Temperature dependence of the heat capacity of a short-circuited TGS crystal measured by adiabatic calorimetry. (After Strukov 1964.)

however, the polarization and dielectric-constant variation are in good agreement with phenomenological theory right up to the phase transition. The difference between experiment and the predictions of eqn (5.6.3) has been attributed by Strukov to short-range polarization fluctuations (Burgess 1958) which are not taken into account in the power-series expansion of the free energy. These fluctuations may be taken into account to a degree by including a term proportional to $(\nabla \mathbf{P})^2$ in the free-energy expansion (Ginzburg 1960) and will be discussed in detail in Chapter 11. They predict, for uniaxial dipolar interactions, a logarithmic divergence of $c^{X,E}$ near T_c and therefore also account for the tail observed in the excess specific heat above T_c, which seems to show a logarithmic variation with temperature over a range from T_c to $T_c + 2°C$.

Some attempts have been made to fit these deviations of specific heat from Devonshire form to Ginzburg-type logarithmic forms, and in this way estimates have been obtained for the spatial extent of these local fluctuations. At least in TGS and its isomorphs Strukov et al. find that the fluctuations involve essentially only nearest neighbours. This contrasts with estimates of correlations of many tens of unit cell lengths made for oxygen octahedron ferroelectrics from optical- and electron-scattering data (see for example Wemple, DiDomenico, and Jayaraman 1969). The reason for the difference is not known. One may even question whether it is real since modern theories of critical phenomena (see Chapter 11) show that even the Ginzburg approach breaks down in fluctuation-dominated (true critical) regions, and that the specific-heat anomaly, for example, should more properly be fitted to a form involving the cube root of a logarithm (see eqn (11.2.21)). Since analyses of this type have not yet been made for ferroelectrics a proper understanding of the critical specific-heat effects may not yet have been made with reference to experimental observations.

In KDP, which has a first-order (but nearly second-order) phase transition, there is also some evidence of a logarithmic dependence of excess specific heat above T_c as shown in Fig. 5.17 (Reese 1969 b). Below T_c the specific heat seems to follow a power law of the form $(T_c - T)^{-1/2}$ over an interval of about one degree. However, some caution must be exercised in interpreting data of this type since a slight broadening of the phase transition due to crystal inhomogeneities can confuse the interpretation and in particular can contribute to a tail above T_c. It is clear from Fig. 5.15 that polarization fluctuations do not appear to have such a marked effect on the specific heat of $LiTaO_3$.

First-order transitions generally have much sharper anomalies of $c^{X,E}$ at T_c. A good example is ammonium dihydrogen phosphate (ADP) shown in Fig. 5.18 which undergoes an order–disorder antiferroelectric transition near 150 K. For this material $\Delta Q = 165$ cal mol^{-1} and $\Delta S = 1 \cdot 10$ cal mol^{-1} K^{-1}. Most of the entropy change in ADP is the configurational

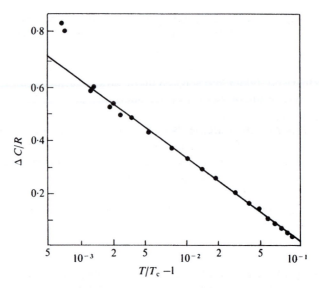

Fig. 5.17. Experimental measurements (closed circles) of the anomalous heat capacity of KDP above the ferroelectric Curie temperature. The background lattice contribution has been subtracted, which may give a systematic uncertainty of absolute magnitude as large as $0.1 R$. (After Reese 1969 b.)

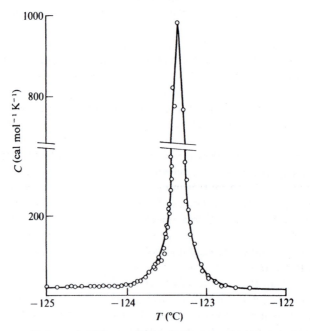

Fig. 5.18. The specific heat of ADP near the Curie temperature. (After Amin and Strukov 1970.)

entropy associated with the ordering of the H-bond dipoles. Nagamiya (1952) calculated for ADP that this H-bond configurational entropy is $0 \cdot 8 \, \mathrm{cal \, mol^{-1} \, K^{-1}}$. Of the remainder Genin, O'Reilly, and Tsang (1968) estimated that $\sim 0 \cdot 2 \, \mathrm{cal \, mol^{-1} \, K^{-1}}$ was due to changes of the vibrational entropy of the NH_4 ion on passing through the transition. It appears that only small changes of entropy are associated with configurational changes of the NH_4 group (see § 6.4.2).

The first reported measurements of low-temperature specific heats for common paraelectrics, ferroelectrics, and antiferroelectrics have recently been given by Lawless (1976). Excellent fits to the data are obtained by adding low-frequency Einstein terms to the normal Debye term. The ferroelectrics measured showed an additional feature at the lowest temperatures: a $T^{3/2}$ contribution which was tentatively assigned to domain-wall contributions.

5.7. Thermal expansion, elastic constants, and piezoelectricity

So far in this chapter we have been primarily concerned with the various compliances which relate the dielectric and thermal properties of ferroelectric crystals, and their anomalous behaviour near a ferroelectric phase transition. To describe these effects the free energy was expanded in terms of electric displacement alone. However, any coupling between electric displacement and elastic strain in the free-energy expansion gives rise to anomalous behaviour of the thermal expansion, piezoelectric, and elastic compliances in the neighbourhood of the Curie temperature owing to the anomalous dielectric behaviour. On the other hand, it may be the elastic properties of the material which trigger the phase transition, in which case the anomalous dielectric behaviour of the material is due to coupling of the displacement with strain.

We have already pointed out that the lowest-order coupling term in the free-energy expansion of a ferroelectric depends on the symmetry of the paraelectric phase. For a centrosymmetric non-polar phase we see from eqn (3.2.30) that the spontaneous strain in the ferroelectric phase is proportional to the square of the electric displacement in the lowest order. In this case the piezoelectric effect in the ferroelectric phase may be considered as electrostriction biased by the spontaneous polarization. On the other hand, for a piezoelectric prototype phase the lowest-order coupling term is bilinear. In this case the spontaneous strain may vary linearly with polarization. These conclusions were well established during early studies of KDP, where the spontaneous strain x_s is linear in P_s, and for $BaTiO_3$, where x_s is quadratic in P_s. These studies are reviewed in texts such as that by Jona and Shirane (1962).

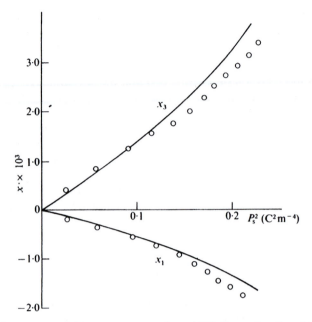

Fig. 5.19. Measurement of the spontaneous strain x of $LiTaO_3$ as a function of P_s^2. The solid curves are the spontaneous strain calculated from the piezoelectric data of Fig. 5.20. (After Iwasaki *et al.* 1970.)

Some more recent studies (Iwasaki *et al.* 1970) of the spontaneous strain in $LiTaO_3$ are shown in Fig. 5.19. It is seen that over a wide temperature range x_s deviates only slightly from a quadratic dependence on P_s^2—although the effect of higher-order coupling terms is clearly evident. A fairly complete set of measurements of the piezoelectric and elastic constants of $LiTaO_3$ are also available and are shown in Fig. 5.20; $LiTaO_3$ has a non-cubic paraelectric phase and below T_c the symmetry is 3m with $P_s = P_3$ along the three-fold axis. Then considering small displacements from the prototype phase we have (since $g_{ij} = 0$)

$$x_i = s_{ij}^D X_j + \tfrac{1}{2} q_{ijk} D_j D_k \qquad (5.7.1)$$

or we can write the displacement in terms of the spontaneous polarization

$$x_i = s_{ij}^D X_j + \tfrac{1}{2} q_{i3k}(P_s^2 + 2P_s \Delta D_k) \qquad (5.7.2)$$

where $q_{i3k} = q_{ik3}$ by symmetry. From eqn (3.2.21) we have

$$x_i = \sum_k \tfrac{1}{2} q_{i3k} P_s^2 + (s_{ij}^D + q_{i3k} P_s d_{kj}) X_j + (q_{i3k} P_s \varepsilon_{kj}) E_j. \qquad (5.7.3)$$

Comparison with eqn (3.2.20) shows that

$$d_{ij}^\dagger = d_{ji} = q_{i3k} P_s \varepsilon_{kj} \quad (i = 1\text{-}6, \ j = 1\text{-}3) \qquad (5.7.4)$$

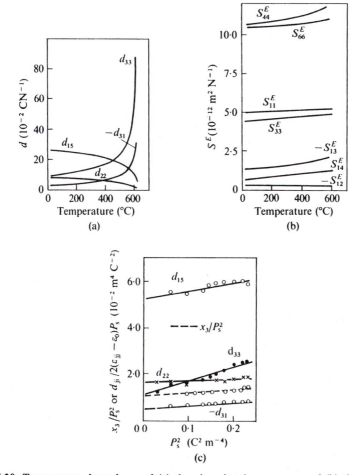

Fig. 5.20. Temperature dependence of (a) the piezoelectric constants, and (b) the elastic compliances of LiTaO₃ as functions of temperature. Graph (c) shows the dependence on temperature of several compliance ratios which theoretically are not expected to show anomalies at T_c. (After Iwasaki *et al.* 1970.)

and

$$s_{ij}^E - s_{ij}^D = q_{i3k}P_s d_{kj} \quad (i, j = 1\text{--}6) \qquad (5.7.5)$$

where the repeated index k implies summation. For many ferroelectrics q_{ijk} are fairly constant over a wide temperature range and the piezoelectric compliances d_{ji} therefore vary near T_c as $P_s\varepsilon_{kj}$, which is as P_s for $j = 1, 2$ or as $P_s\varepsilon_{33}$ for $j = 3$, where ε_{33} is the divergent component of dielectric constant in LiTaO₃. The variation of d_{22}, d_{33}, d_{15}, and d_{31} (which are the only independent non-zero piezoelectric compliances in the class 3m) are shown as a function of temperature in Fig. 5.20 (a). A rapid change of all parameters near T_c^-, with d_{15} and d_{22} going to zero and d_{33} and d_{31} diverging (in general accord with the above), is clearly evident. The ratios $d_{ij}/P_s\varepsilon_{kj}$ on the other hand (Fig. 5.20c) are slowly varying functions of temperature as expected from (5.7.4).

From eqn (5.7.5) the elastic compliance s_{ij}^E varies near T_c as P_sd_{kj} since s_{ij}^D is not expected to show anomalous behaviour in a material in which D is the primary order parameter. It follows (remembering that k in eqns (5.7.4) and (5.7.5) is always summed over its three allowed values 1, 2, 3) that a relationship

$$s_{ij}^E \propto P_s^2 \varepsilon_{33} \qquad (5.7.6)$$

is to be expected for all the elastic compliances near T_c^-. Since $P_s^2 \propto (T_c - T)$ in this limit and $\varepsilon_{33} \propto (T_c - T)^{-1}$, the product on the right-hand side of (5.7.6) is independent of temperature. The slow variation of the LiTaO$_3$ elastic compliances, shown in Fig. 5.20 (b), is consistent with this result. Relationships of this kind can be carried even further by relating the complex elastic constants to the complex dielectric constants via the piezoelectric field induced by the acoustic waves in the crystal. The ultrasonic attenuation can then be related to the dielectric loss. Indeed measurement of ultrasonic attenuation provides a useful technique for studies of relaxation at high frequencies (see Chapter 10).

In the case of LiTaO$_3$ we chose to keep terms linear in x and quadratic in D, and experimental results are reasonably consistent with this approximation over a wide temperature range. However, this is not necessarily the case for other materials. In iron–iodine boracite (Fe$_3$B$_7$O$_{13}$I) for instance the experimental data suggest that the spontaneous strain is not caused by the spontaneous polarization alone (Kobayashi and Mizutani 1970). In this instance the polarization is itself driven by a further coupling to an anti-distortive (zone-doubling) mode which softens at T_c (see § 10.4).

For materials such as KDP and gadolinium molybdate (GMO) which are acentric above the Curie temperature eqns (3.2.18) and (3.2.21) give

$$x_i = s_{ij}^D X_j + g_{i3}^\dagger P_s + g_{ij}^\dagger (d_{jk} X_k + \varepsilon_{jk}^X E_k) \qquad (5.7.7)$$

and by comparison with (3.2.20) we have

$$d_{ki} = d_{ik}^\dagger = g_{ij}^\dagger \varepsilon_{jk}^X \qquad (5.7.8)$$

and

$$s_{ik}^E - s_{ik}^D = g_{ij}^\dagger d_{jk} = g_{ji} g_{lk} \varepsilon_{ij}^X . \qquad (5.7.9)$$

In this case we see that, if the relevant g_{ij} are non-zero and constant, at least some piezoelectric and elastic constants should show a Curie–Weiss type of temperature dependence in the same way as the dielectric constant. This result has been well established experimentally in KDP for s_{66} and d_{36} in terms of a diverging ε_{33} as $T \to T_c^+$ (Jona and Shirane 1962). Gadolinium molybdate on the other hand has the same paraelectric symmetry ($\bar{4}$2m) as KDP but shows quite different dielectric behaviour as shown in Fig. 5.6. In this case there is no clamped dielectric anomaly at the phase transition indicating no softening of a zone-centre optic mode, but there are prominent elastic anomalies (Cross et al. 1968, Höchli 1972). The small dielectric anomaly observed in unclamped crystals is due to piezoelectric coupling to this anomalous elastic behaviour described by eqn (5.7.8) and (5.7.9). The elastic anomalies are in turn driven by a coupling to a soft 'driving' anti-distortive mode (see § 10.3).

Techniques for the measurement of spontaneous strain, piezoelectric, and elastic compliances have all reached a high degree of sophistication and are well documented in the literature (for piezoelectric and elastic measurements see IRE Standards, 1949, 1957, 1958, Mason and Jaffe 1954, Berlincourt 1956, and for precision lattice-parameter measurements see Bond 1960). Most of these compliances can be measured with relative accuracies of about 10^{-5}. Typically, spontaneous strain is measured by X-ray diffraction techniques or by direct measurement of the change in dimensions of a crystal by electrical-capacitance or optical-interference measurements. In the latter methods one end of the crystal is fixed while the separation of the other end from a fixed plane is measured as a function of temperature. Piezoelectric measurements may be made in the same way by measuring the strain created by an applied field. Alternatively the charge developed across a stressed crystal can be measured by one of the pyroelectric methods discussed in § 5.4. The preferred techniques, however, which yield measurements of the elastic and piezoelectric compliances are piezoelectric resonance techniques. A mechanical resonance is excited in a suitably oriented plate or bar of the material with an alternating field by piezoelectric coupling. The resonance frequencies f_r are readily detected by the sharp drop of the crystal impedance at resonance as indicated by eqn (5.2.10) and a sharp increase occurs at the anti-resonance frequencies f_a. These frequencies are directly related to the velocity of sound V_α in the crystal which in turn depends on the effective elastic constant s_α appropriate to the boundary conditions, i.e.

$$V_\alpha = (s_\alpha \rho)^{-1/2} \qquad (5.7.10)$$

where ρ is the crystal density. The piezoelectric constants are evaluated from the electromechanical coupling constants. For instance the coupling constant appropriate to the boundary conditions of eqn (5.2.10) is $k_{31} = d_{31}/(s_{11}^E \varepsilon_{33})^{1/2}$. The open-circuit and short-circuit elastic compliances are related in this case by

$$s_{11}^D = s_{11}^E (1 - k_{31}^2). \qquad (5.7.11)$$

By suitable choice of crystal geometry, s_{11}^D, s_{11}^E, and k_{31} can be measured separately from measurements of f_a and f_r.

The resonance techniques are particularly useful for a preliminary qualitative check of the piezoelectric nature of a material in powder form. The Giebe–Scheibe technique uses the fact that the resonance within individual crystallites can be detected by the change of the impedance of an r.f. capacitor containing the powder. With ferroelectrics, however, negative results are sometimes obtained with a Giebe–Scheibe test if the crystallites are polydomain or if the acoustic Q of the material is low.

Acoustic attenuation can be measured from the width of the piezoelectric resonance, but there are a variety of ultrasonic pulse techniques which are more suitable for direct measurement of acoustic loss (see for instance Litov and Uehling (1970) who used a pulse-echo method for measuring ultrasonic velocity and absorption in DKDP).

The use of thermodynamic relationships to relate the various dielectric and elastic compliances can give some insight as to the nature of the primary order parameter involved in the ferroelectric phase transition, but such interpretations are not unambiguous since the elastic and dielectric measurements only probe the behaviour of optic and acoustic modes near $\mathbf{q} = 0$ of the Brillouin zone (the elastic constants are determined by the square of the slopes of the acoustic branches as $\mathbf{q} \to 0$), while the effects observed at $\mathbf{q} = 0$ may sometimes be secondary effects due to coupling with soft modes at $\mathbf{q} \neq 0$ (as set out above for GMO and iron–iodine boracite). A detailed treatment of coupling to $\mathbf{q} \neq 0$ driving modes is given in Chapter 10.

5.8. High-pressure studies

The temperature dependence of the ferroelectric properties discussed in this chapter has two contributions. Firstly there is a volume effect associated with thermal expansion and secondly there is a pure temperature effect which would be present even if the crystal volume did not change. To separate these effects it is necessary to make measurements as a function of temperature and pressure separately. Several studies have recently been made as a function of pressure which have been crucial in elucidating fundamental aspects of ferroelectricity, and these techniques provide a useful additional handle for the study of the structure sensitivity of ferroelectrics.

Most high-pressure studies are made with hydrostatic pressure since this greatly simplifies the interpretation of results and allows much higher pressures to be applied without destroying the crystals.

Some general conclusions concerning the effect of pressure can be arrived at in a straightforward way by including the lowest-order interaction term between polarization and stress in the power-series expansion of the free energy G_1, i.e.

$$G_1 = F(D) + F(D, X) + F(X). \tag{5.8.1}$$

When hydrostatic pressure is applied to a crystal with a centrosymmetric paraelectric phase and a polarization along one of the crystallographic axes in the ferroelectric phase, the lowest-order interaction term is

$$F(D, X) = \Omega p D^2 \qquad (5.8.2)$$

where $p = X_{11} + X_{22} + X_{33}$ is the hydrostatic pressure and Ω is a function of the electrostrictive compliances. Then, expanding $F(D)$ in the usual fashion

$$G_1 = F(X) + [\tfrac{1}{2}\beta\{T - T_0(0)\} + \Omega p]D^2 + (\gamma/4)D^4 + \ldots \qquad (5.8.3)$$

from which it follows that the reciprocal dielectric constant $\partial^2 G_1/\partial D^2$ follows a Curie–Weiss law above T_c with a pressure-dependent Curie–Weiss temperature

$$T_0(p) = T_0(0) - 2\Omega p/\beta \qquad (5.8.4)$$

and a pressure-independent Curie constant. Alternatively at a constant temperature it is convenient to write the Curie–Weiss law

$$\varepsilon^{-1} = 2\Omega(p - p_0) \qquad (5.8.5)$$

where $p_0 = -\tfrac{1}{2}\beta\{T - T_0(0)\}/\Omega$.

In the same way the other phenomenological relations developed in preceding sections may be applied to crystals under pressure. For instance (compare eqn (3.3.4))

$$P_s^2 = -\frac{\beta\{T - T_0(p)\}}{\gamma}, \qquad P_s \to 0 \qquad (5.8.6)$$

and so on. For a second-order transition the shift of the Curie temperature $T_c = T_0$ is seen to be linear with hydrostatic pressure. For a first-order transition T_c is also found experimentally to be linear in the ferroelectric perovskites, but the difference $T_c - T_0$ tends to decrease with increasing pressure. This suggests that there is a tendency toward second-order characteristics at high pressure. An increase of the peak dielectric constant and a decrease of the polarization discontinuity at T_c observed in $BaTiO_3$ (Samara 1969) are consistent with this suggestion.

The relationships eqns (5.8.4) to (5.8.6) are found to predict the pressure dependence of ferroelectric behaviour of many materials over a wide range of pressure and temperature. A good example is $BaTiO_3$, shown in Fig. 5.21, where the pressure dependence of the dielectric constant is seen to resemble the familiar temperature dependence of ε, i.e. the ferroelectric–paraelectric phase transition can be induced with pressure. The pressure dependence of the Curie constant is weak, indicating that the lowest-order interaction term in the free energy provides an adequate description of the effects. Reducing the Curie temperature in this way has proved useful in $LiH_3(SeO_3)_2$ and its deuterated form since these materials melt below T_c at 1 bar. Application of

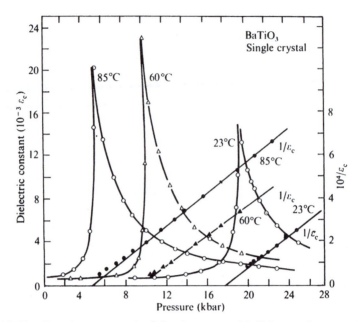

Fig. 5.21. The effect of pressure on the dielectric constant of $BaTiO_3$ at various temperatures. (After Samara 1966.)

pressure allows the ferroelectric transition to be studied by simultaneously decreasing T_c and raising the melting point.

A summary of the pressure dependence of T_c for many materials is listed in Table 5.2. It is seen that T_c may increase or decrease with pressure for both ferroelectric and antiferroelectric materials depending on the details of the ordering mechanism.

It is also worth pointing out that the pressure dependence of T_c is directly related to the discontinuous volume change ΔV and entropy change ΔS for first-order transitions by the Clausius–Clapeyron equation

$$\frac{\partial T_c}{\partial p} = \frac{\Delta V}{\Delta S}. \tag{5.8.7}$$

This relation has been verified experimentally in several materials (Jona and Shirane 1962).

The strong pressure variation of the dielectric constant near T_c predicted by eqn (5.8.5) implies a similarly strong variation of one or more zone-centre optic-mode frequencies. Consider the Lyddane–Sachs–Teller relationship (see § 7.1)

$$\frac{\varepsilon}{\varepsilon_\infty} = \Pi_i \left(\frac{\omega_{Li}^2}{\omega_{Ti}^2} \right) \tag{5.8.8}$$

TABLE 5.2

Values of the transition temperature T_c and its initial pressure derivative for various ferroelectric and antiferroelectric crystals†

Substance	Transition	T_t (K)	dT_t/dp (K kbar^{-1})
BaTiO$_3$	FE–FE	278	−2·8
BaTiO$_3$	FE–PE	390	−4·0 to −6·7
PbTiO$_3$	FE–PE	760	< −8
Ba$_{.05}$Sr$_{.95}$TiO$_3$	FE–PE	40	−7·3
SbSI	FE–PE	293	−37·0
LiH$_3$(SeO$_3$)$_2$	FE–PE	420	−6.0
LiD$_3$(SeO$_3$)$_2$	FE–PE	446	−5.9
NaD$_3$(SeO$_3$)$_2$	FE–PE	258	−3.3
KH$_2$PO$_4$	FE–PE	122	−4.5; −5·5
KD$_2$PO$_4$	FE–PE	221	−2·4; −3·9
TGS	FE–PE	322	+2·6
TGSe	FE–PE	295	+3·7
TGFB	FE–PE	346	+2·5
Rochelle salt	AFE?–FE	255	+3.6
Rochelle salt	FE–PE	297	+11·0
NaNO$_2$	FE–AFE	436	+5·6
NaNO$_2$	AFE–PE	438	+4·9
KNO$_3$	AFE?–FE	∼398	—
KNO$_3$	FE–PE	∼398	+22
PbZrO$_3$	AFE–PE	507	+4.5
PbHfO$_3$	AFE–AFE	433	+5.9
PbHfO$_3$	AFE–AFE	476	+5·0
PbHfO$_3$	AFE–PE	476	+14·0
NH$_4$H$_2$PO$_4$	AFE–PE	151	−3.4
ND$_4$D$_2$PO$_4$	AFE–PE	235	−1.4
Co$_3$B$_7$O$_{13}$I	FE–PE	196	−8.1[a]
RbH$_2$PO$_4$	FE–PE	141	−6·2[b]; −8.2[c]
KH$_2$AsO$_4$	FE–PE	96	−3.3[c]

† Unless otherwise specified the data is taken from Samara 1970 a.

[a] Smutny and Fousek 1970.

[b] Peercy and Samara 1973.

[c] Frenzel, Pietrass, and Hegenbarth 1970.

where ε_∞ is the high-frequency dielectric constant, and ω_{Li} and ω_{Ti} refer respectively to the longitudinal and transverse optical-phonon frequencies. Under the assumption that only one mode $\omega_{Ti} = \omega_s$ varies strongly with pressure then, from eqns (5.8.5) and (5.8.8),

$$\omega_s^2 \propto p - p_0. \tag{5.8.9}$$

In BaTiO$_3$ the pressure variation of the soft optic-mode frequency is supported by direct measurement of the coupled acoustic–optic modes

(Peercy and Samara 1973) by combined Raman–Brillouin scattering. The coupled-mode frequncy $\omega_a(c)$ is given by

$$\omega_a^2(c) = \omega_a^2 - \frac{A^2}{\omega_s^2} \tag{5.8.10}$$

where A^2 is the coupling constant, ω_a is the acoustic-mode frequency and $\omega \sim \omega_a \ll \omega_s$ (Lazay and Fleury 1971). The increase of ω_s with pressure results in an increase of $\omega_a(c)$ due to the piezoelectric coupling between the modes. The pressure variation of the coupling constant A also contributes to the shift of $\omega_a(c)$ so these data do not afford direct verification of (5.8.9). Estimates of $\partial A/\partial p$ from the pressure variation of the sponteneous polarization of $BaTiO_3$ (since the piezoelectric coupling is proportional to P_s) together with calculations of $\partial \omega_s/\partial p$ from the dielectric data ($\partial \varepsilon/\partial p$) are in reasonable agreement with direct measurements of $\omega_a(c)$. (Note: the pressure variation of ω_a is small in $BaTiO_3$.)

The increased stability of the paraelectric phase with pressure of a material like $BaTiO_3$ can be intuitively understood as follows. The frequency ω_s becomes small near T_c because of a cancellation of the short-range interionic forces by long-range Coulomb forces. Decreasing the unit cell volume tends to increase the short-range forces more rapidly than the Coulomb forces so that the difference, which is proportional to ω_s, increases. Thus, a negative value of $\partial T_c/\partial p$ establishes that the associated ferroelectric transition cannot be due to the decreasing volume upon cooling, since decreasing the volume by pressure stabilizes the paraelectric phase. The relative importance of the volume and pure temperature contribution to the variation of the dielectric constant (and hence ω_s) of $KTaO_3$ was recently determined over a wide temperature range (Samara and Morosin 1973). The results are shown in Table 5.3 together with the volume compressibility κ and thermal expansivity β. The crystal $KTaO_3$ is cubic over the entire temperature range, but the soft mode almost becomes unstable at very low temperatures and atmospheric pressure. The pure temperature variation $(\partial \varepsilon/\partial T)_V$ was obtained from the measured parameters via the relation

$$\left(\frac{\partial \ln \varepsilon}{\partial T}\right)_p = -\frac{\beta}{\kappa}\left(\frac{\partial \ln \varepsilon}{\partial p}\right)_T + \left(\frac{\partial \ln \varepsilon}{\partial T}\right)_V. \tag{5.8.11}$$

It is seen that the pure temperature-dependent term in Table 5.3 is dominant in determining the isobaric temperature dependence of ε (and hence ω_s), but at higher temperatures the volume effects become significant. Writing the soft-mode frequency $\omega_s^2(T)$ as the sum of a bare phonon part $\omega_0^2 (=2a_1 - v(0)$ in eqn (2.3.10)) and a thermally averaged anharmonic component (the a_2 terms in eqn (2.3.10)) which includes temperature-dependent and zero-point contributions Samara and Morosin find that the strictly harmonic term $\omega_0 \sim 0$ or is barely imaginary, implying that all the

TABLE 5.3

Value of the static dielectric constant ε of KTaO₃ and its logarithmic pressure and temperature derivatives at different temperatures †

T	ε	β	κ	$\left(\dfrac{\partial \ln \varepsilon}{\partial p}\right)_T$	$\left(\dfrac{\partial \ln \varepsilon}{\partial T}\right)_{p=1\,\text{bar}}$	$=-\left(\dfrac{\beta}{\kappa}\dfrac{\partial \ln \varepsilon}{\partial P}\right)_T$	$+\left(\dfrac{\partial \ln \varepsilon}{\partial T}\right)_V$
(K)		$(10^{-5}\,\text{K}^{-1})$	$(10^{-4}\,\text{kbar}^{-1})$	$(10^{-2}\,\text{kbar}^{-1})$	$(10^{-3}\,\text{K}^{-1})$	$(10^{-3}\,\text{K}^{-1})$	$(10^{-3}\,\text{K}^{-1})$
4·0	3840	~0	4·23	$-39\cdot0\pm0\cdot1$	~$-2\cdot4$	~0	$-2\cdot40$
50·0	1350	0·70	4·28	$-11\cdot85\pm0\cdot1$	$-21\cdot00\pm0\cdot50$	1·78	$-22\cdot78$
77·2	825	0·79	4·30	$-7\cdot24\pm0\cdot05$	$-12\cdot75\pm0\cdot50$	1·33	$-14\cdot08$
180·3	375	1·20	4·39	$-3\cdot17\pm0\cdot05$	$-5\cdot19\pm0\cdot10$	0·87	$-6\cdot06$
248·3	281	1·50	4·45	$-2\cdot30\pm0\cdot05$	$-3\cdot70\pm0\cdot10$	0·78	$-4\cdot48$
300·0	239·0	1·76	4·50	$-1\cdot83\pm0\cdot01$	$-2\cdot70\pm0\cdot10$	0·72	$-3\cdot42$
375·0	199·5	1·97	4·57	$-1\cdot54\pm0\cdot01$	$-2\cdot17\pm0\cdot10$	0·66	$-2\cdot83$
450·0	172·6	2·08	4·63	$-1\cdot36\pm0\cdot02$	$-1\cdot80\pm0\cdot10$	0·61	$-2\cdot41$

After Samara and Morosin 1973.

† The volume expansivity β and compressiblity κ are also given. The isobaric temperature derivatives are separated into pure temperature and pure volume contributions.

stabilization of ω_s at 0 K and 1 bar is provided by zero-point anharmonicities. At high pressure ω_0 becomes real.

The *increase* of Curie temperature with increasing pressure in order-disorder ferroelectrics such as TGS and its isomorphs, Rochelle salt, NaNO₂, and KNO₃ must be attributable to an increase of the dipole interaction energy $J(0)$ which increases as the dipole separation decreases. This behaviour is predicted by the Ising model for a system of fixed dipoles.

Antiferroelectric lead zirconate affords an interesting example of a material which has both a strong temperature-dependent zone centre optic mode, which gives rise to a pronounced dielectric anomaly, and a $\mathbf{q} \neq 0$ mode which becomes unstable (see § 8.2.2). Application of pressure increases T_c and decreases T_0 with the result that the dielectric anomaly at T_c decreases sharply as shown in Fig. 5.22. This experiment shows that decreasing volume softens the $\mathbf{q} \neq 0$ mode, which triggers the antiferroelectric transition while hardening the ferroelectric mode in the same way as the other perovskite oxides.

High-pressure studies have been particularly useful in the study of hydrogen-bonded ferroelectrics in which the proton plays a dominant role in triggering the phase transition. The effect of pressure on the transition temperatures of KH₂PO₄ (KDP), RbH₂PO₄ (RDP), KD₂PO₄ (DKDP), and antiferroelectric NH₄H₂PO₄ (ADP) are shown in Fig. 5.23. Isobaric measurements as a function of temperature are shown for KDP in Fig. 5.24. The most interesting feature of Fig. 5.23 is the complete disappearance of the ferroelectric or antiferroelectric transition at high pressures in all the materials except DKDP where sufficiently high pressures could not be

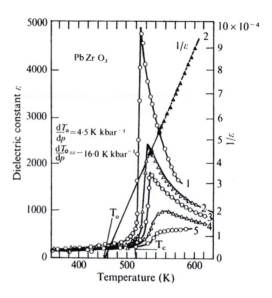

Fig. 5.22. Temperature dependence of the static dielectric constant at different pressures for PbZrO$_3$: curve 1, $p = 1$ bar; curve 2, $p = 1.90$ kbar, curve 3, $p = 3.40$ kbar; curve 4, $p = 7.60$ kbar; curve 5, $p = 11.40$ kbar. In the high-temperature paraelectric phase the Curie–Weiss law is obeyed at each pressure. (After Samara 1970 a.)

achieved. Initially the decrease of T_c with pressure is linear, as with the displacive perovskites, but at high pressures there is a marked departure from linearity with a large increase of magnitude of $\partial T_c / \partial p$.

These features can be qualitatively understood in terms of the tunnel-mode model discussed in § 2.5. According to this model the Curie tempera-ture is given by

$$\tanh \frac{\Delta}{2kT_c} = \frac{2\Delta}{J(0)} \qquad (5.8.12)$$

where $\Delta/\hbar$ is the single-proton tunnel frequency and $J(0)$ is the dipolar interaction (the exchange term in the Ising model). The value of Δ is governed by the details of the double-well potential along the OH—O bond and, since this is the most compressible bond in the crystal, Δ is expected to be a strong function of pressure. The simple classical model of Lippincott and Schroeder (1955), which has been successful in interpreting the behaviour of hydrogen bonds in solids, shows that, with decreasing distance R between the oxygen ions of the OH—O bond, the height of the potential barrier between the two equivalent proton positions decreases and the sensitivity of the proton position $\partial r/\partial R$ increases until the proton becomes symmetrically situated between the oxygens when $R \sim 0.24$ nm. In this situation either ferroelectric or antiferroelectric ordering naturally vanishes.

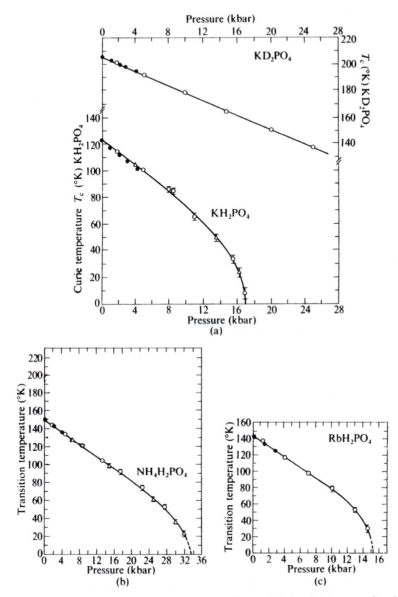

Fig. 5.23. Decrease of the transition temperature of several H-bonded ferro- and antifer-roelectrics with pressure: (a) KDP and DKDP (after Samara 1974); (b) ADP (after Samara 1971); (c) RDP (after Peercy and Samara 1973).

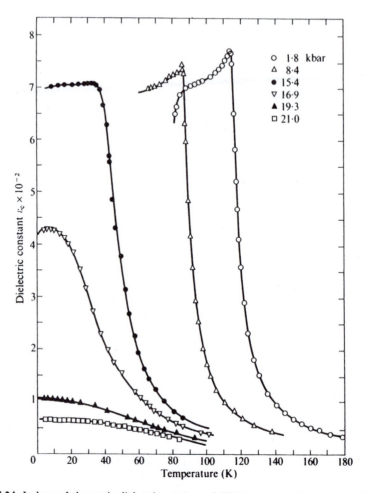

Fig. 5.24. Isobars of the static dielectric constant of KDP measured along the c-axis as a function of temperature. The paraelectric–ferroelectric transition vanishes at pressures >17 kbar. (After Samara 1971.)

However, even before hydrostatic pressure decreases R to this value the proton ordering may vanish if $J(0)$ becomes sufficiently small or if Δ becomes sufficiently large, since eqn (5.8.12) has a non-zero solution only as long as $2\Delta < J(0)$; i.e. the disordering tendency of the tunnel mode is less than the long-range ordering field. The large increase of the slope $\partial T_c/\partial p$ at high pressures observed in Fig. 5.23 can be adequately accounted for with eqn (5.8.12) by a reasonable variation of these parameters, but the fit is not unique.

5.9. The effect of high electric fields

At temperatures close to a ferroelectric phase transition, where the dielectric susceptibility becomes very large, the polarization which can be induced by even moderate external fields can be large. Under these conditions dielectric non-linearities become very important. The lowest-order non-linear dielectric term in the free energy expansion is the term in D^4, i.e.

$$G_1 = \tfrac{1}{2}\beta(T - T_0)D^2 + (\gamma/4)D^4 \qquad (5.9.1)$$

from which we deduce

$$E = \beta(T - T_0)D + \gamma D^3 \qquad (5.9.2)$$

and

$$\varepsilon^{-1} = \beta(T - T_0) + 3\gamma D^2. \qquad (5.9.3)$$

In these equations D includes both the spontaneous and field-induced polarization. We see that the inverse small-signal dielectric constant depends on the square of the electric displacement. Consequently, measurements of the dielectric constant as a function of the applied field provide an indication of the importance of higher-order terms in the expansion (eqn (5.9.1)) and of the temperature dependence of the coefficient γ. In fact in BaTiO$_3$ γ is known to vary linearly with temperature (Drougard, Landauer, and Young 1955) from experiments of this kind. At small fields when the linear (Curie–Weiss) dependence of ε is dominant we find from (5.9.3) that above T_c where $P_s = 0$

$$\varepsilon^{-1} = \beta(T - T_0)\left\{1 + \frac{3\gamma}{\beta^3(T - T_0)^3}E^2\right\} \qquad (5.9.4)$$

and for higher fields, when the first non-linear term is dominant,

$$\varepsilon^{-1} = -\beta(T - T_0) + 3(\gamma E^2)^{1/3}. \qquad (5.9.5)$$

For a first-order phase transition the ferroelectric phase may be induced by an external field at temperatures above, but close to, the Curie temperature. The shift of Curie temperature T_c may be calculated from the Clausius–Clapeyron type of equation

$$\frac{\Delta T_c}{\Delta E} = \frac{\Delta D}{\Delta S} \qquad (5.9.6)$$

where ΔD and ΔS are the discontinuous displacement and entropy change at T_c. Writing $\Delta D = D$ and $\Delta S = \tfrac{1}{2}\beta D^2$ (see § 5.6) in lowest order we find

$$\frac{\Delta T_c}{\Delta E} = \frac{2}{\beta \Delta D}. \qquad (5.9.7)$$

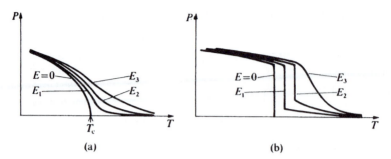

Fig. 5.25. Qualitative behaviour expected for a ferroelectric as a function of electric field for (a) a second-order transition, and (b) a first-order transition. (After Devonshire 1954.)

The anticipated effect of a field on the polarization from simple thermodynamic theory is shown schematically in Fig. 5.25 (b). The dynamic pyroelectric measurements on $BaTiO_3$ shown in Fig. 5.26 provide a convincing demonstration of this behaviour. The shift of the Curie temperature with field is also demonstrated by measurements of the thermal capacity of KDP shown in Fig. 5.27. For $BaTiO_3$ $\partial T_c / \partial E \sim 1 \cdot 4 \times 10^{-3}$ Kcm V^{-1} and for KDP,

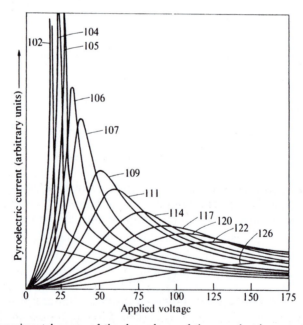

Fig. 5.26. Experimental curves of the dependence of the pyroelectric current response of $BaTiO_3$ (first-order transition) as a function of applied voltage at various temperatures near the ferroelectric–paraelectric Curie temperature T_c. Temperature values in °C are enclosure values only and $T_c \sim 101$°C in zero field. (After Chynoweth 1956 a.)

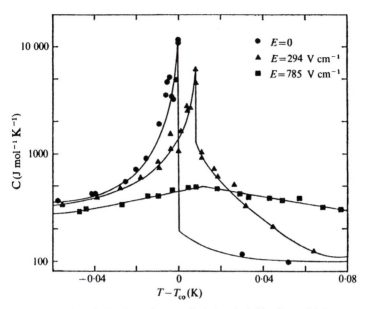

Fig. 5.27. Heat capacity of KDP in various applied electric fields. The solid lines are merely guides to the eye. No latent heat was observed for a field of 785 V/cm^{-1}, but there was evidence of a small latent heat at 294 V cm^{-1}. (After Reese 1969 a.)

$\partial T_c/\partial E \sim 0.47 \times 10^{-4}$ Kcm V^{-1}. In both sets of experimental data it is evident that for applied fields of a certain magnitude the discontinuity at T_c of the ferroelectric properties vanishes as expected.

If a large alternating field is applied to a ferroelectric just above a first-order phase transition a double hysteresis loop is observed in the same way as for antiferroelectric materials when the transition to a ferroelectric phase is induced by a field. In KDP and DKDP double loops have only recently been observed because of the slow response of the polarization to the external field near T_c (Gladkii and Sidnenko 1971, Okada and Sugie 1971) which prevents the detection of double loops at the 60 Hz frequency which is normally used. The crystals require several tens of seconds to equilibrate.

At a second-order phase transition the field dependence of the ferroelectric behaviour is somewhat different. In this case the temperature where the pyroelectric coefficient peaks does not change with applied field; the polarization curve broadens as shown in Fig. 5.25 (a) but the inflection remains at a constant temperature. Nevertheless the peak dielectric constant does shift to higher temperatures in a constant applied field. Differentiating eqns (5.9.2) and (5.9.3) we have

$$\frac{\mathrm{d}(\varepsilon^{-1})}{\mathrm{d}T} = 0 = \beta + 6\gamma D \frac{\mathrm{d}D}{\mathrm{d}T} \tag{5.9.8}$$

and

$$\frac{dE}{dT} = 0 = \{\beta(T-T_0) + 3\gamma D^2\}\frac{dD}{dT} + \beta D \qquad (5.9.9)$$

from which we find that ε^{-1} is an extremum at $T = T_{\text{peak}}$ where

$$(T_{\text{peak}} - T_0) = \frac{3\gamma D^2}{\beta} \qquad (5.9.10)$$

or, substituting in (5.9.2),

$$(T_{\text{peak}} - T_0) = \frac{3}{4}\left(\frac{4\gamma}{\beta^3}\right)^{1/3} E^{2/3}. \qquad (5.9.11)$$

The peak dielectric constant shifts to higher temperatures as $E^{2/3}$. With TGS this field dependence of the dielectric constant is found to be valid for both external bias fields and internal fields owing to dipolar defects (see § 4.3). For the case of alanine-doped TGS the shift of the peak dielectric constant shown in Fig. 4.17 and the depression of the maximum value of ε with the internal bias field E_b determined from the hysteresis loops is in good agreement with the above analysis provided that E is replaced by $E + E_b$.

The large dielectric non-linearity of ferroelectrics near a phase transition implies that the zone-centre optic-mode frequencies are field dependent. If we assume as usual that the longitudinal optic-mode frequencies ω_{Li} and the high-frequency dielectric constant ε_∞ are temperature and field independent and that only one transverse optic mode ω_s is soft, then from the Lyddane–Sachs–Teller relation we have

$$\varepsilon\omega_s^2 = \text{constant.} \qquad (5.9.12)$$

Measurements of field-induced Raman scattering of $KTaO_3$, and $SrTiO_3$ have been reported by Fleury and Worlock (1967) and Worlock and Fleury (1967). There is no first-order Raman effect in the absence of a field because all the long-wavelength phonons have odd parity. Application of a field, however, removes the centre of symmetry and a first-order Raman effect is readily observed. The scattering intensity increases as the external field is increased and in $SrTiO_3$ the soft optic mode was found to shift to higher frequencies. The field and temperature dependence of the soft-mode component polarized parallel to the field is shown in Fig. 5.28. At low fields the shift of ω_s^2 was found to be proportional to E^2 as expected from eqn (5.9.4), while at higher fields the shift of ω_s^2 was proportional to $E^{2/3}$ in agreement with eqn (5.9.5), providing an excellent confirmation of the simple theory.

The coupling of polarization and strain in a ferroelectric results in a qualitatively similar variation of the elastic and piezoelectric compliances with an applied field. The detailed behaviour depends on the crystal

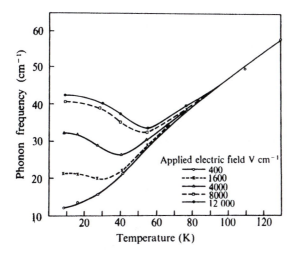

Fig. 5.28. Temperature dependence of the soft-mode frequency of SrTiO₃ for various values of applied electric field. Only the soft-mode component polarized parallel to the electric field is shown. Above 100 K the electric field shift could not be observed. (After Worlock and Fleury 1967.)

symmetry and the nature of the coupling (see Chapter 10). In some instances the effect can be quite marked. For instance in lithium thallium tartrate the large dielectric non-linearity at low temperatures (Fousek, Cross, and Seely 1970) results in a large field dependence of the elastic compliance c_{44} (Sawaguchi and Cross 1971). This compliance can be varied by a factor of 25 with fields of only 10^4 V cm^{-1} leading to a corresponding electrically tunable change of 5:1 of the piezoelectric resonance frequency.

6

STRUCTURAL CRYSTALLOGRAPHY
AND PHASE TRANSITIONS

6.1. X-ray diffraction and static structure determination

6.1.1. General theory of X-ray diffraction

IF ALL ATOMS in a crystal were exactly at their official lattice sites and there were no order–disorder ambiguity as to the position of these sites, then all the measurable physical quantities associated with the crystal would be exactly periodic functions reflecting the symmetry of the unit cell. The particular periodic function to which X-rays respond is that of the local electron density, the $\mathbf{E}$ vector of the X-ray polarizing the atoms which then scatter coherently. If we denote the incident and scattered wavevectors of the X-ray beam by $\mathbf{q}_i$ and $\mathbf{q}_s$ respectively then $\mathbf{Q} = \mathbf{q}_s - \mathbf{q}_i$ is termed the scattering vector. The intensity of scattered radiation is usually expressed by the differential scattering cross-section $d\sigma/d\Omega$ which is the intensity per unit solid angle in the direction of $\mathbf{q}_s$. For a single atom held fixed we write this simply as f_0^2, which is in general a function of scattering vector $\mathbf{Q}$, where f_0 is called the atomic scattering amplitude. The unit of intensity is taken to be that scattered by a single electron under the same conditions.

For the problem of a crystal composed of N unit cells with n atoms per cell, each at its official site $\mathbf{R}_l + \mathbf{R}_b$ (where l denotes cell site and b labels the atoms in a cell), wave theory tells us that if we look in the direction specified by $\mathbf{q}_s$ then the total scattered amplitude A is simply

$$A = \sum_{l=1}^{N} \sum_{b=1}^{n} f_0(b) \exp\{i\mathbf{Q} \cdot (\mathbf{R}_l + \mathbf{R}_b)\}. \tag{6.1.1}$$

The resulting differential cross-section is the square of the modulus of this amplitude and can be written

$$\frac{d\sigma}{d\Omega} = |F_0(\mathbf{Q})|^2 \left| \sum_l \exp(i\mathbf{Q} \cdot \mathbf{R}_l) \right|^2 \tag{6.1.2}$$

where

$$F_0(\mathbf{Q}) = \sum_b f_0(b) \exp(i\mathbf{Q} \cdot \mathbf{R}_b). \tag{6.1.3}$$

The separation of scattering amplitude into a *structure factor* $F_0(\mathbf{Q})$ and a translational *cell factor* is of the greatest utility. For a Bravais lattice in which

$\mathbf{R}_l = \Sigma_i l_i \mathbf{a}_i$ and the scattering vector $\mathbf{Q} = \Sigma_i Q_i \mathbf{a}_i^*$, where $\mathbf{a}_i$ and $\mathbf{a}_i^*$ ($i = 1, 2, 3$) are respectively the basis vectors of the unit cell and the corresponding reciprocal triad, the cell factor becomes

$$\sum_l \exp(i\mathbf{Q} \cdot \mathbf{R}_l) = \sum_{l_1} \sum_{l_2} \sum_{l_3} \exp\{i(Q_1 l_1 + Q_2 l_2 + Q_3 l_3)\} \qquad (6.1.4)$$

and, when N is large, is non-zero only when Q_1, Q_2, and Q_3 are very close to being integral multiples of 2π. This is the condition that $\mathbf{Q}$ should coincide with a reciprocal lattice vector. More quantitatively, the square of the modulus of the structure factor is

$$\left| \sum_l \exp(i\mathbf{Q} \cdot \mathbf{R}_l) \right|^2 = \prod_{i=1}^{3} \frac{\sin^2(\tfrac{1}{2} Q_i L_i)}{\sin^2(\tfrac{1}{2} Q_i)} \qquad (6.1.5)$$

in which L_i are numbers of unit cells along the sides of the block of (macroscopic) crystal, i.e. $N = L_1 L_2 L_3$. For large L_i the right-hand side of eqn (6.1.5) acquires the characteristics of a δ-function at the points $\mathbf{G}$ of the reciprocal lattice and can be written as $N V_{\mathrm{BZ}} \Sigma_G \delta(\mathbf{Q} - \mathbf{G})$ where the δ-function is defined by

$$\delta(\mathbf{Q}) = 0 \text{ for } \mathbf{Q} \neq 0, \qquad \int \delta(\mathbf{Q}) \, \mathrm{d}V_{\mathrm{BZ}} = 1 \qquad (6.1.6)$$

and V_{BZ} is the volume of the Brillouin zone. The final form for the scattering cross-section for the perfect translationally invariant crystal is therefore

$$\frac{\mathrm{d}\sigma}{\mathrm{d}\Omega} = N V_{\mathrm{BZ}} \sum_{\mathbf{G}} |F_0(\mathbf{G})|^2 \delta(\mathbf{Q} - \mathbf{G}). \qquad (6.1.7)$$

The condition $\mathbf{Q} = \mathbf{G}$ or $Q_i = 2\pi n_i$ ($i = 1, 2, 3$), where n_i are arbitrary integers, can be rewritten as

$$\mathbf{Q} \cdot \mathbf{a}_i = 2\pi n_i, \qquad i = 1, 2, 3 \qquad (6.1.8)$$

and these three equations are the so-called Laue equations for X-ray diffraction. The condition $\mathbf{Q} = \mathbf{G}$ itself is called the Bragg condition, and provides a direct determination of the crystal symmetry and of the magnitude of the lattice constants. When experiments are performed as functions of temperature it is also possible to investigate some of the more basic characteristics of structural phase transitions. Thus a ferroelectric transition can be distinguished from an antiferroelectric one very simply by the doubling of the unit cell for the latter. In addition it is possible to judge the order of the phase transition according to whether the parameters are continuous or discontinuous at T_c.

As we have seen there are three Laue equations which have to be satisfied in order that a particular component n_1, n_2, n_3, of the Bragg spectrum may be produced from the crystal. In general this requires either that the incident radiation covers a wide band of wavelengths (Laue method) or that the crystal be capable of orientational motion (Bragg method). We shall not concern ourselves with the details of the possible experimental configurations but simply state that it is in general possible to examine all the three-dimensional spectra, labelling each by its Miller indices $(n_1/m, n_2/m, n_3/m)$ where m, the greatest common divisor of n_1, n_2, n_3, is the 'order' of the spectrum. In general one finds that certain combinations of indices are missing in a crystal specimen and, by systematic study of such absences, it is possible to determine the space group of the crystal. Absences of some Miller diffracted spectra in all orders give clues as to the Bravais lattice. Other absences which refer only to a particular subset of orders indicate screw axes, glide planes, etc.

To obviate the need for Bragg orientation of the crystal one can also use powdered or polycrystalline specimens where the randomness of crystallite orientation always allows for a fraction of the sample to be suitably oriented with respect to the incident beam to enable an arbitrary diffraction line to be observed. On the other hand, the simplicity of the powder (or Debye–Scherrer) method is only achieved at the expense of losing some directional information since the method takes account only of interplanar spacing and, if there are two sets of planes in different directions but with the same separation, then the method will not distinguish them. Also, in low-symmetry structures there is often a serious overlapping between sets of planes whose spacings are closely similar.

Although the angular positions of the spectral lines allow a deduction of the space group, a quantitative measurement of intensity is required to determine the details of how atoms are arranged in the unit cell. Consider the structure factor (eqn (6.1.3)). In the scattering cross-section (eqn. (6.1.7)) it occurs only as $F_0(\mathbf{G})$ where the reciprocal lattice vector $\mathbf{G} = \Sigma_i G_i \mathbf{a}_i^*$ with $G_i = 2\pi n_i$. Writing the positions $\mathbf{R}_b$ within the unit cell as $\mathbf{R}_b = \Sigma_i x_i \mathbf{a}_i$ it follows that one can write a structure amplitude factor

$$F_0(n_1, n_2, n_3) = \sum_b f_0(b) \exp\{2\pi i(n_1 x_1 + n_2 x_2 + n_3 x_3)\}. \qquad (6.1.9)$$

Allowing more generally for a continuous distribution of electron density $\rho(x_1, x_2, x_3)$ and applying the more conventional notation $n_1 = h$, $n_2 = k$, $n_3 = l$ this generalizes to

$$F_{hkl} = \int_0^1 \int_0^1 \int_0^1 \rho(x_1, x_2, x_3) \exp\{2\pi i(hx_1 + kx_2 + lx_3)\}\, dx_1\, dx_2\, dx_3 \qquad (6.1.10)$$

where the integral runs over a unit cell. Since the electron density is a periodic function it can be expressed as a Fourier sum

$$\rho(x_1, x_2, x_3) = \sum_{h'} \sum_{k'} \sum_{l'} C_{h'k'l'} \exp\{2\pi i(h'x_1 + k'x_2 + l'x_3)\}. \qquad (6.1.11)$$

Combining eqns (6.1.10) and (6.1.11) one finds that the electron-density Fourier component C_{hkl} is proportional to the value of the structure amplitude factor F_{hkl}. Thus a knowledge of F_{hkl} is in principle equivalent to a knowledge of the electron distribution in the unit cell.

The major difficulty which arises in paractice is the fact that the experimental intensities, via eqn (6.1.7), measure only the modulus of the scattering amplitude and all information concerning its phase (since it is in general a complex quantity) is lacking. Indeed there is no direct way of finding this phase by experiment and this is a fundamental limitation on the method. For structures which possess a centre of symmetry the difficulty is less general, since the phase is restricted to 0 or π and the lack of knowledge reduces to an uncertainty in sign. There are a number of methods of approach to solving this problem of phase. One is to study an isomorphous series of compounds and to determine which way the F values vary for metal atoms of increasing scattering power (i.e. of increasing atomic number). Those F-values which increase along this series can be taken to have the same sign, the sign of the contribution of the metal atom.

A less powerful but more general method is that due to Patterson (1934). It relies on the fact that the triple summation

$$P(y_1, y_2, y_3) = \sum_h \sum_k \sum_l |F_{hkl}|^2 \exp\{2\pi i(hy_1 + ky_2 + ly_3)\} \quad (6.1.12)$$

is proportional to

$$\int_0^1 \int_0^1 \int_0^1 \rho(x_1, x_2, x_3)\rho(x_1 + y_1, x_2 + y_2, x_3 + y_3)\,dx_1\,dx_2\,dx_3.$$

$$(6.1.13)$$

It follows that large contributions to the Patterson function P give information about interatomic separations. On the other hand, two pairs of atoms with the same separation will combine to give a single peak in P. The usefulness of the Patterson projection depends very much on the atoms present; for example the method gives rather direct evidence if one atom is much heavier than the others since, for this case, the pattern of P is very similar to the actual picture of electron density in the cell. In any event the method is of the utmost value in the early stages of any structural analysis.

The final refinement of structural determination is arrived at by an iterative procedure. Armed with the knowledge acquired from other sources (e.g. Patterson analysis) a trial structure is suggested. From this structure both the magnitudes and phases of the structure factor are calculated. This phase information is then used to construct a Fourier projection from the experimental intensity data. If the assumed model is correct the experimental projection will reproduce the model precisely. Any deviations indicate errors and suggest a refinement of the model. By continuing the process we finally reach a 'best-fit' model which is taken as the actual structure. Inevitably the agreement is less than perfect and the minimum value of the factor

$$\frac{\sum_h \sum_k \sum_l |(|F_0|^2 - |F_c|^2)|}{\sum_h \sum_k \sum_l |F_c|^2} \tag{6.1.14}$$

where F_0 and F_c are respectively the experimental and model values of the structure factor, measures the accuracy of the 'least-square refinement'.

For ease of analysis it is often convenient to attempt a description of the structure in terms of thermally vibrating ions at specific cell positions rather than to attempt a more general electron-density map. In order to describe thermal amplitudes it is necessary to describe deviations $\delta \mathbf{R}_b$ of individual ions from their equilibrium positions $\mathbf{R}_b$. Expressing $\delta \mathbf{R}_b$ in terms of lattice phonons it is then possible to conceive of inelastic contributions to X-ray scattering in addition to the elastic contribution. The former will be discussed in detail in § 6.3. Elastic phonon contributions arise first in second-order perturbation and, allowing for thermal vibrations, we find that the cross-section for Bragg scattering is still given by eqns (6.1.2) and (6.1.3) but where $f_0(b)$ is now replaced by (see for example Glauber 1955, Lipkin 1960)

$$f(b, \mathbf{Q}) = f_0(b) \exp\{- W(b, \mathbf{Q})\} \tag{6.1.15}$$

in which

$$W(b, \mathbf{Q}) = N^{-1} \sum_l \langle \tfrac{1}{2}(\mathbf{Q} \cdot \delta \mathbf{R}_{lb})^2 \rangle \tag{6.1.16}$$

where $\delta \mathbf{R}_{lb}$ labels the displacement of the bth ion in the lth unit cell and N is the number of cells. The factor e^{-W} in (6.1.15) is called the Debye–Waller factor. There does not seem to be any simple relation between the components of the mean-square displacements and the force constants linking the atoms, but it can be shown that the distribution of a nucleus has surfaces of constant probability which are ellipsoids. Thus in a precise least-squares fitting program there can be as many as nine parameters for each atom, three Cartesian co-ordinates plus six thermal parameters specifying the independent components of the 3×3 symmetric tensor $U_{\alpha\beta}(b)$ where

$$W(b, \mathbf{Q}) = \tfrac{1}{2} \sum_\alpha \sum_\beta U_{\alpha\beta}(b) Q_\alpha Q_\beta. \tag{6.1.17}$$

With this degree of complexity it is obviously very useful to have some characterization of a material in advance of X-ray study. Such a characterization for ferroelectrics might include an identification of the polar axis and, if the crystal is transparent in the visible region, the optical anisotropy. Other essentials for good structure determination are the use of high-quality single-domain crystals, an accurate measurement of integrated intensities, and a precise error analysis. The completeness of poling can be determined pyroelectrically or by examination of second-harmonic generation during rotation of the crystal about the polar axis. Twinning can be identified optically and single untwinned portions selected. The final securing of a single-domain crystal is essential if the relative sense of the displacement of the atoms from the higher-symmetry phase is to be unambiguously determined. A multi-domain crystal results in a composite diffraction pattern and, since ferroelectric displacements from the higher-symmetry prototype phase are usually small, the superposition can very easily be interpreted as apparent thermal motion.

Limitations of the X-ray-scattering method for structure determination result from the fact that scattering intensity reflects electron density and that as a result the prominence of a particular atom in the structure of the diffraction pattern is directly proportional to the number of electrons it contains. This circumstance has discouraged the making of detailed X-ray maps for ferroelectrics containing light atoms, particularly hydrogen atoms. The difficulty can largely be overcome by use of neutron-scattering techniques as described in the § 6.2.

6.1.2. Structural classification for ferroelectrics

An excellent account of the modern use of X-ray-scattering analysis for the precise determination of ferroelectric crystal structures has been given by Abrahams and Keve (1971). From a careful scrutiny of ferroelectric atomic arrangements as determined primarily by X-rays they note that ferroelectric materials can usefully be divided into three classes according to the predominant nature of the atomic displacements required to reverse the polarization. The 'one-dimensional' class involves atomic displacements which are all parallel to the polar axis and is conceptually the simplest. In the 'two-dimensional' class polarity reorientation involves atomic displacements in a plane containing the polar axis, the displacements often approximating rotations by atomic groups about an axis perpendicular to that plane. Finally in the 'three-dimensional' class polarity reorientation involves similar magnitude displacements in all three dimensions. This classification involves the static ferroelectric crystallographic configuration and therefore makes no distinction between displacement and order–disorder character. Examples of these three classifications, taken from Abrahams and Keve (1971), are given in Fig. 6.1.

Shared MO_6 octahedra are fundamental to known one-dimensional ferroelectrics, M being one of many of the transition metals. Local dipoles in such octahedra arise from the displacement of the metal atoms from the centres of the octahedra. For one-dimensional ferroelectrics these displacements are along the four-, three-, or two-fold axes of the octahedron which

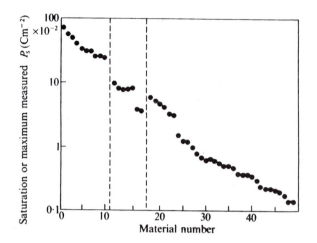

Fig. 6.1. Spontaneous polarization of some ferroelectric materials. Broken lines separate one, two, and three-dimensional classes. Materials are arranged as follows.

1. $LiNbO_3$
2. $PbTiO_3$
3. $LiTaO_3$
4. $Ba_4Na_2Nb_{10}O_{30}$
5. $Ba_{1.3}Sr_{3.7}Nb_{10}O_{30}$
6. $KNbO_3$
7. $Pb(Fe_{\frac{1}{2}}Ta_{\frac{1}{2}})O_3$
8. $BaTiO_3$
9. $SbSI$
10. $Pb(Mg_{1/3}Nb_{2/3})O_3$
11. $BaZnF_4$
12. $BaCoF_4$
13. $BaMgF_4$
14. $BaNiF_4$
15. $NaNO_2$
16. HCl
17. $SC(NH)_2$

18. RbH_2PO_4
19. KH_2AsO_4
20. KH_2PO_4
21. $(NH_2CH_2COOH)_3 \cdot H_2BeF_4$
22. $(NH_2CH_2COOH)_3 \cdot H_2SO_4$
23. $P(CH_3)_4HgBr$
24. $(NH_2CH_2COOH)_2 \cdot HNO_3$
25. $N(CH_3)_4HgI_3$
26. $N(CH_3)_4HgCl_3$
27. $N(CH_3)_4HgBr_3$
28. NH_4HSO_4
29. $RbHSO_4$
30. $NH_2CH_2COOH \cdot AgNO_3$
31. $CaB_3O_4(OH)_3 \cdot H_2O$
32. $C(NH_2)_3Ga(SeO_4)_2 \cdot 6H_2O$
33. $C(NH_2)_3Al(SeO_4)_2 \cdot 6H_2O$
34. $C(NH_2)_3Ga(SO_4)_2 \cdot 6H_2O$

35. $(NH_4)_2SO_4$
36. $C(NH_2)_3Cr(SeO_4)_2 \cdot 6H_2O$
37. $C(NH_2)_3V(SO_4)_2 \cdot 6H_2O$
38. $C(NH_2)_3Cr(SO_4)_2 \cdot 6H_2O$
39. $NaKC_4H_4O_6 \cdot 4H_2O$
40. $C(NH_2)_3Al(SO_4)_2 \cdot 6H_2O$
41. $Ca_2Sr(CH_3CH_2COO)_6$
42. $Sm_2(MoO_4)_3$
43. $LiNH_4C_4H_4O_6 \cdot H_2O$
44. $(NH_4)_2BeF_4$
45. $NaNH_4C_4H_4O_6 \cdot 4H_2O$
46. $Tb_2(MoO_4)_3$
47. $Gd_2(MoO_4)_3$
48. $Eu_2(MoO_4)_3$
49. $LiTlC_4H_4O_6 \cdot H_2O$

are then respectively parallel to the polar axis. The best-known family of this type is perhaps the perovskite family ABO_3, typified by $BaTiO_3$. Both $BaTiO_3$ and $KNbO_3$ display a sequence of ferroelectric phases which correspond to displacements along each of the three allowed directions. A second one-dimensional ferroelectric family is the lithium niobate family, also ABO_3. Currently known examples are $LiNbO_3$, $LiTaO_3$, and possibly $BiFeO_3$. In these the oxygen octahedra all have a three-fold axis parallel to the polar axis. Other one-dimensional families include the ferroelectric members of the tungsten bronzes $A_xB_{10}O_{30}$ and the $SbSI$ family. All of the above will be discussed in some detail in other chapters. It is most likely that many other families of ferroelectrics (perhaps most, if not all, of the oxygen octahedra structures) will eventually be assigned as one-dimensional in the present sense when more structural information becomes available and classification becomes possible.

Characteristic of the class of two-dimensional ferroelectrics is the reversal of polarization by the rotation of atomic groupings about axes normal to mirror planes. The most thoroughly documented examples to date are the members of the barium cobalt fluoride family $BaMF_4$ (M = Mg, Mn, Fe, Co, Ni, Zn). The structure is characterized by infinite sheets of MF_6 octahedra normal to an orthorhombic b-axis. Polarization reversal takes place by rotation of octahedra (by as much as 45°) about axes parallel to c and passing through the M ions. The resulting motion in the ab-plane produces a polarization reversal parallel to the a-axis. Simultaneous with the octahedral rotations, in which neighbouring octahedra rotate in opposite senses, the Ba atom moves along the polar a-axis. Other families of two-dimensional ferroelectrics are not yet as well documented, but it seems likely that the ferroelectric hydrogen halides (HCl, DCl, HBr, DBr), sodium nitrite, and thiourea are all members of the two-dimensional class.

Three-dimensional ferroelectric materials are characterized by crystal structures which are in general more complex than those of the other two classes. The examples presently known all contain one or more of the following structural features: discrete tetrahedral or molelcular ions, hydrogen bonds, or boron–oxygen frameworks. The class is a large one and contains many ferroelastics. Among the better known three-dimensional ferroelectrics are the gadolinium molybdate family, the guanidinium aluminium sulphate hexahydrate (GASH) family, the potassium dihydrogen phosphate (KDP) family, and the boracites. Again, many of these will be discussed in detail later in the book.

The above classification of ferroelectrics according to the character of their polarization reversal is not possible, of course, until the detailed structure of the material is known. On the other hand, the classification appears to be quite significant and demonstrates a number of other correlations among the classes. For example one-dimensional ferroelectrics seem quite generally to have the largest spontaneous polarizations and three-dimensional ones the smallest (Fig. 6.1); also, one-dimensional ferroelectrics seem to be ionic in character, three-dimensional ones molecular, and the two-dimensional group contains both molecular and non-molecular crystals.

A knowledge of crystal structure also allows for a calculation of spontaneous polarization in terms of a point charge or more sophisticated model. Most simply one can try a relation

$$\mathbf{P}_s = \frac{1}{2V} \sum_i Z_i \mathbf{\Delta}_i \qquad (6.1.18)$$

where V is a unit-cell volume, Z_i are effective charges, and $\mathbf{\Delta}_i$ the atomic displacement vectors which measure the change of ion positions on switching. Values of P_s calculated in this way, where either ionic point charges or molecular dipoles are assumed, agree most closely with the measured values for the two-dimensional class. In the one-dimensional class the ionic models give values of P_s which tend to be smaller than experiment indicating, most likely, the importance of a neglected electronic contribution. In the three-dimensional class the small differences between opposing displacement components are often swamped by the experimental uncertainty in the structural parameters.

If and when good electron-density maps become available for ferroelectrics, expressing charge density $\rho(\mathbf{r})$ as a function of position vector $\mathbf{r}$ throughout the unit cell, more quantitative estimates of spontaneous polarization might be envisaged as

$$\mathbf{P}_s = V^{-1} \int_V \mathbf{r}\rho(\mathbf{r}) \, d\mathbf{r}. \qquad (6.1.19)$$

Even with an assumed knowledge of $\rho(\mathbf{r})$, however, great care is required in calculating electric polarization, and eqn (6.1.19) is in fact incorrect in principle (Martin 1974). Spontaneous polarization can be derived formally from the basic definition

$$\mathbf{P}_s = V^{-1} \int_V \mathbf{P}(\mathbf{r}) \, d\mathbf{r} \qquad (6.1.20)$$

where $\mathbf{P}(\mathbf{r})$ is the polarization as a function of position and is related to charge density $\rho(\mathbf{r})$ by

$$\operatorname{div} \mathbf{P}(\mathbf{r}) = \rho(\mathbf{r}). \qquad (6.1.21)$$

Integrating eqn (6.1.20) by parts and using eqn (6.1.21) yields

$$\mathbf{P}_s = V^{-1} \int_V \mathbf{r}\rho(\mathbf{r}) \, d\mathbf{r} + V^{-1} \int_S \mathbf{r}\{\mathbf{n} \cdot \mathbf{P}(\mathbf{r})\} \, dS \qquad (6.1.22)$$

where $\mathbf{n}$ is the outward surface-normal unit vector and S is the surface of the unit cell. The second term represents charge transferred across the cell boundary and is a term well known in electrostatics. In general the magnitude of the two terms in eqn (6.1.22) are both functions of the choice of unit cell; only their sum is invariant. Also, in the general case it is not possible to choose a surface *a priori* for which the surface term vanishes.

Finally, in discussing elastic X-ray diffraction mention should be made of recent efforts to measure the effects of strong externally applied electric fields on the positions and intensities of the Bragg reflections (see for example Chrpa *et al.* 1972). Although such effects are clearly discernible, their interpretation at the quantitative level has not yet been accomplished in materials for which the field-driven perturbation is significant. In principle the method could be of great value in studying ferroelectrics close to their Curie points where response to applied field is very large. Interpretation of the intensity variations is complicated in many cases by the presence of charge carriers (electrons, protons, ions) which either are present in the crystal or migrate to the crystal via the electrodes to become trapped in defect sites. This interferes with charge equilibrium and leads to distortions which cannot be treated as simple stresses of the lattice. Simple stresses of course merely alter the Bragg angles, while individual ion displacements and distortions manifest themselves as changes of intensity. In spite of this separation of effects the shifts are often quite small and their interpretation, even with a valid model, requires extensive technical effort. On the other hand, if the migration effects of charged carriers can be avoided (perhaps by going to lower temperatures) the experiments seem feasible and the rewards (e.g. a study of the nature and strength of the binding forces of the constituent atoms) considerable.

6.2. Elastic neutron scattering and the problems of structure refinement

The process whereby atoms and solids scatter X-rays is an interaction between the electromagnetic radiation field and the electrons of the atom. In this picture it is clear that the prominence of a particular atom is directly proportional to the number of electrons that it contains. It follows that the location of light atoms and, in particular, a study of their thermal motion is particularly difficult by X-ray analysis. A second limitation of the X-ray picture of a structure arises because of possible discrepancies between the position of an atomic nucleus and the centre of gravity of the electron cloud which surrounds it. It is therefore of some advantage to have a technique which locates nuclei and defines internuclear distances in order to supplement the X-ray picture.

Both the above limitations are removed by using a beam of thermal neutrons instead of X-rays for the diffraction experiments. Slow neutrons have other advantages such as the ability to take note of electron spin and thereby to characterize magnetic structures. In addition, since the energy of thermal neutrons is much lower than X-ray energies, the former are a superior probe for measuring energy transfer to phonons and therefore for studying phase-transition dynamics. The process whereby atomic nuclei scatter neutrons is a nuclear one. A single nucleus of spin zero when held fixed has a scattering length a_0 for slow neutrons which corresponds to the atomic scattering factor f_0 for X-rays. However, unlike the latter it is independent of scattering angle and the total scattering cross-section is therefore simply $4\pi a_0^2$. The actual value of the scattering length, though independent of wavelength for the neutron energies considered here, depends on the detailed energy levels in the 'compound nucleus' formed by the initial nucleus and the neutron, and varies widely from element to element and indeed from isotope to isotope. The result is that the overall scattering varies considerably and in a rather haphazard way as we go through the periodic table. In particular the scattering length for hydrogen is only a little below the average value for all nuclei.

When the scattering nucleus has a spin (quantum number I) then there are two different compound nuclei of spin $I \pm \frac{1}{2}$ which can be formed with the neutron, and two distinct scattering lengths a_0^+ and a_0^- are then applicable. The existence of two scattering lengths leads to both coherent and incoherent scattering. If we define the scattering length of the bth nucleus in the lth cell as $a_l(b)$, then the square of the modulus of total scattered amplitude

$$A = \sum_{l=1}^{N} \sum_{b=1}^{n} a_l(b) \exp\{i\mathbf{Q} \cdot (\mathbf{R}_l + \mathbf{R}_b)\} \qquad (6.2.1)$$

can be rewritten as (compare (6.1.2) to (6.1.7))

$$\frac{d\sigma}{d\Omega} = N\left[V_{\text{BZ}} \sum_{\mathbf{G}} |F_0(\mathbf{G})|^2 \delta(\mathbf{Q} - \mathbf{G}) + \sum_{b=1}^{n} \{\langle a^2(b) \rangle - \langle a(b) \rangle^2\} \right] \qquad (6.2.2)$$

where

$$F_0(\mathbf{Q}) = \sum_b \langle a(b) \rangle \exp(i\mathbf{Q} \cdot \mathbf{R}_b) \qquad (6.2.3)$$

and

$$\langle a(b) \rangle = N^{-1} \sum_l a_l(b)$$

$$\langle a^2(b) \rangle = N^{-1} \sum_l a_l^2(b). \qquad (6.2.4)$$

Because of the absence of a phase relationship the incoherent scattering cross-section, given by the second term in eqn (6.2.2), is uniformly distributed. The condition $\mathbf{Q} = \mathbf{G}$ applies only to the coherently scattered part and incoherence is therefore an inconvenience since it allows some of the Bragg intensity to escape and become uniform in $\mathbf{Q}$-space. If a_0^+ and a_0^- are of opposite sign, as for example happens for hydrogen, $\langle a(b) \rangle$ can be quite small and most of the scattering becomes incoherent. Fortunately the situation can be greatly improved by deuteration. If thermal vibrations are allowed, the Debye–Waller modification found for X-rays is again applicable, eqn (6.2.2) remaining valid if each $a_0(b)$ is replaced by

$$a(b, \mathbf{Q}) = a_0(b) \exp\{- W(b, \mathbf{Q})\}. \qquad (6.2.5)$$

In addition to the elastic coherent and incoherent neutron scattering described by eqn (6.2.2) there are also inelastic coherent and inelastic incoherent contributions to scattering. These will be discussed in § 6.4.

The elastic coherent neutron scattering cross-section contains all the information necessary for a static nuclear structure determination. The information thus obtained serves normally to supplement the picture obtained by X-ray diffraction. The analysis by which we extract this information is very similar to that described for X-ray scattering. Thus the square $|F_{hkl}|^2$ of the structure factor is proportional to the intensity scattered from a crystal plane with Miller indices h, k, l, so that a theoretical intensity can be calculated from a model which locates the various atoms in the unit cell. If a fairly accurate X-ray structure determination has already been carried out, only small displacements of the X-ray tentative positions are required to refine the picture. A least-squares refinement analysis is then performed in an analogous fashion to that described in the previous section for X-rays.

Such a least-squares refinement of neutron data, starting with X-ray measurements, has improved our knowledge of the crystallography of many displacement and order–disorder ferroelectrics. For example in $PbTiO_3$, where X-ray diffraction could not determine the signs of the oxygen displacements because of the relatively small scattering lengths of these atoms compared with the massive Pb contribution, a neutron refinement (Shirane, Pepinsky, and Frazer 1955, 1956) settled the question. For the antiferroelectric $PbZrO_3$ neither the signs nor the magnitudes of the oxygen displacements could be explained until a neutron study (Jona, Mazzi, and Pepinsky 1957) was carried out. More recently, however, increasingly powerful computer facilities have made it possible to carry out successful X-ray structural determinations on a wide range of materials. For example, despite the small Li mass a successful structure determination was carried out for $LiNbO_3$ (Abrahams, Hamilton, and Reddy 1966) and later confirmed by neutrons. In isomorphic $LiTaO_3$, however, with the heavier Ta ion, X-rays were not able to fix the Li positions with complete assurance (Abrahams and Bernstein 1967) and a least-squares neutron fit was required to determine both position and temperature factors (Abrahams, Hamilton, and Sequeira 1967).

For cases in which an X-ray structure is not already available the general problem of interpreting the neutron data is very similar to that described for X-rays. The uncertain phase of the scattering amplitude is again the prime difficulty, and Patterson maps again play an important role. Of the neutron analyses of ferroelectrics by Patterson maps and Fourier projections one of the most thorough has been made by Bacon and Pease (1953, 1955) on KH_2PO_4 (KDP). At room temperature (above T_c) it proved impossible to distinguish between a model with strongly anisotropic vibrations of the protons and one in which the protons are randomly distributed in double-minimum potential wells. Below the Curie temperature, however, the protons are clearly found to be ordered with a preference for one or the other ends of the 'double' well.

Despite the recent successes of X-ray and neutron analyses in determining the basic structures of many complex ferroelectrics, such as the tungsten bronzes (Jamieson, Abrahams, and Bernstein 1968, 1969), TGS (Kay and Kleinberg 1973), and beta gadolinium molybdate (Keve, Abrahams, and Bernstein 1971) for example, one must always bear in mind the fact that structure refinement takes place with respect to a particular model and that one can never be sure that some possibly very significant facet of the real structure has escaped detection. The situation is particularly difficult in more complicated materials containing reorienting subgroups. One of the simpler examples of this problems arises for sodium nitrite, $NaNO_2$. Near and above the phase transition in $NaNO_2$ the nitrite group undergoes an order–disorder motion between two positions of local stability. The phase transition inserts a mirror plane normal to the polar axis with the nitrite group disordered on either, but not both, sides of this plane in any one primitive cell. The question arises as to the character of the reorientational motion of the nitrite group, viz. whether it be oscillatory along the direction normal to the mirror plane or rotational about one or other of the symmetry axes within the mirror plane. In any event it is most unlikely that the actual thermally averaged static structure seen by the Bragg neutrons can be approximated by placing the constituent atoms at (even split) positions and allowing only for ellipsoidal thermal vibrations. Additional terms are often added to the Fourier sum for F_{hkl} which describe free or hindered rotations in two or more fold wells, but the different models in general yield discrepancy factors (eqn (6.1.14)) after refinement which do not afford a clear discrimination among the models. It is this fact which makes other methods of probing the statics and dynamics of ferroelectric transitions so important. In this context magnetic studies, as outlined in Chapter 12, are particularly useful. For $NaNO_2$ itself, as in many more complex order–disorder situations, the detailed character of the transition remains uncertain at the time of writing. The most recent neutron diffraction study (Kay 1972) only accents the basic problems: parameter correlations, less than definitive conclusions, and conflict with the findings from other methods (in this case the Raman study of Hartwig, Wiener-Avnear, and Porto 1972).

Even one-dimensional ferroelectrics (in the Abrahams and Keve sense), which should in principle be conceptually the simplest, may present difficulties for the structural crystallographer. The basic problem again is that the X-ray or neutron data have to be fitted to a model which is always less than perfect. If an acceptable discrepancy factor is obtained after refinement using a simple basic model there may be little inclination to pursue possible complexities. A good example here is that prototype of 'displacement' ferroelectrics, $BaTiO_3$. In early experiments good convergence and refinement was obtained using a simple displacement model with isotropic Debye–Waller factors (Frazer, Danner, and Pepinsky 1955). Subsequent allowance for Debye–Waller anisotropy, however, destroyed the convergent behaviour. On increasing the number of structure–factor data the calculation again began to show signs of incipient convergence, but strong correlations developed between two of the parameters, the titanium displacement and a barium Debye–Waller parameter. The correlation is such that the discrepancy factor, or quality of refinement, has an extremum which determines only a linear combination of the correlated parameters (Frazer 1971). By again increasing the number of structure data a marginal convergence was obtained. Finally, in a separate study using both X-ray and neutron refinement (Harada, Pedersen, and Barnea 1970), a confirmation of the structure (at least as far as positional parameters are concerned) was obtained.

But is the $BaTiO_3$ refinement story really complete even yet? In the last year or two there have been indications from diffuse X-ray scattering and from the low-frequency behaviour of the dielectric constant that the actual transition in $BaTiO_3$ may have an order–disorder character. Although this is far from established as a fact, the above structure analysis contains no information which might refute the suggestion since the possibility of a partially disordered model (of the sort proposed by Comes, Lambert and Guinier (1968) and Lambert and Comes (1969) for example) was never allowed for. Such a model, in fact, would probably introduce far more parameter interactions than those encountered in the earlier studies and would almost certainly pose an indeterminate problem in a conventional refinement.

One should not extrapolate from the $BaTiO_3$ experience, however, to conclude that structure determinations of more complex materials must be hopelessly unreliable. The difficulties associated with parameter correlations seem most likely to occur for systems containing atoms which are displaced only slightly from sites of higher symmetry. Unfortunately many ferroelectrics fall into this category and one should perhaps be sceptical of any such results for which the possibility of parameter correlation has not been carefully examined.

The problem of detecting an order–disorder character in a nearly displacive ferroelectric is also not necessarily unsuperable. A recent example of just such a determination has been published for $LiTaO_3$, which had generally been considered to be a displacive ferroelectric prior to a careful neutron investigation of the temperature dependence of the structure on approaching the second-order phase transition (Abrahams et al. 1973). Lithium tantalate is ferroelectric at room temperature with Ta and Li removed in the polar direction from the centrosymmetric positions which they occupy in the prototype configuration (i.e. the high-symmetry parent structure). The phase transition at ≈ 900 K is known to be close to or actually of second order and, as the temperature is raised, the Ta ion is observed to approach the centre of the oxygen framework in a manner which parallels the temperature dependence of spontaneous polarization. The Li atom, on the other hand, appears to remain in its displaced location. Above the Curie temperature an attempted refinement with Li at its centrosymmetric (prototype) site leads to a root-mean-square thermal amplitude of vibration in the polar direction of 0·051 nm. This extremely large value suggests an attempted refinement in terms of a split-Li (double-well or order–disorder) model. The refinement is satisfactory and suggests that the Li is located equally at sites 0·037 nm on either side of centrosymmetry in the paraelectric phase.

The success of this refinement is almost certainly due to the large distance between the double-well sites in this case. For less distant wells the order–disorder character would not be betrayed by anomalously large vibrational amplitudes, and parameter interactions would likely render the order–disorder refinement indeterminate even if other pieces of experimental information suggested such a model. Indeed in the $LiTaO_3$ case itself a theoretical analysis of the dielectric properties (Lines 1972 b) suggests that the local Li potential may have three minima, two at the off-centre sites and one at the centrosymmetric position. Attempts to refine the related model with x per cent Li at centrosymmetry and $(50 - \frac{1}{2}x)$ per cent at each of the off-centre sites resulted in just such excessive parameter interaction and large temperature factors. Part of the problem at least is the oversimplification of the model; a Li ion moving in a triple well is not very well represented by three independent split Li ions at the three minima, particularly when the well depths are comparable to thermal energies as seems likely to be the case in $LiTaO_3$. Nevertheless the order–disorder character of the phase transition in $LiTaO_3$ has been established with reasonable assurance.

All these results indicate that a combination of X-ray and neutron diffraction constitutes an invaluable auxiliary in the understanding of ferroelectricity. However, elastic (or Bragg) scattering informs us only of static equilibrium configurations, and much of the vital character of ferroelectrics is associated with dynamics and non-equilibrium configurations. Information concerning the character of the ferroelectric modes and the nature of the correlations (both static and dynamic) which develop between neighbouring cell desplacements as a phase transition is approached requires a study of inelastic scattering. For X-rays, since the energy of lattice phonons is very small compared with X-ray energies, this involves quasi-elastic scattering. Thermal neutrons, with much smaller energy than X-rays, are much better equipped for a quantitative study of inelastic scattering by phonons. On the other hand, since neutron diffraction requires a high-flux reactor, this technique is not always readily available to crystallographers and we shall discuss the capabilities and limitations of X-ray diffuse (or quasi-elastic) scattering in the following separate section.

However, before moving on to dynamic studies, a few words must be included concerning a most significant development in neutron refinement analysis (Rietveld 1969, Hewat 1973 b) which enables excellent detailed structures to be obtained from powder specimens. Single-crystal neutron-diffraction experiments usually require large single-domain crystals and it is often necessary to maintain an electric field of about 10^4 V cm^{-1} to maintain the single-domain character. Powder diffraction patterns, on the other hand, are very easily obtained. The problem is in the analysis of the resulting complicated pattern, which is merely a function of the scattering angle, all reflections for a given scattering angle being superimposed. Rietveld (1969) has developed a technique which, instead of attempting to measure the structure factor or intensity for each reflection, fits the detailed shape or profile of the whole pattern using a structural model for which atomic positions can be varied continuously. A least-squares refinement is performed with parameters falling into two groups, those describing the diffractometer and those describing the crystal. The calculated profile is then a sum of Gaussian curves whose positions and heights are calculated from the structural parameters and whose width involves the instrumental parameters. Recent modifications of the fitting program even allow for the refinement of anisotropic temperature factors and an overall accuracy can be obtained which approaches that of good single-crystal work. In fact the temperature factors appear to be more precisely obtained from the powder work, primarily because of the lack of any systematic errors (such as those produced by extinction) which can plague the single-crystal experiments.

The profile-refinement technique may prove to be a very significant breakthrough in the structure refinement of moderately complex crystals for which approximate structures are already available. The great advantage is the saving in time and effort but, in addition, the method allows for a much greater freedom for changes of external conditions such as temperature and pressure, and, since a complete set of data can often be collected in a single day (compared with several weeks for typical single-crystal neutron work), the possibility of studying whole series of materials under a variety of conditions is created.

6.3. Diffuse X-ray scattering and static displacement correlations

To this point in the present chapter we have been concerned solely with elastic scattering (or Bragg scattering) which senses atomic motion only via the thermally averaged spatial positions of the independent ions. Formally the latter is a second-order effect arising from the virtual excitation of two phonons with equal and opposite wavevectors and is expressed by the

Debye–Waller factor. We shall now consider what additional information can be gleaned from a study of inelastic (or diffuse) X-ray scattering.

If we allow the ions l, b to be disturbed from their formal positions $\mathbf{R}_l + \mathbf{R}_b$ by a small amount $\mathbf{u}_{lb}$, then the scattering amplitude (eqn (6.1.1)) becomes

$$A = \sum_l \sum_b f_0(b) \exp\{i\mathbf{Q} \cdot (\mathbf{R}_l + \mathbf{R}_b + \mathbf{u}_{lb})\} \tag{6.3.1}$$

and the Debye–Waller effect can be included simply by replacing $f_0(b)$ by $f(b, \mathbf{Q})$ of eqn (6.1.15). Expanding to first order in $\mathbf{u}_{lb}$, and expressing the latter in terms of the phonon branches $\mathbf{u}_j(b, \mathbf{q})$, we find a scattering cross-section for a first-order phonon process to be

$$\frac{d\sigma}{d\Omega} = NV_{\mathrm{BZ}} \sum_{\mathbf{G}} \delta(\mathbf{Q} + \mathbf{q} - \mathbf{G}) \left| \sum_b f(b, \mathbf{Q})\{\mathbf{Q} \cdot \mathbf{u}_j(b, \mathbf{q})\} \exp\{i(\mathbf{Q} + \mathbf{q}) \cdot \mathbf{R}_b\} \right|^2 .$$
$$\tag{6.3.2}$$

Here we have neglected any time-dependent effects because the lattice frequencies are so very much smaller than X-ray frequencies. A proper quantum-mechanical treatment (Laval 1958) would show an energy restriction

$$\hbar c (\mathbf{q}_\mathrm{s} - \mathbf{q}_\mathrm{i}) = \pm \hbar \omega_j(\mathbf{q}) \tag{6.3.3}$$

corresponding to the excitation or absorption of a phonon with energy $\hbar \omega_j(\mathbf{q})$, in addition to the momentum conservation $\mathbf{Q} + \mathbf{q} - \mathbf{G} = 0$ expressed by the δ-function in eqn (6.3.2). More exactly the first-order scattering probabilities should be determined by the 'golden rule' of time-dependent perturbation theory, but the cross-section (eqn (6.3.2)) turns out to be correct if summed over the absorption and emission processes.

Introducing normalized wave amplitudes $\hat{\mathbf{u}}_j(b, \mathbf{q})$ such that

$$\sum_b m_b |\hat{\mathbf{u}}_j(b, \mathbf{q})|^2 = m \tag{6.3.4}$$

where $m = \Sigma_b\, m_b$ is the mass of one unit cell, Cochran (1963) has shown that eqn (6.3.2) can be expressed concisely in terms of the energy $E_j(\mathbf{q})$ of the mode j, $\mathbf{q}$ as

$$\frac{d\sigma}{d\Omega} = \frac{V_{\mathrm{BZ}} E_j(\mathbf{q})}{m \omega_j^2(\mathbf{q})} \sum_{\mathbf{G}} Q^2 |F_j'(\mathbf{Q})|^2\, \delta(\mathbf{Q} + \mathbf{q} - \mathbf{G}) \tag{6.3.5}$$

where the quantity $F_j'(\mathbf{Q})$ is defined by

$$QF_j'(\mathbf{Q}) = \mathbf{Q} \cdot \sum_b \hat{\mathbf{u}}_j(b, \mathbf{q}) f(b, \mathbf{Q}) \exp(i\mathbf{G} \cdot \mathbf{R}_b) \tag{6.3.6}$$

and may be called the structure factor for first-order scattering by analogy with $F_0(\mathbf{Q})$ of eqn (6.1.3), the structure factor for Bragg scattering. Summing over the modes j the total first-order scattering cross-section at wavelength $\mathbf{q}$ is found to be

$$\frac{d\sigma}{d\Omega} = \frac{NQ^2}{m} \sum_j \frac{E_j(\mathbf{q})}{\omega_j^2(\mathbf{q})} |F_j'(\mathbf{Q})|^2 \qquad (6.3.7)$$

where as we have already seen $\mathbf{Q}$ and $\mathbf{q}$ are related by momentum conservation $\mathbf{Q} + \mathbf{q} = \mathbf{G}$. The energy $E_j(\mathbf{q})$ is related to frequency $\omega_j(\mathbf{q})$ by

$$E_j(\mathbf{q}) = \{n_j(\mathbf{q}) + \tfrac{1}{2}\}\hbar\omega_j(\mathbf{q}) \qquad (6.3.8)$$

where $n_j(\mathbf{q})$ for thermal equilibrium is the occupation number of the mode $[\exp\{\hbar\omega_j(\mathbf{q})/kT\} - 1]^{-1}$.

Close to a displacement phase transition one of the modes $j = s$ goes soft and dominates eqn (6.3.7) which, using eqn (6.3.8) with $\hbar\omega_s(\mathbf{q}) \ll kT$, tends towards

$$\frac{d\sigma}{d\Omega} = \frac{NkTQ^2}{m} \{\omega_s(\mathbf{q})\}^{-2} |F_s'(\mathbf{Q})|^2. \qquad (6.3.9)$$

From § 2.3 (e.g. eqn (2.3.17)) we know that for quasi-harmonic motion the reciprocal of the square of the soft-mode frequency is a measure of static wavevector-dependent susceptibility $\chi(\mathbf{q})$. This in turn, via the fluctuation theorem (eqn (2.2.28)), is a measure of static (or equal-time) fluctuations. We obtain only equal-time correlations (i.e. no information concerning true dynamics) because experimentally we are integrating over all frequency by measuring the total intensity without attempting to frequency analyse the scattered beam. This in turn is necessitated by the very small energy of lattice phonons compared with X-ray energies. On the other hand, a knowledge of the temperature and angular dependence of static correlations near a phase transition contains a lot of important information concerning the forces producing the transition (e.g. degree of anisotropy, range, etc.) and allows for a determination of several static critical parameters if the transition is of second order or close to it.

The relationship between diffuse X-ray scattering and static wavevector-dependent susceptibility is, not surprisingly, more general than the quasi-harmonic derivation above would indicate. Consider for example the opposite extreme of an Ising-like order–disorder transition (Yamada and Yamada 1966). Near T_c the lattice motion is dominated by the order–disorder degree of freedom, and we therefore write the electron charge distribution in the lth unit cell as

$$\rho_l(\mathbf{x}) = \langle\rho(\mathbf{x})\rangle + \sigma_l \,\Delta\rho(\mathbf{x}) \qquad (6.3.10)$$

where the origin of $\mathbf{x}$ is taken at the cell centre of symmetry, $\langle\rho(\mathbf{x})\rangle$ is the thermal average distribution, $\Delta\rho(\mathbf{x})$ is an odd function with respect to $x = 0$, and σ_l is an Ising variable with values ± 1. The resulting scattering cross-section is

$$\frac{d\sigma}{d\Omega} = \left| \sum_l f_l(\mathbf{Q}) \exp(i\mathbf{Q} . \mathbf{R}_l) \right|^2 \qquad (6.3.11)$$

where structure factor $f_l(\mathbf{Q})$ is

$$f_l(\mathbf{Q}) = F(\mathbf{Q}) + i\sigma_l F'(\mathbf{Q}) \qquad (6.3.12)$$

with

$$F(\mathbf{Q}) = \int_{\text{cell}} \langle \rho(\mathbf{x}) \rangle \exp(i\mathbf{Q} . \mathbf{x}) \, d\mathbf{x} \qquad (6.3.13)$$

and

$$iF'(\mathbf{Q}) = \int_{\text{cell}} \Delta\rho(\mathbf{x}) \exp(i\mathbf{Q} . \mathbf{x}) \, d\mathbf{x}. \qquad (6.3.14)$$

The simplest formalism is for the disordered phase, for which $\langle \sigma_l \rangle = 0$. For this case the scattering cross-section (eqn (6.3.11)) separates into Bragg and diffuse contributions proportional to $\{F(\mathbf{Q})\}^2 \delta(\mathbf{Q} - \mathbf{G})$ and to $\{F'(\mathbf{Q})\}^2 \sigma(-\mathbf{Q})\sigma(\mathbf{Q})$ respectively, where $\sigma(\mathbf{Q})$ is the Fourier transform of σ_l with respect to the lattice $\mathbf{R}_l$. Thus in thermal equilibrium the X-ray diffuse scattering intensity is proportional to $\langle |\sigma(\mathbf{Q})|^2 \rangle \equiv \langle |\sigma(\mathbf{Q} - \mathbf{G})|^2 \rangle$, which is the intercell (charge-motion) correlation function measuring the equal-time correlations $\langle \sigma_l(t)\sigma_{l'}(t) \rangle$ as a function of range $\mathbf{R}_l - \mathbf{R}_{l'}$ and temperature. Once again, via the fluctuation theorem, it is also a measure of wavevector-dependent static susceptibility.

Away from a phase transition eqn (6.3.7) would indicate that the major contribution to first-order diffuse intensity comes from long-wavelength acoustic modes (which have the smallest frequencies). In fact such a realization can lead to the use of diffuse X-ray scattering for measurement of elastic constants, although the results are generally less accurate than can be obtained using ultrasonics. On approach to a critical point the contribution to diffuse scattering from critical fluctuations increasingly dominates, but the subtraction of the non-critical background can be a problem, particularly at temperatures and wavevectors away from the singularity itself. For ferroelectrics the singularity occurs at reciprocal lattice points $\mathbf{G}$, and X-ray critical scattering is therefore expected to occur close to these points as the Curie temperature is approached. If the long-range dipolar interactions play a significant role, then the critical scattering is expected to be very anisotropic because of the anisotropic nature of the dipolar forces themselves (Cohen and Keffer 1955).

A good example of an Ising-like ferroelectric for which quantitative diffuse X-ray scattering analysis is available is TGS, which is spontaneously polarized parallel to the monoclinic b-axis and exhibits a second-order ferroelectric phase transition at $\approx 49°C$ (see Chapter 9 for details of the structure). Shibuya and Mitsui (1961) made the first observation of X-ray critical scattering in TGS, but no quantitative study of intensity distribution in reciprocal space was made until Fujii and Yamada (1971) undertook a more detailed study. The distribution of dipole moments carried by molecular groups in the unit cell is actually very complicated in TGS but the transition entropy suggests that an Ising dipole approximation, with each dipole at a Bravais lattice point and oriented along the $\pm b$ direction, is a fairly realistic basic model with which to describe the critical phenomena.

From eqns (6.3.11) and (6.3.12) the paraelectric X-ray diffuse scattering due to the fluctuation in the Ising variable is given by

$$\frac{d\sigma}{d\Omega} \propto |F'(\mathbf{Q})|^2 \sum_l \sum_{l'} \langle \sigma_l \sigma_{l'} \rangle \exp\{i\mathbf{q} \cdot (\mathbf{R}_l - \mathbf{R}_{l'})\} \qquad (6.3.15)$$

where $\mathbf{Q}$ is the scattering vector and $\mathbf{q} = \mathbf{G} - \mathbf{Q}$ (with $\mathbf{G}$ a reciprocal lattice vector). In terms of the Fourier transform of the Ising variable this is, as expressed above, equal to $|F'(\mathbf{Q})|^2 \langle \sigma(-\mathbf{q})\sigma(\mathbf{q}) \rangle$ and, via the fluctuation result for an Ising disordered phase, viz.

$$kT\chi(\mathbf{q}) = \langle \sigma(-\mathbf{q})\sigma(\mathbf{q}) \rangle \qquad (6.3.16)$$

(compare eqn (2.2.28)) this becomes

$$\frac{d\sigma}{d\Omega} \propto |F'(\mathbf{Q})|^2 kT\chi(\mathbf{q}). \qquad (6.3.17)$$

In the simple random-phase or correlated effective-field statistical approximations we expect a wavevector dependence of susceptibility of the form (see eqn (2.2.25))

$$\{\chi(\mathbf{q})\}^{-1} = \{\chi(0)\}^{-1} + v(0) - v(\mathbf{q}) \qquad (6.3.18)$$

where $v(\mathbf{q})$ is the Fourier transform of the interaction energy $v_{ll'}$ between the dipoles. At the ferroelectric transition temperature T_c the uniform mode diverges and the distribution of critical scattering around the reciprocal lattice points is best expressed by expanding $v(\mathbf{q})$ around $v(0)$. However, $v(\mathbf{q})$ has a singularity at $\mathbf{q} = 0$ when uniaxial dipolar forces are involved. Separating $v(\mathbf{q})$ into a regular part and a singular one, and assuming for the moment a Curie–Weiss divergence for uniform susceptibility, we find

$$\frac{d\sigma}{d\Omega} \propto kT|F'(\mathbf{Q})|^2 \left\{ k(T - T_c) + \left(\sum_\lambda v_\lambda q_\lambda^2 \right) + C \cos^2\varphi \right\}^{-1} \qquad (6.3.19)$$

where φ is the angle between $\mathbf{q}$ and the direction of polarization, $\sum_\lambda$ runs over $\lambda = x, y, z$, the v_λ are effective interaction parameters, and C is a constant. In a zeroth approximation C is just the Ising-model Curie–Weiss constant (Fujii and Yamada 1971), but more accurately it is modified by the electrostatic interaction between the Ising dipole and the electronic polarization induced by these moments. In other words the Ising system must be considered to be in a polarizable medium. This results in a 'screening' of electrostatic interaction which is in general anisotropic leading to a tensor form for C.

Qualitatively, however, certain properties are immediately obvious from eqn (6.3.19). Because of the last term in the denominator the critical divergence as $T \to T_c$ and $\mathbf{q} \to 0$ occurs only in the $\varphi = \pi/2$ plane (the a^*c^* plane in TGS, where the asterisk denotes a reciprocal vector) if the long-range dipolar forces are significant. Fig. 6.2 shows a typical temperature

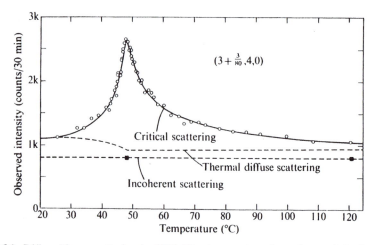

Fig. 6.2. Diffuse X-ray scattering in TGS. The temperature dependence of the intensity observed at the point $(3+3/80, 4, 0)$ with $\mathbf{q} = 3\mathbf{a}^*/80$ is shown by the open circles. The black square shows the intensity observed at the point on the Brillouin-zone boundary. (After Fujii and Yamada 1971.)

dependence of diffuse X-ray intensity close to a reciprocal lattice point in TGS. The total intensity peaks at the Curie temperature but contains incoherent and thermal diffuse contributions in addition to the order-disorder critical component. The total intensity is much weaker than the Bragg peak at the reciprocal lattice point but resolution is good, often better than one part in a hundred of the zone-boundary wavevector. Subtraction of incoherent scattering is easy since it is independent of both temperature and wavenumber. Thermal diffuse scattering from acoustic phonons is parameterized theoretically and the parameters (e.g. Debye–Waller factor) determined experimentally. The separate contributions are shown in Fig. 6.2 and also in Fig. 6.3 where an intensity distribution scan along the b^* axis

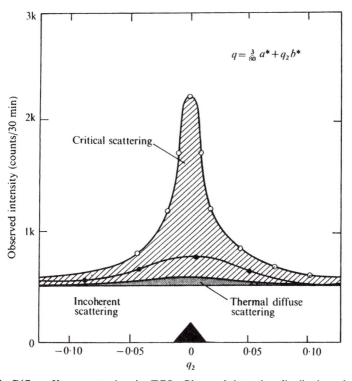

Fig. 6.3. Diffuse X-ray scattering in TGS. Observed intensity distribution along $\mathbf{q} = 3\mathbf{a}^*/80 + q_2\mathbf{b}^*$ at a temperature 0.1 K above T_c (open circles) and at a temperature $T = 25°C$ (some 25 K below T_c). The dotted region shows the part of the thermal diffuse scattering at $T_c + 0.1$ K estimated from the data at 25°C. The hatched region therefore represents the critical scattering at $T_c + 0.1$ K. The black triangle at $q_2 = 0$ shows the size of the resolution function in the direction of the $\mathbf{b}^*$ axis. (After Fujii and Yamada 1971.)

is shown for a temperature $T = T_c + 0.1$ K. Finally in Fig. 6.4 (a) we show the intensity contours of critical scattering in the a^*b^* plane at $T = T_c + 0.1$ K after subtraction of the background. The contour in the b^*c^* plane is qualitatively similar. The theoretical contour calculated from eqn (6.3.19) without screening effects is shown for comparison in Fig. 6.4 (b).

A quantitative analysis of the data for TGS reveals firstly that the linear temperature term in the denominator of eqn (6.3.19) is indeed essentially correct. In critical-exponent terms (see Chapter 11) the intensity divergence in the a^* or c^* directions is a measurement of the correlation-length exponent ν and gives $\nu = 0.502$. A value $\nu = \frac{1}{2}$ is to be expected for long-range forces and $\nu \approx \frac{2}{3}$ for short-range forces. This finding therefore points to the dominance of long-range forces, presumably of dipolar origin, in the critical region. On the other hand a comparison of Figs. 6.4 (a) and (b) indicates that, although the qualitative $\mathbf{q}$-dependence expected from forces

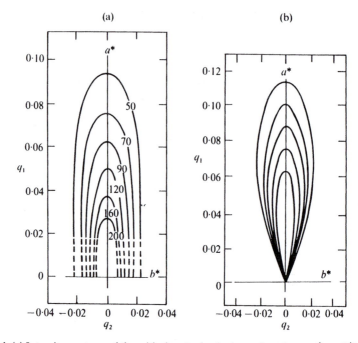

Fig. 6.4. (a) Intensity contours of the critical scattering in the c-plane ($\mathbf{q} = q_1\mathbf{a}^* + q_2\mathbf{b}^*$) around the (3, 4, 0) reflection in TGS at the temperature $T = T_c + 0 \cdot 1$ K (after subtracting the background). The numbers on the contours represent intensity in units of counts per 3 min. (b) The equivalent intensity contours calculated from eqn (6.3.19) without screening. (After Fujii and Yamada 1971.)

of dipolar character is indeed observed, the theoretical pinching into the Bragg point evident in Fig. 6.4 (b) is not observed experimentally. This is in all probability the result of a screening by the electronic response which reduces the influence of the longer-range dipolar forces by adjusting electron density. The reduction is apparently not sufficient to produce short-range critical exponents. It is interesting to compare these results with the equivalent X-ray findings for a non-ferroelectric Ising-model system $NaNO_3$. In $NaNO_3$ the ordering of NO_3^- radicals can also be described as an order–disorder transition (between two stable orientations of each NO_3^-). This time, however, since the NO_3^- radical does not carry an electric dipole moment, electric dipole forces are not expected to be involved. Significantly Terauchi and Yamada (1972) find for this case an intensity distribution which is almost isotropic and which can be explained quantitatively in terms of short-range interactions up to second-neighbour radicals. A measurement of critical exponents very close to the transition temperature was not attempted.

Among the displacement ferroelectrics some of the more extensive diffuse X-ray studies have been performed on the perovskites $BaTiO_3$, $KNbO_3$, and $NaNbO_3$. The intensity distribution from each is very anisotropic and can be related in general to the presence of dipolar forces. In the perovskites, however, the cubic crystal symmetry in the high-temperature phase leads to soft-mode degeneracies and to a much more complex diffuse scattering pattern than for TGS. The situation for the two ferroelectrics $BaTiO_3$ and $KNbO_3$ is qualitatively similar. At high temperatures the cubic phases are paraelectric and on approach to the highest transition temperature diffuse scattering anisotropy develops which is concentrated in {100}-type reciprocal-lattice planes. This indicates the presence of strong displacement correlations in real space along equivalent $\langle 100 \rangle$ cubic axes (Comes *et al.* 1968, 1970). Qualitatively this is just the behaviour to be expected for intercell forces of dominantly dipolar form for which, as a general rule (Lines 1972 a), correlations develop along the incipient polar axis as a ferroelectric transition is approached from the paraelectric side. In the cubic phase of $BaTiO_3$ and $KNbO_3$ this incipient polar direction is a cubic axis and hence three-fold degenerate.

Marked scattering anisotropy is also observed in each of the three ferroelectric phases at lower temperatures and Comes *et al.* have developed an (Ising) model of static disorder in all phases (even the paraelectric cubic phase) to account for the observed anisotropy pattern. However, since the X-ray diffuse spectrum contains no information concerning lattice dynamics it is not possible to assess the static or dynamic nature of the correlated deviations from prototype symmetry from these experiments alone. The static Ising model would imply that the higher-symmetry phases were merely averaged resultants of lower-symmetry unit cells. A displacement model at the opposite extreme would attribute the correlations to quasi-harmonic soft-mode fluctuations about the higher-symmetry thermal equilibrium. The true situation may of course be of intermediate character, e.g. oscillations in a very anharmonic potential or even a shallow double well. Neutron scattering experiments, as we shall see in the following section, seem to support this intermediate picture.

In the soft-mode language of eqn (6.3.9) the X-ray diffuse scattering anisotropy is directly associated with the symmetry and dispersion of the soft mode itself. The many phase transitions of $NaNbO_3$ (§ 8.1.5) give a good example of this. The lowest temperature phase of $NaNbO_3$ below about $-100°C$ is ferroelectric. The room-temperature phase is antiferroelectric and persists up to about 370°C. Between this temperature and 640°C at least four more structural transitions occur with the cubic perovskite phase eventually becoming stable over 640°C. These several transitions have all been indexed recently by means of diffuse X-ray scattering (Ishida and Honjo 1973) and the two highest ones characterized in terms of the condensation of zone-boundary phonons. When a single soft mode dominates the scattering near a phase transition of the displacement type, then it is possible to determine this mode by analysing $F'_s(\mathbf{Q})$ of eqn (6.3.9). If the mode is degenerate this can be taken into account during the analysis. The scattering anisotropy is caused by the dispersion of the soft phonons near the condensation wavevector. In the high-temperature phases of $NaNbO_3$ two zone-boundary modes (corresponding to rotations of the oxygen octahedra) are simultaneously of low frequency. This leads to a complex intensity pattern involving both correlated $\langle 100 \rangle$-type chains and planes which can be understood in terms of the displacements involved in the two low-frequency modes together with the appropriate phase relation of the movement of atoms in successive planes. This combination of correlated chains and planes, first observed in the cubic phase of $NaNbO_3$ by Denoyer, Comes, and Lambert (1970), may be somewhat unusual and is not expected for other perovskite anti-distortive systems like $SrTiO_3$ and $LaAlO_3$ in which only a single zone-boundary mode is soft.

Some antiferroelectric structures are more complicated than a simple up–down arrangement of dipoles (corresponding to a zone-boundary condensation) and exhibit a sinusoidal modulation with a wavelength which may or may not be commensurate with the crystal structure. In such a situation the X-ray diffuse scattering intensity exhibits a maximum at a finite value of $\mathbf{q}$ which measures the pitch of the structural modulation. The side peak is often quite close to the Bragg line and as a result it is often referred to as X-ray satellite scattering. Examples are found in thiourea (Shiozaki 1971) and in sodium nitrite (Yamada, Shibuya, and Hoshino 1963, Yamada and Yamada 1966).

In Fig. 6.5 we show the intensity distribution in $NaNO_2$ (which is an Ising-dipole system) in the *paraelectric* phase at a temperature a few degrees

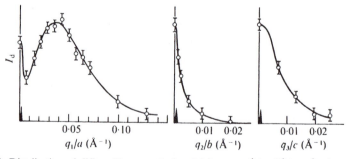

Fig. 6.5. Distribution of diffuse X-ray scattering $I_d(\mathbf{q})$, $\mathbf{q} = q_1 \mathbf{a}^* + q_2 \mathbf{b}^* + q_3 \mathbf{c}^*$, observed at a temperature about 3 K above the transition to the sinusoidal antiferroelectric phase in $NaNO_2$. Small triangles indicate resolution functions and the peak at $q_1 = 0$ includes the contribution of the normal Bragg reflection. The Bravais lattice is orthorhombic with $a = 0.366$ nm, $b = 0.567$ nm, $c = 0.536$ nm. (After Yamada and Yamada 1966.)

above the transition to a sinusoidal antiferroelectric phase. The sharp maximum at the origin corresponds to Bragg reflection and the small triangles indicate the estimated resolution. More detailed results and a detailed discussion of $NaNO_2$ is given in § 9.5, but for the moment we focus attention on the satellite scattering in the a^* direction which represents a sinusoidal correlation of b-component dipoles along the a-axis (the Ising dipole direction is the b-axis, with random hopping to the 'left' or 'right' of the crystallographic ac-plane). The satellite peak position is in fact slightly temperature dependent, moving from $q \approx a^*/8$ at the transition point to $q \approx a^*/5$ in the high-temperature limit, and the peak amplitude increases as the transition is approached. Yamada and Yamada (1966) show that these features are qualitatively explained once again by the peculiar spatial properties of the electric dipolar interaction. Simple dipolar theory predicts a satellite at $q \approx a^*/5$. A more quantitative understanding can be obtained by allowing for additional interactions of a short-range nature and the latter become relatively more important as the transition is approached. One readily verifies that the paraelectric correlations reflect the symmetry of the sinusoidal ordered phase which sets in below the transition.

6.4. Inelastic neutron scattering and dynamic crystallography

If we define $\mathbf{q}_s$ and $\mathbf{q}_i$ to be respectively the scattered and incident wavevectors for neutrons of mass m_n, then, for a coherent scattering process in which $\mathbf{Q} = \mathbf{q}_s - \mathbf{q}_i$ and $\hbar\omega = \hbar^2(q_s^2 - q_i^2)/2m_n$ are respectively the change in momentum and energy of the scattered neutrons, the differential scattering cross-section $d^2\sigma/d\Omega\, d\omega$, i.e. the flux of neutrons into the solid angle $d\Omega$ with energy change between $\hbar\omega$ and $\hbar(\omega + d\omega)$, can be written in the form

$$\frac{d^2\sigma}{d\Omega\, d\omega} = \frac{q_s}{2\pi q_i} S(\mathbf{Q}, \omega) \qquad (6.4.1)$$

where $S(\mathbf{Q}, \omega)$ is called the scattering function and, as was shown by van Hove (1954), is the Fourier transform with respect to both lattice and time of the internuclear correlation function. Specifically

$$S(\mathbf{Q}, \omega) = \iint g(\mathbf{R}, t) \exp\{i(\mathbf{Q} \cdot \mathbf{R} - \omega t)\}\, d\mathbf{R}\, dt \qquad (6.4.2)$$

where the correlation function $g(\mathbf{R}, t)$ is given by

$$g(\mathbf{R}, t) = \int \langle \rho(\mathbf{R}', 0)\rho(\mathbf{R} + \mathbf{R}', t)\rangle\, d\mathbf{R}' \qquad (6.4.3)$$

in which

$$\rho(\mathbf{R}, t) = \sum_i \langle a_i \rangle\, \delta(\mathbf{R} - \mathbf{R}_i(t)) \qquad (6.4.4)$$

with $\langle a_i \rangle$ the coherent scattering length of the ith nucleus. The cross-section integrated over frequency (or equivalently energy) becomes in the elastic limit $q_s \to q_i$

$$\frac{d\sigma}{d\Omega} = \int (2\pi)^{-1} S(\mathbf{Q}, \omega) \, d\omega = S(\mathbf{Q}) \qquad (6.4.5)$$

where $S(\mathbf{Q})$ is therefore the Fourier transform with respect to the lattice of the equal-time (or static) correlation function $g(\mathbf{R}, 0)$. This is the form appropriate for diffuse X-ray scattering.

Following Cochran (1969) we introduce a time-dependent structure factor, or spatial Fourier transform of the scattering density,

$$A(\mathbf{Q}, t) = \sum_i \langle a_i \rangle \exp\{i\mathbf{Q} \cdot \mathbf{R}_i(t)\} \qquad (6.4.6)$$

where i runs over all nuclear sites in the macroscopic lattice. In terms of $A(\mathbf{Q}, t)$ one readily verifies the relationship

$$S(\mathbf{Q}, \omega) = \int \langle A(-\mathbf{Q}, 0) A(\mathbf{Q}, t) \rangle \exp(-i\omega t) \, dt. \qquad (6.4.7)$$

In a translationally invariant lattice we write the general ith site as being the bth site of the lth unit cell, i.e.

$$\mathbf{R}_i(t) = \mathbf{R}_l + \mathbf{R}_b + \mathbf{u}_{lb}(t) \qquad (6.4.8)$$

where $\mathbf{u}_{lb}(t)$ allows for motion about the formal site. Using displacements (eqn (6.4.8)) in eqn (6.4.6) then leads directly to the scattering amplitude (eqn (6.4.7)) and the scattering cross-section (eqn (6.4.1)). For the quasi-harmonic case, in which phonons $\mathbf{u}_j(b, \mathbf{q})$ are introduced with normalized amplitudes $\hat{\mathbf{u}}_j(b, \mathbf{q})$ as given in eqn (6.3.4), an expansion of the exponentials to first order in phonon co-ordinates leads formally (via the introduction of creation and annihilation operators (Cochran 1969) to a cross-section for first-order scattering from phonon j, $\mathbf{q}$ of the form

$$\frac{d^2\sigma}{d\Omega \, d\omega} = \frac{V_{BZ}}{m\omega_j^2(\mathbf{q})} \{E_j(\mathbf{q}) \pm \tfrac{1}{2}\hbar\omega_j(\mathbf{q})\} \, \delta\{\omega \pm \omega_j(\mathbf{q})\}$$

$$\times \sum_{\mathbf{G}} \frac{1}{2} \frac{q_s}{q_i} Q^2 |F_j'(\mathbf{Q})|^2 \, \delta(\mathbf{Q} + \mathbf{q} - \mathbf{G}) \qquad (6.4.9)$$

which is to be compared with eqn (6.3.5). The structure factor $F_j'(\mathbf{Q})$ for first-order scattering is given by eqn (6.3.6) but with $f(b, \mathbf{Q})$ replaced by the coherent neutron scattering length $\langle a(b, \mathbf{Q}) \rangle$.

Equation (6.4.9) contains the energy and momentum conservation laws

$$\mathbf{Q} \equiv \mathbf{q}_s - \mathbf{q}_i = \mathbf{G} - \mathbf{q} \qquad (6.4.10)$$

$$\frac{\hbar^2}{2m_n} (q_s^2 - q_i^2) = \pm\hbar\omega_j(\mathbf{q}) \qquad (6.4.11)$$

where m_n is the mass of the neutron. Separate cross-sections for phonon absorption with neutron energy gain and for phonon emission with neutron energy loss have to be distinguished. In thermal equilibrium $E_j(\mathbf{q})$ in eqn (6.4.9) takes the value of its ensemble average, viz. $\{\hbar\omega_j(\mathbf{q})/2\}\coth\{\hbar\omega_j(\mathbf{q})/2kT\}$. Thus at low temperatures the cross-section for absorption tends to zero (negative sign in eqn (6.4.9)) while that for emission tends to a finite value.

The great advantage of inelastic neutron scattering over light scattering and X-ray scattering is its capability of measuring both the wavelength and time dependence of dynamic correlations (viz. $S(\mathbf{q}, \omega)$). Light scattering, at least in first order, measures essentially $S(0, \omega)$ and X-ray scattering $S(\mathbf{q})$. Infrared and Raman spectra are also limited by optical selection rules. One disadvantage, of course, is the need for a reactor to produce the neutron flux, which limits rather drastically the number of locations able to undertake such experiments. By use of the momentum conservation law to calculate the phonon wavevector $\mathbf{q}$, dispersion relations $\omega_j(\mathbf{q})$ can in principle be measured over the entire Brillouin zone. However, for reasons of expediency measurements are most frequently confined to symmetry directions, such as $\langle 100 \rangle$, $\langle 110 \rangle$, and $\langle 111 \rangle$ in cubic crystals. In hydrogen-bonded ferroelectrics the measurements are more difficult because of the large incoherent cross-sections of the protons. For incoherent scattering any vibration can contribute to scattering in any direction, and the cross-section measures essentially a density of states. The resulting neutron bands tend to be broad and very difficult indeed to interpret with confidence.

Thermal neutrons from water-cooled reactors have energies ranging typically around 30 meV (or 240 cm^{-1}), which corresponds to wavelengths of about 0.38 nm. Thus the neutron beam is capable of undergoing scattering by phonons to give measurable changes in energy as well as wavevector, a property which makes thermal neutrons a unique tool for the study of lattice dynamics. A neutron measurement is in principle simple; a monoenergetic neutron beam strikes a single-crystal sample, and scattered neutrons are detected and analysed as a function of sample orientation and scattering angle. The neutron spectrometer therefore has three parts; a monochromator (usually Bragg scattering from a suitable single-crystal material) to select the monoenergetic neutrons, goniometers to set the sample angles, and an analyser (either a chopper in the scattered beam or a second Bragg-scattering single crystal) to measure the final neutron energy. The use of the second Bragg scatterer for analysing (triple-axis spectrometer) has an advantage that it makes possible any type of scan through energy–wavevector space to be performed; most common is the systematic variation of final energy with scattering vector constant (constant Q-scans) or the variation of $\mathbf{Q}$ for a fixed energy (constant E-scans). Details of instrumental resolution corrections, intensity-measuring techniques, and relevant references can be found in Samuelsen (1971).

For ferroelectrics and antiferroelectrics we are primarily concerned with inelastic neutron scattering from phonons at or near a special symmetry point in the Brillouin zone where a mode softens as a phase transition is approached. The condensing mode may of course interact strongly with other modes and significant anomalies may be present for several branches. One important point, however, is the fact that the neutron scattering cross-section, as the soft-mode energy $\hbar\omega_j$ decreases to values smaller than kT, becomes proportional to $1/\omega_j^2$ (see eqn (6.4.9)). It follows that soft-mode measurements should not be difficult once sufficient knowledge is obtained about the resolution function of the spectrometer (i.e. the spread of neutron energy and wavevector due to finite collimations of the monochromator).

Ferroelectric soft modes are found, if at all, at the Brillouin zone centre and can be either underdamped or overdamped. The perovskite crystals have received to date by far the largest attention in the present context, and as an example we shall discuss here the findings for potassium tantalate niobate (KTN) as given by Yelon et al. (1971), Hewat, Rouse, and Zaccai (1972), and Zaccai and Hewat (1974).

6.4.1. $KTa_{1-x}Nb_xO_3$ and perovskite soft mode dynamics

Two general classes of behaviour seem to typify the perovskite ferroelectrics. In one case (e.g. $BaTiO_3$, $KNbO_3$) the condensing mode is overdamped and extremely anisotropic; in the other (e.g. $SrTiO_3$, $KTaO_3$, $PbTiO_3$) the softening modes tend to be underdamped and less anisotropic, although $SrTiO_3$ and $KTaO_3$ never quite become ferroelectric at atmospheric pressure. Potassium tantalate niobate, which is a mixed crystal $KTa_{1-x}Nb_xO_3$, might therefore be expected to show some properties intermediate between those of its end members $KTaO_3$ and $KNbO_3$. Potassium niobate becomes ferroelectric at $T_c = 708$ K transforming from cubic to a tetragonal phase. At lower temperatures it undergoes further transitions to orthorhombic and finally rhombohedral ferroelectric structures. Potassium tantalate remains cubic to absolute zero, but KTN with $x > 0.05$ exhibits all three phase transitions. With $x \approx 0.4$ the Curie temperature is a little above room temperature. Yelon et al. (1971) have studied a sample with $x = 0.37$ and $T_c \approx 20°C$.

The dispersion relationships measured for $T = 65°C$ (cubic phase) and for $\langle 100 \rangle$ and $\langle 110 \rangle$ directions in the Brillouin zone are shown in Fig. 6.6. The transverse optic (soft) mode is found to be low in energy but overdamped in a $\langle 100 \rangle$ direction. Also, as in $BaTiO_3$ and $KNbO_3$, the optic mode is extremely anisotropic, increasing rapidly for $\mathbf{q}$ in a $\langle 110 \rangle$ direction when mode polarization vector $\mathbf{e}$ is parallel to $\langle 1\bar{1}0 \rangle$. The $\langle 110 \rangle$ optic phonons are quite well defined, much better than in pure $KNbO_3$ for which one can observe only an increased cross-section spread across a range of $\mathbf{q}$-vectors. Also shown in

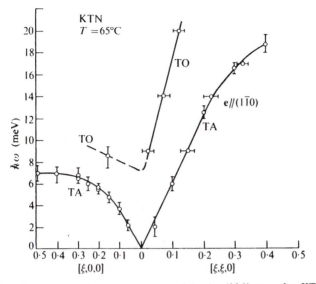

Fig. 6.6. The phonon dispersion curves measured in the $(hk0)$ zone for KTN at 65°C ($T_c \approx 20$°C). Vertical bars denote constant-Q scans and horizontal bars constant-E scans, TO refers to 'transverse optic' TA to 'transverse acoustic' in mode designation, and **e** labels a mode polarization vector. The abscissa is in units of $a^* = 2\pi/a$. (After Yelon *et al* 1971.)

Fig. 6.6 are the transverse acoustic mode dispersion curves and, as Yelon *et al.* (1971) note, they seem to be fairly normal at long wavelengths (i.e. quasi-isotropic). More recently, however, Zaccai and Hewat (1974) have scanned carefully through a {100} plane and have observed a pronounced valley in the dispersion surface of the transverse acoustic mode close to a {100} plane (Fig. 6.7). This finding illustrates the danger of concentrating exclusively upon dispersion curves along directions of high symmetry and illustrates the importance of examining the effects of small changes of wavevector or polarization away from the symmetry line.

After observing the by now expected X-ray diffuse scattering from {100} planes, Zaccai and Hewat studied the diffuse *neutron* scattering from a two-axis neutron diffractometer along several lines crossing such planes and, instead of a single peak on the plane, observed a narrow double peak which could be resolved into two Gaussians. This they were able to interpret in terms of inelastic scattering by a phonon mode with a dispersion which corresponded well to that of the axial transverse acoustic mode shown in Fig. 6.6. Measurement of the eigenvector confirmed the suggestion that the probable mechanism responsible for anisotropic diffuse X-ray scattering (in KTN at least) is an inelastic scattering from the transverse acoustic mode, which is anomalously low in energy when propagating in or very close to {100} planes with polarization normal to the planes. This acoustic-mode

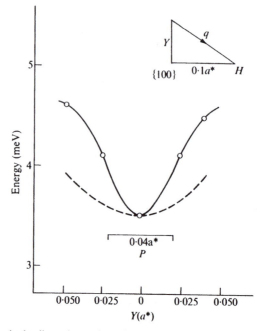

Fig. 6.7. The valley in the dispersion surface of the KTN transverse acoustic mode at $q = 0 \cdot 1a^*$. The broken line is the cross-section of an isotropic dispersion surface $\omega = vq$, P is the projection of the resolution ellipsoid, and H is a Bragg point. (After Zaccai and Hewat 1974.)

anisotropy probably results from a direct interaction with the transverse optic soft mode, the anisotropy of the latter reflecting the symmetry of the long-range electric dipolar contribution to its energy. Close to the zone centre (where its energy is lowest) the soft mode may also contribute directly to diffuse X-ray scattering.

Since the intensity of neutron scattering in a one-phonon process is proportional to the square of the inelastic structure factor $F'_j(\mathbf{Q})$, it is possible to determine the detailed character of the soft mode by finding displacements such that the intensities observed for different $\mathbf{Q}$ but the same $\mathbf{q}$ can be correctly described by $F'_j(\mathbf{Q})$. In KTN the soft mode is found to consist of vibrations of the essentially undistorted oxygen octahedron against the remainder of the cell. In this respect KTN with $x \approx 0 \cdot 4$ is like $KTaO_3$ and less like $KNbO_3$ for which the oxygen octahedron is slightly dynamically distorted. On the other hand, the overdamped character of the {100}-plane soft modes in KTN resembles $KNbO_3$ more than $KTaO_3$. Little study has yet been made of the composition dependence of the properties of KTN, but there are indications that the phase transition may change from first to second order as the concentration of Nb decreases and it is interesting to speculate whether the overdamped or underdamped characteristics of the soft mode are related in any way to the order of the transition.

Although earlier experiments on the so-called less-anisotropic soft-mode perovskite ferroelectrics had failed to show any unusual dynamic anisotropy properties, this has since proved to be due to the restriction of earlier experiments to high-symmetry directions. Significant anisotropy does in fact exist for this group as well and more recent inelastic neutron scattering experiments have revealed the details. For example Comes and Shirane (1972), seeking to understand the anisotropic diffuse X-ray scattering observed in $KTaO_3$, performed a detailed investigation of the anisotropy of the lowest optic and acoustic dispersion branches. Although the gross anisotropy of the soft mode is indeed smaller than in KTN (for example), there is none the less a significant dip in both transverse acoustic and transverse soft optic mode energies close to a {100} plane with polarization along the ⟨100⟩ axis. In addition, unlike KTN, all the modes are well defined in energy (i.e. underdamped). As the temperature is lowered the soft-mode energy decreases in {100} planes, particularly at long wavelengths, but in addition the transverse acoustic mode anisotropy increases, the entire softening of the latter being narrowly restricted to $\mathbf{q}$-vectors in {100} planes. This acoustic anomaly has been explained in terms of an optic–acoustic mode interaction by Axe, Harada, and Shirane (1970).

As was the case for KTN the measured inelastic cross-sections and their anisotropy can be used to help understand the origin of the diffuse scattering. Although a calculation of frequency integration is complicated by the optic–acoustic interaction, such has been accomplished and the resulting agreement with experiment for integrated neutron cross-section is good. Replacing the neutron scattering length by X-ray scattering factors an equivalent calculation shows that the anisotropy of the transverse acoustic branch (which by virtue of its lower energy has a higher cross-section than the soft optic branch) is indeed qualitatively able to account for the diffuse X-ray streaks.

Thus one result of the neutron studies on perovskites seems to be that the anisotropic quasi-elastic X-ray streaks are not usually the result of static deviations from high symmetry, as first proposed by Comes *et al.* (1968). The several-meV width of the diffuse neutron scattering in $KNbO_3$ for example (Nunes, Axe, and Shirane 1971) is conclusive evidence of the dynamic nature of the disorder. In $KTaO_3$ the driving (optic) modes appear to be phonon like, but in many other cases the possibility remains that the modes responsible for the disorder may be of a tunnelling nature. On the other hand, when the relaxation time becomes comparable to the period of the typical phonon the possibility of distinguishing displacive and order–disorder character seems to be lost.

In addition to the zone-centre soft modes of the perovskite ferroelectrics, another class of perovskites such as $SrTiO_3$, $LaAlO_3$, $KMnF_3$, $NaNbO_3$, and $CsPbCl_3$ undergoes phase transitions caused by the condensation of zone-boundary phonons. These also have received extensive study by coherent inelastic neutron study and examples will be discussed in Chapter 8.

6.4.2. KD_2PO_4 and hydrogen-bond dynamics

Among the hydrogen-bonded ferroelectrics and antiferroelectrics much less coherent inelastic neutron work has been performed. This is primarily because of the high incoherent cross-section for hydrogen but is also due, to a significant extent, to the much more complicated crystal structures involved. An example of some of the complexities resulting from a study of this nature can be found in the triple-axis work of Skalyo, Frazer, and Shirane (1970) on KD_2PO_4.

In deuterated KDP the ferroelectric mode occurs at the zone centre, and an analysis of mode symmetry shows that above the Curie point it must be (group theoretically) in a Γ_3 representation. The displacements which occur for a particular phonon mode can in general be described by displacement vectors (one for each of the n atoms in the primitive cell). For KD_2PO_4 there are two molecules per primitive cell and hence 48 modes for each $\mathbf{q}$-vector. The measurement and description of the entire phonon spectrum is just not physically feasible at the present time. However, for a zone-centre mode one can take full advantage of the point-group symmetry of the crystal and the Γ_3 mode in question can be shown group theoretically to have just seven component basis vectors (i.e. it can at most be a linear combination of seven basic modes). Experimentally this mode is found to be overdamped. Quasi-elastic measurements in the region of the (303) Bragg point (for which the intensity was greatest) are shown in profile in Fig. 6.8 and contoured in Fig. 6.9. The now familiar ferroelectric-type anisotropy with scattering dominantly normal to the incipient polar c-axis is again observed. A mode analysis for relative displacements in the ferroelectric mode has also been accomplished with some heretofore unexpected results including a large distortional motion of the oxygen framework and a sizable motion of the deuterium ions along the ferroelectric axis in phase with the phosphorus. Both these results are surprising in the sense that the static neutron results of Bacon and Pease (1953, 1955) on KH_2PO_4 at $T < T_c$ admit to no oxygen-framework distortion of the character found by Skalyo $et\ al.$ and also indicate a hydrogen motion out of phase with P. Although it is possible that these differences result from the fact that the static studies were carried out on the undeuterated crystal and that the low-temperature-phase positional parameters in KD_2PO_4 may be significantly different from those in KH_2PO_4, this has not yet been demonstrated. A detailed discussion of KDP-type crystals follows in Chapter 9.

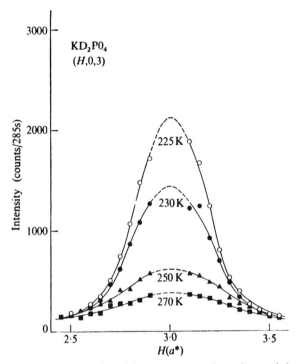

Fig. 6.8. Scan at zero energy transfer of the temperature dependence of the overdamped ferroelectric branch phonon measured in the [010] zone near (303) for KD_2PO_4. The Curie temperature is ≈ 220 K. (After Skalyo et al. 1970.)

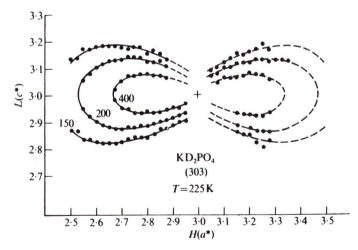

Fig. 6.9. Intensity in the [010] zone near (303) observed for zero energy transfer in KD_2PO_4 at 225 K. (After Skalyo et al. 1970.)

While coherent scattering is the richest source of lattice-dynamical information by far, the intense incoherent scattering from hydrogen has led many observers to study incoherent neutron cross-sections for hydrogen-bonded compounds. Since there is no wavevector conservation in this case the intensity peaks reflect solely the total density of states, and the problem of correct line assignment is difficult and sometimes rather speculative. Atomic or isotopic substitution can be informative in this respect, but a direct comparison with, for example, infrared spectra is perhaps even better. Once the peaks have been associated with confidence to librations or rotations of particular groups in the crystal, their study as functions of temperature through the phase transition can provide information concerning any change in freedom of motion etc. A recent example of such a study is that of Dimic *et al.* (1973) on the NH_4 motion in a number of hydrogen-bonded antiferroelectrics. The paper really serves more to emphasize the limitations of the method than to elucidate many details of the dynamics of the phase transitions. Distinct peaks in neutron intensity are found and assigned with the aid of deuteration and supplementary infrared measurements. A spectrum of low-lying frequencies is associated with acoustic vibrations and higher-frequency ones with the modes of local motion of the NH_4^+ ion. However, little change is observed in passing from the paraelectric to the antiferroelectric phase, and the major finding from the incoherent measurements alone is a rather negative one, namely that the librational and torsional motions of the NH_4 groups are not significantly altered by the onset of antiferroelectric order. The narrow linewidth of the bands is interpreted as indicating a significantly hindered rotational motion of NH_4 throughout, but little of the nature of the critical dynamics is disclosed. It seems clear that the really exciting results concerning transition dynamics will come eventually from neutron coherent scattering which, at least in principle, can always be separated from the incoherent scattering components.

7

STUDY OF SOFT MODES USING INFRARED
AND OPTIC TECHNIQUES

7.1. The dielectric function and linear response

I**T HAS LONG** been recognized that lattice vibrations which involve a
fluctuating electric dipole have quite different properties from non-polar
vibrations. Most importantly they couple directly via the dipole moment to
the radiation field in the crystal to form mixed phonon–photon modes which
have a characteristic dispersion relation and are known as polaritons. They
have frequencies which depend in general upon wavevector **q** and on their
polarization, and they can be readily explored by direct absorption of
infrared radiation or by the inelastic scattering of light. We have seen earlier
that it is just such a polar mode which softens and triggers a ferroelectric or
antiferroelectric phase transition and which has a complex frequency going
to zero for some specific wavevector $\mathbf{q}_0$ at a second-order transition. It
follows that the direct coupling of these modes with incident electromagnetic radiation provides a very useful tool for the study of their thermal
development.

The actual mechanism by which a soft mode becomes unstable has been
discussed at some length in Chapter 2. In this chapter we shall neglect the
driving anharmonicities and simply think of the soft mode, and indeed any
other polar mode, as a damped harmonic oscillator with possibly
temperature-dependent descriptive parameters. Neglected in Chapter 2,
but now to be included, are the effects of electronic contributions to electric
dipole moment. For the present at least these will be represented by a single
damped harmonic oscillator describing electron-shell vibrations about the
ionic cores. The restoring forces for ionic and electronic motion are in
general of comparable magnitude, but the much smaller electronic mass
causes the electronic oscillations to occur at a much higher frequency than
the ionic resonant frequencies.

In the lth cell of a crystal lattice we can now write the equation of motion
for the jth polar mode in the damped harmonic form

$$\ddot{\xi}_{j,l} + \Omega_j^2 \xi_{j,l} + \gamma_j \dot{\xi}_{j,l} = Z_j E_l \qquad (7.1.1)$$

where $\Omega_j^2 \xi_{j,l}$ is the restoring force, $\gamma_j \dot{\xi}_{j,l}$ a damping force of the usual
velocity-dependent type, Z_j an effective charge coefficient, and E_l the
effective Maxwell field at the lth cell site. In addition we can write an
analogous equation (which we shall label with a subscript $j = \infty$) to represent

the electronic oscillator. We have at this stage neglected all explicit coupling between modes and, for additional mathematical simplicity, we shall assume a high-symmetry situation for which the response (or susceptibility) tensor is diagonal along the axes of symmetry and for which, therefore, a simple scalar formalism can be maintained so long as we recognize that each set of parameters defining a polar mode can take on up to three different values corresponding to the three principal directions.

In this approximation the instantaneous value of polarization P_l (the electric moment per unit volume at the lth cell) can be written

$$P_l = \sum_j \frac{z_j \xi_{j,l} + z_\infty \xi_{\infty,l}}{v} \tag{7.1.2}$$

where v is the volume of a primitive cell and where the coefficients z_j are also effective charges, but differ in general from Z_j as the result of local field effects (see for example Born and Huang 1954, Lines 1969 a). Looking for plane-wave solutions

$$P_l; E_l; \xi_{j,l}; \xi_{\infty,l} \sim \exp\{i(\omega t - \mathbf{q} \cdot \mathbf{l})\} \tag{7.1.3}$$

we write

$$P_l = P \exp\{i(\omega t - \mathbf{q} \cdot \mathbf{l})\} \tag{7.1.4}$$

and corresponding equations for the other variables defining amplitudes E, ξ_j, ξ_∞. Substituting into eqn (7.1.1) the amplitudes ξ_j (including ξ_∞) are obtained. Inserting these into eqn (7.1.2) gives

$$P = \left\{ \sum_j \frac{(z_j Z_j / v)}{\Omega_j^2 - \omega^2 + i\omega\gamma_j} + \frac{(z_\infty Z_\infty / v)}{\Omega_\infty^2 - \omega^2 + i\omega\gamma_\infty} \right\} E. \tag{7.1.5}$$

This equation represents the dynamic linear response to the driving field E. The resulting linear dielectric function ε is now obtained from the defining equation

$$\varepsilon_0 E + P = \varepsilon \varepsilon_0 E \tag{7.1.6}$$

which leads to

$$\varepsilon = 1 + \sum_j \left\{ \frac{S_j \Omega_j^2}{\Omega_j^2 - \omega^2 + i\omega\gamma_j} + \frac{S_\infty \Omega_\infty^2}{\Omega_\infty^2 - \omega^2 + i\omega\gamma_\infty} \right\} \tag{7.1.7}$$

in which the parameters $S_j = z_j Z_j (\Omega_j^2 v \varepsilon_0)^{-1}$ and the electronic equivalent $S_\infty = z_\infty Z_\infty (\Omega_\infty^2 v \varepsilon_0)^{-1}$ are introduced as convenient dimensionless measures of mode strength.

For work at infrared frequencies $\omega \ll \Omega_\infty$ the electronic term in eqn (7.1.7) is approximately frequency independent and is normally replaced by the constant limiting value S_∞. This in turn is conventionally included with the vacuum term to define the high-frequency dielectric function $\varepsilon_\infty = 1 + S_\infty$, which is the value which the dielectric constant approaches at frequencies well above the ionic resonances but well below Ω_∞. The linear dielectric function in this frequency range now becomes

$$\varepsilon = \varepsilon_\infty + \sum_j \frac{S_j \Omega_j^2}{\Omega_j^2 - \omega^2 + i\omega\gamma_j} \qquad (7.1.8)$$

where j runs over all the polar modes which couple (i.e. have non-zero mode strength) to the Maxwell field for the particular principal direction considered.

Following Barker (1964, 1967) we now solve for free vibrations; that is consider *all* the equations of motion, including Maxwell's equations, in the absence of free charges and currents. The latter are

$$\nabla . \mathbf{D} = \nabla . \mathbf{H} \equiv \mathrm{div}\,\mathbf{D} = \mathrm{div}\,\mathbf{H} = 0$$

$$\nabla \wedge \mathbf{E} \equiv \mathrm{curl}\,\mathbf{E} = -\mu_0 \dot{\mathbf{H}}$$

$$\nabla \wedge \mathbf{H} \equiv \mathrm{curl}\,\mathbf{H} = \dot{\mathbf{D}}$$

$$\mathbf{D} = \varepsilon_0 \mathbf{E} + \mathbf{P} = \varepsilon\varepsilon_0 \mathbf{E} \qquad (7.1.9)$$

in conventional notation, and the first equation div $\mathbf{D} = 0$ has two sets of roots according to whether $\varepsilon = 0$ or div $\mathbf{E} = 0$. Let us first study the roots of $\varepsilon = 0$. Multiplying through eqn (7.1.8) by all the resonant denominators we consider the roots in the form

$$\varepsilon \prod_{j=1}^{n} (\Omega_j^2 - \omega^2 + i\omega\gamma_j) = \Omega_1^2 \Omega_2^2 \dots \Omega_n^2 \{S_1 + S_2 + \dots + S_n + \varepsilon_\infty\}$$

$$+ i\omega(\dots) - \omega^2(\dots) - \dots + (-1)^n \omega^{2n} \varepsilon_\infty.$$

$$(7.1.10)$$

Thus $\varepsilon = 0$ has $2n$ roots $\omega_{L1}, \omega_{L2}, \dots, \omega_{L2n}$ which are easily seen to occur in pairs, ω_L with its negative complex conjugate $-\omega_L^*$. Writing the right-hand side of eqn (7.1.10) in factored form and making use of the way the roots occur in pairs we have

$$\varepsilon \prod_{j=1}^{n} (\Omega_j^2 - \omega^2 + i\omega\gamma_j)$$

$$= (-1)^n \varepsilon_\infty (\omega - \omega_{L1})(\omega - \omega_{L2}) \dots (\omega - \omega_{L2n})$$

$$= (-1)^n \varepsilon_\infty \{\omega^2 - |\omega_{L1}|^2 - 2i\omega\,\mathrm{Im}(\omega_{L1})\} \dots \{\omega^2 - |\omega_{Ln}|^2 - 2i\omega\,\mathrm{Im}(\omega_{Ln})\}$$

$$(7.1.11)$$

where Im refers to the imaginary part of the (in general) complex roots. A number of sum rules can now be obtained by equating coefficients of equivalent terms in eqns (7.1.10) and (7.1.11). The most important of these (Cochran and Cowley 1962) is that obtained by equating the frequency-independent terms, namely

$$\frac{|\omega_{L1}|^2 |\omega_{L2}|^2 \dots |\omega_{Ln}|^2}{\Omega_1^2 \Omega_2^2 \dots \Omega_n^2} = \frac{\varepsilon_\infty + \Sigma_j S_j}{\varepsilon_\infty} = \frac{\varepsilon(0)}{\varepsilon_\infty} \qquad (7.1.12)$$

where $\varepsilon(0) = \varepsilon_\infty + \Sigma_j S_j$ is the dielectric constant at very low frequencies from eqn (7.1.8). This is the Lyddane–Sachs–Teller relation connecting the absolute value of ω_{Lj}, which we shall see below to be the longitudinal mode frequencies, with the force-constant parameters Ω_j. In the absence of damping Ω_j will actually be the transverse mode frequencies and the ω_{Lj} will be real. Additional discussion of the other sum rules is given by Barker (1964). To deduce the nature of the solutions we rewrite Maxwell's equations (7.1.9) assuming plane-wave solutions $\sim \exp\{i(\mathbf{q}.\mathbf{l} - \omega t)\}$. Since $\nabla = i\mathbf{q}$ for this case these become

$$\mathbf{q}.\mathbf{D} = \mathbf{q}.\mathbf{H} = 0$$

$$\mathbf{q} \wedge \mathbf{E} = \omega \mu_0 \mathbf{H}$$

$$\mathbf{q} \wedge \mathbf{H} = -\omega \mathbf{D}$$

$$\mathbf{D} = \varepsilon \varepsilon_0 \mathbf{E} \qquad (7.1.13)$$

where the symbol $\wedge$ implies a vector product. Thus $\varepsilon = 0$ implies $\mathbf{D} = 0$, $\mathbf{H} = 0$, and $\mathbf{q} \wedge \mathbf{E} = 0$. It follows that $\mathbf{E}$ and $\mathbf{q}$ are parallel to each other and also, via eqn (7.1.5), to $\mathbf{P}$. The solutions are therefore longitudinal modes with frequencies independent of wavevector (flat dispersion curves).

The other possible set of solutions to the equations of motion have div $\mathbf{E} = 0$. For plane-wave solutions this implies, from eqn (7.1.13), that $\mathbf{q}$, $\mathbf{E}$, and $\mathbf{H}$ form a right-handed system of orthogonal vectors in the given order. The wave equation for these transverse modes follows from the simultaneous solution of $\mathbf{q} \wedge \mathbf{E} = \omega \mu_0 \mathbf{H}$ and $\mathbf{q} \wedge \mathbf{H} = -\omega \mathbf{D}$, viz of

$$\nabla^2 \mathbf{E} = -\frac{\omega^2}{c^2} \varepsilon \mathbf{E}. \qquad (7.1.14)$$

This equation describes a wave motion with dispersion

$$q^2 c^2 = \varepsilon \omega^2 \qquad (7.1.15)$$

where ε is given by eqn (7.1.8) and where c is the velocity of light *in vacuo*. These modes are mixed phonon–photon modes and their quanta are referred to as *polaritons*. The general character of the dispersion is most easily envisaged for the case of negligible damping; it is qualitatively sketched in Fig. 7.1(a) for the case of three undamped phonon modes. The curves show regions of photon-like behaviour at very high and very low frequencies (with slopes determined by limiting high- and low-frequency dielectric constants respectively) but join continuously to horizontal phonon-like curves at intermediate frequencies. We note also that there are gaps (where the vibrations are completely damped) between pairs of transverse and longtudinal phonon frequencies. Born and Huang (1954) show that near the transverse phonon frequencies energy is carried primarily by the elastic part of the field, while at very high frequencies, of course, the ions cannot take part in the vibrations to an appreciable extent, owing to their large inertia,

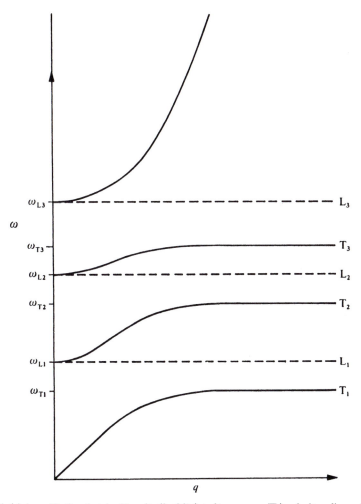

Fig. 7.1. (a) A qualitative sketch of longitudinal (L_i) and transverse (T_i) polariton dispersion for the case of three undamped modes $i = 1, 2, 3$; ω_{Li} labels longitudinal mode frequency and ω_{Ti} transverse mode frequency in the elastic limit.

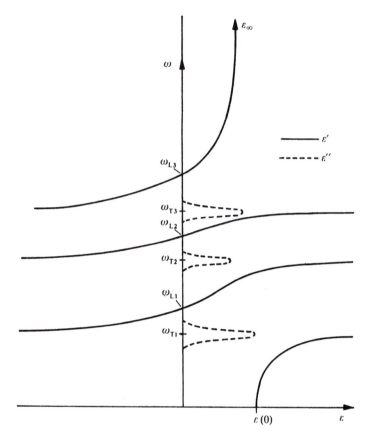

Fig. 7.1. (b) The real and imaginary parts of the dielectric function $\varepsilon(\omega) = \varepsilon' + i\varepsilon''$ for this same three-mode situation. Some damping has now been included to prevent ε'' from shrinking to a series of δ-functions.

and the solutions are essentially radiative. The corresponding curves for the dielectric function are shown in Fig. 7.1 (b). A more realistic spectrum is shown in Fig. 7.2 where we sketch the transverse solutions for the three classical modes which fit the dielectric behaviour of $SrTiO_3$. It is plotted for real frequencies and again shows the high- and low-frequency photon-like behaviour, but now there are no completely non-propagating 'gaps' and transverse vibration modes exist at all frequencies although there are still regions where the modes are highly damped.

We define the transverse phonon frequencies ω_{Tj} as the poles of the dielectric constant. Thus, from eqn (7.1.8) it follows that

$$\omega_{Tj} = \pm(\Omega_j^2 - \tfrac{1}{4}\gamma_j^2)^{1/2} + \tfrac{1}{2}i\gamma_j. \qquad (7.1.16)$$

It is important to realize that the modes for which photon–phonon coupling

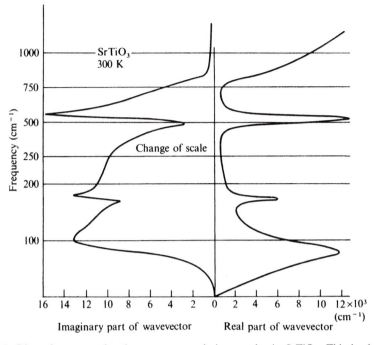

Fig. 7.2. Dispersion curves for the transverse polariton modes in $SrTiO_3$. This is also a three-mode situation and the curves (plotted for real frequency) should be compared with the idealized model situation of Fig. 7.1. (After Barker 1967.)

effects are significant (i.e. polariton modes) cover only a very small region of the Brillouin zone near the zone centre. Indeed, Fig. 7.2 includes a region extending only 10^{-4} of the way to the zone boundary. Even in the absence of damping (Fig. 7.1 (a)) the polariton modes approach the bare phonon frequencies very rapidly as the wavevector increases out from the zone centre, and it is for this reason that throughout the present section we have neglected completely any **q**-dependence of the bare phonon frequencies, representing them essentially by their long-wavelength limiting values. In fact the resolution available in neutron spectroscopy is insufficient to detect any polariton effects at all. On the other hand, we now see from eqns (7.1.12) and (7.1.16) that in the presence of damping it is not the frequencies representing the poles of the response (i.e. transverse phonon frequencies) which enter the Lyddane–Sachs–Teller relation, and recognition of damping and polariton effects are very important if correct results are to be obtained in this context. For the interpretation of infrared absorption and reflection experiments an appreciation of polariton effects is therefore essential. Also, as we shall see later in this chapter, if the infrared active modes are also

Raman active, then the bending down of the dispersion curves near $\mathbf{q} = 0$ (Fig. 7.1 (a)) is also directly detectable by forward Raman scattering.

Perhaps the most serious simplification which has so far been included in the theory of this section, and one which certainly requires further examination when attempts are made to interpret actual spectra, is the assumption of a frequency-independent damping parameter. Since the physical origin of the damping is the coupling of the phonon in question to other phonons, it is easy to conceive of the situation where there is a high density of these coupled phonons at one or more particular frequencies. In such cases we should expect the damping parameter to peak at the latter frequencies to reflect the enhanced leaking of energy from the primary mode near these resonances.

In certain paraelectric and ferroelectric crystals this variation of damping parameter with frequency is very important. An extensive treatment of such effects in $SrTiO_3$, $BaTiO_3$ and $KTaO_3$ has been given by Barker and Hopfield (1964). For example, if the primary response has a resonance near Ω_1 while the damping parameter is peaked near Ω_2, then a transmission experiment might well reveal the existence of a weak combination band mode near Ω_2 in addition to the response of the main mode near Ω_1. Such a two-peaked response is distinctly different from that obtained by the simple addition of two separate responses from independent constant-damping oscillators. It can often be adequately described by assuming a damping parameter $\Gamma_1(\omega)$ of the frequency-dependent form

$$-i\omega\Gamma_1(\omega) \sim \frac{\Omega_1^2\Omega_2^2}{\Omega_2^2 - \omega^2 + i\omega\gamma_2} \qquad (7.1.17)$$

with γ_2 independent of frequency. In addition the transverse mode parameter to be included in the Lyddane–Sachs–Teller relation is no longer Ω_1^2 but must now be computed as the effective restoring force in the low-frequency limit.

If the description of dynamic response in terms of damped harmonic-oscillator response is not valid, as is often the case in order–disorder systems at low frequencies, it is necessary to generalize the Lyddane–Sachs–Teller relationship to allow for poles and zeros in the dielectric function of a more general form. This problem has been discussed by Chaves and Porto (1973) who remove the harmonic-oscillator restriction and give a derivation based simply on causality and energy conservation. When allowing for the appearance of zeros and poles on the imaginary frequency axis, two very distinct situations must be considered separately. In the first one two symmetrical poles go to meet each other at that axis (e.g. Fig. 2.4); this is typical of damped harmonic-oscillator behaviour. For such a case eqn (7.1.12) holds, although the zeros and infinities of ε are no longer necessarily associated with decoupled resonances or normal modes in the conventional sense. If an unpaired pure imaginary pole exists in addition to the pairs of symmetric poles then a single imaginary zero must exist also. Writing the symmetric poles in the form eqn (7.1.16) with subscript T

$$\omega_{Tj} = \pm(\Omega_{Tj}^2 - \tfrac{1}{4}\gamma_{Tj}^2)^{1/2} + \tfrac{1}{2}i\gamma_{Tj} \qquad (7.1.18)$$

and the symmetric zeros in the same fashion but with subscript L, Chaves and Porto deduce a generalized Lyddane–Sachs–Teller relation

$$\frac{\varepsilon(\omega)}{\varepsilon_\infty} = \frac{i\gamma_{Ld} + \omega}{i\gamma_{Td} + \omega} \prod_j \frac{\Omega_{Lj}^2 + i\omega\gamma_{Lj} - \omega^2}{\Omega_{Tj}^2 + i\omega\gamma_{Tj} - \omega^2} \qquad (7.1.19)$$

where γ_{Td} and γ_{Ld} are respectively the relaxation frequencies associated with the unpaired pole and zero. For ferroelectrics of the order–disorder kind the frequency γ_{Td} is probably in the microwave region with γ_{Ld} much higher. At very low frequencies the phonon contribution to ε becomes frequency independent and eqn (7.1.19) can be rewritten as

$$\frac{\varepsilon(\omega)}{\varepsilon_1(\infty)} = \frac{(\gamma_{Ld}/\gamma_{Td}) - i\omega/\gamma_{Td}}{1 - i\omega/\gamma_{Td}} \qquad (7.1.20)$$

where $\varepsilon_1(\infty)$ takes account of electronic plus phonon contributions. For frequencies $\omega \ll \gamma_{Ld}$ this equation simplifies further to

$$\varepsilon(\omega) = \frac{\varepsilon(0)}{1 - i\omega/\gamma_{Td}} \qquad (7.1.21)$$

which has the same form as the Debye eqn (2.4.21) and is known to describe low-frequency response for many order–disorder ferroelectrics.

In the absence of an unpaired pole eqn (7.1.19) reduces to

$$\frac{\varepsilon(\omega)}{\varepsilon_\infty} = \prod_j \frac{\Omega_{Lj}^2 + i\omega\gamma_{Lj} - \omega^2}{\Omega_{Tj}^2 + i\omega\gamma_{Tj} - \omega^2} \qquad (7.1.22)$$

a frequency-dependent extension of the basic Lyddane–Sachs–Teller form (eqn (7.1.12)).

7.2. Infrared spectra

Ionic vibrational modes which involve an oscillating electric dipole are termed optically active since these displacements can couple directly to electromagnetic radiation via the $-\mathbf{P}.\mathbf{E}$ term in the Hamiltonian. This interaction produces a resonant absorption when the frequency of the radiation approximates that of the relevant long-wavelength lattice vibration. Thus by analysing the infrared absorption or reflectivity one can obtain detailed information concerning the frequency and damping of the infrared active optic phonons (or more accurately polaritons). A new importance was given to the study of infrared spectra by the realization that, at least for displacive systems, ferroelectric phase transitions were accompanied by the softening of just such a phonon mode. In particular, via the Lyddane–Sachs–Teller relation (eqn (7.1.12)), the divergence of static dielectric constant $\varepsilon(0)$ can be associated with a vanishing force-constant parameter Ω_j and, via eqn (7.1.16), with a resulting softening of the corresponding transverse phonon frequency. On the other hand, it is important to recognize that the lattice-dynamical theory leading to these predictions is couched in the language of damped harmonic vibrations and that the resulting dispersion analysis is limited to ferroelectrics for which the dielectric response is wholly attributable to optic modes. For order–disorder ferroelectrics with relaxation soft-mode behaviour the propagating modes identified in an infrared study will certainly not account for the temperature dependence of $\varepsilon(0)$ as is immediately evident from the generalized Lyddane–Sachs–Teller form (eqn (7.1.19)).

Since the interpretation of infrared spectra is simplest for displacive systems, we shall discuss first a few representative results for ferroelectric and near-ferroelectric perovskites which can be understood in terms of the theoretical development leading to eqn (7.1.12). In perovskites one normally measures reflectivity, since the scattering is so severe in single crystals as to make any quantitative study of absorption bands in the long-wavelength transmission spectrum impossible. Although strontium titanate becomes ferroelectric only under stress (see § 8.1), the free crystal does exhibit a Curie–Weiss-like susceptibility at higher temperatures with a Weiss temperature of about 35 K. At lower temperatures the susceptibility deviates from the simple law and remains finite down to absolute zero, although it attains large values indicative of a substantial mode softening in the lattice-dynamical picture. Its spectrum is more convenient to measure than that of $BaTiO_3$, because the soft-mode remains underdamped whereas that of $BaTiO_3$ is overdamped. It follows that for $BaTiO_3$ one must consider the shape of the lattice absorption rather than just the peak of the mode in testing for temperature shifts.

In Fig. 7.3 we show the real and imaginary parts of the dielectric function for $SrTiO_3$ at the two temperatures 300 and 85 K. The peak in the imaginary part at low frequency is markedly temperature dependent and corresponds to the resonance of the lowest-frequency transverse optic vibration. This mode (Barker 1966 a) is observed to have a frequency which varies with temperature as $\{\varepsilon(0)\}^{-1/2}$ and hence to be consistent with eqn (7.1.12) if the other transverse and longitudinal mode frequencies are essentially temperature independent. Strontium titanate has in all six polar optic mode frequencies (three longitudinal and three transverse) and the other five are indeed found to be temperature independent to at least 1 or 2 per cent. They can also be seen in Fig. 7.3 as peaks to the imaginary part of ε (the transverse modes ω_T) or in the imaginary part of $1/\varepsilon$ (the longitudinal modes ω_L).

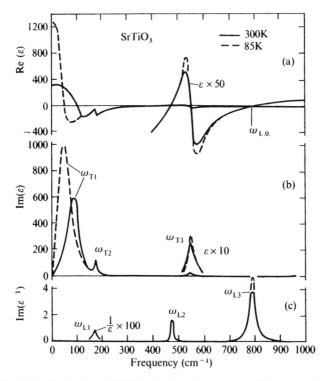

Fig. 7.3. The dielectric function of SrTiO$_3$ for two temperatures. The curves result from a Kramers–Kronig analysis of reflectivity data and hence include finite linewidth effects (compare Fig. 7.2). (After Barker 1967.)

The dielectric constant is obtained from the measured reflectivity R by use of a Kramers–Kronig analysis (Spitzer *et al.* 1962). From $r = R^{1/2}$ one obtains the real and imaginary parts of the refractive index $n + ik$ through the relations

$$r\,e^{i\theta} = \frac{n - 1 - ik}{n + 1 - ik} \qquad (7.2.1)$$

where θ is given by the Kramers–Kronig equation

$$\theta(\omega) = \frac{2\omega}{\pi} \int_0^\infty \frac{\ln\{r(\omega')\}\,d\omega'}{\omega'^2 - \omega^2}. \qquad (7.2.2)$$

From n and k the real and imaginary parts of the dielectric constant follow respectively as

$$\varepsilon' = n^2 - k^2, \qquad \varepsilon'' = 2nk. \qquad (7.2.3)$$

It is immediately evident that the resulting dielectric function is only reliable if $r(\omega)$ has no prominent features at frequencies below those used in the reflectivity experiments, since eqn (7.2.2) requires an extrapolation to zero frequency. This obviously creates difficulties in order–disorder materials.

The association of the peaks in ε'' and zeros in ε' with the transverse and longitudinal mode frequencies is valid as an approximation only if the corresponding modes are weakly damped. In general it is better to proceed by using a lattice-mode model such as that of the previous section leading to an expression (eqn (7.1.8)) for dielectric constant ε as a function of sets of independent mode parameters, one set for each infrared active mode. An analysis of reflectivity

$$R = \left| \frac{\varepsilon^{1/2} - 1}{\varepsilon^{1/2} + 1} \right|^2 \qquad (7.2.4)$$

using such a model proves to be quite successful even for highly damped systems like $BaTiO_3$. In fact the infrared spectrum of $BaTiO_3$ provided the first clear indication that the soft mode in $BaTiO_3$ is overdamped (Spitzer *et al.* 1962). In $SrTiO_3$ the agreement between the two methods is quite close (Barker 1966 a) on account of the relatively small damping parameters. Even here, however, a significant effect is played by damping, and a really good fit to the reflectivity requires the extension of the independent damped oscillator model to include a significant interaction between the modes during the damping process. Fortunately, at least for $SrTiO_3$, the phonon frequencies themselves (as opposed to the detailed frequency dependence of response) turn out to be relatively unaffected by the damping interaction and are therefore quite accurately obtained from the independent oscillator fit.

The coupling of various lattice vibrational modes, which manifests itself as a frequency dependence of the mode parameters in the independent oscillator model, can be quite severe in more strongly damped situations. Barium titanate for example has both a large and frequency-dependent damping coefficient associated with its soft mode, and Ballantyne (1964) has noted that the peak in ε'' certainly does not shift with temperature fast enough to agree with the low-frequency dielectric constant via the Lyddane–Sachs–Teller relationship (eqn (7.1.12)). The higher-frequency infrared modes are weakly damped and essentially temperature independent (Barker 1966 a) so that a test of eqn (7.1.12) for $BaTiO_3$ is possible if an adequate description can be found for the soft mode itself. No simple independent oscillator fit is good over the entire far-infrared frequency range, but for frequencies up to twice that of the soft-mode restoring-force parameter Ω_j ($j =$ soft mode) an approximate fit can be obtained. For this fit the damping parameter is from four to six times the restoring-force parameter and hence, from eqn (7.1.16), the soft-mode frequency is imaginary. The parameter Ω_j which enters the Lyddane–Sachs–Teller relationship is quite real, however, and is found to vary with temperature at least approximately as $\{ r(0) \}^{-1/2}$ as expected from eqn (7.1.12). The important point is that for this overdamped situation the peak in ε'' occurs at a frequency at least three times lower than Ω_j and is therefore quite misleading if used as an approximate measure of restoring force in the simple Lyddane–Sachs–Teller formalism. The entire theoretical picture can be improved significantly by allowing for lattice mode interactions (Barker and Hopfield 1964) leading to frequency-dependent mode parameters although, as mentioned in the previous section, some additional thought is then necessary concerning the manner in which such interactions manifest themselves in the Lyddane–Sachs–Teller relationship.

Infrared reflectivity measurements have been carried out on many other displacive systems including $KTaO_3$ (which, like $SrTiO_3$, is a near-ferroelectric as $T \to 0$ K under zero stress), mixed $KTaO_3$–$KNbO_3$ crystals (which are ferroelectric with a Curie temperature which depends approximately linearly on Nb concentration), and $LiNbO_3$. The measurements are used not only for testing the temperature dependence of soft modes but also for examining the form of the mode splittings which occur on transition from a higher- to a lower-symmetry phase. Thus, for example, the optically active perovskite cubic modes are triply degenerate and split in accordance with group-theoretical predictions on transition to lower-temperature ferro-electric phases. Also the possible existence of additional modes in mixed crystals can be explored in addition to the concentration dependence of the fundamental modes of the end members.

Finally, when the infrared active modes have all been located and their mode strengths determined, eqn (7.1.8) can then be applied in the zero-frequency limit, viz.

$$\varepsilon(0) = \varepsilon_\infty + \sum_j S_j \qquad (7.2.5)$$

to verify or question the displacive nature of a ferroelectric transition. If the transition is caused solely by instabilities in optical vibrations, the sum of the mode strengths and the high-frequency dielectric constant should be equal to the static dielectric constant. Such a test has been applied, for example, for $LiNbO_3$ (Barker and Loudon 1967) to conclude that it is indeed a displacive-type ferroelectric. It is important in this check to recognize that the static dielectric constant in eqn (7.2.5) refers to the clamped configuration, so that in a free sample the 'static' measurement should actually be made at frequencies below all the phonon modes but above any piezoelectric (electromechanical) resonances. The situation for $BaTiO_3$ seems to be not so clear (Chaves and Porto 1973, Burns 1973) with $\varepsilon_\infty + \Sigma_j S_j$ possibly falling short of the clamped d.c. value of dielectric constant and thereby suggesting that $BaTiO_3$ may possess ionic motion in addition to its phonons, a possibility first proposed by Comes et al. (1968).

For order–disorder ferroelectrics with negligible tunnelling characteristics (such as TGS and $NaNO_2$ for example) the relationship (eqn (7.2.5)) seems quite definitely to be violated (Barnoski and Ballantyne 1968, Barker and Tinkham 1963). This implication of the absence of a soft lattice vibrational mode is confirmed by Raman spectroscopy (Hartwig *et al.* 1972) and is in accord with the double-well potential picture of the order–disorder transition given in Chapter 2. In this picture the phonon mode corresponding to oscillation within the wells undergoes no drastic anomaly at the Curie point and its frequency response maintains a peak near the corresponding phonon frequency right through the transition. In addition, however, the response is expected to develop a Debye-like relaxation anomaly at low frequencies corresponding to the thermally activated potential-barrier hopping. This also is confirmed by frequency-dependent dielectric measurements (Hatta 1970, Luther and Muser 1970). Similar behaviour has recently been observed by infrared and microwave techniques in KNO_3 (Hill and Mohan 1971, Chen and Chernow 1967). These systems should better obey a Lyddane–Sachs–Teller relationship of the generalized form (eqn (7.1.19)), although no detailed analyses along these lines seem yet to have been attempted.

The situation for the KDP-type hydrogen-bonded ferroelectrics is less well understood. There have been many conflicting reports on the character of the low-frequency vibrations over the years, particularly in KH_2PO_4 and KD_2PO_4. There must, of course, be a region of intense absorption in these systems in the far-infrared and microwave regions to account for the extremely large values of $\varepsilon(0)$ close to the phase transitions. The question is whether these are propagating modes, such as soft phonons or pseudo-spinwave excitations (possibly highly damped), or whether they are pure relaxation (i.e. non-propagating) modes. The basic problem is that the low-frequency modes are certainly very highly damped, and that in such a case an equally good fit to experiment can often be found either in terms of a Debye relaxation with relaxation time τ or of an overdamped oscillator with damping constant γ and frequency Ω restricted only by the relation $\gamma/\Omega^2 = \tau$. Better success at unravelling the detailed character of the KDP transitions has come through Raman and Brillouin scattering studies and will be discussed later.

Of greater value for complicated crystal structures is an analysis of the infrared spectra at higher frequencies, features of which can often be identified with the internal vibrations of a characteristic subgroup within the lattice. This is usually accomplished by comparing spectra in a series of isomorphous crystals. Lower frequencies can then sometimes be assigned to oscillations of the subgroup as a whole. Examination of the splitting of the higher-frequency lines at a phase transition can often provide useful information concerning the details of the transition itself. A particularly good example is the manner in which the spectrum of the phosphate group in the KDP series of crystals is affected by the various phase transitions which occur in the different members. By analysing the molecular vibrations of the PO_4 grouping making use of group-character tables one can predict the manner in which the free-ion spectra will split when placed in the surroundings representing different subgroups. In particular a different splitting will characterize a ferroelectric phase transition, such as occurs for example in KH_2PO_4, from that which characterizes an antiferroelectric one like that in $NH_4H_2PO_4$ (Wiener-Avnear, Levin, and Pelah 1966). Another example of the splitting of an ion-subgroup spectrum associated with a reduction of lattice symmetry at a ferroelectric phase transition is that of the NH_3^+ spectrum in triglycine sulphate (Galanov and Kislovskii 1965). In addition to the line splitting in KH_2PO_4 some recent single-crystal reflectivity experiments by Levin, Pelah, and Wiener-Avnear (1973) have noted that the PO_4 spectral lines are much broader above T_c than below. This they attribute to the interaction between the phosphate modes and the protonic low-frequency mode (the latter being at frequencies below the 250 cm^{-1} limit of the experiment). Below T_c the proton double well is assumed to become asymmetric removing the resonance condition responsible for the broadening.

Besides line splittings attributed to symmetry reduction and to tunnelling other spectral features have been advanced as evidence for or against subgroup ordering. On the whole, however, the various interpretations tend to be somewhat speculative, and relatively little which is likely to remain unquestioned has yet been learned about ferroelectric phase transitions in H-bonded materials from these efforts. The problem is that the critical dispersion range for H-bonded and order–disorder systems really only starts below about 150 cm^{-1} even in the most favourable situations. For deuterated materials the critical dispersion may not begin much above a few gigahertz. It follows that Raman scattering or microwave techniques are more suited for the study of phase-transition dynamics in these materials. To some extent this is even true for displacive systems since, as we shall discuss in Chapter 11, a long-wavelength dispersion often accompanies even underdamped soft-mode phase transitions in the critical region itself.

One of the most interesting recent developments in the field of infrared spectroscopy has involved the observation of surface polaritons (Ritchie 1973). Surface phonons are a consequence of the finite extent of the crystal, and optical surface phonons have frequencies between the corresponding transverse and longitudinal bulk modes (see Ruppin and Englman 1970 for an introduction). In a crystal whose size is small compared with the wavelength of the relevant phonon only radiative surface modes exist and these can be detected by infrared absorption (Martin 1970). In large crystals with flat surfaces only non-radiative surface modes exist which interact with electromagnetic radiation to produce surface polaritons with pronounced dispersion at long wavelengths. Experimentally these surface polaritons have been observed by low-energy electron spectroscopy or, more recently, by infrared spectroscopy using the technique of attenuated total reflection (ATR).

A surface polariton is characterized by an electric field which oscillates in time and varies sinusoidally in directions along the surface but falls off exponentially on *both* sides of the surface. In the attenuated total reflection method (Otto 1968) a prism is placed close to the dielectric surface (separated by a small air gap) and infrared radiation is internally reflected within the prism at the air-gap interface. The electric-field pattern on the air side is called an evanescent wave and, if the gap is small enough, it couples directly to the component of surface polariton in the gap. In this situation the infrared beam does not undergo total reflection, because some of the energy is used to excite the surface polariton and hence one observes a dip or attenuation in the total reflection at the frequency of the surface polariton whose wavevector matches that of the evanescent wave.

Experimental results for the surface-polariton dispersion of the upper two branches in $SrTiO_3$ are shown in Fig. 7.4. Theoretically the dispersion is

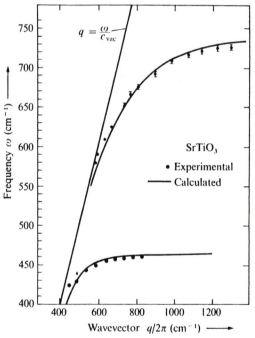

Fig. 7.4. The upper two branches of surface-polariton dispersion of $SrTiO_3$ at room temperature. The experimental points are measured by ATR (see text) and the full curve is calculated from eqn (7.2.6). (After Fischer, Bauerle, and Buckel 1974.)

expected to be given by

$$q(\omega) = \frac{\omega}{c} \left\{ \frac{\varepsilon'(\omega)\varepsilon_G}{\varepsilon'(\omega)+\varepsilon_G} \right\}^{1/2} \tag{7.2.6}$$

where ε' is the real part of the dielectric function of the crystal, ε_G is the dielectric constant of the medium of the gap, and c is the velocity of light *in vacuo* (Ruppin 1970). Using $\varepsilon_G = 1$ for the gap medium (air) and calculating ε' from a Kramers–Kronig analysis of the reflectivity data taken on the same crystal specimen, the fit of eqn (7.2.6) with experiment is quite good (Fig. 7.4). The most interesting question from the point of view of ferroelectricity, viz. how does the lowest surface-polariton-mode dispersion branch (which belongs to the soft ferroelectric mode) vary with temperature as the ferroelectric bulk mode softs, seems not yet to have been answered.

7.3. Raman spectroscopy

7.3.1. Basic concepts

The Raman effect measures the frequency shift of inelastically scattered light produced by phonon excitations. These spectra are in general richer in detail concerning the symmetry type and location in the Brillouin zone of the scattering phonon than are the corresponding infrared resonant spectra and contain, in particular, multi-phonon as well as single-phonon spectral lines. The selection rules for the Raman spectra are in general different from those for the equivalent infrared lines so that the two methods are to a considerable extent complementary for the study of lattice vibrations.

The crystal can modify the incident light wave by taking a small fraction of its energy and reradiating it at a different frequency, with changed polarization, and in a more or less arbitrary direction. This scattering implies the existence of an electric polarization in the crystal with frequency and wavevector different from that of the incident light. This can be accounted for by allowing the electronic polarizability to be modulated by the mechanical disturbances in the crystal lattice. Such disturbances may result from acoustic phonons, optic phonons, or simply static faults and crystal defects, and in a truly quantitative theoretical treatment it is necessary to recognize that the scattering occurs at a particular point within the crystal, and that the generation of radiation and its propagation to the surface should be calculated with due regard for crystalline anisotropy, surface refection, etc. (Lax and Nelson 1974). The modulation is the result of a direct coupling between the electronic and ionic co-ordinates $\xi_{\infty,l}$ and $\xi_{i,l}$ of eqn (7.1.2), a coupling which has been neglected in the linear theory of § 7.1. The detailed consequences of such a coupling have been discussed by a number of authors (e.g. Barker and Loudon 1972) with specific reference to Raman scattering. The basic mechanism can be understood by examining the consequences of expanding the electronic polarizability (or equivalently the high-frequency dielectric constant) as a function of the normal modes of the crystal lattice:

$$\varepsilon_{mn}(\infty) = \varepsilon_{mn}^{(0)}(\infty) + \varepsilon_{mn}^{(1)}(\infty) + \ldots \quad (7.3.1)$$

where $\varepsilon_{mn}^{(1)}(\infty)$, but not $\varepsilon_{mn}^{(0)}(\infty)$, is dependent on unit-cell position and time co-ordinates l and t. When the sample is traversed by a plane monochromatic light source

$$\mathbf{E}_l = \mathbf{E}\,\exp\{i(\mathbf{q}_0 \cdot \mathbf{l} - \omega_0 t)\} \quad (7.3.2)$$

a polarization $\mathbf{P}$ results where

$$P_m = \{\varepsilon_{mn}(\infty) - 1\}\varepsilon_0 E_n = \{\varepsilon_{mn}^{(0)}(\infty) - 1\}\varepsilon_0 E_n + \varepsilon_{mn}^{(1)}(\infty)\varepsilon_0 E_n$$
$$= P_m^{(0)} + P_m^{(1)}. \quad (7.3.3)$$

The second term $P_m^{(1)}$ is the fluctuation contribution and the scattered radiation field $\mathbf{E}_s$ can be computed from the source distribution $P_m^{(1)} = \varepsilon_{mn}^{(1)}(\infty)\varepsilon_0 E_n$. Expanding $\varepsilon_{mn}^{(1)}(\infty)$ in terms of the lattice phonons

$$\varepsilon_{mn}^{(1)}(\infty) = \sum_j \sum_{\mathbf{q}} C_{mn}(\mathbf{q}, j)\xi_j(\mathbf{q})\,\exp[i\{\mathbf{q} \cdot \mathbf{l} - \omega(\mathbf{q}, j)t\}] \quad (7.3.4)$$

where $\omega(\mathbf{q}, j)$ is the dispersion of the jth phonon mode, it is evident that the incident optical plane wave produces scattered components

$$\mathbf{E}_s = \mathbf{E}\,\exp\{i(\mathbf{q}_s \cdot \mathbf{l} - \omega_s t)\} \quad (7.3.5)$$

where

$$\mathbf{q}_s = \mathbf{q}_0 \pm \mathbf{q}, \qquad \omega_s = \omega_0 \pm \omega(\mathbf{q}, j). \quad (7.3.6)$$

The latter phase-matching conditions are of course equivalent to a conservation of energy and momentum in a particle picture of the same interaction. Thus in this picture two radiated waves will be observed in any particular direction and will have frequencies shifted from the incident frequency by $\pm\omega(\mathbf{q}, j)$. The scattered intensity of each component is proportional to the square of the resulting polarization

$$I(\mathbf{q}) \propto \langle |C(\mathbf{q}, j)\xi_j(\mathbf{q})|^2 \rangle E^2. \quad (7.3.7)$$

Detailed derivations of intensity, including the correct factors of proportionality, have been given by Born and Huang (1954), Loudon (1964), and Cummins (1969). Quantum mechanically the intensities of the down- and up-shifted Raman lines (referred to respectively as the Stokes and anti-Stokes lines) are unequal in the ratio $\exp\{-\hbar\omega(\mathbf{q}, j)/kT\}$. This difference becomes large at low temperatures, the physical reason being that the excitation of the anti-Stokes line requires the *absorption* of a lattice phonon

which, of course, to be absorbed must initially have been thermally populated. The Stokes line, on the other hand, corresponds to the *creation* of a
phonon and remains finite at $T = 0$.

Thus far we have not allowed for finite lifetime effects and accordingly the
above equations define sharp excitations. An account of line broadening is
most simply obtained in terms of a frequency-independent damping. The
spectral response or scattering function $g(\mathbf{q}, \omega)$, which is explored in more
detail in Chapter 11, is in general proportional to the imaginary part of the
dynamic susceptibility (via the fluctuation theorem) and hence, for a
damped harmonic-oscillator form (eqn (2.1.33)), can be written in a normalized form

$$F = \frac{\Gamma\Omega^3}{(\Omega^2 - \omega^2)^2 + \omega^2\Gamma^2} \tag{7.3.8}$$

where we have written the frequency $\omega(\mathbf{q}, j)$ simply as Ω and Γ is the
damping constant. This shape function is shown for a series of values of
relative damping Γ/Ω in Fig. 7.5. The curves all have the same area (equal to
$\frac{1}{2}\pi$) and show the shape to be expected for the Raman spectra in this
approximation. When $\Gamma \ll \Omega$ the lineshapes are Lorentzian and centred

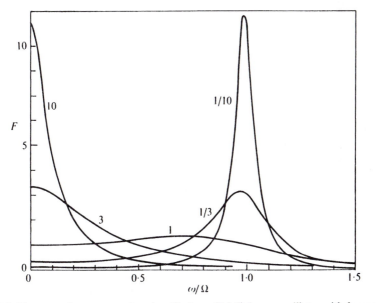

Fig. 7.5. The scattering response function F of eqn (7.3.8) for an oscillator with frequency-
independent damping. Strictly $F(\omega)$ is a valid lineshape only in the classical high-temperature
limit. At lower temperatures the Stokes and anti-Stokes spectra are distorted by quantum-
mechanical Bose–Einstein factors. The curves are labelled by the ratio Γ/Ω of damping to
frequency fundamental.

approximately at $\omega = \Omega$ with Γ equal to the full width at half maximum. As Γ/Ω increases the finite frequency peak becomes less prominent and finally disappears into a central peak which, in the limit $\Gamma \gg \Omega$, is indistinguishable from that of a relaxation process with relaxation time Γ/Ω^2. Near a displacive phase transition the soft-mode frequency is expected to vary with temperature closely as $\Omega^2 = T - T_{\rm c}$, and therefore in a spectrum arising from such a lattice mode one should see temperature-dependent Raman lines with positions converging towards a central peak as the transition temperature is approached.

It is now necessary to consider a little more closely some of the tacit assumptions which have gone into the discussion presented to this point. Firstly we should recognize that in a fully quantum picture the scattering should be viewed as a third-order process (Loudon 1964) in which an incident photon excites a virtual electron–hole pair, after which the electron is scattered by a phonon and finally falls back into the hole to emit the scattered photon. Even within the semi-classical picture we should realize that the validity of an expansion of electronic polarizability in terms of the lattice normal modes depends on the Born–Oppenheimer (or adiabatic) separation of electronic and nuclear motions as discussed in § 1.2. In this picture the electronic structure can adjust instantaneously to the nuclear configuration at every instant of time. Even more importantly we have also assumed the Born approximation for scattering in which the higher-order terms in the expansion (eqns (7.3.1) and (7.3.4)), which correspond to multiple phonon scattering, are neglected.

In the Born approximation the Raman peaks occur at the frequencies of the excited phonons and, because of the conservation of momentum and the small wavevector of light (on the scale of the Brillouin zone), these are essentially zone-centre phonons or polaritons. Higher-order spectra involving more than one phonon are not so restricted and are therefore much richer in detail and in information concerning the symmetry and location in the Brillouin zone of the phonons involved. Furthermore, in addition to the scattering from finite-energy propagating modes in various orders there is in general a quasi-elastic component from static strains, impurities, or flaws. These contribute to the central (or Rayleigh) peak in the spectrum and can easily obscure the details of any critical soft-mode dynamics. On the other hand, interesting contributions to central peaks should result from the polarization fluctuations themselves in the case of an order–disorder transition without significant tunnelling. For this situation one anticipates a soft mode of pure relaxation form from eqn (2.4.21) and the temperature development of the central peak now contains, at least in principle, important information concerning the critical dynamics of the transition.

The criterion which determines whether or not a particular phonon appears in the first-order Raman spectrum is that it should transform, under the symmetry operations of the point group of the lattice, in the same way as do the components of the suceptibility tensor. The reason that we can ignore the frequency and wavevector dependence of the scattering phonon in deducing the selection rules is as follows. The phonon time dependence of oscillation can be neglected because the nuclei move so slowly on the scale of optical vibrations that they are effectively stationary. The argument for ignoring the spatial dependence is that the momentum-conservation law (eqn (7.3.6)) restricts the scattering phonon wavelength to be greater than or of the order of an optical wavelength. The latter is characteristically about a thousand times a unit cell dimension and, correspondingly, the pattern of distortions in neighbouring unit cells is essentially identical.

It is particularly simple to establish the absence of certain first-order Raman spectra in centrosymmetric crystals. These are the modes with displacement amplitudes which reverse in sign under the inversion symmetry operation, since the dielectric constant must of course be invariant under inversion in such a crystal. The optic phonon modes in the cubic perovskite phase are good examples. Such modes, however, are not necessarily infrared inactive since they may indeed generate an electric dipole. On the other hand, there is a rule of crystal physics which states that in a centrosymmetric crystal no phonon can be both infrared and Raman active.

In more general circumstances it is necessary to examine the transformation properties of all the unit-cell lattice vibrations under the operations of the point group. Raman activity occurs for a phonon if the irreducible representation of the point group according to which the relevant phonon transforms is contained in the representation formed by the components of the electronic susceptibility tensor. The irreducible representations by which the components of the electronic susceptibility (or polarizability) tensor transform have been listed by Herzberg (1945), Loudon (1964), and others for all 32 point groups. We reproduce the results in Appendix B. In addition, Petzelt and Dvorak (1976) have recently used group-theoretical techniques to derive general formulae for all strength changes and frequency shifts at a wide range of structural transitions.

Information of this kind is of great value in interpreting Raman spectra, although a correct identification of Raman lines is not always easy because of the presence of multi-phonon scattering processes in addition to the one-phonon lines. In fact Raman spectra are often very complicated and the early literature in ferroelectrics is full of doubtful line assignments. Symmetry identification is usually obtained by examination of the polarization properties of incident and scattered radiation, and difficulties encountered in separating the one-phonon from the multi-phonon components can sometimes be lessened by making use of complementary assignments made in infrared or neutron spectroscopy on the same material.

Second-order spectra are made up from modes contributing terms of the order $\xi_i(\mathbf{q})\xi_i(\mathbf{q}')$ to the electronic polarizability. The conservation of momentum then requires that $\mathbf{q} + \mathbf{q}'$ be essentially zero on the scale of the Brillouin zone. It follows that modes of all wavevectors $\mathbf{q} = -\mathbf{q}'$ are able to contribute and that the two-phonon spectrum is therefore continuous, although it does have peaks corresponding to phonon density-of-states peaks and can lead to spurious one-phonon line assignments.

Much of the experimental work described in the following section has been performed on the perovskites, which have both potentially unstable zone-centre modes and zone-corner (unit-cell doubling) modes in the cubic phase. In this phase, however, the optic modes are all Raman inactive and neutron spectroscopy is better able to study the critical dynamics. Below the transition point it is a simple but powerful result of group theory that (at least for a second-order phase change) there will always be a zone-centre totally symmetric component of the soft mode, regardless of the symmetry or wavevector of the soft mode above the transition (Birman 1973). The totally symmetric mode has a contribution to $\varepsilon_{mn}^{(1)}(\infty)$ of eqn (7.3.1) which is of the same form as the unperturbed polarizability (i.e. its motion does not change the symmetry of the unit cell) and it is therefore always Raman active. In other words the soft mode is always Raman active below T_c and it follows that the dynamics of *any* structural transition of second order can be studied by Raman spectroscopy at least for $T < T_c$. Inelastic neutron scattering can probe both sides of the transition but the much better frequency resolution available with laser optical techniques makes Raman spectroscopy an attractive alternative in many cases. Also, even in high-symmetry phases with Raman inactive soft modes, the technique of Raman spectroscopy can still have its uses, since the application of quite a small electric field is able to lift the Raman forbiddenness sufficiently to allow observation of one-phonon lines. In fact such experiments (e.g. Fleury and Worlock 1968) were among the first truly quantitative studies of soft modes to be made.

7.3.2. Some experimental studies

In a typical light-scattering experiment intense collimated light from a laser is focused on a crystal and scattered light is collected, collimated, and analysed. Light passing through the analysing spectrometer is converted to a series of current pulses by a photomultiplier and the pulses counted or integrated. The use of light scattering for the study of soft modes actually goes right back to 1940 when Raman and Nedungadi (1940) observed the softening of a totally symmetric optic phonon on approach to the $\alpha-\beta$ phase transition in quartz. This early intuitive understanding of a soft mode, subsequently clarified enormously by the lattice-dynamical theory of Cochran (1960), led to a search for similar phenomena in ferroelectrics. Early results for $BaTiO_3$ were confusing and the soft mode associated with the cubic–tetragonal phase change could not be seen. This, it has since become apparent, was not any result of shortcomings in the Raman technique but due rather to an unfortunate choice of crystal. In fact $BaTiO_3$ is complicated by extreme anisotropy in q-space and very short phonon lifetimes to such a degree that questions concerning the true nature of the critical dynamics (i.e. whether the transition should be described in terms of an order–disorder diffusive relaxation or as a grossly overdamped harmonic soft mode) are still raised (e.g. Fontana and Lambert 1972). The resulting Raman spectrum (DiDomenico, Wemple, Porto and Bauman 1968, Burns 1974) is then a broad peak centred throughout on zero frequency and this basic character of the $BaTiO_3$ soft mode is confirmed by neutron scattering (Harada, Axe, and Shirane 1971).

Much simpler in their soft-mode behaviour are the $BaTiO_3$ isomorphs $SrTiO_3$, $KTaO_3$, and $PbTiO_3$. The first two of these never in fact become ferroelectric at zero stress but do exhibit significant zone-centre mode softening at low temperatures. As mentioned above the cubic perovskite soft modes are normally Raman inactive, but the Raman forbiddenness can be lifted by application of a small electric field. In fact both $KTaO_3$ and $SrTiO_3$ (Fleury and Worlock 1968) provide excellent Raman studies of mode softening with the zone-centre soft modes remaining underdamped to the lowest temperatures and with frequencies in accord with the Lyddane–Sachs–Teller expectation. An advantage of studying soft modes in the paraelectric configuration is that the theoretical complication of a temperature-dependent order parameter is avoided and the anharmonic interactions producing the mode softening can be examined in their purest form. Another attractive feature of the zone-centre soft modes in $KTaO_3$ and $SrTiO_3$ is their almost negligible interaction with other modes in this context, couplings which in many other soft-mode systems lead to lineshape asymmetries, level repulsions, and many other complexities which can at times lead to mistaken conclusions regarding the physics of the phase changes in question. Such comparative simplicity makes $KTaO_3$ and $SrTiO_3$ excellent candidates for attempting a quantitative understanding of mode softening in terms of a detailed microscopic model of lattice anharmonicity. Lead titanate (Burns and Scott 1970, 1973 a) also appears to possess close to a textbook example of a soft ferroelectric mode, this time actually leading to a (zero-stress) phase transition. For this case the scattering was observed in the absence of an applied field; all the modes are sharp and underdamped below T_c, obey the expected selection rules, and in particular disappear above T_c (Fig. 7.6). A damped harmonic-oscillator model with frequency-independent damping appears adequate to explain the observed lineshapes.

One of the difficulties of interpreting the Raman spectra of less well-behaved systems is that not only may the lines be broad and indistinct, but they may not always appear to obey the appropriate selection rules. This can

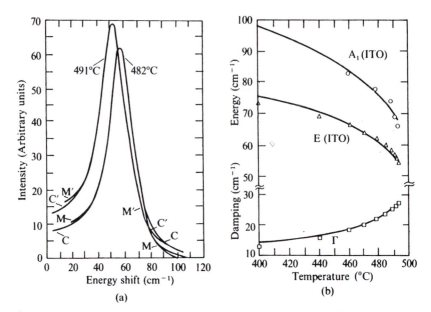

Fig. 7.6. (a) Experimentally observed (M) and calculated (C) Raman lineshapes for the lowest-energy (TO) E mode in $PbTiO_3$ at two temperatures below the first-order ferroelectric transition at $T_c \approx 493°C$. (b) Experimental values of the soft A_1 and E-mode frequencies (and the E-mode damping parameter Γ) as functions of temperature from the same Raman experiments. (After Burns and Scott 1973 a.)

occur whenever impurities, faults, imperfections, or, most importantly, the simple statistical randomness of a double-well potential removes the strict translational invariance of the lattice. Thus, for example, Johnston and Kaminow (1968) have noted a relaxation of the selection rules in $LiNbO_3$ and $LiTaO_3$ which they attribute to grown-in dislocations or defects, while Hartwig et al. (1972) attribute a similar effect above T_c in $NaNO_2$ to the double-well statistical disorder. Either effect might also produce the appearance of 'forbidden' lines in, for example, a cubic perovskite phase. An interesting clarification of the very confused and often conflicting interpretations of Raman and infrared studies on $BaTiO_3$ was provided by Burns and Scott (1971) by examining Raman scattering in the mixed ferroelectric $Pb_{1-x}Ba_xTiO_3$. Since the relevant modes are all narrow and well defined when $x = 0$, a study of the development of the system as a function of increasing x reveals the broadening and converging of the lines (Fig. 7.7) and, most importantly, allows the correct interpretation of the Raman spectrum of the end member $BaTiO_3$ ($x = 1$) to be determined. It is interesting that the soft E-mode decreases in frequency as a function of x but remains underdamped to $x = 0.8$. It is therefore most unlikely that the overdamped character of $BaTiO_3$ can be attributed to impurities and,

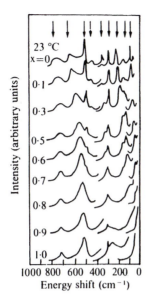

Fig. 7.7. The 23°C ($T < T_c$) powder Raman experimental results for various x in $Pb_{1-x}Ba_xTiO_3$ (the x values are given by the numbers on the curves). The curves are broken at $\sim 350\ cm^{-1}$ and sometimes at $\sim 100\ cm^{-1}$ because of gain changes. The arrows on the top indicate the single-crystal $PbTiO_3$ modes. (After Burns and Scott 1971.)

indeed, the physical reason for this basic difference between the two types ($x = 0$ and $x = 1$) of soft mode remains rather obscure.

As mentioned above the perovskite lattice is also potentially unstable with respect to zone-boundary phonons. The most closely studied of these antiferrodistortive transitions is the 105 K transition in $SrTiO_3$. Above the transition the mode wavevector is too large to allow study by optical techniques but inelastic neutron scattering (Shirane and Yamada 1969, Cowley, Buyers, and Dolling 1969) has verified the instability of the triply degenerate zone corner $(\frac{1}{2}, \frac{1}{2}, \frac{1}{2})$ phonon as T decreases towards $T_c \approx 105$ K. Below T_c the unit cell doubles in size and the soft mode is now at the centre of the new Brillouin zone. An accompanying tetragonal distortion lifts the triple degeneracy below T_c and the resulting A_{1g} singlet and E_g doublet modes are both Raman active and are readily separated by polarization selection rules (Fleury, Scott, and Worlock 1968). The Raman spectra at 88 K are shown in Fig. 7.8 and we note the unequal amplitudes of the Stokes and anti-Stokes lines as expected for a quantum crystal.

In hydrogen-bonded ferroelectrics the anharmonicity associated with the hydrogen-bond double well typically yields overhamped features. Indeed only very recently (Peercy 1973) has the propagating nature of these modes been established beyond question. Early Raman studies on KH_2PO_4 were

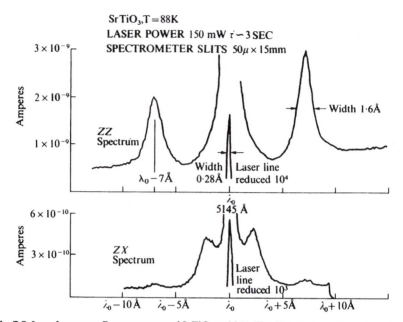

Fig. 7.8. Low-frequency Raman spectra of $SrTiO_3$ at 88 K. The *ZZ* spectrum shows the soft A_1 mode while the *ZX* spectrum shows the soft E mode. (After Worlock and Olsen 1971.)

interpreted in terms of a simple overdamped soft mode (Kaminow and Damen 1968). However, more recent work (also using the damped harmonic-oscillator language) suggests that a simple oscillator model with temperature- and frequency-independent damping is inadequate to describe the soft-mode condensation in many hydrogen-bonded compounds. Cowley *et al.* (1971), working with CsH_2AsO_4 and KH_2AsO_4, suggest that there is evidence for a strong coupling of the soft mode with another optic phonon of like symmetry, and this interpretation has received support in the tunnel-mode formalism as well (Moore and Williams 1972). Any reservations concerning the use of a propagating-mode formalism for these systems were dispelled by Peercy (1973) who studied the pressure dependence of the coupled-mode spectra in KDP. The soft mode, which is overdamped at atmospheric pressure, becomes underdamped for pressures exceeding 6 kbar (Fig. 7.9). The fact that at high pressures the observed response actually peaks at non-zero frequencies indicates that the correct description must be in terms of propagating modes and not of purely diffusive motion.

Allowing for mode coupling in a propagating-lattice or tunnel-mode formalism, a decoupled soft mode can be defined which is normally over-damped in the hydrogen-bonded structures. However, this bare phonon

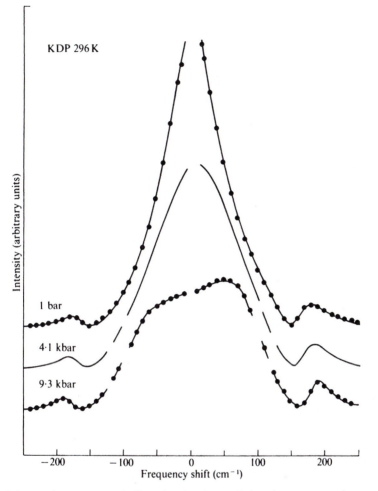

Fig. 7.9. Spectra for B_2 phonons in KDP showing the coupled-mode spectra at pressures of 1, 4·1, and 9·3 kbar. The solid curves are experimental data and the points show calculated curves for the best least-squares fit to theory. (After Peercy 1973.)

frequency still does not exhibit the expected approach to zero as $T \to T_c$. Coombs and Cowley (1973) attribute this anomalous behaviour to a direct coupling of the soft mode with fluctuations in the phonon density. Such a coupling can occur in any piezoelectric lattice structure and manifests itself by lending a frequency dependence to the damping factor (see also Chapter 11). A possible alternative explanation has been offered by Moore and Williams (1972) which does not require a piezoelectric environment so that it is proper at this time to consider the matter less than wholly resolved.

Interest is now mounting rapidly in the Raman spectra of randomly disordered systems. An important consequence of static disorder is the breakdown of wavevector conservation which can have a marked effect on light scattering and optical absorption. As a result of this q-conservation breakdown it becomes possible for a mode anywhere in the Brillouin zone to be first-order Raman active unless prevented by symmetry. By this criterion almost all points in the Brillouin zone (with the exception of a few high-symmetry points) are Raman active and the resulting Raman spectrum tends to reflect the one-phonon density of states. Evidence for just such behaviour has been found in NH_4Cl (Loveluck and Sokoloff 1973) which is not ferroelectric but undergoes an order–disorder transition at 243 K to a phase in which the force constants between ammonium ions are disordered because there are two equally probable orientations of the NH_4^+ ion within a unit cell. In the ferroelectric context Burns and Scott (1973 b) have studied several highly disordered ferroelectrics with disorder due to vacancies or random site occupancy. Evidence for local disorder is clearly manifest in the Raman measurements. For example $Pb_3MgNb_2O_9$ is statistically cubic above T_c, having the ABO_3 perovskite structure, but with the B site randomly occupied by Mg and Nb atoms. In the absence of this randomness (i.e. for a genuine cubic perovskite) no first-order Raman emission is allowed by symmetry. For paraelectric $Pb_3MgNb_2O_9$ a first-order spectrum is found (as evidenced by its Bose–Einstein temperature dependence) but it exhibits no obvious ferroelectric critical behaviour as the transition temperature is approached. This is perhaps not unexpected since the density of states near the zone centre is probably very small. More generally in the ordered phase one expects contributions both from the ordered and disordered sources.

The observation of polariton effects, or in other words the wavevector dependence of transverse optic phonons, by Raman techniques requires the examination of forward scattering. With forward scattering the phonon wavevector is clearly small compared with that of the incident and scattered photon and therefore, at some forward angle, the wavevector approaches the value (typically 10^3–10^4 cm^{-1}) for which the photon and phonon frequencies are resonant. Polariton effects have been studied in a number of cubic and uniaxial crystals. We shall illustrate with an example of the latter, $LiNbO_3$ (Fig. 7.10), which shows the measured wavevector dependence of the lowest-frequency A_1 mode (which is also the soft mode) and one of the higher-frequency modes of the same symmetry. According to theory (see Fig. 7.1 (a)) the former should have a polariton dispersion which tends to zero at $q = 0$, while the latter should tend to a finite frequency corresponding to that of a longitudinal mode. The experimental points obviously are in accord with such an interpetation and the extrapolation of the soft-mode dispersion to a linear form at the origin gives an estimate for static dielectric

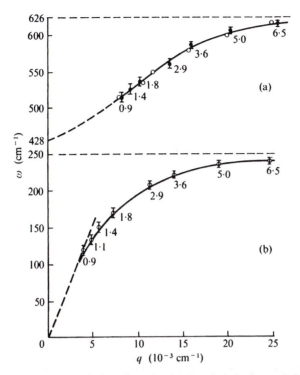

Fig. 7.10. Room-temperature polariton dispersion for the soft A_1 lattice mode in LiNbO$_3$ and for one higher-frequency like-symmetry mode as measured by forward Raman scattering. The scattering angle (in degrees) inside the crystal labels the experimental points. (After Mavrin, Abramovich, and Sterin 1972.)

constant $\varepsilon(0) = 26 \pm 4$ in satisfying agreement with the values deduced from electrical measurements. In general polariton effects are most marked for modes with large oscillator strength, and the polariton effects in the Stokes and anti-Stokes lines are expected to be essentially the same.

Finally, one should emphasize perhaps that all the Raman theory and experimental observation discussed to this point has involved light scattering as the result of a modulation of electronic polarizability. In principle there should also be a corresponding effect resulting from a modulation of ionic polarizability. Thus in the normal Raman effect the incident photon excites intermediate electronic states which are modulated by a thermally generated phonon. For a pure ionic Raman effect the intermediate states are not electronic but vibrational. More generally the intermediate states can be mixed (Martin and Genzel 1974). In fact neither mixed nor pure ionic Raman scattering has yet been observed, the corresponding Raman efficiency being below the limits of the present generation of infrared detectors. On the other hand, calculations do suggest that resonant ionic frequency mixing (with the Raman-active phonon generated by infrared radiation rather than thermal population) may well be observable with present-day equipment.

7.3.3. Rayleigh scattering and critical opalescence

In general any light scattering which is peaked at the frequency of the incident radiation is termed Rayleigh scattering. In addition to the Rayleigh scattering associated with the first-order Raman spectrum at phase transitions (as discussed in the previous sections) additional contributions can arise from any static impurities, dislocations, or other faults in the crystal. Yet another most interesting contribution can arise from the quadratic term in the expansion of the optical dielectric constant in terms of polarization. Let us for simplicity consider merely a symbolic expansion for uniform (i.e. $\mathbf{q} = 0$) polarization as follows

$$\varepsilon = \text{constant} + \left(\frac{\partial \varepsilon}{\partial P}\right)P + \frac{1}{2}\left(\frac{\partial^2 \varepsilon}{\partial P^2}\right)P^2 + \ldots \tag{7.3.9}$$

where the linear term gives rise to one-phonon Raman scattering and, in particular, is absent for a centrosymmetric phase (since ε is then a symmetric tensor and P is a vector).

Consider small increments $\Delta\varepsilon$ in ε and ΔP in P. Let us write $\Delta\varepsilon = (\Delta\varepsilon)_1 + (\Delta\varepsilon)_2 + \ldots$ where the subscripts refer to one-phonon, two-phonon, etc. contributions to the increase in dielectric constant. From eqn (7.3.9) it follows that $(\Delta\varepsilon)_1 \propto \Delta P$, with a coefficient which vanishes for centrosymmetric structures, and $(\Delta\varepsilon)_2 \propto P\Delta P$. As a consequence

$$\langle(\Delta\varepsilon)_1^2\rangle \propto \langle(\Delta P)^2\rangle \tag{7.3.10}$$

and

$$\langle(\Delta\varepsilon)_2^2\rangle \propto P^2\langle(\Delta P)^2\rangle \tag{7.3.11}$$

where the angular brackets refer to thermal or ensemble averages.

Near a second-order phase transition the polarization (or more generally the order parameter) varies with $T_c - T$ as $P^2 \propto (T_c - T)^{2\beta}$ where β is a critical exponent (see Chapter 11) which, in simple effective-field theory, is equal to one-half. On the other hand, via the fluctuation theorem (eqn (2.2.28)), $\langle(\Delta P)^2\rangle$ is proportional to the susceptibility χ which near a critical point varies as $(T_c - T)^{-\gamma}$ with $\gamma = 1$ in simple effective-field theory. It follows that in the limit of long wavelengths the first-order contribution diverges and the second-order contribution (which varies as $(T_c - T)^{2\beta-\gamma}$) remains finite, at least in the mean-field approximation.

Many years ago it was pointed out by Ginzburg (1955) that the second-order contribution also diverges (even in the mean-field scheme) for phase transitions which are on the borderline between second and first order. In the Devonshire formalism for ferroelectrics such a situation occurs when the coefficient of the fourth-order term in the free-energy expansion is zero. At such a point, which is called a Curie critical point in the older literature and a tricritical point in more recent terminology, $P \propto (T_c - T)^{1/4}$ and $\chi \propto (T_c - T)^{-1}$ in Devonshire theory (Grindlay 1970, p. 197). A more careful argument should of course be followed for a quantitative analysis, allowing in particular for wavevector dependence and an adequate representation of the disordered phase (Ginzburg and Levanyuk 1958, Ginzburg 1962) for which $P = 0$ in eqn (7.3.11).

The divergence of Rayleigh scattering intensity which is expected from the above divergence of dielectric fluctuations is termed 'critical opalescence'. We see that it may result from one-phonon scattering in piezoelectric structures, but that it must result if at all from multi-phonon processes in centrosymmetric phases. In ferroelectrics, for which critical exponents seem to be close to the mean-field predictions, the multi-phonon opalescence would seem to require a close to tricritical phase transition. For anti-distortive transitions the $T > T_c$ critical fluctuations are not at the Brillouin zone centre and hence do not contribute a first-order spectrum. Any critical opalescence in such a case must be multi-phonon in origin, and, since $2\beta - \gamma$ is in general less than zero even for ordinary second-order anti-distortive transitions (implying a breakdown of the mean-field approximation, see Chapter 11), second-order transitions of this type are good candidates for the observation of critical opalescence.

The experimental search for critical opalescence in solids has had a long and rather frustrating history. An increase in light scattering near second-order transitions has been reported on several occasions (Yakovlev, Velichkina, and Mikheeva 1956, Fritz and Cummins 1972, Steigmeier, Harbeke, and Wehner 1971), but an unambiguous interpretation of the data is exceptionally difficult owing to the presence of scattering from domains, deformations, dislocations, and other static imperfections (Ginzburg and Levanyuk 1974). It has also been suggested that many structural transitions thought to be of second order may actually take place via a discontinuous intermediate stage in which two phases coexist over a small temperature range. This situation could also give rise to an anomalous increase in Rayleigh scattering in the transition region (Bartis 1973). One of the best-documented examples to date for the observation of critical opalescence in solids is that of Steigmeier, Auderset, and Harbeke (1973) at the 105 K anti-distortive transition in $SrTiO_3$. Since in this case the soft mode above T_c is at the zone boundary, the opalescence (if indeed it is of dynamic origin) must be multi-phonon or, in other words, arise from the direct coupling of light with phonon-density (i.e. temperature) fluctuations. However, at the time of writing no definitive observations have been made of the frequency width of any central peaks and the possibility remains open that this feature is not dynamic in origin.† A thorough search for central peak dynamics using high-resolution light scattering was recently carried out by Lagakos and Cummins (1974) at the ferroelectric transition in KDP, but without success. It appears that the central peak width is certainly less than $\sim 10^9$ Hz, but below this range no direct spectroscopic evidence yet exists.

Finally we note that although temperature fluctuations normally obey the law of heat diffusion leading to a spectrum with a quasi-elastic central peak, under special conditions (Enns and Haering 1966) temperature fluctuations can propagate in the form of weakly damped waves (second sound). In this case the multi-phonon Rayleigh component should split into a doublet, but efforts to observe such an effect seem to be thwarted by the fact that in the second-sound regime the characteristic relaxation times are so long that energy and momentum transfer are restricted to extremely small values.

† See, however, the footnote on page 403.

8

OXYGEN OCTAHEDRA

8.1. The ferroelectric perovskites

A VERY important group of ferroelectrics is that known as the perovskites from the mineral perovskite $CaTiO_3$ (which itself is actually a distorted perovskite structure). The perfect perovskite structure is an extremely simple one with general formula ABO_3, where A is a monovalent or divalent metal and B is a tetra- or pentavalent one (see Fig. 8.1). It is cubic, with the A

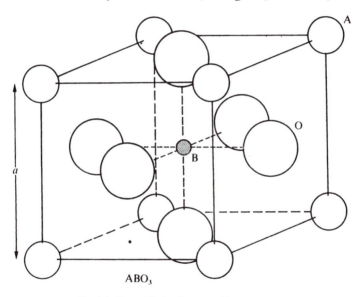

Fig. 8.1. The cubic ABO_3 perovskite structure.

atoms at the cube corners, B atoms at the body centres, and the oxygens at the face centres. The structure can also be regarded as a set of BO_6 octahedra arranged in a simple cubic pattern and linked together by shared oxygen atoms, with the A atoms occupying the spaces between.

The first ferroelectric perovskite to be discovered was $BaTiO_3$. This was a particularly significant event since prior to the discovery only complex hydrogen-bonded crystals were known to exhibit ferroelectricity and the presence of hydrogen had been considered an essential for the phenomenon. Of equal importance was the extreme simplicity of the structure, which gave physicists for the first time an opportunity to study the onset of ferroelectricity from a simple highly symmetric prototype phase. In fact the enormous rise of interest in ferroelectrics in the 1950's was spearheaded by an intense study of the properties of $BaTiO_3$. In more recent years it has become apparent that nature was perhaps a little unkind in giving physicists $BaTiO_3$ as a ferroelectric standard bearer since it has proved to be far from the simple prototype displacement ferroelectric which for so long it promised to be. In fact the extreme anisotropy and overdamped character of its soft phonon modes near the paraelectric–ferroelectric phase transition have led to suspicions that at least an element of order–disorder character may yet be present in the transition. At very least the motion is grossly anharmonic and any description in terms of the quasi-harmonic language of displacive theory is accordingly open to question.

Much simpler in their ferroelectric soft-mode behaviour are the near ferroelectrics $KTaO_3$ and $SrTiO_3$ and the actual ferroelectric $PbTiO_3$. For these the mode damping and anisotropy are less extreme and the displacive picture more realistic. Lead titanate, in addition, has but a single ferroelectric phase, whereas $BaTiO_3$ (and also $KNbO_3$) undergoes successive transitions to three different ferroelectric phases as the temperature is lowered. On the other hand, $BaTiO_3$ and $KNbO_3$ are by no means among the more complex of the perovskite ferroelectrics. The great fascination of the perovskite structure is that it readily undergoes structural transitions involving non-polar phonons (associated with different types of tilts of the oxygen octahedra) in addition to both ferro- and antiferroelectric transitions. Accordingly some perovskite ferroelectrics, such as $NaNbO_3$ for example, can exhibit a large number of phase transitions, some ferroelectric, some antiferroelectric, and some structural. All involve only small distortions from the ideal cubic structure and are therefore appealing objects for experimental and theoretical study.

The O_6 group in particular can be thought of as a hard unit in the sense that it is little distorted from a regular octahedron even when other distortions of the structure are more considerable. In the cubic phase the O_6 octahedra are parallel, but the bond angles at their corners are soft and tilting is easy. The simplest tilts are those in which all octahedra tilt in alternating fashion about the same axis. There are three such possibilities: (i) a two-fold symmetry (diad-axis) tilt, (ii) a three-fold (triad-axis) tilt, and (iii) a four-fold (tetrad-axis) tilt. If the tilts are small it is possible to think of them as made up of component tilts about the three tetrad axes (Megaw 1968). The triad–axis tilt then requires three equal non-zero components and the diad-axis tilt two equal non-zero components. One can, however, have unequal tilt components and many other arrangements are possible. The off-centring of the B-cation is a relatively independent process that can occur in any structure built from the hard octahedra. It is this off-centering which leads to the presence of dipoles and to ferroelectric and antiferroelectric behaviour.

Whereas phase transitions involving tilts alone are almost certainly precipitated by short-range forces, ferroelectric phases are stabilized by a competition between long-range (dipolar) and short-range interactions, and it is still a rather open question as to whether the tendency of the perovskite lattice structure to support ferroelectricity results from a particularly favourable geometry (as suggested by the large Lorentz correction to internal dipolar field originally calculated by Slater (1950)) or whether it is an intrinsic property of the B-cation in the local oxygen-cage environment. The latter explanation is suggested by the tendency of BO_6 groups to support ferroelectricity in varied structural situations with increasing valency and decreasing size of the B-cation relative to the oxygen cage. Even if the intrinsic effect is significant the dipolar forces still play an essential role, as witnessed by the long-range character of the ferroelectric critical phenomena and the quasi-dipolar anisotropy of the soft-mode dispersion.

Early microscopic theories of ferroelectricity in the perovskites were based on Slater's concept of the 'rattling' Ti ion in $BaTiO_3$. The motion of the B-cation is singled out as primarily responsible for the effect and a potential energy for the displaced ion of cubic symmetry is assumed (Slater 1950). More recent theories concentrate on the soft-mode picture set out in Chapter 2. The basic Hamiltonian is just that of eqn (1.2.8) except that for the cubic environment the soft mode is triply degenerate owing to equivalent cubic axes. Thus a basic set of normal-mode co-ordinates is selected which describes motion along each of the three cube axes. As a result the soft-mode momentum and displacement variables (i.e. π_l and ξ_l of eqn (1.2.8)) are now construed as vectors and the local potential $V(\xi_l)$ must be chosen to reflect the cubic symmetry of the basic perovskite structure.

The most recent theory of this kind is by Pytte (1972 b). It is formally a self-consistent phonon treatment with equations of motion linearized by making use of the fluctuation theorem, but for actual calculations the mean-field approximation is adopted. Terms are also included in the effective Hamiltonian which couple the soft-mode co-ordinates with elastic strain and long-wavelength acoustic phonons via a local strain tensor. The model Hamiltonian is found to describe a first- or second-order phase transition depending on the strength of the coupling of soft mode with strain. It can also describe transitions to tetragonal, trigonal, and orthorhombic phases depending on the choice of model parameters. Since the actual calculation was performed in the effective-field approximation no wavevector dependence of the modes is derived. In order to do this it is necessary to have a much more detailed knowledge of the interion interactions and soft-mode displacements. Efforts to determine the detailed wavevector dependence of lattice modes (soft or otherwise) are usually restricted to a harmonic approximation and therefore do not attempt any realistic description of temperature development. One such example is that of Hüller (1969) for BaTiO$_3$. A very recent effort to go beyond the harmonic approximation has been made by Bruce and Cowley (1973) primarily for SrTiO$_3$ but with implications for BaTiO$_3$.

Before any quantitative anharmonic calculations can be made it is first necessary to have a realistic model for the frequencies and eigenvectors of the harmonic modes throughout the Brillouin zone. This information is included succinctly in the phenomenological potential functions $V(\xi_l)$ and $v_{ll'}$, of eqn (1.2.8) for the soft mode, but the aim of a more fundamental microscopic theory must be to derive (and parameterize) these potentials from a basic ionic model. The simplest such model is the rigid-ion model (Kellermann 1940) which neglects the electronic polarizability of the ions. More sophisticated models include the deformation of ions by separating each ion into a core and an electronic shell coupled harmonically together. The resulting shell models, the rigid shell model (e.g. Dick and Overhauser 1958, Woods, Cochran, and Brockhouse 1960) and the deformable shell model (e.g. Schröder 1966) enable a fairly quantitative description of harmonic vibrations to be obtained in many cases. For SrTiO$_3$ detailed shell-model parameters were derived by Stirling (1972) using extensive inelastic neutron scattering data. Starting from this base Bruce and Cowley (1973) parameterize the lowest-order anharmonicities and determine them by fitting certain anharmonic properties. The results support the concept of rigid oxygen octahedra with short-range forces dominated by A–O and B–O interactions, and give reasonable agreement with the temperature-dependent frequency and linewidth of both zone-centre and zone-boundary soft modes in SrTiO$_3$. An attempt to extend the work to BaTiO$_3$ was less successful.

As emphasized by Bruce and Cowley it is difficult to envisage any significant improvements of the model without a considerable increase in the complexity of the numerical calculations. In the light of the relative simplicity of the perovskite structure this would seem to suggest that most theoretical work in the foreseeable future will come, via the effective Hamiltonian approach, i.e. eqn (1.2.8), concentrating on the phenomenological description of a specific soft mode and perhaps its interaction with a few other properties of interest. The self-consistent treatment of several phonon branches via a specific microscopic model of interacting cores and shells would therefore seem to be prohibitively complex for most displacement ferroelectrics at the present time.

8.1.1. Barium titanate

BaTiO$_3$ has the prototype cubic perovskite structure (point group m3m) above 120°C. Below 120°C it transforms successively to three ferroelectric phases: first to 4mm tetragonal, then to mm orthorhombic at about 5°C, and finally to a 3m trigonal phase below $-90°C$. The polar axis in the three ferroelectric phases is [001], [011], and [111] respectively. All three transitions are of first order and the temperature dependence of the dielectric constant shows discontinuities at the transitions and peak values as high as 10^4. Above $T_c = 120°C$ the dielectric constant follows a Curie–Weiss law $\varepsilon = C/(T-T_0)$ with $T_0 < T_c$. An excellent account of the detailed dielectric, optic, and elastic properties of $BaTiO_3$ can be found in Jona and Shirane (1962) to which we refer the reader for details.

$BaTiO_3$ was long considered to be the prototype of the perovskite ferroelectrics and the latter as the simplest of the various ferroelectric types. Most static and dynamic structural interest has centred on the highest-temperature (cubic-to-tetragonal) transition. In the tetragonal phase the Ti and O ions move relative to Ba at the origin from their cubic positions, (viz. Ti at $(\frac{1}{2}, \frac{1}{2}, \frac{1}{2})$ and three oxygens at $(\frac{1}{2}, \frac{1}{2}, 0)$, $(\frac{1}{2}, 0, \frac{1}{2})$, and $(0, \frac{1}{2}, \frac{1}{2})$), to Ti at $(\frac{1}{2}, \frac{1}{2}, \frac{1}{2}+dz_{Ti})$, OI at $(\frac{1}{2}, \frac{1}{2}, dz_{OI})$, and OII at $(\frac{1}{2}, 0, \frac{1}{2}+dz_{OII})$ and $(0, \frac{1}{2}, \frac{1}{2}-dz_{OII})$, where $dz_{Ti} = 5$ pm (pm $\equiv$ picometres), $dz_{OI} = -9$ pm, and $dz_{OII} = -6$ pm (Harada *et al.* 1970). This much at least seems fairly definite for the statics. But is it? The structure has been solved from elastic neutron and X-ray data by assuming a model in which the ions vibrate quasi-harmonically about well-defined local positions. The possibility of a partially disordered model with atomic thermal distribution not necessarily single peaked has not been considered and may well, in any case, lead to an indeterminate problem (see Chapter 6). On the other hand, there is evidence from X-ray diffuse scattering, infrared absorption, Raman spectroscopy, and inelastic neutron scattering (all discussed in the relevant sections of this book) that close to the phase transition the soft mode is grossly overdamped to the extent that a double-well characteristic certainly cannot be discounted.

The most detailed study to date is the inelastic neutron scattering work of Yamada, Shirane, and Linz (1969), Shirane *et al.* (1970), and Harada *et al.* (1971). Strong quasi-elastic scattering in the paraelectric phase is grossly anisotropic with a form characteristic of dipolar-dominated interactions. An energy analysis of the scattering cross-section shows the soft polarization fluctuation to be of a relaxation type which can be explained equally well in terms of an overdamped phonon mode or by a pseudo-spin tunnelling mode. The several meV width of the diffuse scattering, however, rules out the possibility of static disorder of the type propounded by Comes *et al.* (1968) to explain the diffuse X-ray data. When analysed as an overdamped soft phonon for $T > T_c$ the real part of the frequency varies as $\omega^2(T) = A(T-T_0)$, and, as pointed out by Cochran (1968), the coefficient A can be related theoretically to the Curie constant and gives a calculated value quite consistent with that found experimentally from susceptibility. The equivalent calculation for a tunnel model yields a Curie constant which is much smaller than observed. The inference is that the motion is not in deep double wells but more likely is of large amplitude in a grossly anharmonic and possibly shallow double-well potential. A very recent confirmation of the phonon character of the soft mode has come from a measurement in the tetragonal phase of the polariton wave-vector dependence of the soft mode (Heiman and Ushioda 1974, Tominaga and Nakamura 1974) by forward Raman scattering. More and more, however, we are becoming aware that the terms 'displacive' and 'order–disorder' really only describe limiting behaviour, and that some physical systems like $BaTiO_3$ may be distributed more or less between these extremes.

The orthorhombic phase of $BaTiO_3$ was structurally solved by Shirane, Danner, and Pepinsky (1957) with major displacements dx parallel to an [011] direction of the pseudo-cubic cell as follows: $dx_{Ti} = 6$ pm, $dx_{OI} = -6$ pm, and $dx_{OII} = -7$ pm relative to the Ba at (000). Study of the lowest-temperature rhombohedral phase has been much complicated by experimental difficulties, with large single-domain crystal specimens generally unavailable for analysis. Recently (Hewat 1974) it has been found possible to extract fairly quantitative structural information from powder specimens by use of a new profile-refinement technique (see § 6.2). The major distortions from cubic symmetry are parallel to the polar axis [111] and relative to Ba at the origin are 9 pm and −9 pm for Ti and oxygen respectively. The oxygen octahedra are slightly distorted in each of the ferroelectric phases.

When the displacements are known in the various ordered phases it becomes possible to calculate spontaneous polarization directly in terms of effective charges on the displaced ions (e.g. eqn (6.1.18)). These effective charges must include contributions from ion distortions (i.e. electronic contributions) but can be estimated in many cases by combining high-frequency dielectric constant and infrared mode-strength measurements (Axe 1967, Hewat 1973 a). They tend to be larger than the formal ionic charges and do in fact lead to fairly good estimates for spontaneous polarization in the ferroelectric perovskites.

8.1.2. Potassium niobate

$KNbO_3$ is in many ways qualitatively similar to $BaTiO_3$. It undergoes the same series of ferroelectric transitions in the same sequence as the temperature is lowered. The cubic–tetragonal transition is at $T = 435°C$, the tetragonal–orthorhombic at 225°C, and the orthohombic–rhombohedral at −10°C. As in $BaTiO_3$ the transitions are all of first order and hysteresis effects are readily observed. The static structure determination in each of the ferroelectric phases has been carried out by Hewat (1973 b) using the powder profile-refinement technique. The oxygen framework retains a rigid closely octahedral form throughout and the displacements from cubic perovskite are very dominantly parallel to the polar axis in each phase. With K fixed at the origin (000), for reference, the relative displacements are as follows:

Phase	Displacement	Nb	OI	OII
Tetragonal	Parallel to [001]	8 pm	−10 pm	−9 pm
Orthorhombic	Parallel to [011]	8 pm	−13 pm	−12 pm
Rhombohedral	Parallel to [111]	9 pm	−13 pm	−13 pm

The oxygen octahedron distortion is considerably less than in $BaTiO_3$ and an essentially rigid oxygen cage is displaced relative to the reference K ion at each of the phase transitions. In the soft-mode picture the tetragonal phase is produced by the condensation of a soft transverse optic mode at the zone centre and polarized along [001]. The transition stabilizes the other two ([100] and [010]) of the originally degenerate triplet of cubic soft modes until, at 225°C, one of these also condenses giving a net polarization along [011]. Finally at the rhombohedral transition the third soft mode condenses to produce a net polarization along [111]. An analogous picture obviously applies to $BaTiO_3$.

A particularly interesting finding from the neutron refinement is a very marked anisotropy in the thermal vibrations of the oxygen ions. The atoms appear to have almost twice the mean-square displacement in the plane of the prototype cube face as they have perpendicular to this plane. This type of anisotropy has also been found in $K(Ta, Nb)O_3$ (Abrahams and Bernstein 1974) and can be attributed to low-frequency rotations of the rigid oxygen octahedra. These correspond to zone-boundary optic modes which are known to be soft in perovskite-like symmetry, and which, indeed, actually condense to produce superlattice structures in $NaNbO_3$ and many non-ferroelectric perovskites.

A study of the highest temperature cubic–tetragonal phase transition has also been performed with inelastic neutrons. Nunes et al. (1971) observe a dynamic behaviour very similar to that found for $BaTiO_3$. In particular, diffuse neutron scattering is found to be concentrated only in planes for which the polarization vector is parallel to a cubic axis. In these planes the modes are soft and overdamped and indicate the presence of significant dynamic disorder. Away from these planes the modes become well defined with much increased energy. The soft-mode frequency is found to be of Cochran form ($\omega^2 \propto T - T_0$) with a Curie–Weiss temperature $T_0 \approx 370°C$ in good agreement with the value obtained from dielectric measurements (see Jona and Shirane 1962). In the orthorhombic phase damping seems less severe (Currat et al. 1974) and well-defined phonon excitations have been observed in the soft plane.

8.1.3. Potassium tantalate niobate (KTN)

The system of solid solutions $KTa_{1-x}Nb_xO_3$ is ferroelectric for $x \gtrsim 0.05$ and exhibits the same three ferroelectric phases as $BaTiO_3$ and $KNbO_3$. The transition temperature varies fairly linearly with composition (Jona and Shirane 1962) with a cubic–tetragonal transition in the region of room temperature when $x \approx 0.37$. Some static and dynamic structure determinations for cubic and tetragonal phases have been reported by Yelon et al. (1971), Hewat et al (1972), and Zaccai and Hewat (1974), and we have already discussed many of their findings in Chapter 6. As with $BaTiO_3$ and $KNbO_3$ the soft mode is found to be overdamped and highly anisotropic, and to represent essentially a vibration of the oxygen cage against the remainder of the cell. The oxygen octahedra are found (at least for $x = 0.44$) to be essentially undistorted in the tetragonal phase. A significant interaction between soft optic and transverse acoustic modes has been measured in the cubic phase producing an anisotropy of acoustic dispersion which seems to be primarily responsible for the observed anisotropy of diffuse X-ray scattering. Little composition-dependence study has yet been undertaken for KTN except for dielectric properties (Triebwasser 1959) which indicate that the paraelectric–tetragonal phase transition may change from first to second order as the concentration of tantalum increases.

8.1.4. Lead titanate

In the light of the findings that $BaTiO_3$ and $KNbO_3$ are very much more complicated in their dynamic behaviour than was envisaged in the original concept of a displacement ferroelectric, it is particularly interesting to discover that the ferroelectric perovskite $PbTiO_3$ behaves more closely in the expected manner. In fact it appears to be a textbook example of a displacive ferroelectric transition. It undergoes a first-order transition at 493°C from cubic perovskite to a tetragonal ferroelectric phase isomorphic with tetragonal $BaTiO_3$, but no additional transitions are as yet definitely established. Raman measurements (Burns and Scott 1970, 1973 b) show that (a) the selection rules below the Curie point T_c are properly obeyed for all modes, (b) the modes are all sharp and underdamped, and (c) the modes disappear abruptly above T_c as they should for the cubic phase (see Fig. 7.6). The behaviour contrasts with that of $BaTiO_3$ (Quittet and Lambert 1973) where the Raman spectrum is broad peak centred throughout at zero frequency and persists into the nominally cubic phase. The reasons why $PbTiO_3$ should be well behaved in the above sense are not understood at the present time.

Even in its static crystallography the $PbTiO_3$ situation is simpler than its $BaTiO_3$ and $KNbO_3$ isomorphs. This is because the parameters associated with the polar phase (such as spontaneous polarization and ionic shifts) are considerably larger, which means that the difficulties associated with pseudosymmetric crystal-structure determination (see § 6.2) can be avoided. The room-temperature crystal structure of tetragonal $PbTiO_3$ was determined by Shirane *et al.* (1956) with displacements parallel to the polar axis (relative to the Pb ion at the origin) of $dz_{Ti} = 17$ pm and $dz_{OI} = dz_{OII} = 47$ pm. We note that the oxygen octahedra suffer no distortion in going to the ferroelectric phase, and that, unlike $BaTiO_3$ and $KNbO_3$, the oxygens and B-cations are shifted in the same direction in $PbTiO_3$ relative to the A-cations. The large ionic shifts in $PbTiO_3$ lead to a particularly large room-temperature spontaneous polarization (variously measured as $57 \, \mu C \, cm^{-2}$ and $75 \, \mu C \, cm^{-2}$ by Remeika and Glass (1970) and by Gavrilyachenko *et al.* (1970)) more than double that of the corresponding phase of $BaTiO_3$ and $KNbO_3$.

The dynamics of the phase transition have been investigated by inelastic neutron scattering (Shirane *et al.* 1970). The soft [001]-polarized transverse optic mode is underdamped for all except the very longest wavelengths, and is observed to have a (001)-plane dispersion $(\hbar\omega)^2 = (\hbar\omega_0)^2 + \alpha q^2$ where the zone-centre mode softens with temperature in the expected manner, viz. $\omega_0^2 \propto (T - T_0)$ with a Curie–Weiss temperature $T_0 = 440°C$. The detailed dispersion for polarization in hard directions has not yet been reported, but one presumes that the characteristic dipolar-type anisotropy is present. Because of the underdamped nature of the soft modes lead titanate is the first example of a perovskite ferroelectric for which the symmetry of the soft-mode motion can be compared with the static structure in the tetragonal phase. A comparison of calculated soft-mode intensities at several zone centres with experimental observation leads to the conclusion that the tetragonal ionic shifts (in a centre-of-mass co-ordinate frame) do indeed correspond quite closely to the soft-mode motion.

Despite the basic simplicity of the $PbTiO_3$ phase transition, ferroelectric studies of bulk properties have been hampered by the excessive electrical conductivity of the crystals caused most likely by lead deficiency. As a result the basic thermodynamic and dielectric parameters describing the transition are still uncertain. Recently, Remeika and Glass (1970) have grown $PbTiO_3$ crystals doped with a small amount of U^{3+} to charge compensate the lead deficiency and have succeeded in raising the room-temperature resistivity to $10^{10} \, \Omega \, cm$. The results of their subsequent measurements of dielectric constant, specific heat, and temperature derivative of spontaneous polarization as functions of temperature (all measured on the same crystal specimen) are shown in Fig. 8.2. All quantities show the abrupt discontinuities typical of first-order transitions. Above the Curie temperature the dielectric constant follows a Curie–Weiss law with Curie constant $C = 4\cdot 1 \times 10^5$ K and Curie–Weiss temperature $T_0 = 449°C$. A thermal hysteresis of about $10°C$ is typically observed. Reproducibility of measurement on different crystal

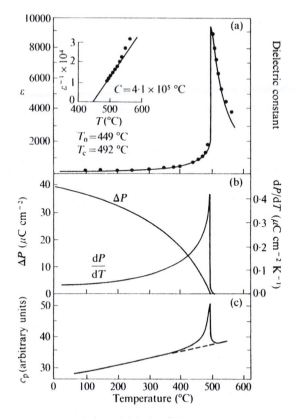

Fig. 8.2. The temperature variation of (a) the dielectric constant ε, (b) the pyroelectric coefficient dP/dT and the change in polarization $\Delta P = \int_{T_c}^{T} (dP/dT)\, dT$, and (c) the specific heat c_p, of PbTiO$_3$. (After Remeika and Glass 1970.)

specimens was good only up to 400°C. Near the phase transition the peak value of dielectric constant varied by up to a factor of 3 while reproducibility of specific heat and dP/dT was only within a factor of 2. The value of the spontaneous polarization discontinuity at T_c was not measured directly, but a value of $17\,\mu\text{C cm}^{-2}$ was estimated indirectly from dP/dT and room-temperature polarization-reversal measurements. Thus, although the phase transition in PbTiO$_3$ seems dynamically to be the simplest yet observed in the perovskites, crystal growth and reproducibility problems still hamper the development of a really quantitative test of our understanding of ferroelectricity for this system.

8.1.5. Sodium niobate

Sodium niobate (NaNbO$_3$) is to date the most complex perovskite ferroelectric known, and in a book of this nature we shall be forced to focus on only a few of its interesting features. The high-temperature phase above about 640°C is simple prototype cubic perovskite as in the other ferroelectric perovskites discussed above. Below 640°C, however, a whole series of structural transitions ensues as the temperature is lowered culminating in a monoclinic ferroelectric phase below about −100°C. The exact number of phases stable at atmospheric pressure is still not absolutely certain but well-documented transitions occur at 640, 575, 520, 480, 370, and −100°C (see for example Avogadro *et al.* 1973). The higher transitions are to nonpolar phases and result from a softening of zone-boundary phonon modes of the oxygen octahedron tilting type. Transitions of this sort are very common among the non-ferroelectric perovskites such as $LaAlO_3$. $KMnF_3$, and $SrTiO_3$ and will be discussed further in the following section. The phases of greatest interest from a dielectric point of view are the two lowest-temperature ones, the room-temperature orthorhombic phase and the low-temperature ferroelectric one.

The room-temperature phase is nonpolar in the absence of applied electric field, but a ferroelectric state (distinct from the low-temperature ferroelectric phase) can be induced by the application of strong fields (Wood, Miller, and Remeika 1962). This phase is therefore antiferroelectric in the classic sense. Its unit cell contains eight formula units, a' and c' being along the original cubic [101] directions and b' along [010] but with a spacing approximately four times the original cubic dimension. Displacements of Nb occur parallel to an a' direction with phase $(+ + − − + + − −)$ along the b' axis (Vousden 1951, Sakowski–Cowley, Lukaszewicz, and Megaw (1969). An applied field parallel to a' then produces an orthorhombic ferroelectric phase by flipping half the dipoles through 180°.

The onset of the natural ferroelectric phase at −100°C is unlike the onset of ferroelectricity in the other perovskites discussed above since it involves an order–order transition from antiferroelectric to ferroelectric. The transition is of first order and there are very large thermal hysteresis effects. The new monoclinic unit cell contains four formula units with the b' spacing now approximately twice the original cubic dimension and the spontaneous polarization parallel to b'.

Much of the more recent experimental work on $NaNbO_3$ has centred on a study of the higher-temperature phases by elastic and diffuse X-ray scattering (Lefkowitz, Lukaszewicz, and Megaw 1966, Denoyer *et al.* 1970, Glazer and Megaw 1972, Ahtee, Glazer and Megaw 1972, Ishida and Honjo 1973, Glazer and Ishida 1974). As a result the origin of the anisotropic diffuse X-ray scattering seems now to be at least qualitatively understood in terms of the soft modes (zone centre and zone boundary) present. In particular the two highest-temperature phase transitions appear to be caused by the condensation of zone-boundary non-polar (i.e. oxygen-cage tilting) modes. At 640°C the transition is caused by the condensing of an M_3 mode at the $(110)\pi/a$ or M-point of the cubic Brillouin zone. This mode involves octahedral rotation about [001] as shown in Fig. 8.3. At this same transition a triply degenerate Γ_{25}-mode (see also Fig. 8.3) at the

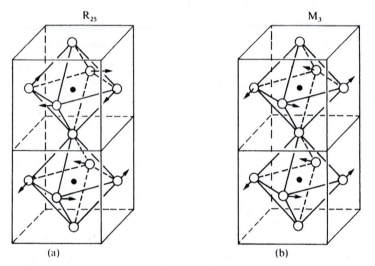

Fig. 8.3. Displacement patterns of the O ions of cubic perovskite ABO_3 for the modes (a) R_{25} and (b) M_3. The corner A-ions are omitted for simplicity. (After Fujii *et al.* 1974.)

$(111)\pi/a$ or R-point of the Brillouin zone splits into an A_{1g}-singlet and an E_g-doublet, and it is the latter doublet which condenses at the 575°C transition. The M_3- *and* Γ_{25}-modes referred to here play an important role in many of the zone-doubling transitions of non-ferroelectric perovskites as discussed in the following section, but only in $NaNbO_3$ do we find such a close interplay of structural, ferroelectric, and genuine antiferroelectric transitions. The phase transition at 520°C is less well investigated, but it appears to involve only oxygen octahedral tilting since an off-centre motion of Nb atoms (which is responsible for ferro- and antiferroelectricity) does not take place until 480°C. The 480°C to 370°C phase is probably antiferroelectric (Avogadro *et al.* 1973). The origin of the diffuse X-ray scattering is also less well understood in the lower-temperature phases and the possibility of a static stacking-fault disorder has been proposed by Darlington (1971).

8.1.6. Potassium tantalate

Potassium tantalate at zero stress retains the prototype cubic perovskite symmetry down to the very lowest temperatures although it is just barely stable against a condensation of the ferroelectric mode at 4 K. The earliest work (Hulm, Matthias, and Long 1950) and some more recent work (Demurov and Venevtsev 1971) has given evidence for a ferroelectric transition at about 10 K in the form of a peak in static susceptibility. However, there is doubt as to whether this transition is an intrinsic property of pure $KTaO_3$ since no transition has been recorded down to helium temperatures in other studies (e.g. Axe *et al.* 1970, Abel 1971, Lowndes and Rastogi 1973, Samara and Morosin 1973). The dielectric constant rises to large values as $T \to 0$ but appears to saturate at a value of about 4×10^3 at helium temperatures. Recently Uwe *et al.* (1973) have observed the onset of ferroelectricity on application of stress at 4.2 K. The application of a uniaxial [100] stress causes the dielectric constant to increase by an order of magnitude and then peak at a stress of order 5×10^8 Nm^{-2}. For larger stress a remanent polarization is produced and ferroelectric hysteresis is observed. The static response at atmospheric pressure can be fitted at higher temperatures (30–300 K) to a Curie–Weiss law of the form $\varepsilon = B + C(T - T_0)^{-1}$ where $B \approx 48$, $C \approx 5 \times 10^4$ K, and $T_0 \approx 13$ K. Below about 30 K the susceptibility falls below the Curie–Weiss curve and follows quite well the form

$$\varepsilon = B + C\{\tfrac{1}{2}T_1 \coth(T_1/2T) - T_0\}^{-1} \qquad (8.1.1)$$

derived by Barrett (1952), with $T_1 \approx 55$ K. In Barrett's theory, which is a quantized version of Slater's (1950) classical theory, the deviations from Curie form result from the quantized nature of the ionic motion of the 'rattling' Ta ion with the lowest quantum level having an energy kT_1. The theory, however, is a mean-field one and neglects any dispersion of the lattice modes within the Brillouin zone. An analytic calculation in the self-consistent phonon approximation seems not yet to have been performed, although Silverman and Joseph (1963) have considered the problem in a perturbation lattice-dynamic treatment. Although the latter is not self-consistent, Samara and Morosin (1973) report that the rather complex expression given by Silverman and Joseph will also fit the experimental susceptibility data over the entire temperature range.

The temperature dependence of soft-mode frequency ω_s has been reported by several groups and the findings (Fig. 8.4) support very nicely the simplest form of the Lyddane–Sachs–Teller relationship, viz. $\omega_s^2 \propto \varepsilon^{-1}$. We note in particular that it deviates significantly from the classic Cochran form for which ω_s^2 is linear with temperature. The form of this deviation both at high and low temperatures has been explained qualitatively within self-consistent phonon theory (Gillis and Koehler 1971) by numerical solution for a model NaCl lattice with long-range and short-range forces. In the self-consistent treatment the renormalized soft-mode frequency (see § 2.3) contains contributions from the harmonic potential and from ensemble-averaged anharmonic contributions. Samara and Morosin (1973) conclude from an analysis of the experimental form of ω_s^2 that the harmonic contribution in $KTaO_3$ is zero or slightly negative. If this is true then the $KTaO_3$ system is stabilized in its cubic phase at low temperatures solely by

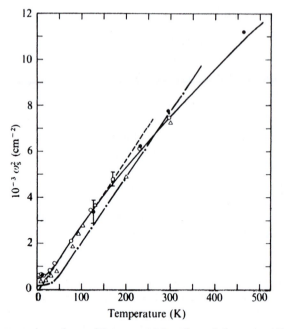

Fig. 8.4. Temperature dependence of the square of the soft-mode frequency of $KTaO_3$ at 1 bar comparing directly measured response (symbols; ○ Shirane *et al.* 1967, ● Perry and McNelly 1967, △ Fleury and Worlock 1968) with that deduced from the dielectric-constant data (solid curve $1·81 \times 10^6/\varepsilon$). Results for $SrTiO_3$ (—•—) are shown for comparison. (After Samara and Morosin 1973.)

zero-point anharmonicity. This very delicate balance is also illustrated by an extremely large pressure dependence of soft-mode frequency (about 20 per cent per kbar at 4 K), the strictly harmonic contribution becoming real at high pressure (see also § 5.8).

In $KTaO_3$ the long-wavelength soft transverse optic modes remain well defined at all temperatures and have been observed in infrared reflectivity (Perry and McNelly 1967), by field-induced Raman scattering (Fleury and Worlock 1968), and by neutron scattering (Shirane, Nathans, and Minkiewicz 1967, Axe *et al.* 1970, Comes and Shirane 1972). The characteristic anisotropy of soft-mode dispersion, with lowest energies confined to modes of ⟨100⟩-type polarization, is again observed together with a rather strong optic–acoustic mode interaction. Coupling to acoustic phenomena will be discussed in Chapter 10, but qualitatively the mode interaction causes a dip in the acoustic branch (Fig. 10.2) and, using a single coupling parameter between the two branches, one can explain not only the shape of the dispersion curve but also the strong intensity anomalies observed for both branches as a function of temperature.

8.1.7. Strontium titanate

In its incipient ferroelectric behaviour at low temperatures $SrTiO_3$ is very much like $KTaO_3$. Unlike $KTaO_3$, however, it also possesses non-ferroelectric structural phase transitions involving oxygen-cage tilting, the most celebrated of which occurs (at atmospheric pressure) at about 105 K. The latter, involving the condensation of a zone-boundary mode, has been the object of very detailed experimental investigation over the last several years and will be discussed further in the next section. For the present we concentrate upon the quasi-instability of the zone-centre ferroelectric mode at low temperatures.

At high temperatures the dielectric response follows a Curie–Weiss law suggesting a ferroelectric phase transition at about 35–40 K. At lower temperatures, however, Barrett-like deviations from linearity occur, and the dielectric constant never peaks at atmospheric pressure but continues to rise down to helium temperatures, finally attaining values in excess of 3×10^4 (Lowndes and Rastogi 1973). The quantitative fit to Barrett's relationship (eqn (8.1.1)) is less good for $SrTiO_3$ than $KTaO_3$ and possibly reflects a greater importance of lattice-mode dispersion in the somewhat less stable titanate. Like $KTaO_3$ the stability of the cubic phase is increased by isotropic pressure (Lowndes and Rastogi 1973) but decreased by uniaxial stress (Burke and Pressley 1971). In the latter circumstance a ferroelectric transition is produced at helium temperatures by a [100] uniaxial stress of about $1 \cdot 0 \times 10^8$ Nm^{-2} compared with an equivalent value some five times larger in $KTaO_3$ (the difference reflecting primarily the higher zero-stress susceptibility of $SrTiO_3$).

Because of the existence of at least one zone-boundary structural transition in $SrTiO_3$ its detailed single-domain response is more complicated than $KTaO_3$ and shows anisotropy and hysteresis. Single-domain crystals can be produced by cooling under an applied stress, and the crystals retain single-domain form when the stress is removed until raised again in temperature above the structural Curie point. Sakudo and Unoki (1971) have performed careful single-crystal measurements of dielectric constant and observe a distinct tetragonal anisotropy of response below about 110 K. A further lowering of symmetry to orthorhombic (or lower) was reported below about 65 K, but the existence of this additional transition, at which the [110] and [1$\bar{1}$0] axes become no longer equivalent, is still questioned by some observers. Several other references relevant to a discussion of this question can be found in the paper by Sakudo and Unoki.

The temperature dependence of the zone-centre 'ferroelectric' mode at low temperatures has been observed in $SrTiO_3$ by field-induced Raman spectroscopy (Fleury and Worlock 1968). The energy decreases by more than an order of magnitude as the temperature is lowered from ~ 300 K anu the mode remains underdamped throughout. The Lyddane–Sachs–Teller relationship seems to be well verified not only as regards the temperature dependence but also the absolute value of amplitude as given in eqn (7.1.12) in terms of the frequencies of the other modes. Fleury and Worlock also find the Lyddane–Sachs–Teller relationship to be valid in strong biasing fields so long as the dielectric constants are interpreted as small-signal incremental responses. Away from the zone centre quasi-elastic neutron scattering measurements (Cowley 1964, Yamada and Shirane 1969) reveal a strong anisotropy of dispersion with quasi-dipolar symmetry favouring the cubic axes as incipient ferroelectric axes. As in $KTaO_3$ an interaction between the soft optic mode and the transverse acoustic branch is evident at low temperatures leading to a softening of the acoustic branch.

8.2. Antiferroelectric and cell-doubling perovskites

The óccurrence of ferroelectricity in perovskite crystals has been correlated in the previous section with the existence of a soft polar $q = 0$ zone-centre optic mode of lattic vibration. It therefore seems only natural to anticipate that perovskite antiferroelectrics will arise via the condensation of a soft polar optic phonon with a non-zero value of wavevector q. Unfortunately this has proved at best to be an oversimplification and, although a number of structural transitions are known to occur via the simple condensation of a zone-boundary lattice mode, none of these to date involves polar phonons nor accordingly exhibits any dielectric anomaly. The perovskite antiferroelectrics, like $PbZrO_3$, $PbHfO_3$, and $NaNbO_3$, all seem to possess structures of a more complicated character produced by phase transitions of first order involving several lattice modes.

The idea that crystals could double their primitive unit cells in a continuous way by means of a soft zone-boundary phonon was contained in one of Cochran's early discussions of ADP (Cochran 1961). It was not until 1968, however, that a Raman study established the existence of such behaviour in a real structural transition—the 105 K second-order displacive transition in $SrTiO_3$ (Fleury et al. 1968). Although this transition is not antiferroelectric (nor even antipolar), it does represent an example of the simplest conceivable anti-distortive transition, a cell-doubling transition, and as such has been the object of rather intense experimental study in the last several years. A similar transition involving polar phonons would (if it were found) be the prototype two-sublattice antipolar (and possibly antiferroelectric) transition first envisaged by Kittel (1951) and Cross (1956). For this reason, coupled with the fact that an understanding of the simple cell-doubling transition provides a valuable introduction to the more complicated and less well understood perovskite antiferroelectrics themselves, we discuss first in this section zone-boundary-mode condensation, which has now been observed in several perovskites including $SrTiO_3$, $KMnF_3$, $LaAlO_3$, and $PrAlO_3$.

8.2.1. Zone-boundary lattice-mode condensation

Early calculations by Cowley (1964) of the lattice dynamics of the cubic perovskite structure revealed potentially low-frequency zone-boundary optic phonon modes such as the Γ_{25} mode at the $\mathbf{q} = (111)\pi/a$ or R-point of the Brillouin zone and the M_3 mode at the $\mathbf{q} = (110)\pi/a$ or M-point. These modes, shown in Fig. 8.3, involve only the rotations of oxygen (or fluorine in $KMnF_3$) octahedra around [100] axes and are therefore non-polar. The Γ_{25}-mode at the R-corner of the Brillouin zone is triply degenerate, and phase transitions associated with it can therefore occur in different ways corresponding to the condensation of different linear combinations of degenerate modes. In fact the cubic transitions in $SrTiO_3$ and $KMnF_3$, on the one hand, and of $LaAlO_3$ and $PrAlO_3$, on the other, exemplify different cases.

Using the basic set of modes corresponding to rotations about the three cubic axes (one of which is shown in Fig. 8.3) the Γ_{25} transitions at 105 K in $SrTiO_3$ and 186 K in $KMnF_3$ correspond to the condensation of just one of the Γ_{25}-modes, thereby producing an ordered phase of tetragonal symmetry with two formula units per primitive cell. The octahedral rotation angle φ about the tetragonal [001]-axis then becomes the order parameter for the staggered or anti-distortive transition. More exactly the transition is only approximately described by a rotation since the oxygen ions (or fluorines in $KMnF_3$) actually remain in the faces of each cube as the temperature is lowered below the transition temperature. In $LaAlO_3$ and $PrAlO_3$ the cubic high-temperature structure undergoes a trigonal distortion to a rhombo-hedral structure described by a staggered rotation of oxygen octahedra about a cube diagonal. Again the rotational description is only approximate but it can be conceived as resulting from the condensation of a linear combination of all three (cubic axis) Γ_{25} R-point modes. Again the primitive cell of the lattice is doubled but this time the angular order parameter describes staggered rotations about a [111] axis. A third group-theoretically possible condensation of the Γ_{25} R-point modes, that of a linear combination of two of the modes, would produce an orthorhombic ordered phase, but it has not yet been observed in nature.

Direct observation of a zone-boundary condensation was first made (using inelastic neutron scattering) by Shirane and Yamada (1969) for $SrTiO_3$. In $SrTiO_3$, $LaAlO_3$, and $PrAlO_3$ the Γ_{25}-transition is of second order and the order parameter φ and soft-mode frequency $\omega(R)$ follow an approximate mean-field temperature dependence $\varphi \propto \omega(R) \propto (T_c - T)^{1/2}$ except very close to T_c itself, where critical effects have been reported (e.g. Müller and Berlinger 1971, Kjems et al. 1973; see Chapter 11). Below the transition in both the trigonal and tetragonal cases the degenerate Γ_{25} cubic mode is split into two modes depending upon whether the polarization vector is perpendicular to (doublet) or parallel to (singlet) the rotational axis.

The simplicity of the soft-mode eigenvectors in these transitions, i.e. the fact that only the oxygen ions move, has enabled Pytte and Feder (1969) and Feder and Pytte (1970) to construct a lattice Hamiltonian containing only a small number of temperature independent parameters. Although the formal statistical solution is given in a self-consistent phonon approximation, comparison with experiment for $SrTiO_3$ and $LaAlO_3$ was for simplicity carried through only at the mean-field level. Nevertheless a very encouraging agreement with experiment is obtained, particularly when allowance is

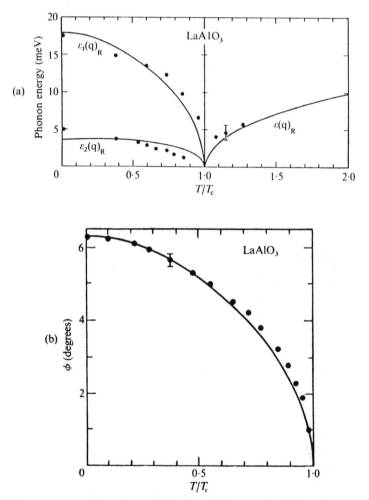

Fig. 8.5. The temperature dependence of (a) soft-mode frequency and (b) order parameter ϕ for the anti-distortive transition in $LaAlO_3$. Solid curves are theoretical (after Feder and Pytte 1970) with experimental data (solid circles) from Müller *et al* (1968) for ϕ and from Axe, Shirane, and Müller (1969) for phonon energy.

made for interactions with strain. The theoretical parameters are greatly overdetermined by the available data on $SrTiO_3$ and to a lesser degree on $LaAlO_3$, and, with the exception of genuine critical effects very close indeed to T_c, the theory is able to account for all aspects of the transition dynamics. The comparison of theory with experiment for order parameter and soft-mode frequencies in $LaAlO_3$ (as given by Feder and Pytte 1970) is shown in Fig. 8.5.

KMnF$_3$ and PrAlO$_3$ are structurally more complex than $SrTiO_3$ and $LaAlO_3$ in that they exhibit additional structural and magnetic phase transitions below their cubic transitions. Firstly, KMnF$_3$ exhibits a cubic–tetragonal transition (at 186 K) which is marginally of first order. Then, as a function of decreasing temperature, a second first-order structural transition takes place at about 91 K with the crystal transforming to a monoclinic structure. This 91 K transition is known to involve the instability of the M_3 phonon (Fig. 8.3) with $q = (110)(\pi/a)$. Finally, slightly below the lower structural transition, at 88 K, an antiferromagnetic spin order is established. The two soft structural modes, Γ_{25} at the R-point and M_3 at the M-point, are very simply related as one can see from Fig. 8.3. The former corresponds to opposite octahedral rotations in successive layers while the latter represents a rotation in the same direction in all layers. If we consider the line connecting the R- and M-points in the Brillouin zone the phase of rotation of adjacent octahedra changes from zero to 180°. Experimentally Gesi *et al.* (1972) find the RM energy branch to be essentially flat at room temperature, decreasing in energy uniformly on cooling, with the R-point mode condensing at 186 K and the M-point mode at 91 K. This implies a negligible coupling between rotational planes and a strong coupling within them for the RM rotations and also accounts for an extremely anisotropic rod-like diffuse X-ray scattering observed by Comes *et al.* (1971). Since the instantaneous correlation of rotational fluctuation is dominated by these low-frequency modes the full system becomes essentially a two-dimensional arrangement of staggered in-plane rotations (Fig. 8.6).

PrAlO$_3$ also exhibits additional phase transitions below the Γ_{25} R-point condensation but of a more complex nature. The cubic–rhombohedral second-order structural transition at 1320 K is, in PrAlO$_3$, followed in turn by a first-order transition to an orthorhombic structure at 205 K and a second-order transition at 151 K which at completion leaves the crystal essentially tetragonal. The characteristic feature of the series of transitions is that upon completion of each the aluminate octahedra are rotated in a staggered sense about [111], [101], and [001] perovskite axes respectively. The Γ_{25} transition is purely lattice dynamical and analogous to the equivalent transition in $LaAlO_3$, but it has been postulated (Harley *et al.* 1973, Birgeneau *et al.* 1974) that the lower two are not pure lattice-dynamical optic mode condensations but are dependent upon the character of the

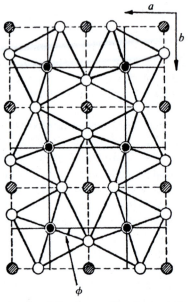

Fig. 8.6. The order parameter ϕ and the staggered in-plane rotational distortion defining the soft mode in $SrTiO_3$ and $KMnF_3$. (After Unoki and Sakudo 1967.)

electronic levels of the orbitally unquenched $(4f^2)$ Pr^{3+} ion. The lower transitions are then the result of a strong coupling, via the electric field gradient at the Pr site, between the electronic states of the Pr^{3+} ion (excitons) and the AlO_6 rotary modes (Γ_{25} R-point phonons). In essence there is a competition between lattice energy, which below the Γ_{25} condensation at 1320 K prefers the [111] rotation axis, and a potential Jahn–Teller splitting of the lowest electronic level (an E_g orbital doublet in the rhombohedral phase) of the Pr^{3+} ion. At high temperatures, when all the crystal-field levels are equally populated, the electronic energy does not play a role and purely lattice-dynamical considerations determine the crystal structure. At lower temperatures, for which only the lowest few electronic levels are populated, the crystal can lower its total free energy by transforming to an orthorhombic structure, gaining electronic energy in excess of the cost in lattice energy. Finally, at 151 K, the crystal begins to transform continuously from orthorhombic to tetragonal in order to effect the largest possible gain in electronic energy. Co-operative Jahn–Teller transitions (Elliot *et al.* 1972, Pytte 1971 b, 1973) have recently attracted much attention, and the 151 K transition in $PrAlO_3$, being of second order, provides a good example of the anticipated critical softening. It involves the coupling of the singlet exciton mode of the $PrAlO_3$ orthorhombic phase with both optic and acoustic phonons of like symmetry. The actual soft mode is the acoustic mode which is pushed down by the softening exciton–phonon mode.

Although the details of these more complex phase transitions are not of direct concern in the context of antiferroelectricity, their presence serves to illustrate the peculiar softness of the perovskite lattice structure with respect to zone-boundary as well as zone-centre transitions. In this respect the Γ_{25} R-point phonon and the M_3 M-point phonon are only two of the more significant candidates for mode softening. In $NaNbO_3$, as discussed in the previous section, at least seven different structural phases exist as a function of temperature with at least six phase transitions, all of which may be essentially pure lattice instabilities. In $NaNbO_3$ the highest-temperature (cubic–tetragonal) transition is produced by the condensation of the M_3 symmetry M-point phonon (Ishida and Honjo 1973) and the second transition by the Γ_{25} R-point condensation (or more accurately by condensation of the E_g doublet component of the singlet and doublet modes into which Γ_{25} splits at the cubic–tetragonal transition). This is the reverse of the situation in $KMnF_3$, but is a sequence which has recently been duplicated in yet another perovskite, $CsPbCl_3$ (Fujii et al. 1974).

Unlike the oxygen-tilting transitions which have no polar characteristics and which are usually second order (or near second order) in a thermodynamic sense, antiferroelectric transitions in the ABO_3 perovskites require the off-centering of the B-cation and usually are markedly first order. The latter displacement would seem to be a relatively independent process and, as discussed in the previous section, can loosely be correlated with the size and valence of the B cation. The relative stability of a ferroelectric or antiferroelectric structure in cases where B-cation off-centering is favoured is presumably largely determined by dipolar energy considerations. Although in principle antiferroelectric transitions have nothing to do with oxygen framework tilts, and simple zone-doubling antiferroelectric transitions to centric ordered phases are conceptually possible, the known perovskite antiferroelectrics are not simple two-sublattice structures and do in general couple their antiferroelectricity with the oxygen-cage rotational motion.

8.2.2. Lead zirconate

The best-known and most frequently studied case of antiferroelectricity in the perovskites is that of $PbZrO_3$. At high temperatures it possesses the prototype cubic perovskite structure but on decrease of temperature it exhibits a marked dielectric anomaly at about 230°C. The high-temperature phase is paraelectric with the dielectric constant following a Curie–Weiss law with Curie constant $C \approx 1 \cdot 6 \times 10^{-5}$ K and a Weiss temperature $T_o \approx 190$°C. Although the general character of this dielectric anomaly appears similar to that in a typical first-order ferroelectric perovskite like $BaTiO_3$, dielectric hysteresis is not generally observed in the low-temperature phase, and X-ray and neutron diffraction experiments

(Sawaguchi, Maniwa, and Hoshino 1951, Jona *et al.* 1957) have established that the room-temperature structure is antiferroelectric.

Early experiments on ceramic samples of $PbZrO_3$ indicated that the phase transition at 230°C was the only structural transition to occur as a function of temperature in zero field, but that application of an external electric field below T_c induced a transition to a rhombohedral ferroelectric phase (probably isomorphous with the rhombohedral ferroelectric phase of $BaTiO_3$). This implied that the antiferroelectric and ferroelectric phases were very closely equal in free energy, conforming nicely with the classic picture of antiferroelectricity. However, although the antiferroelectric nature of the room-temperature phase seems well founded, recent work on high-purity ceramic and single-crystal specimens has questioned the simple paraelectric–antiferroelectric character of the transition at 230°C (Tennery 1966, Goulpeau 1966, Scott and Burns 1972 a). It now seems possible that the rhombohedral ferroelectric phase may actually be stable in zero field within a narrow intermediate temperature range between the antiferroelectric and paraelectric phases. The interposed ferroelectric phase has been observed to be stable over a temperature range of order 10–25 degC below $T_c \approx 230°C$ (Scott and Burns 1972 a), but the range of stability depends very sensitively upon stoichiometry and growth conditions of the samples so that the situation for pure stoichiometric single-crystal $PbZrO_3$ remains uncertain. The existence of the intermediate ferroelectric phase is well established in certain solid solutions containing $PbZrO_3$ (e.g. $PbZr_{1-x}Ti_xO_4$, see Jona and Shirane 1962) but its temperature range of stability appears to extrapolate very close to zero as $x \to 0$.

In spite of this uncertainty concerning the form of the transition or transitions, the antiferroelectric structure at room temperature seems well established to an accuracy of order 1 pm (Jona *et al.* 1957). The structure is orthorhombic (Fig. 8.7), the unit cell containing eight formula units, and is derived from the cubic perovskite form predominantly by antiparallel displacements of the lead ions along one of the original [110] directions, which we now call the a-axis. This primary distortion is shown schematically in Fig. 8.8. In addition, however, smaller distortions are present, the oxygen atoms being displaced antiparallel to each other in the (001) c-plane (Fig. 8.7) and an uncompensated resultant oxygen displacement existing along the [001] or c-axis. As a consequence this c-axis is polar, although it has not yet proved possible to switch the associated polarization and establish ferroelectricity in this direction. If such could be accomplished the term ferrielectric (i.e. possessing both ferroelectric and antiferroelectric characteristics) might seem appropriate in loose analogy with ferrimagnetism.

The complicated perovskite distortion depicted in Fig. 8.7 is easier to comprehend when the associated displacements from cubic form are resolved in terms of normal-mode displacements. Each such set of modes is

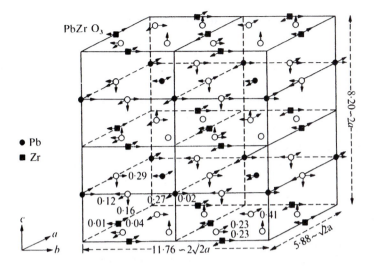

Fig. 8.7. The crystal structure of PbZrO₃ at room temperature, with distances given in ångströms (=0·1 nm). (After Cochran and Zia 1968.)

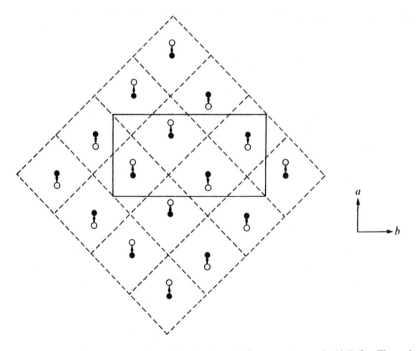

Fig. 8.8. A schematic representation of the basic antipolar arrangement in PbZrO₃. The unit cell is outlined.

distinguished by its wavevector $\mathbf{q}$ and the symmetry of the displacements. The detailed nature of the symmetry restrictions and the conventional symbols used to denote them have been tabulated for the perovskite system by Cowley (1964). Cochran and Zia (1968) find that the room-temperature structure of Fig. 8.7 is made up of mode displacements of four kinds, a Γ_{25} R-point oxygen rotation (familiar from the last section), a zone-centre (Γ_{15}) polar mode, a $(\frac{1}{2}\frac{1}{2}0)(\pi/a)\Sigma_3$ mode, and a $(110)(\pi/a)M'_5$ mode. The Σ_3 mode contains the large a-axis Pb displacements of Fig. 8.8 and defines the basic antiferroelectric character of the phase, while the Γ_{15} component describes the polar displacement in the c-direction. The zone-centre Γ_{15} representation is triply degenerate (transforming like a normal vector), and these modes are just those responsible for ferroelectricity in the conventional perovskite ferroelectrics like $BaTiO_3$.

The transition from cubic perovskite to antiferroelectricity in $PbZrO_3$ therefore involves at least four modes. Since, from the Landau theory of phase transitions, a second-order transition can only involve the condensation of a single mode, it follows that this transition is necessarily of first order (Haas 1965). This is in accord with experimental observations for those $PbZrO_3$ samples which do not have the interposed ferroelectric phase complication. In fact, the $PbZrO_3$ transition may involve more than four modes, since Miller and Kwok (1967) have proved an interesting theorem which enables the prediction of all allowed wavevectors which can contribute to a complicated phase transition. They show that the group of wavevectors which occurs must be one of certain subgroups of the group G, the latter being defined as those reciprocal lattice vectors of the lower-symmetry structure (e.g. the low-temperature zirconate phase) which occur in the first Brillouin zone of the higher-symmetry structure (cubic perovskite). In fact for $PbZrO_3$ the theorem predicts two more modes than are actually found in Cochran and Zia's analysis of the actual structure. The apparent absence of some modes means that the corresponding frequencies are too high and/or the corresponding anharmonic coefficients too small for the associated displacements to be detectable.

No detailed soft-mode measurements have yet been reported for $PbZrO_3$, nor indeed for any other perovskite antiferroelectric. Even near first-order transitions some modes may well exhibit significant temperature anomalies and we can hope to single these out as primarily responsible for (or driving) the transition. In $PbZrO_3$ the dielectric anomaly points to the fact that a zone-centre mode is certainly soft. However, without further information it is difficult to predict the temperature dependence of the other modes or to say whether one bears a larger temperature dependence than the others. Conceptually it is simplest to suppose that only two modes $\omega_1(0)$ and $\omega_2(\mathbf{q})$ are essentially involved, and that $\omega_1^2(0) \propto T - T_0$ and $\omega_2^2(\mathbf{q}) \propto T - T_c$ (where T_c is the transition temperature and T_0 the Curie–Weiss temperature) expresses the approximate temperature dependence of their frequencies, but it seems doubtful whether such a simple picture can represent the situation for $PbZrO_3$ very precisely.

There are some indications that significant mode softening may take place in $PbZrO_3$ at places other than the zone centre, but these are rather indirect. For example, Jain, Shringi, and Sharma (1970) observe a particularly large anomaly in the Mössbauer fraction near T_c. This conforms qualitatively with expectations (Dvorak 1966) for a system in which mode softening takes place over a particularly large fraction of the Brillouin zone and may be suggestive of (though certainly not direct evidence for) the presence of more than one soft mode (see § 12.4). Pressure-dependent measurements by Samara (1970 b) are also relevant. On increase of pressure P the temperature derivative dT_c/dP is observed to be positive while dT_0/dP is negative. Interpreted in terms of lattice dynamics this establishes that the mode or modes responsible for the antiferroelectric transition at T_c are quite different from, and relatively independent of, those zone-centre soft modes responsible for the dielectric anomaly. One further interesting feature observed by Samara is that the $PbZrO_3$ samples exhibiting the narrow intermediate ferroelectric phase at atmospheric pressure quickly have it 'squeezed out' as pressure is increased. As a result all Samara's samples could be made to exhibit a clear first-order paraelectric–antiferroelectric phase transition above 1 kbar.

A very large amount of work has been performed on solid solutions containing $PbZrO_3$. Of particular interest are the $PbZr_{1-x}Ti_xO_3$ ceramics which exhibit strong electromechanical activity and are accordingly of some technical importance (see § 15.2.3). We shall only touch upon their properties here to the extent that they teach us anything about the $x \to 0$ limit. For $0 \leqslant x \leqslant 0.42$ the paraelectric phase transforms first to a rhombohedral ferroelectric phase FE_1 and then to either the zirconate antiferroelectric structure (if $x \lesssim 0.06$) or to a second rhombohedral ferroelectric phase FE_2. The complete phase diagram is given in Fig. 15.8. The ferroelectric phases FE_1 and FE_2 have been determined by X-ray and neutron diffraction (Michel *et al.* 1969) and, while the lower-temperature phase FE_2 contains a Γ_{25} oxygen tilting in addition to Γ_{15} parallel displacements of the cations along [111] directions, the FE_1 phase appears to have no oxygen tilting, leaving only the [111] cation displacements relative to the oxygen framework. This points to a condensation of a zone-centre Γ_{15} mode at the paraelectric–FE_1 transition and to the simple condensation of an R-point Γ_{25} mode at the order–order FE_1–FE_2 transition. With a single mode condensing at each phase change the associated transitions could be (but are not necessarily) second order. Recent measurements by Clarke and Glazer (1974) of the strain anomaly at the paraelectric–FE_1 transition suggest that it is indeed of second order. If confirmed this will represent the first example of a second-order ferroelectric transition found in the perovskites. From the point of view of understanding $PbZrO_3$, however, the FE_1–FE_2 transition is of more relevance. It suggests that the R-point Γ_{25} mode is soft in the FE_1 phase and, since FE_1 is the marginally stable phase interposed between the paraelectric and antiferroelectric phases near the $x = 0$ zirconate limit, this in turn suggests that this same Γ_{25} mode plays a very significant (and possibly driving) role in the antiferroelectric transition despite the fact that it is not itself antiferroelectric.

Other antiferroelectric perovskites are known, such as $PbHfO_3$ (Shirane and Pepinsky 1953), which has two antiferroelectric phases and a third at high pressures (Samara 1970 b), and $NaNbO_3$, which has two antiferroelectric phases, one ferroelectric phase, and several other structural phases (see § 8.1.5). To this point, however, rather little work of direct relevance to antiferroelectricity has been published on either, and the structural details, except for the room-temperature phase of $NaNbO_3$, are not precisely known.

8.3. Lithium tantalate and lithium niobate

Lithium niobate ($LiNbO_3$) and lithium tantalate ($LiTaO_3$) have in recent years been the object of intensive study because of their large optic non-linearities and corresponding device potential in the field of electro-optic modulators, parametric oscillators, harmonic generators, etc. Both compounds are ferroelectric at room temperature and may be grown in the form of large optic-quality single crystals. They are uniaxial at all temperatures, with only a single structural phase transition (paraelectric–ferroelectric), and exhibit second-order (or close to second order) phase transitions. On the debit side, however, as regards obtaining an understanding of the ferroelectric mechanism in these isomorphic systems, is the fact that their Curie temperatures are very high (620°C for $LiTaO_3$ and ≈ 1200°C for $LiNbO_3$) placing the region of most intense structural activity in an inconvenient temperature range, particularly for the niobate which melts only about 50 degC above its very elevated Curie point. Also, with two formula units per primitive cell (10 atoms), and consequently 27 optic modes of lattice vibration, the detailed lattice-dynamical picture is more complex than for the simple perovskites.

The important simplification which makes $LiTaO_3$ and $LiNbO_3$ attractive subjects for analysis is that the ferroelectric soft mode is non-degenerate both above and below T_c, so that, unlike the situation in the cubic perovskites, the displacement Hamiltonian (eqn (1.2.8)) can be used with a simple scalar variable ξ_l to describe the basic features of the ferroelectric transition and to assess the nature of the local potential function $V(\xi_l)$ in each. Thus, to the extent that the soft mode dominates the ferroelectricity, we expect $LiNbO_3$ and $LiTaO_3$ to be examples of the simplest possible ferroelectric transition, viz. the condensation of a single-polar non-degenerate mode. We do not know in advance, of course, the extent to which this mode might be simple in a quasi-harmonic sense.

Although $LiNbO_3$ and $LiTaO_3$ do not have the perovskite structure they are ABO_3 lattices with oxygen octahedra. Fig. 8.9 shows the basic room-temperature structure consisting of a sequence of distorted oxygen

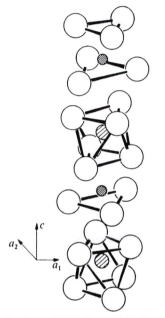

Fig. 8.9. Room-temperature structure of $LiNbO_3$ and $LiTaO_3$. The smallest hatched circles represent Li ions, the intermediate hatched circles Nb or Ta, and the large open circles O ions. The polar axis is the c-axis. (After Abrahams *et al.* 1973.)

octahedra joined by their faces along a trigonal polar c-axis. Within the oxygen cages, as one progresses along the polar axis, the cations appear in the sequence Nb(Ta), vacancy, Li, Nb(Ta), vacancy, Li, ..., and the room-temperature positions of the cations within the oxygen cages are known precisely from X-ray and neutron structure determinations (Abrahams, Reddy, and Bernstein 1966, Abrahams, Hamilton, and Reddy 1966, Abrahams and Bernstein 1967, Abrahams, Hamilton, and Sequeira 1967). Unlike the perovskite ferroelectrics the relative offset of cations from possible non-polar positions is quite large (implying a large room-temperature spontaneous polarization which is indeed observed, $\approx 50 \,\mu C \, cm^{-2}$ in the tantalate and $\approx 70 \,\mu C \, cm^{-2}$ in the niobate), and the form of the high-temperature paraelectric phase is not immediately apparent from the room-temperature observations. The earliest attempt at high-temperature structural determination was a polycrystalline X-ray diffraction study of $LiNbO_3$ (Abrahams, Levinstein, and Reddy 1966) and was only partly successful, since no reliable results could be obtained at or above the Curie point ($T_c = 1210°C$ for the sample under study) owing to loss of oxygen from the sample and a tendency of the $LiNbO_3$ to react with the platinum holder at these elevated temperatures. Little or no change in the room-temperature co-ordinates was observed up to 1000°C but, although the Li position could not be resolved at the higher temperatures, there was a definite tendency for the Nb to approach its oxygen-cage centre on approaching the phase-transition region $T \approx 1200°C$. This strongly suggests a centrosymmetric non-polar phase in which the Li atoms move into their nearest oxygen planes and the Nb atoms are midway between adjacent oxygen planes.

At room temperature for $LiNbO_3$ the Nb c-coordinate is 0·09 nm from its nearest oxygen layer and 0·141 nm from its next-nearest oxygen layer, and the corresponding values for Li are 0·071 nm and 0·160 nm. This implies therefore that (using a centre-of-mass co-ordinate representation) the magnitudes of the respective shifts between T_c and room temperature in $LiNbO_3$ are $x(Li) = 0·051$ nm, $x(O) = -0·02$ nm, and $x(Nb) = 0·006$ nm. In this same picture the equivalent shifts for $LiTaO_3$ are $x(Li) = 0·043$ nm, $x(O) = -0·017$ nm, and $x(Ta) = 0·003$ nm. In spite of the less than definitive study of the structural change at T_c in $LiNbO_3$ the equivalent study for $LiTaO_3$ (which with a lower Curie point and higher melting point than $LiNbO_3$ promised more favourable experimental circumstances) was not performed for a number of years. Finally, however, Abrahams $et al.$ (1973) reported the results of a neutron scattering study of single-crystal $LiTaO_3$ both below and above the Curie point. The results held a surprise since, although the Ta ion did seem to move to the centre of its oxygen cage on approaching the Curie point, the Li ion did not enter its centrosymmetric oxygen plane but took positions ±0·037 nm on either side of this plane with equal probability.

Dielectric, thermal, and pyroelectric measurements on $LiNbO_3$ and $LiTaO_3$ establish that the phase transitions are of second, or very close to second, order. The data are particularly convincing for $LiTaO_3$ (Glass 1968) where the familiar second-order classical relationships for spontaneous polarization $P_s \propto (T_c - T)^{1/2}$, specific heat, and dielectric constant $\varepsilon \propto |T - T_c|^{-1}$ were observed close to the Curie point (see Figs 5.5 and 5.15). All dielectric and thermal properties close to T_c were adequately accounted for in terms of a simple thermodynamic theory with Gibbs free energy expanded as a power series in polarization up to terms of sixth order. On the other hand, such a simple expansion with temperature-independent fourth- and sixth-order coefficients was not capable of describing the experimental data over a wider temperature range.

The observed Curie constants for $LiNbO_3$ and $LiTaO_3$ are of order 10^5 K and are typical for systems of broadly displacive character (order–disorder ferroelectrics such as TGS, $NaNO_2$, and KDP have Curie constants up to two orders of magnitude smaller). In line with this picture direct measurements of long-wavelength lattice-mode frequencies by infrared reflection and by Raman techniques (Barker and Loudon 1967, Johnston and Kaminow 1968, Barker, Ballman, and Ditzenberger 1970) clearly show that, of the many observable optic modes, only one A_1 symmetry mode in each system is strongly temperature dependent and seems to be condensing as the Curie point is approached. This A_1 mode describes ionic motion along the polar axis (although individual oxygen ions may possess a transverse motion which averages to zero over the unit cell) and its frequency appears to follow the classic temperature dependence $\omega^2 \propto T_c - T$ over a wide temperature range on approach to the ferroelectric transition from below. Since the soft mode crosses many approximately temperature-independent modes of E-symmetry (motion perpendicular to the polar axis) as its frequency decreases, it has not yet proved possible to follow it in detail or to study its precise frequency spectrum on very close approach to the critical point. In fact the Raman work by Johnston and Kaminow (1968) on both materials was many times too low in resolution to resolve the Brillouin intensities for example (which are expected to increase near T_c as the result of piezoelectric coupling between optic and acoustic phonons) from the soft-optic-mode frequency spectrum in the critical region. They did, however, observe a maximum in the quasi-elastic scattering intensity within their resolution limitations at the Curie point. These early studies of lattice dynamics in $LiNbO_3$ and $LiTaO_3$ have been followed over the years by more detailed and accurate experiments. Directional polariton-dispersion measurements have revealed that some of the original line assignments were incorrect (Claus *et al.* 1972, Winter and Claus 1972, Posledovich *et al.* 1973) but most of the $4A_1 + 9E$ optically active $\mathbf{q} \to 0$ modes have now been located, at least in $LiNbO_3$, with reasonable assurance at room temperature. Even some information concerning three of the five optically inactive A_2 symmetry modes has recently been obtained for $LiNbO_3$ from inelastic neutron scattering experiments (Chowdhury *et al.* 1974), but all these measurements have so far been confined to room temperature so that little additional information concerning critical dynamics in the $LiNbO_3$ and $LiTaO_3$ structures has been forthcoming to date.

8.3.1. 'Local-mode' analysis

Some clues as to the possible degree of anharmonicity present in $LiNbO_3$ and $LiTaO_3$ can be found from a simple theoretical analysis of the existing static and dynamic experimental data. The comparative wealth of data (e.g. bulk properties such as the temperature dependence of spontaneous polarization, dielectric constant, and specific heat, dynamic information such as soft-mode strength and frequency, detailed structural information, etc.) coupled with the possibility of a theoretical description in terms of a single scalar soft-mode variable presents an excellent opportunity for testing the internal self-consistency of the basic displacement Hamiltonian formulation (eqn (1.2.8)). Expanding the local potential function $V(\xi_l)$ as a power series in even orders of the displacement variable ξ_l to sixth order

$$V(\xi_l) = \tfrac{1}{2}\Omega_0^2 \xi_l^2 + A\xi_l^4 + B\xi_l^6 \qquad (8.3.1)$$

Lines (1969 a, b, 1970 a, b) has shown that the problem is indeed comfortably overdetermined for both $LiNbO_3$ and $LiTaO_3$ and that self-consistency, using completely temperature-independent microscopic parameters, is surprisingly good even over a fairly wide temperature range. In a correlated effective-field statistical approximation the effective lth-cell Hamiltonian near T_c becomes (Lines 1972 b)

$$\mathcal{H}(l) = \tfrac{1}{2}\pi_l^2 + \tfrac{1}{2}\omega_0^2 \xi_l^2 + A\xi_l^4 + B\xi_l^6 - C\xi_l\langle \xi_l \rangle \qquad (8.3.2)$$

where π_l is the momentum operator conjugate to the scalar displacement amplitude ξ_l (see § 1.2). In a simple mean-field theory one could write $\omega_0^2 = \Omega_0^2$ and assume the intercell forces to be entirely included in the coefficient C. Allowing for intercell correlations (§ 2.2) the primarily dipolar intercell forces contribute both to the linear and quadratic terms in eqn (8.3.2). The important findings for both $LiTaO_3$ and $LiNbO_3$ are that $\omega_0^2 \ll \Omega_0^2$ (correlation effects are extremely important) and that the quartic anharmonicity coefficient A is negative. In fact for $LiTaO_3$, using the overall 'best-fit' values for the microscopic parameters ω_0, A, and B, one finds an effective potential $\tfrac{1}{2}\omega_0^2 \xi_l^2 + A\xi_l^4 + B\xi_l^6$ governing local ion motion in the paramagnetic phase ($\langle \xi_l \rangle = 0$) which has local side minima at $\xi_l \approx \pm 0.25$ amu$^{1/2}$ nm in addition to the central minimum at $\xi_l = 0$. These minima may be quite shallow but, within the accuracy to which the relevant microscopic parameters have been determined, could be as deep as the central minimum itself. In terms of the probability distribution of ionic displacement this implies, in particular, side maxima for the Li ions at a distance $\approx \pm 0.04$ nm from the centrosymmetric position and is in general accord with the observation of Abrahams et al. (1973) of a 'split lithium' above the Curie point at positions ± 0.037 nm from the centrosymmetric site. Within a single-mode picture side maxima in distribution probability are also predicted for the heavy ions but the splittings involved are much smaller. The theoretical picture is one of extremely anharmonic motion along the c-axis (the implied potential barrier between central and side minima is probably less than $\sim \tfrac{1}{2}kT_c$) with the largest amplitude at the Li site. It is very unlikely that such a local motion could give anything other than extremely anharmonic long-wavelength critical dynamics, and Penna,

Chaves, and Porto (1976) have now confirmed that the soft mode in $LiTaO_3$ is indeed of diffusive character. Some measurements of ^7Li n.m.r. through the transition region have also been made (e.g. Slotfeldt-Ellingsen and Pedersen 1974) but no reliable interpretation has yet been given.

The situation for $LiNbO_3$ seems to be less extreme than for $LiTaO_3$ and the 'best-fit' microscopic parameters for this case do not give rise to a triple-minimum local potential. Nevertheless the anharmonicity is still severe and again grossly anharmonic critical dynamics seems to be likely. We note in passing that the negative quartic coefficient A in eqn (8.3.2) for $LiTaO_3$ and $LiNbO_3$ does not imply a first-order phase transition (as would an equivalent term in a free-energy expansion in powers of polarization). The corresponding free-energy expansion can in principle be obtained by direct integration from eqn (8.3.2) and its coefficients are in general temperature dependent (i.e. temperature-independent microscopic parameters lead to temperature-dependent free-energy parameters). For this reason a microscopic expansion like eqn (8.3.1) is able to perform acceptably over a much wider temperature range than would an equivalent free-energy expansion with temperature-independent quartic and sixth-order coefficients.

Although the single-soft-mode model appears to work well for $LiTaO_3$ and $LiNbO_3$, one must expect other modes to contribute significantly at points far from the Curie point. Obviously all response measurements perpendicular to the polar axis are dominated by the E modes so that a single-soft-mode analysis is at best restricted to polar-axis response. Even here, however, one might hope to detect non-soft-mode contributions to dielectric response at, for example, very low temperatures. One promising place to look is at the temperature dependence of spontaneous polarization. Since the soft-mode frequency rises to ≈ 220 cm^{-1} ($LiTaO_3$) and ≈ 250 cm^{-1} ($LiNbO_3$) as $T \to 0$ (Johnston and Kaminow 1968), contributions to the temperature derivative dP_s/dT of spontaneous polarization from this mode might be expected to contain respectively a Boltzman factor $\exp(-320/T)$ or $\exp(-360/T)$, with T in degrees K, and to be negligible compared with (for example) power-law acoustic mode effects at low enough temperatures.

8.3.2. Low-temperature polarization

Accurate measurements of spontaneous polarization for ferroelectrics or indeed for pyroelectrics have been very few at low temperatures. Early observations were conflicting and attempts to fit low-temperature data to a simple power law in temperature T varied in their findings from a T^2 law to a T^4 law. The theoretical situation was equally confused with some calculations carried out solely within a Debye configuration (Born 1945) and others entirely within an Einstein anharmonic-oscillator framework (Garrett 1968). Some light has recently been shed on the problem as the result of some careful measurements on $LiTaO_3$ and $LiNbO_3$ by Glass and Lines (1976). Observations from above room temperature down to 5 K were made using the dynamic technique of Chynoweth (1956 a) which records the current response of the crystals to absorbed infrared radiation. The method (see § 5.4) measures the ratio of pyroelectric coefficient dP_s/dT to thermal capacity $\mathscr{C}$, and proves to be a far more accurate relative measure of the low-temperature variation of spontaneous polarization than conventional static techniques such as those involving direct charge integration. Results for $LiTaO_3$ are shown in Fig. 8.10 and have been performed at frequencies

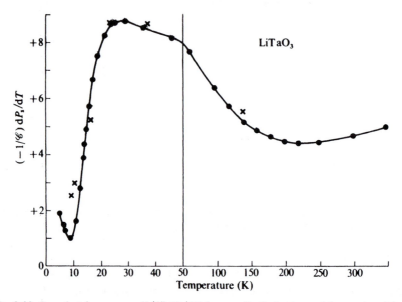

Fig. 8.10. Pyroelectric response $(1/\mathscr{C})dP_s/dT$ for an effectively clamped (crosses) and free (filled circles) crystal of $LiTaO_3$. Only relative measurements are significant although an approximate calibration of the ordinate can be made from the independently measured room-temperature values $dP_s/dT \approx -18 \times 10^{-3} \mu C\,cm^{-2}\,K^{-1}$ and $\mathscr{C} \approx 24\,cal\,mol^{-1}\,K^{-1}$. (After Glass and Lines 1976.)

both below (free crystal) and above (clamped crystal) the fundamental mechanical resonances of the crystal. Results for $LiNbO_3$ are qualitatively similar with a response peak near 30 K and a rounded minimum near 200 K. At low temperatures, for which the heat capacity $\mathscr{C}$ follows a T^3 law, a constant value of response $(1/\mathscr{C})\,dP_s/dT$ therefore corresponds to a T^4 variation of polarization deviation $\Delta P_s = P_s(0) - P_s(T)$. It is immediately evident that no such simple power law explains the observations.

Theoretically Glass and Lines have used the self-consistent phonon theory of § 2.3 to determine the temperature dependence of soft-mode displacement $\langle \xi \rangle$ at very low temperatures. They find a soft-mode contribution to polarization deviation of the form

$$\langle \xi \rangle_{T=0} - \langle \xi \rangle_T \propto \left\langle \{\Omega_q(T)\}^{-1} \exp\left\{-\frac{\hbar\Omega_q(T)}{kT}\right\}\right\rangle_q \qquad (8.3.3)$$

where $\Omega_q(T)$ is the frequency of the mode on the soft branch with wavevector q at temperature T and $\langle \ldots \rangle_q$ is an average over the first Brillouin zone. At low temperatures the T dependence of the soft-mode dispersion is negligible and we can replace $\Omega_q(T)$ by $\Omega_q(0)$ without serious error in eqn (8.3.3). Assuming a mode dispersion of quadratic form

$$\Omega_q(0) = \Omega_{min} + \alpha_x q_x^2 + \alpha_y q_y^2 + \alpha_z q_z^2 \qquad (8.3.4)$$

near the bottom of the band at $q \approx q_{min}$, the form (8.3.3) can be integrated analytically after differentiation with respect to T (neglecting zone-boundary effects) to give a soft-mode contribution

$$\frac{dP_s}{dT} = Kt^{-1/2} \exp\left(-\frac{1}{t}\right) \qquad (8.3.5)$$

in which $t = kT/\hbar\Omega_{min}$. The exponential term is the anticipated Boltzman factor. However, since the energy minimum does not necessarily fall at $q = 0$, Ω_{min} is not necessarily equal to the Raman or infrared measured 'soft-mode' frequency. The $t^{-1/2}$ factor is dependent on the mode dispersion and would, for example, become t^{-2} if mode dispersion were negligible.

At temperatures low enough for a T^3 specific heat law to be valid we therefore anticipate a soft-mode contribution to dynamic pyroelectric response $(1/\mathscr{C}) \, dP_s/dT$ of the form $Kt^{-7/2} \exp(-1/t)$. The fit of this expression to the data for both $LiNbO_3$ and $LiTaO_3$ is shown in Fig. 8.11. The fit is quantitative below about 30 K and indicates values $\hbar\Omega_{min} \approx 63$ cm^{-1} for $LiTaO_3$ and ≈ 67 cm^{-1} for $LiNbO_3$. These energies seem very low to be associated with the low-temperature soft A_1 mode (no A_1 energies below about 230 cm^{-1} were observed for c–axis dispersion in $LiNbO_3$ by Chowdhury et al. (1974), which are the only $q \neq 0$ data yet available) and we must perhaps not exclude other possible explanations. For example, the E mode of local-mode theory does not in fact retain its E-symmetry group-theoretical character at short wavelengths (except for q parallel to the c-axis) in a more rigorous lattice dynamical description (Devine and Peckham 1971). Also Johnston and Kaminow (1968) find evidence that even at long wavelengths crystal imperfections are sufficient to cause a breakdown of E-mode selection rules. It is therefore conceivable that a formal E mode of local-mode theory might be able to influence polarization in a dynamic fashion (i.e. in a manner which does not involve piezoelectric coupling to strain) and that the optic mode contribution at 60–70 cm^{-1} seen so clearly in

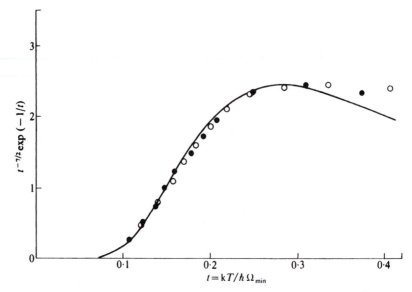

Fig. 8.11. Fit of theory eqn. (8.3.5) and experiment for pyroelectric response $\mathscr{C}^{-1}\,dP_s/dT$ at low temperatures in LiTaO$_3$ (open circles) and LiNbO$_3$ (closed circles) where energy gap $\hbar\Omega_{min} \approx 63$ cm^{-1} for the former and ≈ 67 cm^{-1} for the latter. (After Glass and Lines 1976.)

the pyroelectric data might correspondingly arise through a thermal population of an E-mode branch. Until complete dispersion data become available the question must remain open.†

Finally, since a shift of the baseline in Fig. 8.11 is equivalent to the presence of a T^4 term in polarization deviation and all our data (both clamped and unclamped) have been normalized in amplitude at the low-temperature response peak, it is not possible to exclude the existence of small T^4 contributions to unclamped polarization from the data in Fig. 8.11. In fact such terms are expected to be present theoretically and arise from the thermal population of acoustic modes and a piezoelectric coupling to the spontaneous polarization. There is in fact some evidence for their existence at very low temperatures in LiNbO$_3$ (Glass and Lines 1976), but above about 10 K they are dominated by the lowest optic mode contribution of eqn (8.3.5). Since experimental difficulties increase rapidly at very low temperatures it has not yet proved possible to carry out accurate measurements in a region where the acoustic effect dominates. In addition, in regions where piezoelectric and thermal expansion effects are dominant very careful consideration must be given to the effective crystal constraints and to the relationship between the measured quantity (involving flow of charge from a

† A recent light-scattering study of LiTaO$_3$ by Penna, Chaves, Andrade, and Porto (1976) finds the lowest E mode at 69 cm^{-1}

surface) and the polarization, which by definition is the electric moment per unit (i.e. temperature-independent) volume.

It has also been suggested theoretically (Born 1945) that acoustic modes can contribute directly even to clamped pyroelectricity via second-order terms in an expansion of polarization in phonon variables. This term contributes as T^2 to polarization deviations ΔP_s (but see Szigeti (1975) who disputes this finding) and therefore should enter the $(1/\mathscr{C}) \, dP_s/dT$ plot of Fig. 8.10 as a T^{-2} contribution. It is possible that the very-low-temperature data ($T < 10$ K) seen in Fig. 8.10, but omitted from the theoretical fit in Fig. 8.11, could be indicative of this effect, although such an identification is speculatory at the present time.

8.3.3. Stoichiometry

Detailed characterization of the physical properties of both $LiNbO_3$ and $LiTaO_3$ is complicated by crystal-chemistry considerations. Early optical and dielectric studies of $LiNbO_3$ (Fay, Alford, and Dess 1968, Bergman et al. 1968) and $LiTaO_3$ (Ballman et al. 1967) indicated that the properties of these crystals, such as the Curie temperature and the optical birefringence, depend on the melt composition from which they are grown. Subsequent crystal-growth and X-ray diffraction studies of the phase diagram of the $Li_2O–Nb_2O_5$ system by Lerner, Legras, and Dumas (1968) showed that this dependence results from the existence of solid solutions which incorporate both excess Li_2O and Nb_2O_5. They established that the congruent melting point of $LiNbO_3$ occurs at 48·6 mol per cent Li_2O with the result that crystals grown from a melt of stoichiometric composition deviate significantly from stoichiometry. Confirmation of these results by Carruthers et al. (1971) led to the high-temperature phase diagram shown in Fig. 8.12. To briefly explain this diagram, consider a crystal grown from a stoichiometric melt at point A on the liquidus. The first composition to solidify is the composition at point B on the solidus at about 49 mol per cent Li_2O, leaving the melt richer in Li_2O. Eventually, when the melt composition is about 58 mol per cent rich in Li_2O a crystal composition close to stoichiometric can be obtained, but the large difference between the melt and solid compositions at this point leads to great difficulty in obtaining homogeneous crystals. On the other hand, crystals of uniform composition and high optical quality can be obtained from a melt of 48·6 mol per cent Li_2O since both the crystal and melt have the same composition at all times during growth.

A further complication arises at lower temperatures where the limits of solid solubility decrease. At 1000°C it is seen from Fig. 8.12 that the solubility limits are 46 per cent to a little over 50 per cent. Each side of these limits a new phase is precipitated. At lower temperatures the width of this solubility range decreases, the studies of Scott and Burns (1972 b) indicating that the solubility range may only be about $\frac{1}{2}$ per cent each side of the

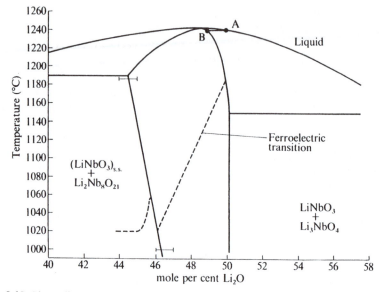

Fig. 8.12. Phase diagram of the Li_2O–Nb_2O_5 system (for explanation see the text). (After Carruthers *et al.* 1971.)

stoichiometric composition for $T \leqslant 600°$ C. Since it was necessary to anneal the crystals for three days to establish equilibrium at 600°C, it is relatively easy to quench a non-equilibrium solid solution such as the congruent melting composition from the higher temperature. Nevertheless, any extended low-temperature heat treatment of non-stoichiometric crystals outside the solubility limits at lower temperatures could result in precipitation and reduced optical quality.

In $LiTaO_3$ the situation appears to be similar, but a complete phase diagram of the Li_2O–Ta_2O_5 has not yet been reported because of the higher melting point and the serious loss of Li_2O from $LiTaO_3$ above 1200°C. However, it has been established that the congruent melting composition is approximately 49 mol per cent Li_2O and the range of solid solubility extends from about 46 per cent to 50·4 per cent Li_2O at 1200°C (Barns and Carruthers 1970). Qualitatively, the ability of the $LiTaO_3$ and $LiNbO_3$ structures to accommodate a fairly large excess of Ta/Nb ions but only a small excess of Li ions is a result of the large Ta—O and Nb—O bond strengths compared with the weaker Li—O bonds.

It is appropriate at this point to discuss briefly some of the methods by which the phase diagrams of these systems have been established, since the techniques reflect both the effect of stoichiometry variations on the physical properties of the material as well as the usefulness of the physical properties in studying crystal chemistry. The usual approach is to measure first the properties of carefully prepared ceramics of known composition and then to use these data as a calibration with which to determine the composition of single crystals grown from various melt compositions. In $LiNbO_3$ and $LiTaO_3$ several parameters are relatively sensitive to the crystal composition so that each of these can be used independently. A particularly convenient calibration is the ferroelectric Curie temperature itself, since this is readily measured in both ceramic and single-crystalline states. For instance a range of Curie temperatures from about 1020 to 1185°C is measured in $LiNbO_3$ as the solid composition is changed from 46 to 50 mol per cent Li_2O as shown by the broken line in Fig. 8.12. In $LiTaO_3$ the range of Curie temperature is from 510 to 690°C over a similar range. This fractional variation of T_c is very much greater than that of the lattice parameters or crystal density with stoichiometry, as shown in Table 8.1, and allows much more precise determination of the composition than conventional X-ray or dilatometric techniques.

TABLE 8.1

(a)	T_c	Δn	a	c	ρ
$LiNbO_3$	1190[a]	−0·105[a]	5·1478[c]	13·867[c]	4·652[c]
$LiTaO_3$	675[b]	−0·002[a]	5·1517[b]	13·778[b]	7·428[b]
(b)	$\partial T_c/\partial C$	$\partial(\Delta n)/\partial C$	$\partial a/\partial C$	$\partial c/\partial C$	$\partial \rho/\partial C$
$LiNbO_3$	+43[a]	−0·016[a]	−0·0013[c]	−0·005[c]	−0·0033[c]
$LiTaO_3$	+80[b]	−0·01[a]	−0·004[b]	−0·01[b]	−0·030[b]

(a) The Curie temperature T_c(°C), optical birefringence $\Delta n = n_e - n_0$, lattice parameters a, c (Å), and the measure density ρ (g cm^{-3}) for stoichiometric $LiNbO_3$ and $LiTaO_3$. For stoichiometric $LiNbO_3$, T_c is above the melting point and the value listed is extrapolated from non-stoichiometric compositions; ρ is also extrapolated.
(b) The variation of T_c, Δn, a, c, and ρ with crystal composition $C = Li_2O/(Li_2O + M_2O_5)$ mol per cent, $M = Nb$, Ta. The data for $LiTaO_3$ are not so linear as for $LiNbO_3$ over the range $C = 46$–50 per cent and the values listed are estimated values at $C = 50$ per cent.
[a] Carruthers et al. 1971.
[b] Barns and Carruthers 1970.
[c] Lerner et al. 1968.

Other sensitive probes of the stoichiometry are the nuclear magnetic resonance spectrum (Peterson and Carruthers 1969) and the Raman spectrum of the materials (Scott and Burns 1972 b), both of which are readily measured on powders and crystals. Because of the large quadrupole moment of the niobium nucleus the surrounding defects drastically broaden the n.m.r. absorption line. Experimental results are shown in Fig. 8.13. The linewidth provides a direct measure of the random defect density and hence a measure of the stoichiometry. Outside the solid-solubility range no further

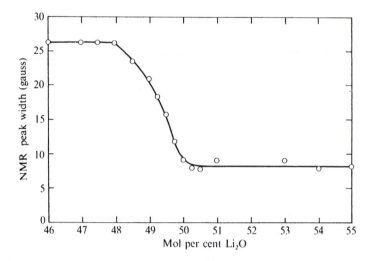

Fig. 8.13. Half-widths of the low-field derivative ^{93}Nb NMR peak for lithium niobate as a function of composition. The width remains constant, as expected, outside the single-phase solid-solution region. (After Peterson and Carruthers 1969.)

change of the linewidth occurs, since this line then only probes the equivalent niobium ions of the LiNbO$_3$ phase (see Fig. 8.12).

The use of Raman spectroscopy as a probe similarly relies on the broadening of Raman-active modes which occurs when the translational symmetry of the lattice is reduced in off-stoichiometric crystals. The experimental measurements are shown in Fig. 8.14 and may be compared directly with those of Fig. 8.13. The results clearly demonstrate the decrease of the solid-solubility range as the powders are equilibrated at lower temperatures. Since the Raman lines of the neighbouring phases do not overlap that of the LiNbO$_3$ phase used as a probe, the linewidth remains constant outside the solubility range.

The defect structure of the LiNbO$_3$ and LiTaO$_3$ lattice which results from the Li$_2$O deficiency remains unsolved. Transparent crystals of high optical quality can be grown from congruent melts despite the fact that 3 per cent of the LiNbO$_3$ units and 2 per cent of the LiTaO$_3$ units have no Li ion. The absence of any paramagnetism or optical absorption associated with the defect structure rules out charge compensation of the Li deficiency by free carriers or simple trapped-electron centres. This is to be contrasted with the perovskite oxides such as BaTiO$_3$, SrTiO$_3$, and K(Ta, Nb)O$_3$ where departures from stoichiometry are frequently associated with colour centres and increased electrical conductivity. The simultaneous increase of both density and unit-cell volume in Li$_2$O-deficient regions of both materials suggests that at least part of the excess Nb ions are situated at additional sites in the unit cell. Indeed, in the n.m.r. spectrum (Peterson and Carnivale 1972) of

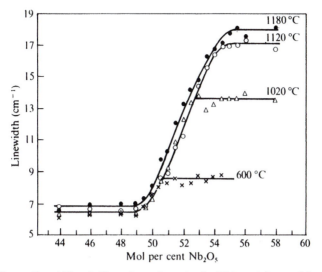

Fig. 8.14. Raman linewidth at half-maximum intensity for lithium niobate solid solution as a function of composition for various quenching temperatures. (After Scott and Burns 1972b.)

off-stoichiometric $LiNbO_3$ a second type of niobium centre appears accounting for up to 6 percent of the total niobiums. However, n.m.r. also shows a new Li site occupied by about 16 per cent of the Li ions (Peterson and Carruthers 1969). These data have not yet been described in terms of any single defect model.

The increase of T_c in both $LiNbO_3$ and $LiTaO_3$ with Li_2O content (as indicated in Table 8.1) is consistent with simple arguments based on the major contributions of the Li and Nb/Ta ions to the ferroelectric behaviour of $LiNbO_3$ and $LiTaO_3$ (Nassau and Lines 1970). The major bonding effects are associated with the highly covalent $Nb(Ta)-O$ bonds, while the major mass effects are associated with the light Li ions. An increasing average bond strength (i.e. 'stiffer' local potential V of eqn (1.2.8)), which results from an increasing Nb density and a decreasing Li density, would thus be expected to decrease the Curie temperature. In the same context one would expect the greater bond strength of the $Ta-O$ bond to give rise to a lower Curie temperature as Nb_2O_5 is replaced by Ta_2O_5 in the $LiNbO_3$ structure. Indeed, T_c decreases linearly throughout the entire solid-solution range between the $LiTaO_3$ and $LiNbO_3$ end members.

In concluding this section it should be pointed out that certain difficulties in the growth of $LiNbO_3$ as large crystals of uniform optical properties may be overcome by the addition of MgO to the melt (Nassau 1967). MgO addition increases T_c as expected from the preceding arguments and thereby compensates in this respect for the loss of Li_2O. By appropriate choice of the constituents in the melt high-optical-quality crystals can be grown with a range of properties so that parameters such as the optical birefringence and phase-matching temperature for parametric interactions can be selected for specific applications (Bridenbaugh *et al.* 1970).

8.4. Tungsten-bronze-type structures

Next to the perovskite oxides the largest family of ABO_3-type oxygen-octahedra ferroelectrics crystallize with structures closely related to the tetragonal tungsten bronzes $K_x WO_3$ and $Na_x WO_3$ described by Magneli (1949). The basic oxygen octahedral framework is shown in Fig. 8.15. The

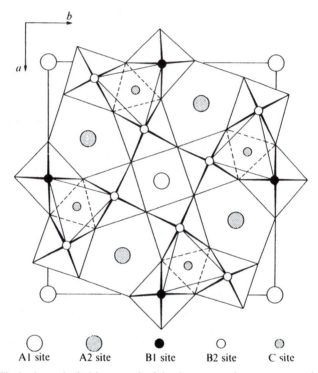

| A1 site | A2 site | B1 site | B2 site | C site |

Fig. 8.15. The basic octahedral framework of the the tungsten bronze structure looking down the tetragonal c-axis. The A-type cations can be accommodated in any or all of the three different types of 'interstitial' site labelled A1, A2, and C. The B-type cations are located at the octahedron centres (B1, B2 in the figure). (After Jamieson *et al.* 1968.)

tetragonal unit cell consists of 10 BO_6 octahedra linked by their corners in such a manner as to form three different types of tunnels running right through the structure parallel to the c-axis. The unit cell is only one octahedron high (~ 0.4 nm) in the c-direction with an $a = b$ dimension of typically 1.25 nm ($= \sqrt{(10)}c$). The long chains of oxygen octahedra along the c-axis resemble those in the perovskites, while normal to this axis the structure consists of slightly puckered sheets of oxygen atoms. The A-type cations enter the structure in the interstitial tunnels in a variety of ways depending on the particular composition. The arrangement provides space for up to four cations in nine-co-ordinated trigonal A2 sites, two cations in somewhat smaller 12-co-ordinated A1 sites, and four cations in the relatively small three-co-ordinated planar C-sites as shown in the figure. There are, in addition, two different B-cation sites which are labeled B1 and B2 in Fig. 8.15.

Only two simple ferroelectric compounds have been discovered with this basic structure (Francombe and Lewis 1958), namely lead metaniobate $PbNb_2O_6$ (Goodman 1953) and lead metatantalate $PbTa_2O_6$ (Smolenskii and Agranovskaya 1954), where the lead atoms are located only in the A1 and A2 sites between the NbO_6 or TaO_6 octahedra. Both of these materials have small orthorhombic distortions from the prototype tetragonal unit cell. $PbNb_2O_6$ becomes tetragonal at the Curie temperature $T_c \approx 575°C$ but $PbTa_2O_6$ remains orthorhombic throughout. As with $LiNbO_3$ and $LiTaO_3$ solid solutions the substitution of tantalum ions for niobium ions in $PbNb_2O_6$ forms a continuous solid solution, lowering the Curie temperature from 575 to 260°C between end members. Goodman (1957) also found that the ferroelectric behaviour of these materials could be considerably modified if lead ions are partially replaced by Mg, Ca, Sr, and Ba ions, a discovery which soon led to a comprehensive study of a large number of alkaline earth niobate solid solutions (Francombe 1960).

Only five out of the six available A sites of the tungsten bronze structure are occupied by lead ions in $Pb(Nb/Ta)_2O_6$, so that the structure is to some extent random even in these simple compounds. Furthermore, both the tantalate and the niobate are only thermodynamically stable at high temperatures (1250°C in the niobate, 1150°C in the tantalate), and the corresponding ferroelectric tungsten bronze phases are only obtained by quenching crystals rapidly to room temperature from these high temperatures. Even the original tungsten bronzes, Na_xWO_3 and K_xWO_3, are only stable off-stoichiometry (Dickens and Whittingham 1968), that is for $x < 1$. These compounds are metallic since the cation deficiency is charge compensated by free carriers. In fact the name 'bronze' described the metallic lustre of these compounds. The ferroelectric tungsten bronzes which are known to be stable at room temperature are all solid solutions of at least two components such as $x(A1)BO_3 + y(A2)BO_3$ (where A1 and A2 here label two different A cations and are not necessarily associated with the A1 and A2 cell sites of Fig. 8.15) where neither component material itself has a stable tungsten bronze structure at room temperature. All this evidence suggests that the structure is only stable when there is a certain degree of disorder, and we shall discuss below how this disorder affects the ferroelectric properties of the materials.

Various subtleties of the structure depend on the particular composition and the crystal growth. Generally speaking the paraelectric phase has tetragonal symmetry, but below the Curie temperature both tetragonal distortion and orthorhombic distortion from the paraelectric structure may occur. The orthorhombic cell has dimensions approximately $1 \cdot 75 \times 1 \cdot 75 \times 0 \cdot 8$ (nm) where

$$a_{\text{orthorhombic}} \approx b_{\text{orthorhombic}} \approx \sqrt{2}\, a_{\text{tetragonal}}$$

and

$$c_{\text{orthorhombic}} \approx 2c_{\text{tetragonal}}.$$

In most cases the spontaneous polarization appears along the c-axis, but $PbNb_2O_6$ is an exception to this rule with the polar axis perpendicular to c.

The interest in tungsten bronze ferroelectrics was renewed in the 1960's because of the large optical non-linearities of these materials. Attention centred on solid solutions of alkali and alkaline earth niobates from which transparent crystals could be grown with a variety of ferroelectric properties depending on the specific cations introduced into the structure. The general composition may be considered to be close to one of the following formulae:

(a) $$(Al)_x(A2)_{5-x}Nb_{10}O_{30}$$

if both A1 and A2 are alkaline earth ions,

(b) $$(A1)_{4+x}(A2)_{2-2x}Nb_{10}O_{30}$$

if A1 is an alkaline earth and A2 is an alkali, and

(c) $$(A1)_{6-x}(A2)_{4+x}Nb_{10}O_{30}$$

if both A1 and A2 are alkali ions. The range of values of x depends on the width of the tungsten bronze solid-solution region. The actual compositions may be more complicated than these three simple solid solutions since the niobium stoichiometry can also vary in these equations. However, when the niobates and tantalates are off-stoichiometry the cation deficiency or excess does not usually give rise to metallic conductivity as in the non-stoichiometric tungstates. Even with wide departures from stoichiometry insulating and transparent crystals may be grown suggesting that, as in $LiNbO_3$ and $LiTaO_3$, charge compensation takes place by ionic rearrangement.

Probably the best-known and most widely studied examples of each of the three categories above are

$$Sr_{5-x}Ba_xNb_{10}O_{30} \qquad \text{(SBN)}$$

$$Ba_{4+x}Na_{2-2x}Nb_{10}O_{30} \qquad \text{(BNN)}$$

and

$$K_{6-x-y}Li_{4+x}Nb_{10+y}O_{30} \quad (KLN)$$

in which ferroelectricity was established by Francombe (1960) for SBN, Rubin, Van Uitert, and Levinstein (1967) for BNN, and Van Uitert $et\ al.$ (1967) for KLN. In SBN the unit cell contains five formula units (10 NbO_6 octahedra) with only five alkaline earth cations to fill six interstitial A1 and A2 sites. Both these ions are too large to enter the small C sites. The structure is thus incompletely filled and a certain degree of randomness is expected. In BNN the A1 and A2 sites are completely filled and the C sites are empty, while in KLN all the A1, A2, and C sites are expected to be filled, with the small Li ions in the C sites (Van Uitert $et\ al.$ 1968). These expectations are more or less borne out by the detailed structure measurements of Jamieson $et\ al.$ (1968, 1969) and Abrahams, Jamieson, and Bernstein (1971) on specific compositions of each of these three compounds. Table 8.2 summarizes the site occupancy of these three materials. It is clear

TABLE 8.2

Site occupancy of some ferroelectric tungsten-bronze-type structures

Formula	A1	A2	C	B1	B2
$Ba_xSr_{5-x}Nb_{10}O_{30}$ $(x \approx 1\cdot38)$	$\approx 1\cdot64$ Sr	$\approx 1\cdot38$ Ba $\approx 2\cdot01$ Sr	0	2 Nb	8 Nb
$Ba_{4+x}Na_{2-2x}Nb_{10}O_{30}$ $(x = 0\cdot13)$	1·74 Na 0·13 Ba	4 Ba	0	2 Nb	8 Nb
$K_{6-x-y}Li_{4+x}Nb_{10+y}O_{30}$ $(x = 0\cdot07, y = 0\cdot23)$	1·75 K 0·25 Li	3·95 K 0·05 Li	3·77 Li 0·23 Nb	2 Nb	8 Nb

After Abrahams $et\ al.$ 1971.

that in SBN ($x \approx 1\cdot38$) the Ba ions prefer the larger site as may be expected from its larger atomic radius, while the Sr ions are randomly distributed over the remaining A1 and A2 sites. In the case of BNN, as before, the Ba prefers the A2 sites and the Na prefers the A1 site so that for $x = 0$ one would expect the structure to be completely ordered (i.e. translationally invariant). In the case of KLN the excess Nb ions in the composition displace Li ions from the C sites to the A1, A2 sites. Again one would expect the structure to become ordered for $x = y = 0$ with Li on C sites and K ions filling A1 and A2 sites. However, studies of the phase equilibria of the K_2O–Li_2O–Nb_2O_5 ternary system (Scott $et\ al.$ 1970) shows that the composition $x = y = 0$ is not stable within the tungsten bronze structure. The structure is only stabilized in the presence of excess Nb so that complete ordering of this compound is not possible.

SBN, on the other hand, has a wide range of solid solubility along the $SrNb_2O_6$–$BaNb_2O_6$ binary join, $1 < x < 4$ (Ballman and Brown 1967, Francombe 1960), and to a lesser extent to 1 per cent on the niobium-rich side and 4 per cent on the Sr + Ba-rich side of this join (Carruthers and Grasso 1970). For $x > 2 \cdot 5$ the structure possesses an observable orthorhombic distortion but below this composition the structure is tetragonal. If the A1 site continues to be occupied only by Sr ions as the Ba/Sr ratio changes, then the entropy of distribution of the cations becomes a minimum at $x = 3 \cdot 3$. Then only Sr ions occupy the A1 sites and only Ba ions occupy the A2 sites. The ferroelectric properties of the material seem to reflect the disorder in the structure as shown in Fig. 8.16. The ferroelectric phase transition is

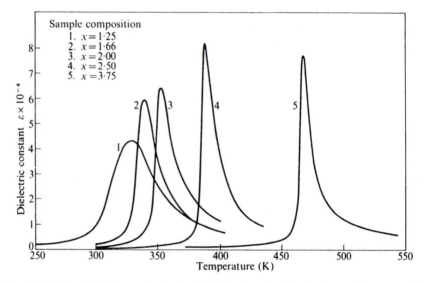

Fig. 8.16. The dielectric constant of single-domain strontium barium niobate crystals $Sr_{5-x}Ba_xNb_{10}O_{30}$ at 1 kHz as a function of material composition x and temperature. (After Glass 1969 a.)

diffuse and increases from about 330 K for $x = 1 \cdot 25$ to about 469 K at $x = 3 \cdot 75$ with a marked sharpening of the associated dielectric anomaly as x increases (see Fig. 8.16). Similar sharpening is observed in other measurements of the temperature-dependent ferroelectric properties—such as the dielectric loss, the specific heat, the pyroelectric coefficient (Glass 1969 a), and the refractive index (Venturini *et al.* 1968). A rather similar broadening of the phase transition is observed in the temperature dependence of the dielectric constant of KLN as the Nb_2O_5 composition of the crystal is increased from 51·6 to 55 mol per cent (Scott *et al.* 1970). From the earlier arguments this broadening corresponds to the increasing disorder of the structure.

The range of the tungsten bronze field of the $BaO-Na_2O-Nb_2O_5$ system is considerably narrower than SBN or KLN, although the Curie temperature can be varied over a range of about 100 degC by varying the niobia content of the melt from 50 to 57 per cent. Below 53 mol per cent the crystals are orthorhombic, but above this concentration the crystals appear optically uniaxial at room temperature suggesting a tetragonal symmetry (Ballman, Kurtz, and Brown 1971). This range includes the ordered 'stoichiometric' compound $Ba_4Na_2Nb_{10}O_{30}$. Dielectric studies on annealed crystals of this material show that the ferroelectric transition at ~585°C is fairly sharp as expected from an ordered structure. Burns and O'Kane (1969) have noted, however, that the Curie temperature of this compound can vary from 585°C in crystals annealed at 800°C to 552°C in crystals quenched from 1300°C. It is tempting to assume that there is some disorder in the cation distribution which can be quenched in at lower temperatures and which affects the ferroelectric behaviour of the crystals in this way.

Because of the complicated structure little has yet been learned about the fundamental nature of the phase transition of tungsten bronzes from scattering experiments. However, these materials are important from a practical point of view, both because of their exceptionally large non-linear polarizability and because of the large range of compositions and ferroelectric properties available for applications. For most of the tungsten bronzes crystal growth and preparation has proved to be a difficult problem. The phase diagrams are complicated and the growth of crystals with non-uniform composition leads to marked variations of the optical and electric properties within a crystal. In the case of orthorhombic structures such as BNN, as-grown crystals form orthorhombic twins at room temperature. For optical applications these twins must be removed by the application of a compressive stress normal to the c-axis during the poling procedure (Singh et al. 1970). In BNN the tetragonal–orthorhombic transition, which is extrinsically ferroelastic and intrinsically anti-distortive (Toledano 1975), occurs below the Curie point near 300°C. Coupling between the ferroelectric and ferroelastic behaviour is not prohibited from a structural point of view but no such coupling has been observed.

8.4.1. Diffuse phase transitions

Broad phase transitions of the kind which are observed in tungsten bronze ferroelectrics were first discovered in solid solutions of the perovskites $BaTiO_3-BaSnO_3$ (Smolenskii and Isupov 1954). Early studies revealed that piezoelectric activity could be observed at temperatures well above the dielectric constant maxima, in direct contrast with experience in simple perovskites which become cubic at temperatures above the dielectric constant peak. Early studies of diffuse phase transitions have recently been reviewed by Smolenskii (1970). It is now evident that broadening of the phase transition is a very common occurrence in solid solutions and other disordered structures. The ferroelectric–paraelectric transition in these systems is characterized not by an abrupt structural transition, but rather by a gradual, diffuse transition which occurs over a temperature range—usually referred to as the Curie range. At temperatures well below this range the materials belong to a pyroelectric crystal class and from most points of view behave as normal ferroelectrics, while at temperatures well above the Curie range the crystals show no ferroelectric behaviour. However, within the Curie range the crystals exhibit unusual dielectric and ferroelectric properties. The temperature at which the dielectric constant peaks depends on the frequency of measurement, and this does not generally coincide with the peak of the dielectric loss, as in normal ferroelectrics, nor with the peak of the pyroelectric coefficient. The spontaneous polarization and other ferroelectric properties such as the specific-heat anomaly, the optical absorption edge, the refractive index, and the electro-optic properties, vary slowly throughout the Curie range rather than exhibit the abrupt changes normally

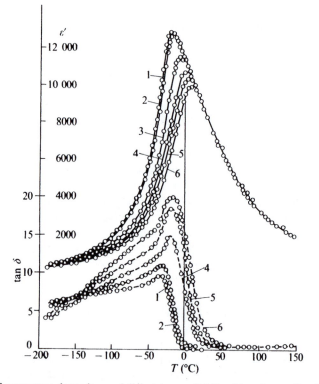

Fig. 8.17. Temperature dependence of dielectric permittivity ε' (continuous lines) and of tan $\delta \equiv \varepsilon''/\varepsilon'$ (broken curves) of a polycrystal sample of $Pb_3MgNb_2O_9$ in weak fields at differemt frequencies: curve 1, 0·4; curve 2, 1; curve 3, 45; curve 4, 450; curve 5, 1500; curve 6, 4500 kHz. (After Smolenskii *et al.* 1960.)

expected at a ferroelectric transition. One system which has been extensively studied (Smolenskii *et al.* 1960) is $Pb_3MgNb_2O_9$ which has a particularly diffuse transition extending over more than 100 degC around room
temperature as shown in Fig. 8.17. The actual shape of the transition
depends on the preparation procedure (Verbitskaya *et al.* 1971). The
absence of superlattice lines in X-ray studies of this material suggests that
the defect structure is random. The frequency dependence of the dielectric
constant, shown in Fig. 8.17, is typical of diffuse transitions. The dielectric
relaxation occurs at low frequencies on the low-temperature side of the
maximum with the result that the dielectric constant maximum shifts to
higher temperatures with increasing frequency.

A similar, but even more pronounced, effect is observed in SBN as shown
in Fig. 8.18. The dispersion curves are very similar to those expected for
normal Debye relaxation with the real part of the dielectric constant peaking
at zero frequency. These data were taken on large (>1 mm) crystals with
mechanical resonance frequencies near 1 MHz, but the dispersion cannot be
attributed to the effects of clamping. It is curious that at very high frequencies practically no dielectric constant anomaly is observed at all—similar to
what would be expected from an extrinsic ferroelectric transition (see
§ 10.2). Throughout the Curie region the dielectric properties are extremely
non-linear. The dependence of the dielectric constant and the pyroelectric
coefficient of SBN on external applied fields, shown in Fig. 8.19, is fairly
typical. The external field has a much larger effect on the position of the
dielectric constant maximum and affects the magnitude of the dielectric
properties over a much wider temperature range than in normal ferroelectrics discussed in § 5.9. As a result of these non-linearities, dielectric
hysteresis loops usually give erroneous values for the spontaneous polarization throughout the Curie region.

As indicated earlier in the section, the broadening of the phase transition
can be attributed to structural disorder and compositional fluctuations in
solid solutions. Systems which are translationally invariant (have perfect
structural order) necessarily have sharp phase transitions, even though
above the Curie temperature some information about the low-temperature
phase may be contained within critically small volumes (the correlation
volume) at an instant of time; the time or ensemble average will be
characteristic only of the paraelectric phase. However, when translational
invariance is destroyed as a result of compositional disorder, broadening of
the phase transition may occur depending on the microscopic details of the
disorder. For instance, in annealed systems where impurities or vacancies
can diffuse freely and reach equilibrium with respect to the other degrees of
freedom of the system a sharp phase transition is expected (Lubensky 1975).
On the other hand, in quenched systems in which the impurities and
vacancies are frozen in fixed positions in the lattice the broadening of the

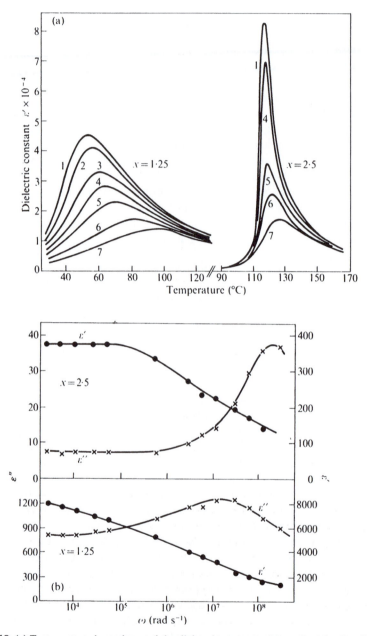

Fig. 8.18. (a) Temperature dependence of the dielectric constant of $Sr_{5-x}Ba_xNb_{10}O_{30}$ (SBN) measured at various frequencies f: curve 1, 50 Hz; curve 2, 500 Hz; curve 3, 5000 Hz; curve 4, 0·05 MHz; curve 5, 0·5 MHz; curve 6, 5 MHz; curve 7, 20 MHz. (b) The variation of the real and imaginary parts ε' and ε'' of the dielectric constant as a function of radian frequency ω at 296 K for the same crystals. (After Glass 1969 a.)

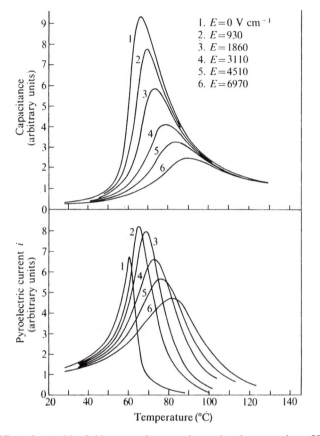

Fig. 8.19. Effect of a d.c. bias field on capacitance and pyroelectric current in an SBN sample. (After Glass 1969 a.)

phase transition is expected if the randomness is not homogeneous (that is, if significant clustering of impurities occurs). If the randomness is homogeneous, a sharp transition can occur even in systems with frozen impurities although critical exponents may be modified from their equivalent 'pure' values. This situation, for the case of dipolar forces, has been analysed by Aharony (1976). There is no experimental information on the details of the disorder in systems with diffuse phase transitions. Most of the diffuse phase transitions studied in oxide ferroelectrics have occurred at low temperatures (<400°C) where the disorder is likely to be frozen. From the theoretical considerations just outlined one would not expect to find diffuse transitions at high temperatures where the defects are more mobile in the lattice. Indeed, sharp transitions are observed in $LiNbO_3$ and $LiTaO_3$ despite fairly large deviations from stoichiometry.

Most of the experimental observations of unusual ferroelectric behaviour within the Curie range can be described at least qualitatively by arguments in which the crystal is represented as a collection of small regions of conventional ferroelectric behaviour where the individual regions have dimensions which are large compared with the correlation length (see § 11.1), but where the normal modes and Lorentz parameters vary from one region to another throughout the system. We have already seen how the ferroelectric Curie temperature can vary significantly with small composition variations in the perovskite and tungsten bronze ferroelectrics. The macroscopic properties of such a mixed system will then be some average over the crystal volume. Any temperature within the Curie region corresponds to the Curie temperature of a small fraction of the total volume while some microregions are above their Curie temperature and others are below. It is obvious that the macroscopic dielectric and ferroelectric behaviour will be a broad envelope of the behaviour of individual regions. The width of the Curie region depends both on the range of the composition fluctuations and on the sensitivity of the Curie temperature (or other properties) to the composition changes. At the Curie temperature the properties of a normal ferroelectric are very sensitive to external fields and pressures, so it is not surprising that the average properties of a mixed system vary significantly with applied field throughout the entire Curie region as shown in Fig. 8.19. The shift of the peak dielectric constant and pyroelectric coefficient to higher temperatures with field would, for a conventional Devonshire ferroelectric, imply that the transition is of first order, but for a material with a diffuse transition this conclusion cannot be drawn.

A temperature-dependent dielectric relaxation is also predicted from such a simple model. At temperatures close to the Curie temperature of the individual microregions, dielectric relaxation occurs at low frequencies as the ferroelectric mode softens. Furthermore, polarization reversal and domain-wall motion below T_c will contribute to the dielectric constant at low frequencies more than at high frequencies. The temperature difference between the peak dielectric constant and the peak dielectric loss is an automatic consequence of the Kramers–Kronig relations when there is a temperature-dependent relaxation.

The real situation is certainly much more complicated than this simple model would suggest, but such a simple picture is helpful in obtaining an intuitive understanding of ferroelectric behaviour within the Curie region. The individual microregions are not, of course, independent since they can interact with each other via depolarization fields, elastic strains, and thermal fluctuations. Attempts to set out a quantitative statistical analysis for the behaviour of such complex systems are as yet rather tentative and have provided only a partial understanding of the system primarily because of the rather drastic simplifications which have been found necessary to obtain solution of the problem. Nevertheless, some success has been achieved in the interpretation of experimental data. A summary of the approach has been given by Yurkevich and Rolov (1975).

While the equilibrium properties of disordered systems can be qualitatively understood in terms of the thermodynamic relationships developed in Chapters 3 and 5 averaged in some statistically significant manner, the dynamic nature of the diffuse ferroelectric transition remains something of a mystery. Disordered systems cannot be described by a set of normal modes with well-defined wavevectors $\mathbf{q}$. The disorder which destroys the translational symmetry is clearly manifest in Raman scattering experiments, with intense first-order Raman scattering being observed above the Curie range of a mixed perovskite (Burns and Scott 1973 b), for example, where a perfect perovskite is cubic above T_c and first-order Raman scattering is forbidden by symmetry. More significant, however, is the fact that no soft modes are observed in the Raman spectra of a large number of mixed ferroelectrics which can account for the large dielectric constant peak observed at low frequencies. Indeed, from the Raman spectra of several crystals with the tungsten bronze structure the low-frequency dielectric constant calculated from the Lyddane–Sachs–Teller relationship (eqn (7.1.12)) is found to be approximately independent of temperature (Burns 1972).

If the behaviour of the low-frequency dielectric constant is due to one of the normal optic modes of the structure (broadened by disorder), then the mode must lie in the frequency range below 10 cm^{-1} which is inaccessible to Raman scattering studies. This possibility is consistent with the dielectric data of Fig. 8.18 which show considerable dispersion of dielectric constant between 50 Hz and 2×10^7 Hz—in fact at the higher frequency the dielectric constant is relatively temperature independent in the transition region. However, this same dielectric behaviour is more characteristic of a Debye relaxation, possibly associated with defect hopping over barriers. The latter description cannot alone be responsible for the entire phenomenon since it would not account for the existence of the structural (i.e. co-operative) transformation or of the appearance of a spontaneous polarization which is much greater than could be obtained from a set of isolated relaxing dipolar defects in the small electric fields required to polarize the crystals. Some defect-coupling mechanism would be essential, and it has been suggested that the spectrum arises from a Debye-type hopping of defects or impurities over a distribution of barriers in regions of large local field connected with the polarization of a higher-frequency soft lattice mode. The polar lattice-mode coupling can then cause a considerable enhancement of the hopping polarization (Burns and Burstein 1974, Barker 1976). In general, of course, mode softening in a peak centred at zero frequency may or may not correspond to any of the modes obtained within a harmonic approximation

of lattice dynamics (Barker 1976). Since a diffuse transition appears to
'sharpen up' as the structural disorder is decreased, the most promising
picture of the general dynamics is perhaps one of a marginally softening
lattice mode which, on softening, interacts increasingly with the defects. This
interaction would produce a frequency-dependent relaxation which has the
effect of slowing and eventually stopping the lattice mode softening and the
transferring of mode strength to a central diffusive peak. The quantitative
effects of such a frequency-dependent relaxation are discussed in § 11.5 in a
different context. Resulting 'central peaks' are a common feature of many
basically displacive transitions even in non-random structures, although in
the latter cases the actual mode softening of the driving lattice mode is
usually quite pronounced, and the appearance of the central peak and arrest
of the mode softening is observed only very close to the sharp transition
temperature; the physical origin of the frequency-dependent damping in
these systems may have no connection with the existence of impurities. In
systems with major structural disorder it is likely that the coupling to
hopping defects is so strong that the lattice-mode softening is essentially
prevented altogether. It then becomes difficult to say whether the more
useful basic 'picture' is one of a soft lattice mode grossly relaxed by polar
defects, or of defects interacting co-operatively via an interaction carried by
a polar lattice mode.

9

ORDER-DISORDER FERROELECTRICS

9.1. KDP-type ferroelectrics (theory)

IN §§2.3 AND 2.5 we demonstrated the manner in which ionic motion in a double-well local potential can be cast theoretically in the language of co-operative magnetism. In this modern approach, as exemplified by the work of Moore and Williams (1972), the potential minima referred to are with respect to a generalized displacement co-ordinate describing the motion of all ions in a unit cell. The more traditional approach for KH_2PO_4 and its isomorphs (Slater 1941, Takagi 1948, Kobayashi 1968) has been to concentrate on the specific motion of the hydrogens (which is of largest amplitude) and to include the motion of the other ions, if at all, in terms of a specific coupling mechanism. Thus, for example, any tunnelling characteristics in the more generalized scheme involve all the atoms (although, of course, the displacements of some atoms may be considerably greater than others), whereas the tunnelling character in the more traditional approach is specifically reserved for the hydrogens.

KH_2PO_4 (KDP) crystallizes in a tetragonal structure above 123 K with a non-centrosymmetric point group $\bar{4}2m$. Below T_c an orthorhombic phase exists (point group mm) which is ferroelectric with the spontaneous polarization mainly due to the displacement of K, P, and O ions in the direction of the polar c-axis. The structure has been observed by X-ray (Frazer and Pepinsky 1953) and by neutron diffraction experiments (Bacon and Pease 1953, 1955) in both phases. The structure consists of two interpenetrating body-centred lattices of PO_4 tetrahedra and two interpenetrating body-centred K lattices, the phosphate and potassium lattices being separated along the c-axis (Fig. 9.1). Every PO_4 is linked to four other PO_4 groups by hydrogen bonds which lie very nearly perpendicular to the ferroelectric c-axis. The linkage is such that there is a hydrogen bond between one upper oxygen of one PO_4 group and a lower oxygen of a neighbouring group. As revealed by the neutron studies only two hydrogens are normally located nearest any one PO_4 group giving rise to a formal ionic configuration $K^+(H_2PO_4)^-$.

Bacon and Pease showed that the distribution of proton density shifted from the centre of the bond above T_c to a localized asymmetric position below T_c. The direction of proton shift was found to reverse when the sense of the spontaneous polarization is reversed. Fig. 9.2 (a) shows schematically the complete ion movement associated with a direction of polarization along the $+c$-axis and Fig. 9.2 (b) depicts the situation for the opposite polarization (Tokunaga 1970). Above T_c the structure of molecular units is as shown

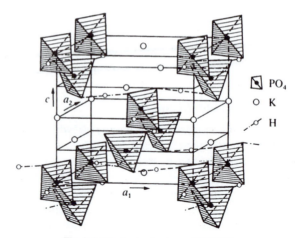

Fig. 9.1. The basic structure of KH_2PO_4.

in Fig. 9.2 (c) with no net polarization. If a rigid order–disorder picture is accepted this last figure is most simply interpreted as the averaged combination of Figs. 9.2 (a) and (b), each atom transferring stochastically between the two 'ordered' configurations. In a displacive picture, on the other hand, each atom would vibrate about an equilibrium configuration (though possibly in a highly anharmonic fashion), the observed elongated probability densities now being accounted for by vibrational anisotropy.

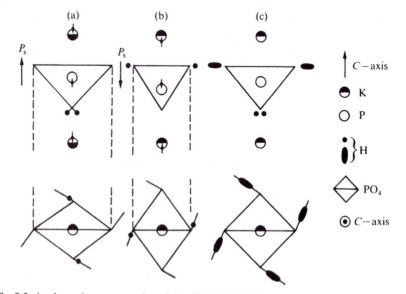

Fig. 9.2. A schematic representation of the dipole unit in KDP: (a) the polarization along the c-axis, (b) along the $-c$-axis, and (c) the paraelectric phase. (After Tokunaga 1970.)

Although the structure of KDP is quite complicated, 48 unit vectors being required to form a basis for the lattice-dynamical problem (i.e. two formula units per primitive cell), a great simplification occurs if we concentrate on the long-wavelength limit for which we expect ferroelectric anomalies to manifest themselves. In fact there are only six modes in this limit which (group-theoretically) are capable of producing an electric polarization parallel to the c-axis.

The frequencies and displacement vectors of these six low-lying modes have been computed by Fujiwara (1970) in a quasi-harmonic approximation. Significantly he finds that the lowest-frequency (or soft) mode possesses a considerable motion of the P, K, and O ions parallel to the c-axis, but that in the normal c-plane the motion is completely dominated by the protons moving parallel to the hydrogen bonds. The displacement amplitude variable ξ in § 2.5 for example therefore refers in the KDP context to a motion of this general form involving all the unit-cell atoms. If the local potential is of double-well form it again refers to all the atoms in the unit cell, there being two stable atomic configurations above T_c with the system as likely to be found in one as the other.

Although it is difficult to justify the above local-mode picture in the KDP structure for large values of the wavevector, it should have great merit for temperatures close to T_c where long-wavelength excitations dominate. In particular the displacement of all the ions observed by neutron diffraction below T_c receives fundamental attention within the model from the outset. There are difficulties, however. The first, envisaged most easily in the order–disorder limit, is that many of the proton configurations corresponding to H_2PO_4 groups (e.g. configurations 3 to 6 of Table 9.1) are excluded completely. Such groupings in KDP itself have an activation energy $\lesssim kT_c$ and are important. They appear in the local-mode picture as strong interactions between the soft mode and other low-lying polar modes, and clearly complicate an otherwise very simple representation. A separate problem concerns the form to be used for near-neighbour intercell forces; one usually assumes a simple bilinear representation but such may not always be adequate. Long-range dipolar interactions between c-axis dipoles are, on the other hand, clearly of bilinear form and are very simply included.

Historically the earliest theoretical models for KDP (Slater 1941, Takagi 1948) followed the idea that the dielectric properties were essentially determined by the proton configuration. Each proton is considered to have two equilibrium positions along a bond connecting two PO_4 groups. Labelling them $+$ and $-$, a projection of the network of hydrogen bonds in a c-plane can be sketched as shown in Fig. 9.3. There are $2^4 = 16$ possible arrangements of the four hydrogens surrounding any PO_4 group, but only six of these, corresponding to H_2PO_4 groups with two hydrogens attached, represent low energies. Two of these (Fig. 9.2 (a) and (b)) represent states of

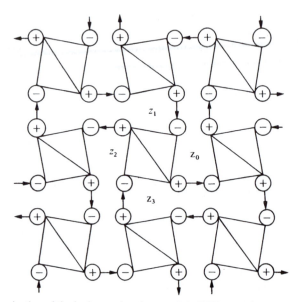

Fig. 9.3. A projection of the hydrogen-bond network in KDP on a plane perpendicular to the polar c-axis. The symbols $+$ and $-$ denote the eigenvalues of the Ising operators S^z, and the arrows display the phase relation of the proton collective mode. (After Tokunaga 1970.)

lowest energy (say zero) and appear as $(+ + + +)$ and $(- - - -)$ in the notation of Fig. 9.3. The other four, viz, $(+ - - +), (+ - + -), (- + - +),$ $(- + + -)$, represent states with a higher energy E_1. Neglect of all other proton configurations constitutes what is known as the 'ice constraint' and was the original Slater approximation. Extending the picture to include the entire 16 arrangements leads to the energy scheme shown in Table 9.1.

A rather interesting point in connection with the energy levels of Table 9.1 is that they require four particle interactions of the form

$$\mathcal{H} = - \sum_{i,j,k,l} J_{ijkl} S_i^z S_j^z S_k^z S_l^z \qquad (9.1.1)$$

for their 'Ising' description (S^z being an Ising variable with eigenvalues $+\frac{1}{2}$ and $-\frac{1}{2}$). Tokunaga and Matsubara (1966) and Blinc and Zeks (1972) have demonstrated that eqn (9.1.1) can approximately be replaced by a sum of pair interactions

$$\mathcal{H} = -\frac{1}{2} \sum_{i,j} J_{ij} S_i^z S_j^z \qquad (9.1.2)$$

if certain conditions are fulfilled. For example writing

$$J_{12} = J_{34} = U$$

$$J_{13} = J_{24} = J_{23} = J_{14} = V \qquad (9.1.3)$$

TABLE 9.1

Energy levels of the 16 proton configurations around a PO$_4$ group in KDP according to the Slater–Takagi model and to the pseudo-Ising model of eqn (9.1.4)

Configuration†		$S_1^z\, S_2^z\, S_3^z\, S_4^z$	Slater–Takagi	Pseudo–Ising
1		$+\;+\;+\;+$	0	$-(2U+4V)/4$
2		$-\;-\;-\;-$	0	$-(2U+4V)/4$
3		$+\;-\;-\;+$	E_1	$U/2$
4		$+\;-\;+\;-$	E_1	$U/2$
5		$-\;+\;-\;+$	E_1	$U/2$
6		$-\;+\;+\;-$	E_1	$U/2$
7		$+\;+\;+\;-$	E_2	0
8		$+\;+\;-\;+$	E_2	0
9		$+\;-\;+\;+$	E_2	0
10		$-\;+\;+\;+$	E_2	0
11		$-\;-\;-\;+$	E_2	0
12		$-\;-\;+\;-$	E_2	0
13		$-\;+\;-\;-$	E_2	0
14		$+\;-\;-\;-$	E_2	0
15		$-\;-\;+\;+$	E_3	$-(2U-4V)/4$
16		$+\;+\;-\;-$	E_3	$-(2U-4V)/4$

the Ising Hamiltonian of a cluster of four protons around a PO$_4$ group becomes

$$\mathcal{H}_4 = -U(S_1^z S_2^z + S_3^z S_4^z) - V(S_1^z S_3^z + S_2^z S_4^z + S_2^z S_3^z + S_1^z S_4^z) \qquad (9.1.4)$$

the energy levels of which are also shown in Table 9.1. It is evident that the

Slater–Takagi energy levels and the pseudo-Ising ones are identical only if

$$U = -2E_2 + 2E_1, \qquad V = 2E_2 - E_1 \tag{9.1.5}$$

and

$$E_3 = 4E_2 - 2E_1. \tag{9.1.6}$$

The last equation is a restriction not required in the full Slater–Takagi theory but which must be satisfied if an exact separation of the four-body interactions into two-body terms should be possible. If $E_3 \gg E_2 \gg E_1$ (the most likely physical situation) then it is easy to see that the restriction (eqn (9.1.6)) is for most purposes a good approximation and Hamiltonian eqn (9.1.2) an acceptable form. Nevertheless it is important to recognize that eqn (9.1.2) is *not* a conventional Ising problem since i and j do not define a Bravais lattice.

The Slater KDP model is not exactly soluble in three dimensions but has been exactly solved, even in the presence of an external electric field, for two-dimensional arrays. Both the exact solutions and approximate three-dimensional solutions have very unusual features quite unlike the ordinary magnetic Ising model. For example the model undergoes a first-order transition at T_c with the polarization being 'perfect' (equal to its $T = 0$ value) for all $T < T_c$. Both ferroelectric and antiferroelectric models can be set up in the present context. Those containing only the six $(H_2PO_4)^-$ configurations are termed 'ice models', but extensions to include configurations 15 and 16 of Table 9.1 (the eight-vertex problem) and to include all 16 configurations of Table 9.1 (the 16-vertex problem) have also been the subject of much extensive mathematical research relating to the special counting problems posed by their statistical solution (an excellent review has recently been given by Lieb and Wu 1972).

Although of great interest as a theoretical model, the above protonic description of the KDP transition is unphysical in that it ignores the important long-range dipolar forces and the possibility of tunnelling. In many ways the former presents the more difficult problem, because the basic elementary dipoles of Figs. 9.2 (a) and (b) involve primarily the c-axis motion of K, P, and O and there is not necessarily a rigid and instantaneous correlation between protonic configuration and dipole moment. The main contribution to polarization comes from the deformation of the $KH_2^{3+} - PO_4^{3-}$ complex along the c-axis. The completely ordered proton system corresponds to a saturation polarization $P_0 = N\mu$, where N is the total number of KH_2PO_4 molecules and μ is the magnitude of an elementary molecular dipole along the c-axis. One simple method of relating spontaneous polarization to proton disorder is to write it in the form $\mu \Sigma_i S_i^z$ where the summation is over the $2N$ proton co-ordinates. As a natural extension of the idea the motion of an individual dipole μ_j is described by

$$\mu_j = \tfrac{1}{2}\mu \sum_i{}' S_i^z \tag{9.1.7}$$

where the primed summation is over the four hydrogen bonds of the jth PO_4 group. This assumption implies that an elementary dipole instantaneously follows the protonic configuration. The long-range dipole interaction can now be formally expressed in terms of the elementary moments and, using eqn (9.1.7), transformed to the pseudo-Ising form (eqn (9.1.2)). In particular, they are readily included in a Slater–Takagi model as a local effective field.

Once we have relaxed the ice constraint (which removes the unusual character of the Slater phase transition referred to above) and included dipolar forces, many of the actual features of KDP type ferroelectric transitions can be explained. In particular, transitions of first or second order can be generated (Silsbee, Uehling, and Schmidt 1964, Blinc and Svetina 1966, Vaks and Zinenko 1973) with realistic transition entropies. One obviously rather restrictive feature of the model, however, is the rigid response coupled with the unphysical requirement from eqn (9.1.7) that the elementary dipole moment becomes just $\frac{1}{2}\mu$ in the E_2 levels of Table 9.1. This restriction can be lifted by describing the dipole system by a separate Ising or phonon co-ordinate representation which is coupled in some simple manner to the protonic system. The Ising-dipole picture was first used by Tokunaga and Matsubara (1966), while a model assuming a quasi-harmonic dipolar representation has been developed by Villain and Stamenkovic (1966) and by Kobayashi (1968). In this latter (coupled proton–lattice) model the motion of the heavy ions is included in a familiar simple phonon form

$$\mathscr{H}_{\text{lattice}} = \tfrac{1}{2}\sum_{\mathbf{q}} \{\pi'(\mathbf{q})\pi'(-\mathbf{q}) + \Omega^2(\mathbf{q})\xi'(\mathbf{q})\xi'(-\mathbf{q})\} \qquad (9.1.8)$$

together with a bilinear interaction between 'spins' and phonons

$$\mathscr{H}_{\text{proton-lattice}} = \sum_{\mathbf{q}}\sum_{i} -F_i(\mathbf{q})S_i^z\xi'(-\mathbf{q}) \qquad (9.1.9)$$

so that the complete representative Hamiltonian becomes

$$\mathscr{H} = \mathscr{H}_{\text{proton}} + \mathscr{H}_{\text{proton-lattice}} + \mathscr{H}_{\text{lattice}} \qquad (9.1.10)$$

in which the first term on the right-hand side is eqn (9.1.2) and the lattice co-ordinate ξ' describes a local motion of (only) the heavy ions, with the symmetry of the static distortion which sets in below T_c.

In this coupled model the order–disorder proton motion is arbitrarily strongly coupled to the optic phonon describing $K–PO_4$ lattice vibrations along the c axis. The coupling itself is most simply pictured as an interaction of the O—H——O bond electric dipole moments with the electric field created by the polar lattice displacements. In the simplest mathematical picture of Kobayashi a single proton per unit cell is assumed leading to a pair of coupled proton–lattice modes. A more accurate model for the KDP structure requires four protons per unit cell (Nettleton 1971 b) leading to four spinwave branches each coupled to an optic phonon branch of the appropriate symmetry. In fact, since there are several phonon branches in the KDP structure, corresponding to each of the relevant symmetries, even the Nettleton model may be too simple. As the proton system approaches its own transition temperature one of the coupled proton–lattice modes becomes unstable causing a spontaneous polarization to set in along the c-axis.

A spectacular feature of the phase transition in KDP compounds is an extremely large isotope effect (on replacing hydrogen by deuterium) in the Curie temperature, which is raised by a factor of nearly 2. Similar isotope effects occur in the shape of the polarization curves (Bantle 1942) and in the saturation polarization P_0. Regarding the latter, Ymry *et al.* (1967) find that P_0 in the deuterated material is some 20 per cent larger than in the undeuterated case. The stimulus for creating the tunnelling models of KDP ferroelectrics was the fact that the order–disorder model, whether protonic, proton–lattice, or local mode, could not convincingly explain the large isotope shift of T_c. The fundamental idea of the tunnelling model was introduced by Blinc (1960) in the protonic context, each proton on the corresponding hydrogen bond making a tunnelling motion between two equilibrium positions. When the deuteron with its heavier mass is substituted for the proton, it reduces very significantly the frequency of the tunnelling motion and affects both the statics and dynamics of the phase transition very significantly.

The manner in which a tunnelling motion modifies the basic Ising-like order–disorder Hamiltonian has been demonstrated, in the local-mode context, in § 2.5. In exact analogy with eqn (2.5.4) the tunnelling term in the protonic or proton–lattice models appears in the proton Hamiltonian eqn (9.1.2) modifying the latter to

$$\mathscr{H}_{\text{proton}} = -\sum_i \Delta S_i^x - \tfrac{1}{2}\sum_{i,j} J_{ij} S_i^z S_j^z \qquad (9.1.11)$$

with Δ defined by eqn (2.5.5), but where now the variable i refers to an individual double-well proton motion rather than to a general local-mode configurational vector.

Let us briefly consider a random-phase solution of the coupled proton–lattice Hamiltonian eqn (9.1.10) including tunnelling. First we consider the statics. Replacing spin operators by their mean-field averages and assuming for simplicity a single pseudo-spin per (Bravais) cell we can express the part of the Hamiltonian depending on phonon co-ordinates as

$$\mathscr{H}(\text{phonon}) = \tfrac{1}{2}\sum_q \{\pi'(\mathbf{q})\pi'(-\mathbf{q}) + \Omega^2(\mathbf{q})\xi'(\mathbf{q})\xi'(-\mathbf{q})\}$$
$$- N^{1/2}\sum_q \langle S^z \rangle F(\mathbf{q})\xi'(-\mathbf{q})\delta_{q,0} \qquad (9.1.12)$$

where we have defined the Fourier transform $F(\mathbf{q})$ by writing

$$\sum_i F_i(\mathbf{q})S_i^z = F(\mathbf{q})S^z(\mathbf{q}). \tag{9.1.13}$$

Thus, the effective field of the spin system produces a displacement of the lattice oscillators. The average displacement obtained by minimizing (9.1.12) with respect to $\xi'(-\mathbf{q})$ is

$$\langle \xi'(\mathbf{q}) \rangle = N^{1/2} \left\{ \frac{F(0)}{\Omega^2(0)} \right\} \langle S^z \rangle \delta_{\mathbf{q},0} \tag{9.1.14}$$

and relates the ordering of protons $\langle S^z \rangle$ to the displacement of the heavy ions $\langle \xi'(0) \rangle$, which is responsible for most of the spontaneous polarization.

The effective-field spin-Hamiltonian for protonic motion, on the other hand, becomes

$$\mathscr{H}(\text{spin}) = -\Delta S^x - J(0)S^z \langle S^z \rangle - N^{1/2}F(0)\langle \xi'(0) \rangle S^z. \tag{9.1.15}$$

Using (9.1.14) the phonon co-ordinate can be eliminated to give

$$\mathscr{H}(\text{spin}) = -\Delta S^x - \{J(0) + NF^2(0)\Omega^{-2}(0)\}S^z \langle S^z \rangle. \tag{9.1.16}$$

Thus as far as the static properties are concerned the whole effect is a renormalization of the spin-spin interaction from $J(0)$ to $J(0) + NF^2(0)\Omega^{-2}(0)$ which now contains an indirect phonon-mediated part in addition to the direct part. Proton-phonon interactions can thereby produce a phase transition even in the absence of the direct spin-spin interaction.

All the effective field results discussed in § 2.5 can now be recovered if the renormalized interaction is used in place of the bare spin-spin term. Thus to this approximation the equilibrium properties of order-disorder or tunnel systems are not changed in any essential way by the proton-phonon coupling, and equivalent critical behaviour will be obtained in proton-lattice and local-mode representations.

A simple random-phase solution for the dynamics analogous to that discussed in § 2.5.2 for the 'bare' system produces eigenfrequencies in the disordered phase of the form (Kobayashi 1968)

$$2\omega_\pm^2(\mathbf{q}) = \Omega^2(\mathbf{q}) + \Omega_B^2(\mathbf{q}) \pm \left[\{\Omega^2(\mathbf{q}) - \Omega_B^2(\mathbf{q})\}^2 + 2N\Delta|F(\mathbf{q})|^2 \tanh\left(\frac{\Delta}{2kT}\right) \right]^{1/2} \tag{9.1.17}$$

in which $\Omega_B(\mathbf{q})$ is the RPA soft-mode frequency of the bare system as given in eqn (2.5.31). The ω_- mode describes an in-phase motion of the pseudo-spin system and the lattice, whereas the ω_+ mode describes a motion of the two systems with opposite phase. The stability limit is determined by the highest temperature at which an $\omega_-(\mathbf{q})$ approaches zero, defining the critical wavevector $\mathbf{q} = \mathbf{q}_0$. This condition

$$2\Delta = \{J(\mathbf{q}) + N|F(\mathbf{q})|^2\Omega^{-2}(\mathbf{q})\}\tanh\left(\frac{\Delta}{2kT_c}\right) \qquad (9.1.18)$$

is again identical to the one obtained for a bare system (compare eqn (2.5.17) for the ferroelectric case $\mathbf{q}_0 = 0$) if only $J(\mathbf{q})$ is replaced by $J(\mathbf{q}) + N|F(\mathbf{q})|^2\Omega^{-2}(\mathbf{q})$. The critical value of $\mathbf{q}$ is determined by maximizing this latter quantity as a function of $\mathbf{q}$. Since it is this critical value $\mathbf{q} = \mathbf{q}_0$ which determines the structure of the ordered phase, the proton–lattice model allows us the possibility of explaining, for example, the antiferroelectricity of $NH_4H_2PO_4$ as compared with the ferroelectricity of KH_2PO_4 in terms of a difference in phonon spectra which moves the relevant maximum from the centre to the edge of a Brillouin zone.

In piezoelectric crystals like the KDP group coupling between the soft mode and the transverse acoustic phonon branch is also allowed. On approaching T_c from above, the ferroelectric soft-mode branch crosses the acoustic branch, resulting in mode mixing and level repulsion (Brody and Cummins 1968, Dvorak 1970) which causes the velocity of the acoustic mode (i.e. the slope of the dispersion curve at the origin) to vanish before the unmixed ω_- mode would condense on its own. The problem can be treated in an analogous fashion to that used above for proton–phonon coupling, viz. by including a bilinear soft-mode–acoustic-phonon coupling term and solving the resulting equations of motion in a simple random-phase approximation.

The large isotope effect on T_c in KDP ferroelectrics is (see eqn (2.5.17)) qualitatively explained by the tunnelling concept. The isotope effect of P_0 is also anticipated in this model because deuteration leads to a larger probability density of a deuteron (or the generalized local mode) at the plus site (say) when $T = 0$ K. However, simple random-phase statistical solutions of the tunnelling problem (e.g. § 2.5, see also Pytte and Thomas 1968, Konsin and Kristoffel 1972) give rise to propagating 'spin-wave' dynamics with an *undamped* frequency softening at T_c. The typically overdamped or Debye-like relaxation observed in actual KDP systems therefore receives no explanation in the simplest picture. Theoretically situations involving large damping are notoriously difficult to handle. The difficulties stem from the extremely large interaction of the elementary excitations of the system. Clearly, techniques based on simple decoupling methods, which give in lowest order undamped 'spin waves', seem to be far too crude as an approach to the KDP problem proper. In recent years various techniques have been developed to cope with highly damped systems (Kawasaki 1967, 1968, Resibois and de Leener 1966, 1969, Blume and Hubbard 1970, Hubbard 1971). The basic idea is that a time-dependent fluctuating field rather than an ensemble-averaged static one plays the role of generating the 'unperturbed' energy levels. Details are mathematically too complex to warrant discussion in a general text of this nature, but the method has been used (Moore and Williams 1972) to discuss the KDP tunnel-mode problem, at least in the disordered phase. Recalling from eqns (2.5.30) and (2.5.31) that for $T > T_c$ the simple random-phase tunnelling solution consists of a pair of undamped S^z fluctuation modes and a single zero-frequency S^x mode, the more accurate solutions show response in the z-direction to be of damped harmonic character, tending towards diffusive form as T_c is approached, and response in the x-direction to be diffusive for low frequencies at all (paraelectric) temperatures. Even this approximation, however, breaks down in the critical region, and the theoretical problem of the tunnel-mode Hamiltonian remains a topic of very active interest.

In the coupled proton–lattice model (with tunnelling) the damping term is often included in a phenomenological fashion (Kobayashi 1968, Silverman 1970). It can then be adjusted as required to give underdamped or overdamped behaviour. In particular, with a simple frequency-independent damping term, the soft mode always assumes a diffusive character close enough to the Curie temperature and can indeed represent the typically overdamped motion of hydrogen-bonded ferroelectrics in a fairly realistic fashion. The most significant difference between the proton–lattice model and the simplest local-mode picture is seen in the dynamics, where the former always predicts at least a pair of temperature-dependent modes while the latter has only one. In the proton-lattice picture the 'in-phase' soft mode is always paired with an 'out-of-phase' proton–lattice mode which also has a temperature-dependent energy but which remains 'hard' at the phase transition. If experimentally one sees only a single soft mode then the explanation within the model is that the coupling is 'tight'. If a second (hard) temperature-dependent mode is seen then the coupling is presumably more relaxed. This latter circumstance can also be explained in the local-mode picture if the model is embellished to include direct coupling to a higher-energy phonon branch. In this case (Moore and Williams 1972) the mathematical expression of the two models becomes formally very similar and differences primarily interpretational.

In summary, what can be said about the differences, advantages, disadvantages, etc. of the various theoretical approaches to the KDP problem? The local-mode and proton–lattice models can really be grouped together as an attempt to obtain simple running-wave solutions. These are ideally suited for the interpretation of scattering or indeed any wavevector-dependent phenomena. The difference is that in its simplest form the local-mode picture involves but a single configurational variable, and therefore assumes a rigid coupling of all unit-cell atoms as regards the phase and character (displacive, tunnel, order–disorder, etc.) of motion, whereas the proton–lattice model allows an extra degree of freedom to enable the description of situations in which the character of the protonic and heavy-ion motions is different. Most importantly, however, both theories ignore the intracell motion which corresponds to excited Slater levels unless complicated couplings to other lattice modes are introduced. These not only complicate the theories unduly but introduce vast numbers of coupling parameters about which little is known *a priori*. Finally, both the running-wave methods can be formally justified only at long wavelengths since they fail to take into account any wavevector dependence of the exact eigenvectors.

The great advantage of the Slater–Takagi model (suitably embellished to include tunnelling and long-range interactions, e.g. Vaks and Zinenko 1973) is that it is able to take account of the detailed short range *intra* cell proton–proton interactions in a simple physical approximation. Since the character of the phase transition seems to be quite sensitive to this feature of the KDP structure (for example the sharpness of the transition), its neglect in the simple running-wave solutions may be of some serious consequence. The basic weakness of the model is that it is as yet unsolved in its general form by any method which takes advantage of translational symmetry. The simple 'cluster' solutions of Blinc and Svetina (1966) take some account of short-range correlations but have not been cast in a form convenient for the discussion of wavevector-dependent response. They are therefore of limited use in interpreting those dynamical characteristics at long wavelengths which are now measurable by resonance and scattering techniques and which presumably contain such detailed information concerning the fundamental nature of the ferroelectric transition in the KDP structure.

9.2. KDP-type ferroelectrics (experiment)

9.2.1. KH_2PO_4 (KDP) and KD_2PO_4 (DKDP)

Ferroelectrics with the basic KH_2PO_4 crystal structure have attracted an enormous amount of interest over several decades. From an experimental standpoint the reasons are practical; large single crystals can be grown easily from water solution and they are in general of high optical quality. As a result they yield useful technological devices such as high-speed electro-optic modulators. Theoretically, as we have seen in the previous section, the occurrence of ferroelectricity in these materials is associated in a large degree with proton tunnelling, but a strong coupling to the heavy ions is necessarily involved in an essential way.

One of the major problems in interpreting the wealth of dynamical experiments on KDP and DKDP themselves has, until quite recently, been the paucity of detailed structural knowledge. The static structure determination for KDP by Bacon and Pease (1955) long remained the only careful study, but it was unable to distinguish whether the hydrogen nuclei were making large anharmonic vibrations about the O—H—O bond centre point in the disordered phase or whether they were disordered in two wells. For many years no static structure determination was attempted for the deuterated crystal at all. Recently the situation has become clearer. Incoherent elastic and coherent inelastic neutron scattering techniques have been used to study the proton motion and distribution in KDP and the deuteron motion in DKDP. Plesser and Stiller (1969) found the proton distribution in KDP at room temperature to be concentrated at sites of order $d = 0.04$ nm apart, with the line joining the sites being inclined to the c-plane (where c is the ferroelectric axis) at $\theta = 6 \pm 3°$. Wallace, Cochran, and Stringfellow (1972), using the dynamical neutron data of Skalyo $et\,al.$ (1970), find a value $\theta \approx 22°$ for DKDP. Most recently, Nelmes, Eiriksson, and Rouse (1972) have reanalysed the data collected by Bacon and Pease using more modern least-squares refinement techniques and also report room-temperature neutron data on single-crystal DKDP. They conclude that it is almost certain that both KDP and DKDP have double-minimum wells. In addition they find that the inter-minimum distance is larger ($d = 0\cdot044$ nm) in the deuterated material than in KDP itself ($d = 0\cdot034$ nm), and that the line joining the minima is indeed inclined to the c-plane and to the oxygen–oxygen direction at a greater angle in DKDP than in KDP. It follows that isotopic effects on the local potential are somewhat more significant than had commonly been assumed.

The effects of deuteration on the static ferroelectric properties were also for many years rather ill defined. The one well-known, very dramatic effect was on the transition temperature which increases from 123 K in KDP to 223 K in DKDP. Deuteration effects on other properties, such as the paraelectric Curie constant C (i.e. dielectric constant $\varepsilon = C/(T - T_0)$), saturation polarization P_0, etc., were generally assumed to be small primarily because of the wide variation of data among the authors and the lack of a definite trend. That this assumption was erroneous has recently become apparent and has been particularly clearly demonstrated by Samara (1973) by a careful study of $K(H_{1-x}D_x)_2PO_4$ in crystals with nominal values of $x = 0, 0\cdot35, 0\cdot80$, and $0\cdot98$. His results are shown in Table 9.2 and indicate significant increases in both C and P_0 upon deuteration. The very large increase in coercive field on deuteration points to a very marked change in domain-wall dynamics in going from KDP to DKDP. In fact the most striking isotope effect of all (Bjorkstam and Oettel 1966) is a reduction of six orders of magnitudes in domain-wall mobility on deuteration. With the now recognized substantial dependence of double-well separation and P_0 on deuteration it is no longer quite so self-evident that a consistent explanation of the large T_c isotope effect requires the concept of tunnelling, since the interaction J_{ij} of eqn (9.1.2) may now be substantially different in the two cases. Other experimental differences, resulting from calorimetric measurements, particularly in the vicinity of the phase transition, have been reported

TABLE 9.2

Summary of the effects of deuteration in a series $K(H_{1-x}D_x)_2PO_4$ of KDP-type crystals

x	$T_c(K)$	$T_0(K)$	$C(K)$	P_0 $(\mu C\,cm^{-2})$	E_s $(kV\,cm^{-1})$	E_c $(V\,cm^{-1})$
0	$122\cdot5\pm0\cdot2$	$122\cdot5\pm0\cdot1$	2910 ± 50	$4\cdot95\pm0\cdot07$	$\geqslant5$	160
0·35	$159\cdot0\pm0\cdot2$	$158\cdot9\pm0\cdot1$	3450 ± 50	$5\cdot40\pm0\cdot07$	$\geqslant10$	380
0·80	$204\cdot8\pm0\cdot2$	$204\cdot0\pm0\cdot2$	3880 ± 50	$6\cdot05\pm0\cdot07$	$\geqslant20$	1560
0·98	$219\cdot8\pm0\cdot1$	$218\cdot3\pm0\cdot1$	4020 ± 50	$6\cdot21\pm0\cdot07$	$\geqslant32$	3380

These data are taken from Samara (1973).

T_c is the transition temperature, T_0 the extrapolated Curie–Weiss temperature, C the Curie constant, P_0 the saturation polarization, E_s the field required to produce saturation and E_c the coercive field. P_0 is measured at 30–40 K below T_c.

For $x = 0$ the measured P_0 was still increasing slightly with decreasing T. The true saturation value for KDP may be as high as $5.10\ \mu C\,cm^{-2}$.

by Reese (1969 a) and Benepe and Reese (1971). Both KDP and DKDP exhibit first-order phase transitions (although that of KDP is only just first order) and show no clear evidence of critical phenomena (i.e. are well interpreted by a Devonshire-type free energy), except possibly in specific heat (Fig. 5.17).

Recent attempts to understand the isotope effects (Kopsky 1971, Holakovsky, Brezina, and Pacherova 1972, and Vaks and Zinenko 1973) are in essential agreement that a tunnelling energy $\Delta \sim 200$–300 K (a measure $\hbar\Omega = \frac{1}{2}\Delta$ is often used in the literature) is present in KDP but that tunnelling is not a significant factor in DKDP. The most complete calculation (by Vaks and Zinenko) uses the Blinc–Svetina cluster model and includes both long-range dipolar forces and tunnelling. It concludes that the first Slater energy level (E_1 of Table 9.1) is at an energy of about 100 K, that the second level E_2 is at an energy of about 1000 K (see also Table 9.3), and that the physically reasonable assertion

$$\frac{(E_1)_H}{(E_1)_D} \approx \frac{(E_2)_H}{(E_2)_D} \approx \frac{(P_0^2)_H}{(P_0^2)_D} \approx \frac{(d^2)_H}{(d^2)_D} \qquad (9.2.1)$$

seems consistent with the data. Only part of the T_c and P_0 isotopic shifts now depend on tunnelling, in fact less than half the T_c shift and perhaps only a small part of the P_0 shift, P_0 being essentially determined by eqn (9.2.1). The most difficult static property to understand self-consistently within the complete scheme is the first-order character of the phase transition for KDP. Experimentally the transition is barely first-order with $T_c - T_0$ only of order 0·05 K (Strukov *et al.* 1972), and Vaks and Zinenko suggest that it may be precipitated by striction, i.e. a coupling to elastic forces hitherto ignored in the theory. (Effects of striction within a Kobayshi model have recently been studied by Nettleton (1974, 1975).) The variation of the first-order character of the phase transition on deuteration is shown in Fig. 9.4. KDP therefore

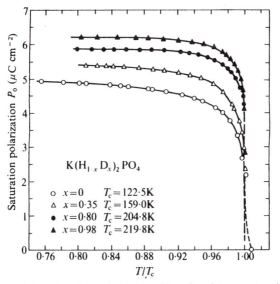

Fig. 9.4. Variation of the saturation polarization with reduced temperature in mixed KDP–DKDP crystals. The indicated transition temperatures were determined from dielectric constant data. (After Samara 1973.)

seems to be very close to a Curie critical or tricritical point (i.e. a point at which a phase transition changes from second to first order). Near such a point one might expect to observe critical opalescence in light scattering (§ 7.3.3), but efforts to detect such a phenomenon in KDP have thus far failed (Lyons, Mockler, and O'Sullivan 1973).

Efforts to interpret isotope effects using running-wave methods in general have most difficulty in explaining the sharpness of the phase transitions (Houston and Bolton 1971), since this property in particular seems to be very sensitive to the details of the near-neighbour proton–proton force characteristics which are not correctly included in these models. Nevertheless, other isotopic effects can be explained within the proton–lattice model (for example) and suggest (Cochran 1969, Pak 1973, Peercy 1974) that the coupling between protons and phonons is strong and is primarily responsible for the lattice instability, i.e. that the $F(0)$ term in eqn (9.1.16) is very significant. This idea of a phonon-induced spin coupling has been questioned by Moore and Williams (1972), but it would seem to be consistent with the notion that the long-range dipolar forces (between c-axis lattice dipoles) are playing an important role, as witnessed by the critical neutron

scattering of Figs 6.8 and 6.9. The concept is also not inconsistent with the
'cluster' finding of Vaks and Zinenko that inter-cluster interactions are of
considerable magnitude.

Support for the importance of proton tunnelling in KDP comes from the
high-pressure dielectric constant measurements of Samara (1971). Samara
has investigated the effects of temperature and pressure on the dielectric
properties of KDP using oriented single-crystal samples. Curves of c-axis
dielectric constant *versus* temperature for a series of pressures are shown in
Fig. 5.24. As the pressure is increased the transition temperature T_c
decreases and eventually vanishes at a pressure of 17 kbar. Blinc, Svetina,
and Zeks (1972) have pointed out that this behavior is in accord with a
tunnelling model, since the variation of external pressure is perhaps the one
way of changing the ratio of the tunnelling to the longitudinal (dipolar) field
externally (see § 5.8). Since the distance between the two potential minima
along a hydrogen bond decreases with increasing presure, the value of the
tunnelling parameter Δ should increase and the dipolar field $J(0)$ decrease.
Within the model the transition temperature, given by eqn (2.5.17),
decreases to zero when $\Delta = \frac{1}{2}J(0)$. The theory also explains at least qualitatively the shape of the T_c *versus* pressure plot which approaches an infinite
slope at $T_c \to 0$. As Blinc, Svetina, and Zeks note, it is certainly not necessary
for the proton double well to change over to single-well form in order to lose
the ferroelectricity. In conformity with this general picture the effect of
pressure on DKDP is less than on KDP (Samara 1970 a).

To this point we have concerned ourselves with the equilibrium properties. A more detailed understanding of the microscopics should be available
from an interpretation of dynamic measurements. Temperature-dependent
modes have been sought in a number of optical experiments. Infrared
measurements on KDP by Barker and Tinkham (1963) showed a relaxation
behaviour of the dielectric response at relatively low frequencies. Kaminow
and Damen (1968), using Raman scattering, detected a heavily damped
mode but found that it could be well described with a resonant-frequency
softening as $\{(T - T_c)/T\}^{1/2}$. This interpretation has not gone unquestioned,
but the fact that this mode is indeed of propagating character was finally
confirmed by Peercy (1973) who studied it as a function of pressure (see Fig.
7.9). Thus, at least above the phase transition, the approach to T_c is
characterized by a broad response over $\sim 150 \text{ cm}^{-1}$, which is centred (at
amospheric pressure) at zero frequency, but with width decreasing and
intensity increasing as $T \to T_c^+$ in qualitative agreement with a soft-mode
model. A quantitative analysis shows, however, that the spectrum is complicated by an interaction of this overdamped soft mode with another mode of
the same symmetry lying at about 180 cm^{-1}. Representing both modes by
simple damped oscillators (She *et al.* 1972), and including a simple coupling,
the spectra can be well interpreted in terms of a soft and hard pair of modes

reminiscent of Kobayashi's proton–phonon model. In fact Peercy and Samara (1973) have noted that in the limit of negligible damping the coupled-mode response and the Kobayashi finding (eqn (9.1.17)) can be written in an identical form. In any event the coupled-mode spectra yield directly a pair of uncoupled frequencies and an interaction term. She *et al.* (1972) find that the uncoupled tunnel mode, which in this picture is the bare-proton mode, does not go soft until extrapolated to $T'_c \approx 30$ K (compared with the actual $T_c = 123$ K). Since T_c is approximately proportional to the effective 'exchange, which from the last section is $J(0) + NF^2(0)\Omega^{-2}(0)$, while T'_c is proportional to $J(0)$, the ratio of the direct to coupled energies is of order 1 to 3 in this estimate and the great importance of the proton–lattice coupling $F(0)$ is apparent.

Below T_c in KDP the Raman scattering measurements (C. M. Wilson 1971), Shigenari and Takagi 1971) give first a broad zero-centred spectrum but at lower temperatures ($T_c - T \sim 20$ K) show well-defined inelastic side peaks in addition to the quasi-elastic scattering. This behaviour (Blinc, Lavrencic, *et al.* 1973) is characteristic of the tunnel-mode theory as given in § 2.5 with a pair of propagating 'transverse' spin-wave branches framing a diffusive 'longitudinal' one. This interpretation has yet to be thoroughly tested in a quantitative fashion, but the disappearance of the side peaks upon deuteration (Blinc, Lavrencic, *et al*, 1973) lends support to the picture. The fact that the soft mode is overdamped in KDP when $T > T_c$ and underdamped when $T < T_c$ would then be explained by the much larger real frequency component below T_c (by virtue of the $J(0)\langle S^z \rangle_0$ term in eqn (2.5.31)). Note that for a tunnelling energy $\Delta \to 0$, as anticipated physically for DKDP, the equivalent 'transverse' modes do not contribute to polarization fluctuations since their eigenvectors have no components in the direction of the spontaneous polarization (§ 2.5). The anticipated spectrum in this limit is then of relaxation character (from the longitudinal mode) which is as observed. Whether there are mode-interaction complications below T_c similar to those existing above T_c is an interesting question, and recent pressure-dependent Raman measurements by Peercy (1975 a) suggest that there are.

The full story of the KDP ferroelectric transition has still not been told, however, since on approach to T_c from above the coupled proton–lattice soft mode ω_- of eqn (9.1.17) itself interacts piezoelectrically with xy-shear acoustic modes (Pytte 1971 a). The soft proton–lattice mode pushes the acoustic phonon frequency with which it couples to zero at a temperature a little higher (123 K) than that at which the former mode would extrapolate to zero (116 K). Thus the actual KDP transition is then manifested by the vanishing of an elastic constant (actually c_{66}) which can be observed by ultrasonic or Brillouin techniques (see Chapter 10). Details for KDP (Fig. 9.5) have been obtained by Garland and Novotny (1969) and by Brody and

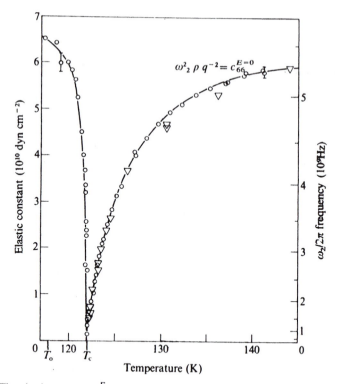

Fig. 9.5. The elastic constant c_{66}^E as a function of temperature in the neighbourhood of the ferroelectric transition in KDP. Triangles denote ultrasonic measurements after Garland and Novotny (1969) and circles denote Brillouin shifts after Brody and Cummins (1968).

Cummins (1968, 1969). Also very close to the transition a quasi-elastic component with dramatically increasing intensity appears in the dynamic response (Lagakos and Cummins 1974) (Fig. 11.8). Such a component seems to be a rather common feature of critical scattering near structural transitions (see §§ 7.3 and 11.5) and is possibly 'critical opalescence' connected with a difference between adiabatic and isothermal response, the mode appearing when the periodic time of oscillation of the soft mode becomes less than the time required for local temperature equilibrium to be established. Other mechanisms, however, are possible and indeed, since the width of the central peak has still to be resolved, it may yet be of a static nature and result simply from defects or impurities.

For DKDP the most detailed studies yet available (Paul *et al.* 1970, Skalyo *et al.* 1970, Reese, Fritz, and Cummins 1973) all point to a very narrow soft mode centred on zero frequency (critical dispersion range $1–10 \text{ cm}^{-1}$). Ultrasonic studies by Litov and Uehling (1968, 1970) suggest a soft mode of Debye form with a polarization reciprocal relaxation time

$\tau^{-1} \approx 4 \times 10^9 (T - T_c) \, s^{-1} \, K^{-1}$. The mode is nicely accessible to Brillouin scattering experiments (Reese *et al.* 1973, see Fig. 10.3) which confirm the ultrasonic findings. Very close to T_c there is a strong interaction with acoustic vibrations similar to the situation discussed for KDP, except that in DKDP the soft ferroelectric mode falls in the same frequency region as the acoustic modes and both can be observed simultaneously in the Brillouin experiments.

Because of the smaller incoherent scattering background inelastic coherent neutron scattering experiments are more easily performed for DKDP than for KDP itself. Skalyo *et al.* (1970) find quasi-elastic scattering dominantly in a plane normal to the polar c-axis and of a character which points to the importance of dipolar forces between c-axis dipoles (see Figs 6.8 and 6.9). A mode analysis to determine the displacement vectors in the soft mode showed surprising features, including a large distortional motion of the oxygen tetrahedra, but the analysis was based on KDP structural parameters (since those for DKDP were not available at the time of the Skalyo *et al.* work) and there are indications that the recently determined differences in static structure may influence this finding (Nelmes *et al.* 1972).

Until very recently all the essential dielectric information concerning the phase transition in KDP had been assumed contained in the response parallel to the polar axis. As a result the subject of transverse susceptibility has been largely ignored. However, it now appears (Havlin, Litov, and Uehling 1974) that the transverse susceptibility also provides evidence of the phase transition and that it can be used to provide a numerical value for the tunnelling energy. The important point to recognize is that, in contrast to the case of an applied field along the ferroelectric c-axis, for which all hydrogen bonds are equivalent and the physical state is adequately described by a single ordering parameter, a transverse applied field (say along the a- or b-axis) sees in the ordered state a chain of hydrogen bonds with hydrogen ions alternately on the 'left' or 'right' of their bonds (Fig. 9.3). The basic tunnel-mode formalism can still be used if this antiferroelectric characteristic of the transverse response is recognized, and leads to the prediction of a cusp-like singularity in the transverse susceptibility at T_c, Fig. 9.6. Comparison with experiment for KDP is good and leads to a tunnelling energy $\Delta \sim 200$ K in general accord with estimates from other measurements. The equivalent findings for DKDP have been given by Havlin *et al.* (1976). It has also been suggested (Litov and Havlin 1974) that, since KDP is piezoelectric in the transverse direction at all temperatures, the antiferroelectric-like character of the transverse dielectric response is directly responsible for an analogous anomaly in the transverse (d_{14}) piezoelectric constant.

Since KDP and DKDP both exhibit first-order phase transitions and, being ferroelectric, are expected to have very narrow critical regimes near T_c

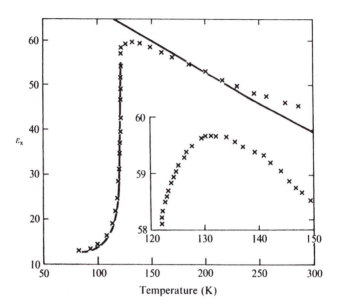

Fig. 9.6. The transverse dielectric constant ε_x of KDP as a function of temperature near the ferroelectric T_c. A more detailed plot of the temperature range immediately above $T_c \approx 122$ K is shown inset. The solid curve represents a theoretical fit to the data using a cluster approximation. (After Havlin, Litov, and Sompolinsky 1975.)

even in a second-order limit (see Chapter 11), observation of non-Landau behaviour near T_c is expected to be difficult. Even in KDP, for which the first-order character is very slight, most thermodynamic measurements conform with a simple Devonshire-like description (Benepe and Reese 1971). There is some evidence (Reese 1969 b, Garber and Smolenko 1973) that non-Landau behaviour can be seen in the specific heat for KDP within a few hundredths of a degree of the transition, but the effects of crystal imperfections etc. in this region are not well documented nor understood and the evidence must as yet be deemed inconclusive. One interesting point, however, is that the specific heat at constant polarization (as well as that at constant field) undergoes an anomaly at T_c suggesting that a portion of the calorimetric anomaly in KDP is not directly connected with the polarization discontinuity.

One should perhaps caution the reader that the applicability of Landau or Devonshire-like free energies does not necessarily imply a numerical accuracy of mean-field theory. This is particularly true for systems (like ferroelectrics) in which long-range dipolar forces play a role. In such cases theories like the correlated effective-field approximation of § 2.2, which incorporate a measure of spontaneous fluctuations, differ very significantly in their numerical predictions from mean-field theory but still (with only logarithmic corrections) give rise to mean-field critical exponents (Lines 1972 a). In particular Cochran (1969) has noted a serious numerical failure of some specific mean-field relationships when applied to KDP and DKDP.

9.2.2. Other KDP ferroelectric isomorphs

Substitution of potassium with rubidium or caesium, or of phosphorus with arsenic, produces crystals isomorphous with KDP which in general undergo ferroelectric phase transitions of the KDP type. In each case a large increase in Curie temperature is brought about by deuteration, and it seems clear that the basic mechanism responsible must be broadly similar to that discussed for KDP and DKDP.

RbH_2PO_4 (RDP) has a ferroelectric transition at $T_c \sim 145$ K and there is some evidence that the phase transition is of second order (Strukov et al. 1973 b) in both hydrogenous and deuterated forms. Even if true, however, no accurate measurements of critical exponents are yet available, although there is some indication of a logarithmically diverging specific heat. The basic character of the transition is little affected by deuteration but there is a marked increase in T_c (to ~ 223 K) and smaller increases in saturation polarization ($5\cdot3$–$6\cdot3$ μC cm^{-2}) and Curie constant (3000–3800 K). A close analogy with the KDP and DKDP data in Table 9.2 is apparent.

Since RDP has a larger compressibility than KDP it is better suited for the investigation of pressure-dependent effects, and the first pressure-dependent Raman measurements for KDP-type materials have been reported (see Fig. 5.23) for RbH_2PO_4 by Peercy and Samara (1973). As with KDP the transition temperature T_c decreases with increasing pressure and eventually vanishes (for RDP with a critical pressure of about 15 kbar). Interpreting the Raman spectrum in terms of a coupled-mode representation Peercy (1974) used the Kobayashi proton–lattice analogy to evaluate the values and pressure dependence of all the pertinent model parameters. He was able to demonstrate that with these parameters the Kobayashi theory is able to account for the temperature and pressure dependence of the dielectric properties in a self-consistent fashion. He calculates, in particular, a tunnelling parameter $\Delta = 157$ cm^{-1} and an effective proton–proton interaction (including the phonon coupling, see eqn (9.1.16)) of 473 cm^{-1}. One additional point of interest is that for pressures in excess of the critical pressure, for which the material remains paraelectric down to absolute zero, strong deviations from Curie–Weiss response are found below about 50 K. These are reminiscent of those found for the incipient ferroelectrics $SrTiO_3$ and $KTaO_3$ and probably have a similar explanation, associated with quantum lattice effects, as first suggested by Barrett (1952) (see eqn (8.1.1)).

Rather extensive Raman work has also been reported on the KDP-type arsenates. In fact the work of Katiyar, Ryan, and Scott (1971) on CsH_2AsO_4

(CDA) and KH_2AsO_4 (KDA) was the first to stress the importance of a coupled-mode representation for understanding the low-frequency Raman spectra in KDP-type materials. Noting that the soft mode had an unusual spectral shape which could not be described by a single anharmonic-oscillator form (nor as a Debye relaxation spectrum), they were able to find an explanation in terms of an anti-resonant interference between the ferroelectric mode and a low-frequency B_2 phonon. Representing both by damped oscillator expressions they expressed the coupled-mode response by

$$\chi''(\omega) = \text{Im} \sum_{i,j} P_i P_j G_{ij}(\omega) \qquad (9.2.2)$$

where P_i is the strength of the mode i and $G_{ij}(\omega)$ are the solutions to the coupled equations

$$\begin{pmatrix} \omega_i^2 - \omega^2 + i\omega\Gamma_i & \omega_{ij}^2 + i\omega\Gamma_{ij} \\ \omega_{ij}^2 + i\omega\Gamma_{ij} & \omega_j^2 - \omega^2 + i\omega\Gamma_j \end{pmatrix} \begin{pmatrix} G_{ii} & G_{ij} \\ G_{ij} & G_{jj} \end{pmatrix} = \begin{pmatrix} 1 & 0 \\ 0 & 1 \end{pmatrix}. \qquad (9.2.3)$$

Although eqn (9.2.3) enables an excellent fit to be obtained to the low-frequency Raman spectrum it is unfortunately overdetermined (corresponding to an arbitrary choice of phase for the intereacting modes) with the two limiting cases occurring for real coupling ($\Gamma_{ij} = 0$) and imaginary coupling ($\omega_{ij} = 0$). With real coupling and zero damping $\Gamma_i = \Gamma_j = 0$ we recover Kobayashi's solutions (9.1.17), Where ω_i and ω_j are identified as the proton and lattice frequencies $\Omega_B(\mathbf{q})$ and $\Omega(\mathbf{q})$ respectively. Much attention has been paid (e.g. Cowley *et al.* 1971, Lowndes, Tornberg, and Leung 1974) to the experimental values obtained for ω_i and ω_j in the real-coupling and imaginary-coupling limits. The values differ for the two limits, although both indicate qualitatively that one frequency is essentially independent of temperature while the other tends to vary as $(T - T_c')^{1/2}$ where T_c' is less than the Curie temperature T_c, very significantly so in many KDP-type systems with the choice of real coupling. The attention given to ω_i and ω_j seems to us misplaced, the important frequencies being those at which the coupled response diverges, namely the solutions of

$$\begin{vmatrix} \omega_i^2 - \omega^2 + i\omega\Gamma_i & \omega_{ij}^2 + i\omega\Gamma_{ij} \\ \omega_{ij}^2 + i\omega\Gamma_{ij} & \omega_j^2 - \omega^2 + i\omega\Gamma_j \end{vmatrix} = 0. \qquad (9.2.4)$$

In the case of negligible damping and real coupling these solutions are just the $\omega_\pm$ of the Kobayashi model. More generally they are complex, and the detailed nature of these modes and their dependence if any on the choice of phase of the coupling term has yet to be fully explored.

The most detailed light scattering measurements on KDP-isomorphs to date are those of Lowndes *et al.* (1974). These authors have reported measurements of the temperature dependence of the Raman active modes (both above and below T_c) in KDA, RDA, and CDA and also in their deuterated isomorphs. A thorough examination of these spectra has resulted in the identification of internal $(AsO_4)^{3-}$ modes and the O—H—O valence vibration bands as well as the collective soft modes. The internal arsenate modes are similar to those observed in solution and in other compounds, while the O—H—O vibrations exist as broad bands in the 1000–3000 cm^{-1} region. The latter are lowered by about 25 per cent on deuteration. The breadth of many of the spectral features diminishes markedly in the ordered phase suggesting that the linewidth is related to proton or deuteron disorder above T_c. Finally there is some evidence, particularly in the deuterated compounds, that the ferroelectric mode couples to more than one optic phonon so that a two-coupled-oscillator fit can at best describe these spectra in an approximate fashion.

According to Blinc, Burgar, and Levstik (1973), RDA, KDA, and CDA all possess ferroelectric phase transitions of first order. Thermodynamic properties have been measured by Fairall and Reese (1972, 1974) and by Strukov et al. (1973 a, b). In general the substitution P^{5+} to As^{5+} intensifies the first-order characteristics of the phase transitions, and the properties are well explained by the tunnelling 'cluster' theory of Vaks and Zinenko (1973) but with tunnelling frequencies which are considerably lower than in the analogous phosphate crystals. In addition all the crystals studied show a clear tendency to decrease their degree of first-order character on increasing the radius of the univalent ion $(K^+ \rightarrow Rb^+ \rightarrow Cs^+)$. These characteristics and other relevant data for this series of ferroelectrics and their deuterated isomorphs are given in Table 9.3.

TABLE 9.3

Curie temperature T_c, *proton configurational energy levels* E_1 *and* E_2,† *tunnelling parameter* Δ, *and degree of first order character* $P(T_c)/P(0)$ *of a series of KDP isomorphs*

	KDA	RDA	CDA	KDP	RDP	DKDA	DCDA	DKDP
T_c(K)	96	110	146	122	147	161	190	220
E_1/kT_c	0·361	0·405	0·455	0·494	0·510	0·345	0·440	0·513
E_2/kT_c	4·34	4·20	3·97	5·80	4·43	4·26	3·90	4·90
Δ/kT_c	0·50	0·60	0·40	1·80	1·0	0	0	0
$P(T_c)/P(0)$	0·862	0·755	0·625	0·365	0·300	0·922	0·701	0·724

Data from Fairall and Reese 1975.
† See Table 9.1.

9.3. $(NH_4)H_2PO_4$: a KDP-type antiferroelectric

The phase transition in ammonium dihydrogen phosphate (ADP) has been known for almost 40 years, although for a long time it was thought to be an ammonium rotation transition and basically different from the ferroelectric KDP-type transition. It was Nagamiya (1952) who recognized the transition as antiferroelectric and who demonstrated that many of its characteristics could be understood in terms of a Slater-type ferroelectric model but with the energy E_1 of Table 9.1 negative. He also proposed a pattern of proton ordering below T_c which produces an antiparallel arrangement of electric dipoles in an a- (or b-) axis direction. An ordered phase of precisely this nature has since been confirmed for ADP (Hewat 1973 c).

Above its antiferroelectric transition point ADP is isomorphic with KDP. The uniform static susceptibility is of Curie–Weiss form both parallel and perpendicular to the c-axis but with a *negative* Curie–Weiss temperature in both directions. At a temperature $T_c \approx 150$ K a sharp first-order transition to the antipolarized state takes place accompanied by pronounced thermal hysteresis and a discontinuous drop in both dielectric susceptibilities. The crystals invariably shatter on entry into the ordered phase because different domains distort in different directions (i.e. either the a- or b-directions can become antiferroelectric axes below T_c). This fact has precluded the taking of precise dielectric or electromechanical measurements in the ordered phase. The final determination of the structure of the ordered phase by Hewat (1973 b) was made on a powdered specimen using the neutron powder profile refinement technique recently perfected by Rietveld (1969).

The crystal structure as deduced by Hewat (a deuterated sample was actually used for the neutron experiment) is most simply described by referring to the deuteron positions. Table 9.4 gives the co-ordinates for the two equivalent deuteron sites above T_c (when each is occupied for half the time), and in the low-temperature phase where we find that only one of these alternative positions is occupied. We notice that unlike KDP, where two 'upper' or two 'lower' proton sites (with respect to the c-axis) are filled for an arbitrary H_2PO_4 group in the ordered phase, ordered ADP has one 'upper' and one 'lower' site filled. Figure 9.7 shows how these positions are taken up to produce a perfectly ordered arrangement of bonds with electric dipoles in c-planes and arranged in an antiferroelectric pattern. This is precisely the ordering proposed back in 1952 by Nagamiya. The phosphate group at $(\frac{1}{2}\frac{1}{2}\frac{1}{2})$, which is equivalent to that at (000) in the high-temperature phase, becomes inequivalent in the low-temperature phase. These phosphate groups, and the ammonium groups too, are slightly distorted by the proton (or deuteron) ordering (the ammonium groups are displaced as well), but the resulting dipole moments cancel exactly and the ordered phase is indeed antipolar.

Given this structure an explanation in terms of a Slater model with negative E_1 looks immediately promising. Such a negative E_1 certainly favours an ordering of the ADP type over KDP ferroelectricity. A closer look, however, shows that the above assumption alone cannot predict the ADP-type transition (Ishibashi, Ohya, and Takagi 1972). If the internal

TABLE 9.4

Fractional co-ordinates of the deuterium atoms on the hydrogen bonds in deuterated ADP

	a = b = 7·502 Å, c = 7·520 Å, T = 22°C Paraelectric phase				a = 7·507 Å, b = 7·529 Å, c = 7·445 Å, T = 77·4 K Antiferroelectric phase			
	X	Y	Z	Occupation	X	Y	Z	Occupation
D_1	0·277(3)	0·144(1)	0·140(4)	$\frac{1}{2}$	0·275(3)	0·144(3)	0·134(3)	1
D_2	0·223	0·144	0·110	$\frac{1}{2}$				0
D_3	−0·144	0·277	−0·140	$\frac{1}{2}$	−0·146	+0·277	−0·145	1
D_4	−0·144	0·223	−0·110	$\frac{1}{2}$				0

From Hewat 1973 b.

In the low-temperature phase ($T_c = 150$ K) these atoms are ordered on one of the two sites over which they are distributed in the high-temperature phase.

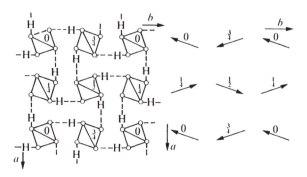

Fig. 9.7. The ordering of the protons on the hydrogen bonds, and the resulting antiferroelectric dipole arrangement, of $NH_4H_2PO_4$ in the antiferroelectric phase. The fractions $0, \frac{1}{4}, \frac{1}{2}, \frac{3}{4}$, denote the c-axis co-ordinates of the PO_4 groups. (After Nagamiya 1952.)

energy consists only of the orientational energy of $(H_2PO_4)^-$ radicals and if the 'ice' constraint is imposed (i.e. only one proton per hydrogen bond and only two protons attached to each PO_4 group), then both ferroelectric and antiferroelectric dipolar arrangements are equally possible in the c-plane under the assumption of negative E_1. In other words, if there is no additional energy which selects one ordered state over another, several perfect-order states are possible for ADP at 0 K within the ice model. According to Ishibashi *et al.* (1972) the inclusion of the dipolar interactions is sufficient to stabilize the ADP structure of Fig. 9.7. Including these interactions as an effective field in a Slater cluster environment, the theory predicts a first-order phase transition to an ordered state in which the sublattice polarization is saturated at all $T < T_c$. Unfortunately the degree of sophistication for antiferroelectric KDP-type theories has not yet progressed to incorporate either proton tunnelling or proton–lattice coupling, although it seems evident from the effect of deuteration on T_c (T_c increases from about 150 K to about 235 K on deuteration in ADP) that tunnelling is just as important in ADP as it is in KDP. Extension of theory in this manner would certainly seem to be feasible at this time but progress has been much slower in the antiferroelectric context (both experimentally and theoretically) than for the analogous ferroelectric problem.

Part of the reason for this slow progress in KDP-type antiferroelectrics is the fact that the soft mode responsible for the transition occurs at the edge of the Brillouin zone as T_c is approached from above and is therefore not accessible to light-scattering experimentation. Although a ferroelectric mode shows some sign of softening too, an understanding of the mechanism of the transition obviously requires a knowledge of short-wavelength dynamics, the details of which are readily accessible only to inelastic neutron study. In the ordered phase, where the soft mode would appear at the centre of the reduced Brillouin zone, the situation is experimentally difficult because of the tendency of crystals to shatter on ordering. Nevertheless, some information can be gleaned from light scattering and both infrared and Raman measurements have been reported.

Infrared absorption spectra for ADP (and for the isomorphous arsenate which is also antiferroelectric) have been studied by Wiener-Avnear, Levin and Pelah (1970) between 250 and 4000 cm^{-1}. Above T_c there is a close resemblance between the spectrum of ADP and KDP except, of course, for the additional ammonium peaks in ADP. This close similarity suggests that the same proton–lattice coupling postulated for KDP (between tunnelling protons and PO$_4$ clusters) also exists in the ammonium salt causing a similar broadening of most of the PO$_4$ peaks. On crossing T_c or upon deuteration a marked narrowing of these peaks occurs in both KDP and ADP, again pointing to similar proton–lattice coupling. Below T_c, however, the spectra of the two salts are very different, extra lines appearing in ADP owing to the destruction of the two-fold symmetry about the c-axis in the antiferroelectric structure (see Fig. 9.7).

The low-frequency Raman spectra of ADP have been reported by Ryan, Katiyar, and Taylor (1972) and by Broberg et al. (1972). In the high-temperature phase there are B$_2$- and E-symmetry Raman-active proton modes which are expected to interact with like-symmetry lattice modes in both KDP and ADP. In KDP it is the B$_2$ mode which is soft and overdamped and which precipitates the phase transition to the ferroelectric state. Although, as mentioned above, neither zone-centre mode is expected to be critically soft in ADP, the temperature-dependent behaviour of these long-wavelength modes is of interest since they dominate the contributions to dielectric constant in the crystal. The observed B$_2$ spectrum was analysed in terms of the interacting-mode model (eqn (9.2.2)) and the (bare) tunnel mode ω_i, Γ_i was found to be overdamped. Both the square of the frequency ω_i^2 and the static susceptibility $\chi(B_2) = \Sigma_{i,j} P_i P_j G_{ij}(0)$ are found to depend linearly on temperature, and the susceptibility, in particular, compares reasonably well with the temperature-dependent microwave data of Kaminow (1965). The E-mode analysis is more complicated since the bare proton mode apparently couples with several optic phonons and a two-mode analysis is correspondingly crude. Despite this difficulty a fairly reliable estimate has been obtained for the relaxation time $\tau = \Gamma_i/\omega_i^2$ of the overdamped protonic E mode and shown to vary with temperature as $T/\tau \propto (T - T_0)$ with $T_0 \approx 80$ K. The antiferroelectric transition at about 150 K takes place long before either the B$_2$ or E zone-centre modes approach their condensation.

An ordered phase with the character of Fig. 9.7 would, in a tunnel or proton–lattice formalism, be produced by the condensation of an antiferroelectric mode at the zone boundary along the c^* axis (called the Z-point) where c^* is the reciprocal lattice vector in the c-direction. This mode has actually been observed by quasi-elastic neutron scattering for the deuterated material DADP (Meister *et al.* 1969). While an analysis of the detailed nature of the soft mode has yet to be made, the quasi-elastic scattering near the Z-point confirms the existence of a soft mode at that point and can be understood in terms of an overdamped mode with roughly temperature-independent damping constant and a frequency

$$\omega_A^2(\mathbf{q}, T) \propto \{T - T_0 + f(\Delta\mathbf{q})\}; \qquad \Delta\mathbf{q} = \mathbf{q}_Z - \mathbf{q} \qquad (9.3.1)$$

where $\mathbf{q}_Z$ is the $\mathbf{q}$-vector at the Z-point of the Brillouin zone. A quantitative analysis shows that $T_0 \approx 195$ K (T_c for DADP is 235 K) and that

$$f(\Delta\mathbf{q}) = (T - T_0)(\Delta\mathbf{q} \cdot \mathbf{A} \cdot \Delta\mathbf{q} + \ldots) \qquad (9.3.2)$$

in which $A_{xx} \approx 16 A_{zz}$ and $A_{zx} = 0$. It follows that low-frequency excitations are concentrated along the z-axis (which is the c-axis), which means that the dominant quasi-elastic scattering is also in this direction, but that the strongly correlated motion and antiferroelectric fluctuations are dominantly in the ab-plane. This is exactly the opposite situation from KDP (or indeed most other uniaxial ferroelectrics) for which quasi-elastic scattering occurs in c-planes with polarization fluctuations along the c-axis. The departure of T_0 from T_c is not unexpected since the phase transition in DADP is strongly of first order. A determination of the actual atomic movements in the antiferroelectric mode would obviously be of great value, but the problem involves the determination of 20 amplitude parameters and a solution has not yet been obtained.

Other experimental methods, such as incoherent neutron scattering, and Mossbauer and electron paramagnetic resonance (e.p.r) of transition-metal impurities, have added general support to the coupled proton–lattice picture without being able to improve the quantitative picture. Pressure-dependent measurements (e.g. Samara 1971) show that ADP and DADP behave qualitatively like KDP and DKDP under pressure with T_c decreasing as a function of increasing pressure and eventually going to zero (Fig. 5.23). Again this can be semi-quantitatively understood in terms of a tunnelling mechanism. Perhaps the most convincing or direct demonstration of the existence of strong proton–lattice coupling has been given for the arsenates KDA and ADA (where the latter, $NH_4H_2AsO_4$, is a KDA-type antiferroelectric) by comparing the temperature dependence of the ^{75}As and proton hyperfine structures as functions of temperature using ENDOR (electron–nuclear double resonance) and e.p.r. techniques (Dalal and McDowell 1972). An analysis of the spectra makes possible a determination of the characteristic times associated with the particular types of motion involved and shows that at high temperatures the motion of the arsenic nuclei is clearly governed by different processes from that of the protons. However, as the temperature is lowered the characteristic times of the

arsenic and proton motions gradually merge, indicating an increasing correlation of the two motions and hence presumably a strong proton–lattice
coupling (see Fig. 12.1).

It would therefore appear that the basic principles involved in the
production of an antiferroelectric phase in the KDP antiferroelectrics are
broadly understood. Development will presumably follow rather closely the
progress made for the isomorphic ferroelectrics, e.g. in terms of mode
interactions and tunnelling concepts. One must not, of course, presume that
surprises will necessarily be absent, and indeed one has already surfaced in
the recent e.s.r. observations of Lamotte, Gaillard, and Constantinescu
(1972) who find that in the disordered state of ADA there are more
'up–down' configurations of H_2AsO_4 radicals than 'sideways' configurations. This implies that the configurational energy E_1 in ADA is positive and
suggests that a negative E_1 is therefore not a necessary requirement for
antiferroelectricity of the Fig. 9.7 type. That this is indeed the case has been
established theoretically by Ishibashi, Ohya, and Takugi (1974) who show
that a strong dipolar interaction along the a- or b-axis can still make the
antiferroelectric configuration energetically stable as long as a critica
positive value of E_1 is not exceeded.

9.4. Triglycine sulphate (TGS)

The ferroelectric nature of triglycine sulphate, $(NH_2CH_2COOH)_3H_2SO_4$,
usually abbreviated TGS, was discovered by Matthias, Miller, and Remeika
(1956). Despite its complex chemical and crystallographic form it has
become the object of active experimental research for two reasons. Firstly it
is one of very few ferroelectrics known to exhibit a second-order phase
transition and hence to offer possibilities for the observation of genuine
critical (i.e. fluctuation-dominated) phenomena very close to T_c. Secondly, it
is an order–disorder ferroelectric, is uniaxial and, unlike the KDP-
ferroelectrics, has dielectric properties which are not grossly affected by
deuteration. All this suggests that in spite of its structural complexity the
ferroelectric character may be basically simple and describable in terms of a
theoretical model with small if not negligible tunnelling.

Above the Curie temperature $T_c = 49°C$ the crystal has monoclinic symmetry and belongs to the centrosymmetric crystal class $2/m$. Below T_c the
mirror plane disappears and the crystal belongs to the polar point group 2 of
the monoclinic system. The polar axis is along the monoclinic (two-fold
symmetry) b-axis. Large crystals are easily grown from water solution. The
first detailed investigation of crystal structure was carried out by Hoshino *et
al.* (1959) using X-ray diffraction (Fig. 4.16). Subsequent neutron scattering
work (Kay and Kleinberg 1973) and X-ray scattering studies (Itoh and
Mitsui 1973) have confirmed many of the qualitative features of the earlier

measurements but have refined the structure much further and have also investigated temperature dependence.

In order to understand the structure it is first necessary to discuss briefly the crystal-chemistry characteristics of the glycine molecule. This molecular group NH_2CH_2COOH commonly crystallizes in two different forms. One is a structure in which the two carbons and the two oxygens are approximately coplanar, while the nitrogen atom is significantly displaced out of this plane; the other is a structure in which all the carbon, nitrogen, and oxygen atoms are close to planar. In TGS, which has three glycine groups, two (designated II and III in Fig. 4.16) are quasi-planar and one (designated I) is non-planar. The two planar groups II and III are arranged almost perpendicular to the polar b-axis and are mirror images. They are connected by a hydrogen bond between oxygen atoms (say O_{II} and O_{III}) which are some 0·25 nm apart. There are other hydrogen bonds in the detailed structure (see the original papers for their probable arrangement), but the glycine II—glycine III bond is particularly significant and has been postulated by some to drive the ferroelectric transition. The recent neutron study by Kay and Kleinberg finds the O_{II}—H—O_{III} proton to be in a double-well potential with $\sim 0·03$ nm between the minima (which is comparable with the situation in KDP). The X-ray measurements of Itoh and Mitsui, on the other hand, find no pronounced double-minimum character to the charge density between O_{II} and O_{III}.

The order–disorder character of the transition at 49°C has been questioned over the years, but the overwhelming weight of physical evidence indicates an order–disorder transition. For example the order–disorder model is consistent with the magnitude of the Curie–Weiss constant and the transition entropy (see Jona and Shirane 1962), and the dielectric relaxation is of Debye form. Perhaps the most convincing evidence is from the X-ray measurements of Itoh and Mitsui who find without question that the glycine I molecule is 'split' above the Curie temperature. Above T_c the b-plane through the glycine I molecule is a plane of mirror symmetry, and Itoh and Mitsui find that, even if a refinement of X-ray scattering is started assuming a planar glycine I in the mirror plane, the structure refines to show that all the oxygen, carbon, and nitrogen atoms occupy double-minimum sites straddling the mirror plane. This means that the mirror plane above T_c refers only to the averaged structure. Since glycine I is in fact grossly non-planar as regards the nitrogen atom, the distance between the double nitrogen minima is large (almost 0·1 nm) although the other glycine I splitting of heavy atoms is of order 0·01–0·03 nm.

Chemically two protons from the H_2SO_4 group are more properly associated with the glycines and the compound is often written as glycine di-glycinium sulphate $(NH_2CH_2COOH)(NH_3CH_2COOH)_2^+ . SO_4^{2-}$. Thus, for example, glycine I is more correctly a glycinium ion. The other proton can be thought of as the O_{II}—H—O_{III} proton and resonates between glycine II and glycine III, at least in the paraelectric phase. It is evident that there are two areas of marked asymmetry in the crystal, the first (according to neutron scattering) at the glycine II—glycine III hydrogen bond and the second at the glycine I location, particularly the nitrogen site. A comparison of the average amplitudes of motion for the heavy atoms near T_c shows by far the greatest amplitude for the glycine I nitrogen in the b-direction. This in turn points to the possibility of a driving role for the nitrogen double-well instability in the mechanism of the ferroelectric transition. Although the hydrogen bond may still play an important role, the smallness of the effects of deuteration on the dielectric properties (e.g. T_c is increased by only about 3 per cent in TGS compared with about 80 per cent for KDP) suggests that the role may be less than a dominant one and that perhaps we should look to the dynamics of the glycine I nitrogen atom as a more likely mechanism.

The attractive feature of ferroelectricity in TGS is its second-order uniaxial transition which follows almost perfectly the simple macroscopic Devonshire predictions for static properties. Thus (see for example Jona and Shirane 1962) a quantitative understanding of the static dielectric properties can be obtained close to T_c in terms of a Devonshire (or Landau) free-energy expansion in powers of the polarization (as set out in Chapter 3) and the relevant macroscopic parameters can be determined. On the microscopic level the fact that mean-field (or Landau) critical exponents are obtained (e.g. a Curie–Weiss susceptibility and a spontaneous polarization $P^2 \propto T_c - T$) points to the dominance of long-range forces but does not tell us immediately about the form of the local potential function. Gonzalo (1970) has examined the consequences of using a mean-field Ising model to explain both the temperature and field dependence of static dielectric properties near T_c. Within this theory the equation of state can be written in terms of T_c, the saturation polarization P_0, and the saturation internal field E_0, as

$$p = \tanh\left(\frac{p+e}{t}\right) \qquad (9.4.1)$$

in which $t = T/T_c$, $p = P/P_0$, and $e = E/E_0$. Agreement near T_c is good only if p and e are replaced by $0 \cdot 139p$ and $0 \cdot 450e$ respectively. Equation (9.4.1) is more restrictive than Devonshire theory in that P_0 and E_0 are not adjustable parameters but are fixed within the model. Moreover, the microscopic theory attempts to describe dielectric properties over the entire temperature range. Some improvement in the Ising picture can be achieved by recognizing that the total polarization is made up of an induced (i.e. electronic) part as well as a dipolar part (Gonzalo 1974), but the observed low-temperature flattening of spontaneous polarization is still not in accord with mean-field behaviour. Tello and Hernandez (1973) have looked for an explanation by allowing tunnelling within the basic Ising scheme (following

Blinc and Svetina 1966), but this explanation must be treated with reservation since equivalent adjustments in polarization curves (see for example Lines 1969a) can be made by allowing the double-well local potential to deviate from Ising δ-function form. In addition there is no evidence for significant tunnelling in the dynamic properties of TGS, and one must also remember that the dielectric properties are only likely to be dominated by the single 'soft' degree of freedom represented by an Ising, local-mode, or tunnel formalism close to the phase transition itself. Indeed one would be surprised if such a simple model which neglects all other degrees of freedom were quantitative over a large temperature range.

As far as static properties are concerned great interest has also centred around the possibility of observing deviations from mean-field behaviour extremely close to the second-order transition (see Figs 11.1, 11.2, and 11.3). Such critical behaviour is anticipated in a uniaxial ferroelectric, even in the presence of long-range dipolar forces, but takes a subtle form involving logarithmic terms (see Chapter 11). The most pronounced deviation from mean-field behaviour is expected for specific heat, which is finite but discontinuous in Landau (e.g. Devonshire) theory but diverges as $|\ln(|T - T_c|/T_c)|^{1/3}$ in a more accurate theoretical estimate. Experimentally for TGS mean-field behaviour seems to hold at least for $|(T - T_c)/T_c| > 10^{-4}$. Deviations from mean field have been observed closer to T_c, but the effects seem often to be dependent on crystal quality and there is hesitancy to ascribe them in any quantitative sense as critical effects. The best evidence to date for a critical divergence of specific heat in TGS comes from a measurement of thermal expansion (Schurmann et al. 1973) which theoretically is expected to exhibit the same critical anomaly as specific heat. However, no claim has yet been made of a fit to the precise theoretical form above. More details of the efforts to measure ferroelectric critical phenomena in TGS are given in Chapter 11.

To this point we have mentioned only uniform static properties. A somewhat more detailed picture of the statics of the TGS transition can be obtained by study of the wavevector dependence of static fluctuations by means of X-ray scattering (Shibuya and Mitsui 1961, Fujii and Yamada 1971, Pura and Przedmojski 1973). The early X-ray work of Shibuya and Mitsui observed only the temperature dependence of diffuse intensities and compared them with the temperature dependence of the dielectric constant. The later studies observed details of the intensity distribution in reciprocal space. Fujii and Yamada find a very anisotropic intensity distribution with scattering primarily in the plane normal to the polar b-axis (Fig. 6.4). A more detailed discussion has been given in § 6.3. In short the scattering resembles qualitatively that expected for a uniaxial system in which long-range dipolar forces are playing an important role and for which the correlation length (defining the scattering divergence at T_c) diverges with a mean-field critical exponent. Nevertheless there is evidence that the dipolar forces are screened in some way (the theoretical 'pinching' into the Bragg point evident in Fig. 6.4 (b) is missing from the experimental distribution), probably by the electronic response which readjusts electron density in such a way as to cut off the long-range tail from the dipolar distribution. In particular Pura and Przedmojski (1973) have examined closely the diffuse X-ray scattering close to the b-axis and found a small increase in scattering very close to this axis. Following Kocinski and Wojtczak (1973) they suggest that an explanation can be found solely in terms of short-range interactions if an allowance is made for the asymmetry of the crystal structure. Since physically there can be no doubt of the presence of dipolar forces it seems more likely that the answer lies somewhere between the two extremes, i.e. dipolar forces are present (and primarily responsible for the mean-field critical behaviour) but may be screened to the extent that they do not completely dominate the short-range interactions at small $\mathbf{q}$ nor produce non-analytic behaviour as the wavevector $\mathbf{q} \to 0$.

Investigation of the dynamics of polar motion near the phase transition in TGS has had a rather confused history. Below T_c, as with all ferroelectrics, the dynamic properties are dominated by domain motion. Above T_c, on the other hand, the interpretation should be much simpler. The first measurements of the frequency dependence of dielectric response above T_c were made by Hill and Ichiki (1962) at frequencies up to $\sim 10^{11}$ Hz (Fig. 5.8). Both the frequency and temperature dependence of the complex dielectric constant were measured with the conclusion that TGS relaxation was Debye-like but not with a single relaxation time. A Gaussian distribution of relaxation times was found at each temperature and explained in terms of clusters of different sizes in which dipoles tend to align themselves. More recently (Barker 1967, Mansingh and Lim 1972) the reality of this polydispersive behaviour has been questioned, and new measurements by Luther and Muser (1969, 1970) and by Unruh and Wahl (1972) have confirmed that the correct relaxation is much more closely monodispersive as in eqn (2.4.16). A very small region of polydispersive behaviour was found within a small fraction of a degree of T_c (and may be explicable in terms of surface effects) but otherwise a single Debye relaxation term seemed adequate. The Debye form (eqn (2.4.16)) in which the relaxation time τ varies with temperature as

$$\tau = A(T - T_c)^{-1} \qquad (9.4.2)$$

in mean-field approximation measures also the depth of the potential double well through the constant A. The latter is independent of temperature in the Ising (infinite well depth) limit but if the barrier to thermal hopping is of finite energy E then (see § 2.4)

$$A \sim 2\pi \frac{\hbar}{k} \exp\left(\frac{E}{kT}\right). \qquad (9.4.3)$$

Experimental measurements of dielectric dispersion, as reported by Mansingh and Lim (1972) and by Luther (1972) point to a value E only of the order or a little greater than kT_c. This finding is supported by the ultrasonic measurements of O'Brien and Litovitz (1964) who report $A = 2.25 \times 10^{-10}$ Ks. It follows that any theoretical model based on an Ising scheme is likely to be no more than qualitative for TGS. The Debye form supports the notion of stochastic (thermally activated) barrier hopping rather than dynamical tunnelling as the mechanism for establishing equilibrium between the double wells of the local potential. Measurements at higher Brillouin frequencies (Gammon and Cummins 1966) find a rather smaller value ($\sim 3 \times 10^{-11}$ Ks) for A which suggests that there may be a frequency-dependent relaxation rate. Possible origins of frequency-dependent damping are discussed in § 11.5.

Early infrared measurements by Barker and Tinkham (1963) had already pointed to a non-resonant response, but further infrared and Raman studies have yet to yield much which is qualitatively new concerning the transition dynamics. Most efforts to study the dynamics in detail have thus far come from magnetic resonance experiments. These include measurements of deuteron or ^{14}N quadrupole splitting and electron spin resonance of glycine radicals (formed by gamma irradiation) or of doped transition metal ions. The results have been controversial on the whole with some (e.g. Bjorkstam 1967, Kato and Abe 1972) favouring a displacive transition and others (e.g. Blinc, Pintar, and Zupancic 1967, Blinc *et al.* 1971, Nishimura and Hashimoto 1973) claiming evidence for an order–disorder character.

With the hindsight provided by the recent X-ray data of Itoh and Mitsui (1973) it now seems possible to reconcile most of the seemingly conflicting evidence and to form at least a qualitative picture of the phase-transition dynamics as follows. Glycine II and glycine III are chemically equivalent above T_c as far as quadrupole resonance frequencies are concerned. This points to a fast exchange of the glycinium–glycine roles of these two ions owing to proton transfer in the symmetric double well of the hydrogen bond. The character of this motion appears to be resonant (Brosowski *et al.* 1974) pointing to a tunnelling mechanism. Below T_c glycine II and glycine III become inequivalent as the proton minima develop unequal depths. Glycine I, above T_c, in all probability describes a thermally activated hopping motion between equivalent potential wells on either side of the b mirror plane. This motion, which carries most of the polarization, is therefore of order–disorder character. More accurately the glycine I motion and the glycine II—glycine III hydrogen bond motion must be coupled via the detailed hydrogen-bonded structure of the complete unit cell and coupled modes can probably be defined. However, theoretical development along these lines has not yet developed and it remains to be seen whether one or the other of these motions can in any realistic sense be considered as driving the other to the transition.

Many of the earlier difficulties of reconciling the magnetic resonance data with an order–disorder picture stemmed from efforts to interpret the results in terms of the older X-ray diffraction data of Hoshino et al. (1959) which claimed that glycine II was non-planar and glycine III planar. Bjorkstam (1967), on the other hand, measured the ND_3^+ deuteron resonance in glycine I and still claimed evidence for a displacement character at T_c. This evidence was a failure to observe the critical slowing down of the order–diorder relaxation (or hopping) time which would be expected to broaden the quadrupole lines when it approached the time ($\sim 10^{-6}$ s) characteristic of quadrupole resonance. Blinc et al. (1971) later explained this by noting that a relaxation time as long as 10^{-6} s would not be expected until $T - T_c \lesssim 10^{-2} - 10^{-3}$ K and would therefore escape detection with Bjorkstam's temperature resolution of order 0·1 K.

A few experiments have been reported on TGS isomorphs such as the selenate and the fluorberyllate but hardly enough to add much to our detailed understanding of the phase transition. The general static dielectric properties of these isomorphs are seemingly as simple and well behaved as for TGS itself, with one possibly interesting exception. Peshikov (1972) has reported that the selenate transition changes from second to first order on deuteration, and also that for the undeuterated crystal the line of second-order transitions changes continuously into one of first order under hydro-static compression, the critical pressure being about 7·5 kbar. Deuterated TGS is still a second-order ferroelectric and no change in the order of transition in undeuterated TGS occurs at least up to 23 kbar pressure. The selenate finding suggests the possibility of an experimental study of a ferroelectric tricritical point and, for example, the possible observation of critical opalescence in this material under pressure.

9.5. Sodium nitrite (NaNO₂)

Sodium nitrite, $NaNO_2$, has excited extensive scientific investigation since the discovery of its phase transition by Sawada et al. (1958). The prime incentive has been the fact that $NaNO_2$ possesses the simplest crystal structure of any typical order–disorder ferroelectric yet characterized in detail. Understanding its phase transition in a quantitative fashion would therefore provide extremely valuable groundwork for subsequent efforts to unravel more complicated situations. A complication is that the simple ferroelectric ordered phase is separated from the disordered paraelectric phase by a sinusoidal antiferroelectric phase which is stable over only a very narrow temperature range of order 1 K. On the plus side, however, is the fact that the existence of this interposed phase and the details of its sinusoidal polar variation contain rather specific information about the relative importance of short-range and dipolar forces in $NaNO_2$.

$NaNO_2$ is ferroelectric at room temperature with a structure belonging to a body-centred orthorhombic group (space group $Im2m$–C_{2v}^{20}) with unit cell dimensions $a = 0.356$ nm, $b = 0.556$ nm, and $c = 0.538$ nm. The orthorhombic unit cell of this ferroelectric phase is shown in Fig. 9.8. The

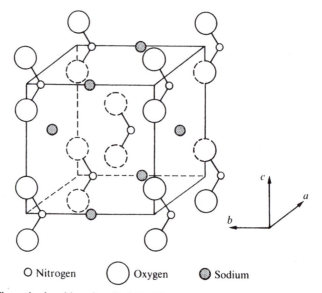

O Nitrogen   ◯ Oxygen   ◉ Sodium

Fig. 9.8. The orthorhombic unit cell of $NaNO_2$ in the low-temperature ferroelectric phase. (After Sakurai, Cowley, and Dolling 1970.)

primitive cell has only one formula unit with but four atoms and 12 normal modes of vibration for any wavevector **q**. Three of these are internal NO_2 vibrations with high frequency ($\sim 10^3$ cm^{-1}), while the other nine involve translational motion of Na and NO_2 ions together with librational motions of the latter. The structure is simple enough that each of these modes can be studied through the phase transition and their behaviour correlated with the onset of lattice instability.

In the ferroelectric phase the NO_2 ions are aligned with the bc-plane in such a way that their electric dipole moments are all pointing along the (positive) b-axis. At $T = T_c \approx 163°C$ a first-order transition takes place to an antiferroelectric phase in which the average dipole moment along the b-axis, for any plane of nitrite ions normal to the a-axis, displays a sinusoidal modulation along the a-axis (Yamada *et al.* 1963, Yamada and Yamada 1966). This sinusoidal ordering has a period of about $8a$, although the precise period may decrease slightly with increasing temperature; thus, if the average dipole moment is a maximum along $+b$ for one (100)-plane of ions, then it is approximately maximum along $-b$ for a (100)-plane at a distance $4a$ along the a-direction. This ordering creates satellite Bragg peaks on either side of the regular reflections in X-ray and neutron experiments (e.g. Fig. 6.5) and these satellite peaks merge into broad diffuse critical scattering as the temperature approaches a (probably) second-order transition at $T_N \approx 164 \cdot 5°C$ at which the system becomes paraelectric with a centrosymmetric structure (space group Immm–D_{2h}^{25}).

The most accurate structure determination in the ferroelectric and paraelectric phases to date is the neutron work of Kay (1972). Neutron diffraction data were taken at 150°C (ferroelectric) and at 185° and 225°C (paraelectric) in an effort to define the mechanism of the transition. In the ferroelectric phase $x = 0$ and $z = 0$ are mirror planes (where x, y, z running between 0 and 1 define the orthorhombic cell axes of Fig. 9.8 along a, b, c respectively) with the nitrite ion in the $x = 0$ plane and the two oxygen atoms related by the mirror plane at $z = 0$. The phase transition inserts a mirror plane perpendicular to the b-axis at $y = 0$. The nitrite and sodium ions are disordered on either, but not both, sides of this plane in any one primitive cell (Fig. 9.9). The point of contention is the manner in which the nitrite ion

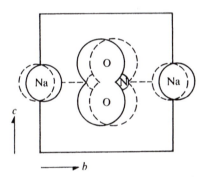

Fig. 9.9. The order–disorder arrangement for the molecular unit in $NaNO_2$. (After Yamada *et al.* 1963.)

moves between its two configurations. Since the internal O—N—O bending frequency remains high throughout (~ 800 cm^{-1}) a motion confined to the bc-plane is immediately ruled out. Kay tries several models for refining the elastic neutron data and finds a best fit in the ferroelectric region which, in addition to the obverse and reverse structures of Fig. 9.9, requires a fraction of the molecules to be rotating in two- or four-fold shallow wells about the c-axis. No evidence for free or hindered rotation was discernible in the paraelectric regime (i.e. refinement was not improved by adding the extra degree of freedom to the obverse–reverse model). The refined atomic positions in the obverse configuration are

$$Na(0, 0·54, 0), N(0, 0·07, 0), O(0, -0·05, 0·19).$$

Since the above structure determination is model dependent and contains no direct dynamic information, its predictions concerning rotational motion are only suggestive at best. Dynamic information, however, can be obtained from single-crystal coherent inelastic neutron scattering (Sakurai *et al.* 1970). Although a simple structure in relation to other order–disorder ferroelectrics, $NaNO_2$ is nevertheless perhaps the most complex material for which a full inelastic neutron investigation of vibrational modes has been undertaken. Treating the NO_2 group as a rigid unit (i.e. ignoring the three high-frequency intranitrite modes) the nine other phonon branches have been carefully observed with the finding that none of them has an anomalously low frequency in the region of the transition. The phase transition therefore does not result from any instability of the crystal against a normal mode describing small displacements of the atoms. The measurements of critical scattering show that the changes in ferroelectric fluctuations occur far more slowly than any of the phonon vibrations and involve large displacements of all the atoms. (This situation, in terms of a simple double-well local-mode model, has been discussed in general terms in § 2.4). Study in detail of such slow long-wavelength fluctuations is more accurately pursued by other techniques.

Studies of low-frequency ($\leqslant 10^9$ Hz) dielectric dispersion and loss have been reported by several workers, most recently by Hatta (1968, 1970). The response in the paraelectric phase is of simple Debye form and monodispersive to within about a half degree of the transition temperature T_N. The relaxation rate shows a critical slowing down of form (9.4.2) with a potential barrier E between the potential minima of order $7kT_N$, according to Yamada, Fujii, and Hatta (1968). Even at high temperatures well in excess of T_N where intercell correlations are small the nitrite flipping frequency is less than $10^{11}\,s^{-1}$. Since the lowest optic phonon frequency is of order 5×10^{12} Hz (Axe 1968, Barnoski and Ballantyne 1968), it is evident that all phonon frequencies are large compared with the flipping frequency at all temperatures.

The picture which emerges is one of an almost classic order–disorder transition involving the flipping of nitrite ions coupled, though probably not rigidly, to Na^+ ion motion (from the crystal structure one would expect the sodium motion to be governed by the relative orientation of at least two nitrite groups). The fact that the phase transition is not directly related to a softening of any phonon or tunnel mode has been confirmed by many observations including infrared reflectivity (Barnoski and Ballantyne 1968) and Raman spectroscopy (Hartwig *et al.* 1972). Above the phase transition there are in general more resonances than allowed group theoretically by the D_{2h}^{25} space-group symmetry. This is simply another illustration of the fact that the higher symmetry of the paraelectric phase is only of a statistical nature with a flipping time between wells which is long compared with the period of the optic phonon frequencies. A particularly interesting finding of the Raman study by Hartwig *et al.* is the presence in the high-temperature phase of lines which cannot be explained even allowing for a simple obverse–reverse nitrite disorder. The authors interpret the presence of these anomalous lines as evidence of a large torsional motion of the nitrite ion about the a-axis (see Fig. 9.8). This finding conflicts with that of Kay (1972) from neutron diffraction and of Singh and Singh (1974) from a ^{14}N quadrupole resonance study, both of which claim evidence in favour of a rotational flipping about the c-axis. There are in fact numerous claims in the literature of evidence in favour of flipping about either the a-axis or the c-axis (Singh and Singh cite several of them), but, with no obvious concensus, the matter must be left open at the present time. It is, of course, not impossible (nor perhaps even unlikely) that both mechanisms could be playing a significant role.

The most detailed information concerning the form of the *inter*cell potential function has come from X-ray scattering (Yamada *et al.* 1963, Yamada and Yamada 1966) and has already been discussed in § 6.3 on diffuse X-ray scattering. The nature of the X-ray diffuse scattering above the second-order transition T_N to the antiferroelectric phase has been carefully probed (e.g. Fig. 6.5), and its distribution in reciprocal space a^*, b^*, c^* and its temperature dependence analysed. Taking a model in which the NO_2 and Na are rigidly coupled within a primitive cell and are restricted to 'up' and 'down' Ising positions along the b-axis, Yamada and Yamada find that an interaction potential of pure long-range dipolar form can immediately account for most qualitative features of the intensity distribution. Since the lattice structure is very simple the dipole sum is readily performed and exhibits a minimum in its Fourier transform when $\mathbf{q} = \pm \mathbf{a}^*/5$. This is very significant and explains the appearance at T_N of the antiferroelectric configuration with a sinusoidal polarization variation in the a-direction. The experimental scattering peak corresponding to the real potential minimum actually occurs closer to $a^*/8$ near T_N moving towards $a^*/5$ in the high-temperature limit. In a more detailed analysis Yamada and Yamada include near-neighbour interactions from the a, b, and body-diagonal directions. They find a good, but not perfect, agreement with experiment when the body-diagonal interaction is ferroelectric and the other two antiferroelectric, all three showing a small temperature dependence (possibly resulting from thermal expansion). Scattering in general is heavily concentrated in the b-plane (i.e. a^*c^*-plane), a normal distribution for dipolar forces, and the absolute potential minimum at $a^*/5$ is only about 7 K lower than the 'ferroelectric' saddle at the origin. The competition between the antiferroelectric and ferroelectric structures is therefore qualitatively understandable, although the precise manner in which the various interactions trigger the first-order transition to the polar phase remains uncertain.

Phenomenological theory, including coupling to strain (e.g. Gesi 1965), can represent the macroscopic data fairly well, but neglects fluctuation effects and requires a number of *ad hoc* assumptions concerning the temperature dependence of the phenomenological parameters. Microscopic statistical theories are still at a very rudimentary stage and no effort seems yet to have been made to pursue a local-mode model in detail. Existing calculations (e.g. Flugge and Meyenn 1972, Bonilla, Holz, and Rutt 1974) are for static properties only and rest on very specific assumptions concerning polarizability effects, the validity of point-charge and point-dipole approximations, the importance and nature of the intracell Na^+–NO_2^- coupling, and the neglect of intercell correlations (i.e. the use of mean-field theory). The severe nature of the many approximations combined with the rather modest successes of the various theories makes it difficult to claim that much in the way of microscopic insight has resulted from them to date.

Although there remains some doubt as to whether the antiferroelectric–paraelectric transition is truly of second order there is no doubt that the loss of ferroelectricity at T_c is first order. Hamano (1973) finds a clear discontinuous jump of some $5 \cdot 8 \, \mu C \, cm^{-2}$ at T_c and estimates that the low temperature saturation polarization may be close to $12 \, \mu C \, cm^{-2}$. The dielectric constant in the polar phase follows a Curie–Weiss law over a fairly wide range and thermal hysteresis is readily observed at T_c. Efforts at measuring static critical exponents have not been very convincing because of the first-order nature of the ferroelectric transition although Hatta (1970) claims some slight deviations from Curie–Weiss form in the paralectric regime. One of the more careful efforts to measure static exponents on $NaNO_2$ was made by Hatta and Ikushima (1973) for specific heat near the T_N point. They find the transition to be of second order but the measured divergence of specific heat, with exponents $\alpha = 0 \cdot 38$, $\alpha' = 0 \cdot 18$, is very difficult to understand from any theoretical standpoint (see Chapter 11). It is clear that much remains unresolved at the quantitative level even among the static properties. In fact (see, for example, Gesi 1969) slight anomalies are observed in static elastic and dielectric properties at temperatures other than T_c and T_N. Although the situation is not yet clear it seems that there may be other subtle phase transitions possessed by the simple $NaNO_2$ structure (particularly if pressure is included as a variable (Rapoport 1966)). One very interesting possibility (Hoshino and Motegi 1967) is that a second polar phase may exist in a very small temperature interval between the room-temperature ferroelectric phase and the sinusoidal antiferroelectric phase.

This phase, according to Hoshino and Montegi's X-ray diffraction study, has a sinusoidal variation of polarization in the a-direction of a character similar to that which would result from adding a constant b-axis polarization to the sinusoidal antiferroelectric structure.

10

COUPLING TO THE SOFT MODE

10.1. Acoustic anomalies near ferroelectric transitions

IT IS NOW well established that near the temperature of a second-order structural phase transition, and in particular near the Curie point of a ferroelectric, anomalously strong ultrasonic absorption occurs. The increase arises from a coupling between strain and polarization (for ferroelectrics) and is attributable to at least two absorption mechanisms, one via a piezoelectric coupling (which is linear in both strain and polarization) and the other via electrostrictive interactions (which are linear in strain and quadratic in polarization). For a ferroelectric with a piezoelectric high-temperature phase both mechanisms operate on both sides of the phase transition. However, for a ferroelectric with a centrosymmetric high-temperature phase (such as that described by our basic model Hamiltonian eqn (1.2.8)) only the electrostrictive term is active above T_c. This remains active as the result of polarization fluctuations $\langle \Delta P_l \, \Delta P_{l'} \rangle \neq 0$.

In this section we shall consider the perturbation produced on an acoustic travelling wave by those interactions with a soft optic (ferroelectric) lattice variable. The resulting anomalies affect both the real and imaginary parts of the acoustic frequency and are observable respectively as shifts in acoustic velocity (or equivalently elastic moduli) and in ultrasonic attenuation. By observing either as a function of frequency in a piezoelectric phase one can obtain a measure of relaxation time τ for fluctuations from equilibrium which occur in the polarization or order parameter characterizing the transition. This time τ diverges as the Curie point is approached, and in favourable cases some comparisons are possible with τ measurements from other types of experiment.

10.1.1. Anomalies in sound velocity and elastic constants

In the theoretical development of Chapter 2 we pursued a model Hamiltonian which concentrated explicitly on the motion of a single ferroelectric or soft-mode variable ξ. The model neglected any possible coupling terms to other lattice variables such as strain and acoustic modes, or electronic co-ordinates, and therefore carried no information concerning the existence of critical anomalies in other than the soft-mode response. In general the existence of coupling terms, whether in low or high order, leads to characteristic anomalies in other observables which are essentially driven by the primary soft mode. One of the most prominent of these extrinsic anomalies associated with ferroelectrics (and indeed with many non-polar structural transitions as well) occurs as the result of direct coupling between the optic soft mode and the acoustic modes and gives rise to critical variations of sound velocity and elastic constants with temperature on passing through the transition temperature.

Let us consider explicitly the coupling of the soft optic mode to strain. The model Hamiltonian then consists of three parts

$$\mathcal{H} = \mathcal{H}_{\text{opt}} + \mathcal{H}_{\text{el}} + \mathcal{H}_{\text{int}} \tag{10.1.1}$$

where $\mathcal{H}_{\text{opt}}$ is the soft-mode Hamiltonian eqn (1.2.8), viz.

$$\mathcal{H}_{\text{opt}} = \sum_l \{\tfrac{1}{2}\pi_l^2 + V(\xi_l)\} - \tfrac{1}{2}\sum_l \sum_{l'} v_{ll'}\xi_l\xi_{l'} \tag{10.1.2}$$

$\mathcal{H}_{\text{el}}$ is the elastic Hamiltonian, and $\mathcal{H}_{\text{int}}$ is a low-order interaction term coupling the two. To terms of harmonic order we can express the elastic Hamiltonian in the form

$$\mathcal{H}_{\text{el}} = \sum_l \{\tfrac{1}{2}M\dot{\mathbf{u}}_l^2 + \tfrac{1}{2}\sum_i \sum_j c_{ij}x_{i,l}x_{j,l}\} \tag{10.1.3}$$

in which $\mathbf{u}_l$ is the (acoustic) displacement variable for the lth cell, M is the cell mass, c_{ij} are the elastic constants, and $x_{i,l}$ the localized strain variables (in Voigt notation, see eqn (3.1.4)) in the lth cell. The interaction term of lowest order is the electrostrictive one of form

$$\mathcal{H}_{\text{int}} = \sum_l G_i x_{i,l}\xi_l^2 \tag{10.1.4}$$

if we assume for the moment a centrosymmetric prototype paraelectric phase. If the prototype phase is of higher than axial symmetry then the soft-mode variable ξ_l ceases to be scalar and a more complex subscript formalism must be developed. An example of this for cubic symmetry has been given by Feder and Pytte (1970) with reference to the perovskites. Coupling of the soft mode to other mode variables (say ξ_l') is also readily included in a more general formalism by introducing an interaction term

$$\mathcal{H}_{\text{int}}' = \sum_l G'\xi_l'^2\xi_l^2 \tag{10.1.5}$$

which also can be formally generalized for high-symmetry situations.

Considering eqn (10.1.1), distortion from the prototype symmetry can be described by non-vanishing expectation values of ξ_l and $x_{i,l}$. We set

$$\xi_l = \langle\xi\rangle_0 + \eta_l$$
$$x_{i,l} = \langle x_i\rangle_0 + u_{i,l} \tag{10.1.6}$$

where the angular brackets denote thermal averages and η_l and $u_{i,l}$ fluctuations about these values. Substituting eqn (10.1.6) into eqn (10.1.4) yields terms which describe the coupling of optic and acoustic phonon modes:

$$\mathcal{H}_{\text{int}} = \sum_l 2G_i \langle \xi \rangle_0 u_{i,l} \eta_l + \ldots \qquad (10.1.7)$$

The fluctuations $u_{i,l}$ about the average values of local strain $\langle x_i \rangle_0$ may be expressed in terms of the normal-mode co-ordinates $Q(\mu, \mathbf{q})$ of acoustic phonons according to

$$u_{jk,l} = \tfrac{1}{2}iN^{-1/2} \sum_\mu \sum_{\mathbf{q}} \{q_j e_k(\mu\mathbf{q}) + q_k e_j(\mu\mathbf{q})\} Q(\mu\mathbf{q}) \exp(-i\mathbf{q} \cdot \mathbf{l})$$
$$(10.1.8)$$

where μ labels one of the three acoustic branches, $\mathbf{e}(\mu\mathbf{q})$ is the polarization vector for this mode at wavevector $\mathbf{q}$, and we have reverted from the Voigt notation to conventional tensor formalism for strain. Introducing running waves

$$\eta(\mathbf{q}) = N^{-1/2} \sum_l \eta_l \exp(i\mathbf{q} \cdot \mathbf{l}) \qquad (10.1.9)$$

the soft-mode interaction term in eqn (10.1.7) can be expressed for small $\mathbf{q}$ as

$$\mathcal{H}_{\text{int}} = 2i\langle \xi \rangle_0 \sum_\mu \sum_{\mathbf{q}} qG(\mu)Q(\mu\mathbf{q})\eta(-\mathbf{q}) + \ldots \qquad (10.1.10)$$

where the detailed form of the coupling factor $G(\mu)$ may be determined by substitution of eqns (10.1.8) and (10.1.9) into eqn (10.1.7) for any particularly geometry of interest. The important features of eqn (10.1.10) are (a) the linear dependence on the (ferroelectric) order parameter $\langle \xi \rangle_0$, and (b) the linear dependence on wavevector magnitude q. The former implies that a lowest-order bilinear interaction term is absent above the Curie point (a result of the choice of a centrosymmetric prototype optic potential), and the latter establishes that even for $T < T_c$ the interaction will not perturb the soft mode at $\mathbf{q} = 0$ itself although it will affect the optic mode dispersion on approach to $\mathbf{q} = 0$.

Following Pytte (1970 a) we now consider sound propagation along a high-symmetry direction for which only one acoustic mode couples to the soft mode. If for simplicity we assume an approximation for optic and acoustic motion with uncoupled modes represented by damped harmonic oscillators, with frequencies $\Omega_{\text{opt}}(\mu\mathbf{q})$ and $\Omega_{\text{ac}}(\mu\mathbf{q})$ and damping constants $\Gamma_{\text{opt}}(\mu\mathbf{q})$ and $\Gamma_{\text{ac}}(\mu\mathbf{q})$ respectively, then the introduction of the bilinear coupling term (10.1.10) to the motion perturbs the frequencies according to the equation

$$\{\omega^2 - \Omega_{ac}^2(\mu\mathbf{q}) + i\omega\Gamma_{ac}(\mu\mathbf{q})\}\{\omega^2 - \Omega_{opt}^2(\mu, -\mathbf{q}) + i\omega\Gamma_{opt}(\mu, -\mathbf{q})\}$$
$$= \frac{4q^2\langle\xi\rangle_0^2 G^2(\mu)}{M}. \qquad (10.1.11)$$

Solving for perturbed acoustic frequency in the small-q limit, assuming the acoustic frequency small compared with the optic frequency, we find (Pytte 1970 a)

$$\omega = \omega_\mu(1 - F_\mu c_\mu^{-2}) + i(\tfrac{1}{2}\Gamma_\mu + q^2 F_\mu\tau) \qquad (10.1.12)$$

in which we have written the unperturbed long-wavelength acoustic frequency as $\omega_\mu + \tfrac{1}{2}i\Gamma_\mu$ (i.e. $\omega_\mu^2 = \Omega_{ac}^2 - \tfrac{1}{4}\Gamma_{ac}^2$ and $\Gamma_\mu = \Gamma_{ac}$) and where

$$\tau = \Gamma_{opt}/\Omega_{opt}^2 \qquad (10.1.13)$$

$$F_\mu = \frac{2}{M}\left(\frac{\langle\xi\rangle_0}{\Omega_{opt}}\right)^2 \frac{G^2(\mu)}{1 + \omega_\mu^2\tau^2} \qquad (10.1.14)$$

and c_μ is the sound velocity of the pertinent acoustic mode ($\omega_\mu = c_\mu q$).

If the soft-mode damping Γ_{opt} is neglected we find from eqn (10.1.12) a finite discontinuity in ω and hence in sound velocity at T_c since both $\langle\xi\rangle_0^2$ and Ω_{opt}^2 vary as $T_c - T$ in eqn (10.1.14). The effect of the $(1 + \omega_\mu^2\tau^2)^{-1}$ factor is to smooth out this discontinuity. Using the relationship between sound frequency and elastic constant (sound velocity is proportional to the square root of the ratio of an elastic modulus to the mass density) the corresponding changes in the frequency-dependent elastic constants are readily obtained. The formal results apply to any phase transition involving a non-degenerate $\mathbf{q} = 0$ soft optic mode where a bilinear coupling of strain to the soft mode is present in the ordered phase but absent on symmetry grounds in the high-temperature phase. Extension to cover cases where the optic mode condensation occurs at $\mathbf{q} = \mathbf{q}_0 \neq 0$ is trivial, and in fact the final result (eqn (10.1.12)) is unchanged in this context. Extension to cover the cases of degenerate soft modes and/or piezoelectric high-temperature phases is straightforward and has received attention in the literature (see, for example, Pytte 1971 a).

Much experimental work to date has been performed on the perovskite structure, which is centrosymmetric in the high-temperature phase but possesses a degenerate soft mode. In this situation there are three distinct coupling parameters G_{11}, G_{12}, and G_{44} allowed by symmetry. For a non-polar structural transition, such as the zone-boundary $SrTiO_3$ transition, the predicted changes in elastic constants are (Pytte 1971 a)

$$\Delta c_{11}(\omega) = \Delta c_{12}(\omega) = -G_{12}^2 D_3(\omega)$$

$$\Delta c_{13}(\omega) = -G_{11}G_{12}D_3(\omega)$$

$$\Delta c_{33}(\omega) = -G_{11}^2 D_3(\omega) \qquad (10.1.15)$$

$$\Delta c_{44}(\omega) = -G_{44}^2 D_1(\omega)$$

$$\Delta c_{66}(\omega) = 0$$

in which

$$D_\lambda(\omega) = \frac{4\rho}{M}\left\{\frac{\langle\xi\rangle_0}{\Omega_{\mathrm{opt}}(\lambda)}\right\}^2 (1+\omega^2\tau_\lambda^2)^{-1} \qquad (10.1.16)$$

where ρ is the density of the crystal and λ labels the components of the three-fold degenerate soft mode (i.e. the rotations of octahedral units about the λ axis). Experimental measurements of the sound-velocity anomaly proportional to $(C_{44})^{1/2}$, as measured in SrTiO$_3$ by Laubereau and Zurek (1970) for two different frequencies, are shown in Fig. 10.1. The

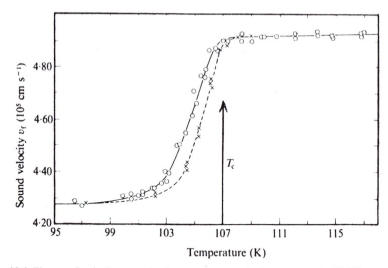

Fig. 10.1. The sound-velocity anomaly of transverse acoustic phonons near the 105 K structural transition in SrTiO$_3$ propagating in a [100] direction for two different frequencies. As measured by Brillouin scattering by Laubereau and Zurek 1970.) (After Pytte 1971 a.)

frequency dependence can be ascribed to a $(1+\omega^2\tau^2)^{-1}$ factor with $\tau = 1\cdot6\times10^{-11}(T_c-T)^{-1}$ s K.

For a *ferroelectric* transition in the perovskite structure the situation (eqn (10.1.15)) is complicated a little because we must distinguish between transverse and longitudinal optic mode frequencies, which are no longer degenerate even at $\mathbf{q} = 0$ because of the presence and non-analytic form of the dipolar forces. The soft mode in the cubic phase, which is triply degenerate in the $SrTiO_3$ structural case, splits into a doubly degenerate transverse mode and a singlet non-degenerate longitudinal mode. In the tetragonal ferroelectric phase pure longitudinal and transverse modes can only be defined along the c-axis ($\lambda = 3$) and in the c-plane ($\lambda = 1, 2$). As a result the elastic anomalies (eqn (10.1.15)) become modified as follows:

$$\Delta c_{11} = \Delta c_{12} = -G_{12}^2 D_3^T$$

$$\Delta c_{33} = -G_{11}^2 D_3^L$$

$$\Delta c_{44} = -G_{44}^2 D_1^{L,T} \qquad (10.1.17)$$

$$\Delta c_{66} = 0$$

where L and T refer to longitudinal and transverse modes respectively. The change in c_{44} can have two values depending on the propagation and polarization vectors considered (Shirane, Axe, Harada, and Linz 1970).

For structures in which the bilinear interaction term does not vanish in the high-temperature phase (e.g. systems with piezoelectric high-temperature phases) expressions of the form (eqns (10.1.12) and (10.1.15)) are obtained both below and above T_c except that the factor $\langle \xi \rangle_0^2$ does not appear in the coupling term (eqns (10.1.14)) or (10.1.16)). It follows that F_μ and $D_\lambda(\omega)$ now diverge as the soft optic frequency tends to zero and that the soft mode accordingly pushes the acoustic frequency with which it couples all the way to zero. The result is the condensation of an acoustic mode which occurs in general at a temperature higher than that at which the uncoupled optic mode would otherwise condense. An example of this has been seen in KDP (Fig. 9.5) both by Brillouin scattering and by ultrasonics (Garland and Novotny 1969, Brody and Cummins 1968, 1974). In this case the driving mode is a coupled pseudo-spin optic-phonon mode and the phase transition takes place about 4 K above the extrapolated zero of the driving mode. Also, despite the heavily damped form of the ferroelectric driving mode, the acoustic mode is only slightly damped by the interaction and appears as a well-defined Brillouin peak in the Raman experiments.

To this point we have neglected components higher than bilinear in η_l and $u_{i,l}$ arising from the electrostrictive term (eqn (10.1.4)). Although the bilinear term vanishes in a centrosymmetric phase the total effect of eqn (10.1.4) does not, and higher-order components (involving static fluctuations) can still perturb the acoustic velocity (Tani and Tsuda 1969, Dvorak 1971 d) and in addition can significantly perturb the acoustic mode dispersion at order q^3 (Axe et al. 1970). The centrosymmetric ($T > T_c$) acoustic anomaly is too small to be seen in $SrTiO_3$ (Fig. 10.1), but has been observed ultrasonically in $BaTiO_3$ for which the relevant acousto-optic coupling is greater (Kashida *et al.* 1973). It takes the form of a critical decrease of sound velocity proportional to $(T - T_0)^{-\zeta}$, where T_0 is the Weiss temperature and exponent $\zeta \approx \frac{1}{2}$. The q^3 perturbation of acoustic dispersion has been observed for many perovskites (e.g. for $KTaO_3$ by Axe *et al.* 1970 and for $PbTiO_3$ by Shirane, Axe, Harada, and Remeika 1970) and is quite

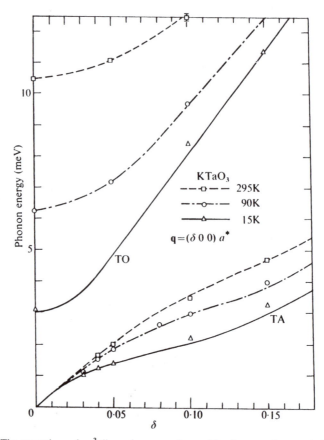

Fig. 10.2. The acoustic mode q^3 dispersion anomaly resulting from an electrostrictive coupling to the soft (ferroelectric) optic mode in $KTaO_3$ as the latter condenses towards $T_c \approx 0$ K. The limiting acoustic mode slope at $q = 0$ is approximately temperature independent and in agreement with ultrasonic velocity measurments. (After Axe *et al.* 1970.)

pronounced (see Fig. 10.2). In addition we must remember that the interaction (eqn (10.1.4)) is only one of many possible interaction forms of higher than second order. Additional effects might also result from the third-order term quadratic in strain and linear in soft-mode variable or even from fourth-order terms quadratic in both strain and optic variables.

Although microscopic theories developed from model Hamiltonians are from a fundamental point of view more satisfying than phenomenology, it is only fair to point out that quite successful approaches to the problem of coupled long-wavelength acoustic and optic modes were developed phenomenologically before the microscopic picture presented above was first set out. The types of phenomenological theory commonly used are based on power-series expansions of crystalline energy density. They had their origins in the work of Landau and Khalatnikov (1954) and of Sannikov (1962), and have been pursued with increasing degrees of sophistication by Levanyuk (1965), Dvorak (1967, 1968, 1971 d), Fleury and Lazay (1971), and Reese *et al.* (1973). The phenomenology begins by writing an energy density in terms of strain and displacement (or polarization) including a piezoelectric coupling between them. The relevant harmonic coefficients involve elastic, polarizability, and piezoelectric tensors respectively. Equations of motion can be formally derived via the Lagrangian and are conventionally restricted to a coupled harmonic oscillator form with damping added phenomenologically. Solving for the perturbed (i.e. coupled) frequencies the results are equivalent to those obtained from the microscopic approach. The big difference, however, is that in the macroscopic formalism a temperature dependence must be inserted into the polarizability and piezoelectric (and possibly even the elastic) coefficients in a purely phenomenological fashion whereas in the microscopic Hamiltonian formalism it can be calculated explicitly (at least in principle) via statistical mechanics. Thus while the macroscopic theory is very useful in relating experimentally observed quantities, it is not able to calculate a temperature variation explicitly.

By including driving terms coupling to the elastic variable, polarization variable, or both, the forced coupled acoustic–optic equations of motion can also be formally solved for Raman and Brillouin scattering response. Fits of phenomenological theory with measured scattering then determine the macroscopic energy coefficients involved and can be compared with the predictions resulting from using the same phenomenology to interpret static properties. In this way the internal self-consistency of different energy expansions may be compared (Brody and Cummins 1974).

Although the major optic response and acoustic response are usually in quite different frequency ranges, close to a phase transition the soft optic mode softens to the extent that both modes can sometimes be observed in the same Brillouin experiment. A particularly clear example of this has been observed in DKDP by Reese *et al.* (1973). Fig. 10.3 shows a coupled acoustic–optic response in DKDP at various temperatures approaching the Curie point. Near room temperature only the soft acoustic mode components are discernible along with small parasitic Rayleigh scattering in the centre of the spectrum. At this temperature the overdamped or Debye-like optic mode contributes only a small effectively constant background. As the temperature is lowered towards T_c the Brillouin components increase in intensity and decrease in frequency, while the broad soft ferroelectric mode becomes clearly discernible within about 10 K of T_c eventually 'washing out'

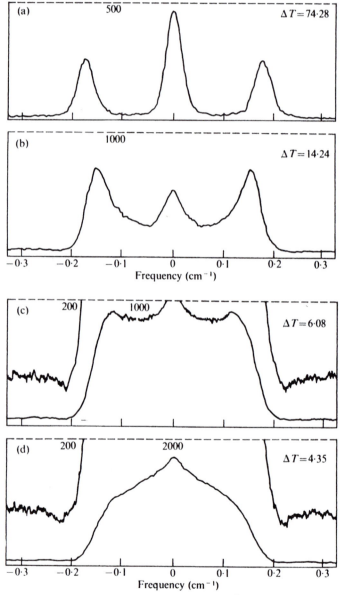

[*Figure continued on facing page*]

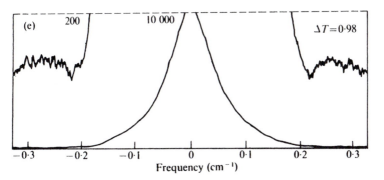

Fig. 10.3. Five typical Brillouin spectra showing the coupling of ferroelectric and acoustic soft modes in DKDP; $\Delta T = T - T_e$ and the appropriate count-rate scales are shown on each spectrum. (After Reese *et al.* 1973.)

the Brillouin components completely within about 6 K of T_c. In undeuterated KDP, by contrast, the soft-mode frequency is very much higher and the Brillouin peaks remain clear to within a 0·1 K of the Curie point, the undeuterated soft mode still appearing as a constant background on the Brillouin frequency scale of Fig. 10.3 at this temperature (Brody and Cummins 1968, 1969).

Another interesting case is the observation by Fleury and Lazay (1971) of the coupling of a transverse acoustic mode to the ferroelectric mode in BaTiO₃ in the piezoelectric tetragonal phase below the Curie point. Since the high-temperature phase is not piezoelectric, the acoustic frequency is only perturbed in a minor fashion but the lineshape anomalies are quite pronounced and include interference dips so that the spectrum is much more complicated than the simple sum of Lorentzians.

10.1.2. *Ultrasonic attenuation*

In addition to the shift in acoustic frequency given by the real part of the perturbed solution (eqn (10.1.12)) there is also an associated linewidth anomaly given by the imaginary part. The latter manifests itself also as a resonant contribution to ultrasonic attenuation. The amplitude absorption coefficient α_μ is defined as the imaginary part of the complex propagation vector $\mathbf{q}$ (i.e. where polarization, strain, etc. vary as $\exp \mathrm{i}(\omega t - \mathbf{q} \cdot \mathbf{l})$) and is therefore given by the imaginary part of the mode frequency ω (or energy if the latter is expressed in units of $\hbar$) divided by the sound velocity c_μ. It follows immediately from eqn (10.1.12) that the contribution of the *resonant* interaction to α_μ is $q^2 F_\mu \tau / c_\mu$ or

$$\alpha_\mu^{\mathrm{res}} = \left\{ 2G^2(\mu) \frac{\langle \xi \rangle_0^2}{M c_\mu^3} \right\} \Omega_{\mathrm{opt}}^{-2} \frac{\omega_\mu^2 \tau}{1 + \omega_\mu^2 \tau^2}. \qquad (10.1.18)$$

This contribution vanishes above T_c where the order parameter $\langle\xi\rangle_0$ is zero. Again this is a consequence of our choice of a centrosymmetric prototype phase. In this case the high-temperature phase absorption then results from multi-phonon processes arising from terms higher than bilinear in the variables. Thus, for example, with $\langle\xi\rangle_0$ and $\langle x_i\rangle_0$ both zero in eqn (10.1.6) the electrostrictive interaction Hamiltonian eqn (10.1.4) Fourier transforms to a form

$$\mathcal{H}_{int} = \sum_{\mu}\sum_{q}\sum_{q'} g(\mu\mathbf{q}\mathbf{q}')Q(\mu\mathbf{q})\eta(\mathbf{q}'-\mathbf{q})\eta(-\mathbf{q}'). \qquad (10.1.19)$$

Expanding $Q(\mu\mathbf{q})$ and $\eta(\mathbf{q})$ in terms of creation and annihilation operators we find that eqn (10.1.19) describes processes involving three phonons, e.g. the splitting of an acoustic phonon into two optic phonons, the fusion of two optic phonons into an acoustic phonon, and the scattering of an acoustic phonon with the absorption and emission of optic phonons. An analysis of the contribution of these scattering processes to acoustic attenuation predicts a singular behaviour but with different critical exponents depending upon whether the soft optic mode is underdamped or overdamped near the critical region. For the overdamped case, and for $\omega_\mu \ll \Gamma_{opt}$, Pytte (1970 a) finds a *scattering* contribution to absorption coefficient $\alpha_\mu^{scatt} \propto \omega_\mu^2 |T - T_c|^{-3/2}$. For an underdamped soft mode the equivalent finding takes the form $\alpha_\mu^{scatt} \propto \omega_\mu^2 |T - T_c|^{-1/2}$. It must be emphasized, however, that these exponents have been derived assuming an isotropic soft-mode frequency and a mean field soft-mode behaviour. The *resonant* contribution to absorption (eqn (10.1.18)) peaks below the Curie point when $\omega_\mu\tau \approx 1$ and then goes to zero as $T \to T_c^-$. Note, however, that when $\omega_\mu^2\tau^2 \ll 1$ we have $\alpha_\mu^{res} \propto \omega_\mu^2\tau$, which from eqn (10.1.13) shows a critical divergence proportional to $(T_c - T)^{-1}$ as $T \to T_c^-$.

As with the case of the sound-velocity anomaly the resonant contribution to sound absorption can also be deduced phenomenologically. The qualitative form

$$\alpha_\mu^{res} = A\frac{\omega_\mu^2\tau}{1+\omega_\mu^2\tau^2} \qquad (10.1.20)$$

is confirmed, with the polarization (order-parameter) dependence of the amplitude factor A disappearing for piezoelectric prototype systems. The basic Landau and Khalatnikov formalism has been extended by Levanyuk (1965), Dvorak (1971 d), and Levanyuk and Shchedrina (1972) to allow for the inclusion of fluctuation (scattering) terms. Pytte's findings for α_μ^{scatt} can be reconfirmed for an isotropic soft mode, but the temperature exponents are sensitive to the dimensionality and character of the fluctuations. For example, the temperature exponents are found to be -1 and $-\frac{1}{2}$ respectively for the case of overdamped uniaxial ferroelectrics with centrosymmetric and piezoelectric non-polar phases.

In comparing these theoretical predictions with experiment many difficulties present themselves. One expects, for example, a critical attenuation which is dependent both on the polarization and direction of propagation of the sound wave. It is also expected to be sensitive to the detailed character of the polarization fluctuations (many structural transitions have quasi-two-dimensional fluctuations and uniaxial ferroelectrics generally have quasi-linear fluctuations). Experimentally the reproducibility of results between different crystal specimens has proved to be a problem. This may be due to scattering from inhomogeneities or to the presence of domain scattering. One must also be concerned about the uniqueness of interpretation as regards critical exponents. It is often possible to accommodate a range of temperature exponents when fitting over less than a decade in $|T - T_c|/T_c$. The problem is compounded when an effort is made to determine both the frequency and temperature exponents. Finally, none of the theoretical estimates for critical exponents made above is expected to be valid if the system in question enters a truly 'critical' or fluctuation-dominated region very close to a Curie point. The shortcomings of both macroscopic and simple microscopic schemes are well known in critical theory (see Chapter 11). The breakdown is expected to be worst for structural transitions in which the forces involved are of short range, although the effect may still be pronounced for high-symmetry dipolar-dominated situations. An effort to go beyond simple Landau-exponent theories in the estimate of ultrasonic critical attenuation has recently been made by Schwabl (1973) who suggests that in certain situations a cross-over from one set of exponents further from T_c to a second set close to T_c may be observable.

If the simple theory leading to eqn (10.1.18) is valid anywhere it should be for uniaxial ferroelectrics for which the polarization fluctuations are relatively suppressed and true critical effects are expected to be minimal. For $\omega_\mu^2 \tau^2 \ll 1$ we anticipate $\alpha_\mu^{res} \propto \omega_\mu^2/|T - T_c|$ in the piezoelectric phase, and this relationship has been verified for TGS and its isomorphs below T_c (Minaeva and Levanyuk 1965, Minaeva et al. 1967, Minaeva, Strukov, and Varnstorff 1968) (see Fig. 10.4). A similar $(T - T_c)^{-1}$ divergence of absorption is also observed on approach to T_c from the centrosymmetric high-temperature phase and is presumably due to fluctuation scattering in agreement with theory for an overdamped uniaxial ferroelectric (Levanyuk and Shchedrina 1972). Additional verification of simple resonant theory has been obtained for Rochelle salt which is piezoelectric in both ordered and disordered phases (Baranskii et al. 1962). However, the predicted absorption is seen only for special directions of propagation (generally perpendicular to the polar direction), and this is readily understandable in terms of the anisotropy of dipolar forces which only allows for a soft mode with $\mathbf{q}$ perpendicular to the ferroelectric axis.

For structural transitions like those in $SrTiO_3$ and $KMnF_3$ which involve zone-boundary-mode condensation the agreement with theory is much less

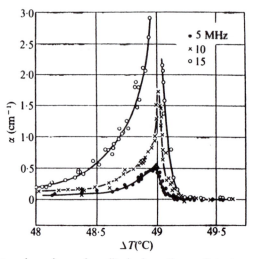

Fig. 10.4. Temperature dependence of amplitude absorption coefficient α near the ferroelectric Curie temperature in TGS, for longitudinal ultrasonic waves at frequencies of 5 (solid circles), 10 (crosses), and 15 (open circles) MHz. (After Minaeva and Levanyuk 1965.)

convincing, presumably because of the onset of genuine critical effects. In KMnF$_3$ a Γ_{25} optic phonon mode goes soft at the Brillouin zone corner (R-point) at 185 K. Ultrasonic attenuation near this soft-mode transition has been very carefully measured by Domb, Mihalisin, and Skalyo (1973) who find that the critical attenuation diverges in the high-temperature cubic phase as $\omega^x (T - T_c)^{-y}$, where $x = 1 \cdot 18 \pm 0 \cdot 03$ and $y = 1 \cdot 30 \pm 0 \cdot 04$, over a two-decade region in temperature $0 \cdot 001 < (T - T_c)/T_c < 0 \cdot 1$ and a one-decade variation of frequency 30–270 MHz. Other, less-detailed experimental findings (e.g. Furukawa, Fujimori, and Hirakawa 1970, Courdille and Dumas 1971) support the general conclusion that the frequency exponent is a little larger than unity and the temperature exponent $\sim 1 \cdot 3$, although there is some suggestion that the exponents may vary slightly with the sound-propagation direction. No cross-over effects such as predicted by Schwabl were observed. The most disturbing feature is the frequency exponent which differs very markedly from the quadratic variation of extant theories, although this experimental conclusion has been questioned by Hatta, Matsuda, and Sawada (1974) who claim that a frequency dispersion of the form (eqn (10.1.20)) *is* able to fit their data in a satisfactory manner. Additional experimental evidence is available for other zone-boundary transitions, in particular the 105 K transition in SrTiO$_3$, but the findings are again not always consistent. The important question, namely whether true critical (fluctuation dominated as opposed to Landau fluctuation) phenomena are being observed in the zone-boundary cases, has not yet been convincingly answered.

Finally, since the anharmonicity of the lattice is considerable in ferroelectrics, we should expect that elastic constants of higher order than second would also vary very sharply in the vicinity of a ferroelectric transition. Much of the early work on non-linear acoustic effects was performed on ordinary piezoelectric semiconductors, but it is now apparent that ferroelectrics operating near their Curie temperatures are much better materials for observing acoustic non-linearities. Work in this field is not yet well developed, but experiments on the behaviour of third-order elastic moduli are now beginning to take place using harmonic generation and frequency mixing and some results on non-linear anomalies have already appeared for $BaTiO_3$, SbSI, and TGS (see for example Serdobol'skaya and Kuak Tkhi Tam 1972). A simple theoretical analysis of ultrasonic harmonic generation has also recently been given (Spector 1974).

10.2. Extrinsic (improper) ferroelectrics and ferroelastics

The simplest macroscopic or phenomenological description of a phase transition can be given in terms of a single order parameter, say θ_1, by defining a suitable free-energy function F as a power series

$$F = \tfrac{1}{2}\alpha\theta_1^2 + \tfrac{1}{4}\gamma\theta_1^4 + \tfrac{1}{6}\delta\theta_1^6 + \ldots \qquad (10.2.1)$$

in this variable. An example is the series eqn (3.3.1) for the elastic Gibbs free energy G_1 in terms of electric displacement D. The coefficients α, γ, δ are in general temperature dependent. More generally one can allow for additional macroscopic parameters θ_p, $p > 1$, and write F as the sum of power series like eqn (10.2.1). Contours of constant F can then be considered plotted in a multi-dimensional space for which the axes are the θ_p's. As the temperature of the system is lowered towards a phase transition there are two basic possibilities. If the curvature $\partial^2 F/\partial \theta_p^2$ becomes zero at the origin for one order parameter ($p = 1$) while F remains greater than zero at all other points, then we have the condition for a second-order transition involving a mode for which a (complex) frequency has become zero at the transition temperature. The possibility of this occurring simultaneously for two or more modes which are not degenerate by symmetry is infinitesimal, and we conclude that a second-order transition involves a single (but possibly degenerate) soft mode. If, on the other hand, the free energy develops a minimum away from the origin of the configuration space and this minimum falls to $F = 0$ before the development of a zero curvature at the origin, then we have the condition for a first-order transition. This may involve a single mode (which may well have fallen considerably in frequency, though not to zero) or it may involve many modes. A first-order transition involving many modes is most likely to occur when the modes involved are fairly soft (low in frequency), since a large positive quadratic term will usually ensure that a mode of high frequency cannot be significantly involved. Examples of first-order transitions involving many modes have already been described for $PbZrO_3$ and $NaNbO_3$ in Chapter 8.

These arguments, however, are incomplete since we have not given adequate thought to the effects produced by the existence of coupling terms between the variables. Including such terms, and in particular those involving a coupling to the soft-mode variable θ_1 at a second-order phase transition, one can deduce that one or more order parameters which correspond to modes of any frequency can be *induced* by this coupling in spite of the fact that the transition is of second order. Consider for example a simple two-variable situation with a free energy defined by

$$F = \tfrac{1}{2}\alpha_1\theta_1^2 + \tfrac{1}{4}\gamma_1\theta_1^4 + \tfrac{1}{6}\delta_1\theta_1^6 + \tfrac{1}{2}\alpha_2\theta_2^2 + C\theta_1^2\theta_2. \qquad (10.2.2)$$

Writing $\alpha_1 = A(T - T_c)$ with A a positive constant and assuming all other coefficients (including the coupling C) to be independent of temperature, we have

$$\frac{\partial F}{\partial \theta_2} = \alpha_2\theta_2 + C\theta_1^2 \qquad (10.2.3)$$

which is zero when $\theta_2 = -(C/\alpha_2)\theta_1^2$. Substituting in eqn (10.2.2) gives

$$F = \tfrac{1}{2}\alpha_1\theta_1^2 + \tfrac{1}{4}\gamma_1'\theta_1^4 + \tfrac{1}{6}\delta_1\theta_1^6 \qquad (10.2.4)$$

in which

$$\gamma_1' = \gamma_1 - 2\frac{C^2}{\alpha_2}. \qquad (10.2.5)$$

Finally from (10.2.4) we find the equilibrium equation

$$\frac{\partial F}{\partial \theta_1} = \alpha_1\theta_1 + \gamma_1'\theta_1^3 + \delta_1\theta_1^5 = 0. \qquad (10.2.6)$$

With α_1 linear in $T - T_c$ this represents a second-order phase transition at $T = T_c$ if $\gamma_1' > 0$, with $\theta_1 \propto (T_c - T)^{1/2}$ and therefore $\theta_2 \propto T_c - T$, in the low-temperature phase. If $\gamma_1' < 0$ the transition occurs at $T > T_c$ and is of first order. We therefore have two non-zero order parameters appearing below a second-order transition when $\gamma_1' > 0$. These order parameters are nevertheless of quite different characters, and we can consider one to be primary (θ_1) and the other to be secondary (θ_2), induced by the condensation of the primary mode. To make this clear consider the field variables E_1 and E_2 corresponding to the order parameters θ_1 and θ_2. They are defined by

$$E_1 = \left(\frac{\partial F}{\partial \theta_1}\right)_{\theta_2, T} = \alpha_1\theta_1 + \gamma_1\theta_1^3 + \delta_1\theta_1^5 + 2C\theta_1\theta_2 \qquad (10.2.7)$$

$$E_2 = \left(\frac{\partial F}{\partial \theta_2}\right)_{\theta_1,T} = \alpha_2\theta_2 + C\theta_1^2. \qquad (10.2.8)$$

The associated reciprocal responses (or permittivities) follow as

$$\left(\frac{\partial E_1}{\partial \theta_1}\right)_{\theta_2,T} = \alpha_1 + 3\gamma_1\theta_1^2 + 5\delta_1\theta_1^4 + 2C\theta_2 \qquad (10.2.9)$$

$$\left(\frac{\partial E_2}{\partial \theta_2}\right)_{\theta_1,T} = \alpha_2. \qquad (10.2.10)$$

It is now clear that, whereas the parameter θ_1 response $(\partial E_1/\partial\theta_1)^{-1}$ diverges as $T \to T_c$ ($\theta_1 \to 0$, $\theta_2 \to 0$) by virtue of the proportionality of α_1 to $T - T_c$, the θ_2 response remains quite finite and indeed, within our present model, is completely temperature independent.

In general, therefore, we expect a single-order parameter to have a divergent response at a second-order phase transition, and we shall designate this order parameter as the primary one. We shall refer to the transition as being *intrinsically* of the character θ_1. Any additional induced order parameters (or secondary order parameters) will be referred to as *extrinsic* order parameters. Correspondingly the (complex) frequency associated with the primary or intrinsic mode goes to zero at a second-order transition while the frequencies of the secondary or extrinsic modes remain finite.

As an example, the cubic–tetragonal transition in $BaTiO_3$ can be discussed in terms of eqn (10.2.2) if θ_1 is the polarization and θ_2 the strain. In this case the coefficient γ_1' is negative and the transition is of first order, but a softening overdamped mode is still clearly in evidence associated with fluctuating polarization and the primary and secondary characteristics of polarization and strain are quite clearly defined. It follows that $BaTiO_3$ is an intrinsic ferroelectric and an extrinsic ferroelastic, where we define ferroelasticity (§ 1.1.2) as the possession of a spontaneous strain which can be reoriented either experimentally by application of stress or conceptually in terms of small atomic displacements required by the structure in changing orientations (the concept of ferroelasticity was introduced by Aizu (1969)). A second example is the structural transition in $SrTiO_3$. Here the mode corresponding to the primary order parameter is an anti-distortive mode with wavevector $\mathbf{q}$ at the R-point or $(111)(\pi/a)$ corner of the Brillouin zone. The transition is actually of second order and a coupling to a secondary strain order parameter results in the appearance of spontaneous strain below T_c. The $SrTiO_3$ transition is therefore intrinsically of anti-distortive structural character but is extrinsically ferroelastic.

Having defined the concept of extrinsic and intrinsic order parameter, the interesting question from a ferroelectric point of view is whether ferroelectrics exist which are polar only in the extrinsic sense. The possibility of such a situation was first touched upon by Indenbom (1960) and was studied from a free-energy or Landau point of view by Levanyuk and Sannikov (1968). In the symbolism of eqn (10.2.2) it requires the extrinsic order parameter to be a polarization component (say P_i) and the intrinsic order parameter θ_1 to be some as yet unspecified property. However, the simple representation (eqn (10.2.2)) is as yet too restricted to describe a realistic situation. In general the polarization P and the intrinsic variable θ_1 are not scalars, and the primary interaction term is not necessarily of second order in the latter and first order in the former. In order to determine the basic allowed form for a realistic free energy it is necessary to consider the transformation properties of both variables under the transformations belonging to the symmetry group of the prototype phase.

If θ_1 is scalar, for example, cross-terms of the type $\theta_1 P_i$ are allowed only if θ_1 transforms like P_i, since none of the terms contained in the free-energy expansion can be changed by any of the symmetry-group transformations. Such a situation can occur, and will be discussed at the end of this section, but does not lead to what are thought of as typical extrinsic characteristics. Terms of the type $\theta_1^2 P_i$ are admissible only for pyroelectric prototypes since θ_1^2 is necessarily invariant. Since we are interested in transitions from non-polar to polar phases this form of coupling is excluded. Terms $\theta_1 P_i P_j$ are allowed if θ_1 transforms like $P_i P_j$, and terms $\theta_1^2 P_i P_j$ are always admissible for any structure since one can always construct invariants out of $P_i P_j$ (e.g. $P_x^2 + P_y^2$ and P_z^2 for a uniaxial crystal). It follows that the simple form (eqn (10.2.2)) with $\theta_2 = P_i$ cannot be written to represent a non-polar high-temperature phase if θ_1 is a scalar. One can of course investigate the consequences of coupling forms like $\theta_1^2 P_i P_j$ (see, for example, Holakovsky 1973) and some interesting polar characteristics result. However, we shall concentrate on conditions which allow for the existence of a lower-order coupling linear in polarization P_i. This can occur when the intrinsic order parameter θ_1 is a two-component vector, say $\theta_1 = (\eta_x, \eta_y)$, when cross-terms of the form $\eta_x \eta_y P_i$ or $(\eta_x^2 - \eta_y^2) P_i$ are allowed by symmetry in non-polar high-temperature prototype phases if $\eta_x \eta_y$ or $\eta_x^2 - \eta_y^2$ transform like P_i. To illustrate this we consider specifically a simple crystal structure which allows a cross term $\eta_x \eta_y P_i$ (Cochran 1971).

As a very simple possible situation we imagine the ZnS structure (Fig. 10.5) elongated a little along the c-axis (or z-direction) to produce a tetragonal crystal as our high-temperature prototype phase. We suppose that a transverse mode with wavevector $q = 2\pi/c$ along [001] is soft. (This mode is at a Brillouin zone boundary since the lattice is not primitive.) This mode is doubly degenerate in the c-plane and to simplify the picture as much as possible we assume force constants which leave half the atoms (shaded in Fig. 10.5) undisplaced in this mode. The variables η_x and η_y then represent displacements of the open-circle atoms along the [100] and [010] directions

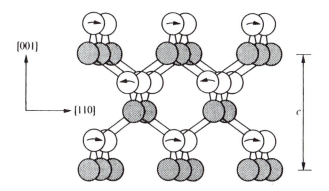

Fig. 10.5. The zincblende structure viewed in a direction close to the [110] direction; the overlapping circles represent atoms lying on a [110] line. For additional details see text. (After Cochran 1971.)

respectively. The essential terms in the free energy for the extrinsic ferroelectric effect under consideration are

$$F = \tfrac{1}{2}\alpha_1(\eta_x^2 + \eta_y^2) + \tfrac{1}{4}\gamma_1(\eta_x^4 + \eta_y^4) + \tfrac{1}{2}\varepsilon_1\eta_x^2\eta_y^2 + \tfrac{1}{2}\alpha_2 P_z^2 + C\eta_x\eta_y P_z \quad (10.2.11)$$

where P_z is the component of spontaneous polarization in the c-direction. We assume, in the simplest Landau approximation, that the intrinsic quadratic coefficient α_1 is proportional to $T - T_c$ and that all other parameters are independent of temperature.

The manner in which the final cross-term arises physically within the crystal structure of Fig. 10.5 is very easy to grasp. When the soft-mode motion has a large amplitude with displacements directed along [110], the anharmonic forces result in a secondary displacement of these atoms along the $-c$ direction as indicated by the curvature of the arrows in Fig. 10.5. If, however, the primary mode displacements are out of the plane of the diagram (along $[1\bar{1}0]$) then by symmetry the secondary displacements are along $+c$. In other words displacements $\pm\eta_x$, $\pm\eta_y$ induce a polarization of the opposite sign to $\pm\eta_x$, $\mp\eta_y$.

Using eqn (10.2.11) the condition $\partial F/\partial P_z = 0$ gives

$$P_z = -\frac{C}{\alpha_2}\eta_x\eta_y \quad (10.2.12)$$

and substitution into eqn (10.2.11) leads to

$$F = \tfrac{1}{2}\alpha_1(\eta_x^2 + \eta_y^2) + \tfrac{1}{4}\gamma_1(\eta_x^4 + \eta_y^4) + \frac{1}{2}\left(\varepsilon_1 - \frac{C^2}{\alpha_2}\right)\eta_x^2\eta_y^2. \quad (10.2.13)$$

This expression can lead to either a first- or a second-order transition depending on the relative values of the parameters involved. We shall concentrate on second-order situations as follows. Using eqn (10.2.13) the equilibrium configuration is given by

$$\frac{\partial F}{\partial \eta_x} = \alpha_1 \eta_x + \gamma_1 \eta_x^3 + \left(\varepsilon_1 - \frac{C^2}{\alpha_2}\right) \eta_x \eta_y^2 = 0 \qquad (10.2.14)$$

$$\frac{\partial F}{\partial \eta_y} = \alpha_1 \eta_y + \gamma_1 \eta_y^3 + \left(\varepsilon_1 - \frac{C^2}{\alpha_2}\right) \eta_y \eta_x^2 = 0. \qquad (10.2.15)$$

Solutions of these equations exist with the direction of anti-distortion along an axis ($\eta_x \neq 0$, $\eta_y = 0$; $\eta_y \neq 0$, $\eta_x = 0$) or along a face diagonal ($\eta_x = \pm\eta_y \neq 0$). The latter solution is relevant for our purposes and is

$$\eta_x^2 = \eta_y^2 = -\alpha_1 \left(\gamma_1 + \varepsilon_1 - \frac{C^2}{\alpha_2}\right)^{-1}. \qquad (10.2.16)$$

Writing $\alpha_1 = A(T - T_c)$ with A positive in the usual Landau manner we see immediately that eqn (10.2.16) defines a second-order phase transition if $\gamma_1 + \varepsilon_1 - C^2/\alpha_2$ is positive. The primary or intrinsic order parameter $\eta_x = \pm\eta_y$ is proportional to $(T_c - T)^{1/2}$ below T_c, while the extrinsic polarization P_z follows from eqn (10.2.12) and is proportional to $\pm(T_c - T)$. That the polarization is truly an extrinsic order parameter is verified by calculating the isothermal polar response (permittivity) near T_c. It is

$$\left(\frac{\partial^2 F}{\partial P_z^2}\right)^{-1}_{\eta_x, \eta_y, T} = \alpha_2^{-1} \qquad (10.2.17)$$

and within the model is completely temperature independent. The model therefore defines an extrinsic ferroelectric which is intrinsically anti-ferrodistortive.

Although no extrinsic ferroelectric as simple in character as that outlined above has yet been discovered in nature there is no longer any doubt that extrinsic ferroelectrics in the more general sense do exist. Axe, Dorner, and Shirane (1971, 1972) have now established beyond question that terbium molybdate $Tb_2(MoO_4)_3$ is such a system. However, the coupling is more complicated than that outlined for the ZnS structure above, since P_z and (η_x, η_y) are not directly coupled through a term $P_z\eta_x\eta_y$ but indirectly via two terms involving $P_z x_s$ and $\eta_x\eta_y x_s$, where x_s is a shear strain. An analogous situation also occurs in isomorphic gadolinium molybdate $Gd_2(MoO_4)_3$ and we shall discuss the details in the following section.

Meanwhile it is interesting to note (Cochran 1971) that an expression for free energy analogous to eqn (10.2.11) should apply for the antiferroelectric $(NH_4)H_2PO_4$ (ADP). However, this provides an example of the situation where the direction of antipolarization is in practice along an axis ($\eta_x = 0$ or

$\eta_y = 0$). It follows from eqn (10.2.12) that the induced spontaneous polarization is zero. It is also of some interest to note that an expression for F of the form (eqn (10.2.11)) can also be used to describe the simultaneous development of ferromagnetism and ferroelectricity. If for example the variables η_x and η_y correspond to components of magnetization then the model describes an extrinsic ferroelectric driven by a primary ferromagnetic mode. Nickel iodine boracite $Ni_3B_7O_{13}I$ (Ascher *et al.* 1966) appears to be a material for which a coupling between ferromagnetism and ferroelectricity is involved, although whether the coupling is direct or via strain is not yet clear. A summary of the extent of theoretical understanding of boracite phase transitions at the time of writing can be obtained from Dvorak (1974).

The general conclusion is that a second-order transition can involve only a single primary variable, by which we mean that only a single variable has a response which diverges at the transition temperature T_c. Nevertheless, via a coupling mechanism, one or many other non-zero order parameters may appear below T_c and tend to zero as $T \rightarrow T_c^-$. These secondary or extrinsic order parameters are *induced* by the primary instability and their response functions do not exhibit typical second-order singularities at T_c. In particular, spontaneous polarization can occur below T_c as a secondary order parameter induced by any of a number of possible primary characteristics. We refer to the resulting ferroelectrics as extrinsic ferroelectrics. They are also often referred to as 'improper' ferroelectrics in the literature.

Strictly speaking (Dvorak 1974) the terms intrinsic and extrinsic (or proper and improper) should be used only to describe a particular transition and not a material, since it is quite possible for the same material to display both extrinsic and intrinsic ferroelectric phase transitions. In addition, even at the same transition it is possible for one component of polarization to be the intrinsic parameter while other components are extrinsic in the sense that they acquire non-zero values solely through an interaction with the former. The full range of possibilities can include some very complex situations which are perhaps not yet fully categorized. Aizu (1972 a) has made an effort to bring some order to a very complex situation by defining an 'index of faintness' n equal to the power of intrinsic variable which couples to the extrinsic variable under consideration. Thus, for example, free energy (eqn (10.2.11)) describes an extrinsic ferroelectric with index of faintness $n = 2$.

Extrinsic ferroelectrics with index of faintness $n \geq 2$ possess what have come to be considered as typical extrinsic properties such as a lack of critical divergence of clamped and free dielectric constant at T_c. The situation with $n = 1$ is rather different and will now be considered specifically. In this case a simple free-energy expression can be written of the form

$$F = \tfrac{1}{2}\alpha_1\theta_1^2 + \tfrac{1}{4}\gamma_1\theta_1^4 + \tfrac{1}{6}\delta_1\theta_1^6 + \tfrac{1}{2}\alpha_2\theta_2^2 + C\theta_1\theta_2 \qquad (10.2.18)$$

where for an extrinsic ferroelectric the variable θ_2 corresponds to a polarization component. Defining field variables

$$E_1 = \left(\frac{\partial F}{\partial \theta_1}\right)_{\theta_2,T} = \alpha_1 \theta_1 + \gamma_1 \theta_1^3 + \delta_1 \theta_1^5 + C\theta_2 \qquad (10.2.19)$$

$$E_2 = \left(\frac{\partial F}{\partial \theta_2}\right)_{\theta_1,T} = \alpha_2 \theta_2 + C\theta_1 \qquad (10.2.20)$$

it follows immediately that the reciprocal dielectric response at constant θ_1 (e.g. clamped response) is

$$\left(\frac{\partial E_2}{\partial \theta_2}\right)_{\theta_1,T} = \chi_{\theta_1,T}^{-1} \doteq \alpha_2 \qquad (10.2.21)$$

and is temperature independent in the usual scheme. It follows that eqn (10.2.18) certainly does not describe an intrinsic ferroelectric. However, let us now calculate the dielectric response at constant field E_1 in the disordered phase (for which the calculation is easiest, i.e. $E_1 \propto \theta_1 \propto \theta_2$ small). Calculating θ_1 from eqn (10.2.19) and substituting in eqn (10.2.20) leads to

$$\left(\frac{\partial E_2}{\partial \theta_2}\right)_{E_1,T} = \chi_{E_1,T}^{-1} = \alpha_2 - \frac{C^2}{\alpha_1}. \qquad (10.2.22)$$

Writing $\alpha_1 = \beta(T - T_0)$ this can be recast in the form

$$\chi_{E_1,T} = \alpha_2^{-1} + \frac{C^2 \alpha_2^{-2}}{\beta(T - T_c)} \qquad (10.2.23)$$

in which $T_c = T_0 + C^2 \alpha_2^{-1}/\beta$. This 'free' response shows a divergence when $T \to T_c^+$ and is quite unlike the calculated free response for the $n = 2$ case which, from eqns (10.2.7) and (10.2.8), is readily shown to be equal to the clamped response α_2^{-1} in the zero-field limit as $T \to T_c^+$ and hence to undergo no anomaly in the disordered phase.

If the $n = 1$ coupling constant C is large the free susceptibility (eqn (10.2.23)) follows a Curie–Weiss law over a wide temperature range above T_c. An example of this case is KDP for which the intrinsic order parameter θ_1 is an ordering of protons between double-well sites. If C is small the free susceptibility follows a Curie–Weiss law only very close to T_c. Away from T_c it is essentially temperature independent. Ammonium sulphate, $(NH_4)_2SO_4$, appears to be an example of the latter (Kobayashi, Enomoto, and Sato 1972). Recently it has been shown that the elastic, electromechanical, and optic properties of ammonium sulphate all conform to the picture of an extrinsic ferroelectric with index of faintness $n = 1$ (Ikeda et al. 1973, Sawada, Takagi, and Ishibashi 1973, Petzelt, Grigas, and Mayerova 1974). It has been suggested that the intrinsic variable θ_1 represents a rotational non-polar mode in this case which weakly couples with the polarization, although its true nature may be more complex (Jain and Bist 1974).

Extrinsic ferroelectrics with $n = 1$ are therefore rather special cases. They have been termed pseudo-proper or pseudo-intrinsic by some authors. Owing to the bilinear interaction the crystal actually becomes unstable with respect to a mixed mode. This mode contains a large polarization contribution in KDP and a small contribution from polarization in ammonium sulphate. Neither $n = 1$ nor $n \geqslant 2$ extrinsic ferroelectrics seem likely to be rarities in nature and examples of each are accumulating rapidly. Among the $n \geqslant 2$ category we can now include, in addition to boracites and rare-earth molybdates, langbeinites (Dvorak 1972 a), ammonium Rochelle salt (Sawada and Takagi 1972), diammonium dicadmium sulphate (Aizu 1972 b), and rubidium trihydrogen selenate (Dvorak 1972 b), while other likely candidates include ammonium fluoberyllate, sodium trihydrogen selenate, thiourea, and bismuth titanate (Dvorak 1974).

10.3. Gadolinium molybdate and its isomorphs

Gadolinium molybdate $Gd_2(MoO_4)_3$ and the isostructural molybdates of Sm, Eu, Tb, and Dy were observed by Borchardt and Bierstedt (1967) to undergo ferroelectric transitions in the temperature range $150 < T_c < 190°C$. Subsequent investigations of the dielectric, optical, and mechanical properties of these materials revealed many unusual features (Cross, Fouskova, and Cummins 1968, Cummins 1970, Keve, Abrahams, Nassau, and Glass 1970). Firstly, despite the onset of ferroelectricity below T_c, no dielectric anomaly exists at all for the clamped response, and only a very minor anomaly below T_c occurs for the free-crystal permittivity (Fig. 5.6). Similarly, although a spontaneous strain appears at T_c, no marked softening of elastic response occurs above T_c. However, significant anomalies in some elastic constants are seen below T_c (Höchli 1972, Itoh and Nakamura 1973, 1974). The transitions are in general of first order, and in the ordered phase the existence of states $\pm P_z$ of spontaneous polarization (which are readily switched by application of an applied electric field) are coupled with two mechanical configurations described by a shear strain $\pm x_{xy}$. The states of opposite strain can be switched by application of mechanical stress and it is clear that the ordered phase is ferroelastic as well as ferroelectric. In fact these two properties are so closely coupled in these materials that P_z and x_{xy} always change simultaneously and can hence both be switched by either the application of applied stress or electric field (Smith and Burns 1969, Kumada, Yumoto, and Ashida 1970).

The existence of a marked elastic anomaly in the ordered phase of gadolinium molybdate (GMO) and the essential absence of an associated dielectric anomaly led Cross *et al.* (1968) initially to suggest that the phase

transition is driven by an elastic instability, with the spontaneous polarization resulting as an incidental but necessary consequence of the resulting
strain via the piezoelectric effect (i.e. in the terminology of the last section
that GMO is intrinsically ferroelastic and extrinsically ferroelectric). The
explanation is not satisfactory, however, since an intrinsic ferroelastic
necessarily exhibits a critical divergence of elastic strain as $T \rightarrow T_c^+$ and such a
decrease in stiffness is completely absent on the high-temperature side in the
GMO-type transitions.

The correct basic character of these transitions was first deduced by Pytte
(1970 b) based on X-ray data which indicated that a unit-cell doubling was
involved. He suggested that the transition was driven by a zone-boundary
structural mode and that these rare-earth molybdates are therefore intrinsically antiferrodistortive and extrinsically both ferroelastic and ferroelectric.
Independently Levanyuk and Sannikov (1970) and Aizu (1971) came to the
same conclusion. The matter was finally settled beyond a doubt by an
inelastic neutron study of terbium molybdate, which was chosen over the
more extensively studied GMO because of its more favourable neutron
scattering properties. The neutron measurements (Axe et $al.$ 1971, 1972)
established that a lattice vibrational mode at the Brillouin-zone M-point,
$(110)(\pi/a)$, condenses at the transition. The mode is doubly degenerate,
corresponding to the xy-degeneracy of the tetragonal high-temperature
phase, with a frequency $\omega_M \propto (T - T_0)^{1/2}$ condensing towards the Weiss
temperature $T_0 \approx 149°C$. The actual transition in terbium molybdate is of
first order and takes place at $T_c \approx 159°C$ before complete softening has
occurred.

Although the basic properties of the ferroelastic–ferroelectric rare-earth
molybdates are expected to be closely similar, most work has been performed on GMO. The high-temperature phase β-GMO which undergoes
transformation to the ferroelectric–ferroelastic phase at $T_c \approx 160°C$ is actually only a metastable phase although it transforms to the absolutely stable
α-phase imperceptibly slowly below about 600°C. A precise crystal-
structure determination has been performed by Keve, Abrahams, and
Bernstein (1971) (see also Jeitschko 1972) who find that the ordered phase is
orthorhombic, though pseudo-tetragonal ($a = 1·039$ nm, $b = 1·042$ nm,
$c = 1·070$ nm), with four formula units per unit cell and a polar point group
mm2. Above T_c it becomes tetragonal (with a reduction of unit-cell volume
by a factor of 2) and possesses a point group $\bar{4}2m$ which is non-polar but is
piezoelectric. The structure is made up of three independent sets of
$(MoO_4)^{2-}$ tetrahedra forming successive layers along the c-axis. One set is
oriented with apex pointing along $+c$, one with apex along $-c$, and the third
with a pseudo-$\bar{4}$ axis parallel to c (see Fig. 10.6). The Gd^{3+} ions occupy the
approximate centres of GdO_7 polyhedra with average Gd—O distance of
0·233 nm. The average Mo—O distance is about 0·175 nm.

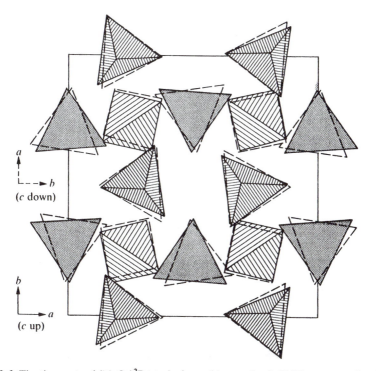

Fig. 10.6. The three sets of $(MoO_4)^{2-}$ tetrahedra making up the β-GMO structure, showing their orientations in the two ferroelastic–ferroelectric configurations. (After Keve, Abrahams, and Bernstein 1971.)

In the low-temperature phase the polar axis is parallel to c and the room-temperature value of spontaneous polarization is small (in the range 0.17–$0.20\,\mu\mathrm{C\,cm}^{-2}$) and arises from the mutual displacements of Gd^{3+} and $(MoO_4)^{2-}$ sublattices, with components arising from independent Gd ions nearly cancelling. In the high-temperature phase this cancellation is complete, but an antiparallel arrangement of electric dipoles persists and the tetragonal phase can perhaps be considered antipolar although it is not antiferroelectric in the sense of being switchable to a ferroelectric by application of an external electric field. In the orthorhombic phase a compressive stress applied along the b-axis causes a simultaneous interchange of the a- and b-axes together with a reversal of electric polarity. The two effects crystallographically are due to a single process involving displacements Δ of all atoms. The fact that the orthorhombic phase is pseudotetragonal (with $b-a$ being only some 0.3 per cent of a or b) results in all atoms being associated in pairs according to

$$(x_1, y_1, z_1) = (y_2, \tfrac{1}{2} - x_2, 1 - z_2) + \Delta \qquad (10.3.1)$$

where the vector $\boldsymbol{\Delta}$ is not in general in any high-symmetry direction and is small $(0{\cdot}046\,\text{nm} \geqslant |\Delta| \geqslant 0{\cdot}001\,\text{nm}$ in GMO). The displacements of the oxygen tetrahedra on domain reversal are shown in Fig. 10.6. The effect is readily observable experimentally by optic methods since the optic plane in (010) is discontinuously rotated by 90° as pressure is applied along [010]. The application of an electric field along [001] produces a similar rotation of the optic figure as a and b are interchanged.

In the low-temperature phase there are four formula units to the unit cell (68 atoms) leading to 204 branches of the phonon spectrum. In the high-temperature prototype phase these figures are reduced by a factor of 2, the basic halving of the unit-cell volume on entering the non-polar phase being depicted in Fig. 10.7. The 102 phonon branches in the non-polar phase still

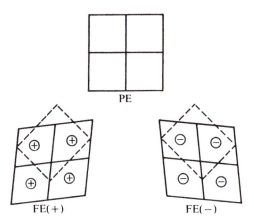

Fig. 10.7. Schematic drawing for the two possible GMO configurations in the ferroelectric phase FE originating from the paraelectric phase PE. The solid lines describe unit cells of PE and the broken lines those of FE projected onto the xy-plane. The $(+)$ and $(-)$ indicate the directions of spontaneous polarization along z.

represent an immensely complicated dynamical problem. Further simplification results by considering the MoO_4 groups as rigid, reducing the degrees of freedom to 48, and finally by using group-theoretical techniques to take advantage of restrictions produced by the nature of the crystal symmetry (Dvorak 1971 a, Petzelt and Dvorak 1971). Even in simple terms it is clear that neither elastic nor $q = 0$ optic phonon polar instabilities can explain a cell-doubling characteristic at T_c. The reduction in translational symmetry necessary to explain Fig. 10.7 must involve displacements modulated with the wavevector $\mathbf{q}_M = (110)(\pi/a')$ where we write the high-temperature prototype cell dimensions (a', a', c) with $a' \approx a/\sqrt{2} \approx 0{\cdot}73\,\text{nm}$. Group theoretically it transpires that there are only three possible physical representations at the M-point and that they are all doubly degenerate (owing essentially to the equivalence of the x- and y-axes). We need not concern ourselves here with the possible physical displacements involved but merely designate the soft M-point zone-boundary mode as (η_1, η_2) or $\eta(\gamma_1, \gamma_2)$ with $\gamma_1^2 + \gamma_2^2 = 1$. The order parameter η can be thought of as the thermal average of the critical phonon amplitude and the γ_1 and γ_2 as direction cosines.

It is now evident that the appearance of spontaneous strain and electric moment below T_c must arise in GMO and its isomorphs via a coupling to the soft anti-distortive mode $\eta(\gamma_1, \gamma_2)$. In the co-ordinate system (xyz) of the high-temperature tetragonal phase, the lowest-order terms in a power-series expansion of free energy $F(T, P, \eta, x)$ which are compatible with the symmetry change occurring at T_c have been given by Dvorak (1971 b). Following Axe et al. (1971, 1972) we shall here restrict ourselves only to those terms which are necessary for an understanding of the principal features of the GMO transition. Accordingly we write

$$F = F(\boldsymbol{\eta}) + F(\mathbf{P}) + F(\mathbf{x}) + F(\mathbf{P}, \mathbf{x}) + F(\boldsymbol{\eta}, \mathbf{x}) \qquad (10.3.2)$$

where

$$F(\boldsymbol{\eta}) = \tfrac{1}{2}\omega_M^2 \eta^2 + \tfrac{1}{4}\eta^4 \sum_\alpha V_\alpha f_\alpha^{(4)}(\gamma_i) + \tfrac{1}{6}\eta^6 \sum_\alpha W_\alpha^{(6)}(\gamma_i) \qquad (10.3.3)$$

$$F(\mathbf{P}) = \tfrac{1}{2}\chi_{33}^{-1} P_z^2 \qquad (10.3.4)$$

$$F(\mathbf{x}) = \tfrac{1}{2}c_{66}^P x_{xy}^2 \qquad (10.3.5)$$

$$F(\mathbf{P}, \mathbf{x}) = a_{36} P_z x_{xy} \qquad (10.3.6)$$

$$F(\boldsymbol{\eta}, \mathbf{x}) = \tfrac{1}{2}x_{xy}\eta^2 \sum_\alpha g_{66,\alpha} f_\alpha^{(2)}(\gamma_i). \qquad (10.3.7)$$

Here c_{66}^P is the elastic constant at constant polarization, χ_{33} is the clamped dielectric susceptibility, a_{36} is a piezoelectric coefficient, and ω_M is the frequency of the soft mode at the M-point. As is usual in simple Landau theory we shall assume $\omega_M^2 = \beta(T - T_0)$ with $\beta > 0$ and the other parameters to be basically temperature independent. Since the transition is of first order we have had to include terms up to sixth order in $F(\boldsymbol{\eta})$, and we anticipate that the Weiss temperature T_0 will be somewhat below the actual transition temperature T_c. The symbol $f^{(n)}(\gamma_i)$ in eqns (10.3.3) and (10.3.7) represents a homogeneous function of direction cosines γ_1 and γ_2 of order n, the form of which is dictated by the requirement of the invariance of F under all symmetry operations of the prototype group. The explicit forms are given by Axe, et al. (1972). The direct interaction $F(\mathbf{P}, \boldsymbol{\eta})$ between polarization and soft mode is not absent on symmetry grounds but is established as very small in the rare-earth molybdates by the essential absence of an anomaly in the clamped dielectric constant. We neglect its effects although a more complete treatment has been given by Dvorak (1971 b).

Using the equilibrium conditions

$$\frac{\partial F}{\partial P_z} = 0 \quad \text{and} \quad \frac{\partial F}{\partial x_{xy}} = 0$$

we find

$$P_z = -a_{36}\chi_{33}x_{xy} \tag{10.3.8}$$

and

$$x_{xy} = -\tfrac{1}{2}\frac{\eta^2}{c_{66}^E}\sum_\alpha g_{66,\alpha}f_\alpha^{(2)}(\gamma_i) \tag{10.3.9}$$

in which $c_{66}^E = c_{66}^P - a_{36}^2\chi_{33}$. Thus P_z is proportional to strain which is in turn proportional to the square of the intrinsic order parameter η^2. The primary experimental features—that the polarization changes when the strain is switched and that the clamped crystal exhibits no dielectric anomaly—are immediately explained. Using eqns (10.3.8) and (10.3.9) to eliminate P_z and x_{xy} from the free energy F we find

$$F = \tfrac{1}{2}\omega_M^2\eta^2 + \tfrac{1}{4}\eta^4\sum_\alpha B_\alpha f_\alpha^{(4)}(\gamma_i) + \tfrac{1}{6}\eta^6\sum_\alpha W_\alpha f_\alpha^{(6)}(\gamma_i) \tag{10.3.10}$$

which differs from $F(\eta)$ of eqn (10.3.3) only in a change of V_α to B_α, where B_α and V_α can be simply related via terms in g_{66} and c_{66}^E (Axe et al. 1972). Writing $\gamma_1 = \cos\varphi$ and $\gamma_2 = \sin\varphi$ it is possible to minimize eqn (10.3.10) with respect to the angle φ to obtain

$$\tan(4\varphi) = 4\left\{\frac{B_3 + W_3\eta^2}{B_1 - B_2 + (W_1 - W_2)\eta^2}\right\} \tag{10.3.11}$$

which is in general temperature dependent through η^2. In this respect the rare-earth molybdates differ from most other well-studied materials undergoing displacive transitions in which the composition of the spontaneous displacements (expressed in terms of degenerate soft-mode eigenvectors) is usually fixed by symmetry considerations.

A careful search for possible small temperature-dependent φ changes requires precision structural determination as a function of temperature and has not yet been performed. The existing data of Axe et al., however, are essentially consistent with a temperature-independent φ which could result from eqn (10.3.11) in any one of three ways: (a) the isotropy of fourth-order terms $B_1 - B_2 = B_3 = 0$; (b) the isotropy of sixth-order terms $W_1 - W_2 = W_3 = 0$; (c) the equivalence of fourth- and sixth-order anisotropy $B_3/(B_1 - B_2) = W_3/(W_1 - W_2)$. Assuming for simplicity $\gamma_1 = 1$, $\gamma_2 = 0$, the free energy (10.3.10) takes on the old familiar form of (3.3.1) namely

$$F = \tfrac{1}{2}\alpha\eta^2 + \tfrac{1}{4}\gamma\eta^4 + \tfrac{1}{6}\delta\eta^6 \tag{10.3.12}$$

where $\alpha = \omega_M^2 = \beta(T - T_0)$, $\gamma = B_1$ and $\delta = W_1$. Using the results of § 3.3.2 we find a first-order transition when $\gamma < 0$ at the transition temperature

$$T_c = T_0 + \frac{3}{16} \frac{\gamma^2}{\beta\delta} \qquad (10.3.13)$$

at which the value of the spontaneous intrinsic order parameter is η_0 where

$$\eta_0^2 = \frac{-3\gamma}{4\delta} \qquad (10.3.14)$$

and the corresponding reciprocal response κ as $T \to T_c^+$ is

$$\kappa = \beta(T - T_0) = \kappa_0 + \beta(T - T_c), \qquad T > T_c \qquad (10.3.15)$$

where

$$\kappa_0 = \beta(T_c - T_0) = \frac{3\gamma^2}{16\delta}. \qquad (10.3.16)$$

Using eqns (10.3.14) to (10.3.16) we can transform the equilibrium condition $\partial F/\partial \eta = 0$ for $T < T_c$ to the form

$$\eta^2 = \tfrac{1}{3}\eta_0^2[2 + \{4 - 3(T - T_0)(T_c - T_0)^{-1}\}^{1/2}]. \qquad (10.3.17)$$

If static displacements are small then the intensity of the superlattice (neutron) reflection is proportional to η^2. In addition the proportionality of P_z to x_{xy} from eqn (10.3.8) and of x_{xy} to η^2 from eqn (10.3.9) suggests that for $T < T_c$ within Landau theory the superlattice intensity, spontaneous polarization, and spontaneous strain (i.e. angular distortion) should all be proportional to that same temperature function given in eqn (10.3.17). The resulting fits of this theory to the measurements of Cummins (1970) for polarization and shear angle and of Axe et al. (1972) for Bragg neutron intensity are shown in Fig. 10.8. We see that the three quantities do have the same temperature dependence near the transition temperature and are described by eqn (10.3.17) quite closely. The results, in which the Weiss temperature is an adjustable, point to a 'best-fit' value $T_0 \approx 155°C$ which is some 6 degC higher than that determined directly from the soft-mode condensation $\omega_M^2 \propto (T - T_0)$. This may indicate oversimplifications made in the Landau calculation itself (e.g. neglect of the temperature dependence of φ), or may imply a breakdown of Landau theory due to the onset of fluctuation-dominated motion very close to the transition. The theory can be extended to calculate elastic-constant anomalies and the free-crystal dielectric response by considering the coupled equations of motion describing fluctuations about the equilibrium values given above.

Because of the absence of direct soft mode to P_z coupling the GMO phase transition is simply the piezoelectric analogy of the perovskite zone-boundary transitions. The absence of this coupling, however, is only a

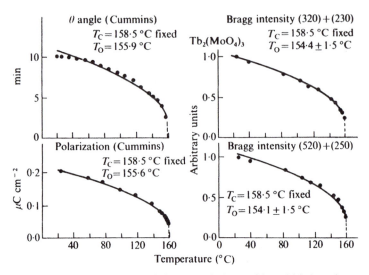

Fig. 10.8. Temperature dependence of three physical quantities which have been used to characterize the phase transition in rare-earth molybdates. All have been fitted (solid curves) using eqn (10.3.17). The good fit shows that the spontaneous strain and spontaneous polarization are both proportional to the square of the zone-boundary soft-mode displacements. (After Dorner, Axe, and Shirane 1972.)

quantitative approximation and has no fundamental significance for the origin of the transition. Actually a very small change in clamped dielectric response χ_{33} has recently been detected by more sensitive far-infrared measurements (Petzelt and Dvorak 1971). In other systems with extrinsic ferroelastic and ferroelectric properties one must in general anticipate an anomaly in clamped dielectric constant as well as in the elastic constants at constant polarization.

A microscopic justification for the above phenomenology has been given by Boyer and Hardy (1973) who have carried out a microscopic lattice-dynamical calculation using a simple point-ion model with long-range coulombic and short-range non-coulombic forces, the latter being determined from the balance of long- and short-range forces necessary to produce the static equilibrium conditions. For complex structures of low symmetry like GMO the equilibrium conditions are such as to define uniquely the first derivatives of all interactions which are likely to be important. The method proves to be quite successful and calculated zone-centre frequencies reproduce quite closely the basic features of the observed Raman spectrum (Ullman *et al.* 1973). The theoretical dispersion curves show a pronounced softening of two phonon branches which become doubly degenerate at the Brillouin-zone M-point.

One final feature of extrinsic ferroelectric soft-mode behaviour, which has been demonstrated first for GMO but is probably common to all such ferroelectrics, is the temperature dependence of the soft-mode oscillator strength on approach to T_c from the low-temperature side. First we note that the soft mode is always a zone-centre mode in the Brillouin zone of the ordered structure. Intrinsic ferroelectric transitions of the displacement kind are marked by the condensation of an infrared active mode whose dimensionless oscillator strength S_i, as defined in § 7.1, is proportional to the inverse square of its frequency Ω_i. For this reason a quantity $S_i' = S_i \Omega_i^2$, which is expected to be essentially temperature independent in this instance, is also commonly used as a measure of mode 'strength' in the literature. From eqn (7.1.8) the mode contributes an amount

$$\Delta\varepsilon(\omega) = \frac{S_i'}{\Omega_i^2 - \omega^2 + i\gamma_i\omega} \qquad (10.3.18)$$

to the dielectric function, which diverges as $\Omega_i^{-2} \propto |T - T_c|^{-1}$ in the static limit as the Curie temperature T_c is approached from either side.

In the extrinsic case the soft mode is a zone-boundary mode above T_c and contributes nothing to the dielectric anomaly. Below T_c Dvorak and Petzelt (1971 a) have deduced that the now zone-centre soft mode acquires a 'strength' S_i' which goes to zero as $T \to T_c^-$ in the form

$$S_i' \propto P_z \propto \Omega_i^2. \qquad (10.3.19)$$

Substituting eqn (10.3.19) into eqn (10.3.18) establishes that the extrinsic soft mode contribution to the dielectric constant in the ordered phase should be constant, dropping discontinuously to zero at the transition.

The far-infrared transmission spectra of GMO (Petzelt 1971) confirm these effects clearly. The doubly degenerate M-point mode of the tetragonal phase splits into two zone-centre A_1 modes in the ordered phase. Although the splitting becomes indistinct experimentally as $T \to T_c^-$ owing to critically diverging linewidths (Khmel'nitskii and Shneerson 1973), the decrease in intensity is clear and the approximate constancy of the implied contribution to static dielectric constant established. The magnitude of the latter is very small $\Delta\varepsilon(0) \approx 0 \cdot 03$ in GMO and, indeed, was not detected earlier by standard dielectric techniques. It is believed that a similar intensity lowering should be observable in Raman scattering as well although the quantitative law may be different. Raman measurements (Fleury 1970, Ullman *et al.* 1973) have confirmed the presence of the zone-centre soft mode in the ordered phase, but attempts to delineate the temperature dependence of line intensities have not yet been successful.

10.4. Classification of ferro-materials

With the rapidly increasing interest in the study of phase transitions in general and of structural phase transitions in particular, the need for a simple representation of the dielectric, elastic, magnetic, and symmetry properties involved becomes acute. Aizu (1965, 1966 b) and Shuvalov (1970) have recognized and enumerated the various possible ferroelectric–paraelectric phase transitions in terms of their point groups, and Aizu (1969) has extended the categorization to include the concept of ferroelasticity.

For discontinuous (first-order) type ferroelectric transitions there is no necessary symmetry compatibility between the high- and low-temperature phases. However, for all continuous (or quasi-continuous) transitions there are symmetry restrictions which can be probed by group theory via a Landau free-energy analysis. A concise description of the method is given by Blinc and Zeks (1974), and here it will suffice to state that all the possible ferroelectric states derivable from a particular higher-symmetry configuration by quasi-continuous phase transition can be enunciated, together with the number of possible ferroelectric domains (with different but equivalent orientations of spontaneous polarization), for each lower-symmetry phase. All the possible ferroelectric point groups which can result from the 32 paraelectric point groups are tabulated in Appendix C. The equivalent findings for collinear antiferroelectrics (e.g. excluding spiral structures such as that found in $NaNO_2$) are given in Appendix D. A simplified but more general representation which includes magnetic as well as dielectric and elastic properties of materials has been introduced by Keve and Abrahams (1970). Its simplicity, and the fact that it can include a recognition of coupling between properties, makes the latter particularly suitable for discussion in the context of coupled phase transitions.

Each phase of a crystal is represented by a triplet of symbols followed by the point group of that phase. The first symbol refers to the dielectric state of the crystal as follows.

F for actual or potential ferroelectric in which the spontaneous polarization can be reversed or reoriented either experimentally or conceptually (i.e. including pyroelectric).

A for antiferroelectric, characterized by the existence of ordered local dipoles in the structure but a net zero polarization (i.e. including antipolar).

π for piezoelectric, encompassing all piezoelectric materials except those characterized by F or A.

P for paraelectric, characterized by zero net polarization and containing no ordered local dipoles in the structure.

The second symbol refers to the elastic state of the crystal as follows.

F for ferroelastic, characterized by the existence of a net spontaneous strain $\mathbf{x}$ which can be reoriented either experimentally or conceptually.

A for antiferroelastic, defined by the existence of a single stable mechanical state in which the unit cell contains an equal number of opposite local spontaneous strains with a zero resultant.

P   for paraelastic, encompassing all other crystals with a single stable mechanical state and zero net strain.

The third refers to the magnetic state in the normally accepted terminology as follows.

F   for ferromagnetic, characterized as possessing a non-zero net magnetic moment (i.e. including ferrimagnetism, weak ferromagnetism, etc.).
A   for antiferromagnetic.
P   for lack of magnetic order (i.e. paramagnetic or diamagnetic).

In ferrophases the possibility of any pair of ferroelectric, ferroelastic, and ferromagnetic order parameters being coupled always exists. When this occurs the strongly coupled properties are denoted by a subscript c on both property symbols. The possibility that all three order parameters are strongly coupled in a ferroelectric, ferroelastic, ferromagnetic phase cannot be discounted, and, indeed, such appears to occur in the boracites.

In order to familiarize ourselves with the Keve–Abrahams representation we now use it to discuss some familiar transitions from earlier chapters and then to discuss some new situations. Emphasis in this book is obviously given to ferroelectric transitions and most of the examples are taken from Abrahams and Keve (1971) or Abrahams (1971).

Consider first the series of transitions which take place as a function of decreasing temperature in $BaTiO_3$. At high temperatures $BaTiO_3$ has the prototype cubic perovskite structure, point group m3m; it contains neither dipole moments nor local strains and hence, since no paramagnetic ions are involved, it must be PPP. Below $T \approx 120°C$ it transforms to a tetragonal structure with point group 4mm and acquires a spontaneous polarization along the c-axis. The difference in length between the new a- ($=b$) and c-axes is only ≈ 1 per cent ($c > a$) and the c-axis can be transformed ferroelastically into an a-axis (with corresponding change in the direction of spontaneous polarization) by the application of uniaxial pressure. The tetragonal phase is therefore F_cF_cP. At about 5°C a further transition to an orthorhombic ferroelectric phase occurs (point group mm2). The resulting difference in the new b- and c-dimensions is again very small. Atomic displacements of only a few hundredths of a nanometre are necessary to interchange the orthorhombic b- and c-axes and produce a 90° reorientation of the polar axis. This suggests that the mm2 phase is also ferroelastic and fully coupled to the ferroelectricity (F_cF_cP). Finally, below $-90°C$, $BaTiO_3$ becomes rhombohedral (point group 3m). Spontaneous polarization is along the (cubic) prototype body diagonal, and the very small distortion angle from 90° suggests that this phase is also ferroelastic with the possibility of reorienting the polar axis to another body diagonal by uniaxial compression along the polar axis. The sequence of $BaTiO_3$ phase changes is therefore succinctly described in the Keve–Abrahams notion by

$$BaTiO_3 : PPP(m3m)F_cF_cP(4mm)F_cF_cP(mm2)F_cF_cP(3m). \quad (10.4.1)$$

Consider now the case of KDP. Its high-temperature phase is tetragonal with a piezoelectric point group $\bar{4}2m$. Since the lattice constants $a = b \approx 1\cdot05$ nm and $c \approx 0.69$ nm there is no question of ferroelastic behaviour. This phase is therefore πPP. At about 122 K a phase transition takes place to an orthorhombic phase which is spontaneously polar. In the orthorhombic phase (point group mm2) the difference between the new a and b unit-cell dimensions is small ($\approx 0\cdot01$ nm) suggesting the possibility of ferroelasticity. In the ordered phase two independent oxygen atoms at $x_1y_1z_1$ and $x_2y_2z_2$ are associated by the relationship

$$(x_1y_1z_1) = (y_2, 1-x_2, 1-z_2) + \Delta \qquad (10.4.2)$$

where $|\Delta|$ is only $0\cdot015$ nm. The potassium is at the special position $(00z)$ and, under the transformation (eqn (10.4.2)), moves to $(0, 0, 1-z)$ through a displacement of $0\cdot022$ nm. The hydrogen is linked to the oxygen at $x_1y_1z_1$ and is displaced with that atom by the same $|\Delta|$. Thus, reversal of polarization necessarily interchanges the a and b-axes (and also reverses the sense of b). The relationship (eqn (10.4.2)) conversely requires that the application of a mechanical stress along [100] will have the identical effect. It follows that the ordered phase is F_cF_cP and that we describe the KDP crystal as

$$\text{KDP: } \pi PP(\bar{4}2m)F_cF_cP(mm2). \qquad (10.4.3)$$

One readily verifies that gadolinium molybdate (GMO) has the same representation

$$\text{GMO: } \pi PP(\bar{4}2m)F_cF_cP(mm2). \qquad (10.4.4)$$

Order parameters are most likely to be strongly coupled, as in the above examples, when they occur together at the same transition. An example of a ferroelectric ferroelastic for which the coupling is essentially absent can be found in the tungsten bronze barium sodium niobate. The high-temperature phase is tetragonal (point group 4/mmm) and PPP. As the temperature is decreased a ferroelectric phase transition occurs first at which the structure remains tetragonal (new point group 4mm) and hence acquires no ferroelasticity. At a lower temperature, however, a second transition occurs, this time to an orthohombic ferroelectric phase (point group mm2) which is only slightly strained. The a- and b-axes can be interchanged by application of mechanical stress but no measurable effect is produced on the spontaneous polarization by the ferroelastic reversal. We therefore represent barium sodium niobate as

$$\text{Ba}_2\text{NaNb}_5\text{O}_{15}: \text{PPP}(4/mmm)\text{FPP}(4mm)\text{FFP}(mm2). \qquad (10.4.5)$$

Zero coupling, of course, naturally occurs in the Keve–Abrahams sense when a phase involves but a single order parameter. For ferroelectrics this is a not uncommon situation and most frequently occurs when the high- and low-temperature phases involve the same crystal system (e.g. tetragonal, as for the tungsten bronzes, or trigonal, as for $LiNbO_3$ and $LiTaO_3$). Barium strontium niobate, for example, has three phases all within the tetragonal crystal system

$$Ba_x Sr_{5-x} Nb_{10}O_{30}: PPP(4/mmm)\pi PP(\bar{4}2m)FPP(4mm). \quad (10.4.6)$$

Both uncoupled magnetic and coupled ferroelastic–magnetic systems are common in nature and, although not strictly relevant in a book on ferroelectricity, will be briefly mentioned here for completeness. Simple magnetic systems occur frequently for which the phase transition involves no change in crystal system. Examples are the linear-chain antiferromagnets

$$CsMnCl_3 \cdot 2H_2O: PPP(mmm)PPA(mmm) \qquad (10.4.7)$$

$$(CH_3)_4NMnCl_3: PPP(6/m)PPA(6/m). \qquad (10.4.8)$$

Coupled elastic and magnetic properties, on the other hand, are found in

$$K_2NiF_4: PPP(4/mmm)PF_cA_c(mmm) \qquad (10.4.9)$$

$$KFeF_3: PPP(m3m)PF_cA_c(3m) \qquad (10.4.10)$$

$$Mn_3O_4: PPP(4/mmm)PF_cF_c(m). \qquad (10.4.11)$$

Again, however, as for ferroelectric–ferroelastic phases, the coexistence of two order parameters does not necessitate a strong coupling between them. Weak or zero coupling is likely in ferroelastic ferromagnets (or antiferromagnets) if the two order parameters appear at different transitions. An example is

$$RbFeF_4: PPP(4/mmm)PFP(mmm)PFA(mmm). \qquad (10.4.12)$$

From a ferroelectric point of view a very important class of materials would be that involving the coexistence of ferroelectricity and ferromagnetism, and, in particular, examples in which a coupling between the properties is significant. Substances possessing simultaneous ferroelectricity and ferromagnetism were predicted as early as 1958 by Smolenskii and Joffe. The search for examples, however, has not met with much success and so far only two classes of material have been found with coexisting ferroelectricity and ferromagnetism. Firstly there are some solid solutions of the perovskite type which were first investigated by Smolenskii *et al.* (1961), and secondly there are the ferroelectric magnetic boracites. At least some of the latter (Ascher *et al.* 1966) exhibit simultaneous non-zero ferroelectric and ferromagnetic order parameters in low-temperature phases, but unfortunately detailed information concerning these phases is to date rather scanty and conflicting.

The scarcity of ferroelectric ferromagnets in nature must at least partly be due to the preponderance of antiferromagnetic over ferromagnetic exchange interactions in insulators and semiconductors. Thus the coexistence of ferroelectricity and antiferromagnetism is likely to be far more common than the coexistence of ferroelectricity and ferromagnetism and although no intensive search has been made, classes of ferroelectric antiferromagnets are rapidly being discovered. One is the $BaMF_4$ family (with M a 3d transition-metal ion) typified by $BaCoF_4$ which is orthorhombic (mm2) and already ferroelectric in its high-temperature phase. The spontaneous polarization $P_s \approx 8 \cdot 0 \, \mu C \, cm^{-2}$ has been switched at room temperature (DiDomenico et al. 1969, Keve, Abrahams, and Bernstein 1970), but the ferroelectric Curie point lies above the melting point so that the prototype phase (i.e. the theoretical highest possible symmetry phase of PPP form) is in this case hypothetical. In the orthorhombic phase there is no close equivalence of axes and no ferroelasticity is involved. At low temperatures an antiferromagnetic ordering occurs while the nuclear structure remains essentially unchanged. It has been predicted by Keve, Abrahams, and Bernstein (1970) that the magnetic phase is accompanied either by a slight lowering in symmetry to a ferroelastic monoclinic phase or by a retention of orthorhombic symmetry but in a cell larger than the magnetic monoclinic cell. The latter situation would correspond to an antiferroelastic state. These alternatives, including the hypothetical prototype, are

$$BaCoF_4: \quad \begin{matrix} PPP(mmm)FPP(mm2)FFA(2) \\ PPP(mmm)FPP(mm2)FAA(mm2). \end{matrix} \qquad (10.4.13)$$

From a device point of view the most intriguing prospect is to find a ferroelectric ferromagnet with strong interproperty coupling. The search to date has been on the whole disappointing, but one group of materials, namely the boracites, at least holds promise. The boracites are rather complex crystals formed by $M_3B_7O_{13}X$ where M is one of the divalent metals of the 3d series and X is a halogen. Recently it has been found that the boracite structure forms also in some cases with X equal to OH, S, Se, or Te and with M equal to monovalent Li. Although the detailed behaviour naturally varies with M and X, the halogen boracites nearly all exhibit transitions from a high-temperature $\pi PP(\bar{4}3m)$ piezoelectric cubic phase to one or more lower-symmetry phases which are ferroelectric. The first transition takes place in general from the cubic phase to an mm2 orthorhombic phase and involves a doubling of the unit-cell volume. The orthorhombic phase is both ferroelectric and ferroelastic with the two properties closely coupled. In addition, in view of the cell doubling, neither spontaneous strain nor spontaneous polarization can be the primary order parameter. The situation is reminiscent of GMO and must involve an intrinsic $(110)(\pi/a)$

zone-boundary phonon mode which couples to the elastic and ferroelectric parameters. The details of the coupling may well vary from one boracite to another but what little information is available at the time of writing (a review by Nelmes (1974) summarizes most of the existing data and includes extensive references) points to the existence of relatively large dielectric anomalies and quite small elastic anomalies, the reverse of the GMO situation, suggesting perhaps a dominant direct coupling of polarization to the driving zone-boundary mode and possibly a piezoelectrically driven elastic anomaly. In any event the transition is written in Keve–Abrahams notation $\pi PP(\bar{4}3m)F_cF_cP(mm2)$ and describes an extrinsic ferroelectric and extrinsic ferroelastic situation.

The form of the free energy requires that the cubic–orthorhombic transition be of first order (Dvorak and Petzelt 1971 b) and this is confirmed by measurement in a number of cases (e.g. Kobayashi, Schmid, and Ascher 1968). The introduction of terms coupling order parameters into the free energy accounts for most of the observed 'anomalous' dielectric and elastic behaviour (Dvorak 1971 c, 1974) although some details remain to be explained. Thus the phenomenological approach to this highest-temperature phase change seems reasonably successful although there has as yet been no direct observation of the soft zone-boundary mode responsible for the instability.

From a magnetic point of view the lower-temperature transitions are the relevant ones, and unfortunately very little accurate structural information is yet available concerning them. Investigation of the temperature dependence of the magnetic properties of Fe, Ni, Co, and Cu halogen boracites by Schmid, Rieder, and Ascher (1965) and by Quèzel and Schmid (1968) suggested that they order in a basically antiferromagnetic fashion, but with the sublattice magnetizations in the ferroelectric lattice slightly canted to produce a resultant ferromagnetic moment. The most exciting finding was that of Ascher et al. (1966) who, for nickel iodine boracite, demonstrated a direct coupling between the ferromagnetic component M_s and the spontaneous polarization P_s in the magnetically ordered low-temperature phase. M_s and P_s are mutually perpendicular and P_s can be reversed from [001] to [00$\bar{1}$] by switching M_s through 90° from [± 1, ± 1, 0] to [± 1, ∓ 1, 0], and vice versa M_s can be switched by reversing P_s. Since the ferroelectric and ferroelastic properties of the higher-temperature phase are also coupled, it seems highly likely that the magnetic phase of this boracite (at least) is a fully coupled $F_cF_cF_c$ phase.

Despite the obvious interest of these materials neither the complete sequence of ferroelectric nor of magnetic transitions is clear at the time of writing even in nickel iodine boracite, which has perhaps attracted most interest. The very low-temperature (6 K) neutron scattering measurements by von Wartburg (1974) show the canted antiferromagnetic character of the magnetic phase quite clearly, but with the spontaneous ferromagnetic moment M_s parallel rather than perpendicular to P_s. This points to the existence of at least two magnetic phases. Even the magnetic properties of the high temperature cubic phase are not well understood, with the rounded susceptibility maxima which occur there being sometimes interpreted as evidence for the existence of an antiferromagnetic transition but more likely being due to simple short-range magnetic-correlation effects. In the light of this immense confusion the time is not ripe for a detailed discussion of the ferromagnetic boracites. Progress in this area will require much more accurate knowledge of all the crystal structures involved, and there are already indications that they may be many and varied (Kobayashi, Sato, and Schmid 1972). In this respect it is encouraging to note that a concentrated programme of crystal-structural studies of the halogen boracites is presently being launched at the University of Edinburgh (Nelmes 1974, Nelmes and Thornley 1975).

11

CRITICAL PHENOMENA

11.1. Landau theory

IN RECENT years considerable attention has been focused on the so-called critical phenomena which occur very close to a second-order phase transition. This interest is centred on the nature of the thermodynamic singularities and long-range spatial correlations which develop at a critical point, and on developing an understanding of the basic simplicity and apparent close similarity of critical behaviour in widely different physical contexts. Thus, for example, a ferromagnet near its Curie point behaves quite similarly to a liquid near its critical point, and a superconducting transition is not very different from a second-order ferroelectric one. In this book we shall naturally be primarily concerned with the ferroelectric critical point; nevertheless, in view of the rather general character of most critical theories we shall first consider the more general context and later focus upon the precise character of the ferroelectric second-order transition. In this way we can fit the ferroelectric transition into the more general picture and glean a little information concerning the nature of the forces responsible for ferroelectricity.

Obvious questions which arise at the outset involve the degree to which different phase transitions are alike and in what respects they differ. The simplest view of phenomena near a critical point, attributable in its general form to Landau (1937 a, b, 1965), is of a universal character and therefore attributes certain common characteristics to all phase transitions. Although the theory does not agree quantitatively with general experimental observations *very* close to a critical point, it does provide much qualitative insight and also points to the origin of the breakdown of the simple theory as the critical point is approached.

For a mathematical description of a phase transition it is essential to introduce that quantity which determines the manner in which the less-symmetric phase departs from the configurations of the more-symmetric phase, i.e. the order parameter. In the absence of an applied field the order parameter is adjudged to have the value zero in the more-symmetric phase. The continuity of the change of phase for a second-order transition then requires that this order parameter can take on arbitrarily small values near the critical point.

Let us denote a local order parameter $\langle \xi(\mathbf{r}) \rangle$, using angular brackets to indicate that it is in fact an ensemble average of some dynamic variable $\xi(\mathbf{r})$.

We shall find it instructive to consider the situation in the presence of a field $h(\mathbf{r})$, which is in general a function of spatial position $\mathbf{r}$ and which is directly coupled to $\langle \xi(\mathbf{r}) \rangle$. Thus, for example, in ferroelectrics $h(\mathbf{r})$ is the Maxwell field and $\langle \xi(\mathbf{r}) \rangle$ the local displacement, while in antiferroelectrics $h(\mathbf{r})$ is the staggered field and $\langle \xi(\mathbf{r}) \rangle$ the sublattice displacement. For infinitesimal fields $h(\mathbf{r})$ it would perhaps seem reasonable to expand the appropriate Gibbs free energy about its value at the critical point. We write (Kadanoff *et al.* 1967) a free energy

$$G = \int g(\mathbf{r}) \, d\mathbf{r} \tag{11.1.1}$$

where

$$g(\mathbf{r}) = g_0(\mathbf{r}) - h(\mathbf{r})\langle \xi(\mathbf{r}) \rangle + \frac{a}{2} \langle \xi(\mathbf{r}) \rangle^2 + \frac{c}{4} \langle \xi(\mathbf{r}) \rangle^4$$

$$+ \mu \{ \nabla \langle \xi(\mathbf{r}) \rangle \cdot \nabla \langle \xi(\mathbf{r}) \rangle \}. \tag{11.1.2}$$

The first term $g_0(\mathbf{r})$ is the free energy per unit volume in the high-symmetry phase, the second is the field energy, and the last term serves to damp out the spatial variations in local order parameter. The coefficients can in general be functions of temperature and of other constraints appropriate to the free energy G. Odd terms in $\langle \xi(\mathbf{r}) \rangle$, other than the field term, have been omitted on the assumption that the form of the forces producing the transition does not change when the sign of $\langle \xi(\mathbf{r}) \rangle$ is reversed.

Minimizing G as a function of $\langle \xi(\mathbf{r}) \rangle$ determines the equilibrium situation in the form

$$\{ a + c\langle \xi(\mathbf{r}) \rangle^2 - 2\mu \nabla^2 \} \langle \xi(\mathbf{r}) \rangle = h(\mathbf{r}). \tag{11.1.3}$$

When the applied field is position independent $(h(\mathbf{r}) = h)$ we assume that the order parameter is also independent of $\mathbf{r}$. This is the assumption that spontaneous fluctuations are small. Writing $\langle \xi(\mathbf{r}) \rangle = M$ for this case we find, from (11.1.3), the equation

$$(a + cM^2)M = h. \tag{11.1.4}$$

In particular, when $h = 0$ we have two solutions $M = 0$ and $M^2 = -a/c$. The former minimizes G when $a > 0$ and the latter when $a < 0$. Since it is necessary for M to vanish above the critical temperature T_c we require $a > 0$ for $T > T_c$ and $a < 0$ for $T < T_c$. Landau chooses the simplest form which accomplishes this purpose, viz.

$$a = b(T - T_c) \tag{11.1.5}$$

where b is a positive constant. It follows that the order parameter within Landau theory varies with temperature as

$$M = \left(\frac{b}{c} \right)^{1/2} (T_c - T)^{1/2}. \tag{11.1.6}$$

It also follows immediately from eqn (11.1.4) that the uniform susceptibility $\chi = dM/dh$ in zero field is given by

$$\chi^{-1} = b(T - T_c), \qquad T > T_c \tag{11.1.7}$$

and

$$\chi^{-1} = 2b(T_c - T), \qquad T < T_c. \tag{11.1.8}$$

Finally, also from eqn (11.1.4), when $T = T_c$ ($a = 0$) we find

$$M = (h/c)^{1/3}. \tag{11.1.9}$$

Now let us consider eqn (11.1.3) for the case of *non-uniform* field. The formal statistical definition of order parameter $\langle \xi(\mathbf{r}) \rangle$ is

$$\langle \xi(\mathbf{r}) \rangle = \frac{\mathrm{Tr}(\xi(\mathbf{r}) \exp\{-(1/kT)\{\mathcal{H}_0 - \int \xi(\mathbf{r}')h(\mathbf{r}')\,d\mathbf{r}'\}\})}{\mathrm{Tr}(\exp[-(1/kT)\{\mathcal{H}_0 - \int \xi(\mathbf{r}')h(\mathbf{r}')\,d\mathbf{r}'\}])} \tag{11.1.10}$$

in which $\mathcal{H}_0$ is the macroscopic-system Hamiltonian in the absence of applied field. By direct classical differentiation (i.e. ignoring any problems of non-commuting $\mathcal{H}_0$ and $\xi(\mathbf{r})$) we obtain

$$\delta\langle \xi(\mathbf{r}) \rangle = \frac{1}{kT} \int g(\mathbf{r}, \mathbf{r}')\,\delta h(\mathbf{r}')\,d\mathbf{r}' \tag{11.1.11}$$

where $g(\mathbf{r}, \mathbf{r}')$ is called the static two-site correlation function (or simply correlation function) and is given by

$$g(\mathbf{r}, \mathbf{r}') = \langle \{\xi(\mathbf{r}) - \langle \xi(\mathbf{r}) \rangle\}\{\xi(\mathbf{r}') - \langle \xi(\mathbf{r}') \rangle\} \rangle. \tag{11.1.12}$$

Thus $g(\mathbf{r}, \mathbf{r}')$ is a measure of fluctuations and, since we shall be concerned with correlations at long range, i.e. $|\mathbf{r} - \mathbf{r}'| \gg$ lattice spacing, quantum fluctuations are indeed insignificant and the classical formalism valid. Calculating the first-order change in eqn (11.1.3) as $\langle \xi(\mathbf{r}) \rangle$ and $h(\mathbf{r})$ undergo increments $\delta\langle \xi(\mathbf{r}) \rangle$ and $\delta h(\mathbf{r})$, and using (11.1.11), we find

$$\{a + 3c\langle \xi(\mathbf{r}) \rangle^2 - 2\mu \nabla^2\}g(\mathbf{r}, \mathbf{r}') = kT\delta(\mathbf{r} - \mathbf{r}'). \tag{11.1.13}$$

In the limit of zero applied field, using (11.1.5) and (11.1.6), this equation has a solution for spontaneous fluctuations in the form

$$g(\mathbf{r}, \mathbf{r}') = \frac{kT}{8\pi\mu} \frac{\exp(-|\mathbf{r} - \mathbf{r}'|/\xi')}{|\mathbf{r} - \mathbf{r}'|} \tag{11.1.14}$$

in which ξ' is a *correlation length* given by

$$\xi' = \left(\frac{2\mu}{b}\right)^{1/2}(T - T_c)^{-1/2} \quad \text{for } T > T_c \tag{11.1.15}$$

and

$$\xi' = \left(\frac{\mu}{.}\right)^{1/2}(T_c - T)^{-1/2} \quad \text{for } T < T_c. \tag{11.1.16}$$

Comparing the results set out above with those of § 3.3 it is immediately apparent that the Devonshire theory of ferroelectricity is a particular example of Landau theory; so also is the mean field theory of § 2.1, although the latter is more restrictive in specifying a particular value for T_c. The general theory is capable of making predictions about the static (i.e. equilibrium) properties near a critical point. We note in particular that the correlation length ξ' diverges when $|T - T_c| \to 0$ indicating that fluctuations become of very great spatial range near a critical point. Finally at T_c itself the correlations fall off with distance only as a power law, viz. $|\mathbf{r} - \mathbf{r}'|^{-1}$. This is a result which is valid in Landau theory only in three dimensions since eqn (11.1.14) is the solution of eqn (11.1.13) only for dimensionality $d = 3$. In the general case of d dimensions the Landau solution to eqn (11.1.13) predicts a drop-off as $|\mathbf{r} - \mathbf{r}'|^{-(d-2)}$ at $T = T_c$. One immediately senses trouble in two dimensions.

Qualitatively, however, the idea that correlations or fluctuations increase as a critical point is approached and that many observables (such as susceptibility and correlation length diverge in a simple power-law fashion as $T \to T_c$ seems to be valid. These power-law exponents are referred to as *critical exponents*, and the more common ones are defined in Table 11.1 together with their conventional symbols and Landau values. In a wide variety of critical phenomena (e.g. liquid helium, superconductivity,

TABLE 11.1

Critical exponents

Property	Definition	Exponent	Landau	Ising $(d = 2)$	Ising $(d = 3)$
Order parameter	$\sim (T_c - T)^\beta$	β	$\frac{1}{2}$	$\frac{1}{8}$	$\approx \frac{5}{16}$
Susceptibility for $T - T_c > 0$	$\sim (T - T_c)^{-\gamma}$	γ	1	$\approx \frac{7}{4}$	$\approx \frac{5}{4}$
Susceptibility for $T - T_c < 0$	$\sim (T_c - T)^{-\gamma'}$	γ'	1	$\approx \frac{7}{4}$	$\approx \frac{5}{4} \frac{21}{16}$
Order parameter *at* T_c as function of field h	$\sim h^{1/\delta}$	δ	3	≈ 15	≈ 5
Correlation length for $T - T_c > 0$	$\sim (T - T_c)^{-\nu}$	ν	$\frac{1}{2}$	1	$\approx 0{\cdot}64$
Correlation length for $T - T_c < 0$	$\sim (T_c - T)^{-\nu'}$	ν'	$\frac{1}{2}$	1	?
Pair correlation function at T_c	$\sim |\mathbf{r} - \mathbf{r}'|^{-(d-2+\eta)}$	η	0	$\frac{1}{4}$	$\approx 0{\cdot}04$
Specific heat for $T - T_c > 0$	$\sim (T - T_c)^{-\alpha}$	α	0(disc)†	0(log)†	$\approx \frac{1}{8}$
Specific heat for $T - T_c < 0$	$\sim (T_c - T)^{-\alpha'}$	α'	0(disc)	0(log)	$\approx \frac{1}{8} \frac{1}{16}$

† (disc) refers to a finite discontinuity and (log) to a logarithmic divergence. Ising values are for nearest neighbour force models.

magnetism, liquid–gas, structural, etc.) simple power-law dependences defining critical exponents are generally observed to be valid close enough to the critical point. The exponents themselves, on the other hand, are very often quite different from the Landau predictions. Moreover, one very famous many-body system, the two-dimensional ($d = 2$) Ising model with nearest-neighbour forces, can be solved exactly for many of its critical properties (see for example Stanley 1971 a) and exhibits critical exponents very different indeed from the Landau predictions. Its three-dimensional counterpart, the $d = 3$ Ising model, although it has not received exact solution for any of its properties, has yielded to intense numerical attack to the extent that most of its critical exponents are known quite accurately. They also differ from the Landau estimates but not so markedly as do the $d = 2$ exponents. (The Landau exponents and those of the two Ising models are shown in Table 11.1.) It would appear that Landau theory becomes a better approximation as dimensionality increases and there is, indeed, evidence that Landau theory is actually correct for $d > 4$ (Abe 1972, Wilson and Fisher 1972), small comfort though this may be to experimentalists locked in a three-dimensional world.

In any event it is evident that the Landau approach fails to provide a valid *general* theory of static critical phenomena. Since all the Landau conclusions are consequences of the simple assumption that the free energy is expandable as a power series in the order parameter and $T - T_c$, this statement just cannot be generally true. For example, the Gibbs potential in zero field for the $d = 2$ Ising model can actually be written for $T > T_c$ as

$$G(T) = G(T_c) + a(T - T_c) + b(T - T_c)^2 \ln(T - T_c) + \dots. \quad (11.1.17)$$

It follows that the specific heat, given by the second derivative with respect to temperature, has a logarithmic divergence as $T \to T_c$. On the other hand, one cannot prove that the Landau assumption is necessarily always false, even in three dimensions. Is it perhaps possible that it is essentially correct for certain particular cases? Most ferroelectrics, for example, certainly do not seem to exhibit a marked divergence in critical specific heat (although we note that even a cusp-like singularity at T_c would make for Landau breakdown at one higher order than in eqn (11.1.17)) and they often have many Landau-like critical characteristics. The answer appears to concern the range of the forces responsible for the transition. The exact Ising solutions referred to above involved only nearest-neighbour forces. In an opposite extreme a model with *equal* pair interactions between *all* constituents of a macroscopic ensemble is also exactly solvable and does show Landau behaviour (Kittel and Shore 1965). This is equivalent to a system with forces of infinite range. More generally, for forces which fall off with distance at long range as $|\mathbf{r} - \mathbf{r}'|^{-(d + \sigma)}$ the most accurate analyses presently available (see for example Fisher, Ma, and Nickel 1972) suggest that Landau behaviour occurs in d dimensions when $\sigma < d/2$, i.e. a fall-off more slowly than $|\mathbf{r} - \mathbf{r}'|^{-9/2}$ in three dimensions.

In ferroelectrics, regardless of the significance of short-range forces, one expects electric dipolar interactions to be present and to dominate at long range. Since dipolar forces fall off as $1/|\mathbf{r} - \mathbf{r}'|^3$ in three dimensions can we therefore not anticipate Landau critical behaviour at least for ferroelectrics? Unfortunately we cannot. The enormous complexity of dipolar forces results from their peculiar *angular* dependence and the results above were obtained assuming interactions dependent *only* on range. Actually, as we shall see later in this chapter, a true critical region with non-Landau behaviour does exist for the case of dipolar forces, although its extent, as measured by the ratio $(T - T_c)/T_c$ over which the true limiting behaviour is found, is often much smaller than that characteristic of transitions dominated by short-range forces.

Outside the true critical region the Landau approximation is expected to be valid and Ginzburg (1960) has presented an argument which uses Landau theory to predict its own range of validity. The fundamental approximation used in the solution of eqn (11.1.3) for the case of uniform field was that fluctuations in the local order parameter were negligibly small. On the other hand, in the limit of zero field, fluctuations (eqn (11.1.14)) were obtained which actually diverged in range in the limit $T \to T_c$. It follows that Landau theory is basically inconsistent near a critical point since it first assumes fluctuations to be small and later calculates them to be all important. In essence Landau theory is valid as long as the fluctuations are small in some sense, i.e. T is not too close to T_c. Ginzburg used the criterion for $T < T_c$ that fluctuations in the order parameter over distances comparable to the correlation length must be small in comparison with the order parameter itself; that is

$$g(\mathbf{r}, \mathbf{r}') \ll M^2 \text{ when } |\mathbf{r} - \mathbf{r}'| \sim \xi'. \qquad (11.1.18)$$

Using (11.1.6) and (11.1.14) for zero field and $T < T_c$ the condition becomes

$$\frac{kT_c}{8\pi e \mu \xi'(T)} \ll \frac{b}{c}(T_c - T). \qquad (11.1.19)$$

This can be written in terms of more physcial quantities by using eqn (11.1.16) for correlation length and then extrapolating this expression to $T = 0$ to define a zero-temperature correlation length $\xi'(0)$ in terms of which

$$\xi'(T) = \xi'(0)|\varepsilon|^{-1/2} \qquad (11.1.20)$$

where $\varepsilon = (T - T_c)/T_c$ and $\xi'(0) = (\mu/bT_c)^{1/2}$.

The jump in heat capacity per unit volume at T_c and zero field predicted by Landau theory is

$$\Delta C = \frac{b^2 T_c}{2c} \qquad (11.1.21)$$

and in terms of $\xi'(0)$ *and* ΔC the Ginzburg criterion becomes

$$\left(\frac{16\pi e \Delta C}{k}\right)^{-1} \{\xi'(0)\}^{-3} \ll |\varepsilon|^{1/2}. \qquad (11.1.22)$$

Although the numerical factor is not to be taken seriously, the important result, that the temperature extent of the fluctuation-dominated (or critical) region near a second-order phase transition varies as $\{\xi'(0)\}^{-6}$, where $\xi'(0)$ is a measure of the range of the forces involved, has been verified many times (both for $T > T_c$ and $T < T_c$) by more careful reasoning (e.g. Vaks, Larkin, and Pikin 1967, Amit, Bergman, and Ymry 1973). Hence, as the range of zero-temperature correlations becomes longer and longer, the Landau theory becomes better and better. The concept of Landau breakdown and the Ginzburg criterion can also be extended for use with first-order transitions (Benguigui 1975 b), when it is found that deviations from Landau behaviour are to be expected even here in certain circumstances.

11.2. Modern theories of critical phenomena

The last several years have seen an explosion of activity in the field of critical phenomena in general. This enormous interest has resulted from the realization that the universal static phenomenological theory of Landau is incorrect for most real systems in the limit of an arbitrarily close approach to a critical point. Theoretical insight began with the development of exact high-temperature and low-temperature series expansions for simple model systems. The first several terms in various thermodynamic properties of interest can be obtained exactly but with rapidly increasing labour for each subsequent term. The prediction of critical behaviour then is made possible by developing extrapolation techniques for estimating the limiting behaviour of a power series from a knowledge of only its first several terms. There is nothing exact in the science of extrapolation, of course, but in many cases very smoothly varying characteristics of the series can be found from which rather precise extrapolation predictions can be made. The fact that the method works admirably for those few models where exact solutions are available (e.g. the $d = 2$ Ising model, spherical model, etc., see Stanley 1971 a) gives one confidence in its predictions.

Nothing can be proved, unfortunately, by series extrapolation, and there remains a dearth of rigorous theorems concerning critical phenomena. Some proofs, leading mostly to critical exponent inequalities, have been obtained thermodynamically, and the intriguing thing about this is that many of these inequalities seem to be satisfied experimentally and by extrapolation procedures as equalities. No one has yet proved that any of these inequalities are in fact equalities, but a new phenomenological theory which suggests that they are has been intriguingly successful. This new approach, which has been called *scaling theory*, involves making a simple assumption concerning the basic form of the thermodynamic potential. The unproven assumption is that sufficiently close to a critical point the singular part of the thermodynamic potential, say the Gibbs potential (eqn (3.2.4)) for zero strain, can be written as a generalized homogeneous function of temperature (or more precisely of $\varepsilon = (T - T_c)/T_c$) and field. This is another way of saying that there exist real numbers a and b for which

$$G(\lambda^a \varepsilon, \lambda^b E) = \lambda G(\varepsilon, E) \qquad (11.2.1)$$

for any value of λ. The scaling hypothesis does not specify a and b but, as we shall indicate below, all the thermodynamic critical exponents turn out to be directly related to a and b if eqn (11.2.1) is true. In this case only two critical exponents can be taken as independent and a relationship can be written between any three. These relationships are referred to as *scaling laws*. Thus, scaling theory does not determine any critical exponent numerically but, if any two are known fairly accurately either from experiment or theory, it does predict the values of all the others.

In order to demonstrate the manner in which eqn (11.2.1) leads to scaling laws let us differentiate eqn (11.2.1) with respect to field E at constant temperature. We find

$$\lambda^b \frac{\partial G(\lambda^a \varepsilon, \lambda^b E)}{\partial(\lambda^b E)} = \lambda \frac{\partial G(\varepsilon, E)}{\partial E}. \qquad (11.2.2)$$

However, since $(\partial G/\partial E)_{X,T} = -D$ from eqn (3.2.5), this can be written as

$$\lambda^b D(\lambda^a \varepsilon, \lambda^b E) = \lambda D(\varepsilon, E) \qquad (11.2.3)$$

and in particular, for $E = 0$, as

$$D(\varepsilon, 0) = \lambda^{b-1} D(\lambda^a \varepsilon, 0). \qquad (11.2.4)$$

This equation is valid for all λ by reason of the homogeneity assumption (eqn (11.2.1)). Putting $\lambda = (-1/\varepsilon)^{1/a}$ we obtain

$$D(\varepsilon, 0) = (-\varepsilon)^{(1-b)/a} D(-1, 0) \qquad (11.2.5)$$

where $D(-1, 0)$ is just a constant. Since the definition of the critical exponent β in Table 11.1 is $D(\varepsilon, 0) \sim (-\varepsilon)^\beta$ it follows that

$$\beta = (1-b)/a. \qquad (11.2.6)$$

In addition, putting $T = T_c$ ($\varepsilon = 0$) in eqn (11.2.3) and setting $\lambda = E^{-1/b}$ gives

$$D(0, E) = E^{(1-b)/b} D(0, 1) \qquad (11.2.7)$$

and hence, letting $E \to 0$ and using the definition of critical exponent δ from Table 11.1, we find

$$\delta = b/(1-b). \qquad (11.2.8)$$

A third critical exponent can be obtained from $(\partial^2 G/\partial E^2)_{X,T}$, which is the isothermal dielectric constant at constant (zero) stress. Near a critical point this is equal to the isothermal susceptibility $\chi_T(\varepsilon, E)$ and by direct differentiation from eqn (11.2.2) we find

$$\lambda^{2b}\chi_T(\lambda^a\varepsilon, \lambda^b E) = \lambda\chi_T(\varepsilon, E). \qquad (11.2.9)$$

Putting $E = 0$ and $\lambda = (-\varepsilon)^{-1/a}$ gives

$$\chi_T(\varepsilon, 0) = (-\varepsilon)^{(1-2b)/a}\chi_T(-1, 0). \qquad (11.2.10)$$

Using the definition of exponent γ', viz. $\chi_T(\varepsilon, 0) \sim (-\varepsilon)^{-\gamma'}$, provides another relationship

$$\gamma' = (2b - 1)/a. \qquad (11.2.11)$$

Since we now have three equations (eqns (11.2.6), (11.2.8) and (11.2.11)) relating three critical exponents to the parameters a and b we can eliminate the latter between them to obtain a scaling law, namely

$$\gamma' = \beta(\delta - 1). \qquad (11.2.12)$$

In like manner a whole host of other scaling laws can be derived relating the static critical exponents. A few of the more common ones are listed in Table 11.2.

TABLE 11.2

Some relations between critical exponents predicted by the scaling hypothesis

$$\alpha + 2\beta + \gamma = 2$$
$$\alpha' + 2\beta + \gamma' = 2$$
$$\alpha = \alpha'$$
$$\gamma = \gamma'$$
$$\alpha + \beta(\delta + 1) = 2$$
$$\gamma(\delta + 1) = (2 - \alpha)(\delta - 1)$$
$$\beta\delta = \beta + \gamma$$
$$(2 - \alpha - \gamma)\delta = 2 - \alpha + \gamma$$
$$d\nu = 2 - \alpha$$
$$(2 - \eta)\nu = \gamma$$
$$2 - \eta = \frac{d(\delta - 1)}{\delta + 1} = \frac{d\gamma'}{(2\beta + \gamma')}$$
$$d - 2 + \eta = \frac{2d}{(\delta + 1)} = \frac{2d\beta}{(2 - \alpha)} = 2\beta/\nu$$
$$\nu = \nu'$$

Additional relationships, this time involving the correlation exponents ν, ν', and η, can be generated if the pair correlation function is assumed to be a generalized homogeneous function of ε and $|\mathbf{r}-\mathbf{r}'|$. These relationships, some of which are also shown in Table 11.2, are somewhat less general than those resulting from the thermodynamic potential homogeneity since they involve lattice dimensionality d in most cases. They are also demonstrably false in the Landau limit—that is for long-range forces or high dimensionality—although they are exact for the $d = 2$ Ising model. The thermodynamic potential scaling laws are seemingly consistent with all experiments and model calculations to date, but evidence for the breakdown of correlation-function scaling, even in three dimensions, is quite strong, although the extent of the breakdown for $d = 3$ is very small.

Attempts to provide an intuitive physical picture which leads to the homogeneity hypothesis have appeared in the literature in several forms (e.g. Kadanoff et al. 1967, Griffiths 1967, Halperin and Hohenberg 1967, 1969). Although none of them is rigorous they do tend to render the assumptions plausible on physical grounds. On the other hand, the great weakness in the whole approach is the failure of the method to make specific predictions of numerical exponent values. There is ample evidence that different systems do have different exponents, and one would like to know the important characteristics upon which the critical exponents depend. There is a general feeling that many of the microscopic details of a system should not be involved, since, very close to a critical point, the correlated motion is of quasi-macroscopic dimensions and could hardly be responsive to anything but the grossest features of the relevant Hamiltonian such as symmetry class, range of interaction, dimensionality, etc. Thus the number of basically different types of critical behaviour (or classes of universality as they are now called) may be relatively few. This idea, or principle of universality (Griffiths 1970), has received much support from series-expansion analysis on model systems but, at this moment in time, remains a hypothesis.

A most exciting development in critical theory, which lends support to the concepts of universality and scaling and moreover enables numerical estimates of specific critical exponents to be obtained, is the 'renormalization-group' procedure of K. G. Wilson (1971). The renormalization group is a method of dealing with problems characterized by involving a very large number of degrees of freedom and is not in fact limited to critical phenomena. Under normal circumstances the 10^{23} or so degrees of freedom of a macroscopic sample can be reduced enormously by constructing a microscopic representative sample of it. Thus a liquid of some 1000 atoms, say, would behave (as regards energy per unit volume, density, etc.) very much the same as a liquid with 10^{23} atoms. How far can this reduction in size be carried without qualitatively changing the macroscopic properties? This limiting size in effect defines the correlation length ξ' of the problem.

Far from a critical point this correlation length may only be one or two lattice spacings and small 'cluster' theories can be developed to represent bulk properties. On the other hand, critical phenomena are an example of the opposite extreme for which the correlation length becomes very large indeed. For such situations we are not interested in fluctuations with wavelength smaller than the correlation length and, since we expect the characteristics of long-wavelength fluctuations to be independent of the microscopic details, it should therefore be possible to obtain an effective Hamiltonian with these irrelevant details removed. The philosophy is akin to that of hydrodynamics where one introduces concepts like density $\rho(x)$ which represent an average over the original microscopic degrees of freedom; all microscopic fluctuations are eliminated, the price being that $\rho(x)$ is supposed to describe only quasi-macroscopic fluctuations.

The new effective Hamiltonian is easy to determine in principle by simply integrating out the irrelevant variables, namely those variables with wavevector $\mathbf{q}$ large compared with a cut-off wavevector Λ. Thus we write a new distribution function

$$\exp\left\{\frac{-\mathcal{H}(\Lambda)}{kT}\right\} \equiv \prod_{i,q>\Lambda} \int \exp\left(\frac{-\mathcal{H}_{\text{micro}}}{kT}\right) d\varphi_{iq} \qquad (11.2.13)$$

where the multiple integral is taken over all microscopic variables φ_{iq} for which q is larger than Λ. The cut-off Λ is taken to be larger than the reciprocal correlation length $(\xi')^{-1}$ in order to leave the φ_{iq}'s in the immediate vicinity of $q \sim 1/\xi'$ unintegrated.

To study critical behaviour using $\mathcal{H}(\Lambda)$, however, is not an immediate advantage since $\mathcal{H}(\Lambda)$ is almost certainly more complicated in form than the original microscopic Hamiltonian. The particular characteristic of a critical point which makes the procedure useful is that as $T \to T_c$ the correlation length diverges so that one may subsequently perform additional transformations eliminating more and more degrees of freedom. Thus, for example, each subsequent 'renormalization' might decrease the value of cut-off Λ by a factor s. So one has a transformation, say τ, which is to be applied repeatedly to define new effective Hamiltonians:

$$\tau\mathcal{H}(\Lambda) = \mathcal{H}\left(\frac{\Lambda}{s}\right)$$

$$\tau\mathcal{H}\left(\frac{\Lambda}{s}\right) = \mathcal{H}\left(\frac{\Lambda}{s^2}\right)$$

$$\tau\mathcal{H}\left(\frac{\Lambda}{s^2}\right) = \mathcal{H}\left(\frac{\Lambda}{s^3}\right), \text{ etc.} \qquad (11.2.14)$$

In the limit of a critical point we continue this set of operations (the renormalization group) to a fixed-point solution, namely an interaction

Hamiltonian $\mathcal{H}^*$ satisfying

$$\tau\mathcal{H}^* = \mathcal{H}^*. \tag{11.2.15}$$

This is something like an eigenvalue equation and the possible types of critical behaviour (classes of universality) are determined by the possible fixed points of τ. If a particular system has a microscopic Hamiltonian $\mathcal{H}$, one has to construct the sequence (eqn (11.2.14)) from $\mathcal{H}$ in order to find which particular fixed point gives the limit of that particular sequence. The mathematical complexity of the method is such that we shall not discuss a specific example in detail in this book but merely focus on the findings most relevant for an understanding of ferroelectric and structural phase transitions. An excellent introduction to the renormalization-group method has been given by Ma (1973).

A major result of the renormalization-group approach is that for short-range forces the Landau critical exponents are valid for dimensionality $d > 4$. For $d < 4$ the critical exponents are most readily calculated as a power series in $\varepsilon = 4 - d$, where dimensionality is allowed to be a continuous variable; exponents in three dimensions then correspond to $\varepsilon = 1$. At the time of writing the theory is still in its infancy and is only just reaching the point where it can compete numerically with series-extrapolation results for many simple isotropic $d = 3$ interaction Hamiltonians. On the other hand, for more complicated situations, such as (for example) the addition of cubic anisotropy or of isotropic dipolar interactions to a simple isotropic short-range interaction Hamiltonian, the method has already yielded valuable information (Aharony 1973 a, b; Aharony and Fisher 1973). The findings for dipolar interactions are particularly relevant for ferroelectrics, and for cases of high structural symmetry (e.g. perovskites in the paraelectric phase) are quite surprising. Since dipolar intereactions are of long range they might be expected to bring the critical behaviour closer to Landau form. However, at least to first order in $4 - d$, the renormalization-group procedure suggests that the opposite is true with small shifts of critical exponents away from Landau form (Table 11.3). As the critical point is approached it is possible that a cross-over region exists in which the critical behaviour changes from that characterizing normal isotropic short-range interactions to that characterizing dipolar behaviour and described by a new fixed point of the renormalization-group equations. If, on the other hand, the dipolar interactions are strong the short-range critical behaviour may not develop at all. In the dipolar regime the correlation function loses its isotropic form and develops an anisotropy $\langle \xi^\alpha(-\mathbf{q})\xi^\beta(\mathbf{q})\rangle \sim \{\delta_{\alpha\beta} - (q^\alpha q^\beta/q^2)\}$ where α, β label orthogonal directions x, y, z.

The situation for a *uniaxial* ferroelectric is somewhat easier to envisage. The qualitative development of uniaxial dipolar fluctuations can be simply described in the correlated effective-field theory of § 2.2. In a paraelectric

TABLE 11.3

Exponent	Landau	Renormalization-group calculations (Aharony and Fisher 1973, Bruce and Aharony 1974)		Short range force $d = 3$ series expansion estimates (taken from Stanley 1971 b)	
		Isotropic short range	Isotropic dipolar	Planar	Heisenberg
η	0	0·0208	0·0231	(~ 0)	$(\sim 0·03)$?
2ν	1	1·375	1·384	$\sim \frac{4}{3}$	$\sim 1·40$
γ	1	1·365	1·372	$\sim \frac{4}{3}$	$\sim 1·38$
α	0	$-0·125$	$-0·135$	(~ 0)?	$\sim -0·1$
δ	3	4·458	4·454	$(\sim 5·0)$?	$\sim 5·0$
β	$\frac{1}{2}$	0·380	0·381	$(\sim \frac{1}{3})$?	$\sim 0·38$

Renormalization estimates are correct to second order in $\varepsilon = 4 - d$ with $d = 3$ (d, dimensionality). More recent estimates of the critical exponents for the planar model have been given by Ferer, Moore, and Wortis (1973).

phase and for classical motion the static susceptibility $\chi(\mathbf{q})$ is directly related to static fluctuations $\langle \xi(-\mathbf{q})\xi(\mathbf{q}) \rangle$ via the fluctuation-dissipation theorem (eqn (2.2.28))

$$kT\chi(\mathbf{q}) = \langle \xi(-\mathbf{q})\xi(\mathbf{q}) \rangle \qquad (11.2.16)$$

where now the lth-cell co-ordinate ξ_l describes uniaxial motion in the (say) z-direction and hence $\chi(\mathbf{q})$ is more exactly the Fourier transform of the diagonal (zz) component of the susceptibility tensor. Using eqn (2.2.25) and Fourier transforming eqn (11.2.16) back to real space enables us to express the spatial correlations between the instantaneous values of ξ_l at cells $\mathbf{l}$ and $\mathbf{l} + \mathbf{R}$ as

$$\langle \xi_l \xi_{l+\mathbf{R}} \rangle = \frac{kT}{N} \sum_{\mathbf{q}} \exp(i\mathbf{q} \cdot \mathbf{R})/\{\{\chi(0)\}^{-1} + v(0) - v(\mathbf{q})\}. \qquad (11.2.17)$$

In the long-wavelength limit the uniaxial dipolar interaction potential $v(0) - v(\mathbf{q})$ takes on the qualitative form (Cohen and Keffer 1955)

$$v(0) - v(\mathbf{q}) \sim \frac{q_z^2}{q^2} + bq^2 \qquad (11.2.18)$$

where b is a structure-dependent parameter. Inserting eqn (11.2.18) into eqn (11.2.17) and performing the spatial integration now allows us to examine the spatial correlations as $T \to T_c$ (i.e. $\{\chi(0)\}^{-1} \to 0$). Lines (1972 a) finds that in the limit $R \to \infty$ the longitudinal and transverse correlations both vary as R^{-3} but with the amplitude of the longitudinal correlations increasingly dominant over that of transverse correlations as the Curie point is approached. This indicates that the polarization correlations are enormously concentrated in the uniaxial (or incipient polar) z-direction as the phase transition is approached from above. The entire effect, of course, is produced by the anisotropic q_z^2/q^2 term in eqn (11.2.18), which raises the energy of soft-mode excitations propagating near the z-direction and thereby inhibits their thermal population relative to x- and y-propagating excitations.

Thermodynamic critical behaviour for the uniaxial dipolar model in the correlated effective-field scheme is of Landau form except for logarithmic corrections. This prediction is in fact confirmed by more careful analysis (Larkin and Khmel'nitskii 1969, Aharony 1973 c, Nattermann 1975) including renormalization-group calculations, although the detailed form of the logarithmic corrections is not given correctly by the approximate (correlated effective-field) method. Among the more important findings of the more accurate calculations are the following: the uniform susceptibility $\chi(0)$ diverges as

$$\chi(0) \sim \varepsilon^{-1}|\ln \varepsilon|^{1/3} \tag{11.2.19}$$

the correlation length ξ' as

$$\xi' \sim \varepsilon^{-1/2}|\ln \varepsilon|^{1/6} \tag{11.2.20}$$

and the specific heat c as

$$c \sim |\ln \varepsilon|^{1/3} \tag{11.2.21}$$

where $\varepsilon = |T - T_c|/T_c$. The susceptibility and correlation length diverge more rapidly than in phenomenological theory, while the specific heat shows a logarithmic divergence in place of the phenomenological finite discontinuity at T_c. Finally (Larkin and Khmel'nitskii 1969) the spontaneous polarization P_s goes to zero as $T \to T_c$ from below according to the relation

$$P_s \sim \varepsilon^{1/2}|\ln \varepsilon|^{1/3}. \tag{11.2.22}$$

Critical singularities of additional properties for uniaxial ferroelectrics, such as free energy expansion coefficients, Debye–Waller factors, and electrical resistivity, also show logarithmic deviations from mean field and have been recently given by Binder et al. (1976).

The situation for antiferroelectrics is naturally quite different since dipolar interactions are not of long range (in the normal sense of producing measurable effects, e.g. depolarizing fields, at infinity) in an antiferroelectric configuration. In agreement with physical expectations Aharony (1973 d) finds that the main parameters characterizing dipolar interactions do indeed become irrelevant for the description of antiferroelectric critical phenomena and that the critical exponents, in particular, retain their short-range values. However, the decay of dipolar effects as the critical point is approached is slow, and thus the possibility of observing their existence is not necessarily beyond the bounds of possibility for antiferroelectrics. If critical exponents do indeed retain their short-range values they should correspond numerically to the $d = 3$ Ising exponents (Table 11.1) if the critical ionic motion is approximately constrained to one direction, to the so-called 'planar' exponents (Table 11.3) if the motion is restricted to two dimensions, and to the Heisenberg exponents (Table 11.3) if the critical motion is quasi-isotropic. Although the planar and Heisenberg exponents are still not known accurately enough to attach much relevance to the small differences between them, they both differ very markedly from the phenomenological exponents (Table 11.1) and therefore predict quite non-Landau behaviour for antiferroelectrics inside the critical or fluctuation-dominated region.

Finally, a section on the theoretical aspects of ferroelectric and antiferroelectric critical phenomena would not be complete without mention of the exact solutions which have been obtained for some two-dimensional ferroelectric and antiferroelectric models. Although of limited practical application these models, which are related to the $d = 2$ Ising model, are important since they add to the very small number of phase-transition models which are exactly soluble and because they have a number of very surprising and unusual critical properties. The interested reader is referred to the reviews by Lieb (1969) and Lieb and Wu (1972).

11.3. Experimental observation of static critical phenomena

Observation of critical phenomena in structural phase transitions is particularly difficult because of the often strong coupling of the order parameter with both uniform and local strains. Internal strains within monocrystal samples can easily spread the phase transition over several degrees, and the coupling to uniform strains (lattice compressibility) can affect the simple power-law exponents which might otherwise be relevant for a rigid structure (Fisher 1968, Aharony 1973 e). In addition uncertainties are sometimes large owing to the presence of 'rounding', which is the deviation of experimental findings from a power law when the critical temperature is approached too closely. This effect is often caused by the limitation of the correlation length by impurities, defects, etc., and may obscure the true limiting behaviour altogether (Balagurov and Vaks 1973). The experimental problems are particularly severe for ferroelectrics where the critical temperature range is usually very small and the expected deviations of critical from classical behaviour often subtle (e.g. logarithmic corrections). The situation is also complicated by the scarcity of ferroelectrics with unquestioned second-order phase transitions. At the time of writing there is no well-studied example of a ferroelectric displacive transition of second order.

11.3.1. Ferroelectrics

The most thoroughly studied second-order ferroelectric phase transition is that of triglycine sulphate (TGS). Many authors have measured the static dielectric and thermal properties of TGS in recent years, and Landau critical behaviour is invariably observed for $\varepsilon = |T - T_c|/T_c$ greater than $\sim 10^{-4}$. Deviations from mean-field behaviour are usually observed for smaller values of ε but their form and interpretation still seems obscure. Before the theoretical result that a uniaxial dipolar transition like that of TGS should give rise only to logarithmic departures from Landau behaviour was well known, these deviations were sometimes interpreted as actual non-Landau critical effects (Blinc, Burgar and Levstik 1970). Now they are more often given without comment and may well be merely rounding effects of the sort referred to above. Some of the more recent (and presumably most accurate) static measurements on TGS are shown in Figs. 11.1, 11.2, and 11.3. Rounding in the best samples is not observed until ε is of order 10^{-5}, and the width of the rounded temperature region appears to depend on annealing conditions (Deguchi and Nakamura 1972 b). The non-divergence of the extrapolated Landau dielectric constant at T_c in Fig. 11.2 may possibly be related to the existence of surface layers (Sekido and Mitsui 1967).

From the theoretical results of § 11.2 it is apparent that the best chance of observing a true critical region for a uniaxial dipolar configuration is for the

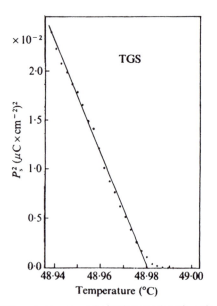

Fig. 11.1. The square of the spontaneous polarization as a function of temperature for TGS in the close vicinity of the Curie point. (After Deguchi and Nakamura 1972 a.)

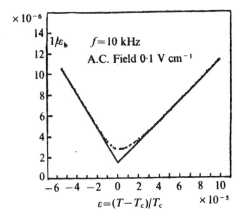

Fig. 11.2. The reciprocal dielectric constant for TGS (in the polar direction) in the vicinity of the Curie point as a function of reduced temperature ε. (After Nakamura *et al.* 1970.)

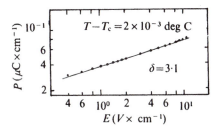

Fig. 11.3. Induced polarization in TGS as a function of electric field E in the vicinity of the Curie point. (After Deguchi and Nakamura 1972 a.)

specific heat, which diverges (albeit only logarithmically) as $T \to T_c$ in critical theory, but is always finite (suffering a finite discontinuity at T_c) in Landau approximation. However, even here the evidence for a truly critical regime is somewhat inconclusive although Grindlay (1965) has interpreted the experimental results of Strukov (1964, 1966 b) as showing some evidence of such a singularity. Additional evidence in favour of the critical divergence comes from a measurement of the coefficient of thermal expansion (Schurmann *et al.* 1973) which should exhibit the same critical anomaly as the specific heat (Janovec 1966). On the other hand, a distinct cross-over from classical to critical behaviour has not been convincingly demonstrated, and the precise theoretical form predicted by eqn (11.2.21) has not yet been verified for a uniaxial ferroelectric.

Another tool which has been used to look for static critical phenomena in TGS is diffuse X-ray scattering (Fujii and Yamada 1971). Near T_c an increase in quasi-elastic intensity is produced by scattering from the long-wavelength low-frequency critical fluctuations of electron density distribution. The scattering measures essentially the spatial Fourier transform of the correlation function (see § 6.3) which, for isotropic short-range forces and small $\mathbf{q}$, can be written

$$g(\mathbf{q}) = \langle \xi(-\mathbf{q})\xi(\mathbf{q}) \rangle = \frac{kT}{(\xi')^{-2} + |\mathbf{q}|^2} \tag{11.3.1}$$

where ξ' is the correlation length defined in Table 11.1 and eqn (11.1.4). As $\mathbf{q} \to 0$ and $T \to T_c$ the scattering diverges as $(\xi')^2$ and hence provides a measure of the critical exponent ν. In principle eqn (11.3.1) presupposes that the critical exponent $\eta = 0$ (Fisher 1964), and it should strictly be modified to allow $g(\mathbf{q})$ at $T = T_c$ to vary as $|\mathbf{q}|^{-2+\eta}$ i.e. with correlations in real space going as $|\mathbf{r} - \mathbf{r}'|^{-(d-2+\eta)}$ as in Table 11.1. Theoretically in three dimensions $(d = 3)$ the exponent η is very small and no experimental evidence for a non-zero value has yet emerged in connection with structural transitions (although it has in other critical contexts, e.g. magnetism). For uniaxial dipolar forces the correlations $g(\mathbf{r}, \mathbf{r}')$ do not decay exponentially at large distance $|\mathbf{r} - \mathbf{r}'|$ and the conventional definition of correlation length (eqn (11.1.14)) is not appropriate (Lines 1972 a). The critical scattering is highly anisotropic and diverges near T_c only for wavevectors perpendicular to the polar axis, a result which is suggested even in the approximate statistical theory leading to eqns (11.2.17) and (11.2.18). For such a situation (Aharony 1973 c) a more general definition of correlation length can be cast directly in terms of the correlation transform $g(\mathbf{q})$ as

$$(\xi')^2 = \{g(0)\}^{-1} \left\{ \frac{\partial g(q_x, q_y, 0)}{\partial (q_x^2)} \right\}_{q_x = q_y = 0} \tag{11.3.2}$$

where z is the polar axis, so that critical scattering experiments measure the correlation exponent directly. For TGS Fujii and Yamada (1971) report $\xi' \sim (T - T_c)^{-\nu}$ with $\nu = 0.50$ (the Landau value). It follows that all the critical exponents yet measured for TGS are classical to within experimental error. This is not really surprising theoretically for the uniaxial dipolar situation, but whether the true critical region has actually been approached is uncertain. With the exception of specific heat experimental findings have not been carefully analysed to see whether the theoretical predictions (eqns (11.2.19) to (11.2.22)) can yield a better description of the data than the simple Landau forms. It is interesting to note, however, that just such evidence is beginning to appear for the analogous magnetic situation (e.g. the uniaxial dipolar ferromagnets $GdCl_3$ (Kotzler and Scheithe 1973) and $LiTbF_4$ (Ahlers, Kornblit, and Guggenheim 1975) which supports the non-Landau forms. Finally, since long-range dipolar forces tend to be screened in a dielectric (particularly in a narrow-band-gap situation) by their interaction with the electronic polarization, there is still some doubt as to the degree of their dominance over short-range interactions in the ferroelectric context (Kocinski and Wojtczak 1973, Pura and Przedmojski 1973), and it is perhaps still possible that the observed Landau-like critical behaviour reflects a dominance of long- but finite-range interactions in ferroelectrics like TGS.

Many other observations of classical static critical behaviour have been recorded in the ferroelectric literature (e.g. Figs. 5.5, 5.14, 5.15, 5.16) with the only real arguments concerning the possible observation of genuine critical phenomena occurring with respect to specific-heat measurements (Fig. 5.17). Several quite careful measurements, for example, have been made on KH_2PO_4 and its isomorphs (e.g. Reese and May 1967, 1968) and attempts made to assess the form of the apparently diverging specific heat. However, since these transitions (with the possible exception of RbH_2PO_4 (Amin and Strukov 1968)) are actually just of first order, the efforts to fit the specific-heat anomalies to particular theoretical forms are hindered by the imprecise knowledge of T_c itself (which results from the presence of thermal hysteresis (Reese 1969 a)). One should also remark that since $\lim_{\alpha \to 0}(|T - T_c|^{-\alpha} - 1)\alpha^{-1} = -\ln(|T - T_c|)$ it is extremely difficult to distinguish between a small α power divergence, such as is predicted theoretically for the $d = 3$ Ising model with short-range forces (Table 11.1) for example, and a logarithmic divergence. It is also interesting to note that Cowley (1976) has recently suggested that for ferroelectrics with a piezoelectric non-polar phase (like KDP) mean-field critical behaviour may actually occur even in the limit $T \to T_c$, since the piezoelectric coupling suppresses fluctuations beyond that caused by the uniaxial macroscopic field alone.

It is unfortunate that none of the nominally pure simple perovskite ferroelectrics seem to exhibit a second-order phase transition, since theory (Table 11.3) predicts quite definite non-classical divergences for dipolar-driven transitions of high symmetry, such as the cubic paraelectric–ferroelectric perovskite situation. However, some interesting preliminary work on the mixed perovskite $PbZr_{0.9}Ti_{0.1}O_3$ by Clarke and Glazer (1974) indicates that it may have just such a transition. This solid solution has the cubic perovskite structure at high temperatures and transforms to a ferroelectric phase at 269°C. The transition is well defined and the lack of thermal hysteresis suggests that it represents a genuine critical point. Little attention has yet been paid to this possible rare example of a second-order ferroelectric perovskite transition but work is beginning, and Clarke and Glazer claim to have detected critical fluctuations in the order parameter by measuring the temperature dependence of spontaneous strain in the ordered phase. Assuming this to be proportional to the square of the order parameter for small distortions from centrosymmetry (i.e. using eqn

(3.2.30)) they find a critical exponent $\beta \approx 0 \cdot 33$, which is to be compared with the Landau value of 0.5. Although an indirect measurement, this work suggests that $PbZr_{1-x}Ti_xO_3$ should be studied more closely and attempts made to observe the polarization and dielectric anomalies directly.

Although the study of high-symmetry dipolar critical phenomena in ferroelectrics has barely begun, analogous work in a ferromagnetic context has been progressing at a slightly faster rate. High-symmetry insulator ferromagnets are also not common but some (e.g. EuO and EuS) do exist, and a recent careful analysis of specific-heat data for EuO has been interpreted as showing a cross-over from Heisenberg to isotropic dipolar critical behaviour near T_c (Salamon 1973), although the interpretation may not go unquestioned. This magnetic situation, though, is rather different from its high-symmetry ferroelectric analogue. The magnetic dipolar forces are small compared with near-neighbour exchange forces in EuO, and the initial deviations from Landau behaviour are of Heisenberg rather than dipolar form making the rather delicate cross-over from Heisenberg to isotropic dipolar critical behaviour (Table 11.3) difficult to determine precisely. A high-symmetry ferroelectric might perhaps be expected to exhibit a direct Landau to isotropic dipolar cross-over which should be much easier to demonstrate experimentally when a good example is found.

11.3.2. Antiferrodistortive transitions

Since the electric dipolar forces are not effectively of long range in an antiferrodistortive configuration, a second-order phase transition from a high-symmetry dielectric phase to such an ordered structure is expected to reflect more the nature of the non-dipolar short-range forces and hence to have a critical region of considerably greater extent than in ferroelectrics (Aharony 1973 d). In addition even for anisotropic systems the critical exponents should be markedly non-Landau. It is therefore perhaps not surprising that the first unequivocal observation of fluctuation-dominated static critical phenomena for structural transitions was observed for such antiferrodistortive systems.

Since the order parameter and divergent susceptibility are 'staggered' quantities for this case and do not couple to a uniform electric field, conventional dielectric studies are not able to probe the nature of these critical divergencies. Indeed, except for true antiferroelectrics, there will be no pronounced dielectric anomaly associated with such transitions at all. In this situation some of the most detailed static investigations have been performed using magnetic techniques (see Chapter 12). The best-documented experimental systems are the perovskites $SrTiO_3$ and $LaAlO_3$ which both undergo second-order displacive phase transitions to antiferrodistortive phases as discussed in § 8.2. In $SrTiO_3$ a rotation of alternate TiO_6 octahedra occurs about a cubic axis, while in $LaAlO_3$ a similar rotation of alternate AlO_6 octahedra takes place around a body diagonal, each case defining an angular order parameter φ. The angular variable increases from zero as T falls below the antiferrodistortive transition temperature T_c, thereby defining the critical exponent β according to $\varphi \propto (T_c - T)^\beta$.

An analysis of the electron paramagnetic resonance (e.p.r.) spectrum of Fe^{3+} substitutional impurities (for Ti^{4+} or Al^{3+}) provides a measure of local symmetry and hence of φ. For $SrTiO_3$, in which the substitution produces a charge misfit, even more precise data can be obtained from Fe^{3+}–vacancy pair spectra, and the agreement of the two methods is excellent (Müller and Berlinger 1971) showing the expected power law form $\varphi \propto (T_c - T)^\beta$ with exponent β very close to $0\cdot33$ for both $SrTiO_3$ and $LaAlO_3$. The experimental data for $SrTiO_3$ are shown in Fig. 11.4. Landau behaviour ($\beta = 0\cdot5$) is

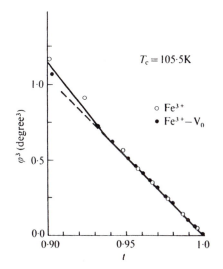

Fig. 11.4. The cube of the rotational parameter φ as a function of reduced temperature $t = T/T_c$ in $SrTiO_3$. (After Müller and Berlinger 1971.)

observed for both systems far from T_c with a cross-over to critical behaviour when $0\cdot1 > \varepsilon > 0\cdot05$. Using the Ginzburg relationship (eqn (11.1.22)) with the measured classical specific-heat jump yields an estimate for zero-temperature correlation length of order 1–2 nm which is to be compared with a distance of about 0·8 nm between equivalent octahedral units. It therefore seems evident that short-range interactions are indeed dominating the critical properties in these systems.

Additional information has been obtained by von Waldkirch et al. (1972) and von Waldkirch, Müller, and Berlinger (1973) from an analysis of the temperature-dependent linewidth of the Fe^{3+}–vacancy resonance in $SrTiO_3$

when $T > T_c$. The measurement provides an estimate of the correlation-length exponent ν and also indicates the degree of anisotropy associated with the critical static fluctuations. The findings show that the fluctuations are quite layer like, with correlation length within (001) planes being about 40 times larger than between-plane correlations. The critical exponent is $\nu = 0 \cdot 63 \pm 0 \cdot 07$ and there is no indication (i.e. outside of experimental error) that exponent η is non-zero. The layer-like nature of the fluctuations seems to be a rather common feature for the zone-boundary structural phase transition in perovskite-related compounds (i.e. octahedral tilting) and produces very significant anisotropic effects in X-ray measurements. Thus two-dimensional correlations are found in the critical anisotropic X-ray diffusive scattering data of Denoyer *et al.* (1970, 1971) on $NaNbO_3$, which undergoes an analogous transition to $SrTiO_3$. These contrast with the chain-like correlations observed in the perovskite *ferroelectrics* (Comes *et al.* 1968, 1970) and presumably reflect the difference in symmetry of the forces responsible for ordering in the two cases.

Supporting data on static critical phenomena in $SrTiO_3$ and $LaAlO_3$ have been obtained from elastic and quasi-elastic neutron scattering experiments with a confirmation of non-classical exponents and anisotropic correlations. Thus, for example, the intensity of Bragg reflection peaks which become permitted below the transition temperature are a measure of the square of the order parameter. An observation by Plakhty and Cochran (1968) on $LaAlO_3$ yields $\beta^{-1} = 3 \cdot 0 \pm 0 \cdot 5$, while for $SrTiO_3$ a more accurate estimate $\beta = 0 \cdot 34$ has been made by Riste *et al.* (1971). The latter authors also measure the diffuse integrated intensity of scattering, which is proportional to the static correlation (eqn (11.3.1)), and find correlation exponents $\nu = 0 \cdot 76 \pm 0 \cdot 08$ $(T > T_c)$ and $\nu' = 0 \cdot 55 \pm 0 \cdot 09$ $(T < T_c)$.

It is clear that a fluctuation-dominated region can be observed in these anti-distortive materials, but the detailed numerical values of sufficient critical exponents are not yet known to test scaling theory in this context. The most accurately known exponent is β and it has a value $\approx \frac{1}{3}$ which is somewhat lower than that expected for short-range isotropic forces (Table 11.3). Although the addition of cubic anisotropy may possibly stabilize a new 'cubic' fixed point in the renormalization-group terminology (Aharony 1973 a, Cowley and Bruce 1973) and precipitate a new set of 'cubic' critical exponents, a more likely explanation of the discrepancy has been given by Aharony and Bruce (1974), who point out that the experiments are performed on samples which are shaped in such a way as to transform into a monodomain low-temperature phase. This indicates the presence of an ordering mechanism which breaks the cubic symmetry (e.g. a strain field) and would lead to a cross-over from the Heisenberg or cubic critical behaviour expected for a free sample to an Ising-like behaviour. The experimental value $\beta \approx \frac{1}{3}$ is quite consistent with this suggestion if we presume that the experiments were performed in the cross-over region in which the effective value of β changes from a Heisenberg value of $0\cdot 38$ to an Ising value of $\frac{5}{16} \approx 0\cdot 31$ (Table 11.1).

A rapidly increasing number of both ferrodistortive and antiferrodistortive transitions are now being studied to at least some degree, and there is every indication that the rather spotty picture, as it now stands, concerning static critical phenomena at structural transitions will rapidly become more complete. To date, however, the experimental emphasis has been much more on phase-transition dynamics (see § 11.5). This is understandable in the sense that there is a greater diversity of dynamic behaviour, much of which still evades physical understanding. Static critical phenomena tend to be less dramatic, and careful quantitative tests of scaling and renormalization relationships are usually more easily conducted in other contexts such as liquid–gas transitions and magnetism where the problems alluded to at the beginning of this section are less severe.

11.4. Dynamic scaling and the soft mode

In the various microscopic statistical models of Chapter 2 explicit forms were obtained for dynamic susceptibility—e.g. eqns (2.1.33), (2.2.36), (2.3.19), (2.4.16), and (2.5.34)—which enabled the concept of soft modes to be introduced as the (in general complex) frequency poles of this collective response function. Just as the static or equilibrium predictions of such approximate theories necessarily become invalid as a critical point is approached arbitrarily closely, so must the dynamic predictions (such as the form of the temperature dependence of soft-mode frequency for example) break down also. It is therefore natural to ask whether the static critical theories of § 11.2 can be extended to include dynamic phenomena.

In contrast to equilibrium properties near critical points the dynamic behaviour of such systems has until recently received only qualitative theoretical treatment. A major reason has been that the time-dependent properties of many-body systems are by their very nature much more diverse and system dependent than their static counterparts. Nevertheless, since the dynamic correlation functions are directly amenable to experiment it is obviously very important to understand them, and a number of significant advances have been made in recent years which have provided some rather precise experimental tests of our understanding of dynamic critical phenomena. Although the first major successes occurred in the fields of magnetism and liquid helium, more recent work on structural phase transitions has revealed the cross-over from Landau to critical dynamic behaviour in this context also.

A dynamic scaling hypothesis was formulated first by Ferrell and co-workers (Ferrell *et al.* 1967, 1968) for the study of the λ-point of He^4. The concept was later reformulated and generalized by Halperin and Hohenberg (1967, 1969) in the language of correlation functions. The classical correlation function $g(\mathbf{r}, \mathbf{r}')$ of eqn (11.1.12) is a static, or equal-time, correlation comparing the displacements $\xi(\mathbf{r})$ and $\xi(\mathbf{r}')$ in cells $\mathbf{r}$ and $\mathbf{r}'$ at the *same* time. This concept can be generalized to a dynamic correlation function $g(\mathbf{r}, t)$ defined by

$$g(\mathbf{r}, t) = \langle\{\xi(\mathbf{r}, t) - \langle\xi(\mathbf{r}, t)\rangle\}\{\xi(0, 0) - \langle\xi(0, 0)\rangle\}\rangle \qquad (11.4.1)$$

from which one can define a frequency and wavevector-dependent Fourier transform $g(\mathbf{q}, \omega)$ from

$$g(\mathbf{r}, t) = \int (2\pi)^{-3}\, \mathrm{d}\mathbf{q} \int (2\pi)^{-1} \exp\{i(\mathbf{q} \cdot \mathbf{r} - \omega t)\} g(\mathbf{q}, \omega)\, \mathrm{d}\omega. \qquad (11.4.2)$$

The static correlation $g(\mathbf{r}, t = 0)$ and its Fourier transform $g(\mathbf{q})$ have a critical form at and close to a critical point T_c, which has already been defined in § 11.1 in terms of static critical exponents and a correlation length ξ' (which is not to be confused with the primitive-cell displacement variable $\xi(\mathbf{r})$).

In the limit of long wavelengths $q \ll \xi'^{-1}$ the form of $g(\mathbf{q}, \omega)$ can often be determined by macroscopic theory. However, as the critical point is approached the correlation length diverges and one would expect that the range of $\mathbf{q}$-values for which the macroscopic form is valid would itself tend to zero. It is the central tenet of the scaling hypothesis that close to T_c the correlation length ξ' is the only important temperature-dependent length in the problem. The behaviour of the correlation function is therefore determined essentially by the ratio of q to $1/\xi'$ when q^{-1} and ξ' are both large compared with primitive-cell dimensions. Following Halperin and Hohenberg (1967, 1969), we illustrate this by a plot of q against ξ'^{-1} (Fig. 11.5). The origin of the plot is the critical point itself. Three asymptotic regions are identified in the q, ξ'^{-1} plane: the shaded region I, corresponding to $q \ll 1/\xi'$, $T < T_c$, is the macroscopic or hydrodynamic region in the ordered phase, region II is the so-called critical region which describes phenomena occurring over distances small compared to ξ' but large compared to all other lengths in the problem, and region III is the macroscopic region for $T > T_c$.

The static correlation function $g(\mathbf{q})$ is singular at the origin but finite elsewhere. The *static* scaling relationships of Table 11.2 can be derived from an assumption concerning the form of $g(\mathbf{q})$, specifically that $g(\mathbf{q})$ varies smoothly throughout the (q, ξ'^{-1}) plane and is described by a single function of $q\xi'$ which is essentially determined by its behaviour in the three asymptotic regions I, II, and III. A *dynamic* scaling hypothesis can now be formulated in similar fashion by focusing on the soft modes $\omega(\mathbf{q})$ which dominate the frequency spectrum of $g(\mathbf{q}, \omega)$ and therefore the dynamic response (i.e. susceptibility) $\chi(\mathbf{q}, \omega)$ to which it is directly related through the

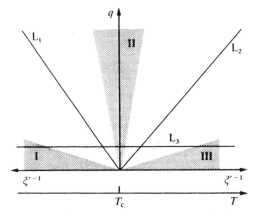

Fig. 11.5. The quasi-macroscopic domain of wavevector q and coherence length ξ'. In the three shaded regions the correlation functions have different characteristic behaviours. These regions are defined by $q\xi' \ll 1$, $T < T_c$, $q\xi' \gg 1$, $T \approx T_c$, and $q\xi' \ll 1$, $T > T_c$ respectively. The asymptotic forms for these regions merge when extrapolated to the lines L_1 or L_2 ($q\xi' = 1$ for $T < T_c$ and $T > T_c$ respectively). An experiment done at constant q (line L_3) passes through all three regions as the temperature is varied. (After Halperin and Hohenberg 1969.)

fluctuation-dissipation theorem

$$g(\mathbf{q}, \omega) = \hbar \coth\left(\frac{\hbar\omega}{2kT}\right)|\text{Im}\{\chi(\mathbf{q}, \omega)\}| \qquad (11.4.3)$$

where $\text{Im}(x)$ signifies the imaginary part of x, and in the classical limit the right-hand side coefficient reduces to $2kT/\omega$.

Noting that the static correlation $g(\mathbf{r}, 0)$ and its lattice Fourier transform $g(\mathbf{q})$ are related by

$$g(\mathbf{r}, 0) = \int (2\pi)^{-3} \exp(i\mathbf{q}.\mathbf{r})g(\mathbf{q}) \, d\mathbf{q} \qquad (11.4.4)$$

it is easy to show that the dynamic correlation function $g(\mathbf{q}, \omega)$ can always be recast in the form

$$g(\mathbf{q}, \omega) = 2\pi g(\mathbf{q})\{\omega(\mathbf{q})\}^{-1} f_\mathbf{q}\left\{\frac{\omega}{\omega(\mathbf{q})}\right\} \qquad (11.4.5)$$

where eqn (11.4.4) implies that

$$\int_{-\infty}^{\infty} f_\mathbf{q}(x) \, dx = 1. \qquad (11.4.6)$$

The function f describes the frequency distribution normalized to unity and will be single peaked at the origin for diffusive motion and double peaked near $\omega = \pm\omega(\mathbf{q})$ for a lightly damped propagating harmonic motion. The *characteristic frequency* $\omega(\mathbf{q})$ is more generally determined by the constraint

$$\int_{-1}^{1} f_{\mathbf{q}}(x)\,\mathrm{d}x = \tfrac{1}{2} \tag{11.4.7}$$

which puts the centre of gravity of the positive or negative distribution at $x = 1$, which is $\omega = \omega(\mathbf{q})$. That this conforms in the usual concept of soft-mode frequency in the limit of quasi-harmonic motion is readily verified from the mean-field response (eqn (2.1.33)) for the damped harmonic oscillator. Using eqns (2.1.33) and (11.4.3) gives the mean-field classical dynamic correlation function in the form

$$g(\mathbf{q}, \omega) = \frac{2kT\chi(\mathbf{q}, 0)\Omega^2(\mathbf{q})\Gamma}{\{\Omega^2(\mathbf{q}) - \omega^2\}^2 + \Gamma^2\omega^2} \tag{11.4.8}$$

where $\Omega(\mathbf{q})$, from § 2.1, is the soft-mode frequency and Γ is a damping constant. Using the static form of the fluctuation-dissipation theorem, i.e. eqn (2.2.28), in the classical limit, viz.

$$g(\mathbf{q}) = kT\chi(\mathbf{q}, 0) \tag{11.4.9}$$

we can reduce eqn (11.4.8) to the form eqn (11.4.5) where

$$\{\omega(\mathbf{q})\}^{-1}f_{\mathbf{q}}\left\{\frac{\omega}{\omega(\mathbf{q})}\right\} = \frac{\Omega^2(\mathbf{q})\Gamma\pi^{-1}}{\{\Omega^2(\mathbf{q}) - \omega^2\}^2 + \Gamma^2\omega^2}. \tag{11.4.10}$$

It is now a matter of simple algebra to confirm that the constraint (eqn (11.4.7)), which now takes the explicit form

$$\int_{-\omega(\mathbf{q})}^{\omega(\mathbf{q})} \frac{\Omega^2(\mathbf{q})\Gamma\pi^{-1}\,\mathrm{d}\omega}{\{\Omega^2(\mathbf{q}) - \omega^2\}^2 + \Gamma^2\omega^2} = \frac{1}{2} \tag{11.4.11}$$

requires us to identify $\omega(\mathbf{q})$ with $\Omega(\mathbf{q})$ as $\Gamma \to 0$.

The dynamic scaling hypothesis states that *at* $T = T_c$ the characteristic frequency $\omega_c(q)$ takes the form

$$\omega_c(q) = \text{constant} \times q^z, \qquad q \equiv |\mathbf{q}| \tag{11.4.12}$$

where z is an unspecified exponent and that the shape function f is a smooth function of its argument. For temperatures *near to* T_c the dynamic scaling hypothesis requires that the characteristic frequency is a homogeneous function of q and the inverse correlation length ξ'^{-1}, i.e.

$$\omega(q) = \text{constant} \times q^z F\!\left(\frac{q}{\xi'^{-1}}\right) \tag{11.4.13}$$

where F depends on q *only* through the ratio q/ξ'^{-1}. In the hydrodynamic (or macroscopic) regions $\omega(q)$ typically varies as $q^x (q \to 0)$ where x has the value 0, 1, or 2. It follows that the hydrodynamic characteristic frequency can be written as

$$\omega(q) = (\xi')^{-z}(q\xi')^x. \qquad (11.4.14)$$

Thus, knowing $\omega(q)$ in the non-critical regions enables statements to be made about the q-dependence of characteristic frequency in the critical regime.

The dynamic scaling hypotheses can be thought of as a generalization of Wilson's renormalization-group ideas to dynamics. Thus, if one defines a renormalization group for the equations of motion of a system in which explicit consideration of short-wavelength fluctuations are successively eliminated and if the equations approach a fixed point under the renormalization transformation, then the dynamic scaling hypotheses generally hold. Explicit calculations of this type have been carried out (e.g. Halperin, Hohenberg, and Ma 1974) and, in a majority of cases, dynamic scaling properties confirmed. In general, when $\xi(\mathbf{r}, t)$ in eqn (11.4.1) refers to the order parameter itself, dynamic scaling does appear to hold quite well in those systems for which it has been experimentally tested (particularly, for solids, in a magnetic context). Unfortunately the situation for displacive and order–disorder structural transitions at the time of writing is much less well documented, and precise tests of scaling in the ferroelectric and antiferro-electric (or anti-distortive) contexts have still to be performed.

The prime shortcoming of dynamic scaling theory is its failure to say anything specific about the shape function f in the critical region. Thus, for example, no predictions are made concerning the diffuse or propagating character of these critical modes. Nor is it possible to predict numerical exponents for transport coefficients such as thermal conductivity or electrical conductivity from scaling arguments alone. On the other hand, the concept of universality, via the renormalization group, should still be valid in the dynamic context. As in the case of static critical phenomena universality does not mean that *all* systems have the same critical behaviour, but that there exist distinct classes of systems with identical critical behaviour. The classes for static and dynamic phenomena are not identical, however, and several dynamic classes may correspond to the same static class. This is another way of saying that time-dependent properties of many-body systems are more diverse and Hamiltonian dependent than static ones. In fact we already have had an example of this from the mean-field theories of Chapter 2. Although all mean-field theories exhibit the same static critical behaviour, we have found in §§ 2.1 and 2.4 respectively that *propagating* soft modes have a (mean-field) critical frequency which goes as $(T - T_c)^{1/2}$ as $T \to T_c^+$, whereas a pure diffusive critical mode (such as that of the kinetic Ising model of § 2.4) has a (mean-field) critical frequency going like $(T - T_c)^1$. This is equivalent to $x = 0$, and $z = 1$ (propagating) or $z = 2$ (diffuse), in eqn (11.4.14).

Attempts to go beyond scaling theory in order to calculate transport coefficients and to assess the critical shape function f have taken a number of forms (see Stanley 1971 a, b). Methods starting from the formal microscopic equations of motion can be developed by decoupling procedures or diagrammatic expansions in an algebraic fashion or can be computed numerically using Monte Carlo procedures. In many cases the dynamic scaling properties have been established explicitly. The most detailed work to date has been performed on the kinetic Ising model of § 2.4. Some rigorous inequalities have been obtained (Halperin 1973), and particularly interesting is the finding for the critical exponent Δ of characteristic frequency at the instability wavevector $\mathbf{q}_0$ ($\mathbf{q}_0 = 0$ for ferroic systems), which is defined by

$$\omega(\mathbf{q}_0) \sim |T - T_c|^{\Delta}. \qquad (11.4.15)$$

Halperin finds $\Delta \geqslant \gamma$, which is $\Delta \gtrsim 1\cdot25$ in three dimensions (Table 11.1). This immediately rules out the mean-field result $\Delta = 1$. Equivalent relationships for the the propagating-mode situation are not yet available, but it is probable that the all-important feature determining the dynamic universality class is the propagating or diffuse nature of the critical fluctuations and that, in view of the tendency of damped harmonic (i.e. propagating) modes to become wholly diffusive on approach to T_c (e.g. Fig. 7.5), the true limiting behaviour may always be characteristic of the diffusive limit in structural transitions. This does not, of course, rule out the possible existence of an intermediate 'critical' regime, between the classical and limiting regimes, with exponents characteristic of the propagating-mode universality class, nor does it imply a universal dynamic critical behaviour for structural transitions since many different universality classes of static phenomena (e.g. long-range forces, short-range forces, uniaxial symmetry, planar symmetry, cubic symmetry, etc.) are possible. In particular, for situations where the disordered phase is of higher lattice symmetry than the ordered phase there may be several low-frequency dynamic modes below T_c which converge to produce a degenerate soft mode above T_c. Whereas the order parameter normally couples only to the lowest of these modes, other variables may couple to the higher-frequency 'soft' modes in varying proportions and may have correlation functions with several different characteristic frequencies or with frequencies varying with different exponents z in different regions of temperature and wavevector.

11.5. Experimental observation of critical dynamics

Despite the obvious importance of the structure of a crystalline solid in determining its physical properties, the subject of phase transitions between crystal structures for a long time received much less experimental study than did many other examples of critical phenomena. Only recently have detailed experiments, using neutron and light scattering techniques and magnetic resonance probes, revealed the similarities and differences between structural transitions and other critical phenomena. A very important step in this understanding is an appreciation of the soft-mode concept as set out in Chapter 2, the idea that associated with each structural transition is a dynamic motion, the complex frequency of which goes to zero at a second-order transition. This critical slowing down can, of course, take place in different forms ranging from the propagating extreme, for which the real part of the frequency dominates throughout the slowing-down process, through significantly damped and tunnel-mode dynamics, to the wholly diffusive behaviour of the order–disorder limit. In this section we shall discuss only those observations and possible explanations which suggest non-classical dynamic behaviour or, more exactly perhaps, behaviour which cannot be understood in terms of the simple microscopic and phenomenological theories as set out in Chapters 2 and 3. The more conventional behaviour usually observed at all but the smallest values of $|T - T_c|/T_c$ has been discussed in earlier chapters.

11.5.1. Displacement transitions

In the absence of good examples of second-order displacement ferroelectrics most experimental work concerning the critical dynamics of displacement structural transitions has to date involved antiferrodistortive instabilities, most notably the 105 K transition in $SrTiO_3$. In $SrTiO_3$ (§ 8.2) a second-order transition is triggered by the softening of a triply degenerate zone-corner mode as T approaches T_c from above. Below T_c this mode splits into singly (A_{1g}) and doubly (E_g) degenerate modes which increase in frequency and diverge as T is lowered below T_c. Far from T_c, on both sides, the modes are easily measurable by neutron scattering (e.g. Shirane and Yamada 1969, Cowley et al 1969) or, below T_c by Raman spectroscopy (Fleury et al. 1968), and their temperature dependence is understandable in terms of simple microscopic theory (Pytte and Feder 1969, Feder and Pytte 1970).

Closer to T_c the situation is complicated by the appearance of a zone-centre diffusive mode in addition to the propagating soft mode (Riste et al. 1971). The first demonstration of this phenomenon came from the observation of constant wavevector energy (i.e. frequency) scans of inelastic neutron scattering cross-section shown in Fig. 11.6. As the Curie temperature is approached from above the soft modes disappear into the diffusive mode as the system enters the critical regime. This obviously makes the measurement of soft-mode frequency within the critical regime at best (i.e. if the diffusive peak is simply an impurity phenomenon) extremely difficult; more likely it calls into question the simple interpretation of the critical dynamics as the simple softening of a damped propagating soft mode.

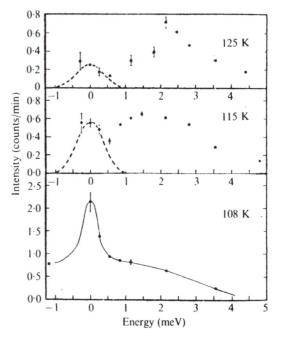

Fig. 11.6. Energy scans at $Q = 0$ for the inelastic scattering of neutrons at a Brillouin zone corner (R-point) for $SrTiO_3$ on approach to its antidistortive phase transition at $T_c \approx 105$ K. The broken curves indicate instrumental resolution as seen from incoherent background. (After Riste *et al.* 1971.)

Although the literature does contain at least one claim (Steigmeier and Auderset 1973) to have followed the softening A_{1g} mode below T_c right into the critical region by Raman study (with an estimate $\Delta \approx \frac{1}{3}$ for the critical frequency exponent of eqn (11.4.15) differing from the classical value $\Delta = \frac{1}{2}$ expected for propagating critical motion), the general feeling is that the matter is far from settled. For example, it now appears that a relaxation central peak may be a fairly common feature of second-order structural transitions and may indeed be an essential part of the transition dynamics which must be understood before limiting exponents can be reliably deduced. Such a diffusive component has also been observed in $LaAlO_3$, $KMnF_3$, and Nb_3Sn, the latter, near its cubic–tetragonal transition at 45 K, being a particularly striking example (Shirane and Axe 1971). In this case the soft mode is a transverse acoustic phonon and we show the neutron inelastic scattering energy scans at constant Q in Fig. 11.7 as the critical temperature is approached from above. Light scattering spectra in SbSI and $LiTaO_3$ also give some indication of this sort of behaviour (Steigmeier *et al.* 1971, Johnston and Kaminow 1968).

The quantity directly measured by an inelastic scattering experiment (providing the particle involved, e.g. neutron, photon, etc., couples linearly

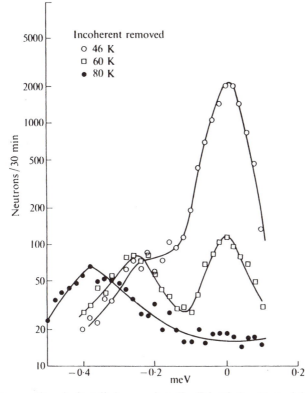

Fig. 11.7. Cross-section of a $(\zeta, \zeta, 0)$ shear mode as $T \to T_c$ in Nb$_3$Sn with $\zeta = 0.02$ as measured at $(2-\zeta, 1+\zeta, 0)$. The incoherent scattering background has been removed and negative energy corresponds to neutron energy gain. The phase transition occurs when $T \approx 45$ K.

to the order–parameter displacement variable $\xi(\mathbf{r}, t))$ is the dynamic correlation function $g(\mathbf{q}, \omega)$, the Fourier transform with respect to lattice and time of $g(\mathbf{r}, t)$ in eqn (11.4.1). In the scattering literature this quantity is more usually written $S(\mathbf{q}, \omega)$ and called the dynamic scattering function or spectral function. We note in addition that it is directly related to dynamic response (or susceptibility $\chi(\mathbf{q}, \omega)$ via the fluctuation theorem (eqn (11.4.3)). Dynamic response also appears in the theoretical many-body literature under another name, the retarded Green's function. In Chapter 2 we calculated dynamic response in various microscopic model systems. For damped harmonic motion, for example, the result (eqn (2.1.33)) leads via the fluctuation theorem to the scattering function

$$g(\mathbf{q}, \omega) = \hbar \coth\left(\frac{\hbar\omega}{2kT}\right) \frac{\chi(\mathbf{q}, 0)\Omega^2(\mathbf{q})\Gamma\omega}{\{\Omega^2(\mathbf{q}) - \omega^2\}^2 + \Gamma^2\omega^2} \qquad (11.5.1)$$

where $\chi(\mathbf{q}, 0)$ is the static susceptibility, $\Omega(\mathbf{q})$ the soft-mode frequency for zero damping, and Γ a damping constant.

When $\Omega(\mathbf{q}) \gg \Gamma$ this consists of a pair of Lorentzian peaks of width Γ at $\omega = \pm\Omega(\mathbf{q})$ with relative intensities $I(+)/I(-) = \exp\{\hbar\Omega(\mathbf{q})/kT\}$. For a fixed damping constant Γ the spectral shape changes gradually as $\Omega(\mathbf{q})$ decreases to values less than Γ until it goes over to a single central peak whose line*width* $\sim\Omega^2(\mathbf{q})/\Gamma$ is then more properly referred to as the soft-mode frequency (see Fig. 7.5). In no circumstances, however, can the form (eqn (11.5.1)) give rise to pronounced side peaks and a central diffusive peak together, which is the characteristic of Figs. 11.6 and 11.7 and of so many of the displacement structural spectra.

A possible explanation may involve the interaction of the soft mode with other excitations. Shirane and Axe (1971) were the first to show that the three-peak data could be fitted using a damped harmonic-oscillator response with a *frequency-dependent* damping factor. In terms of susceptibility one writes

$$\chi^{-1}(\omega) = \Omega^2 - \omega^2 + i\omega\Gamma - \frac{\delta^2\gamma}{(\gamma - i\omega)} \qquad (11.5.2)$$

which describes, in addition to the usual damped oscillator response, the effect of a coupling to some relaxation process with relaxation time $1/\gamma$. In eqn (11.5.2) the uncoupled soft-mode frequency is written Ω and a different low-frequency response ($\omega < \Omega$, $\omega < \Gamma$) is obtained depending on the relative values of ω and γ. In fact we find

$$\chi^{-1} \sim \Omega^2, \qquad \omega \gg \gamma \qquad (11.5.3)$$

$$\chi^{-1} \sim \Omega^2 - \delta^2, \qquad \omega \ll \gamma. \qquad (11.5.4)$$

The observation of the three-peaked response triggered a host of possible microscopic explanations for the frequency-dependent damping term. In general these involve a second- or higher-order coupling of the soft mode to other modes, such as a coupling to local energy-density fluctuations for example (Cowley 1970, Feder 1971, 1973, Enz 1974). In this picture the central mode appears when the periodic time of oscillation of the soft mode becomes comparable to or less than the time required for local temperature equilibrium to be established, in other words, when adiabatic and isothermal phonon responses begin to differ. At the transition temperature the inverse isothermal susceptibility vanishes (by its definition as the second derivative of free energy with respect to the order parameter) and the central mode therefore increasingly dominates as the transition is approached. In terms of eqn (11.5.2) the width of the central component (which is proportional to $\Omega^2 - \delta^2$) goes to zero as $T \to T_c$ with Ω^2 and δ^2 remaining finite. If δ and γ are temperature independent and Ω^2 varies as some positive power ot $T - T_c$, then the soft-mode doublet (positive and negative frequency resonances) will begin to collapse as $T \to T_c^+$ but will 'hang up' at $\omega = \pm\delta$ and begin to transfer its strength to the singularly narrowing central peak. Some evidence of just this sort of behaviour has been found in $SrTiO_3$ by Shapiro *et al.* (1972). However, although singular behaviour of quasi-elastic scattering *intensity* has often been observed, no temperature dependence of the central-peak *linewidth* has yet been measured; the available neutron and optic energy resolutions have permitted only setting an upper limit to the central-peak width. At the time of writing the smallest claimed value for this upper limit is that of Darlington, Fitzgerald, and O'Connor (1975) for $SrTiO_3$ who claim that the greater portion of the central opalescence is 'elastic' within $\pm 2 \times 10^7$ Hz. The possibility therefore remains open that this feature is not dynamic in origin or at least that it is not fundamental to the critical dynamics of the transition.[†]

If the three-peaked critical response is indeed a common characteristic of so-called displacive structural transitions then it becomes clear that a critical region characteristic of propagating excitations may be something of a rarity. Nevertheless, the eventual dominance of diffusive dynamics does not preclude the possibility of the existence of a non-classical propagating region, such as that possibly seen by Steigmeier and Auderset (1973), if the propagating excitations in the range outside the immediate vicinity of the transition are long lived and if the fluctuation-dominated temperature range is large. In the ferroelectric context one promising place to look is in the narrow-band-gap materials to be described in Chapter 14. For these the electronic screening of the dipolar energy tail is expected to be much more complete than in the wide-band-gap materials, such as the perovskites, and the ferroelectric transitions may accordingly be driven by short-range forces and display correspondingly a wider critical temperature range. Indeed there are already indications from Raman scattering experiments on SbSI that the frequency exponent Δ of eqn (11.4.15) is different from its expected classical propagating value of 0.5. The results are discussed in more detail in § 14.3, but the theoretical interpretation of the findings is not yet settled since there is evidence of mode-interaction effects and the phase transition is actually marginally of first order.

11.5.2. *Order–disorder transitions*

It is apparent from the above discussion that the mere characterization of groups of structural transitions as displacive or order–disorder is in itself very difficult. We have already expressed reservations as to whether a *limiting* critical regime characterized by propagating modes exists for any of these transitions. Nevertheless some transitions do possess well-defined

---

[†] The first central-peak width and lineshape measurements have now been reported by Fleury and Lyons (1976) near the displacive ferroelectric transition in lead germanate, $Pb_5Ge_3O_{11}$. The width is $\sim 7 \times 10^9$ Hz and the shape complex, suggesting complicated interactions between the central peak, soft mode, and acoustic phonons.

propagating modes which soften in the characteristic Cochran (1960) fashion in the Landau regime. These we have chosen to refer to as displacive, even if, as in BaTiO$_3$, KNbO$_3$, etc., the Landau modes are overdamped. There is now convincing evidence, however, that the Landau regime in some hydrogen-bonded systems is also correctly characterized in terms of heavily damped propagating modes (Peercy 1973). Indeed under pressure the associated 'soft mode' can even be made underdamped. The detailed character of the critical fluctuations is not yet known for any of these systems, and it is therefore in a rather traditional sense that we include the KDP family, for example, with the presumably more characteristic members of the order–disorder family such as NaNO$_2$, TGS, and others for which the Curie temperature does not depend (or depends very little) on the isotopic mass of the ferroelectric dipole.

In general the theoretical behaviour which one might expect from the simplest fundamental order–disorder transition is a dynamic response of pure relaxation form (eqns (2.4.16), (2.4.17)). In the Landau regime the detailed temperature dependence of the basic hopping time τ_0 (which can be related perhaps to a Boltzmann factor involving the height of the relevant potential barrier (Mason 1947)) will be significant, but in the critical region itself τ_0 is effectively a constant and the critical slowing down is characterized by a relaxation time

$$\tau \sim \tau_0 \left\{ \frac{T - T_c}{T_c} \right\}^{-\Delta} \qquad (11.5.5)$$

which diverges with exponent Δ, the latter having the numerical value unity in simple effective-field theory (eqn (2.4.19)).

Although behaviour broadly characteristic of critical slowing down has been observed by many workers on several order–disorder materials (see Mitsui, Nakamura, and Tokunaga (1973) for references), much of the earlier work, primarily performed by using conventional frequency-dependent dielectric measurements, cannot be fitted quantitatively to a single temperature-dependent relaxation-time model. Certain so-called polydispersive analyses have been attempted on the basis of a formula proposed by Cole and Cole (1941), but there is doubt as to whether such polydispersive characteristics are representative of the behaviour of pure (i.e. translationally invariant) crystal lattices. Mansingh and Lim (1972) suggest that the earlier polydispersive findings are more likely the result of an uncertainty in the measurements of dielectric loss. Reanalysing earlier data they find a monodispersive process consistent with virtually all the results on order–disorder ferroelectrics available to date. Monodispersive or Debye-type response in the long-wavelength limit has also been reported in some more recent experiments (e.g. NaNO$_2$ by Hatta 1970, TGS by Luther and Muser 1970) and a non-classical value of frequency exponent Δ claimed by Hatta.

More recent experiments using Raman and Brillouin light scattering experiments have probed both the wavelength- and frequency-dependent structure of dielectric response and have revealed a number of important subtleties which modify the simple order–disorder or tunnel-mode picture of such transitions and which must be properly understood before any trustworthy critical exponents can be derived from the data. The most detailed work has so far been performed on the KDP family of crystals. For the undeuterated crystals at least one must allow for tunnelling, and the simplest possible interpretation would be expected in terms of the pseudo-spin model of § 2.5. The simple random–phase theoretical treatment of the model given in that section must be improved, of course, in order to bridge the gap between microscopic formalism and dynamic scaling. Progress in this direction has been made (Moore and Williams 1972, Stinchcombe 1973), but it is becoming increasingly clear that the major difficulty preventing a reliable interpretation of critical dynamics for ferroelectric and antiferroelectric order–disorder transitions is the oversimplicity of the tunnel model itself (or of the kinetic Ising model in the zero tunnelling limit).

One major complication appears to be a coupling between the softening tunnel or Debye mode and another optic phonon (possibly its 'Kobayashi' partner, see § 9.2.2) at higher frequencies. This coupling seriously affects the dynamics to the extent that $\mathbf{q}$-dependent experimental data taken near T_c cannot in general be fitted to a simple damped harmonic-oscillator or Debye-relaxation response (Katiyar et al. 1971, Scott and Wilson 1972). When this coupling is properly represented it is possible to recognize one of the modes as the soft mode, but there are additional complexities of a rather general nature. The frequency of the soft mode, though varying with temperature in an approximately classical fashion (with the square of the frequency linear in temperature), extrapolates to zero at a 'Curie' temperature which is often not the same as the actual 'clamped' transition temperature. It has been suggested that this results from a coupling of the soft mode with phonon-density fluctuations (Coombs and Cowley 1973, Young and Elliott 1974) which introduces frequency-dependent damping characteristics in a manner similar to that described for displacive transitions. This should lead to a central peak in Raman response and such, indeed, has recently been observed for KDP by Lagakos and Cummins (1974) and is shown in Fig. 11.8. In this case the observation of the soft mode (by Raman techniques) and of the central peak (by Brillouin techniques) requires a three-mode coupled analysis since the piezoelectric nature of the lattice leads to a direct coupling to acoustic phonons, in addition to the inclusion of a relaxing self-energy term $\delta^2/(1 - i\omega\tau)$. Problems remain, however, since for KDP the difference between the condensation limit of the soft mode and the clamped Curie temperature is zero within experimental error and the width of the Rayleigh peak so small that no dynamic characteristics can be

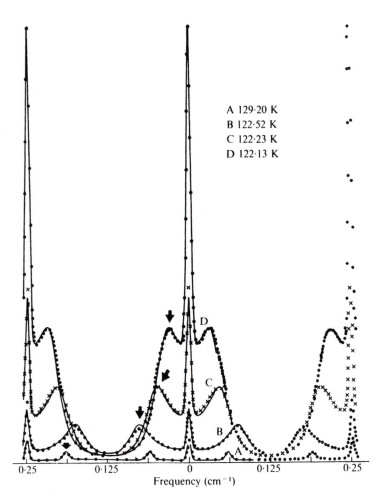

Fig. 11.8. Brillouin spectra of KDP at temperatures approaching the Curie point at $T_c \approx 122$ K. The arrows indicate the positions of the anti-Stokes Brillouin peaks of the order centred at zero. The solid curves are obtained by a fit to a theoretical model analysis. (After Lagakos and Cummins 1974.)

seen (leaving the question of its origin open). Indeed, if produced by the phonon-fluctuation mechanism leading to (11.5.2), then the characteristic frequency γ must be much smaller than that of a 'typical' acoustic phonon and could indicate an importance of long-wavelength acoustic fluctuations. On the other hand, the possibility at this writing still remains (as in the displacive case) that the central peak is not of dynamic origin at all. In fact, more recent work on CsH_2AsO_4 by Lagakos and Cummins (1975) has found no evidence for a central peak there at all, and has explained the previously reported gap between T_c and the extrapolated soft-mode condensation on a particularly large difference between the free and clamped transition temperatures in this material.

All this suggests that precise experimental tests of dynamic scaling concepts are not likely to come early in the field of tunnel-mode or relaxation-mode ferroelectrics. A more promising context for experimental exploration of structural critical disorder and its relationship to the kinetic Ising model would seem to be in the field of second-order phase transitions in binary alloys. Some such transitions (β-brass, Fe_3Al, Ni_3Mn) are thought to be excellent examples of Ising systems and static experimental results in the critical regime are generally very close to Ising expectations, even supporting a minor scaling violation for the $d = 3$ static Ising model. The hope is therefore that the corresponding critical dynamics will reflect the properties of the dynamic Ising model discussed in § 2.4. Initial studies on Ni_3Mn by Collins and Teh (1973) give an Ising relaxation time $\tau \sim (T_c - T)^{-\Delta}$ with $\Delta = 1 \cdot 04 \pm 0 \cdot 09$. This is close to the Landau prediction $\Delta = 1$ and in violation of the rigorous inequality $\Delta \geqslant \gamma \approx 1 \cdot 25$ of Halperin (1973). It therefore in a sense poses more questions than it settles. On the other hand, there seems to be little doubt that the field of critical dynamics of solids in general is about to expand from the realm of magnetic transitions, for which it is already well established, to the less well-trodden ground of structural transitions and spatial order.

12

MAGNETIC AND OPTIC STUDIES OF LOCAL ENVIRONMENT

12.1. Introduction

EXPERIMENTAL methods discussed in the earlier chapters are basically of two types, those which measure a wavevector-dependent response or scattering function of some kind and those which measure bulk properties (i.e. the summed response). In simple-crystal structures, and particularly for quasi-harmonic response, there is a theoretical advantage in obtaining direct measurements in reciprocal space rather than real space since quasi-harmonic propagating lattice waves are approximate eigenstates of the lattice-dynamical system. In grossly anharmonic or complicated structures, on the other hand, there is much less advantage and studies in real space (i.e. of local environment and local dynamics) can be of particular value. In addition, for the study of lattice defects an investigation of local environment is obviously of fundamental importance.

In this chapter we discuss the various methods available for the measurement of local crystal symmetry and of local vibrational motion in ferroelectric systems. A basic disadvantage of many of these techniques is that they make use of properties of ions, such as paramagnetism and nuclear quadrupole moment, which are not possessed by the majority of ions which make up the more common ferroelectrics and antiferroelectrics. This implies a necessity of doping, which in turn can modify the local environment under study and lessen the ability to investigate properties of the nominally pure material. On the other hand, the methods also have advantages including a great sensitivity and the ability to probe in detail the often very important roles played by impurities and defects in ferroelectrics.

One measures in general the perturbing effect of the local ligand field $V(\mathbf{r})$ on the electronic or nuclear energy levels of the atom under study. As a result one obtains information of three basic kinds. The first concerns the symmetry of the local potential $V(\mathbf{r})$ and provides a very sensitive indicator of small changes in symmetry, particularly those associated with subtle phase changes. The second concerns the magnitude of $V(\mathbf{r})$ or of some of its spatial derivatives and, through a simple model representation of the local environment which gives rise to the potential, leads to information concerning the character (e.g. degree of covalency) of the chemical bonds formed between the local atom and its neighbours. The third involves the motion of the local atom in the crystal field and possibly even some information concerning the dynamics of the ligand ions as well.

Formally the ligand-field potential $V(\mathbf{r})$ can be expanded as a series in spherical harmonics and usually only a few low-order terms are required for an adequate description. In ferroelectrics, for example, it is quite common for a fairly high local symmetry in a paraelectric environment (say cubic, involving spherical harmonics of order four) to be reduced to a lower symmetry (say tetragonal or trigonal, involving spherical harmonics of order two) at the onset of ferroelectricity. The largest effects occur when the local ion contains an unfilled electronic shell and an unquenched orbital angular momentum. For such cases the angular momentum $\mathbf{L}$ couples directly to the crystal field to perturb the free-ion energy levels very significantly. These systems often exhibit local transition frequencies in the optic range, and the line splittings and shifts resulting from a change in $V(\mathbf{r})$ can be observed by optic techniques. In ions with an unfilled electronic shell but quenched orbital angular momentum the interaction between the local ion and $V(\mathbf{r})$ is much less direct and relatively small splittings of spin levels S occur via spin–orbit coupling. In the ground state these crystal-field effects can be observed by electron paramagnetic resonance. Finally, for ions with filled electronic shells the crystal-field potential $V(\mathbf{r})$ can still be sensed through its perturbation of nuclear energy levels if the nuclear charge is sufficiently asymmetric (nuclear spin quantum number $I \geqslant 1$). The resulting splitting of a ground nuclear energy level can be measured by nuclear magnetic resonance techniques and of an excited nuclear level by Mössbauer spectroscopy.

12.2. Nuclear magnetic resonance

12.2.1. Basic concepts

Magnetic resonance absorption in a quite general context provides a very sensitive indicator of changes in local environment within a crystal lattice. For example, since the complexity of a given spectrum is a multiple of the number of inequivalent sites, resonance experiments can provide for extremely sensitive detection of subtle phase changes and are a valuable supplement to X-ray and neutron diffraction in this sense. Atomic displacements involved in ferroelectric and antiferroelectric transitions are often small and difficult to study by diffraction methods, but the small static atomic displacements generally cause large changes in internal fields and field gradients which are easily detected by magnetic resonance techniques. In addition the resonance lineshapes contain information concerning the dynamics of the transition and on the lifetime and character of the elementary excitations involved. Since magnetic atoms are not common constituents of ferroelectric materials, *electronic* resonance must usually be performed on ions introduced in small concentration into the material under investigation. *Nuclear* magnetic resonance, on the other hand, can often be performed on at least some atoms in the nominally pure material since many constituent nuclei possess non-zero nuclear magnetic moments.

Nuclei with a non-zero spin $\mathbf{I}$ possess a magnetic dipole moment $\boldsymbol{\mu}_N = \gamma_N \hbar \mathbf{I}$, where γ_N is the nuclear gyromagnetic ratio. If the nuclear spin quantum number has a value $I \geq 1$ the nucleus also possesses a non-zero electric quadrupole moment Q defined by

$$eQ = \int (3z^2 - \mathbf{r}^2)\rho(\mathbf{r})\, d^3\mathbf{r} \qquad (12.2.1)$$

where e is the unit of electronic charge and $e\rho(\mathbf{r})$ is the nuclear charge density with an axis z of axial symmetry. In nuclear magnetic resonance (n.m.r.) and nuclear quadrupole resonance (n.q.r.) experiments one studies the interaction of $\boldsymbol{\mu}_N$ and Q respectively with the local magnetic field and the local electric field gradient at the nuclear site. It is evident that the more detailed information is likely to result from a study of nuclei possessing both dipole and quadrupole moments, and it is here that we meet the first difficulty in that neither of the most common constituent nuclei of ferroelectrics (in terms of which isomorphic trends could most easily be detected) possesses a quadrupole moment. Oxygen (i.e. ^{16}O) with $I = 0$ has neither a dipole nor quadrupole moment and hydrogen (^1H) with $I = \frac{1}{2}$ has only a dipole moment. The limitation for hydrogen-bonded materials is unfortunate (although proton resonance experiments can provide some useful information) since it is the order–disorder proton which undergoes perhaps the most fascinating motion associated with ferroelectric phase transitions, namely the H-bond hopping–tunnelling motion between double-well local-potential minima. A partial alleviation of the difficulty can be achieved by deuteration since the deuteron, with nuclear spin $I = 1$, has a quadrupole moment. However, with its heavier mass the deuteron almost certainly tunnels far less readily than the proton, and results obtained for deuterated materials cannot be taken as representing very closely the equivalent proton dynamics.

In oxygen-octahedra ferroelectrics one does not usually anticipate that any one type of atom will have a critical motion which is fundamentally different in character from the others (although such a possibility cannot be excluded) and most crystals do contain nuclei other than oxygen which possess a quadrupole moment. An example is the ^7Li nucleus ($I = \frac{3}{2}$) in $LiTaO_3$ which moreover, being light, has a much greater amplitude of motion than the oxygen ions and is therefore a more interesting object for study. In general, however, magnetic resonance has been used less in oxygen-octahedra crystals than in a hydrogen-bonded context because the absence of hydrogen and the relatively simple crystal symmetries of the former (e.g. the perovskites) provide examples which are far more attractive candidates for analysis by scattering techniques, and scattering methods provide (at least ideally) much more complete information than magnetic resonance which analyses only the local environment of the relevant nucleus. In principle a fairly complete understanding of ferroelectric dynamics might be obtainable by measuring all possible magnetic resonances in a particular structure, but in practice this usually proves to be beyond the capabilities of a single experimenter and a coherent picture more often evolves, if at all, from an analysis of the overlapping efforts of many workers.

We write the relevant Hamiltonian for a given nucleus

$$\mathcal{H} = \mathcal{H}_D + \mathcal{H}_Q \qquad (12.2.2)$$

as the sum of a magnetic dipole term $\mathcal{H}_D$ and a quadrupole term $\mathcal{H}_Q$. The former is expressible in the familiar form $-\mu_N \cdot \mathbf{H}_{loc}$ which measures the interaction of the nuclear magnetic moment μ_N with the internal magnetic field $\mathbf{H}_{loc}$ at the nuclear site. This field is in general the sum of static and dynamic applied external fields and the local effective field due to all other magnetic dipoles in the sample. The quadrupole contribution $\mathcal{H}_Q$ takes a form which can be deduced by considering the coupling energy between the electronic potential $V(\mathbf{r})$ at the nucleus and the nuclear charge $e\rho(\mathbf{r})$, viz.

$$e \int \rho(\mathbf{r}) V(\mathbf{r}) \, d^3\mathbf{r}. \qquad (12.2.3)$$

More rigorously $V(\mathbf{r})$ is the potential inside the nucleus due to external charges and we have neglected any penetration of electronic charge into the nucleus. Expanding this potential as

$$V(\mathbf{r}) = V(0) + \sum_i x_i \frac{\partial V}{\partial x_i} + \frac{1}{2} \sum_{i,j} x_i x_j \frac{\partial^2 V}{\partial x_i \, \partial x_j} + \cdots \qquad (12.2.4)$$

where x_i ($i = 1, 2, 3$) defines a Cartesian co-ordinate system, and introducing eqn (12.2.4) into eqn (12.2.3), we can expand the coupling energy. The zeroth-order term is simply a constant and, since nuclei have no permanent electric dipole (if, as seems well established experimentally, stationary nuclear states have well-defined parities), the first-order term vanishes. Thus the first term of any importance experimentally is the second-order, or electric quadrupole, term

$$\mathcal{H}_Q = \frac{1}{2} e \int \sum_{i,j} \frac{\partial^2 V}{\partial x_i \, \partial x_j} \rho(\mathbf{r}) x_i x_j \, d^3\mathbf{r} \qquad (12.2.5)$$

where the derivative is measured at the nuclear origin $\mathbf{r} = 0$. Using the fact that the nuclear charge density can be expressed in the form

$$e\rho(\mathbf{r}) = \left\langle \Psi_N \left| \sum_p e\delta(\mathbf{r} - \mathbf{r}_p) \right| \Psi_N \right\rangle \qquad (12.2.6)$$

where Σ_p runs over the nuclear proton positions $\mathbf{r}_p$, and that $V(\mathbf{r})$ is a solution of Laplace's equation $\nabla^2 V(\mathbf{r}) = 0$, we can rewrite (12.2.5) as

$$\mathcal{H}_Q = \frac{e}{6} \sum_{i,j} \frac{\partial^2 V}{\partial x_i \, \partial x_j} \left\langle \Psi_N \left| \sum_p (3x_p^i x_p^j - r_p^2 \delta_{ij}) \right| \Psi_N \right\rangle. \qquad (12.2.7)$$

For a stationary nuclear state with spin I there is a fundamental theorem of quantum mechanics (the Wigner–Eckart theorem) which establishes the proportionality of a traceless tensor involving co-ordinate variables, like $3x_p^i x_p^j - r_p^2 \delta_{ij}$, and a closely related symmetrized operator involving analogous angular-momentum variables. The detailed relationship for this case is

$$\sum_p (3x_p^i x_p^j - r_p^2 \delta_{ij}) = \frac{Q}{I(2I-1)} \left[\frac{3}{2}(I_i I_j + I_j I_i) - I(I+1)\delta_{ij} \right] \qquad (12.2.8)$$

where Q is the quadrupole moment, and leads to a quadrupole Hamiltonian

$$\mathcal{H}_Q = eQ[6I(2I-1)]^{-1} \sum_{i,j} \frac{\partial^2 V}{\partial x_i \, \partial x_j} \left[\frac{3}{2}(I_i I_j + I_j I_i) - I(I+1)\delta_{ij} \right]. \qquad (12.2.9)$$

Choosing a co-ordinate system (x, y, z) for which the second potential derivative is diagonal we define components V_{xx}, V_{yy}, V_{zz} which necessarily satisfy the Laplace equation $V_{xx} + V_{yy} + V_{zz} = 0$. Choosing $V_{zz} \geqslant V_{xx} \geqslant V_{yy}$ the electric field gradient (e.f.g) can now be expressed quite generally in terms of the two parameters

$$eq \equiv V_{zz}, \qquad \eta \equiv \frac{V_{xx} - V_{yy}}{V_{zz}} \qquad (12.2.10)$$

in terms of which the quadrupole Hamiltonian simplifies to

$$\mathcal{H}_Q = e^2 Qq\{4I(2I-1)\}^{-1}\{3I_z^2 - I(I+1) + \eta(I_x^2 - I_y^2)\}. \qquad (12.2.11)$$

The case $\eta = 0$ corresponds to axial symmetry of the nuclear surroundings and, as a result of the definition (eqn (12.2.10)), the parameter η is always between zero and unity. For a cubic or tetrahedral environment $V_{xx} = V_{yy} = V_{zz} = 0$ and the quadrupole coupling vanishes.

The Hamiltonian for a particular nucleus now looks like

$$\mathcal{H} = -\boldsymbol{\mu}_N \cdot (\mathbf{H}_Z + \mathbf{H}_{RF} + \mathbf{H}_{DD}) + \mathcal{H}_Q \qquad (12.2.12)$$

where $\mathbf{H}_Z = (0, 0, H_0)$ is a static applied magnetic field in a laboratory Z-direction, $\mathbf{H}_{RF} = (H_1 \cos \omega t, 0, 0)$ is a small radiofrequency applied field normal to Z, and $\mathbf{H}_{DD}$ is a time- and space-varying local field due to all magnetic dipoles in the sample (except the one at the site under study). With $\boldsymbol{\mu}_N = \gamma \hbar \mathbf{I}$ the eigenvalues of $-\boldsymbol{\mu}_N \cdot \mathbf{H}_Z$ are the Zeeman levels

$$E_M = -M\hbar\omega_L \qquad (12.2.13)$$

where M can take values $-I, -I+1, \ldots, I$, and $\omega_L = \gamma H_0$ is the Larmor frequency. Transitions between these equidistant states $M \to M \pm 1$ can be induced by H_{RF} provided that the resonance condition $\omega = \omega_L$ is fulfilled. Thus, in the absence of $\mathbf{H}_{DD}$ and $\mathcal{H}_Q$ the n.m.r. spectrum would be but a single infinitely sharp resonance line $\omega = \omega_L$.

The time-averaged part of the magnetic dipole interaction field $\mathbf{H}_{DD}$ (which is usually small compared with H_0 in n.m.r. experiments) shifts the resonance to $\omega = \gamma(H_0 + \langle H_{DD} \rangle)$ since $\langle H_{DD} \rangle$ is in the direction of the static applied field. The dynamic part of $\mathbf{H}_{DD}$ provides a relaxation mechanism for the resonance line and therefore defines a linewidth. It follows that through $\mathbf{H}_{DD}$ both the position and shape of the n.m.r. absorption line become a function of crystal structure and the state of the motion. By observing changes in position and linewidth as the crystal is rotated and as the temperature is changed one can detect ferroelectric transitions and gain information about crystal structure and transition dynamics. Unfortunately the information, via H_{DD}, is rather indirect and usually requires elaborate model calculations.

The quadrupole Hamiltonian $\mathcal{H}_Q$ takes its simplest non-trivial form when the asymmetry parameter $\eta = 0$. In this case the energy levels of $\mathcal{H}_Q$ are

$$E_Q = e^2 Qq \{4I(2I-1)\}^{-1} \{3M^2 - I(I+1)\}. \tag{12.2.14}$$

These states are not equidistant but transitions between them can again be produced by a small field H_{RF} with the selection rule $M \to M \pm 1$ as before. In the absence of an applied static field H_0 there is a single resonance frequency if $I = 1$ or $\frac{3}{2}$ but a more complicated spectrum for larger values of I. A more normal experimental situation for nuclear quadrupole study is that in which a large applied static magnetic field H_0 produces a Zeeman splitting which is much larger than the quadrupole term. In this case the nuclear spins are quantized along Z rather than along the principal axis z of the e.f.g. tensor. The resulting eigenvalue problem can be solved by perturbation theory to give energy levels

$$E_M = E_M^{(0)} + E_M^{(1)} \tag{12.2.15}$$

where $E_M^{(0)}$ is the Larmor term which is now perturbed by

$$E_M^{(1)} = e^2 Qq \{8I(2I-1)\}^{-1} (3\cos^2\theta - 1)\{3M^2 - I(I+1)\} \tag{12.2.16}$$

in which θ is the angle between Z and z. As a result the single Zeeman line of the pure magnetic-dipole case is split into (in general) $2I$ resonances. A generalization for the case of asymmetric field gradients $\eta \neq 0$ is straightforward. A measurement of this spectrum about three mutually orthogonal crystal orientations now enables a determination of the principal axes x, y, z of the e.f.g. tensor and of the quadrupole coupling constant Qq and the asymmetry parameter η (Volkoff, Petch, and Smellie 1952). Since one can correlate the axes x, y, z directly with structural features of the lattice, quadrupole or quadrupole perturbed magnetic resonance gives rather direct information concerning lattice structure, particularly if used in conjunction with X-ray or neutron diffraction measurements. Usually an onset of structural ordering results in an increase in the number of resonance lines since there is a distinct set of levels for each group of inequivalent nuclei and, by counting the number of lines, the size of the unit cell can be determined. For cases of pseudosymmetry an equivalent determination by diffraction techniques may be difficult. Also, if a rotating molecule is frozen into one of several possible orientations below T_c, the e.f.g components appropriate to these become averaged with the onset of rotation, causing a coalescence of lines which can be used to identify the motion which becomes unfrozen at T_c. In an analogous fashion the merging of lines known from prior X-ray or neutron work to result from double-well (i.e. H-bond) deuterons can be taken as evidence for the onset of thermal hopping or quantum tunnelling. In these sort of contexts it is usually necessary to have auxiliary X-ray or neutron diffraction information (e.g. to identify the atomic subgroups involved in the ordering or motion) in order to be able to extract the maximum amount of useful information from the magnetic resonance studies.

The most direct and quantitative information about local *dynamics* comes from a measurement of spin–lattice relaxation time (Rigamonti 1967, Bonera, Borsa, and Rigamonti 1970, Blinc, Zumer, and Lahajnar 1970). The establishment of thermal equilibrium between the perturbed nuclear spin system and the lattice takes place in two steps. Firstly the magnetic dipolar interactions between the spins quickly bring about an internal equilibrium to the spin system in which the Boltzmann distribution of spin eigenlevels defines a 'spin temperature' T_s. The time involved, $T_2 \sim (\gamma H_{DD})^{-1}$, determines the width of the resonance lines and, in general, is very short compared with a second characteristic time T_1, the longitudinal or spin–lattice relaxation time, which governs the rate at which the spin temperature T_s approaches the lattice temperature T. Throughout this relaxation the spin system is described essentially by a single constant $\beta = 1/kT_s$ and the approach towards equilibrium with the lattice is described by

$$\frac{d\beta}{dt} = -(1/T_1)(\beta - \beta_0) \qquad (12.2.17)$$

where $\beta_0 = 1/kT$. Because of Curie's law this equation at high static applied field coincides with the usual phenomenological definition of longitudinal relaxation time in magnetism (see for example Abragam 1961) which is through the equation

$$\frac{dM_Z}{dt} = -\frac{1}{T_1}(M_Z - M_0) \tag{12.2.18}$$

where M_0 is the equilibrium value of the magnetization. Experimentally T_1 is measured by observing directly the decay of the Zeeman-level population (which is proportional to the signal intensity) after the system has been perturbed. The various experimental techniques which can be pursued to this end have been outlined by Abragam (1961).

The spin–lattice relaxation time T_1 is a macroscopic transport parameter measuring the rate of return to equilibrium of the expectation value of nuclear spin magnetization along the direction Z of the static applied field H_Z. Theoretically it is calculated by using time-dependent perturbation theory to examine the effect of the fluctuating part of the total Hamiltonian $h(t) = \mathcal{H}_D + \mathcal{H}_Q - \langle \mathcal{H}_D + \mathcal{H}_Q \rangle$ on the stationary states $|\alpha_i\rangle$ of the time-averaged total Hamiltonian. In this manner the spin transition probability between levels k and m is found to be

$$W_{km} = \hbar^{-2} J_{km}(\omega_{km}) \tag{12.219}$$

where $\hbar\omega_{km}$ is the energy separation of the levels k and m, and $J_{km}(\omega)$ is the Fourier transform with respect to time of the autocorrelation function

$$G_{km}(t) = \overline{\langle \alpha_k | h(0) | \alpha_m \rangle \langle \alpha_m | h(t) | \alpha_k \rangle} \tag{12.2.20}$$

where the bar indicates an ensemble average over the lattice variables. If, as is often an adequate approximation, $h(t)$ is regarded as a sum of products of spin and lattice operators (where only the lattice operators are assumed to have an explicit time dependence) and near T_c the motion is dominated by a critical or 'soft' displacement variable $\xi_i(t)$, then the lowest-order term in eqn (12.2.19) becomes proportional to the Fourier transform with respect to time (sometimes called the spectral density) of the autocorrelation function for order-parameter fluctuations, i.e.

$$W_{km} \propto \langle |\Delta\xi_i|^2 \rangle_{\omega km} = \int_{-\infty}^{\infty} \sum_i \langle \Delta\xi_i(0)\, \Delta\xi_i(t) \rangle \exp(-i\omega_{km}t)\, dt$$

$$= \sum_q \int_{-\infty}^{\infty} \langle \Delta\xi(\mathbf{q}, 0)\, \Delta\xi(-\mathbf{q}, t) \rangle \exp(-i\omega_{km}t)\, dt \tag{12.2.21}$$

where $\Delta\xi = \xi - \langle\xi\rangle$ and where we have now reverted to the normal symbolic representation $\langle \ldots \rangle$ for ensemble averages.

The problem of calculating W_{km} (or equivalently T_1) therefore reduces theoretically in lowest order to the evaluation of the autocorrelation function for the order parameter. A very simple example might be given in terms

of an independent-site (non-interacting) kinetic Ising model (see § 2.4) with
stochastic hopping probability $(2\tau_0)^{-1}$. This system has a dynamic susceptibility $\chi_s(\omega)$ given by (2.4.21) as

$$\chi_s(\omega) = \frac{\chi_s(0)}{1 - i\omega\tau_0}.$$ (12.2.22)

The classical fluctuation–dissipation theorem (or Nyquist theorem) establishes a close relationship between $\chi_s(\omega)$ and the corresponding fluctuation
spectrum in the form

$$\langle |\Delta\xi_i|^2 \rangle_\omega = \frac{2kT}{\omega} \operatorname{Im} \chi_s(\omega).$$ (12.2.23)

It follows from eqns (12.2.21) to (12.2.23) that

$$T_1^{-1} \propto \frac{\tau_0}{1 + (\omega\tau_0)^2}.$$ (12.2.24)

For an interacting system the more general classical form of Nyquist's
theorem is

$$\int_{-\infty}^{\infty} \langle \Delta\xi(\mathbf{q}, 0) \, \Delta\xi(-\mathbf{q}, t) \rangle \exp(-i\omega t) \, dt = \frac{2kT}{\omega} \operatorname{Im} \chi(\mathbf{q}, \omega)$$ (12.2.25)

and for the interacting kinetic Ising model we find correspondingly

$$T_1^{-1} \propto \sum_{\mathbf{q}} \frac{\chi(\mathbf{q}, 0)\tau(\mathbf{q})}{1 + \{\omega\tau(\mathbf{q})\}^2}$$ (12.2.26)

in which $\tau(\mathbf{q})$ is related to the 'non-interacting' intrabond jump time τ_0
by eqn (2.4.17). In normal experimental circumstances $\omega\tau(\mathbf{q}) \ll 1$, and
$T_1^{-1} \propto \Sigma_{\mathbf{q}} \tau(\mathbf{q})\chi(\mathbf{q}, 0)$ which exhibits critical behaviour near T_c (see Fig. 12.2)
though only of logarithmic form for an anisotropic dipolar situation. Very
close indeed to a ferroelectric T_c the condition $\omega\tau(\mathbf{q}) \ll 1$ ceases to hold as
$\mathbf{q} \to 0$ (curtailing the limiting singularity) and some frequency dependence of
T_1 is expected. By comparing the observed and theoretical temperature
dependence of T_1 the characteristic jump time τ_0 of quasi-Ising systems such
as H-bond deuterons can be determined. For example Blinc, Stepisnik, et al.
(1971) find for DKDP

$$T > T_c; \qquad \tau_0 = (4 \times 10^{-10})/T \text{ s}$$ (12.2.27)
$$T < T_c; \qquad \tau_0 = (1 \cdot 6 \times 10^{-9})/T \text{ s}$$

with temperature T measured in K. The former value is intermediate
between that obtained from dielectric relaxation (Hill and Ichiki 1962) and
from ultrasonic absorption (Litov and Uehling 1968) in paraelectric DKDP.

12.2.2. Some experimental findings

At the time of writing the most extensively studied material from the standpoint of magnetic resonance is DKDP. For each separate deuteron environment one expects two quadrupole perturbed deuteron magnetic resonance lines with a weakly temperature-dependent splitting. In DKDP below the ferroelectric transition temperature T_c four lines are seen with two separate splittings indicating two deuteron sites with different e.f.g. tensors (Bjorkstam 1967, Fig. 12.1). For each site the largest component

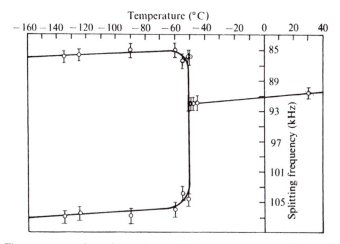

Fig. 12.1. The temperature dependence of x-bond deuteron quadrupole splitting in KD_2PO_4 at $\theta_x = -50°$, where z is the polar axis, x and y are essentially parallel to the z-plane H-bonds, and θ_x is an angular measure of the direction of applied field (being a rotation about the x-axis measured from the y-axis with positive values increasing the angle made with the polar z-axis). (After Bjorkstam 1967.)

(V_{zz}) of e.f.g. is along the O—D axis while the next largest (V_{xx}) is normal to the local P—D—O plane (which is different for the two ends of an H-bond in the KDP structure, see Fig. 9.1). Such a result had been predicted by Chiba (1964) and suggests that in less well-documented structures n.m.r. may be of rather general use in determining bond directions. Above T_c only a single splitting is observed for which the e.f.g. tensor is almost exactly the average for the two equilibrium H-bond deuterium sites. This requires a dynamic hopping or tunnelling mechanism for the deuterons in the paraelectric phase with a fluctuation time short compared with a quadrupole period ($\sim 10^{-5}$ s). Within a half degree or so of T_c both high- and low-temperature spectra coexist establishing the first-order nature of the DKDP transition. Quantitatively the difference between the two splittings in the ordered phase can be related to the temperature dependence of deuteron ordering (see for example Blinc 1968) or, what should be essentially equivalent in the KDP structure, to the temperature dependence of spontaneous polarization P_s. This correspondence has been confirmed experimentally for DKDP and suggests that at least in certain systems n.m.r. quadrupole splitting can be used as a measure of dP_s/dT. The method is likely to be of particular value in antiferroelectric structures for which it could provide a measure of *sublattice* polarization, a property which has yet to be measured by any other means. Finally, although the contribution to T_1 from *intra*bond deuteron motion dominates the spin–lattice relaxation near T_c (Fig. 12.2), there are other

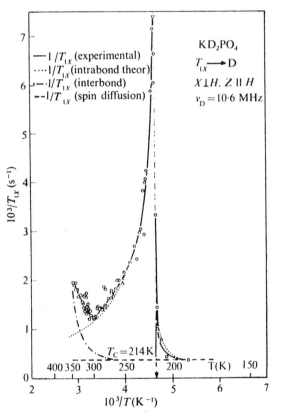

Fig. 12.2. The temperature dependence of the deuteron spin–lattice relaxation rate in KD_2PO_4; the solid curve is drawn through the experimental points as a guide to the eye, the dotted curve is the theoretical contribution from intrabond relaxation, and the dot–dashed and broken curves are respectively the corresponding contributions from interbond and spin–diffusion relaxation. (After Blinc, Stepisnik *et al.* 1971.)

contributions which become more important in other temperature ranges. Each of these can be determined separately and one, the deuteron *inter*bond contribution (which dominates at high temperatures where the intrabond motion becomes too fast to be effective in producing spin relaxation), is particularly interesting since it can be related to the electrical conductivity (Schmidt and Uehling 1962). The interbond jump time is of order 0.4 s at room temperature, and the motion indicates an activation energy of 0.58 eV which is to be compared with a probable intrabond jump rate as given by eqn (12.2.27) and an activation energy of less than 0.1 eV.

A somewhat less direct but nevertheless valuable method of obtaining dynamic information from magnetic resonance experiments is to analyse the temperature dependence of the zero-field n.q.r. spectrum itself. The latter, from eqn (12.2.14), is proportional to q (i.e. V_{zz}) and temperature dependent since the e.f.g. at a particular nuclear site is a function of the motion of its local environment, both zero point and thermal. As molecules execute (say) torsional oscillations (librations) in a crystal, the principal axes of the e.f.g. vary and the amplitude of the motion, which is temperature dependent, lends a temperature dependence to the thermally averaged e.f.g. and hence to the quadrupole coupling constant $e^2 Qq$. The quantitative aspects of this sort of situation have been discussed by Bayer (1951) and Kushida (1955), and the theory has been applied successfully in a number of ferroelectric contexts. Thus, for example, Singh and Singh (1974) have obtained an understanding of the temperature dependence of the two ^{14}N ($I = 1$) nuclear quadrupole frequencies $(3 \pm \eta)(e^2 Qq/4\hbar)$ in $NaNO_2$ by allowing for the effects of the NO_2 librational modes on the e.f.g. at the nitrogen site. The fit with experiment is encouraging (although some deviation is found very close to the phase transition where the amplitude of motion becomes very large) since the parameters involved, for example librational frequencies and corresponding moments of inertia etc., are not adjustables but are known from other experiments. An interesting sideline, not yet given much theoretical consideration, is the observation by Singh and Singh of the strong correlation between the temperature dependence of the n.q.r. frequencies and the corresponding change in optical birefringence. A similar observation in a different material had been made earlier by Schempp, Peterson, and Carruthers (1970) who studied the ^{93}Nb n.q.r. in $LiNbO_3$. The latter authors also used Bayer–Kushida theory to explain the temperature dependence of the n.q.r., but the experiment was carried out for $T \ll T_c$ where many modes are contributing comparable amounts to the effect and analysis was made in terms of a single (weighted-average) Einstein oscillator. Nevertheless, the fit to experiment in this temperature range was good and the resulting Einstein frequency agreed very well with that required to fit the corresponding specific-heat curve. A more general theory, in situations where the local motion is more complex than simple torsional oscillations, should be cast in terms of an expansion of V_{zz} in phonon co-ordinates, particularly, for a displacive system near T_c, involving the soft mode (Scott and Worlock 1973).

Although electric-field gradients are very difficult to calculate from first principles, as indicated in the last section, simple point-charge calculations can provide at least some semi-quantitative estimates of covalency in some cases. In this vein Peterson, Bridenbaugh, and Green (1967) and Peterson and Bridenbaugh (1968) have deduced that point charges very markedly reduced from their formal valence values are necessary to explain the magnitude of the ^{7}Li resonance in $LiNbO_3$ and $LiTaO_3$. Assuming a lithium point charge $e(Li) = 1$ and charge conservation they deduce values $e(Nb) = 1 \cdot 59$ and $e(Ta) = 1 \cdot 21$ to be compared with the formal valence charges $e(Nb) = e(Ta) = 5$. These reduced values are in general agreement with an independent estimate from a statistical analysis of bulk properties (Lines 1970 b).

Even at a qualitative level some interesting correlations can be made. Thus, for example, the variation of quadrupole coupling constant eQq with the length of the shorter $O_1 - D$ bond in a deuterated hydrogen bridge is sensitive enough to allow discrimination between symmetric and asymmetric double wells in many cases. Also (Peterson and Carruthers 1969) a careful monitoring of n.m.r. lineshape and linewidth in solid-solution crystals can make n.m.r. a sensitive and accurate probe of stoichiometry and characterization of the composition of melt-grown crystals (see § 8.3). Hopefully it will eventually be possible to analyse these lineshapes theoretically and to suggest definite models for the defect structures as a result.

Since the proton has $I = \frac{1}{2}$ and no quadrupole moment the information obtained by studying proton resonance is more limited than that which can be obtained from deuteron or more general quadrupole resonance. Spin-relaxation times T_1 can still be measured, since the theory leading to eqn (12.2.26) is still valid when $\mathcal{H}_Q = 0$, but the remaining relaxation mechanism (produced by the lattice-vibrational modulation of H_{DD}) is far less efficient than quadrupole coupling, and extraneous competing mechanisms such as spin diffusion to magnetic impurities often dominate the order-parameter contribution. Even so it has proved quite possible to subtract empirically the non-divergent background relaxation from the critical part in a number of proton resonance experiments near protonic order–disorder ferroelectric transitions. A logarithmic divergence of T_1^{-1} near T_c has been observed in several cases (e.g. Hikita, Sakata, and Tatsuzaki (1973) in dicalcium strontium propionate, see Fig. 12.3) and can be understood from eqn (12.2.26) via eqn (2.4.17) if anisotropic dipolar forces are playing a significant role. More specifically, however, eqn (12.2.26) assumes an order–disorder or overdamped tunnelling dynamics which manifests itself in the frequency dependence of T_1^{-1}. In non-deuterated hydrogen-bond materials the possibility of an underdamped proton-tunnelling motion must always be held as a possibility. For such a

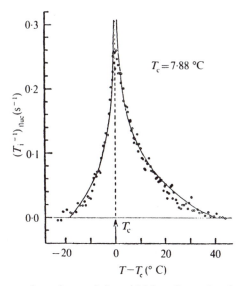

Fig. 12.3. Temperature dependence of the critical or fluctuation-dominated part of the reciprocal spin–lattice relaxation time for protons in dicalcium strontium propionate. The solid curve indicates a fit to a logarithmic divergence. (After Hikita *et al.* 1973.)

situation T_1^{-1} would show a damped harmonic-oscillator frequency dependence of the form

$$T_1^{-1} \propto \sum_{\mathbf{q}} \chi(\mathbf{q}, 0)\Omega^2(\mathbf{q})\Gamma(\mathbf{q})[\{\Omega^2(\mathbf{q}) - \omega^2\}^2 + \omega^2\Gamma^2(\mathbf{q})]^{-1}. \quad (12.2.28)$$

There is some evidence (Brosowski *et al.* 1974) that the glycine II–glycine III proton exchange in TGS may be an example of such a motion.

The other proton resonance observable which is affected by lattice dynamics is the resonance linewidth. By observing changes in linewidth with temperature one can detect phase transitions, and by rotating the crystal with respect to the applied field one can gain information concerning the local crystal structure. Unfortunately the information is rather indirect and is often open to interpretation in several ways. Thus, even though the theory of motional narrowing for resonance linewidths has long been textbook material in a formal sense (see, for example, Abragam 1961), the degree of ordering which occurs at a ferroelectric transition is often insufficient to produce major line broadening, and the linewidth changes which are seen are often subtle and difficult to interpret in an unambiguous fashion. Nevertheless, in favourable circumstances information concerning jump frequencies or rotations of the molecular complex containing the resonating nucleus can be extracted from lineshape and linewidth data and compared, for consistency, with equivalent information obtained from the relaxation experiments.

12.3. Electron paramagnetic resonance

The vast majority of known ferroelectrics do not exhibit electronic paramagnetism and paramagnetic centres must therefore usually be artificially introduced into a ferroelectric lattice if electron paramagnetic resonance (e.p.r.) is to be observed. If such centres are introduced, say by doping, the resulting magnetic levels depend on the local ligand or crystal field at the defect site, and can be probed by applying a static magnetic field H_0 to the sample and inducing transitions between the resulting Zeeman levels by means of a high-frequency field. By varying the orientation of H_0 with respect to the local-crystal-field axes the resulting e.p.r. spectrum reflects the point symmetry of the local environment at the defect site.

The most convenient magnetic ions to use as a probe in the present context are obviously those which perturb the environment the least and which have spin quantum numbers S large enough to show splittings in high as well as low symmetries. Perhaps the best candidates (though many others have been used) are those with half-filled d or f shells, viz. $3d^5$: Cr^+, Mn^{2+}, Fe^{3+} with $S = \frac{5}{2}$, or $4f^7$: Eu^{2+}, Gd^{3+}, Tb^{4+} with $S = \frac{7}{2}$, the final choice depending on the site, size, and valance of the substituted cation. All these ions have orbitally non-degenerate ground states $(L = 0)$ and do not introduce additional distortions into the lattice such as might occur when $L \neq 0$ as a result, for example, of the Jahn–Teller effect (Abragam and Bleaney 1970). The $L = 0$ restriction also implies that the coupling of spin and space co-ordinates is indirect and correspondingly weak. This leads to long spin–lattice relaxation times and enables studies to be made with ease over a very wide temperature range.

With $S > \frac{3}{2}$ cubic as well as axial and rhombic components of local crystal field can be probed, and in this respect at least e.p.r. has surpassed n.m.r. and Mössbauer techniques, which have to date measured only quadrupole splittings (which go to zero in a cubic environment) although in principle higher-order splittings can be observed in the nuclear context if nuclear spin $I > \frac{3}{2}$. On the other hand, there are a number of obvious disadvantages to e.p.r. studies of ferroelectrics. Firstly there is no known negative paramagnetic ion with $L = 0$ which can be used to probe anion sites. Secondly, even with $L = 0$ defects, a valance or size misfit can perturb the environment which is to be probed to a significant degree, and, finally, there is always the necessity of determining correctly the site in which the magnetic impurity resides. This site determination may not be a simple matter in structures with complicated unit cells since one may have to choose between several possible substitutional or interstitial positions. Nevertheless, various techniques are available and many have been discussed in detail in the literature (e.g. Müller 1971).

The splitting of an $L = 0$, spin-S, ground-state level by the local ligand field is traditionally expressed in spin-Hamiltonian formulation. In this formalism the symmetry of the local crystal field is reflected in the symmetry of equivalent spin operators. To terms quadratic in the spin operators the effective Hamiltonian can be written

$$\mathcal{H} = -\mu_B \mathbf{H}_0 \cdot \mathbf{g} \cdot \mathbf{S} + \mathbf{S} \cdot \mathbf{\Lambda} \cdot \mathbf{S} \qquad (12.3.1)$$

where the first term is the Zeeman interaction of electron spin $\mathbf{S}$ with the static applied field $\mathbf{H}_0$ and the second (fine-structure) term represents the interaction between the electron spins via the combined action of spin–orbit coupling and low-symmetry crystal field. In the principal co-ordinate system, x, y, z, the fine-structure term can be expressed in the form

$$\mathbf{S} \cdot \mathbf{\Lambda} \cdot \mathbf{S} = D\{S_z^2 - \tfrac{1}{3}S(S+1)\} + E(S_x^2 - S_y^2) + \text{constant} \qquad (12.3.2)$$

with D representing axial and E rhombic splitting terms. Operators with degree higher than $2S$ can always be omitted since they have zero matrix elements. Since the next-highest terms in the spin Hamiltonian are of fourth order, eqn (12.3.1) is sufficient for all $S \leqslant \tfrac{3}{2}$. For $S \geqslant 2$, however, fourth-order terms should be added, the most important of which is the cubic term

$$B\{S_{x'}^4 + S_{y'}^4 + S_{z'}^4 - \tfrac{1}{5}S(S+1)(3S^2 + 3S - 1)\} \qquad (12.3.3)$$

in which x', y', z' label the (four-fold) quaternary axes and the parameter B is proportional to the fourth derivative of the local-crystal-field potential. In low-symmetry crystals D and E are typically of order 10^{-1}–10^{-2} cm^{-1} (10^9–10^{10} s^{-1}) when $L = 0$, and fourth-order splittings are perhaps an order of magnitude smaller. The advantage of determining local pseudo-symmetries (i.e. x', y', z') in addition to lower-order distortions (x, y, z) is self-evident.

In more formal terms the complete spin Hamiltonian to fourth order is given by

$$\mathcal{H} = -\mu_B \mathbf{H}_0 \cdot \mathbf{g} \cdot \mathbf{S} + \sum_{m=-2}^{2} B_2^m O_2^m + \sum_{m=-4}^{4} B_4^m O_4^m \qquad (12.3.4)$$

where the operators O_l^m are normalized spin operators which transform as corresponding homogeneous Cartesian polynomials. The latter are defined as linear combinations of the tensor operators T_l^m which transform in the same manner as the spherical harmonic polynomials Y_l^m. The operators O_l^m are tabulated in many textbooks on paramagnetic resonance and represent in equivalent spin-operator language the results of expanding the local-crystal-field potential in spherical harmonics. In particular, expressed in the principal-axis co-ordinate system, only two of the second-order operators

O_2^m are non-zero, namely

$$O_2^0 = 3S_z^2 - S(S+1), \qquad O_2^2 = S_x^2 - S_y^2 \qquad (12.3.5)$$

from which the equivalence of eqn (12.3.4) to this order and eqn (12.3.1), eqn (12.3.2) is immediately apparent. For the $4f^7$ ions with spin $\frac{7}{2}$ it is necessary to add sixth-order terms to the spin Hamiltonian since these do not vanish when $S \geqslant 3$.

Electron paramagnetic resonance studies have thus far been pursued more extensively in perovskite and perovskite-like crystals than in hydrogen-bonded ferroelectrics. This is primarily because in the highly symmetric perovskites a lifting of the degeneracy of the spin levels can detect even very small deviations from the cubic paraelectric phase. This in fact was the manner in which the now much studied 105 K zone-doubling transition in $SrTiO_3$ was first detected. The accuracy of the e.p.r. method is quite impressive in such circumstances and the location of the pseudo-quaternary axes in the perovskites can be achieved to better than one-tenth of a degree. A sufficient concentration (~ 10–100 parts per million) of Fe^{3+} and Mn^{2+} impurities is often already present in the nominally pure materials and doping is not necessary. Too large a concentration of defects should be avoided, since magnetic dipolar interactions then cause excessive broadening of the resonance lines and preclude accurate measurement.

Paramagnetic centres can also be produced in ferroelectrics as damage centres induced by irradiation of the crystal with X-rays or γ-rays. This technique has been applied more frequently to H-bonded ferroelectrics. The irradiation normally produces a molecular ion by electron capture and the gross features of the spectrum can be understood in terms of the interaction of the unpaired electron ($S = \frac{1}{2}$) with a neighbouring nuclear spin I. Since for electronic spin $\frac{1}{2}$ the electronic contribution reduces to a Zeeman term the relevant spin Hamiltonian is

$$\mathcal{H} = -\mu_B \mathbf{H}_0 \cdot \mathbf{g} \cdot \mathbf{S} - \mu_N \mathbf{H}_0 \cdot \mathbf{g}_N \cdot \mathbf{I} + \mathbf{S} \cdot \mathbf{A} \cdot \mathbf{I} \qquad (12.3.6)$$

where μ_N, $\mathbf{g}_N$, and A are respectively the nuclear magneton, nuclear g-factor, and hyperfine coupling tensor. The resulting number of resonance lines ($2I+1$) usually allows identification of the nucleus responsible for the splitting. Once such identification is made one can correlate, for example, displacements of ions which result in a multiplication of the number of inequivalent sites with a corresponding multiplication of the number of resonance lines. In addition, a study of the temperature development of the spectrum can often lead to information concerning the motion of the surrounding molecular groups or of the reorientation of the centre itself.

One of the best examples in which e.p.r. in irradiated crystals has really clarified the understanding of a ferroelectric transition is in KH_2AsO_4

(KDA). In gamma- or neutron-irradiated KDA or DKDA a stable paramagnetic centre $(AsO_4)^{4-}$ is formed by electron capture. No fundamental structural change seems to occur on formation of the centre, which is hydrogen-bonded to four other tetrahedral units in the KDP-type crystal. As a result the e.p.r. of the $(AsO_4)^{4-}$ centre is an important microscopic probe for studying several properties of the crystals. The gross features of the spectrum can be understood via eqn (12.3.6) as an interaction of the unpaired electron with the ^{75}As nucleus $(I = \frac{3}{2})$ but, in addition to the four-line ^{75}As hyperfine structure, the spectrum is further split by interactions with the proton nuclear spins on the hydrogen bonds. The latter have been used as a detector of the reorientation frequencies of the H_2AsO_4 ferroelectric dipoles near the paramagnetic centre and, moreover, the range of anticipated frequencies (of order 10^{10} s^{-1}) falls in the range where e.p.r. is ideally suited for investigation.

The possibility of observation of the proton hyperfine structure is a consequence of the fact that the unpaired electron enters an A_1-type molecular orbital on the arsenate ion with a rather large s-character. Thus, when the protons are moving rapidly compared with e.p.r. frequencies the electron sees a spin $\Sigma_i \mathbf{I}_i$ (summing over the four protons) of magnitude 2 and each ^{75}As line is a quintet. If the motion is slow, on the other hand, the electron sees a proton nuclear spin of magnitude 1 arising from the two protons closest to it and each ^{75}As line is a triplet. Experimentally for KDA a cross-over from the quintet to triplet form is observed as a function of decreasing temperature at temperatures well above T_c (Blinc, Cevc, and Schara 1967). No additional anomaly in proton splitting is seen at T_c since the relevant motion on the e.p.r. time scale is already effectively frozen in at temperatures well above T_c. The local motion, of course, is not expected to exhibit the same critical slowing down anticipated for the $q = 0$ ferroelectric mode since modes of all wavelengths contribute to it.

The early e.p.r. work on KDA (which has been summarized by Blinc 1968) concentrated on the protonic perturbation of the ^{75}As levels. More recently Dalal and McDowell (1972) have made a very careful study of both the ^{75}As and protonic hyperfine spectra as functions of temperature taking advantage of the increased accuracy made available by use of ENDOR (electron–nuclear double resonance) techniques. In a completely rigid K–AsO$_4$ framework the ^{75}As hyperfine structure would be expected to reflect the presence of two differently oriented AsO$_4$ tetrahedra in the paraelectric unit cell of these crystals. If, however, the K–AsO$_4$ system were in motion and sufficiently fast, only a single site spectrum would be expected. Experimentally Dalal and McDowell observe the single-site spectrum at high temperatures, but at lower temperatures (though still well above T_c) they find that the ^{75}As spectrum begins to reflect the symmetry of the inequivalent sites. It is therefore apparent that the K–AsO$_4$ (i.e. heavy-ion) framework is in motion with a significantly temperature-dependent frequency. The most important finding, however, comes on comparison of the characteristic times for motion of the protons on the one hand and of the heavy ions on the other (Fig. 12.4). From the figure it is apparent that the

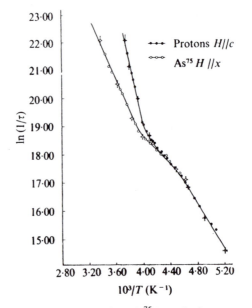

Fig. 12.4. Correlation times for the motion of ^{75}As and of protons calculated from the temperature dependence of ^{75}As and proton hyperfine structure in KH_2AsO_4. (After Dalal and McDowell 1972.)

motion of arsenic nuclei and of the protons are governed by different processes at higher temperatures but by the same process at lower (but still paraelectric) temperatures. This is very important direct evidence for the existence of a coupled proton–lattice motion in the KDP-type structure as T_c is approached. Indications are that a similar coupling is also present in the deuterated material (DKDA) and in the isomorphic antiferroelectric $NH_4H_2AsO_4$ (ADA) and suggests that such a proton–lattice coupled mode may be a rather general feature of KDP-type transition dynamics, thereby lending support to the Kobayashi model of KDP-type ferroelectricity (§ 9.1).

Electron paramagnetic resonance measurements have also provided accurate and valuable information concerning static phenomena close to transition points. Historically the first were those on $BaTiO_3$ (Hornig, Rempel, and Weaver 1959) using Fe^{3+} at Ti^{4+} sites to investigate the cubic–tetragonal (ferroelectric) transition. In most ferroelectrics, however, dielectric anomalies occur near T_c and are readily measured by conventional d.c. or a.c. techniques, so that information obtained from e.p.r. often tends merely to confirm existing results. For example, the axial parameter D of eqn (12.3.2) for Fe^{3+} and Gd^{3+} in the tetragonal phase of $BaTiO_3$ varies with temperature in the same way as the square of the spontaneous polarization (as expected theoretically, see for example Rimai and de Mars (1962)). Also, at least for uniaxial systems, few significant deviations from classical critical exponents are anticipated near ferroelectric Curie points (see Chapter 11), and it is therefore more challenging to use the e.p.r. method, which is capable of considerable accuracy to investigate non-polar structural transitions for which non-classical behaviour is expected if short-range forces dominate the transition. In fact some of the earliest really quantitative observations of non-classical static critical behaviour near second-order structural transitions were obtained in this way for the zone-doubling transitions in the perovskites $SrTiO_3$ and $LaAlO_3$ (Müller and Berlinger 1971).

In $LaAlO_3$ below T_c a rotation of alternate AlO_3 octahedra takes place about a trigonal axis defining an angular order parameter φ (see § 8.2.1). When Fe^{3+} is substituted for Al^{3+} (requiring no charge compensation) the temperature dependence of the axial e.p.r. parameter D reflects the local distortion below T_c and provides a measure of the square of the angular order parameter (since the trigonal structure has a centre of inversion). Experimentally (see, for example, Müller 1971) a relationship $D^{3/2} \propto T_c - T$ is found close to the second-order Curie point T_c implying a critical temperature dependence of order parameter $\varphi \propto (T_c - T)^{1/3}$ with critical exponent $\beta = \frac{1}{3}$ to be compared with the classical exponent $\frac{1}{2}$ (see § 11.3.2). Since for Fe^{3+} the spin Hamiltonian contains cubic as well as axial terms the angular variable φ can actually be determined directly by rotation of a monodomain sample about the domain axis, since for a particular domain there are two Fe^{3+} spectra which are rotated relative to one another by 2φ.

A similar situation has been observed for Fe^{3+} centres in $SrTiO_3$, for which the zone-doubling distortion takes place about a cubic axis, and a critical exponent $\beta \approx \frac{1}{3}$ is again measured for the angular order parameter (Fig. 11.4). For this case, however, charge compensation is required since the Fe^{3+} goes into the Ti^{4+} site. This produces iron–vacancy (Fe^{3+}–V_0) centres (for which an Fe^{3+} iron has a nearest-neighbour oxygen vacancy) in addition to the 'cubic' Fe^{3+} centres. The Fe^{3+}–V_0 e.p.r. spectrum can also be used to determine the temperature dependence of φ (the results being fully consistent with those obtained from Fe^{3+} centres) but, more importantly, they can be used to study critical fluctuations. The Fe^{3+}–V_0 system is strongly axial and its angular variable $\varphi'(T)$ is proportional to but smaller than the order-parameter variable $\varphi(T)$. It follows that fluctuations in $\varphi(T)$ produce a pronounced broadening of the otherwise narrow e.p.r. line. A careful analysis of this broadening (e.g. von Waldkirch et al. 1972, 1973) enables details of static critical fluctuations to be obtained, including the critical exponent for correlation length and the degree of anisotropy of the intercell motion correlations. The latter in $SrTiO_3$ as $T \to T_c^+$ are found to be platelike with correlations within (001)-planes very large compared with the equivalent fluctuations between planes. Below T_c the resonance lines are observed to develop an asymmetry, an effect which has been interpreted by Müller and Berlinger (1972) as resulting from an asymmetry of local fluctuations $\delta\varphi_i$ around the value for which the probability of observing φ_i is a maximum. Such an asymmetry is expected theoretically in an ordered phase (where odd moments of fluctuation are non-zero), but its accessibility to experiment, and the fact that the asymmetry appears to increase in a critical manner near T_c (as observed by Müller and Berlinger), had not been recognized before the e.p.r. experiments were performed.

Some at least of these static critical data should also be measurable using the n.m.r. techniques of the previous section. Thus, for example, the quadrupole coupling constant e^2Qq in $LaAlO_3$ is theoretically quadratic in the order parameter $\varphi(T)$ (Borsa, Crippa, and Derighetti 1971) and its measurement for a quadrupolar nucleus (^{27}Al or ^{139}La) in $LaAlO_3$ as $T \to T_c^-$ should therefore determine the critical temperature dependence of the order parameter. The disadvantage of the use of n.m.r. for this purpose is the smallness of the typical quadrupolar splitting (a few kHz to a few hundred kHz) and the difficulty of resolving the quadrupole structure from the dipolar-broadened central line. In favourable cases, however, some fairly quantitative measurements in the critical region may be possible.

Other recent e.p.r. studies on perovskites have included the observation of critical linewidth broadening for Mn^{2+} in $BaTiO_3$ near the ferroelectric transition temperature (Shapkin, Gromov, and Petrov 1973) and a measurement of the perturbing effect of an applied electric field (i.e. the spin-resonance Stark effect) on the spectrum of Fe^{3+} and Gd^{3+} in $SrTiO_3$ (Unoki and Sakudo 1973, 1974). In the latter work the electric-field-induced shifts, which are quite large at lower temperatures where the dielectric constant becomes large, can be described by adding to the spin Hamiltonian additional terms involving the polarization, viz.

$$\mathscr{H}_P = \sum_{i,j,k} R_{ijk} P_i S_j S_k + \sum_{i,j,k,l} T_{ijkl} P_i P_j S_k S_l + \ldots \qquad (12.3.7)$$

where i, j, k, l each run over three Cartesian co-ordinates and the form of tensors R_{ijk} and T_{ijkl} is largely dictated by symmetry considerations.

12.4. Mössbauer spectroscopy

12.4.1. Theory

The Mössbauer effect, which is the absorption or emission of gamma rays by atoms bound in solids without the excitation of phonons (i.e. the observation of zero-phonon gamma rays), is a relative newcomer to the field of ferroelectric experimentation, and to this point its full potential as a probe of local environment in ferroelectric crystals has probably not been realized. When a gamma ray is emitted from an atom in a solid as the result of the decay of an excited nuclear state the emission is accompanied by the (recoil) transfer of integral multiples of phonon energies to the lattice. However, since the effect is a quantum one, the possibility that no energy is transferred is included in this statement and a fraction of the emitted rays will accordingly be zero-phonon lines. A similar statement is applicable to the scattering of X-rays or neutrons where the distinction between elastic and inelastic processes is perhaps more familiar. If the fraction f of zero-phonon gamma rays can be made large and the environment of the emitting atom is of high symmetry (e.g. no quadrupole splitting), then the resulting source is a very sharp line which can be used to study the resonant absorption by an atom of the same element in the ferroelectric environment of the crystal under study. In the absorber the excited nuclear level contains, through its energy (via an electric monopole interaction with the s-electron cloud—the so-called isomer shift or chemical shift) and possible quadrupole splitting, a considerable amount of information concerning the symmetry and degree of covalency of the local crystal environment relative to that of the standard emitter.

To scan across the energy spectrum of the absorber with a 'Mössbauer spectrometer' it is necessary to have a method of modulating the energy of the source gamma rays. By far the most convenient method is to make use of the Doppler effect by introducing a mechanical motion with precisely controlled velocity. A typical width for a zero-phonon Mössbauer line is 10^{-8} eV (i.e. a few MHz) and to scan across the zero-phonon absorption spectrum requires in general velocities only of order a few mm s^{-1}. To scan typical one-phonon sidebands would take velocities of order 10^6 larger, and for this reason the conventional Mössbauer scan is concerned only with processes in which the quantum state of the lattice remains unchanged unless very soft 'soft' modes with $\omega \sim$ MHz come into range. The quantities of particular interest in observing Mössbauer spectra of transmitted gamma-ray intensity *versus* energy near ferroelectric transitions are the changes of absorption (i.e. Mössbauer fraction f), of energy shift, and of magnetic dipole or electric quadrupole splitting. The Mössbauer fraction f is, by its very definition, of critical importance to the method, and one can show that it contains within it information concerning the local motion of the scattering atom.

The probability per unit time that a plane gamma ray of wavevector $\mathbf{q}$ will induce a nuclear transition with absorption of a quantum of momentum $\hbar\mathbf{q}$ can be obtained from the standard quantum theory of radiation (as given for example in the book by Heitler 1954) by use of time-dependent perturbation theory involving the $\mathbf{p} \cdot \mathbf{A}$ perturbing interaction between the momentum **p** of a nucleon and the vector potential **A** of the photon field. The resulting absorption cross-section (Singwi and Sjölander 1960) for gamma-ray energy E is

$$\sigma_a(E) = \frac{\sigma_0 \Gamma^2}{4} \sum_{m,n} \rho_n \frac{|\langle m| \exp(i\mathbf{q} \cdot \delta\mathbf{r})|n\rangle|^2}{(\Delta + E_m - E_n - E)^2 + (\Gamma/2)^2} \qquad (12.4.1)$$

in which σ_0 is the absorption cross-section at resonance ($E = \Delta$) for the free absorber, ρ_n is the distribution function $\exp(-E_n/kT)/\Sigma_n \exp(-E_n/kT)$, Γ is the full width at half maximum of the excited level, Δ is the energy difference between the ground and excited states of the nucleus, E_n and E_m are respectively the energies of the crystal lattice states $|n\rangle$ and $|m\rangle$, and finally $\delta\mathbf{r}$ is the displacement of the absorbing nucleus from its equilibrium position. The form of (12.4.1) as a sum of the squares of the absolute values of transition amplitudes implicitly assumes that the lattice states are long lived on the scale of a nuclear lifetime (an assumption that requires careful consideration particularly in the context of impure crystal lattices—see for example Lipkin 1964). Rewriting (12.4.1) in the form

$$\sigma_a(E) = \sigma_0 \left(\frac{\Gamma}{2}\right)^2 \sum_{m,n} \rho_n |\langle m| \exp(i\mathbf{q} \cdot \delta\mathbf{r})|n\rangle|^2$$

$$\times \int_{-\infty}^{\infty} \frac{\delta\{\omega - (E_n - E_m)/\hbar\} \, d\omega}{(\Delta - E - \hbar\omega)^2 + (\Gamma/2)^2} \qquad (12.4.2)$$

using the Fourier transform representation for the δ-function, viz.

$$\delta\left\{\omega - \frac{(E_n - E_m)}{\hbar}\right\} = (2\pi)^{-1} \int_{-\infty}^{\infty} \exp(i\omega t) \exp\left\{\frac{i(E_m - E_n)t}{\hbar}\right\} dt \qquad (12.4.3)$$

and introducing Heisenberg (time-dependent) operators

$$\delta\mathbf{r}(t) = \exp\left(\frac{i\mathcal{H}t}{\hbar}\right) \delta\mathbf{r}(0) \exp\left(-\frac{i\mathcal{H}t}{\hbar}\right) \qquad (12.4.4)$$

where $\mathcal{H}$ is the crystal Hamiltonian, one readily transforms (12.4.1) to the form

$$\sigma_a(E) = \sigma_0 \left(\frac{\Gamma}{2}\right)^2 \int_{-\infty}^{\infty} dt \langle \exp\{-i\mathbf{q} \cdot \delta\mathbf{r}(0)\} \exp\{i\mathbf{q} \cdot \delta\mathbf{r}(t)\}\rangle$$

$$\times (2\pi)^{-1} \int_{-\infty}^{\infty} \frac{\exp(i\omega t) \, d\omega}{(\Delta - E - \hbar\omega)^2 + (\Gamma/2)^2} \qquad (12.4.5)$$

in which the notation $\langle \ldots \rangle$ includes both quantum and ensemble averaging.

Integrating explicitly with respect to ω one obtains

$$\sigma_a(E) = \frac{\sigma_0 \Gamma}{4\hbar} \int_{-\infty}^{\infty} dt \, \exp\left\{-\frac{i}{\hbar}(E - \Delta)t\right\} \exp\left(-\frac{\Gamma}{2\hbar}|t|\right)$$

$$\times \langle \exp\{-i\mathbf{q} \cdot \delta\mathbf{r}(0)\} \exp\{i\mathbf{q} \cdot \delta\mathbf{r}(t)\}\rangle. \qquad (12.4.6)$$

To this point no specific assumption has been made concerning the nature of the motion expressed by $\delta\mathbf{r}(t)$. Not surprisingly, however, most detailed study has been performed assuming a harmonic representation. In this case $|m\rangle$ and $|n\rangle$ are harmonic-oscillator states and it can be established that the ensemble average in eqn (12.4.6) can then be related to the well-known Debye–Waller factor in the form (see, for example, Maradudin 1964, 1966)

$$\langle \exp\{-i\mathbf{q} \cdot \delta\mathbf{r}(0)\} \exp\{i\mathbf{q} \cdot \delta\mathbf{r}(t)\}\rangle = \exp(-2W) \exp\langle\{\mathbf{q} \cdot \delta\mathbf{r}(0)\}\{\mathbf{q} \cdot \delta\mathbf{r}(t)\}\rangle$$

$$(12.4.7)$$

in which

$$2W = \langle\{\mathbf{q} \cdot \delta\mathbf{r}(0)\}^2\rangle. \qquad (12.4.8)$$

The final expression for absorption cross-section in the harmonic approximation is

$$\sigma_a(E) = \tfrac{1}{2}\sigma_0\gamma \, \exp(-2W) \int_{-\infty}^{\infty} dt \, \exp(i\omega t - \gamma|t|) \exp\langle\{\mathbf{q} \cdot \delta\mathbf{r}(0)\}\{\mathbf{q} \cdot \delta\mathbf{r}(t)\}\rangle$$

$$(12.4.9)$$

in which $\hbar\omega = \Delta - E$ and $\gamma = \Gamma/2\hbar$. The dependence on crystal dynamics therefore enters through the autocorrelation function $\langle\{\mathbf{q} \cdot \delta\mathbf{r}(0)\}\{\mathbf{q} \cdot \delta\mathbf{r}(t)\}\rangle$ and the Debye–Waller factor. These may be expanded in terms of phonon modes when the normal-mode expansions of $\delta\mathbf{r}(0)$ and $\delta\mathbf{r}(t)$ are introduced. The resulting expression contains not only the Mössbauer term, representing processes in which there is no exchange of quanta between the gamma ray and the lattice, but also a 'one-phonon' term (corresponding to an exchange of one quantum) and 'multi-phonon' terms. The final picture is a sharp Mössbauer line superposed on a broad background of one- and multi-phonon lines (see, for example, Visscher 1960).

The cross-section term for the zero-phonon Mössbauer line itself is easily obtained directly from eqn (12.4.1) by setting $m = n$, viz.

$$\sigma_a^0(E) = \sigma_0\left(\frac{\Gamma}{2}\right)^2 \sum_n \rho_n \frac{|\langle n|\exp(i\mathbf{\acute{q}} \cdot \delta\mathbf{r})|n\rangle|^2}{(\Delta - E)^2 + (\Gamma/2)^2}. \qquad (12.4.10)$$

This can be written simply as

$$\sigma_a^0(E) = \frac{\sigma_0 f(\Gamma/2)^2}{(\Delta - E)^2 + (\Gamma/2)^2} \qquad (12.4.11)$$

where

$$f = \sum_{n} \rho_n |\langle n|\exp(i\mathbf{q} \cdot \delta\mathbf{r})|n\rangle|^2 \qquad (12.4.12)$$

is the Mössbauer fraction by definition, i.e. $\sigma_a^0(\Delta) = \sigma_a f$. In a harmonic approximation $f = e^{-2W}$, the Debye–Waller factor.

The requirement for efficient Mössbauer absorption is therefore that $W \propto \langle\{\mathbf{q} \cdot \delta\mathbf{r}(0)\}^2\rangle$ be small. This condition is most easily satisfied at low temperatures but, because of the existence of zero-point motion, there is also a restriction to small $\mathbf{q}$ or, in other words, gamma rays of low energy (typically $\lesssim 100$ keV). The latter restriction essentially limits the Mössbauer effect to heavy elements, most of which unfortunately do not occur in the more familiar ferroelectrics and antiferroelectrics. It follows that doping is very often necessary, but with the additional drawback that one cannot easily substitute heavy Mössbauer atoms for much lighter atoms without greatly changing the basic character of the local environment, particularly the dynamics. Nevertheless, there are some existing ferroelectrics and antiferroelectrics which do contain Mössbauer nuclei in their nominally pure states, so that the Mössbauer technique is certainly not without promise in the ferroelectric context.

As mentioned earlier in the section the information to be obtained from Mössbauer spectroscopy in ferroelectrics comes from a measurement of energy shift and possible dipole or quadrupole splitting of the observed resonance on the one hand, and a study of the absorption or Mössbauer fraction f on the other. In the earlier equations we have neglected the presence of the spin of the nucleus and the associated effects of magnetic dipole or electric quadrupole splitting. These splittings can be comparable in magnitude with the linewidth but are always very small compared with phonon frequencies. Allowance for these effects simply leads to a replacement of eqn (12.4.6) by a sum of several terms with energies Δ and possibly linewidths Γ which depend on the spin state of the nucleus. With regard to the dipole and quadrupole splittings, an analysis in terms of crystal symmetry and local dynamics can be carried forward in an analogous fashion to that described in the n.m.r. section, except that now, of course, the relevant nuclear spin quantum number I is that associated with the *excited* Mössbauer nuclear level. The Mössbauer method is therefore particularly useful in cases where the ground nuclear level has no quadrupole splitting but the Mössbauer level has. Such a case occurs in the most widely used Mössbauer isotope ^{57}Fe, for which $I = \frac{1}{2}$ in the ground nuclear state and $I = \frac{3}{2}$ in the Mössbauer level. Iron also occurs in some known ferroelectrics, such as potassium ferrocyanide ($K_4Fe(CN)_6 \cdot 3H_2O$) and ferric ammonium sulphate ($NH_4Fe(SO_4)_2 \cdot 12H_2O$), and is one of the lightest Mössbauer nuclei, making it particularly useful in the ferroelectric context.

Isomer shifts in principle contain information about the chemical bonds which bind the Mössbauer nucleus to neighbouring atoms. Their quantitative interpretation remains difficult, however, primarily because of lack of knowledge concerning detailed wavefunctions for electrons in the crystal-lattice environment and the all too common necessity of approximating the latter with their free-ion or slightly embellished free-ion functional forms. Qualitative inferences can sometimes be drawn from isomer-shift anomalies near transition temperatures in terms of the effects of thermal expansion or electrostriction, but the interpretation is further complicated by the existence of a second contribution to energy shift, namely the second-order Doppler shift, which is also temperature dependent and arises from the change in kinetic energy of the nucleus produced by the relativistic mass increase on absorption of the gamma quantum. The latter can be related in theory to the phonon spectrum and is expected to reflect, for example, a soft-mode anomaly or, away from a transition, to provide a measure of Debye temperature. However, one can rarely separate these different contributions to the centre energy shift with assurance, and interpretations accordingly are most speculative.

At the time of writing the study of ferroelectric transitions using Mössbauer spectroscopy is still very much at the pioneer stage and, with the exception of the straightforward interpretation of quadrupole splittings, progress leans more towards developing Mössbauer theory by interpreting transitions already well studied by other techniques than to obtaining new information pertaining to ferroelectricity itself. One of the primary difficulties is that much of the conventional Mössbauer theory as outlined in the earlier part of this section is carried out in a harmonic approximation, whereas critical dynamics, even in most displacive ferroelectrics, is grossly anharmonic. For example, the critical anomaly occurring in the Mössbauer fraction f is almost certainly a function of the degree of anharmonicity associated with the soft mode responsible for the anomaly.

The Mössbauer fraction, which is essentially the area under the Mössbauer absorption curve, is normally expected to decrease as one approaches a displacive ferroelectric transition from either the high- or low-temperature sides as a result of the enhancement of the mean-square amplitude of local motion $\langle \{\delta \mathbf{r}(0)\}^2 \rangle$ produced by the decrease in frequency of the soft mode. The connection between the Mössbauer fraction f and the temperature-dependent mode responsible for displacement ferroelectricity was first suggested by Muzikar, Janovec, and Dvorak (1963) and observed experimentally by Bhide and Multani (1965) using ^{57}Fe in $BaTiO_3$. Even in the harmonic approximation an exact formal analytic expression for $f = e^{-2W}$ is complicated, particularly in impurity Mössbauer situations when contributions to $\langle \{\delta \mathbf{r}(0)\}^2 \rangle$ from local defect modes must also be included (Maradudin 1964, 1966). It requires in particular a complete knowledge of the normal-mode spectrum since all modes contribute to the local motion to some degree. Away from the transition, where all modes are considered to be quasi-harmonic and 'normal', it has been found that $\langle \{\mathbf{q} \cdot \delta\mathbf{r}(0)\}^2 \rangle$ is fairly independent of host masses and can be well approximated by an expression obtained from a fictitious monoatomic crystal having masses equal to that of the Mössbauer atom and force constants equal to those of the original crystal. In this spirit the non-critical behaviour is often given·in a simple Debye approximation and takes the form

$$f(T) = \exp\left[-\left\{\frac{3E^2}{4Mc^2k\Theta_D}\right\}\left\{1 + 4\left(\frac{T}{\Theta_D}\right)^2 \int_0^{\Theta_D/T} \frac{x\,dx}{e^x - 1}\right\}\right] \quad (12.4.13)$$

where E is the gamma-ray energy, M is the mass of the Mössbauer atom, and Θ_D is the Debye temperature (see, for example, Wertheim 1964). The soft-mode anomaly can then be envisaged as superposed on this Debye background (Fig. 12.5). Its detailed form depends on the character of the soft-mode dispersion near the critical point and is rarely discussed in a quantitative sense. Comparison with experiment, such as that carried out by Bhide and Hegde (1972) for ^{57}Fe in PbTiO$_3$, is usually performed by a crude approximation neglecting dispersion effects. In spite of the less than quantitative interpretation there is convincing qualitative evidence for the existence of mode softening in the characteristic f-anomaly near displacive transitions.

One of the few theoretical discussions of the f-anomaly going beyond harmonic theory in the ferroelectric context has been given by Vaks, Galitskii, and Larkin (1968). Their approach is rather illuminating and proceeds directly from the general eqn (12.4.6). For $|E - \Delta| \lesssim \Gamma$, eqn (12.4.6) corresponds essentially to the Mössbauer (zero-phonon) line and the time integral involved can be considered in a rather general context. For typical displacive systems the soft (but not necessarily harmonic) vibrational motion has a characteristic frequency very large compared with $\Gamma \sim 10^6\,\text{s}^{-1}$ (except at temperatures pathologically close to a second-order transition) and the fluctuations within the angular brackets in eqn (12.4.6) can therefore be averaged before the time integral is performed. Direct integration then yields

$$\sigma_a(E) = \frac{\sigma_0 f(\Gamma/2)^2}{(E - \Delta)^2 + (\Gamma/2)^2} \quad (12.4.14)$$

if $|E - \Delta| \lesssim \Gamma$, where the Mössbauer fraction

$$f = |\langle \exp(i\mathbf{q} \cdot \delta\mathbf{r})\rangle|^2. \quad (12.4.15)$$

Order–disorder motion is in general relaxational:

$$\langle \delta r_\alpha(0)\,\delta r_\beta(t)\rangle = \langle \delta r_\alpha(0)\,\delta r_\beta(0)\rangle \exp\left(-\frac{t}{\tau}\right), \qquad t > 0 \quad (12.4.16)$$

(in which α, β label Cartesian co-ordinates) with characteristic relaxation time τ usually greater than or of order $\hbar/\Gamma$. For simplicity we confine ourselves to small $\mathbf{q}$ for which

$$\langle \exp\{-i\mathbf{q} \cdot \delta\mathbf{r}(0)\} \exp\{i\mathbf{q} \cdot \delta\mathbf{r}(t)\}\rangle$$
$$= 1 + \sum_{\alpha,\beta} q_\alpha q_\beta \{\langle \delta r_\alpha(0) \, \delta r_\beta(t)\rangle - \langle \delta r_\alpha(0) \, \delta r_\beta(0)\rangle\}. \quad (12.4.17)$$

Using eqns (12.4.16) and (12.4.17) directly in eqn (12.4.6) yields a Mössbauer cross-section

$$\sigma_a^0(E) = \frac{\sigma_0 \Gamma}{2}\left\{\frac{(\Gamma/2)}{(E-\Delta)^2+(\Gamma/2)^2} + \sum_{\alpha,\beta} q_\alpha q_\beta \langle \delta r_\alpha(0)\, \delta r_\beta(0)\rangle F\right\} \quad (12.4.18)$$

in which

$$F = \frac{(\Gamma/2)+(\hbar/\tau)}{(E-\Delta)^2+\{(\Gamma/2)+(\hbar/\tau)\}^2} - \frac{(\Gamma/2)}{(E-\Delta)^2+(\Gamma/2)^2}. \quad (12.4.19)$$

For the case of long relaxation times $\Gamma\tau \gg \hbar$, $F \to 0$ and no anomaly near T_c is to be expected. In this case the atom does not have time to jump from one potential well to the other during the gamma absorption time. When $\Gamma\tau \sim \hbar$ the lineshape (eqn (12.4.18)) is a superposition of two lines of different widths, and finally, when $\Gamma\tau \ll \hbar$, the first term in F becomes negligible (since $|E-\Delta| \lesssim \Gamma$ for the zero-phonon line) and the cross-section, as in the displacement formulation, can once again be expressed in terms of the static fluctuations of the Mössbauer atom and may exhibit an anomaly at T_c.

12.4.2. Experimental observations

Experimentally the anticipated minimum in f near displacive ferroelectric transitions has been observed in several cases. A particularly pronounced minimum has been observed for ^{57}Fe in PbTiO$_3$ (Fig. 12.5) and an effort to discuss the theoretical situation for this case has recently been given by Samuel and Sundaram (1974). Work on order–disorder systems, and particularly on ferroelectric potassium ferrocyanide trihydrate (K$_4$Fe(CN)$_6$·3H$_2$O, often abbreviated as KFCT) which contains the ^{57}Fe Mössbauer nucleus naturally, has had a more confused history. Hazony, Earls, and Lefkowitz (1968) first measured an f-anomaly in KFCT and reported it to be anomalous in that the Mössbauer fraction exhibited a maximum near T_c. Efforts to repeat the experiment (Gleason and Walker 1969, Clauser 1970) found no anomaly at all. Finally, however, Montano, Shechter, and Shimony (1971) confirmed the f-maximum, but showed that it was a single-crystal property depending on the relative orientation of gamma-ray $\mathbf{q}$ and the crystal axes and suggested that it probably averaged

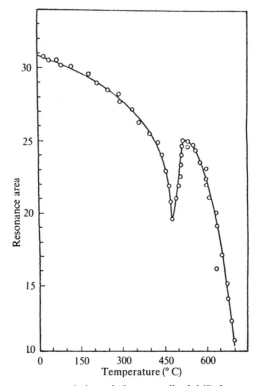

Fig. 12.5. The temperature variation of the normalized Mössbauer resonance area for ferroelectric $PbTiO_3$ showing a well-defined anomaly in the vicinity of T_c. (After Bhide and Hegde 1972.)

to near zero in the powder samples used by Gleason and Walker and by Clauser. KFCT is pseudotetragonal and becomes ferroelectric below $T_c = 248$ K with polar axis in a [101]-direction. The water molecules are in (010)-layers and between each pair of water layers are two layers of $Fe(CN)_6^{4-}$ groups and interspersed potassium ions. It has been suggested that the ferroelectric mechanism in KFCT is an order–disorder transition among the water molecules (Blinc, Brenman, and Waugh 1961) but a detailed understanding is still lacking. Montano *et al.* (1971) find that the Mössbauer line in single-crystal KFCT is slightly quadrupole broadened and that as a result the f-factor is anisotropic. The f-anomaly can now be explained in terms of a gradual reorientation of the electric field gradient (e.f.g.) principal axes through the second-order transition region and has nothing to do with the critical dynamics. Since the f-anomaly is found to be absent when the gamma-ray wavevector $\mathbf{q}$ is parallel to the [010]-direction, it is likely that the e.f.g. reorientation takes place in an (010)-plane and that, in accord with (12.4.19), relaxation times are long.

Work conducted to date on KDP systems is sparse and confusing. Thus, while Brunstein *et al.* (1970) (studying ^{57}Fe in KDP itself) find direct evidence in the Mössbauer line*shape* for the presence, close to T_c, of tunnel-mode sidebands with frequencies of the order of a few MHz, Sastry (1972), conducting the equivalent experiment in isomorphic (but antiferroelectric) ADP, could find no evidence at all that the Mössbauer nucleus was taking part in the critical dynamics. It is possible that ^{57}Fe does not go into equivalent sites in the two isomorphs. Anomalies similar to those discussed for ferroelectrics are of course generally expected for soft-mode antiferroelectrics as well, although the lesser role played by the long-range dipole forces in antiferroelectrics might affect the details. The best-documented Mössbauer study of an antiferroelectric is probably that for $PbZrO_3$. Jain *et al.* (1970) used the Mössbauer isotope ^{119}Sn as an impurity in $PbZrO_3$, and observed a very pronounced minimum in the Mössbauer fraction near T_c together with a sharp change in isomer shift and the expected collapse of the quadrupole splitting on approach to the cubic paraelectric phase. These general results were confirmed by Canner *et al.* (1971) who compared antiferroelectric $PbZrO_3$ with ferroelectric $PbTi_{0.2}Zr_{0.8}O_3$. Neither the isomer shift nor the *f*-anomaly seemed sensitive to the change from antiferroelectric to ferroelectric. In this case, however, the Mössbauer nucleus was ^{57}Fe, which was introduced in both materials as 5 mol per cent $BiFeO_3$. Such a large impurity concentration with random distributions of Pb^{2+} and Bi^{3+} ions on one site and Ti^{4+}, Zr^{4+}, and Fe^{3+} on another locally polarizes the lattice even in the paraelectric phase and results in a remanent e.f.g. and quadrupole splitting above the transition temperature.

Some efforts to deduce the nature of the chemical bonds from isomer shifts and quadrupole splittings have been made, but accuracy is difficult to obtain. Some functional relation should certainly exist between isomer shift and the degree of ionicity of the bonds for a particular nucleus in a sequence of compounds. If such can be established for a series of simple salts for which an ionicity scale is already known, it is then possible to relate a given isomer shift to this curve and to make semi-quantitative estimates of ionicity from this scale. Bhide and Multani (1965) use such a curve, plotted for a series of ferrous and ferric salts, to estimate an ionicity of 60 ± 20 per cent for $BaTiO_3$ from the isomer shift. More frequently estimates of ionicity are made using the quadrupole splitting by comparing the measured electric field gradient with that calculated using a simple point-charge model of local environment. We have already alluded to such a calculation performed for $LiNbO_3$ and $LiTaO_3$ in the n.m.r. section. Again the theory is extremely crude but a surprising degree of self-consistency can be obtained with estimates of ionicity from other sources. Thus an e.f.g. point-charge calculation for $BaTiO_3$ (Bhide and Multani 1965) can account very well for the Mössbauer quadrupole splitting if effective charges are reduced to some 60 per cent of their formal valence values. This agrees not only with the estimate from isomer shift but also with that of Axe (1967) from an interpretation of the mode strengths of polar vibrations. We note (Axe 1967) that these reduced charges refer to the static situation. Dynamic effective charges are quite different (being in general larger than the formal valence values) and reflect the fact that ion-core and shell charges contribute to static and dynamic properties in different ways (i.e. the core and shell are not rigidly coupled). It therefore requires at the very least two effective charges (e.g. core charge and shell charge) in order to describe static and dynamic properties in an approximately self-consistent fashion. More accurately, charge transfer between cations and anions during motion should also be recognized. A recent analysis of the whole concept of effective charges in ionic crystals has been given by Goyal, Agarwal, and Verma (1974).

12.5. Optical spectroscopy

Optical and electron paramagnetic resonance techniques provide complementary information concerning the electronic structure and local environment of impurities and defects in ferroelectric host crystals and consequently defects are excellent probes of the site symmetries and crystal fields within a ferroelectric, provided they do not greatly perturb the host crystal. In addition optical techniques have proved particularly useful for the study of pure ferroelectric materials which are not paramagnetic. For instance a number of optical experiments, including ultraviolet reflectivity (Cardona 1965, Kurtz 1966), electroabsorption, and electroreflectance (Frova and Boddy 1967), have been aimed at relating the optical properties of ferroelectrics to the electronic band structure and the electrical conductivity and at showing how the spontaneous polarization and the ferroelectric phase transition affect the optical and electro-optical properties. Even though a discussion of the electronic band structure may seem out of place in a chapter which is otherwise devoted to studies of localized states and local crystal symmetry, this section on electronic spectroscopy seems nevertheless the most appropriate place for this topic.

Optical studies have also been directed toward understanding how optical excitation of ferroelectrics affects the polarization. In addition to the pyroelectric effect which arises from the excitation of vibrational modes of the lattice a similar polarization change arises from electronic excitation of polar materials.

Much of the motivation for optical studies of ferroelectrics has been the large number of optical applications to which ferroelectrics seem well suited. Most of these applications are discussed in Chapter 16, except for non-linear optics which is the subject of Chapter 13 and ferroelectric lasers which will be briefly discussed in this section.

12.5.1. Electronic band structure

The majority of optical studies of the electronic band structure of ferroelectrics have been concerned with perovskite-type oxides ABO_3 because of the simplification arising from the simple cubic structure of the paraelectric phase. Detailed calculations of the band structure have been made for cubic $SrTiO_3$ (Kahn and Leyendecker 1964) using a linear combination of atomic orbitals (LCAO) as the basis functions. The valence band is found to be derived primarily from oxygen 2p-orbitals while the lowest conduction band is derived primarily from Ti (B cation) d-orbitals. This calculation finds that the conduction band is many valleyed with minima along $\langle 100 \rangle$ directions at the Brillouin zone boundary while the valence band is relatively flat. These predictions for $SrTiO_3$ are also expected to apply for the other perovskites such as $BaTiO_3$ and $KTaO_3$. However, a subsequent calculation (Mattheiss 1972) using the augmented plane-wave method has been combined with the LCAO scheme to determine the band structure of several cubic perovskites. These calculations predict that the conduction-band minima for $SrTiO_3$ and $KTaO_3$ consist of warped bands at the zone centre rather than many valleys near the Brillouin zone boundary as Kahn and Leyendecker proposed. These calculations do suggest that the difference between the direct gap at the zone centre and the indirect gaps is only about $0 \cdot 1$ eV and is very sensitive to small improvements of the theory.

The ferroelectric transition from the cubic to tetragonal phases in the ferroelectric perovskites causes a splitting and shift of the critical points of the cubic band structure. The modified band structure based on the $SrTiO_3$ calculations of Kahn and Leyendecker has been predicted by Brews (1967) for $BaTiO_3$ from the change of overlap of the cation and oxygen orbitals due to the spontaneous polarization. In particular the calculations predict that the upper valence band is not greatly affected by the polarization, yet the conduction-band minima at the zone boundary perpendicular to the polarization are shifted to higher energies with respect to the valleys parallel to the polarization.

These calculations are found to predict correctly the experimental observations of the band splitting in $BaTiO_3$ and $KTa_x Nb_{1-x} O_3$ (KTN) observed by polarized optical absorption on passing from the cubic to tetragonal phase. As shown for $BaTiO_3$ in Fig. 12.6 the absorption edge for light polarized parallel to the polar axis (π polarization) is shifted to higher energies than the perpendicular polarization (σ polarization) (Wemple 1970). These polarizations are consistent with the selection rules for interband transitions determined by Brews.

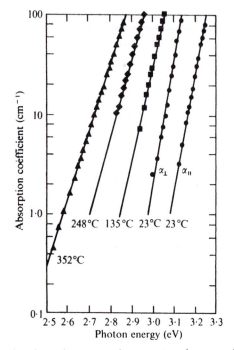

Fig. 12.6. Representative absorption *versus* photon-energy data near the interband edge in BaTiO₃. Below the Curie point at 132°C the two edges correspond to light polarized parallel ($\alpha_{\parallel}$) and perpendicular ($\alpha_{\perp}$) to the tetragonal (polar) c-axis. (After Wemple 1970.)

At the cubic–tetragonal transition in SrTiO₃, on the other hand, the observed optical dichroism is considerably smaller (only 3 meV) and has the opposite sign ($\alpha_{\parallel} > \alpha_{\perp}$) to BaTiO₃ (Müller, Capizzi, and Gränicher 1970, Capizzi and Frova 1971). This splitting is considerably smaller than the 90 meV splitting predicted by Mattheiss (1972) for SrTiO₃ and suggests that the origin of the dichroism is more complicated in SrTiO₃ than that in BaTiO₃. Magnetoresistance and other experimental measurements (Frederikse *et al.* 1967, Frederikse 1969) also support the predictions of a many-valleyed conduction band with zone-boundary minima. However, these experimental data do not rule out the possibility of the model involving warped bands at the zone centre predicted by Mattheiss.

For the perovskite oxides the electronic energy gap is around 3–4 eV which is a convenient region for optical measurements. However, there are experimental complications in determining the exact position of the band edges, because in all the perovskites studied the absorption edges have tails which broaden the edge by several tenths of an eV up to absorption coefficients α of at least 10^3 cm^{-1} (DiDomenico and Wemple 1968). These tails typically have an exponential form normally referred to as Urbach tails

(Urbach 1953), i.e.

$$\alpha = \alpha_0 \exp(\gamma h\nu) \tag{12.5.1}$$

where $h\nu$ is the photon energy. The parameter γ is found to have a temperature dependence of the form

$$\gamma \propto (T - T') \tag{12.5.2}$$

where T' is just a constant obtained by fitting the absorption curve and its physical significance has not been identified. At the present time there is no completely satisfactory explanation for the Urbach tail although it has been well established that both defects (Redfield 1963) and electron–phonon interaction (Mahan 1966) can give rise to broadening of the edge. Wemple (1970) found that the temperature dependence of the Urbach tail could be described by eqn (12.5.1) and (12.5.2) all the way through the ferroelectric phase transition in $BaTiO_3$, so that if electron–phonon interaction played the dominant role in the edge broadening then then the soft transverse optic mode did not make a significant contribution.

For many experiments the actual value of the band gap is not important provided the gap is consistently identified with a particular point on the absorption edge. For instance the entire absorption edge is found to shift in a rigid manner toward higher energies in an applied electric field. Indeed, it has been determined by optical absorption (Harbeke 1963, Gähwiller 1965), electroreflectance, and electro-absorption (Frova and Boddy 1967, Gähwiller 1967) techniques that in several ferroelectrics the field-induced shift is both much larger and opposite in sign to the usual Franz–Keldysh (Franz 1958) effect observed in other insulators and semiconductors. In the paraelectric phase the shift ΔE_g varies quadratically with applied field E and in the ferroelectric phase the shift is linear. There is now considerable evidence for many materials both above and below the Curie temperature that

$$\Delta E_g = \beta \, \Delta P^2 \tag{12.5.3}$$

where below the Curie temperature P includes both the spontaneous and field-induced polarization, i.e. below T_c

$$\Delta E_g = 2\beta P_s \, \Delta P = 2\beta P_s \varepsilon\varepsilon_0 E. \tag{12.5.4}$$

The constant β is a polarization potential analogous to the more familiar deformation potential but with the polarization dyadic $P_i P_j$ replacing the strain tensor x_{ij} (Zook and Casselman 1966). The value of β will be different for each light polarization and for each interband transition. In a general way the energy shift of the nth band gap at the point $\mathbf{q}$ of the Brillouin zone can be described by

$$\Delta E_g^n(\mathbf{q}) = \sum_{ij} \sigma_{ij}^n(\mathbf{q}) P_i P_j \tag{12.5.5}$$

where the σ_{ij} are the appropriate polarization potentials. The constant β defined by eqn (12.5.3) is thus an average of the σ_{ij} coefficients weighted by the appropriate selection rules for each transition of interest and each light polarization (DiDomenico and Wemple 1969). At the optical absorption edge of BaTiO$_3$ Wemple (1970) found that for π-polarized light and the electric field applied along the polar axis

$$\beta_{11} = 1 \cdot 16 \pm 0 \cdot 05 \, \text{eV m}^4 \, \text{C}^{-2}$$

and is independent of temperature. This value of β is consistent with theoretical estimates based on changes of atomic overlap (Brews 1967, Zook and Casselman 1966) and is typical of other perovskite-type oxides. The significance of this will be discussed in the next chapter.

It would be expected from these considerations alone that the temperature dependence of the separation of the two absorption edges observed in π- and σ-polarized light could be predicted from the temperature dependence of the spontaneous polarization P_s and the polarization potentials. It is seen from Fig. 12.7, however, that the temperature dependence of the

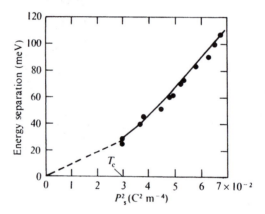

Fig. 12.7. Dependence of band-edge separation in the tetragonal phase of BaTiO$_3$ on the square of the spontaneous polarization P_s^2. The indicated lines serve only to accentuate the non-linear behaviour. (After Wemple 1970.)

band-edge separation in BaTiO$_3$ is not proportional to P_s^2. The additional temperature dependence of the absorption edges has been attributed to spontaneous polarization fluctuations $\langle(\Delta P)^2\rangle = \langle P^2\rangle - P_s^2$ by Wemple (1970). Such fluctuations would contribute to the band-edge shift but would not of course be apparent in static polarization measurements, i.e.

$$\Delta E_g = \beta P_s^2 + \langle \beta_c \rangle \langle (\Delta P)^2 \rangle \tag{12.5.6}$$

where the polarization potential $\langle \beta_c \rangle$ is averaged over the wavevector dependence of the interaction.

An estimate of spatial extent of these fluctuations may be obtained from the classical fluctuation–dissipation theorem (11.4.9) in the form

$$kT\varepsilon = \int [\langle P(0)P(\mathbf{R}) \rangle - P_s^2] d^3\mathbf{R} = \langle (\Delta P)^2 \rangle V_c \qquad (12.5.7)$$

which now defines a 'correlation volume' V_c. The additional term in eqn (12.5.6) can be sufficiently great to account for the experimental results of Fig. 12.7 if $\beta \sim \langle \beta_c \rangle$ and $V_c \sim 10$–10^2 nm^3. The size of this correlation volume and its non-critical behaviour at T_c is consistent with other observations of polarization fluctuations.

12.5.2. Impurity spectra

The spectra of impurities often provide more information about a host crystal than the spectrum of the host material alone. There are several types of impurity ion or other defects which introduce absorption in the spectral region where the host crystal is transparent. The most commonly studied defects are probably the transition–metal ions which have partially filled inner d-shells in the ground state, and the lanthanide and actinide elements which have partially filled f-shells. Absorption in the visible region of the spectrum arises from transitions within the d- or f-shell configurations. Because of the shielding of these orbitals by the outer-shell electrons the states can be described quite accurately in terms of the states of the free ion which are modified by the crystal field at the impurity site due to the sourrounding host ions. Even though d–d or f–f transitions violate the selection rule $\Delta L = \pm 1$ for electric dipole transitions of the free ions, in the crystalline environment they become allowed owing either to a mixing with electron states of opposite parity (such as ligand orbitals) via odd terms in the crystal field or to vibrational–electronic (vibronic) interactions. In addition to these relatively weak transtions more intense charge-transfer transitions are frequently observed in the optical spectrum owing to transfer of an electron to the defect site from a neighbouring host ion or from the defect to a host ion.

As probes of the crystal field and site symmetry the transtition–metal ions are the most useful. This group of ions has been studied in great detail in a wide variety of environments and their behaviour is quite well understood. There are several comprehensive reviews on this subject, such as Ballhausen (1962), for non-ferroelectric materials. Since most of these studies have been involved with an octahedral environment of the impurity ions the existing theories can be applied in a straightforward manner to the oxygen-octahedra ferroelectrics. For instance, for peroviskites in the paraelectric phase the optical spectrum is determined entirely by the splitting of the

free-ion states by the cubic crystal field. In the ferroelectric phase the
spontaneous polarization introduces additional contributions to the crystal
field at the impurity site which can result in a shift or a splitting of the spectral
lines or a change of the intensity, depending on the states involved and the
optical selection rules in the modified environment.

A favourite impurity probe has been the Cr^{3+} ion because the sharp
spectral line (the 'R' line) which appears near 700 nm in an octahedral
environment is very sensitive to small changes of crystal symmetry. A small
tetragonal or trigonal distortion of the octahedral site splits the line into a
doublet, the magnitude of the splitting providing a measure of the distortion.
The temperature dependence of the R-line splitting in $SrTiO_3$ is shown in
Fig. 12.8. Above ≈ 107 K the lines are degenerate in the cubic environment

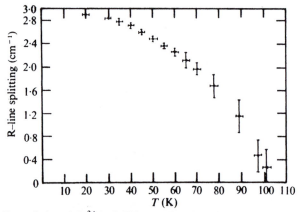

Fig. 12.8. R-line splitting of Cr^{3+} in $SrTiO_3$ as a function of temperature (points with error
bars). (After Stokowski and Schawlow 1969.)

while below this (anti-distortive) temperature the intensities of the two
components observed in polarized light are characteristic of a tetragonal
distortion. In fact these lines have been used as sensitive indicators of the
domain structure of the host crystal (Stowkowski and Schawlow 1969,
Chang 1972). The variation of the R-line frequencies with temperature and
applied electric field in $SrTiO_3$ shows very unusual behaviour. The thermal
shift cannot be explained by the usual interaction with acoustic lattice modes
at low temperatures, the dominant interaction apparently being with the
(ferroelectric) soft transverse optic mode. As a result the R-line shift has a
similar behaviour to the reciprocal dielectric constant, i.e. $\Delta\nu \propto 1/\varepsilon(T)$.
Upon application of an electric field E along the tetragonal axis the R-line
splitting increases in proportion to the square of the induced polarization:

$$\Delta\nu \propto \left\{\frac{\varepsilon(0, T)}{\varepsilon(E, T)} - 1\right\} = AP^2. \qquad (12.5.8)$$

This result would indicate that the frequency of the R-lines depends on the square of the ionic displacements as would be expected from simple perturbation theory.

The influence of polarization fluctuations $\langle (\Delta P)^2 \rangle$ on the R-line width has been observed in methylammonium chrome alum (Morita, Ono, and Murata 1974); an anomalous maximum of the R-line luminescence occurs at the order–disorder ferroelectric phase transition (164 K) which cannot be described by conventional theories of Raman scattering.

Vibronic sidebands of sharp electronic transitions are often observed in the optical spectrum owing to electron–phonon interaction at the impurity-ion site. Phonon modes of all wavevectors contribute to the vibronic structure and symmetry selection rules are much less restrictive as a result than in pure-crystal infrared absorption or Raman scattering experiments. The vibronic structure may thus reflect the phonon spectrum of the host. However, vibronic spectroscopy requires the presence of an impurity and this can modify the vibrational spectrum of the pure host modes or cause the appearance of new local modes. The resulting spectra tend to be rather complicated but nevertheless some progress has been made in their interpretation. For example Kim, Powell, and Wilson (1974) have succeeded in associating most of the intense peaks appearing at the high-energy side of the R-line of Cr^{3+}-doped $SrTiO_3$ with specific low-frequency lattice modes at different points in the Brillouin zone. Unfortunately the temperature-dependent broadening and shift of the more intense R-lines discussed above interferes with an accurate determination of the temperature dependence of the low-lying vibrational sidebands and prevents a precise measurement of mode softening on approach to a phase transtition. Mode softening was evident in the vibronic sidebands of Eu^{3+} lines near 590 nm and 610 nm in $BaTiO_3$ (Weber and Schaufele 1965). In this case the no-phonon line disappears in the cubic phase, yet the vibronic structure remains. The temperature dependence of the sidebands observed with low resolution was found to be roughly consistent with the temperature dependence of the transverse optic mode near $\mathbf{q} = 0$.

Crystal-field theory has also been applied with some success to the boracites in which the transition-metal ions are part of the host-crystal structure and not present as impurities (Pisarev et al. 1969). In the boracites $CoB_7O_{13}X$ and $NiB_7O_{13}X$ where X = Cl, Br, I the transition-metal ions are in sites of low symmetry and the spectra are much more complicated than the known spectra of these ions in an octahedral environment. In spite of this theoretical predictions of the changes of the crystal-field parameters at the Curie temperature, based on a point-charge calculation, were in surprisingly good agreement with experimental observations. With these materials point-charge calculations have the advantage that the positions of the transition-metal ions in the lattice are accurately known from structural determinations. When using impurity-ion probes quantitative predictions of this kind based on point-charge calculations can be quite erroneous if the impurity position in the unit cell is not known exactly. Slight displacements from the host-ion position (for which the impurity substitutes) can greatly affect the calculations.

The 4f-electrons of rare-earth ions are more effectively shielded by the outer electrons than the transition-metal ions and the crystal field has much less influence on the energy levels than spin–orbit coupling. Nevertheless, the spectral lines are generally much sharper as a result so that splittings of the lines due to changes in the crystal field can still be detected. Sm^{3+}-doped $BaTiO_3$ provides an interesting example of a case in which the determination of the host symmetry using impurity-ion probes can be misleading (Makishima et al. 1965). As the temperature is lowered the following changes are observed: (a) at the cubic–tetragonal transition a small splitting of several lines is observed, but no frequency shifts, (b) the spectrum in the orthorhombic phase is identical with that of the tetragonal phase, and (c) the spectrum of the rhombohedral phase again consists of singlets identical to the cubic phase. This unexpected behaviour shows that in the rhombohedral phase the spontaneous polarization has no effect on the spectrum. Similar quasi-cubic behaviour in this phase was noticed during an e.s.r. study of Fe^{3+}-doped $BaTiO_3$ (Sakudo 1963). The small effects of the polarization on the Sm^{3+} spectrum in the tetragonal and orthorhombic phases were attributed to the much larger size of the Sm^{3+} ion than the Ti^{4+} ion for which it substitutes.

12.5.3. Colour centres and polarons

Optical absorption by defects other than impurities can also give rise to intense colouration of crystals. However, optical spectroscopy of colour centres has not received a great deal of attention in ferroelectrics, possibly because the spectra are generally rather broad and relatively insensitive to subtle changes of crystal field. Frequently the appearance of colour centres is associated with increased electrical conductivity. For example the deep blue colouration of $BaTiO_3$ and other perovskite oxides which follows heat treatment in a reducing atmosphere is related to their semiconducting properties. The colouration is due to an intense absorption band peaking near 0.6 eV which extends into the visible part of the spectrum and two other weaker peaks at 2.1 eV and 2.6 eV. The origin of the 0.6 eV band has been the subject of several investigations. The intensity of the absorption is directly related to the concentration of the (n-type) charge carriers regardless of whether the conductivity is due to reduction or to the presence of impurities. The interest in this absorption has been to determine whether it is due to small polarons—in which case it would permit identification of the nature of the conduction mechanism in these crystals—or whether it is due to conventional deep localized states of a defect superimposed on a background of free-carrier absorption according to Drude theory. A band showing similar behaviour is observed in $SrTiO_3$ near $0\cdot2$ eV. In $SrTiO_3$ the shape of the band is in reasonable agreement (Barker 1966 b) with small-polaron theory which predicts (Reik 1963) a frequency dependence of absorption coefficient

$$\alpha(\omega) = \frac{\sigma_0}{nc\varepsilon_0} \frac{\sin(\hbar\omega/2kT)}{\hbar\omega/2kT} \{\exp(-\omega^2\tau^2)\} \qquad (12.5.9)$$

where n is the refractive index, σ_0 is the real part of the low-frequency conductivity, and τ is a scattering time which depends on the phonon frequencies coupling with the electrons. However, the temperature dependence is not in good agreement with this theory. In $BaTiO_3$ the high-energy tail of $\alpha(\omega)$ has a power-law dependence similar to free-carrier absorption (Baer 1966) rather than the exponential dependence required by eqn (12.5.9). The band intensity is consistent with that expected from states associated with colour centres such as oxygen vacancies—for instance a model of an oxygen vacancy with a first ionization energy of $0\cdot025$ eV and a second ionization energy of $0\cdot2$–$0\cdot3$ eV could account for the experimental data (Berglund and Braun 1967). At room temperature all oxygen vacancies would be ionized giving a linear correlation between conductivity and absorption. On the other hand, the bands at $2\cdot1$ eV and $2\cdot6$ eV have been attributed to singly ionized oxygen vacancies (Coufova and Arend 1961).

Electrical transport measurements indicate that the electron mean free path is less than a lattice constant, which would favour small polarons, yet the measurements can equally well be interpreted in terms of a band model without invoking polaron theory (Berglund and Braun 1967). Reflectivity data (Gerthsen et al. 1965) and thermo-e.m.f. measurements and the temperature dependence of the carrier mobility (Bursian et al. 1972) also point to small-polaron conductivity, but at present there is no conclusive evidence for either model and the problem remains unresolved.

Occasionally the colour change of a crystal which has been doped with impurities of a different valence from the host ions may not be due to transitions involving the impurity but to defects which compensate for the extra charge on the impurity. In the previous model explaining the conduction of $BaTiO_3$ in terms of oxygen vacancies these vacancies can be introduced as the charge-compensating defects of trivalent dopants such as La^{3+} ions. Similarly, doping $SrTiO_3$ with Ce^{2+} or Nb^{4+} impurities, which presumably substitute for Sr^{2+} and Ti^{4+} ions respectively, both give rise to absorption bands which also appear in hydrogen-reduced, but undoped, $SrTiO_3$. Presumably the same negatively charged, but unidentified, defect is responsible for the absorption in each of these cases (Rozhdestvenskaya et al. 1970).

12.5.4. Stimulated emission

Additional motivation for the optical study of impurites has been provided by the practical benefits that would be afforded by a ferroelectric laser host. The large optical non-linearities of ferroelectrics would allow second-harmonic generation, parametric oscillation, and modulation within the laser medium itself. Although coherent emission has been observed from several rare-earth ions in ferroelectric hosts (Kaminskii 1972) no practical system has been realized at the present time. The first report (Borchardt and Bierstedt 1966) of stimulated emission from Nd^{3+} ions in $Gd_2(MoO_4)_3$ required an extremely high threshold energy of 940 J with flash-lamp pumping at room temperature. Improved performance was later realized with Nd^{3+}-doped $LiNbO_3$ (Evlanova *et al.* 1967) emitting at $1\cdot0845$ μm in π polarization and $1\cdot0932$ μm in σ polarization owing to transitions between states of the 4f manifold. The only report of frequency multiplication within the laser medium itself is with Tm^{3+}-doped $LiNbO_3$ (Johnson and Ballman 1969). The fundamental frequency at $1\cdot8532$ μm is an ordinary wave and phase matching with the second harmonic at $0\cdot9266$ μm was achieved. However, the birefringence and dispersion of $LiNbO_3$ are such that the phase-matching angle lies at 43° to the polar axis so double refraction resulted in a very short coherence length. A conversion efficiency of about 10^{-6} was achieved.

For efficient lasing crystals of high optical quality are necessary. For instance a dramatic lowering of the threshold energy for lasing in $Gd_2(MoO_4)_3$:Nd was attributed to the use of single-domain crystals Bagdasarov *et al.* 1971). Optically induced changes of refractive index (see § 5.8) due to the intense pump light have proved to be a serious problem for $LiNbO_3$ lasers. The index changes upset the laser cavity and increase the threshold for operation, particularly for π-polarized emission (Kaminskii 1972). Even using a low-wavelength pump of 6 kW cm^{-2} at $0\cdot7525$ μm from a krypton laser, laser operation ceased after about 20 s. It should be pointed out that even if these cumulative optical-index changes could be eliminated the transient polarization changes which accompany optical excitation of ferroelectrics may still limit efficient laser operation.

12.6. Excited-state polarization

Just as the excitation of vibrational modes of a spontaneously polarized material gives rise to a pyroelectric polarization, so does electronic excitation of such a material give rise to a macroscopic polarization change. The acentric character of the host polarizes an impurity ion or defect producing an electric dipole moment which will in general be different in the ground and excited states. In a pyroelectric host this change of moment $\Delta\mu$ has the same sense for all identical impurities so that the resulting polarization change is

$$\Delta P = n \Delta \mu \qquad (12.6.1)$$

where n is the density of impurities in the excited state.

The origin of this polarization may be clarified by considering the simplest case of a one-dimensional asymmetric potential—the triangular potential

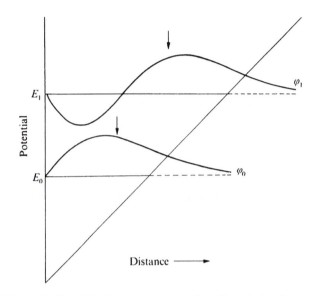

Fig. 12.9. Schematic plot of the lowest two eigenvalues E_i and eigenfunctions φ_i of the Schrödinger equation (12.6.2) for a particle in a triangular potential well.

well shown in Fig. 12.9. By solving Schrödinger's equation for this potential, viz.

$$-\left(\frac{\hbar^2}{2m}\right)\frac{d^2\varphi}{dx^2} = (E + Fx)\varphi \qquad (12.6.2)$$

where F is the electric field describing the sloping wall of the well, the dipole moment associated with each state can be evaluated. The solutions of eqn (12.6.2) for eigenvalues $E = E_i$ are (Gol'dman et al. 1960)

$$\varphi_i = \left(\frac{2mF}{\hbar^2}\right)^{1/3}\left(\frac{1}{\pi F}\right)^{1/2} \mathrm{Ai}(-\gamma_i) \qquad (12.6.3)$$

where the $\mathrm{Ai}(-\gamma_i)$ are the Airy functions of

$$\gamma_i = \left(x + \frac{E_i}{F}\right)\left(\frac{2mF}{\hbar^2}\right)^{1/3} \qquad (12.6.4)$$

Since the left-hand wall of the well is vertical the eigenstates are determined by the condition that $Ai(-\gamma_i) = 0$ at $x = 0$. The eigenfunctions are shown in Fig. 12.9 for the ground and first excited states of the potential. The centres of charge in the two states

$$\bar{x}_i = \frac{\langle \varphi_i | x | \varphi_i \rangle}{\langle \varphi_i | \varphi_i \rangle} \tag{12.6.5}$$

are shown by arrows in the figure. It is clear that electronic excitation to the first excited state gives a change of dipole moment

$$\Delta \mu = e(\bar{x}_1 - \bar{x}_0). \tag{12.6.6}$$

For instance $\Delta \mu \geq 9$ Debye ($\sim 3 \times 10^{-29}$ cm) when $F = 10^8 V\ cm^{-1}$.

It must be emphasized that although there will be a change of dipole moment at any asymmetric potential a macroscopic polarization change is only observed in pyroelectric host crystals. In acentric non-pyroelectric hosts the net dipole moment change per unit volume averages to zero.

In a real crystal the states are considerably more complicated than an isolated set of one-electron states such as Fig. 12.9, and the magnitude of $\Delta \mu$ depends not only on the local field F and the electronic configurations but also on the vibronic interaction and the dipole-induced polarization of the surrounding crystal.

Experimental demonstration of such an electronic polarization change was first made with $LiNbO_3$ and $LiTaO_3$ crystals doped with Cr^{3+} ion impurities which introduce absorption where the host crystals are transparent (Glass and Auston 1972). The Cr^{3+} ions were excited with short optical pulses from the ground $^4A_2$ state to the localized $^4T_2$ state of the $3d^3$ electronic configuration. The macroscopic polarization change ΔP was measured as a function of time by integrating the charge which develops on the crystal faces normal to the polar axis, using the parallel-plate electrode geometry of Fig. 12.10 (a). The duration of the incident pulse $I(t)$ was short compared with the relaxation time τ ($\geq 1\ \mu s$) of the $^4T_2$ state, this relaxation being monitored by simultaneous detection of the $^4T_2$–$^4A_2$ luminescence (Glass 1969 b). Typical results are shown in Fig. 12.10 (b). The polarization change ΔP showed an initial rapid increase due to the change of electronic dipole moment (in addition to some rapid vibronic relaxation in the excited state), followed by a decrease with the electronic relaxation time due to the loss of excited-state polarization. In the absence of any vibrational relaxation ΔP would become zero again when all the Cr^{3+} ions relaxed to their ground state. However, non-radiative vibrational relaxation gives rise to a heating of the host-crystal lattice resulting in a pyroelectric polarization ΔP_∞ which decreases with the thermal relaxation time of the crystal on a much longer time scale (milliseconds).

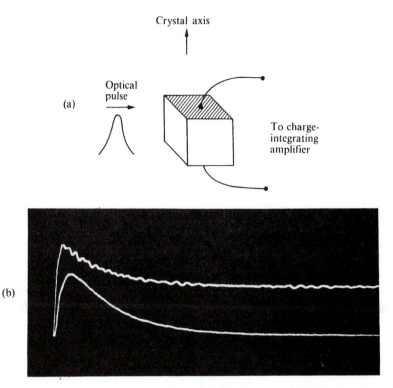

Fig. 12.10. (a) Sketch of the experimental geometry used for measurement of excited-state polarization and (b) the experimental results (oscilloscope traces) showing optical emission (lower trace) and polarization change (upper trace) of LiNbO$_3$: Cr at 173 K. The vertical scale in (b) is arbitrary but the horizontal scale is determined as 5μ s/division. The noise on the upper trace is piezoelectric ringing due to thickness resonances of the crystal. (After Glass and Auston 1972.)

For a two-level system with a relaxation time τ the total polarization change, neglecting piezoelectric effects, is given by the equation

$$\frac{dP(x, t)}{dt} + \frac{1}{\tau} P(x, t) = \frac{dn(x, t)}{dt} \Delta\mu + \frac{p}{\tau} \Delta T(x, t) \qquad (12.6.7)$$

where the first and second terms on the right-hand side are the excited-state dipole and pyroelectric polarization respectively. For uniform illumination of the entire crystal volume this equation can be written

$$\frac{dP(t)}{dt} + \frac{1}{\tau} P(t) = \frac{\alpha\Delta\mu}{h\nu} I(t) + \alpha(1-e)\frac{p}{c_p}\frac{1}{\tau}\int I(t)\,dt \qquad (12.6.3)$$

provided the electronic relaxation time is much shorter than the thermal relaxation of the crystal with its surroundings. The quantities α and e are the absorption coefficient and radiative quantum efficiency respectively and can be obtained by conventional spectroscopy in a separate experiment. The pyroelectric coefficient p and the specific heat c_p are characteristic of the host material and $h\nu$ is the incident photon energy. The extension of eqn (12.6.8) to include more than two states is straightforward. Analysis of the experimental data for chromium-doped $LiNbO_3$ and $LiTaO_3$ gave $\Delta\mu \sim 3$ Debye and 5 Debye respectively for the $^4A_2$-$^4T_2$ transition. The d-orbitals of the Cr^{3+} ion are thus greatly affected by the spontaneous polarization of the host, the ground- and excited-state orbitals being affected in different ways.

In contrast to the d-orbitals of the chromium ions, the f-shell of Nd^{3+} ions in $LiNbO_3$ is only weakly affected by the spontaneous polarization because the orbitals are well shielded by the outer electrons and are relatively insensitive to the local fields of the host material. Only the pyroelectric polarization could be measured in this material because of the small value of $\Delta\mu$.

Eq. (12.6.8) also describes the average polarization change of the crystal volume resulting from non-uniform illumination of the crystal, which would be measured with the experimental arrangement of Fig. 12.10 (a). However, the local polarization change $P(x, t)$ at point x in the crystal will depend on the thermal diffusivity κ_0 of the crystal and the incident intensity pattern. Consider a single Fourier component of the intensity pattern of spatial frequency K $(=2\pi q)$

$$I(x, t) = I(t) \sin(Kx). \qquad (12.6.9)$$

The excited-state polarization for localized excitation reproduces this intensity pattern, but the pyroelectric polarization diffuses to the darker regions of the crystal with a relaxation time $\tau_t = (\kappa_0 K^2)^{-1}$ characteristic of a sinusoidal temperature distribution (Carslaw and Jaeger 1959). In the limiting case $\tau_t \gg \tau$ then eqn (12.6.8) is unchanged by thermal diffusion during electronic relaxation, while if $\tau_t \ll \tau$ then the pyroelectric polarization is uniform throughout the crystal during electronic relaxation. In this latter case the electronic and thermal contributions to the polarization may be determined independently by optical scattering techniques which are sensitive to the spatial frequency of the polarization inhomogeneities (see § 12.8).

It is clear that the excited-state dipole and pyroelectric effects cover complementary time scales, and at least in transition-metal-ion-doped $LiNbO_3$ and $LiTaO_3$ can have comparable magnitudes. Since the response time of the electronic polarization change is determined only be an electronic transition rate, the effect has practical significance for fast optical applications such as the detection of ultra-short optical pulses, optical logic, or the generation of far-infrared radiation. However, from the fundamental point of view the electronic polarization change provides an additional handle for the study of excited-state dipole moments of atoms and molecules and local fields in pyroelectric materials.

12.7. Photovoltaic effects

The polarization change due to localized electronic excitation which we have discussed in the preceding section produces a transient displacement current $J = dP/dt$ but no steady-state current since there is no continuous motion of electronic charge in one direction. With constant illumination of a crystal under open-circuit conditions the depolarization field $E = -\Delta P/\varepsilon_0 \varepsilon$ along the polar axis relaxes with the dielectric relaxation time of the material.

We shall now consider the steady-state electromotive force (e.m.f.) which may be generated when the incident radiation is of sufficiently high energy to create free carriers which move preferentially in one direction in the absence of an external field.

Conventional photovoltaic effects in non-polar materials fall into two general categories. One is the Dember effect whereby the photovoltage develops between the front and back surfaces of a crystal when the front face is illuminated with light of energy greater than the electronic band gap. The light is absorbed as it penetrates the crystal resulting in a gradient of the free-carrier concentrations (both electrons and holes) through the crystal thickness. The free carriers diffuse in this gradient toward the back face of the crystal, the e.m.f. resulting from the different mobilities of the electrons and holes. The second class of photovoltaic effect arises with homogeneous absorption in materials containing macroscopic inhomogeneities which cause band bending—for instance excess free carriers created by the light in a semiconductor close to a p–n junction move in one direction under the influence of space-charge fields in the junction region.

In all of these cases the magnitude of the photovoltage across a single element cannot exceed the value of the electronic energy gap. In pyroelectric materials the spontaneous polarization gives rise to new photovoltaic effects which are not observed in materials with a centre of symmetry and which are not necessarily limited to voltages less than the band gap.

Chynoweth (1956 b) observed steady-state photocurrents in single-crystal $BaTiO_3$ which were closely related to both the magnitude and the sign of the macroscopic polarization. The effect was attributed to internal fields due to space charge near the crystal surfaces. Before discussing how these surface layers affect the photovoltaic behaviour of pyroelectrics we shall extend the discussion of the previous section to show how the excitation of free carriers from localized states can give rise to a volume photovoltaic effect due to the microscopic asymmetry at the absorbing centre.

Consider the asymmetric square-well potential shown in Fig. 12.11, which represents the potential at the absorbing centre. The solution of

Schrödinger's equation for this potential by standard methods (Gol'dman *et al.* 1960) gives bound states within the well and a continuum of free states for energies E greater than V_2. The eigenfunctions for $E < V_2$ and a typical eigenfunction for $V_1 > E > V_2$ are shown in Fig. 12.11 for values of $V_1 = 2$ eV, $V_2 = 1$ eV, and a width of the well $a = 10^{-7}$ cm. Excitation of an electron from the ground state E_0 to the localized state E_1 gives a change of dipole moment $\Delta\mu = 8\cdot6$ Debye as discussed for triangle wells in the previous section. However, on excitation from the ground state to states of energy $V_1 > E > V_2$ the electron is free to move only in one direction as shown in the figure. Electrons moving in the opposite direction are at least partially reflected by the potential barrier, the reflection coefficient depending on the value of V_1 and the width of the barrier. Even when the barrier width is sufficiently narrow to allow significant tunnelling the probability of electron transfer to the left is smaller than to the right of the well in Fig. 12.11. If the defect potential is a small, slowly varying perturbation of the

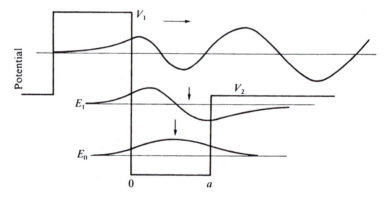

Fig. 12.11. The qualitative form of the bound and free eigensolutions for motion in an asymmetric square well.

host crystal then the free-electron states may be described in terms of the conduction-electron states of the host. Thus optical excitation of free carriers in a pyroelectric host, in which the asymmetry has the same sense at all equivalent defects, gives rise to a net current in one direction. For $E \gg V_1$ electron transfer in both directions becomes equally probable and the net free-carrier current becomes small.

After the electron is scattered its motion becomes isotropic until recombination. If the recombination process involves identical states to the excitation process, then the recombination current density J_r exactly cancels the excitation current density J_e giving no net steady-state current. However, this will not generally be the case since the impurity potential will be modified by the relaxation of the impurity and surroundings following ionization. Alternatively excitation to states $E \gg V_1$ will give negligible current upon excitation $(J_e \sim 0)$, but thermal recombination at the impurity from the bottom of the conduction band can give a recombination current so the net current does not vanish. In either case some coupling to phonons is required. It is clear from these considerations that the photocurrent can have either sign with respect to the polar axis depending on the shape of the potential at the impurity and on the excitation energy.

For an intuitive understanding of the charge-transfer process the microscopic one-dimensional model of a pyroelectric shown in Fig. 12.12 is

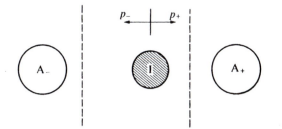

Fig. 12.12. Simple one-dimensional model of a pyroelectric illustrating the charge-transfer process arising from an excitation of an impurity ion I and the unequal probabilities p_+ and p_- of charge transfer to host cations A_+ and A_- respectively (see text).

helpful. It is evident that the probabilities of charge transfer p_+, p_- from the impurity I to the host cations A_+, A_- in the positive and negative directions along the polar axis will differ. Following excitation the charge defect relaxes to a new position changing the overlap with the host cations so that the recombination probabilities p'_+, p'_- differ from p_+ and p_-. The steady-state photocurrent due to the free carriers is

$$J = J_e - J_r = e \frac{\alpha I}{h\nu} \{ l_+(p_+ + p'_+) - l_-(p_- + p'_-) \} \qquad (12.7.1)$$

where α is the absorption coefficient, I the incident intensity, and l_+, l_- are the electron mean free paths in the positive and negative directions. If the mean free path is limited by recombination then the term in the outer parentheses reduces to $(l_+ p_+ - l_- p_-)$.

An additional contribution to the total photocurrent comes from the ionic relaxation following excitation and recombination. Following excitation this current may be written

$$J_i = \frac{\alpha I}{h\nu} (Z_i \Delta l_i) \qquad (12.7.2)$$

where Δl_i is the displacement of the ith ion of charge Z_i and the product $Z_i \Delta l_i$ is summed over all ions. On recombination the displacements Δl_i have, of course, the same magnitude and opposite sign, but the charges Z_i' are not the same since the impurity has an additional charge after recombination. The total photocurrent may therefore be written

$$J = \kappa \alpha I \qquad (12.7.3)$$

where

$$\kappa = (h\nu)^{-1}\{el_+(p_+ + p_+') - el_-(p_- + p_-') + (Z_i - Z_i')\Delta)\Delta l_i\} \qquad (12.7.4)$$

is a constant depending on the host crystal, the absorbing centre, and the incident photon energy $h\nu$. We see from this equation that even if the probabilities of electron transfer in the positive and negative directions at the impurity site are the same (i.e. $V_1 = V_2$ in Fig. 12.11), there is still a net current due to coupling of the charged defect with the lattice. This effect may be present even for transitions between valence and conduction bands of a perfect polar crystal provided there is an appropriate coupling of this excited electronic state with the polar lattice modes.

Under open-circuit conditions this photocurrent charges the crystal capacitance generating an electric field E given by

$$J = \kappa \alpha I + \sigma E \qquad (12.7.5)$$

where σ is the electrical conductivity of the crystal during illumination. The saturation field for $J = 0$ is thus

$$|E_{\text{sat}}| = \frac{\kappa \alpha I}{\sigma}. \qquad (12.7.6)$$

In principle κ itself is a function of E since the microscopic potential at the defect site changes with external field. In the example of Fig. 12.11 κ would reduce to zero for $E = (V_1 - V_2)/a$. In practice the saturation field determined by the conduction current will be much smaller than the microscopic internal field at the impurity site.

Detailed experimental studies of this volume photovoltaic effect were first carried out on $LiNbO_3$ doped with divalent iron impurities (Glass, Von der Linde, and Negran 1974). The absorption spectrum of these crystals shown in Fig. 12.13 shows an absorption band at about 9000 cm^{-1} due to localized d–d transitions, a band centred near 20 000 cm^{-1} which has been attributed to intervalence transfer from the Fe^{2+} ions to next-nearest-neighbour Nb^{5+} ions (Clark *et al.* 1973), and an absorption edge at about 25 000 cm^{-1} due to charge transfer from the oxygen ligands to the impurity. There is some recent experimental evidence (Krätzig and Kurz 1976) that the interpretation of the spectrum is somewhat more complicated than this. The photocurrent along the polar axis of a single-domain crystal was measured with uniform illumination as a function of wavelength under short-circuit conditions, yielding the values of κ plotted in Fig. 12.13. The photocurrent was

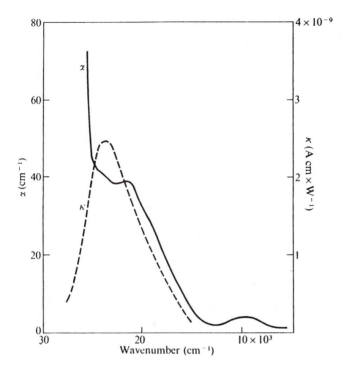

Fig. 12.13. The absorption spectrum of iron-doped LiNbO$_3$ (solid curve) and the measured photocurrent along the polar axis (broken curve) for a single-domain crystal with uniform illumination. (After Glass, Von der Linde, and Negran 1976.)

measured after the initial pyroelectric transient had decayed, and at times long compared with the dielectric relaxation time of the crystal. It is clear from Fig. 12.13 that the photocurrent is greatest for electron transfer to the conduction band (based on Nb orbitals), but decreases at higher energies where absorption due to charge transfer from the ligands (hole creation) is dominant.

Open-circuit saturation voltages in excess of 10 000 V, corresponding to fields of about 10^5 V cm^{-1}, were observed in iron-doped LiNbO$_3$ because of the very low crystal conductivity. For instance with illumination at 20 000 cm^{-1} the photoconductivity is $\sigma \sim 10^{12} \, I \, \Omega$ cm, giving a calculated saturation field of $\sim 10^5$ V cm^{-1} in agreement with experiment. Such anomalously high photovoltages cannot be generated by conventional techniques. The higher conductivity of semiconducting pyroelectrics or efficient photoconductors often makes it difficult to distinguish between this volume effect based on microscopic inhomogeneity and conventional macroscopic effects.

Larger than bandgap steady-state photovoltages have also been observed in polycrystalline materials such as thin films of ZnS and CdTe, striated ZnS crystals, and ferroelectric ceramics. In these materials contribution to the external photovoltage may come from space-charge layers at the grain boundaries in addition to the bulk effect just discussed.

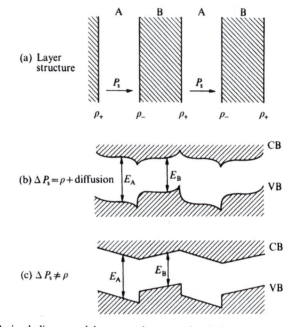

Fig. 12.14. A simple linear model representing a set of grain boundaries in a polycrystalline material: (a) alternating components A and B of polar and non-polar crystallites with space charges ρ_+ and ρ_- at the interfaces compensating for the depolarization fields; (b) the corresponding electronic band structure showing the valence (VB) and conduction (CB) bands (diffusion of the compensating space charge into the crystallites below the interface has been assumed); (c) as (b) but with incomplete compensation of depolarization fields.

Consider the layer structure of Fig. 12.14 which can be taken to represent a set of grain boundaries in a polycrystalline material. If the alternating components A, B are materials containing a centre of symmetry, then the whole structure is symmetrical so that any voltage developing across an AB junction due to space charge at the boundary will be opposed by an equivalent but opposite voltage at a BA junction. The macroscopic voltage across the structure cannot exceed that at one uncompensated junction at the end layer. If either A or B or both are pyroelectric, then the AB and BA junctions are no longer equivalent and a net photovoltage can develop across the entire structure equal to the sum of the voltages developed across each pair of layers in series. The total voltage can exceed the bandgap.

Many of the early studies were made with ZnS crystals which grow as layers of hexagonal and cubic forms of ZnS with the layers perpendicular to the hexagonal axis, similar to the model of Fig. 12.14. The hexagonal ZnS structure is pyroelectric while the cubic form is not. There was some evidence in these crystals that the magnitude of the external photovoltage was correlated with the number of striations (Merz 1958). If we assume then that the A layers have a polarization P_s normal to the layers while the B layers have $P_s = 0$, then at the boundaries div $\mathbf{P} \neq 0$ so that space charge will accumulate at the interface to compensate for the depolarization fields. The space charge is distributed in a layer of finite thickness below the grain boundary, as discussed in § 4.4, and for sufficiently thin layers may extend through the entire layer. The effect on the electronic band structure is to bend the conduction and valence bands downward in the region of excess negative charge at the AB junction and upward in the region of excess positive charge at the BA junction as shown schematically in Fig. 12.14. It has been assumed in this figure that the conduction bands at the centre of both the A and B regions have the same energy. The open-circuit photovoltage for a periodic band structure of the kind shown in the figure has been calculated by Neumark (1962). The maximum photovoltage per AB unit is found to be the difference of the band gaps $E_A - E_B$ of the two regions. The difference of the band gaps in cubic and hexagonal ZnS is approximately 0·1 eV, which is roughly consistent with the experimental determination of the total voltage divided by the number of striations observed under a microscope. Neumark was also able to account for the change of sign of the photovoltage in ZnS with decreasing wavelength. At photon energies $E_A < h\nu < E_B$ optical absorption occurs primarily in hexagonal regions of crystal which results in a photovoltage of opposite sign to when the absorption in A and B regions is comparable.

The preceding arguments do not only apply to alternating layers of differing structures. For instance in ceramics the neighbouring regions have the same structure but different orientation of the polarization, so that div $\mathbf{P} \neq 0$ at the boundary. Because of the anisotropy of the structure the electronic band structure along the axis normal to the grain boundary will also be different in neighbouring grains.

The anomalous photovoltage observed in piezoelectric CdTe (Pensak 1958) presumably has the same origin as that in pyroelectrics, but the polarization is due to strain at the grain boundaries. The piezoelectric polarization then makes AB and BA grain boundaries inequivalent.

12.8. The photorefractive effect

By analogy with the term 'photochromics' which refers to optically induced colour change, the term 'photorefractive' has been adopted to refer to optically induced changes of refractive index which occur in many spontaneously polarized materials. The photorefractive effect was first reported in $LiNbO_3$ and $LiTaO_3$ with focused laser beams in the blue or green regions of the spectrum (Ashkin *et al.* 1966), although unusual effects observed earlier (Peterson *et al.* 1964) can be ascribed to this effect. The index change of the crystal at the focus distorted the wavefront of the transmitted optical beam and for this reason was frequently referred to as 'laser damage', although the origin of the effect, as we shall see, is quite different from the catastrophic irreversible damage which occurs at much higher optical intensities. While the photorefractive effect does provide a serious limitation for some optical applications of pyroelectrics, the effect shows promise for optical memories (see § 16.3). A considerable amount of effort has therefore been devoted to both increasing and decreasing the susceptibility of crystals to these index changes for the various applications.

To account for his early observations of laser-induced index changes in $LiNbO_3$, Chen (1969) proposed that free carriers excited in the illuminated regions of the crystal were displaced along the polar axis of the crystal to trapping sites, the resulting space-charge fields E_i giving rise to the index change Δn_j via the electro-optic effect, i.e. for a linear electro-optic material such as $LiNbO_3$

$$\Delta n_j = \tfrac{1}{2} n_j^3 r_{ji} E_i \qquad (12.8.1)$$

where r_{ji} is the linear electro-optic coefficient. After the light is turned off the index change decreases again as the internal fields relax with the dielectric relaxation time of the material.

This model has also been used successfully to account for similar optically induced index changes observed in several other pyroelectric and ferroelectric materials including single crystals and ceramics. Both the magnitude of the index change and kinetics of the relaxation vary widely from one material to another. Doping crystals with impurities which are readily photoionized by the incident radiation greatly increases the susceptibility of crystals to index changes. In particular the Fe^{2+} absorption near $20\,000\ cm^{-1}$ in $LiNbO_3$ shown in Fig. 12.13 is found to be particularly efficient at increasing the photorefractive sensitivity (defined as the index change per *absorbed* photon density) at this wavelength. The Fe^{2+} ion is photoionized via the reaction

$$Fe^{2+} + h\nu \rightleftharpoons Fe^{3+} + e\,(\text{conduction}) \qquad (12.8.2)$$

both Fe^{2+} and Fe^{3+} valence states being stable in $LiNbO_3$ (Peterson, Glass, and Negran 1971). Indeed the early measurements on nominally undoped $LiNbO_3$ and $LiTaO_3$ can be attributed to Fe^{2+} contamination.

The necessary conditions for the photorefractive effect in an electro-optic host are thus (a) suitable combination of incident wavelength and absorbing centres which are photoionized by the radiation, (b) suitable trapping sites, and (c) free carrier transport to generate the internal fields.

It is possible to satisfy the first requirement either by extrinsic impurity or defect absorption within the band gap of the host crystal, as in the example of Fe^{2+}-doped $LiNbO_3$, or else by intrinsic absorption across the band gap of the host crystal itself. Various mechanisms for single-photon or two-photon absorption are shown schematically in Fig. 12.15. The mobile carriers may

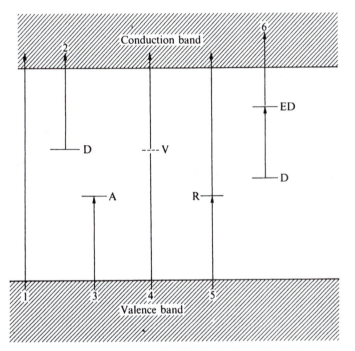

Fig. 12.15. A schematic representation of various mechanisms for single- and two-photon absorption in a host crystal: (1) direct valence band (VB) to conduction band (CB) process; (2) direct excitation of a donor electron to CB; (3) direct excitation of a VB electron to an acceptor; (4) interband two-photon absorption via a virtual intermediate state V; (5) a two-step interband absorption via a real intermediate state R; (6) a two-step absorption from a donor D to CB via a real excited donor level ED.

be electrons or holes. Of course two-photon absorption requires much higher intensities but the energy required to produce a free carrier is just the total transition energy.

In the case of iron-doped $LiNbO_3$ the Fe^{2+} impurities act as electron donors and the Fe^{3+} ions act as electron traps so that space-charge fields are due to a spatial redistribution of the two valence states. In the same way other multivalent impurities such as Cu^{1+}, Cu^{2+}, Mn^{2+}, Mn^{3+} may act as both the sources and traps of photoelectrons (Phillips, Amodei, and Staebler

1972). However, little attention has been paid to the trapping mechanism. Index changes are readily produced in nominally pure $LiNbO_3$ and $K(Ta, Nb)O_3$ by two-photon absorption (Von der Linde, Glass, and Rodgers, 1974) but the nature of the trapping sites has not been determined.

Requirement (c) for the photorefractive process is that free carriers, either electrons or holes (or both), are sufficiently mobile to reach the trapping sites before recombination. The space-charge field E_i generated by the charge displacement is given by the relation

$$E_i = \frac{1}{\varepsilon\varepsilon_0} \int \rho \, dx = \frac{1}{\varepsilon\varepsilon_0} \int J \, dt \qquad (12.8.3)$$

where the current density $J(x)$ is a function of the incident light intensity $I(x)$ and is given by the transport equation

$$J(x) = \sigma(x)(E_i + E_e) + \kappa\alpha I(x) + eD\frac{dn}{dx} + p\frac{dT}{dt} + \Delta\mu\frac{dn}{dt} \qquad (12.8.4)$$

where the first term is electronic conduction in an electric field which is the sum of the internal space-charge field E_i and the external applied field E_e, the second term is the volume photovoltaic current discussed in the last section, the third term is free-carrier diffusion in a concentration gradient dn/dx where D is the diffusion coefficient and e the electronic charge, and the fourth and fifth terms are transient terms due to the pyroelectric effect resulting from localized heating of the crystal by the radition and excited-state polarization of localized excited states (see § 12.6). For steady illumination the last two terms are not usually important but for short-pulse excitation these terms could be dominant. The relative importance of the first three terms in the transport equation depends on experimental conditions and the material characteristics. For instance the diffusion term increases with increasing spatial frequency of the incident radiation, although for iron- and copper-doped $LiNbO_3$ the photovoltaic effect is dominant for all practical spatial frequencies (up to 10^5 lines cm). In these crystals the photovoltaic effect is more important than conductivity for external applied fields up to about 10^4–10^5 V cm^{-1}. On the other hand, in $Sr_{1-x}Ba_xNb_2O_6$, $K(Ta, Nb)O_3$, and PLZT ceramics photoconductive transport is dominant even for small applied fields.

The change of refractive index given by eqn (12.8.1) depends on the total field $E_i + E_e$ at any point in the crystal. However, the external field produces a spatially uniform index change which does not cause any scattering of light so that the index inhomogeneity depends only on the spatially varying internal field E_i. It is seen from eqn (12.8.4) that in a highly conducting crystal there will be no index change since all electric fields vanish. In crystals of small dark conductivity E_i will continue to increase until the local current density in the illuminated region vanishes. For photoconductive transport this occurs when the internal space-charge field is equal (but opposite) to the external field, while for photovoltaic transport $J = 0$ when the space-charge field is equal to the open-circuit saturation photovoltage (see eqn 12.7.4). When the light is turned off the space charge will relax with the dielectric relaxation time $\varepsilon\varepsilon_0/\sigma$ of the material so that the permanence of the index change depends on the dark resistivity. If the electron-trapping states are close to the conduction-band edge the electrons can be thermally liberated and return to their original sites. Alternatively, other conduction mechanisms can relax the internal fields. For instance the index change can be removed by photoconductivity if the crystal is uniformly illuminated under short-circuit conditions with light of suitable wavelength. The rapid relaxation reported in $BaTiO_3$ (Townsend and LaMacchia 1970) can be attributed to the relatively high dark conductivity of this material. In undoped $LiNbO_3$ at room temperature the dark conductivity is extremely small and the index change persists for many months. However, at temperatures around 150°C rapid relaxation of the space-charge fields has been attributed to ionic conductivity (Staebler and Amodei 1972) at these temperatures, so that the ionic space-charge pattern replicates the electronic space-charge pattern but with opposite sign. Subsequent uniform illumination redistributes the electronic charge, but not the ionic charge, so that internal fields are again generated optically which replicate the original intensity pattern.

Johnston (1970) has pointed out that even after all internal fields have relaxed a polarization variation can occur provided div $\mathbf{P} = \rho$. Thus, depending on the manner in which the internal field is reduced to zero, an index change could be present following dielectric relaxation, but this will not necessarily involve the electro-optic coefficients used in eqn (12.8.1). An example of this is afforded by materials of low coercivity such as $Sr_{1-x}Ba_xNb_2O_6$ (Thaxter and Kestigian 1974) and PLZT (Micheron, Mayeux, and Trotier 1974) in which the optically generated fields may be sufficiently great to re-orient locally the spontaneous polarization. An interesting case is when ΔP_i is due to localized excited-state polarization $n\Delta\mu$ (see § 12.6). In this case the depolarization fields must also be taken into account in determining the index change. No space-charge fields arise since no free carriers are generated. For a thin-plate geometry with ΔP_i normal to the plate the depolarization field can generate an index change via the electro-optic effect, while for the thin-plate geometry with ΔP_i in the plane the depolarization fields are negligible.

For quantitative measurement of the photorefractive effect there are two basic techniques. One is to write the index change with a focused laser beam and to measure the birefringence change $\Delta(n_x - n_z)$ with a probe beam which does not affect the index change. One such arrangement is shown in Fig. 12.16 (a). The x and z axes of the crystal are normal to the incident

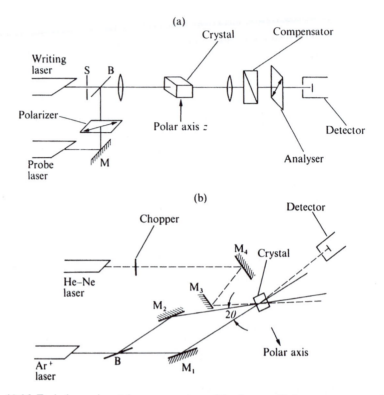

Fig. 12.16. Typical experimental arrangements used for the quantitative measurement of the photorefractive effect. (a) The measurement of changes in birefringence produced by a writing laser is accomplished by use of a probe laser which does not affect the index change. (b) A holographic technique using two equal-intensity laser beams to produce a phase grating in the crystal as the result of induced index changes caused by the optical interference pattern. (After Peterson *et al.* 1971.)

radiation and at 45° to the axes of the polarizer and analyser. If the transmission of the Pockel's cell arrangement is initially set at zero with the compensator then the laser-induced briefringence gives rise to a transmission

$$T = \sin^2\left\{\frac{\pi}{\lambda}\int_0^d \Delta(n_x - n_z)\,\mathrm{d}y\right\} \qquad (12.8.5)$$

where d is the crystal thickness. Birefringence changes of about 10^{-5} can be measured by this technique. The method gives a direct measure of the usefulness of a crystal for non-linear optical applications but suffers from the disadvantage for quantitative measurements that thermally induced index changes may be larger than the photorefractive effect. The holographic technique shown in Fig. 12.16 (b) is generally more accurate for quantitative measurements of the index change (not birefringence as before). Two plane waves of equal intensity intersect in the crystal resulting in an optical interference pattern

$$I(x) = I_0(1 + \cos Kx) \qquad (12.8.6)$$

having a single spatial period $2\pi/K$. This pattern forms a phase grating in the crystal because of the induced index change. The diffraction efficiency η of such a grating, provided Bragg conditions are satisfied is

$$\eta = \exp(-\alpha d) \sin^2\!\left(\frac{\pi \, \Delta nd}{\lambda \, \cos \theta}\right) \qquad (12.8.7)$$

where Δn is the modulation amplitude of the index change and α is the absorption coefficient at the probe wavelength λ. The minimum index change which can be measured is limited by scattering from defects and typically $\Delta n > 10^{-7}$ can be measured. Thermal effects are usually less serious with this technique since the 'thermal grating' relaxes very rapidly at the high spatial frequencies of the optical interference pattern. This technique also has the advantage that the electric fields and polarization changes are well defined along the grating vector K, so that the orientation properties of the photorefractive effect are easily investigated. In this way Chen (1969) was able to identify the electro-optic origin of the effect, since the magnitude and symmetry of the diffraction efficiency η depended on the electro-optic coefficients r_{ij} (see eqns (12.8.7), (12.8.1)).

A useful quantitative characterization of the photorefractive sensitivity S is the change of refractive index produced in the material per unit of absorbed optical energy, i.e.

$$S = \frac{\Delta n}{\alpha W} \qquad (12.8.8)$$

where $W = \int I \, dt$ is the incident energy density. For iron-doped $LiNbO_3$ at a wavelength of 514.5 nm the value of S is $\sim 10^{-5} \, cm^3 J^{-1}$, while by two-photon absorption in $KTa_{0.65}Nb_{0.35}O_3$ a value of S as high as $0.1 \, cm^3 \, J^{-1}$ has been observed (Von der Linde et al. 1974). To gain some insight as to what this sensitivity means in terms of the microscopic processes it is useful to express S in terms of the average distance l travelled per excited photoelectron from the source to the trapping site. The polarization change $\Delta P = Nel$ where e is the electronic charge and $N = \alpha W/h\nu$ is the total number of absorbed photons (excited carriers) per unit volume. Then using eqns (12.8.1) and (12.8.8) we have

$$S_j = \left[\frac{n_j^3 r_{ji}}{2\varepsilon_0(\varepsilon_i - 1)}\right]\!\left(\frac{el}{h\nu}\right). \qquad (12.8.9)$$

Thus S depends on the dielectric properties of the material and the dipole moment created per excited photoelectron. For the $LiNbO_3$: Fe and KTN crystals cited above the value of l correspond to about $10^{-4}\,\mu m$ and $10\,\mu m$ respectively, clearly demonstrating the superior transport properties of KTN.

13

NON-LINEAR OPTICS

13.1 Wave propagation in non-linear dielectrics.

A VARIETY of optical phenomena can be classified under the heading of non-linear optics. Several of these, such as the Kerr and Pockels electro-optic effects, the Zeeman effect, and the photoelastic effect in which a low-frequency electric or magnetic field or mechanical strain is mixed with the optical wave, do not require high optical intensities for their observation and have been known and understood for many years. On the other hand, non-linear phenomena which involve mixing two or more optical waves, such as second-harmonic generation, sum and difference frequency generation, or two-photon absorption, had to await the availability of high-intensity sources—namely lasers. Another large group of optical non-linearities involves the scattering of light from atomic or electronic oscillators or other fluctuations within the optical medium. In this category fall, for instance, second- or higher-order Raman scattering from optic phonons in general (including polariton scattering from infrared active phonons), Brillouin scattering from acoustic phonons, and Rayleigh scattering from entropy or orientation fluctuations.

The purpose of this chapter is not to survey the field of non-linear optics, which has been the subject of several entire texts, but rather to examine the important role played by ferroelectrics in the areas of electro-optics and optical mixing, and to discuss phenomenological and microscopic theories which relate the acentric nature of the host crystal to the non-linear optical susceptibilities.

In Chapter 3 the non-linear polarization of ferroelectrics was described by an expansion of the crystal free energy in increasing powers of electric displacement and mechanical strain. This leads to the relation

$$\mathbf{D} = \mathbf{D}(\text{linear}) + \mathbf{D}(\text{non-linear}) \qquad (13.1.1)$$

where the linear part is

$$D_i(\text{linear}) = \varepsilon_{ij}E_j + d_{ijk}X_{jk} + p_iT \qquad (13.1.2)$$

and the non-linear part may be written as a series expansion in the form

$$D_i(\text{non-linear}) = \chi^{\text{NL}}_{ijk}E_jE_k + \phi_{ijkl}X_{jk}E_l + \xi_{ij}TE_j + \chi^{\text{NL}}_{ijkl}E_jE_kE_l + \dots \qquad (13.1.3)$$

where we have reverted to tensor (rather than Voigt) notation and the repeated-index summation convention is once again adopted. As in the rest of this book magneto-electric effects are ignored since they have very small influence on the behaviour of non-magnetic ferroelectrics. We are primarily concerned only with terms in the expansion in which at least one of the components is an optical-frequency electric field. The lowest-order non-linear coefficients (or compliances) χ_{ijk}^{NL}, ϕ_{ijkl}, and ξ_{ij} are then referred to as the second-order non-linear optical susceptibilities.

At low frequencies the electromagnetic wave nature of the alternating electric field does not have to be considered because the wavelength is much longer than the crystal dimensions normally studied. To see how the non-linear electric displacement affects the propagation of light in a crystal we make use of the wave equation for transparent insulating materials, viz.

$$\nabla^2 E_i + \mu_0 \ddot{D}_i = 0 \tag{13.1.4}$$

where we assume that the magnetic permeability μ_0 of the medium is essentially that of free space, since we are not concerned here with magnetic materials. Taking just the linear electric displacement of (13.1.2) we have

$$\nabla^2 E_i + \varepsilon_0 \mu_0 \ddot{E}_i = -\mu_0\{(\varepsilon_{ij} - \varepsilon_0\delta_{ij})\ddot{E}_j + d_{ijk}\ddot{X}_{jk} + p_i\ddot{T}\} \tag{13.1.5}$$

where the time-varying polarization of the medium now appears as a source term in the equation. It is clear that the piezoelectric and pyroelectric terms will not influence the wave propagation in the medium but they can act as sources of electromagnetic radiation. Since typical frequencies for elastic wave propagation do not exceed 10^9 Hz, the strain term will not be important, even at microwave frequencies. However, radiation has been generated in the microwave region in LiNbO$_3$ and LiTaO$_3$ using the pyroelectric term by changing the crystal temperature sufficiently rapidly (Auston, Glass, and Ballman 1972). In the same way the change of the electronic polarization which accompanies optical absorption (see § 12.6) can act as the source of radiation, but this term has been omitted for simplicity. Both of these terms are unique to pyroelectric materials.

The linear electric polarizability of the medium can conveniently be included with the free-wave term of the left-hand side of eqn (13.1.5) and its effect is to reduce the velocity from $c = (\varepsilon_0\mu_0)^{-1/2}$ to $v_i = (\varepsilon_i\mu_0)^{-1/2}$. The non-linear electric displacement changes not only the phase velocity, but also the frequency of wave propagation in the medium. If the non-linearities are small (which they are in ferroelectrics at optic frequencies), then the non-linear terms can be regarded simply as new distributed source terms which are, to a good approximation, superimposed on the linear solution. The refractive index

$$n_i = \varepsilon_0^{-1/2}\left(\frac{\partial D_i}{\partial E_i}\right)^{1/2} \tag{13.1.6}$$

now depends on the applied field, the stress, and the crystal temperature.

The lowest-order non-linear susceptibilities are referred to as the electro-optic effect (χ_{ijk}^{NL}), the photoelastic effect (ϕ_{ijkl}), and the thermo-optic effect (ξ_{ij}). Although the photoelastic effect is of significant practical importance, since it allows for the interaction of acoustic and optic waves and makes possible the acousto-optic modulation of light, and the thermal term is important because it gives rise to the thermal self-focusing of light beams, neither of these terms will be considered here since the ferroelectric nature of the host is not involved in an essential way. Indeed, the acousto-optic materials with the highest figure of merit are not pyroelectric (Pinnow 1971). Attention will thus be restricted to the expansion of D in terms of the electric field alone:

$$D_i = \varepsilon_{ij} E_j + \chi_{ijk}^{NL} E_j E_k + \chi_{ijkl}^{NL} E_j E_k E_l + \ldots . \qquad (13.1.7)$$

Indirect photoelastic contributions to the electro-optic effect due to the piezoelectric excitation of acoustic waves at low frequencies ($\lesssim 10 \, \text{MHz}$) will be included with the direct effect in the same way as we have treated clamped and unclamped linear dielectric compliances in Chapter 5. The relation between the clamped and unclamped electro-optic susceptibilities is then

$$\chi_{ijk}^{NL,X} = \chi_{ijk}^{NL,x} - \phi_{ijkl}\left(\frac{\varepsilon_{ij}}{d_{ijk}}\right). \qquad (13.1.8)$$

The clamped electro-optic susceptibility contains terms due to both optic mode and electronic non-linearities. At frequencies above all lattice-vibrational frequencies only the non-linear electronic polarizability contributes to the electro-optic effect, and instead of χ_{ijk}^{NL} the conventional notation is d_{ijk} (and should not be confused with the linear compliance d_{ijk} of eqn (13.1.2)).

When inserted into the wave equation it is easily seen that the non-linear terms act as sources of the harmonics of any fundamental wave introduced into the medium. Thus, for instance, for a fundamental frequency ω the source term in E_i^2 has a frequency variation

$$(E_0 \cos \omega t)^2 = E_0^2(\tfrac{1}{2} + \tfrac{1}{2} \cos 2\omega t). \qquad (13.1.9)$$

Likewise the third-order term gives rise to third harmonics and so on. The non-linear terms also mix two or more frequencies simultaneously introduced into the crystal and act as sources at the sum and difference frequencies. In particular, a source term of the form $(E_0 \cos \omega_1 t + E_0 \cos \omega_2 t)^2$ has Fourier components at frequencies $\omega_1 + \omega_2$, $\omega_1 - \omega_2$, $2\omega_1$, $2\omega_2$; and d.c. In general the non-linear susceptibility will be different for each of these processes both because of dispersion and because of the tensorial properties of χ_{ijk}^{NL} requiring transformation according to the symmetry requirements for the crystal. It is obvious from eqn (13.1.7) that χ_{ijk}^{NL} vanishes for any system with a centre of inversion. In fact, the non-vanishing elements are the same as those of the piezoelectric tensor d_{ijk} (or d_{ij} in the Voigt notation of Chapter 3) and are listed in Appendix E.

The importance of ferroelectrics in non-linear optics lies in their combination of three properties. Firstly they lack a centre of inversion in the ferroelectric phase and therefore allow second-order interactions. Secondly their nonlinear susceptibilities are in general large compared with other piezoelectrics, and thirdly the large spontaneous birefringence of ferroelectrics often allows the phase matching of the fundamental and second-harmonic waves (or sum or difference frequencies) which, as we discuss in the following section, optimizes the conversion efficiency from the fundamental to second harmonic.

13.1.1. Phase matching

Including just the lowest-order non-linear term in the wave equation and simplifying the notation by omitting subscripts we have

$$\nabla^2 E + \frac{1}{v^2}E = -\mu_0 \chi^{NL}\frac{\partial^2(E^2)}{\partial t^2}. \qquad (13.1.10)$$

Consider a plane wave of infinite extent in the xy-direction (to avoid diffraction effects) propagating along the z-axis,

$$E(\omega) = E_0 \cos(\omega t - q_1 r) \qquad (13.1.11)$$

with velocity ω/q_1. The non-linear source term at frequency 2ω is then, to a good approximation,

$$\chi^{NL}(2\omega)E_0^2 \cos(2\omega t - 2q_1 z). \qquad (13.1.12)$$

However, the generated wave at frequency 2ω actually travels with a velocity different from the source term because of the non-linearity of the frequency *versus* wavevector characteristic of the material (i.e. dispersion). It can therefore be expressed as

$$E(2\omega) = E_1 \cos(2\omega t - q_2 z) \qquad (13.1.13)$$

with amplitude E_1 proportional to the non-linear susceptibility χ^{NL} but with wavevector $q_2 \neq 2q_1$. This means that the source will generate harmonics which interfere destructively with the propagating wave after a distance of

$$L_c = \frac{\pi}{q_2 - 2q_1} \qquad (13.1.14)$$

which is referred to as the coherence length. If the optical thickness of a non-linear crystal is an even number of coherence lengths then the second-harmonic intensity is zero, while for an odd number of coherence lengths the signal reaches a maximum equal to the second-harmonic intensity generated in the last coherence length, this intensity being proportional to L_c^2. A versatile experimental technique for measuring both the coherence length and the non-linear susceptibility involves rotating the crystal about one of its principal axes with respect to an incident fundamental beam as shown in Fig. 13.1. The second-harmonic intensity varies periodically as the optical path

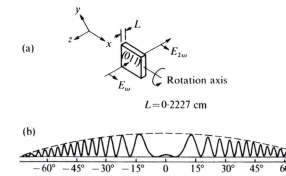

$$L = 0.2227 \text{ cm}$$

Fig. 13.1. Maker fringe spectrum for quartz (011) sample showing (a) slab orientation and (b) the experimentally observed fringe pattern. The broken line is a theoretical fit to the fringe envelope. (After Kurtz 1972.)

length changes (Maker *et al.* 1962). A careful study of these 'Maker fringes' in ADP, KDP, and SiO$_2$ by Jerphagnon and Kurtz (1970) has shown that L_c can be determined with an accuracy of about 1 per cent (with typical values between 1 and 20 μm). Further details of the technique have been reviewed by Kurtz (1972).

An alternative technique for measuring both χ^{NL} and L_c is to translate a wedge prepared from the crystal across the fundamental beam at normal incidence. The second-harmonic intensity changes as the optical path length within the wedge changes. In this case the fundamental beam remains at normal incidence to the wedge and no correction for the change in refractive index with rotation of the crystal is necessary (Wynne and Bloembergen 1969).

One way to prevent cancellation of the second harmonic due to mismatch of the velocities at ω and 2ω is to construct a layer structure where the thickness of each layer is a single coherence length but in which alternating layers are rotated with respect to one another in such a way as to reverse the sign of χ^{NL}. An experiment of this kind was carried out with BaTiO$_3$ (Miller 1964 a) using the periodic 180° domain structure of unpoled crystals.

A better way to avoid phase cancellation is to match the phase velocities at the two frequencies so that L_c becomes very large. In principle this can be done by making use of anomalous dispersion, optical activity, or optical birefringence. Of these, anomalous dispersion is usually accompanied by strong absorption and optical activity is not often large enough to compensate for dispersion. As a result phase matching in homogeneous media has in practice been restricted to birefringent materials (Giordmaine 1962). For phase matching to be possible birefringence must exceed dispersion. A simple example of how the phase velocities at two frequencies can be matched is afforded by KDP which has only two (different) allowed non-vanishing coefficients for second-harmonic generation, namely $\chi_{123}^{NL} = \chi_{231}^{NL}$ and χ_{312}^{NL}, where direction $3(= z)$ is the polar axis. The only non-linear terms are thus

$$D_x(2\omega) = \chi_{123}^{NL} E_y(\omega) E_z(\omega)$$

$$D_y(2\omega) = \chi_{231}^{NL} E_z(\omega) E_x(\omega) \qquad (13.1.15)$$

$$D_z(2\omega) = \chi_{312}^{NL} E_x(\omega) E_y(\omega).$$

If the fundamental is an ordinary wave (i.e. no z-component of $\mathbf{E}(\omega)$) then the second-harmonic wave has extraordinary polarization (Fig. 13.2 (a)). The dispersion of ordinary and extraordinary waves is shown in Fig. 13.2 (b) as the angle θ between the z-axis and the direction of propagation is varied. It can be seen that at $0.6943\,\mu$m, the wavelength of the ruby laser, the incident fundamental and the generated second-harmonic wave are phase matched at about 50^0. Other wavelengths can be phase matched for different angles of propagation in the crystal.

An alternative procedure for phase matching, instead of varying the crystal angle (which has obvious disadvantages for use in tunable laser systems), is to change the crystal birefringence by changing the temperature. In most ferroelectrics the birefringence varies quite rapidly with temperature—particularly close to a phase transition—so that temperature-tuned phase-matched harmonic generation is possible over a fairly wide wavelength range. In practice the optimum condition for phase matching is when the fundamental and second-harmonic waves propagate along a principal axis of the crystal with $\theta = 90°$. With this geometry the phase velocities are relatively insensitive to small variations of the crystal alignment or crystal temperature (the extraordinary index varies as $\cos^2\theta$). This 'non-critical' condition also avoids 'walk-off' of the harmonic and fundamental beam in the crystal due to double refraction. That is, since D and E do not exactly coincide in a birefringent crystal, except along the principal axis, the propagation wavevector $\mathbf{q}$ is not exactly perpendicular to $\mathbf{E}$ nor in precisely the same direction as the energy propagation. This limits the interaction length of the fundamental and second harmonic even under phase-matched conditions.

Very few crystals satisfy this rather special requirement. It can be seen from Fig. 13.2 that, throughout the wavelength range shown, non-critical phase matching is not possible in KDP at room temperature (although it has

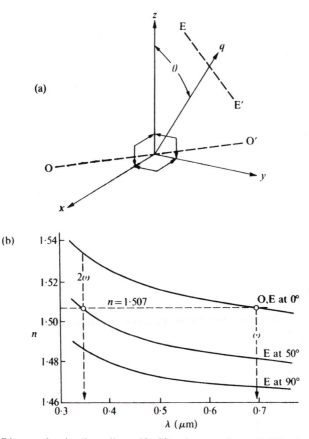

Fig. 13.2. (a) Diagram showing the ordinary (O–O′) and extraordinary (E–E′) polarization of a wave propagating at an angle θ to the optic axis z of uniaxial crystal. O–O′ is in the xy-plane and E–E′ is in the perpendicular plane containing the propagation vector **q**. (b) The index of refraction of KDP as a function of wavelength for O and E polarizations and three different directions of propagation, showing that extraordinary harmonic waves propagating at 50° to the optic axis are phase-matched to the ordinary fundamental wave of a ruby laser. (After Baldwin 1969.)

been achieved at lower temperatures at 514·5 nm). LiNbO$_3$ and Ba$_4$Na$_2$Nb$_{10}$O$_{30}$ can, however, be non-critically phase matched over the wavelength range including the important laser wavelength at 1·06 μm (Miller, Boyd, and Savage 1965, Geusic *et al.* 1968). It is partly for this reason that these crystals have attracted a great deal of attention for non-linear optical applications. An additional reason, however, is the particularly large non-linear susceptibilities possessed by these crystals (Boyd *et al.* 1964, Geusic *et al.* 1968). The phase-matching temperature T_p varies considerably with the stoichiometry of the crystals, particularly in LiNbO$_3$. Here the congruent melting composition (see § 8.3.3) at 48.6 mol per cent Li_2O can be non-critically phase matched for 1.06 μm radiation at about 10°C (Fay *et al* 1968, Bergman *et al*. 1968) while crystals grown from a melt of 54 mol per cent Li_2O are phase matchable at about 180°C at this wavelength. Within this range T_p varies approximately as

$$T_p(°C) = 50\cdot5 + 32(x - 50). \qquad (13.1.16)$$

The addition of MgO to the melt also raises the phase-matching temperature for all Li/Nb stoichiometries, so that T_p can be selected for specific wavelengths by selecting the appropriate crystal composition, although some compositions are difficult to grow as single crystals of high quality. This is a useful property of $LiNbO_3$ because of difficulties with optically induced index changes which distort the optical wavefront of intense visible laser light propagating in the crystals at room temperature. The problem is particularly serious in the green and blue regions of the spectrum—for instance at 0·53 μm, the second harmonic of the 1·06 μm laser. At temperatures above about 120°C the index change begins to anneal out reversibly, so that at temperatures well above this, depending on the photorefractive sensitivity of the crystals, $LiNbO_3$ can be used for non-linear optical applications.

The same general conditions as those outlined for second-harmonic generation also apply for phase matching three-wave interactions when all three frequencies $\omega_1, \omega_2, \omega_3$ are different. The conditions for phase matching may be expressed as the momentum conservation relation

$$q(\omega_3) = q(\omega_1) + q(\omega_2) \qquad (13.1.17)$$

or equivalently

$$n_3\omega_3 = n_1\omega_1 + n_2\omega_2 \qquad (13.1.18)$$

where, or course, for energy conservation we require

$$\omega_3 = \omega_1 + \omega_2. \qquad (13.1.19)$$

From these two equations one readily establishes the relation

$$\frac{n_2 - n_1}{n_3 - n_1} = \frac{\omega_3}{\omega_2}. \qquad (13.1.20)$$

If all of these frequencies are in the optical regime—well above the restrahl yet well below the electronic absorption edge—then dispersion is usually sufficiently small to allow phase matching with a modest birefringence ($\leq 0\cdot1$). However, in difference frequency-mixing experiments ω_1 and ω_2 may be in the optical region and ω_3 below the restrahl in the microwave region. In this case the dispersion across the restrahl is usually very large compared with the optical birefringence because of the ionic contribution to the polarizability at low frequencies. The refractive index n_3 in eqn (13.1.20) can then be replaced by the square root of the clamped dielectric constant. It can be seen that phase matching of this interaction is still possible, however, even for a small birefringence. Thus, if $n_2 - n_1 \sim 10^{-3}$ and $\varepsilon^x \sim 100\varepsilon_0$, then a difference frequency at 1 cm^{-1} in the microwave spectrum can be phase matched with optical frequencies $\sim 10^4 \text{ cm}^{-1}$.

It is important for maximum energy conversion that not only should phase matching be possible but that the non-linear susceptibilities are large in the directions for which phase matching is possible and that the crystal quality should be sufficiently good to allow long interaction lengths.

If a non-linear crystal is twinned, or contains ferroelectric domains, then the phase relationship between the fundamental and second-harmonic waves is destroyed, since in general the phase-matching conditions are different on each side of an interface or domain wall. In fact second-harmonic generation can be used as a measure of crystal quality by measuring the second-harmonic intensity as the crystal temperature (or angle θ) is changed. For a perfect single crystal a single sharp peak in the second-harmonic intensity is observed at the phase-match temperature (or angle), but in imperfect crystals this peak is broadened, or several peaks are observed at different temperatures (or angles). Examples of poor quality and high optical quality $LiNbO_3$ crystals are shown in Fig. 13.3. Broadening of the phase-matching temperature is also observed owing to composition variations, and may occur in mixed, or non-stoichiometric, crystals since the optical birefringence is usually composition dependent.

In highly polycrystalline materials some information can still be obtained concerning the magnitude of the non-linear coefficients and whether the material can be phase matched in single-crystal form. A useful experimental technique for surveying powders of non-linear optic materials has been described by Kurtz and Perry (1968). In this method the second-harmonic intensity $I(2\omega)$ generated by a pulsed laser incident on a thin layer of powder is compared with the intensity generated in a quartz standard of identical geometry. Since there are a large number of oriented grains the second-harmonic signal measures some angular average of the product $(L_c \chi^{NL})^2$. For non–phase-matchable materials it is possible to define an averaged coherence length $\bar{L}_c$ and the harmonic intensity then takes the form

$$I(2\omega) \propto \frac{\bar{L}_c^2 \langle (\chi_{ijk}^{NL})^2 \rangle (n+1)^{-6}}{r_g} \tag{3.1.21}$$

where n is the average refractive index and r_g is the average grain size, which is assumed to be greater than L_c. For phase-matchable materials the angular average is dominated by the large values of L_c near the phase-match angle, and the result can then be recast in the form

$$I(2\omega) \propto \frac{(\chi_{pm}^{NL})^2 \sin \theta_m (n+1)^{-6}}{\sin \rho} \tag{13.1.22}$$

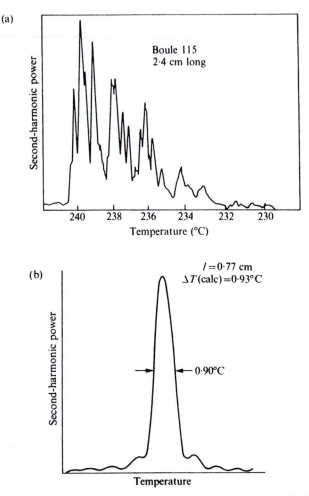

Fig. 13.3. Temperature dependence of the second-harmonic intensity of LiNbO₃ for light propagating normal to the optic axis: (a) an inhomogeneous single crystal and (b) a high-quality crystal. (After Harris 1969 and Nash *et al.* 1970.)

where θ_m is the phase-match angle, ρ is the walk-off angle, and χ_{pm}^{NL} is the non-linear coefficient corresponding to the phase-matched conditions. This is independent of the particle size provided $r_g \gg \pi/(q_1 \sin \rho)$ where q_1 is the wavenumber of the fundamental. A plot of the grain-size dependence of $I(2\omega)$ is shown in Fig. 13.4, including the limit $r_g \ll L_c$. It is clear from this figure that phase-matchable materials and non-phase-matchable materials are easily distinguished from a measurement of $I(2\omega)$ as a function of grain size, and that an estimate of the coherence length and the magnitude of the non-linear susceptibility relative to quartz can be obtained.

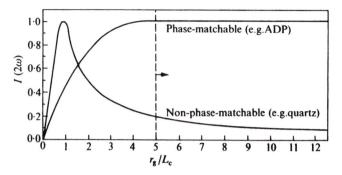

Fig. 13.4. Dependence of the second-harmonic intensity on average particle size r_g for phase-matchable and non-phase-matchable materials. (After Kurtz and Perry 1968.)

13.2. Electro-optic and non-linear optic coefficients

13.2.1. The non-linear susceptibility

The non-linear susceptibilities which determine the magnitude of the coupling between the various electric fields in a solid vary considerably from one material to another, even within materials of similar structural types. This section will be devoted to understanding how the lowest-order non-linear susceptibility χ_{ijk}^{NL} depends on the microscopic and macroscopic properties of materials. Several theories have been developed which assist in determining the origins and magnitude of χ_{ijk}^{NL} and even, to some extent, allow the prediction of the non-linear coefficients from other characteristics of the structures.

The most general comment concerning the non-linear susceptibility is that the number of independent tensor coefficients (determined by crystal symmetry) is the same as the number of independent piezoelectric coefficients (see Appendix E). As with the third-rank piezoelectric d-tensor (see Chapter 3) the two indices jk of the tensor χ_{ijk}^{NL} are interchangeable, and it is often convenient to use the contracted Voigt notation for this pair of indices, viz.

$$
\begin{array}{ccc}
11 \to 1 & 22 \to 2 & 33 \to 3 \\
23 = 32 \to 4 & 31 = 13 \to 5 & 12 = 21 \to 6.
\end{array}
\tag{13.2.1}
$$

Kleinman (1962) has pointed out that if dispersion between the three frequencies $\omega_1, \omega_2, \omega_3$ which are coupled by χ_{ijk}^{NL} is negligible, then all three indices can be permuted leading to a further reduction in the number of independent tensor elements. For example, if all three frequencies are in the transparent region of the optical spectrum of a crystal, as with second-harmonic generation, then, since dispersion is small, the Kleinman permutation relations give to a good approximation

$$\chi_{12} = \chi_{26} \qquad \chi_{13} = \chi_{35} \qquad \chi_{14} = \chi_{25} = \chi_{36}$$
$$\chi_{15} = \chi_{31} \qquad \chi_{16} = \chi_{21} \qquad \chi_{23} = \chi_{34} \qquad \chi_{24} = \chi_{32}$$

$$(13.2.2)$$

using the contracted notation and dropping for the moment the superscript NL. Consider the hexagonal crystal class 6 for which (see Appendix E) there are four independent coefficients allowed by the symmetry: χ_{15}, χ_{33}, χ_{14}, χ_{31}. Permuting the three tensor indices Kleinman's relations (eqn (13.2.2)) show that $\chi_{14} = \chi_{25}$, yet the group symmetry requires that $\chi_{14} = -\chi_{25}$. It follows that $\chi_{14} \approx 0$ for second-harmonic generation. It must be emphasized, however, that since the dispersion is usually non-zero even in the visible spectrum, these relations are only approximations. They are certainly not valid for the electro-optic coefficients because of dispersion across the restrahl, i.e. the low-frequency non-linear coefficients contain contributions from both the ionic and electronic polarizability

$$\chi^{NL}(\text{total}) = \chi^{NL}(\text{ionic}) + \chi^{NL}(\text{electronic}). \qquad (13.2.3)$$

Because of the different historical development of electro-optic effects involving a low-frequency electric field and non-linear optic effects involving three optical frequency fields, it is now conventional to describe the effects by different parameters. The coefficients for optical mixing are usually labelled $d_{ijk}(=\tfrac{1}{2}\chi^{NL}_{ijk})$ and defined by the relation

$$P_i(\omega_1 \pm \omega_2) = \sum_{jk} d_{ijk} E_j(\omega_1) E_k(\omega_2) \qquad (13.2.4)$$

where P_i is the non-linear source polarization in the wave equation, and the $\pm$ on the left-hand side implies taking account of only the relevant (i.e. $\omega_1 + \omega_2$ or $\omega_1 - \omega_2$) component on the right-hand side. For the specific case of second-harmonic generation we have

$$P_i(2\omega) = \sum_{jk} d_{ijk} E_j(\omega) E_k(\omega). \qquad (13.2.5)$$

An expression analogous to eqn (13.2.5) can also be written in terms of the source field $E_i(2\omega)$ inducing the second-order polarization (Miller 1964 b) in the form

$$E_i(2\omega) = \sum_{jk} \delta_{ijk} P_j(\omega) P_k(\omega) \qquad (13.2.6)$$

where the δ_{ijk} are referred to as Miller's delta and are obviously related to the d_{ijk} by the transformation

$$d_{ijk}(2\omega) = \sum_{lmn} \chi^L_{il}(2\omega) \chi^L_{jm}(\omega) \chi^L_{kn}(\omega) \delta_{lmn}(2\omega) \qquad (13.2.7)$$

where the superscript L denotes a linear susceptibility. For crystals other than triclinic or monoclinic this relationship can be written in a principal-axis system in the simplified form

$$d_{ijk}(2\omega) = \chi^L_{ii}(2\omega)\chi^L_{jj}(\omega)\chi^L_{kk}(\omega)\delta_{ijk}(2\omega). \qquad (13.2.8)$$

Miller showed that the δ_{ijk} varied much less from one crystal to another than did the corresponding d_{ijk}. In particular, the dispersion of δ_{ijk} within a given material is small. A theoretical justification for this relative constancy of delta can be found and will be given later in the section.

In contrast with the electronic susceptibility, the low-frequency electro-optic coefficients were originally defined in terms of the dielectric imper-meability tensor $\kappa = \varepsilon^{-1}$. In particular, a linear electro-optic coefficient r_{ijk} is conventionally defined by the relation

$$\Delta\kappa_{ij} = r_{ijk}E_k \qquad (13.2.9)$$

where the frequency of the field E_k lies below the vibrational frequencies of the crystal.

The origin of this definition lies in the fact that κ_{ij} describes a refractive-index ellipsoid called the optical indicatrix. In a principal Cartesian axis system x_1, x_2, x_3 it has the equation

$$\left(\frac{x_1}{n_1}\right)^2 + \left(\frac{x_2}{n_2}\right)^2 + \left(\frac{x_3}{n_3}\right)^2 = 1. \qquad (13.2.10)$$

A plane passing through the origin perpendicular to the direction of light propagation defines an ellipse whose semi-major and semi-minor axes are equal to the principal indices of the crystal for light polarized in those directions. Equation (13.2.9) now describes the distortion of this ellipse by the electric field. The coefficient r_{ijk} can be formally related to the non-linear susceptibility $\chi^{NL}_{ijk}(\omega; \omega, 0)$ via the equations $\kappa = \varepsilon^{-1}$, (13.2.9), and (13.1.3).

A second electro-optic coefficient in common use is that relating the impermeability to the polarization change, and is defined by

$$\Delta\kappa_{ij} = f_{ijk}P_k \qquad (13.2.11)$$

where, of course,

$$f_{ijk} = \frac{r_{ijk}}{\varepsilon_k - \varepsilon_0}. \qquad (13.2.12)$$

This 'f-coefficient' may be conveniently termed a polarization-optic coeffi-cient. In the same way as Miller's delta, this f-coefficient is found to vary over a smaller range than the r_{ijk}.

Higher-order coefficients can naturally be defined by a simple extension of the eqns (13.2.4), (13.2.6), (13.2.9) and (13.2.11) to higher orders in P or E. However, in this chapter we only require the quadratic polarization-optic coefficient g_{ijkl} defined by

$$\Delta\kappa_{ij} = f_{ijk}P_k + g_{ijkl}P_kP_l. \tag{13.2.13}$$

At high frequencies the f-coefficients contain only the electronic contributions to polarizability and can be directly related to the d and δ coefficients as described by Wemple and DiDomenico (1972).

Certain relations between the various linear and non-linear susceptibilities may be derived in ferroelectrics from a purely phenomenological approach by expanding the appropriate free energy of the crystal in terms of the order parameter of the low-temperature phase. From Chapter 3 and Chapter 5 we already have seen that the low-frequency dielectric behaviour of ferroelectrics with a centrosymmetric prototype phase can be described quite well by the expansion (omitting subscripts for clarity)

$$G_1 = \tfrac{1}{2}\alpha D^2 + \tfrac{1}{4}\gamma D^4 + \tfrac{1}{6}\delta D^6 \tag{13.2.14}$$

where the electric displacement D is the primary order parameter measured from the high-temperature phase, and the coefficients γ and δ (the latter not to be confused with Miller's delta) are relatively temperature independent.

Differentiating twice with respect to D leads to an expression for dielectric impermeability (see Chapter 3) in the form

$$\kappa = \varepsilon^{-1} = \alpha + 3\gamma D^2 + 5\delta D^4. \tag{13.2.15}$$

Writing D as the sum of the spontaneous polarization P_s and a field-induced electric displacement D_E, we find

$$\kappa = (\alpha + 3\gamma P_s^2 + 5\delta P_s^4 + \ldots) + (6\gamma P_s + 20\delta P_s^3 + \ldots)D_E$$
$$+ (3\gamma + 30\delta P_s^2 + \ldots)D_E^2 \tag{13.2.16}$$

and so on. At optical frequencies, writing $\varepsilon = n^2\varepsilon_0$, we can write

$$(n^2\varepsilon_0)^{-1} = (n_0^2\varepsilon_o)^{-1} + f'D_E + g'D_E^2 + \ldots \tag{13.2.17}$$

where the coefficients of D_E and D_E^2 are primed since f and g conventionally refer to the coefficients in an expansion in terms of polarization P. The first term in eqn (13.2.17) describes the refractive-index ellipsoid in the ferroelectric phase with no applied field, while the higher-order terms describe the distortion of the index ellipsoid by an applied field. At low frequencies the term $f'D_E$ describes the linear (Pockels) electro-optic effect and the term $g'D_E^2$ describes the quadratic (Kerr) effect. At optical frequencies these terms describe the non-linear optical susceptibilities. Comparing eqns (13.2.16) and (13.2.17) we find to the lowest order

$$\left(\frac{1}{n_0^2}\right)_{T<T_c} - \left(\frac{1}{n_0^2}\right)_{T>T_c} = 3\gamma\varepsilon_0 P_s^2 \tag{13.2.18}$$

and

$$f' = 6\gamma P_s = 2g'P_s. \qquad (13.2.19)$$

These two relationships give the interesting results that the change of refractive index along the polar axis should vary as the square of the spontaneous polarization, i.e.

$$\Delta n = \frac{3\gamma\varepsilon_0 P_s^2 n^3}{2} \qquad (13.2.20)$$

and that the linear electro-optic coefficient (or the coefficient for second-harmonic generation) is to a first approximation equal to the quadratic coefficient biased by the spontaneous polarization. This is equivalent to the result obtained in Chapter 5 that the piezoelectric coefficient in the ferroelectric phase is due to electrostriction in the centrosymmetric phase biased by the polarization.

If the paraelectric phase is not centrosymmetric, then odd powers of D must be included in the free-energy expansion (eqn (13.2.14)). The lowest-order term in Δn then becomes linear in P_s, and the coefficient of D_E in eqn (13.2.16) has a term independent of P_s which describes the non-linear properties of the paraelectric phase.

These simple predictions of Landau theory are generally consistent with experimental observations. For instance, it is well known (e.g. Jona and Shirane 1962) that the spontaneous birefringence of ferroelectrics varies either linearly or quadratically with spontaneous polarization in the ferroelectric phase depending on whether the paraelectric phase is piezoelectric (e.g. KDP) or centrosymmetric (e.g. BaTiO$_3$). Furthermore, the temperature dependence of the non-linear coefficients d_{ijk} of BaTiO$_3$ (Miller 1964 a), LiTaO$_3$ (Glass 1968), and LiNbO$_3$ (Bergman 1976) vary linearly with P_s as expected from eqn (13.3.19) (see for example Fig. 13.5). Simultaneous measurements of the spontaneous birefringence and the second-harmonic intensity of KDP (Vallade 1975) show that the linear susceptibility $\chi_{122}^{L}(\omega)$ is proportional to the non-linear susceptibility $\chi_{311}^{NL}(2\omega)$ as shown in Fig. 13.6 and that they are both proportional to P_s as predicted by phenomenological calculations of this kind. There is now a large body of data on both linear and non-linear optical properties of many ferroelectrics which supports this kind of approach. However, these simple relationships are only valid if the electric displacement (spontaneous polarization) is the primary order parameter of the transition and if the temperature variation of the susceptibilities is dominantly due to variation of P_s. If P_s is not the only important parameter of the transition then these relationships may or may not hold, depending upon the microscopic origins of the polarizability and of the polarization. In the ferroelectric NaNO$_2$, for example, there seems to be no simple relationship between the non-linear susceptibilities and the

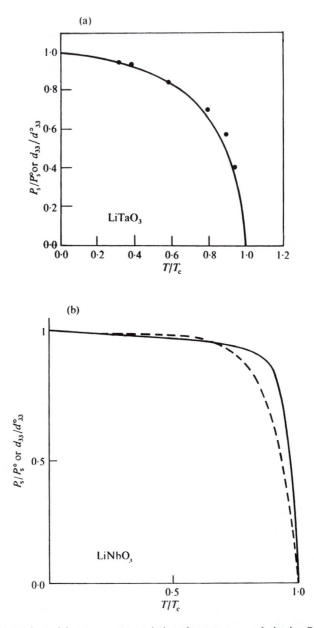

Fig. 13.5. Comparison of the temperature variation of spontaneous polarization P_s with that of the non-linear coefficient d_{33} from second-harmonic generation experiments: (a) for $LiTaO_3$ (after Glass 1968) where the solid curve measures P_s and the filled circles d_{33}; (b) for $LiNbO_3$ (after Bergman 1976), where the solid curve again measures P_s and the d_{33} values are here shown as a broken curve. The reduced values, rather than absolute values, are given for convenience.

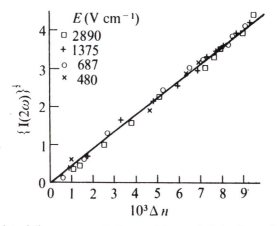

Fig. 13.6. A plot of the square root of second-harmonic intensity against spontaneous birefringence in KDP showing the proportionality between the non-linear coefficient $\chi^{NL}_{311}(2\omega)$ and the linear coefficient $\chi^{L}_{122}(\omega)$. (After Vallade 1975.)

spontaneous polarization, and the different χ^{NL}_{ijk} coefficients differ markedly in their temperature dependence (Vogt 1974).

Since a third-rank tensor such as δ_{ijk} must transform in a well-defined manner under the symmetry operations of the space group to which the crystal belongs, it can be decomposed into irreducible parts (Buckingham and Orr 1967, Jerphagnon 1970). It follows that δ_{ijk} can be expressed as the sum of a vector part $v = \delta_{311} + \delta_{322} + \delta_{333}$ and a so-called septor part s. The exact form of s for the various crystal classes has been tabulated by Jerphagnon. The vector part is only non-zero in crystals which have a unique polar axis while the septor part is present in all acentric crystals. Thus, at a ferroelectric transition from a polar to non-polar phase the vector part disappears to leave only a septor part if the paraelectric phase is piezoelectric. Since v transforms as a vector, Jerphagnon investigated the manner in which this term varies with P_s and found experimentally that v is proportional to P_s for a large number of materials as shown in Fig. 13.7. This result, in fact, is implicit in the phenomenological expansion (eqn (13.2.16)) if full tensor notation is retained.

One of the useful features of the phenomenological approach is that the parameters γ and δ in eqn (13.2.14) do not vary greatly from one material to another. This is sometimes helpful in making rough predictions concerning the behaviour of materials in the ferroelectric phase once the spontaneous polarization is known. However, such predictions are not always reliable. For instance, Miller's δ (which can be related to f') can vary by a factor of 10 or more each side of its average value of about 10^{-6} e.s.u.; it may even be negative. To understand these variations it is necessary to study the microscopic origins of the non-linearities. In the following sections some of the

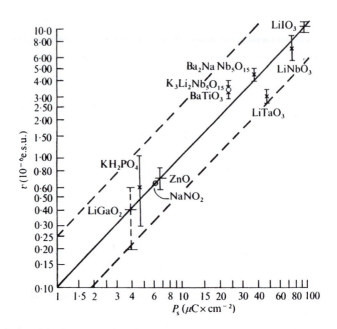

Fig. 13.7. A plot of the Jerphagnon invariant v *versus* spontaneous polarization P_s for a series of materials. (After Kurtz 1972.)

approaches used in describing the non-linear properties of crystals in terms of other macroscopic or microscopic crystalline properties will be discussed.

13.2.2. The anharmonic oscillator

The theories describing the lower-frequency behaviour of ferroelectrics which were developed in the early part of this book assumed that the ionic motion played the dominant role and that any vibronic coupling could be treated in a simple parametric fashion assuming that the valence electrons responded instantaneously to ionic motion. This assumption is often adequate in crystals where the ferroelectric transition is driven by an instability of a lattice-vibrational mode, and can even be expanded (see Chapter 14) to include transformations in which thermal excitation of valence electrons play an important role. For high frequencies, however, the dynamics of the electronic motion becomes of prime importance since the ionic lattice becomes unable to follow the high-frequency fields and only the valence electrons can contribute to the polarizability. In particular, a theory describing the non-linear interaction between low- and high-frequency fields must therefore take into account both the motion of valence electrons and of ions, and requires a model Hamiltonian which includes both electronic and ionic variables as well as pertinent lower-order interactions. The simplest of such models is the anharmonic-oscillator model first used by Bloembergen (1965), and later developed by Kurtz and Robinson (1967) and by Garrett (1968) to include electro-optic effects. The value of the model is that it is sufficiently sophisticated to describe many of the non-linear optical experiments yet sufficiently simple to afford a reasonable intuitive understanding of the microscopic processes involved.

The model then associates a single electronic oscillator p_e, ξ_e with the electronic motion and a single ionic oscillator p_i, ξ_i with the ionic motion and writes a Hamiltonian which includes in the potential term anharmonic contributions to both intra- and inter-oscillator energy to third order as follows:

$$\mathcal{H} = \frac{p_e^2}{2m_e} + \frac{p_i^2}{2m_i} + \tfrac{1}{2}m_e\omega_e^2\xi_e^2 + \tfrac{1}{2}m_i\omega_i^2\xi_i^2$$
$$+ A\xi_i^3 + B\xi_i^2\xi_e + C\xi_i\xi_e^2 + D\xi_e^3 \qquad (13.2.21)$$

where m_e and m_i are respectively the electronic and ionic effective masses, and ω_e and ω_i are the corresponding resonance frequencies. The presence of an applied electric field E adds a term $-E(e_e\xi_e + e_i\xi_i)$ where the quantity in parentheses is the induced dipole moment and e_e and e_i are the relevant effective charges. Following Garrett (1968) an explicit local-effective-field correction to E is not made, any effects of this kind being simply absorbed into the effective-charge-value definitions. The constants A and D refer to the departure from harmonicity of the purely ionic and electronic potentials, with the constants B and C describing the character of the mode interaction.

The lowest-order non-linear terms are cubic since we are assuming a non-centrosymmetric (e.g. pyroelectric) equilibrium configuration. From eqn (13.2.21) we generate the equations of motion

$$\ddot{\xi}_e = -\omega_e^2\xi_e + \frac{e_e}{m_e}E - \frac{3D}{m_e}\xi_e^2 - \frac{2C}{m_e}\xi_e\xi_i - \frac{B}{m_e}\xi_i^2 \qquad (13.2.22)$$

$$\ddot{\xi}_i = -\omega_i^2\xi_i + \frac{e_i}{m_i}E - \frac{C}{m_i}\xi_e^2 - \frac{2B}{m_i}\xi_e\xi_i - \frac{3A}{m_i}\xi_i^2. \qquad (13.2.23)$$

This simple model therefore represents a single $q = 0$ dispersionless mode with the minimum number of parameters to describe linear and non-linear optical effects, including the interaction of low- and high-frequency fields, and even the primary pyroelectric effect. In any specific problem the number of parameters can be reduced by retaining only those terms which are likely to be important.

The simplest case is when all interacting fields have frequencies much greater than ω_i. The ionic displacement is then negligible and the electronic equation of motion simplifies to

$$\ddot{\xi}_e + \omega_e^2 \xi_e + \frac{3D}{m_e} \xi_e^2 = \frac{e_e}{m_e} E. \qquad (13.2.24)$$

The solution of the linear part of this equation in response to a sinusoidal field $E = E_0 e^{-i\omega t} + E_0^* e^{i\omega t}$ is

$$\xi_e^L = \frac{e_e}{m_e} (\omega_e^2 - \omega^2)^{-1} \{E_0 \exp(-i\omega t) + E_0^* \exp(i\omega t)\} \qquad (13.2.25)$$

and the linear susceptibility $N\xi_e^L e_e / \varepsilon_0 E$ follows as

$$\chi_e^L(\omega) = \frac{Ne_e^2}{\varepsilon_0 m_e (\omega_e^2 - \omega^2)} \qquad (13.2.26)$$

where N is the density of electronic oscillators. The first-order non-linear solution of eqn (13.2.24) is

$$\chi_e^{NL}(2\omega; \omega, \omega) = \frac{3DN(e_e^3 / \varepsilon_0 m_e^3)}{(4\omega^2 - \omega_e^2)(\omega^2 - \omega_e^2)^2}. \qquad (13.2.27)$$

This equation can be written in terms of the linear and non-linear susceptibilities in the form

$$\frac{\chi_e^{NL}(2\omega; \omega, \omega)}{\chi_e^L(2\omega)\chi_e^L(\omega)\chi_e^L(\omega)} = \frac{-3\varepsilon_0^2 D}{e_e^3 N^2} \qquad (13.2.28)$$

which is seen, from eqn (13.2.7), to be essentially Miller's delta. All frequency factors have disappeared from the right-hand side of eqn (13.2.28), which accounts for the experimental observations that Miller's delta shows much less dispersion than either the linear or non-linear susceptibilities. It is not clear from this model, however, why Miller's delta should be relatively constant from one material to another. This implies that the ratio D/N^2 does not vary greatly for different materials. It has been noted (Garrett and Robinson 1966) that, if D is estimated by setting the harmonic and anharmonic terms equal in eqn (13.2.24) when ξ_e is equal to one lattice spacing, Miller's delta is indeed the correct order of magnitude, but this does not explain why D/N^2 should not vary greatly.

The second case to consider is the interaction of three fields, one of which has a frequency comparable with (or lower than) the ionic resonance ω_i while the other two have frequencies much greater than ω_i. This situation occurs in an experiment where two optical frequencies ω_2 and ω_3 are mixed via crystal non-linearity to generate a difference frequency $(\omega_3 - \omega_2) \lesssim \omega_i$. The nonlinear susceptibility $\chi^{NL}(\omega_3 - \omega_2; \omega_3, \omega_2)$ can be calculated in a similar manner as before, but the C term describing the electron-phonon coupling must now be included as well as the D term. The linear response at

ω_3 and ω_2 is still overwhelmongly electronic so that the higher-order B interaction term in the potential is not needed. The solution of eqns (13.2.23) and (13.2.24) are then (Garrett 1968)

$$\chi_e^{NL}(\omega_3 - \omega_2; \omega_3, \omega_2) = \frac{3(D/m_e)}{\omega_e^2 - (\omega_3 - \omega_2)^2} F_{\omega_3 - \omega_2}\{(\xi_e^L)^2\} \quad (13.2.29)$$

$$\chi_i^{NL}(\omega_3 - \omega_2; \omega_3, \omega_2) = \frac{(C/m_i)}{\omega_i^2 - (\omega_3 - \omega_2)^2} F_{\omega_3 - \omega_2}\{(\xi_e^L)^2\} \quad (13.2.30)$$

where F_ω is the Fourier component at frequency ω. The total non-linear susceptibility is therefore

$$\chi^{NL}(\omega_3 - \omega_2; \omega_3, \omega_2) = \chi_e^{NL} + \chi_i^{NL}$$

$$= -\frac{2\varepsilon_0^2}{e_e^2 N^2} \chi_e^L(\omega_3)\chi_e^L(\omega_2)\left\{\frac{3D}{e_e}\chi_e^L(\omega_3 - \omega_2) + \frac{C}{e_i}\chi_i^L(\omega_3 - \omega_2)\right\} \quad (13.2.31)$$

which now replaces Miller's rule. The presence of the last term in C shows that the non-linear susceptibility becomes resonant when $\omega_3 - \omega_2 = \omega_i$, but the crystal, of course, is highly absorbing at this frequency.

An expression for the electro-optic susceptibility $\chi^{NL}(\omega; \omega, 0)$ may be calculated in the same way or by interchanging both the indices and the frequencies in eqn (13.2.31), i.e.

$$\chi^{NL}(\omega; \omega, 0) = -\frac{2\varepsilon_0^2}{e_e^2 N^2}\{\chi_e^L(\omega)\}^2\left\{\frac{3D}{e_e}\chi_e^L(0) + \frac{C}{e_i}\chi_i^L(0)\right\}. \quad (13.2.32)$$

Both the optical mixing and electro-optic susceptibilities are expressed as the sum of an electronic part and an ionic part. The electronic part can be calculated directly from the linear electronic susceptibilities (i.e. refractive indices) with the non-linear susceptibility for second-harmonic generation in particular being given by eqn (13.2.28). Kaminow and Johnston (1967) have shown that the ionic contribution can be calculated from the Raman scattering cross-sections and the infrared oscillator strengths of the polar optic modes of the crystal (each mode contributing separately to the non-linear susceptibility). In fact, Kaminow and Johnston find from experimental measurements on $LiNbO_3$ that the electronic contribution accounts for less than 10 per cent of the electro-optic effect (although it is by no means negligible) at room temperature.

From eqn (13.2.32) we can also see that at temperatures approaching a ferroelectric phase transition, when the ionic part of the linear polarizability becomes very much greater than the electronic part, the temperature dependence of the electro-optic susceptibility should be essentially that of the linear susceptibility. This is indeed confirmed experimentally. The dispersion of χ^{NL} should follow that of $(\chi^L)^2$.

The pyroelectric effect can also be determined within the anharmonic-oscillator model, since the change of equilibrium displacement with temperature represents a change of the spontaneous polarization. However, since the displacements are not expressed with respect to the paraelectric phase in the model, the spontaneous polarization itself is not predicted. From the potential energy of the system we readily establish that the equilibrium displacements $\langle \xi_e \rangle$ and $\langle \xi_i \rangle$ are given by the solutions of

$$\frac{\partial \mathcal{H}}{\partial \xi_e} = m_e \omega_e^2 \xi_e + 3D\xi_e^2 + 2C\xi_e \xi_i + B\xi_i^2 = 0$$

(13.2.33)

$$\frac{\partial \mathcal{H}}{\partial \xi_i} = m_i \omega_i^2 \xi_i + C\xi_e^2 + 2B\xi_e \xi_i + 3A\xi_i^2 = 0.$$

If we assume that the electronic oscillator is not thermally excited ($kT \ll \hbar\omega_e$) then we can (classically) set $\langle \xi_e^2 \rangle = 0$. Also, since we presume ξ_e and ξ_i to be essentially uncorrelated, we also have $\langle \xi_e \xi_i \rangle = 0$. With these assumptions it follows from (13.2.33) that

$$\langle \xi_e \rangle = -\frac{B\langle \xi_i^2 \rangle}{m_e \omega_e^2}$$

(13.2.34)

$$\langle \xi_i \rangle = -\frac{3A\langle \xi_i^2 \rangle}{m_i \omega_i^2}.$$

(13.2.35)

The average thermal energy per atom $m_i \omega_i^2 \langle \xi_i^2 \rangle$ may be expressed in terms of the specific heat of the system C_v as

$$\langle \xi_i^2 \rangle = \frac{1}{m_i \omega_i^2} \int C_v \, dT$$

(13.2.36)

so that the pyroelectric coefficient p is given by

$$-p = -\frac{1}{V}\frac{d}{dT}(e_e \langle \xi_e \rangle + e_i \langle \xi_i \rangle) = \frac{C_v}{Vm_i \omega_i^2}\left(\frac{e_e B}{m_e \omega_e^2} + \frac{3e_i A}{m_i \omega_i^2}\right)$$

$$= \frac{C_v \varepsilon_0}{NVm_i \omega_i^2}\left(\frac{B}{e_e}\chi_e^L(0) + \frac{3A}{e_i}\chi_i^L(0)\right)$$

(13.2.37)

where V is the unit-cell volume. The model therefore predicts that the pyroelectric compliance p should vary in the same way as the specific heat. However, since the model takes no account of the existence of acoustic modes, this result is not expected to hold at low temperatures, and nor will it be applicable near T_c where the pyroelectric behaviour is dominated by a temperature-dependent mode. Nevertheless, the ratio p/C_v is in general found to be only slowly varying at temperatures where there is no appreciable mode softening compared with the variation of p or C_v alone.

Comparing eqns (13.2.32) and (13.2.37) it can be seen that a simple relationship exists between the electro-optic susceptibility and the pyroelectric coefficient if the terms in χ_e^L are small compared with the terms in χ_i^L. This is expected to be true for most pyroelectrics. Then we find

$$p = \chi^{NL}(\omega; \omega, 0)\left\{\frac{3ANe_e^2}{2CVm_i\omega_i^2\varepsilon_0(\chi_e^L)^2}\right\}C_v. \qquad (13.2.38)$$

It is found experimentally (Soref 1969) that the quantity in parenthesis in eqn (13.2.38) is remarkably constant from one material to another. Indeed, p is directly proportional to χ^{NL} within a factor of 2 for a wide range of materials covering more than three decades of p. This implies once again that the anharmonicity parameters, or at least the ratios A/C, do not vary greatly from one material to another.

13.2.3. Bond anharmonic polarizability model

Despite its conceptual appeal the anharmonic-oscillator model has the disadvantage that it fails to provide a basis for computing the acentricity parameters A, B, C, D of the interionic potential, nor does it shed any light on the relationship between the different elements of the non-linear susceptibility tensor. Calculation of the linear and non-linear susceptibilities from a general quantum-mechanical approach is too complicated to be useful since a precise knowledge of the electronic wavefunctions of a crystal is not generally available. Electronic contributions to the electro-optic effect have been calculated from approximate ground-state wavefunctions in a few simple systems with moderate success (Robinson 1967, Flytzanis and Ducuing 1969) but there have been no attempts to tackle ferroelectrics from first principles. The theoretical situation becomes even more complicated when ionic motion contributes to the distortion of the electronic charge distribution at low frequencies.

The most useful approaches to describe the non-linear optical properties of complex crystals rely on more phenomenological theories based on physically realistic models. The bond-anharmonicity model is particularly appealing since it is capable of describing the macroscopic properties of crystals by means of a few parameters which have a well-defined physical origin. Furthermore, Levine (1973 a, b) has shown that the non-linear optical coefficients may be calculated from the crystal structure and the linear dielectric properties within the framework of the Phillips–Van Vechten (1969) dielectric theory. The predicted coefficients for second-harmonic generation are in excellent agreement with experiment for a wide variety of materials including complex ferroelectric structures.

The starting point of the bond-anharmonicity model is to resolve the macroscopic susceptibilities into the contributions of individual valence bonds of the structure, assuming that the bond polarizabilities are additive.

This approach has been successfully used for many years to describe the linear susceptibilities of inorganic compounds, and its application to non-linear susceptibilities was proposed by Robinson. One writes, for example, a sum over bonds b of the form

$$d_{ijk} = \sum_b G_{ijk}^b(lmn)\beta_{lmn}^b \qquad (13.2.39)$$

where local-field correction factors have been absorbed into the non-linear bond polarizabilities β and $G_{ijk}^b(lmn)$ are geometric factors relating the bond direction lmn to the crystal axes ijk. Decompositions of this type can be equally well expressed in terms of Miller's delta for the bonds, and in some respects this approach is more useful since the reduced dispersion of δ makes comparison of theory with experiment more straightforward. Furthermore, the δ-tensor is independent of local-field factors (Robinson 1967). Nevertheless the more familiar d_{ijk} and β parameters are retained here.

In simple structures such a microscopic picture sometimes allows relationships between different elements of d_{ijk} to be determined. For instance, in the wurtzite structure the non-linear polarizability arises from the tetrahedral bonds, and one finds (Robinson 1968) that

$$d_{311}/d_{333} = -\tfrac{1}{2} \qquad (13.2.40)$$

in reasonable agreement with experiment for a number of compounds.

The simplification afforded by decomposing the macroscopic properties of the crystal into those of its bonds lies in the fact that useful approximations can be made to reduce the number of independent tensor components β_{lmn} based on physical models. A reasonable simplification is to assume that the bonds are cylindrically symmetrical with non-linear polarizabilities $\beta_{\parallel}$ and $\beta_{\perp}$ parallel and perpendicular to the bond axis. In the zincblende structure, which has only one type of bond in tetrahedral arrangement, the only macroscopic coefficient d_{14} can then be expressed as

$$d_{14} = \frac{8}{3\sqrt{3}}(\beta_{\parallel} - 3\beta_{\perp}). \qquad (13.2.41)$$

In more complex structures there may be several dissimilar bonds so that the number of parameters must be further reduced for such a model to be useful.

The quantum calculations of Flytzanis and Ducuing (1969) and of O'Hare and Hurst (1967) show that the non-linear bond polarizability is generally highly anisotropic with $\beta_{\parallel} \gg \beta_{\perp}$ so that to a reasonable approximation $\beta_{\perp}$ may be neglected (the calculations do depend rather critically on the details of the ground-state wavefunctions but the approximation $\beta_{\perp} \to 0$ will be discussed below). Even in quite complex structures the non-linear coefficients can be described in terms of a few of the more polarizable bonds

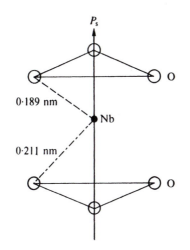

Fig. 13.8. Schematic of the NbO_6 octahedron on $LiNbO_3$, showing the two dissimilar Nb—O bonds.

(Jeggo and Boyd 1970). For instance, in $LiNbO_3$ the polarizability of the Li—O bond is neglected and the non-linear polarizability is ascribed solely to the covalent Nb—O bonds. The NbO_6 octahedron, shown in Fig. 13.8, has two different Nb—O bonds of length 0·189 nm and 0·211 nm for which we define polarizabilities β_1 and β_2 respectively. The three coefficients for second-harmonic generation allowed by the crystal symmetry (3m) and Kleinman's symmetry conditions are

$$d_{311} = \frac{1}{V}(1 \cdot 104\beta_1 - 1 \cdot 106\beta_2)$$

$$d_{333} = \frac{1}{V}(0 \cdot 643\beta_1 - 1 \cdot 796\beta_2) \qquad (13.2.42)$$

$$d_{222} = \frac{1}{V}(0 \cdot 396\beta_1 - 0 \cdot 195\beta_2)$$

from which the betas are overdetermined and the three d-coefficients can be related to each other. For $LiNbO_3$ it is found from d_{311} and d_{222} that

$$\beta_1 = (-51 \pm 7) \times 10^{-28} \, \text{m}^3 \times d_{312}^{\text{KDP}}$$
$$\beta_2 = (-64 \pm 8) \times 10^{-28} \, \text{m}^3 \times d_{312}^{\text{KDP}} \qquad (13.2.43)$$

giving $d_{333} = (70 \pm 9)d_{312}^{\text{KDP}}$ in good agreement with the measured value $d_{333} = (86 \pm 22)d_{312}^{\text{KDP}}$ and thereby justifying the neglect of the Li—O bond $(d_{312}^{\text{KDP}} = 4 \cdot 4 \times 10^{-13} \, \text{MKS})$.

In comparing different ferroelectric niobates and tantalates Jeggo and Boyd found that the absolute magnitude of $\beta_\parallel$ for the Nb—O bonds increased (became increasingly negative) with increasing bond length although, since the range of bond lengths was small, no reliable power dependence could be derived from the data.

It is clear from Fig. 13.8 that if the Nb ion is moved along the threefold axis of the octahedron the macroscopic d-coefficients can vary both because of the change in the angle of the Nb—O bonds and via the change in length of the individual bonds. Since the primary order parameter is the spontaneous polarization in LiNbO$_3$ the displacement of the niobium ion along the polar axis is proportional to P_s, so that as a first approximation this model predicts that the d_{ijk} are also proportional to P_s. The relationship should, for example, be adequate for describing the temperautre or pressure variation of the d_{ijk} in a specific material, or even within a group of structurally and chemically similar materials. However, since this result has already been obtained from the purely phenomenological method of § 13.2.1 it is not really model dependent so that any reasonable parameterization procedure should deduce it. Indeed Bergman (1976) has obtained a good fit to the experimental data for a variety of materials using the two parameters $\beta_\parallel$ and $\beta_\perp$, neglecting the bond-length variation of β, and describing the temperature dependence of d_{ijk} in terms of the rotation of the bonds alone.

The advantage of a more physical approach is its ability to go a step beyond the mere deduction of relationships between tensor coefficients or relative variations of d_{ijk}. Levine has shown that, by using a reasonable model for the bond charge, the non-linear polarizability $\beta_\parallel$ can be calculated with surprising accuracy from the linear polarizability of the bonds. This procedure allows one to predict the non-linear optical properties of a material of known structure without adjustable parameters which must be fitted to measurements of d_{ijk}.

A schematic representation of the bonding region between two atoms A and B is shown in Fig. 13.9. The bond charge q is situated at the covalent radii r_A and r_B. The motion of the charge in response to an applied field E is responsible for all the linear and non-linear polarizability. The bond charge q can be thought of as arising from two sources. Firstly the negative bond charge compensates for the unscreened charge on the atomic cores. The atomic cores are only partially screened by the dielectric constant, the unscreened portion being proportional to ε^{-1}. The second contribution is due to the overlap in the bonding region, which is proportional to the fractional covalency of the bond f_c.

If we define a field dependent bond polarizability

$$\alpha = \alpha_0 + \beta E \tag{13.2.44}$$

where the local-field correction factors have been absorbed into α_0 and β,

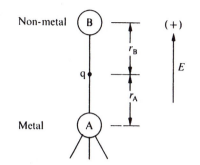

Fig. 13.9. Schematic representation of the bonding region between a metal atom and a non-metal atom showing the bond charge q located at a distance r_A and r_B from the A and B atoms. The posititive electric field (E) direction is also shown as pointing from metal to non-metal (i.e. cation to anion). (After Levine 1973 a.)

then we can express the non-linear polarizability as the field derivative of the linear polarizability

$$\beta = \frac{\partial \alpha}{\partial E}. \qquad (13.2.45)$$

The application of a field displaces the bond charge by an amount Δr given by the linear polarizability

$$\alpha E = q \, \Delta r \qquad (13.2.46)$$

so that (13.2.45) can be rewritten in the form

$$\beta = \frac{\partial \alpha}{\partial r} \frac{\partial r}{\partial E} = \frac{\alpha}{q} \frac{\partial \alpha}{\partial r} \qquad (13.2.47)$$

and the problem of estimating non-linear susceptibility reduces to one of expressing α in terms of the covalent radii r_A and r_B. To do this the anharmonic potential seen by the bond charge must be evaluated. The expressions given by Phillips and Van Vechten (1969) are well suited for this purpose since they give an excellent account of the linear susceptibility and the covalency of the bond. Phillips and Van Vechten's theory starts with pseudopotentials for atoms A and B in a diatomic unit cell and separates the electronic potential into symmetric and antisymmetric parts. These potentials lead to a series of energy bands (depending on the structure) which may be represented by an average covalent energy gap E_h associated with the symmetric potential and an ionic energy gap C associated with the antisymmetric potential. The effective energy gap between the valence and conduction bands is then given by

$$E_g^2 = E_h^2 + C^2. \qquad (13.2.48)$$

A measure of the covalency of the bond can then be defined as

$$f_c^2 = 1 - \frac{C^2}{E_g^2}.$$ (13.2.49)

In particular, for homopolar crystals $C = 0$ and $f_c = 1$.

The linear optical susceptibility is given by

$$\chi_e^L = N_b \alpha = \frac{(\hbar\omega_p)^2}{E_g^2}$$ (13.2.50)

where N_b is the number of bonds per unit volume and where the plasma frequency ω_p is obtained from the number N of valence electrons per unit volume in the form

$$\omega_p^2 = \frac{4\pi N e^2}{m}$$ (13.2.51)

in which e and m are the charge and mass of an electron. Expressing ω_p in terms of N and e rather than the number of bonds per unit volume and the bond charge avoids the introduction of local-field factors.

Now the antisymmetric energy gap C is given by the difference between the screened Coulomb potentials of atoms A and B in the form

$$C^2 = b \left\{ \frac{Z_A}{r_A} - \frac{Z_B}{r_B} \right\} \exp\{-\tfrac{1}{2}k(r_A + r_B)\},$$ (13.2.52)

where the exponential term is the Thomas–Fermi screening factor and the factor b is approximately constant within any crystal class. The covalent energy gap, on the other hand, depends on the average core radius r_c according to

$$E_h^{-2} \propto \{(r_A - r_c)^{2s} + (r_B - r_c)^{2s}\}$$ (13.2.53)

where the exponent $s \approx 2 \cdot 5$. Eqns (13.2.52) and (13.2.53) now describe the dependence of the average energy gaps on the geometry of the bonds. Together with eqns (13.2.48) and (13.2.50) we immediately obtain the dependence of χ_e^L or α on the covalent radii, from which it is possible to calculate β from eqn (13.2.47). We find that the non-linear polarizability β can be written as the sum of two terms, one coming from the C^2 term and one from the E_h^2 term. More specifically

$$\beta = \frac{-\alpha^2}{q E_g^2} \frac{\partial}{\partial r} (E_h^2 + C^2)$$ (13.2.54)

where we have assumed that the bond length $r_A + r_B$ remains constant so that the plasma frequency does not change.

The macroscopic d_{ijk} coefficients are now obtainable by summing β for each bond according to eqn (13.2.39) to give a final form

$$d_{ijk} = \sum_b G_{ijk}(lmn)\{\phi_1^b(r_A^b - r_B^b) + \phi_2^b(C^b)\}_{lmn} \qquad (13.2.55)$$

where the explicit form for the two terms in parenthesis may be obtained by differentiating E_h^2 and C^2 (see Levine 1973 a, b). Eqn (13.2.55) has been written in this way to emphasize the two sources of bond acentricity. The first is due to the difference in atomic sizes and the second to the ionicity of the bond. The quantities entering into ϕ_1 and ϕ_2 are all symmetrical with respect to atoms A and B, but both ϕ_1 and ϕ_2 depend in a complicated way on the bond length $r_A + r_B$.

The sources of acentricity in eqn (13.2.55) show clearly how cancellation effects may lead to small optical non-linearities if they have opposite signs, and how d_{ijk} may have either sign. The first term is only significant when the two atoms have quite different radii, as for example in a bond containing one atom from the first row of the periodic table. If the ionicity term is dominant, then the non-linear susceptibility of the bond varies primarily with the difference of the core charges $Z_A - Z_B$ and with the bond length $r_A + r_B$, which is the nearest-neighbour distance. It is found experimentally that for compounds with the same $Z_A - Z_B$ the non-linear susceptibility increases as the bond length increases in agreement with this model, but no simple power-law dependence of β on the bond length is predicted by eqn (13.2.55).

The most important point about the relationship (eqn (13.2.54)) is that the non-linear susceptibility depends only on those quantities which enter into the linear susceptibility (eqn (13.2.50)). Indeed, all the parameters in the model are macroscopic in that the effects of the structure are included in the pre-screening factor b and the fact that local-field factors are included in β. Once the factor b has been determined from a measurement of the linear susceptibility of a specific simple structure, then the same factor can be used to determine the linear susceptibility of bonds in similar structural arrangements using eqn (13.2.55) and the covalent radii which have been tabulated by Van Vechten (1969). This now allows one to calculate the linear and non-linear susceptibilities of the individual bonds in more complex multi-bond crystals. For instance, the $LiNbO_3$ and $LiTaO_3$ structures are decomposed into Nb—O, Ta—O, and Li—O bonds. The susceptibilities of the Nb—O and Ta—O bonds are then obtained from the long-wavelength refractive indices of Nb_2O_5 and Ta_2O_5, while those of the Li—O bond are obtained by scaling the inverse linear susceptibility of known binary prototype crystals with respect to Z_A-Z_B (see Levine 1973 a, b). It is found that nearly all the linear (83 per cent) and the non-linear (~ 99 per cent) susceptibilities of $LiNbO_3$ may be attributed to the Nb—O bonds (and similarly for $LiTaO_3$ to the Ta—O bonds). In fact for the Nb—O bonds Levine finds $\beta_1 = -47$ and $\beta_2 = -60$ (units of 10^{-28} m^3 d_{312}^{KDP} which may be compared with eqn (13.2.43).

The bond-charge model can also predict d_{36} of KDP with remarkable accuracy. In the acentric paraelectric phase the H—O bonds are disordered and do not contribute to the macroscopic non-linear susceptibility on the average although they do contribute to the linear susceptibility. Since the K^+ ions are expected to contribute a negligible amount, all the non-linearity resides in the P—O bonds. The susceptibility of these bonds is obtained by using the value of the pre-screening factor b of the binary compound SiO_2, since the Si atoms in SiO_2 have the same tetrahedral co-ordination as the P atoms in KDP. Miller's delta calculated using this procedure is in good agreement with experiment:

$$\delta_{36}(calc) = +0.84 \times 10^{-6} \text{ e.s.u}$$

$$\delta_{36}(expt) = +1.18 \times 10^{-6} \text{ e.s.u.}$$

The good agreement in many of these complex crystals shows that one can indeed use the bond non-linearity determined from one crystal in a different compound provided that the two crystals are sufficiently similar. In fact the agreement between the total linear susceptibility of the crystal and the sum of the linear susceptibilities of the individual bonds affords an independent check on the correctness of the procedure.

The non-linear polarizability perpendicular to the bond axis does not appear in the bond-charge model since it is clear from Fig. 13.9 that the largest asymmetry in the bond-charge motion lies along the bond between dissimilar atoms. For an isolated bond the motion perpendicular to the bond axis is symmetric in lowest order so that non-linearity here is a higher-order effect. This is consistent with the quantum-mechanical calculations of Flytzanis and Ducuing and means that in the tetragonal phase of $BaTiO_3$, for example, the four Ti—O bonds which are essentially perpendicular to the polar axis make a negligible contribution to d_{333} because of the unfavourable geometric factors associated with $\beta_{\parallel}$. Essentially all the non-linear polarizability comes from the two Ti—O bonds parallel to P_s, which do not cancel because of their unequal lengths. In fact, in this case the bond-charge model gives good agreement with experiment. If, on the other hand, two parameters $\beta_{\parallel}$ and $\beta_{\perp}$ are used to describe d_{333} without considering the bond-length dependence of $\beta_{\parallel}$, then the main contribution appears to come from $\beta_{\perp}$ of the four bonds 'perpendicular' to P_s and the temperature dependence of d_{333} is then due to the rotation of these bonds with respect to the crystal axes.

The latter kind of parametrization has been used by Bergman and Crane (1975) and is able to account for the non-linear coefficients of a number of compounds with reasonable accuracy. For example, the three compounds $LiNbO_3$, $Ba_2NaNb_5O_{15}$, and $KLiNb_2O_6$ (all basically $Nb—O$ bond compounds) have measured non-linear susceptibilities which can be accounted for with the single pair of values $\beta_\parallel = (82 \pm 7)10^{-28} \, m^3 \times d_{312}^{KDP}$ and $\beta_\perp = (12 \pm 3)10^{-28} \, m^3 \times d_{312}^{KDP}$ neglecting completely any variation of $\beta_\parallel$ with bond length. The iodates also provide a large set of data in a group of materials where the IO_6 configurations are very similar within the group and where the $I—O$ bonds are the primary contributors to the susceptibility. Using six parameters, viz. $\beta_\parallel$ and $\beta_\perp$ for the long $I—O$ bonds (0·29 nm), short $I—O$ bonds (0·19 nm), and the lone pair of iodine valence electrons, which are well overdetermined by the data, Bergman has found that $\beta_\perp$ can be as high as 30 per cent of $\beta_\parallel$. (By using different β for the long and short bonds some bond-length dependence has now been implicitly recognized.)

From purely physical arguments it is hard to determine which of the above approaches is more realistic. In the real situation we know that $\beta_\perp$ is certainly not zero, but we also know from the pressure dependence of the linear susceptibility that the linear and non-linear polarizabilities definitely do depend on bond length. The neglect of one or the other of these effects is therefore a hypothesis, but the numerical successes and predictive aspects of the bond-charge model lend some support to this approach.

13.2.4. The polarization potential

The concept of a polarization potential was introduced earlier in the book (e.g. § 12.5) to describe the Stark-like shift of the electronic energy bands of a ferroelectric due either to an applied field or to a spontaneous polarization, i.e.

$$\Delta E_g^n(\mathbf{q}) = \sum_{ij} \sigma_{ij}^n(\mathbf{q}) P_i P_j \qquad (13.2.56)$$

where the polarization potentials σ_{ij} depend on the particular interband transition n of concern and the polarization P is the total polarization measured from the non-polar phase.

DiDomenico and Wemple (1969) have shown that the electro-optic properties of ferroelectrics may be simply expressed in terms of an effective polarization potential associated with the lowest-energy intraband transitions. The usefulness of the approach stems from the experimental evidence that the ferroelectric oxides, which comprise the largest group of electro-optic materials of practical importance, have similar optical properties near the electronic band edge and that the polarization potential associated with this lowest-energy transition is almost the same for all oxygen-octahedra ferroelectrics.

The similarity of the optical properties lies in the fact that the BO_6 octahedron governs the lower-lying conduction bands and the upper valence bands as discussed in § 12.5. Other ions in the structure contribute to higher-lying conduction states, but these generally have only a small effect on the optical properties of the crystals in the visible region of the spectrum provided the electronic polarizability of the ions is small.

The relationship between the quadratic electro-optic effect and the polarization potential may be derived directly from eqn (13.2.56) and the Sellmeier dispersion relation, which states that in regions of low optical absorption the index of refraction n of a dielectric is given by

$$n^2(\omega) - 1 = \sum_i \frac{f_i}{\omega_i^2 - \omega^2} \qquad (13.2.57)$$

and separates the important interband optical transitions into individual dipole oscillators of strength f_i and frequency ω_i. In the transparent region of ferroelectrics the lowest-energy oscillator $E_g = \hbar\omega_e$ is the largest contributor to the dispersion of the refractive index and all other oscillators can, to a good approximation, be combined to form a single constant term A, i.e.

$$n^2(\omega) - 1 = A + \frac{f_e}{\omega_e^2 - \omega^2} \qquad (13.2.58)$$

where all the parameters can be obtained from measurements of $n(\omega)$. For oxygen-octahedra ferroelectrics the lowest-energy oscillator corresponds to excitations to the B-cation d-electron t_{2g} band, and for many the ultraviolet reflectivity spectra (Cardona 1965, Kurtz 1966) show that the associated energy gap is $E_g \approx (5 \pm 0.5)$ eV. The next-highest conduction band at about 9 eV is also derived from the B-cation d-orbitals (the e_g symmetry band) and can be included in a more general discussion, but the essential arguments are retained by considering only the lowest t_{2g} band in eqn (13.2.58).

Differentiating eqn (13.2.58) and assuming that the oscillator strength remains constant we find

$$\Delta n = \frac{\omega_e f_e \, \Delta E_g}{n \hbar (\omega_e^2 - \omega^2)^2}. \qquad (13.2.59)$$

It is now convenient to introduce the experimentally accessible polarization potential β, as in § 12.5, which is an average of the σ_{ij} weighted by the appropriate selection rules. The parameter will depend on the direction of light polarization with respect to the axes of the oxygen octahedra. For instance, if the polarization P is along the fourfold octahedral axis, then we can write the band-edge shifts observed with light polarized parallel and perpendicular to P in the form

$$\Delta E_\| = \beta_{11} P^2$$
$$\Delta E_\perp = \beta_{12} P^2. \qquad (13.2.60)$$

Alternatively if P is along the threefold axis, then

$$\Delta E_\parallel - \Delta E_\perp = \beta_{44} P^2. \qquad (13.2.61)$$

The β_{ij} defined in this way then correspond to the three independent g-coefficients g_{11}, g_{12}, and g_{44} (in reduced notation) of crystals with O_h symmetry.

From eqns (13.2.59) and (13.2.60) we have, dropping the subscripts,

$$\Delta n = \left\{ \frac{\omega_e f_e \beta}{\hbar n^4 (\omega_e^2 - \omega^2)^2} \right\} n^3 P^2. \qquad (13.2.62)$$

Comparing this expression with the definition of the quadratic electro-optic coefficient

$$n = \tfrac{1}{2} n^3 g P^2 \qquad (13.2.63)$$

we find the relationship

$$g = \frac{2\omega_e f_e \beta}{\hbar n^4 (\omega_e^2 - \omega^2)^2}. \qquad (13.2.64)$$

At frequencies well below ω_e we see that since β, ω_e, f_e, and n are nearly the same for the entire class of oxygen-octahedra ferroelectrics (because of the similar band structure) then we also expect similar g-coefficients. Using typical values for these parameters one finds

$$g_{ij} \approx \beta_{ij}/20 \qquad (13.2.65)$$

where g_{ij} is in m^4/C^2 and β_{ij} is in $eV\, m^4\, C^{-2}$. For the perovskite oxides $g_{11} \approx 0 \cdot 17$, $g_{12} \approx 0 \cdot 04$, and $g_{44} \approx 0 \cdot 12$ ($m^4\, C^{-2}$) giving values $\beta_{11} \approx 3 \cdot 4$, $\beta_{12} \approx 0 \cdot 8$, and $\beta_{44} \approx 2 \cdot 4$ ($eV\, m^4\, C^{-2}$) in good agreement with both theoretical estimates (Brews 1967) and with experimental measurement of field-induced shifts of the band edge.

The quadratic electro-optic coefficients of non-perovskite ferroelectric oxides are in closer agreement with the perovskite values if proper account is taken of the packing density ζ of the oxygen octahedra in the structure. The parameter ζ is defined as the number of octahedra per unit volume normalized so that $\zeta = 1$ in perovskites. It is clear that for a constant dipole moment μ of each octahedron the spontaneous polarization and the linear susceptibility scale linearly with ζ, i.e.

$$P_x = \zeta P_p \qquad (n_x^2 - 1) = A\zeta \qquad (13.2.66)$$

where $A \approx 4$ for oxygen-octahedra ferroelectrics and the subscripts x and p refer to complex oxides and perovskites respectively. Since, from eqn (13.2.63) we have

$$\Delta n_p^2 = n_p^4 g_p P_p^2 \qquad \Delta n_x^2 = n_x^4 g_x P_x^2 \qquad (13.2.67)$$

it follows that

$$\frac{g_p}{g_x} = \zeta\left(\frac{1+A\zeta}{1+A}\right)^2 \sim \zeta^3 \qquad (13.2.68)$$

where the last step follows since $\rho \gtrsim 1$ (and $A \sim 4$) for these materials. A list of the parameters entering into eqn (13.2.68) is shown in Table 13.1 for several oxide ferroelectrics. The relative invariance of the quadratic electro-optic coefficients when corrected for the packing density is immediately evident.

TABLE 13.1

Room-temperature spontaneous polarization and equivalent perovskite quadratic electro-optic coefficients for several oxygen-octahedron ferroelectrics

Material	ζ	P_s	$(g_{11})_p - (g_{12})_p$	$(g_{44})_p$	$(g_{11})_p$	$(g_{12})_p$
LiNbO$_3$	1·2	0·71	0·12	0·11	0·16	0·043
LiTaO$_3$	1·2	0·50	0·14	0·12	0·17	0·03
Ba$_2$NaNb$_5$O$_{15}$	1·03	0·40	0·12	0·12	0·17	0·048
Ba$_{1/2}$Sr$_{1/2}$Nb$_2$O$_6$	1·06	0·25	0·13	—	—	—
BaTiO$_3$	1·0	0·25	0·14	—	—	—

From DiDomenico and Wemple 1969.
All values in MKS units.

The g_{ij} coefficients are simply related to the linear polarization-optic coefficients f as in eqn (13.2.19). For C_{4v} structures (perovskites or tungsten bronzes) the specific relationships are

$$f_{13} = 2g_{12}P_s$$
$$f_{33} = 2g_{11}P_s \qquad (13.2.69)$$
$$f_{51} = g_{44}P_s$$

and for C_{6v} structures the corresponding equations are

$$f_{13} = \tfrac{2}{3}(g_{11} + 2g_{12} - g_{44})P_s$$
$$f_{33} = \tfrac{2}{3}(g_{11} + 2g_{12} + 2g_{44})P_s \qquad (13.2.70)$$
$$f_{51} = \tfrac{2}{3}(g_{11} - g_{12} + \tfrac{1}{2}g_{44})P_s.$$

Since the g_{ij} are relatively invariant, apart from small packing-density factors, we see that large linear polarization-optic coefficients can only be obtained with structures having a large spontaneous polarization.

13.3. Hyper-Rayleigh and hyper-Raman scattering

Any spatial or temporal modulation of the non-linear optical susceptibility χ_{ijk}^{NL} gives rise to second-harmonic scattering. Since the selection rules determining second-harmonic scattering differ from those of ordinary Rayleigh and Raman scattering of fundamental light, supplementary information can be obtained by such measurements. Elastic and inelastic scattering of second-harmonic light is referred to as hyper-Rayleigh and hyper-Raman scattering.

The simplest example, which has already been touched upon in Chapter 4 as a means of observing domains, is the change in sign of χ_{ijk}^{NL} at $180°$ domain walls in ferroelectrics, We can write this as

$$\chi^{NL}(\mathbf{r}) = M(\mathbf{r})\chi_0^{NL} \tag{13.3.1}$$

where $M(r)$ is the modulation function which has the values ± 1 within positive and negatively oriented domains. The scattered harmonic intensity is determined by the Fourier transform $M(\mathbf{q})$ of $M(\mathbf{r})$, i.e.

$$I^{NL} \propto |M(\mathbf{q})|^2 \tag{13.3.2}$$

with

$$\mathbf{q} = \mathbf{q}_2 - 2\mathbf{q}_1. \tag{13.3.3}$$

An example of second-harmonic scattering (Vogt 1974) from domain walls is shown in Fig. 13.10 for $NaNO_2$ which has the domain structure shown in the insert. Since the domain walls are perpendicular to the a-axis, wavevector $\mathbf{q}$ must be parallel to $\mathbf{a}$ if $M(\mathbf{q})$ is to be non-zero. This condition determines the possible directions of $\mathbf{q}_2$. If the fundamental light is propagated along the c-axis, then the maximum intensity of scattered light is expected as shown on each side of $\mathbf{q}_1$. Since there is no discontinuity of linear susceptibility at domain walls, no diffraction of the fundamental light is observed (i.e. the domains are not visible to ordinary light). Thus, information concerning the geometry and distribution of domains in a crystal can be obtained in this way from the angular variation of $I(2\omega)$. As discussed earlier, domains can have a marked effect on collinear second-harmonic generation because of the phase discontinuity at the wall.

In principle, second-harmonic scattering can also be used to study dynamic fluctuations of the structure in the neighbourhood of a ferroelectric phase transition and, indeed, Inoue (1974) has recently reported such critical second-harmonic scattering in $NaNO_2$. Also, for the structural transition in NH_4Cl, Freund (1967) found that the scattered harmonic intensity gave some evidence for pseudo-long-range order in the disordered phase over a temperature range of about $5°C$ above T_c. This may just be evidence of very-long-range correlations above T_c (i.e. of the range of the

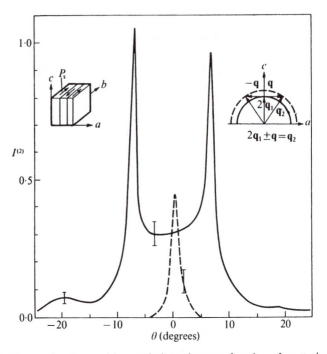

Fig. 13.10. The scattered second-harmonic intensity as a function of scattering angle in multi-domain $NaNO_2$. See inset for domain structure and wavevector configuration. For comparison, the dashed line is the equivalent scattering after elimination of domains by an external field. (After Vogt 1974.)

function $M(\mathbf{r})$), but the dimensions of the correlation volume would have to be greater than about $0 \cdot 1 \ \mu m$ (i.e. greater than $\sim \pi/|q_2 - 2q_1|$).

In order to determine whether the scattering is due to static or dynamic modulation of χ^{NL} a spectral analysis of the scattered harmonic light is necessary. In this way Maker (1970) was able to observe rotational molecular motion in liquids from the spectral broadening of the second-harmonic light.

The hyper-Raman effect involves transitions between vibrational levels of the scattering system and has been observed in liquids and crystals (Terhune, Maker, and Savage 1965, Savage and Maker 1970). This effect depends on the change of bond polarizability $\beta_\parallel$ with the bond length during the vibration. The hyper-Raman effect involves different selection rules from the ordinary Raman effect and can be used to study modes which are normally Raman inactive. This type of experiment could be useful in studying ferroelectric soft modes in, for example, a paraelectric centrosymmetric phase, but experimental difficulties have so far prevented its use in detailed studies of this kind.

14

SEMICONDUCTOR FERROELECTRICS

14.1. Ferroelectric instabilities in narrow-band-gap materials

IN § 1.2 WE DERIVED an effective Hamiltonian for ionic motion in the form

$$\mathcal{H}_{\text{eff}}(\text{ion}) = \sum_i \frac{p_i^2}{2m_i} + U(\mathbf{R}_i, \mathbf{R}_j, \ldots) + E(\mathbf{R}_i, \mathbf{R}_j, \ldots) \qquad (14.1.1)$$

where the first and second terms are respectively the kinetic and potential energies of the lattice of ion cores and $E(\mathbf{R}_i, \mathbf{R}_j, \ldots)$ is the contribution of the valence electrons via the electron–ion (or vibronic) interaction. The latter is derived in the adiabatic approximation which assumes that valence electrons respond essentially instantaneously to a redistribution of ionic co-ordinates $\mathbf{R}_i, \mathbf{R}_j, \ldots$.

In the subsequent development of the basic model Hamiltonian (eqn (1.2.8)) a further very important assumption was made, namely that the electronic contribution $E(\mathbf{R}_i, \mathbf{R}_j, \ldots)$ to effective potential is independent of temperature. This approximation therefore neglects all thermal excitations of electrons from the valence to the conduction band and hence is equivalent to an assumption that the relevant band gap is very large compared with thermal energies. In this limit, which is valid for most ferroelectrics discussed to this point, the valence electrons play only a passive role, although the contribution $E(\mathbf{R}_i, \mathbf{R}_j, \ldots)$ to the resultant effective potential $V(\mathbf{R}_i, \mathbf{R}_j, \ldots) = U(\mathbf{R}_i, \mathbf{R}_j, \ldots) + E(\mathbf{R}_i, \mathbf{R}_j, \ldots)$ may still be essential for the stabilization of ferroelectricity in many materials.

We shall now relax this wide-band-gap restriction and consider the effects of a thermal excitation of electrons on the theory of ferroelectricity in narrow- or intermediate-band-gap materials. Let us return to the fundamental Hamiltonian eqn (1.2.1)

$$\mathcal{H} = \mathcal{H}(\text{ion}) + \mathcal{H}(\text{electron}) + \mathcal{H}(\text{electron–ion}). \qquad (14.1.2)$$

The first term can be expressed in the normal way in terms of ionic momentum and displacement co-ordinates. Let us now concentrate on electronic motion under the influence of $\mathcal{H}(\text{electron})$ and $\mathcal{H}(\text{electron–ion})$ using the adiabatic approximation in which electrons can react instantaneously to ionic motion.

The traditional approach for valence-electron motion is to begin with a one-electron model in which each electron is treated as an independent particle and the effects of interelectron coulomb forces are ignored. In this approximation the independent electrons are assumed to be moving in the effective potential of the ion cores, which is the instantaneous value of $\mathcal{H}$(electron–ion). We shall expand the electron–ion potential about the maximum (ionic) symmetry configuration (which is the prototype lattice) and obtain electronic band states for this highest symmetry condition. For simplicity we consider here just the effects of a uniform static distortion and write accordingly

$$\mathcal{H}(\text{electron–ion}) = W(\xi_0, \mathbf{r}_e) = W(\xi_0^0, \mathbf{r}_e) + \frac{\partial W}{\partial \xi_0}(\xi_0 - \xi_0^0) + \dots \quad (14.1.3)$$

where ξ_0 indicates a uniform displacement of ions with the symmetry of the soft mode, $\mathbf{r}_e$ is the electronic position co-ordinate, and the derivative $\partial W/\partial \xi_0$ is taken about the high-symmetry configuration. Taking the highest-symmetry ionic displacement ξ_0^0 to be zero by definition we can reduce eqn (14.1.3) to

$$W(\xi_0, \mathbf{r}_e) = W_0(\mathbf{r}_e) + W'\xi_0 + \dots \quad (14.1.4)$$

where

$$W_0(\mathbf{r}_e) = W(\xi_0^0, \mathbf{r}_e), \qquad W' = \frac{\partial W}{\partial \xi_0}. \quad (14.1.5)$$

The term $W_0(\mathbf{r}_e)$ is now combined with $\mathcal{H}$(electron) to define a simple one-electron band structure. The other terms in eqn (14.1.4) then describe a mixing and renormalization of nearby valence and conduction bands by the static distortion.

To reduce the theory to its basic mathematical essentials we neglect all wavevector dependencies and consider a model of two non-degenerate electronic bands with dispersionless energies ε_1 and ε_2. At $T = 0$ the lower band ε_1 is fully occupied by the valence electrons while the upper conduction band ε_2 is empty. The resulting eigenvalue problem for vibronically interacting electronic levels then reduces in lowest order to the simple form

$$\begin{pmatrix} \varepsilon_1 + W'_{11}\xi_0 & W'_{12}\xi_0 \\ W'_{12}\xi_0 & \varepsilon_2 + W'_{22}\xi_0 \end{pmatrix} \begin{pmatrix} A \\ B \end{pmatrix} = E \begin{pmatrix} A \\ B \end{pmatrix} \quad (14.1.6)$$

in which the subscripts denote matrix elements in the form

$$W'_{ij} = \langle \Psi_i | W' | \Psi_j \rangle \quad (14.1.7)$$

where Ψ_i and Ψ_j $(i, j = 1, 2)$ are the unperturbed electronic eigenstates for the two electronic levels in the prototype ion configuration.

The uniform (i.e. zero-wavevector) lattice-mode distortion ξ_0 which removes the prototype symmetry centre and leads to a spontaneous polarization is necessarily of odd inversion symmetry. It follows that intraband vibronic matrix elements can be non-zero only in even order (i.e. involving $\xi_0^2, \xi_0^4, \xi_0^6$, etc.) and that as a result $W'_{11} = W'_{22} = 0$. If we assume our two electronic bands to be of opposite parity then W'_{12} is non-zero and the lowest-order vibronic-mixing problem reduces to

$$\begin{pmatrix} \varepsilon_1 - E & W'_{12}\xi_0 \\ W'_{12}\xi_0 & \varepsilon_2 - E \end{pmatrix}\begin{pmatrix} A \\ B \end{pmatrix} = 0 \qquad (14.1.8)$$

and defines perturbed electronic band energies

$$E_\pm = \tfrac{1}{2}(\varepsilon_1 + \varepsilon_2) \pm \tfrac{1}{2}(\Delta^2 + 4W'^2_{12}\xi_0^2)^{1/2} \qquad (14.1.9)$$

in which $\Delta = \varepsilon_2 - \varepsilon_1$ is the energy gap between the unperturbed valence and conduction levels. The free energy for the electronic system now follows as $F = -NkT \ln Z$ where Z, the canonical partition function, is given by $Z = \Sigma \exp(-E/kT)$ with E running in our case over the four possible energies $\Sigma_n f_n \varepsilon_n$ ($n = 1, 2$; $f_n = 0, 1$) of the two-level electronic system. Measuring energies ε_n from the midpoint between the bands we find

$$Z = 2 + \exp\left(\frac{\Delta'}{2kT}\right) + \exp\left(-\frac{\Delta'}{2kT}\right) \qquad (14.1.10)$$

in which $\Delta' = (\Delta^2 + 4W'^2_{12}\xi_0^2)^{1/2}$. The free energy follows immediately as

$$F_{el}(T, \xi_0) = -NkT \ln[2 + 2\cosh\{(2kT)^{-1}(\Delta^2 + 4W'^2_{12}\xi_0^2)^{1/2}\}]. \qquad (14.1.11)$$

In the limit $kT \ll \Delta$ this reduces to $F_{el} = -\tfrac{1}{2}N(\Delta^2 + 4W'^2_{12}\xi_0^2)^{1/2}$, neglecting constant terms, and can be expanded as a (Devonshire) series in even powers of 'polarization' ξ_0. The physically ionic contribution to free energy F_{ion} can also be expanded as a Devonshire series and, as a result, the total free energy $F = F_{ion} + F_{el}$ is also expressible in this form when $kT/\Delta \to 0$. In order to see most clearly the effects produced by electronic thermal excitations we shall now assume that the ionic free energy F_{ion} results from a stable purely harmonic ion core motion, i.e.

$$F_{ion}(\xi_0) = \tfrac{1}{2}N\omega^2\xi_0^2. \qquad (14.1.12)$$

In this approximation no ferroelectric anomaly can possibly result from the ionic terms. The total free energy now takes the form

$$F(T, \xi_0) = \tfrac{1}{2}N\omega^2\xi_0^2 - NkT \ln[2 + 2\cosh\{(2kT)^{-1}(\Delta^2 + 4W'^2_{12}\xi_0^2)^{1/2}\}] \qquad (14.1.13)$$

and can describe a ferroelectric instability if $\partial F(T, \xi_0)/\partial \xi_0 = 0$ has a real solution $\xi_0 \neq 0$. From eqn (14.1.13) by direct differentiation we find

$$\xi_0^2 = \frac{W'^2_{12}}{\omega^4}\tanh^2\{(4kT)^{-1}(\Delta^2 + 4W'^2_{12}\xi_0^2)^{1/2}\} - \frac{\Delta^2}{4W'^2_{12}}. \qquad (14.1.14)$$

As $\Delta/kT \to 0$ (high temperature) we find $\xi_0^2 \to -\Delta^2/4W_{12}'^2$ so that lattice distortion is absent. However, at the low-temperature extreme we find

$$\xi_0^2 = \frac{W_{12}'^2}{\omega^4} - \frac{\Delta^2}{4W_{12}'^2} \tag{14.1.15}$$

which is positive, indicating a real value for ξ_0, when $2W_{12}'^2 > \omega^2 \Delta$.

Clearly when this condition is satisfied the low-temperature stable phase is ferroelectric while the high-temperature one is paraelectric. The ferroelectric instability is induced at a temperature T_c which, by putting $\xi_0 \to 0$ in eqn (14.1.13), can be calculated in the form

$$kT_c = \frac{\Delta}{4}\left\{ \text{arctanh}\left(\frac{\omega^2 \Delta}{2W_{12}'^2}\right) \right\}^{-1} \tag{14.1.16}$$

and approaches zero as $\omega^2 \Delta/2W_{12}'^2$ approaches unity from below. In the light of our assumptions concerning the form of the ionic free energy this instability results solely from a vibronic coupling to the electronic motion. The stronger the electron–phonon interaction parameter W_{12}', the softer the lattice (i.e. the smaller ω^2), and the smaller the energy band gap Δ, the easier the phase transition can arise.

Physically the transition arises as the result of a lowering of the valence-band energy by the electron–phonon interaction. If the resulting lowering of electronic energy is larger than the potential energy of the lattice mode describing the active vibration then a spontaneous distortion results. The existence of thermal excitations between bands is not essential to this mechanism and for wide band gaps (Shukla and Sinha 1966) we can express it in the language of § 1.2 by saying that the ferroelectric instability in this model results from the electronic potential term $E(\mathbf{R}_i, \mathbf{R}_j, \ldots)$ rather than from the physically ionic potential $U(\mathbf{R}_i, \mathbf{R}_j, \ldots)$. In particular, the contribution $E(\mathbf{R}_i, \mathbf{R}_j, \ldots)$ can be expressed explicitly in terms of the electron band and vibronic interaction parameters, and hence the expansion coefficients of the resulting total effective ion potential $V(\mathbf{R}_i, \mathbf{R}_j, \ldots)$, leading to the potentials V and v of eqn (1.2.8), can be separated into their physically electronic and ionic parts. In the case of intermediate- or small-band-gap situations, where electronic excitations are important, $E(\mathbf{R}_i, \mathbf{R}_j, \ldots)$ becomes temperature dependent and an amplification of the basic model (eqn (1.2.8)) along the lines set out above becomes essential for a quantitative theoretical analysis.

Models of this type, allowing for the thermal excitation of valence electrons, were first introduced by Kristoffel and Konsin (1967, 1968, 1969) and by Bersuker and Vekhter (1967, 1969) and have been pursued at length in the Soviet literature (see Kristoffel and Konsin 1973 for references). A thermodynamic or free-energy phenomenology has also been expanded to include the effects of electronic excitations (e.g. Pasynkov 1973), but to obtain a detailed understanding of the associated phenomena it is necessary to work microscopically (and dynamically) and to take account not only of band structures and phonon dispersion but also of possible band degeneracies. If the active bands or vibrations are degenerate, then a number of low-symmetry ordered configurations become possible and the consequences of thermal activation of electrons increases. The microscopic analyses have been formally carried out for both ferroelectrics and antiferroelectrics (see, for example, Bersuker, Vekhter, and Muzalevskii 1974), but quantitative comparisons with experiment are extremely difficult since so many parameters enter even when band structures are fairly well known.

Most contact with experiment to date has been at a qualitative or, at best, semi-quantitative level in an effort to understand the presence of ferroelectricity in certain classes of materials and to account for relative values of T_c within a single class in terms of an assumed dominant electronic origin of the strongly anharmonic potential required. In this work little attention has been paid to the statistical effects introduced by thermal excitations to the conduction band, emphasis being placed on determining the form and magnitude of the effective lattice potential energy $E(\mathbf{R}_i, \mathbf{R}_j, \ldots)$ produced by the vibronic coupling to the filled valence band. The form is usually determined by symmetry, but a measure of the vibronic coupling coefficient W'_{12} can be obtained, for example, by observing the electronic band gap as a function of applied field or of spontaneous polarization below T_c. Thus from eqn (14.1.9) the band gap in the presence of spontaneous 'polarization' ξ_0 is

$$E_+ - E_- = (\Delta^2 + 4W'^2_{12}\xi_0^2)^{1/2} \tag{14.1.17}$$

which, on expanding the square root, can be written as $\Delta + \delta E(\xi_0)$ where

$$\delta E(\xi_0) = \frac{2W'^2_{12}\xi_0^2}{\Delta}. \tag{14.1.18}$$

The quantity $2W'^2_{12}/\Delta$ is therefore directly proportional to the polarization potential defined by Wemple (1970) and is measurable (see § 12.5). In more detail one would have to worry about the tensor nature of the quantity and the effects of polarization fluctuations near the phase transition, but, in general, simple model arguments of this type are sufficient to give order-of-magnitude estimates for the important vibronic coupling parameters. Armed with these it is now possible to ascertain whether electronically produced lattice anharmonicities are alone sufficient to account for a ferroelectric instability and hence to probe the physical origin of the ordering process.

Results of a calculation of this nature for $BaTiO_3$ are reported by Kristoffel and Konsin (1973). They find that, although a compensation occurs between short-range and dipolar ionic interactions, the resulting purely ionic soft-mode 'stiffness' ω^2 of (14.1.12) is positive (though small) and that, even in this wide-band-gap ($\sim$3 eV) material, the ferroelectric instability depends for its very existence upon the presence of vibronic interaction. The compensation of short-range and Coulomb ionic forces, which is the more traditional explanation of ferroelectricity, is nevertheless of major importance, the smallness of ω^2 serving to make possible the achievement of the inequality $2W'^2_{12} > \omega^2\Delta$ even in the presence of the wide band gap.

Work of this kind on genuine intermediate-band-gap ferroelectrics, such as GeTe and SbSI (with band gaps of order $\lesssim 1$ eV), is less well developed at this time, although it seems likely that physically ionic anharmonicities will still play a significant secondary role at these values of Δ and that a wholly electronic, or rather vibronic, theory is likely to be a gross oversimplification. For very small gaps the simple one-electron approximation breaks down, and interactions and correlations between charge carriers become increasingly important. In the simple one-electron theory the influence of the distortion enters only as a change in energy of the electrons. In a more careful dynamical approach, for a situation where electronic thermal excitation effects are large, the lattice distortion appears in the electronic equations of motion as an effective interaction term between electrons. In the zero-band-gap ($\Delta \to 0$) limit the problem becomes similar to that for a metal–nonmetal transition in a half-filled band. The vibronic interaction now splits the band (or separates the two bands) to produce a change from metallic to non-metallic form. This problem has been considered theoretically in some detail (Hallers and Vertogen 1973) and the electron-correlation effects are very significant, although the simple static model seems to be valid for a calculation of equilibrium thermodynamics.

14.2. The effect of impurity carriers on ferroelectric stability

In the context of semiconductor materials the words intrinsic and extrinsic are used to denote the properties of 'pure' and impure or doped crystals. Thus, in an intrinsic semiconductor the carriers (electrons in the conduction band or holes in the valence band) are necessarily produced by thermal excitation alone. Marked intrinsic effects in ferroelectric semiconductors can therefore occur only when the relevant band gap is relatively small (i.e. not too large with respect to thermal energies). This is the situation briefly discussed in the previous section.

In this section we shall neglect intrinsic effects (thereby assuming intermediate to large band gap situations) and concentrate on the effects which carriers introduced by the presence of impurities can have on ferroelectric stability. The resulting carrier can be excess electrons in the conduction band (resulting from the presence of donor impurities), holes in the valence band (from acceptor impurities), or both. In general these impurities also introduce localized energy levels into the energy gap which affect the statistical theory significantly at low temperatures when these levels may be occupied by a significant fraction of the impurity carriers. We shall discuss here, as in § 14.1, a very idealized model which is capable of illustrating only the basic qualitative effects, and in this picture the presence of localized energy levels will be ignored.

In the simplest theory (Kristoffel and Konsin 1971) the one-electron picture of the last section is again used, neglecting electron correlations and replacing the unperturbed electronic bands by simple dispersionless levels with energies ε_1 and ε_2 respectively ($\varepsilon_2 - \varepsilon_1 = \Delta$). In the presence of a vibronic interaction W'_{12} between these bands the resulting perturbed energies are given by eqn (14.1.9) as

$$E_{\pm} = \tfrac{1}{2}(\varepsilon_1 + \varepsilon_2) \pm \tfrac{1}{2}(\Delta^2 + 4W'^2_{12}\xi_0^2)^{1/2} \qquad (14.2.1)$$

where ξ_0 is the static uniform ionic displacement giving rise to the ferroelectric distortion from the prototype centrosymmetric structure.

In this picture the presence of each additional impurity electron from a donor ion adds a potential-energy term E_+, and the presence of each valence-band hole contributes a potential-energy term $-E_-$ (i.e. the removal of a valence-band electron). As a result, if there are N_e electronic carriers of impurity origin and N_h holes, with N_e and N_h both small compared with the total number of energy states N in a band, then the effective potential resulting from electronic origins (neglecting constant terms) can be written

$$E(\xi_0) = \tfrac{1}{2}(N_e + N_h - N)(\Delta^2 + 4W'^2_{12}\xi_0^2)^{1/2}. \qquad (14.2.2)$$

In the intrinsic system the equivalent potential term is

$$E(\xi_0) = -\tfrac{1}{2}N(\Delta^2 + 4W'^2_{12}\xi_0^2)^{1/2} \qquad (14.2.3)$$

so that the contribution solely from the carriers is

$$E_{\text{carrier}} = \tfrac{1}{2}N(f_e + f_h)(\Delta^2 + 4W'^2_{12}\xi_0^2)^{1/2} \qquad (14.2.4)$$

in which $f_e = N_e/N$ and $f_h = N_h/N$ are small numbers. Expanding (14.2.4) for small displacements ξ_0 we find

$$E_{\text{carrier}} = \tfrac{1}{2}N(f_e + f_h)\left(\Delta + \frac{2W'^2_{12}}{\Delta}\xi_0^2 - \frac{2W'^4_{12}}{\Delta^3}\xi_0^4 + \ldots\right). \qquad (14.2.5)$$

We see immediately from this equation that the carrier electrons add to the (harmonic) stiffness of the lattice and reduce the (quartic) anharmonicity. Both these effects make a lattice more resistant to ordering or, in other words, reduce the Curie temperature in systems which are ferroelectric in the intrinsic semiconducting state.

Quantitatively, even with the neglect of electron-correlation effects, the resulting decrease in T_c is difficult to estimate. For example, the existence of donor and acceptor impurity ions in the lattice affects the form of the 'bare' ion–core interactions in addition to producing the electron and hole carrier effects. Also, the degree to which the intrinsic semiconductor ferroelectricity results from electronic or bare-ion anharmonicities remains unspecified. However, as a very simple model let us return to the approximation of § 14.1 for which physically ionic anharmonicity is absent. Combining eqn (14.2.5) with bare ionic energy (eqn (14.1.12)) then allows us to write an effective ionic-plus-carrier contribution to free energy in the form $\frac{1}{2}N\omega'^2\xi_0^2+\ldots$, where

$$\omega'^2 = \omega^2 \left\{1+(f_e+f_h)\frac{2W_{12}'^2}{\Delta\omega^2}\right\}. \tag{14.2.6}$$

If the ferroelectric transition takes place via the vibronic mechanism of the last section, then the result of an effective change of lattice stiffness from ω^2 to ω'^2 can be found by substituting eqn (14.2.6) in eqn (14.1.16) in the form of a reduction in Curie temperature of magnitude

$$T_c(0) - T_c(f) = \frac{4kT_c^2(0)}{\Delta}(f_e+f_h)\cosh^2\left\{\frac{\Delta}{4kT_c(0)}\right\} \tag{14.2.7}$$

where $T_c(0)$ is the Curie temperature for the intrinsic semiconductor and $T_c(f)$ is the Curie temperature in the presence of an extrinsic carrier concentration f_e+f_h.

In a more accurate treatment it is of course necessary to introduce a wavevector dependence of vibronic interaction parameter $W_{12}'(q)$ and of microscopic displacement parameter $\xi(\mathbf{q})$. This is essential for a discussion of dynamics, for example, such as the effect of the carrier concentration on the soft-mode condensation (see for example Kristoffel and Konsin 1971). In this picture the form of the potential (eqn (14.2.2)) is

$$E(\xi) = \frac{1}{2}(N_e+N_h-N)\left\{\Delta^2+\frac{4}{N}\sum_\mathbf{q}W_{12}'^2(\mathbf{q})\xi^2(\mathbf{q})\right\}^{1/2} \tag{14.2.8}$$

and the bare-ionic potential energy

$$E_{ion} = \frac{1}{2}\sum_\mathbf{q}\omega^2(\mathbf{q})\xi^2(\mathbf{q}). \tag{14.2.9}$$

Relevant free energies must then be calculated via the partition function. However, the problem is a complicated one and little progress has yet been achieved towards a quantitative comparison with experiment. In fact experimental results themselves are rather scant in this respect, although some evidence is available (see for example Hirahara and Murakami 1958, and

Grigas, Grigas, and Belyatskas 1967) supporting the general expectation of a decrease in Curie temperature with increasing carrier concentration.

Of the many sweeping simplifications which have gone into the creation of the very simplest model of a semiconductor ferroelectric as used in this and the last section, one in particular, namely the neglect of electron correlations, is deserving of more careful consideration. The important point here, which can be considered aside from the details of band structures etc., is that the itinerant electrons are able to distribute themselves spatially in such a way as to partially screen the Coulomb potential arising from the ionic lattice. This screening effect eliminates the long-range tail of the effective ion–ion dipolar interactions and therefore tends to remove that aspect of the ferroelectric transition in insulators, namely the quasi-Landau critical behaviour leading to mean-field critical exponents, which has discouraged a search for fluctuation-dominated characteristics near ferroelectric transitions.

In the simple ionic theory of ferroelectricity the long-range part of the ionic potential field is assumed to have coulombic form

$$V(\mathbf{r}) = \sum_i \frac{Z_i e}{|\mathbf{r} - \mathbf{R}_i|} \tag{14.2.10}$$

where $Z_i e$ is the charge on the ith ion at position $\mathbf{R}_i$. In the absence of significant electrical conductivity this would seem to be a realistic assumption and leads to a suppression of fluctuation effects compared with an equivalent short-range-potential situation. In particular for a uniaxial symmetry the resulting wavevector dependence (eqn (11.2.18)) with its non-analyticity at $\mathbf{q} = 0$ leads at most to small and subtle logarithmic deviations from mean-field static behaviour. With such a dipole potential the richer critical phenomena associated with short-range forces are suppressed.

The basic effects of ion–electron correlations are well known (e.g. Ziman 1960) and lead to an exponential reduction of the long-range dipolar tail. More precisely the effective long-range potential of a screened static ith ion with bare coulombic potential field (eqn (14.2.10)) is

$$W_i(\mathbf{r}) = \frac{Z_i e}{|\mathbf{r} - \mathbf{R}_i|} \exp(-\lambda |\mathbf{r} - \mathbf{R}_i|) \tag{14.2.11}$$

where the inverse range λ for a classical electron gas is proportional to the electron density and inversely proportional to temperature. The screened interaction between remote static neighbour ions i and j follows as

$$V(\mathbf{R}_i, \mathbf{R}_j) = \frac{Z_i Z_j e^2}{|\mathbf{R}_i - \mathbf{R}_j|} \exp(-\lambda |\mathbf{R}_i - \mathbf{R}_j|). \tag{14.2.12}$$

Although dynamical screening is more generally a function of frequency as well, in the soft-mode limit $\mathbf{q} \to 0$, $\omega \to 0$, $V(\mathbf{R}_i, \mathbf{R}_j)$ of eqn (14.2.12) qualifies as a short-range potential for all $\lambda > 0$. This means that a second-order uniaxial phase transition produced by these interactions has a temperature region $|T - T_c|$ about the Curie temperature for which critical exponents are not of Landau form and for which the motion is truly fluctuation dominated. Although this region may well be small for low concentrations of holes and electrons (small λ), it nevertheless exists and therefore opens up a possible area for the study of fluctuation-dominated phenomena in uniaxial ferroelectrics. Indeed, as we shall describe in the following section, observations of dielectric properties close to T_c in the semiconductor ferroelectric SbSI suggest that such effects may already have been observed.

The second effect of carrier screening is to shift the Curie temperature itself from its value in an equivalent unscreened lattice of ions. The shift can be qualitatively assessed by noting from eqn (2.1.8), for example, that, other things being equal, the Curie temperature is an increasing function of $\Sigma_{j-i} V(\mathbf{R}_i, \mathbf{R}_j)$. It follows that one expects T_c to decrease as λ, or equivalently carrier concentration, increases. A more quantitative estimate of this reduction in Curie temperature by electron–ion screening has been given by Hallers and Caspers (1969).

As a general finding, therefore, we conclude that the presence of carriers in ferroelectrics influences the effective equations for ionic motion in several ways, all of which seem to decrease the stability of the ordered phase. To this point, however, we have restricted our considerations to homogeneous or single-domain states. In some ferroelectric semiconductors (most notably perhaps in SbSI, e.g. Kawada and Ida (1965)) inhomogeneous states have been observed very close to T_c which are made up of alternating layers of paraelectric and ferroelectric phases. An explanation has been offered by Larkin and Khmel'nitskii (1968) in terms of a macroscopic periodic spatial variation in the concentration of carriers. The fact that these carriers can diffuse through the sample and form regions with higher or lower carrier density leads, via the effect on T_c set out above, to a possibility of establishing corresponding regions of paraelectric and ferroelectric nature when the temperature is very close to the Curie point. The positive (negative) carriers move dominantly into positive (negative) space-charge layers which remain paraelectric, while the space charge in these layers then acts to cancel the depolarization field associated with the terminal charges of the intervening polar layers.

Larkin and Khmel'nitskii considered specifically the case (e.g. SbSI) where carriers of sufficient concentration are produced by illumination of the sample (photoconductivity). The dimensions of the periodic layers are then limited by the finite lifetime of the carriers and, in particular, by the distance over which the carriers have time to diffuse during their lifetime. Regions of stability and metastability of the inhomogeneous states were deduced theoretically as a function of carrier concentration, and it was concluded that for carrier densities above a critical value the inhomo-geneous state is stable near T_c. Qualitative agreement of theory with the experimentally observed values of critical concentration ($\sim 10^{17}\,\text{cm}^{-3}$) and of layer periodicity (~ 0.01 cm) was claimed for illuminated SbSI. More recently van den Berg (1972) has formulated a similar theory, leading again to an inhomogeneous stable structure near the ferroelectric phase transition, for the defect structure $Fe_{1-x}S$, which is ferroelectric, for small values of x, below about 140°C.

14.3. Antimony sulphoiodide

The existence of V–VI–VII compounds (where V = Sb, Bi, VI = S, Se, Te, and VII = Cl, Br, I) has been known for over a century, but recent interest stems from the work by Dönges (1950, 1951) which established their basic crystal structure and showed a variation of colour with composition, which pointed to the existence of band gaps in the visible or near infrared. This fact led Nitsche and Merz (1960) to grow single crystals of these materials and to study the photoelectric properties as a function of composition. The crystals are needle like, indicating highly anisotropic properties, and typically show a maximum photocurrent in the 500–800 nm range. The most sensitive are antimony sulphoiodide (SbSI) and antimony sulphobromide (SbSBr), and interest in the series was enhanced further when Fatuzzo et al. (1962) established that SbSI, at least, was also ferroelectric. It has since been established that other members of the series are also ferroelectric (e.g. SbSBr, BiSI, and BiSBr (see Nitsche, Roetschi, and Wild 1964, Pikka and Fridkin 1968)), but by far the most research work to date has been performed on SbSI. With a ferroelectric Curie temperature just above room temperature (≈ 20°C) and a close to second-order transition (at atmospheric pressure; the transition actually becomes truly second order at a pressure $p = 1.4$ kbar, $T_c = 235$ K (Peercy 1975 b) and disappears altogether at $p > \sim 9.5$ kbar (Samara 1975)) it is ideal for the study of the effects of electrical conductivity on ferroelectricity and, vice versa, the effects of a ferroelectric ordering on typical semiconducting phenomena.

The paraelectric structure of SbSI, first determined by Dönges (1950), consists of chains of atoms (related by c-repeat) along the polar c-axis of the orthorhombic structure (point group mmm) as shown in Fig. 14.1. A more detailed structure determination, including the changes in ion position on transition to the ferroelectric phase, has been given by Kikuchi, Oka, and Sawaguchi (1967). In the paraelectric phase at 35°C the orthorhombic unit-cell dimensions are $a = 0.852$ nm, $b = 1.013$ nm, and $c = 0.410$ nm, and clusters of the atoms in the c-axis rods form square-pyramidal S_3I_2 groups with the Sb ion in the centre of the pyramid base. In this high-symmetry phase all atoms lie on mirror planes normal to the c-axis and the crystal is centrosymmetric. There are four formula weights per unit cell and

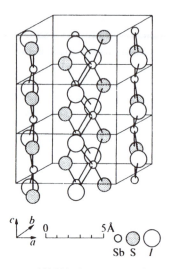

Fig. 14.1. Crystal structure of SbSI in the non-polar phase. (After Dönges 1950.)

the electric dipoles normal to the c-axis, which are formed by each pyramidal complex, cancel by symmetry when summed over the cell. On passage into the polar phase the position parameters normal to the c-axis are substantially unchanged but the Sb and S chains move along the c-axis with respect to the I sites by distances of about 20 pm and 5 pm respectively (at 5°C). This removes the mirror-plane symmetry giving rise to an mm2 (orthorhombic) polar phase with spontaneous polarization along the c-axis. The reported experimental value of spontaneous polarization is 25 μC cm^{-2}.

The large Curie–Weiss constant ($>2 \times 10^5$ K) (Mori and Tamura 1964) suggests that the phase transition is probably of a basically displacive character and such has been confirmed by dynamic measurements as discussed below. However, the X-ray measurements of thermal vibrations in the structure determination show a large c-axis amplitude for Sb ions above T_c which decreases substantially on passage to the ferroelectric phase. This is suggestive of a rather flat potential well or maybe even a shallow double well. In the latter case one would expect the simple soft-mode picture to be complicated very close to T_c in the manner described at the beginning of § 2.4. The small entropy change at T_c (Mori, Tamura, and Sawaguchi 1965) is also in accord with a dominantly displacive picture.

Recently considerable attention has been paid to the lattice dynamics of SbSI. Soft-mode behaviour was actually observed in several optical experiments both by infrared reflection techniques and in several Raman scattering experiments. The soft mode is infrared active in both the high-temperature and the ferroelectric phases but is only Raman active below T_c (Agrawal and Perry 1971). The far-infrared data are not really accurate enough to determine the quantitative temperature dependence of the soft mode, but the soft frequency Ω_s seems to follow a condensation power law with an exponent smaller than the $0 \cdot 5$ value predicted by the RPA theory of (2.1.35) (see for example Petzelt 1969, Sugawara and Nakamura 1972). Raman experiments in the ordered phase give results which vary between $\Omega_s \propto (T_c - T)^{1/4}$ and $\Omega_s \propto (T_c - T)^{1/3}$, but the accurate form is complicated below T_c by the existence of significant interactions between the soft A_1 mode and at least one (and possibly two) other low-frequency optic phonons. The resulting system of coupled oscillators exhibits the level repulsion characteristics typical of interacting modes, and the final limiting form for Ω_s may only materialize quite close to the Curie point itself. Since the transition is actually just of first order, this makes the quantitative deduction of a critical frequency exponent doubly difficult.

The first Raman evidence for mode interaction effects was given by Harbeke, Steigmeier, and Wehner (1970). An interaction of the soft mode with two other low-frequency modes was claimed by Teng, Balkanski, and Massot (1972) who, instead of varying the temperature through T_c, scanned the region of the ferroelectric transition by using hydrostatic pressure at constant temperature. Since the Curie temperature is considerably lowered by application of hydrostatic pressure (Samara 1968), the ferroelectric transition can be induced by pressure when T is kept fixed at a few degrees above T_c. Teng et $al.$ claim that this method is capable of increased accuracy over the more conventional temperature variation and their results for the frequency dependence of the lowest three lattice modes as $T \rightarrow T_c$ are shown in Fig. 14.2. The flat frequency region in the $3°C < T_c - T < 7°C$ was reproducible for several different samples but is not understood.

The most recent studies of lattice dynamics in this context have looked at SbSI in relation to its isomorphs. Thus, for example, Furman, Brafman, and Makovsky (1973) report a Raman study of some $SbSI_x Br_{1-x}$ solid solutions and of SbSBr in particular. The advantages of SbSBr over SbSI for Raman study include a lower Curie temperature ($T_c \approx 39$ K), which leads to much narrower spectral linewidths near the transition, and a larger band gap ($\approx 2 \cdot 2$ eV), which avoids serious absorption and accompanying dichroic effects from the commonly used He–Ne $632 \cdot 8$ nm parent line. Furman et $al.$ also report mode-interaction effects for SbSBr in the ordered phase below 39 K and, using simple interacting anharmonic-oscillator theory, they are able to extract a soft-mode frequency $\Omega_s \propto (T_c - T)^{0 \cdot 30}$. The spectra of the solid solutions are essentially interpolations between the spectra of the two end members with the same number of modes for every value of x. Such 'one-mode' behaviour is qualitatively understandable in the present context by applying a mass-criterion relationship developed, albeit for a simpler case, by Chang and Mitra (1968). Non-classical critical behaviour has also

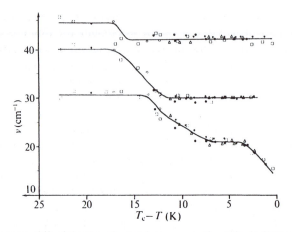

Fig. 14.2. Frequency shift of the three lowest-frequency optic modes in SbSI as a function of temperature near the Curie point. The temperature variation is induced by pressure and the different symbols indicate data taken at different fixed temperatures (except for the open squares which were obtained at constant pressure). (After Teng *et al.* 1972.)

been claimed for static phenomena, most notably for clamped dielectric constant, spontaneous polarization, and specific heat, by Steigmeier and Harbeke (1972), both from their own observations and from a reanalysis of existing data.

In spite of the direct observations of soft-mode dynamics the available evidence is not all in favour of a simple displacive transition. In addition to the soft-mode resonances cited above, some SbSI crystals (at least) have been observed to exhibit a significant further dispersion of relaxation form at long wavelengths (e.g. in the $0 \cdot 1$–$1 \cdot 0$ cm^{-1} region). In addition, the degree to which the measured soft-mode frequency alone can, via the Lyddane–Sachs–Teller relationship, account for the complete static dielectric constant as $T \rightarrow T_c$ seems to be a matter of some controversy. All this may well be evidence in favour of a shallow double-well local potential, the possibility of which has already been mentioned in connection with the X-ray scattering findings. Moreover, a Rayleigh component which increases in intensity near T_c is usually seen in the Raman spectra (e.g. Steigmeier, Harbeke, and Wehner 1971, Steigmeier, Auderset, and Harbeke 1975) and this too could possibly be indicative of an order–disorder component. On the other hand, as we have seen in § 11.5, central peaks of this kind can also be explained in other ways and the evidence therefore remains inconclusive.

Compared with the rather extensive study of phonon-related phenomena in SbSI rather less is known about the details of electronic and optical properties. The problem is that SbSI has a crystal structure which is very complicated compared with the crystals for which accurate band-structure calculations have been performed to date. In addition, a precise determination of the electronic structure requires many good experimental results and at present the experiments are really insufficient for this purpose. Early efforts made use of the extreme crystal anisotropy of the SbSI structure and the weak interactions between the chains to calculate densities of states assuming a one-dimensional model. More recently Nako and Balkanski (1973) have made the first detailed calculation of the complete valence- and electronic-band structure. They use a pseudopotential method but insert pseudopotentials obtained for various other structures in place of the more normal procedure of determining them directly for the crystal in question by comparison with experiment. In spite of the resulting uncertainties a band structure is derived which is able to account for most of the known optical properties. For example the absorption edge (corresponding to direct transitions which are dipole allowed) occurs at a Brillouin zone boundary and is about $1 \cdot 8$ eV for $E \| c$ and $1 \cdot 9$ eV for $E \perp c$ in the paraelectric phase. In the ferroelectric phase the absorption edge increases by $\sim 0 \cdot 1 - 0 \cdot 2$ eV for both polarization, an effect due primarily to the spontaneous polarizations but including a small shift arising from the small change in lattice constants. Below the absorption edge weak indirect transitions can occur and the minimum indirect band gap is about $1 \cdot 4$ eV.

Experimental observations of the absorption edge were first made in detail by Harbeke (1963) who measured the variation of absorption with the direction of the light E-vector (dichroism) and observed the change in absorption as a function of applied electric field. The absorption edge (or direct band gap) increases in energy with applied field, the shift reaching a maximum at the Curie temperature. It is basically understandable in terms of the polarization potential mechanism (eqn (14.1.18)) where the order parameter ξ_0 (or polarization) is induced by the field. More recent observations of reflectivities in both phases up to $3 \cdot 5$ eV (Bercha et al. 1971) have revealed that fine structure exists in the range $2-3 \cdot 5$ eV and that the measured reflectances for $E \| c$ and $E \perp c$ indicate a large anisotropy of the electronic properties. The measured dielectric constant in the range $2-6$ eV for $E \| c$ has also been revealed to have an interesting (three-peaked) structure (Nikiforov and Khasabov 1971). These effects are all at least qualitatively understandable in terms of the band structure calculated by Nako and Balkanski.

In contrast with the semi-quantitative understanding achieved in connection with dielectric and static electronic properties of the single-domain intrinsic SbSI system, the many and varied effects induced by the presence of multi-domains and of impurity and non-equilibrium carriers are at best only qualitatively appreciated. The existence of a very high photosensitivity in SbSI makes it possible to create quite significant concentrations of electronic carriers (whether they are dominantly of electronic or hole character is not yet certain) by illumination. The presence of carriers leads to a screening of internal fields and hence leads to changes in domain structure (the photodomain effect) which in turn results in changes in the shape of electromechanical hysteresis loops, in the temperature hysteresis at T_c, and in a crystal deformation (the photodeformation effect). Additional effects have been observed on switching characteristics and on phase boundaries near T_c. The latter have been referred to in the last section and lead to the existence of alternating ferroelectric and paraelectric phases. These interface boundaries in turn produce an effect referred to as pleochroism, which is the polarization of initially unpolarized light by passage through a crystal containing these interfaces and is probably due to light scattering from the interface walls. A vast amount of study of these properties, collectively referred to as photoferroelectric properties, has been reported primarily in the Soviet literature and some considerable progress has been made toward their interpretation. Much of the recent literature has been surveyed in the Proceedings of the 1972 Rostov on Don Symposium on Semiconducting Ferroelectrics published as volume 6 of the journal *Ferroelectrics* (1973).

Little reliable information is as yet available on transport properties and results by different authors have frequently been in conflict. Thus, for example, linear, sublinear, superlinear, and quadratic volt–ampere characteristics have all been reported for single crystal SbSI under (ostensibly) the same conditions, and there are similar contradictions in data concerning the nature of the conduction carriers and the temperature dependence of conduction current. Some of the most recent findings, including those of electrical-conductivity anomalies near the SbSI Curie point, have been reviewed by Fridkin *et al.* (1973). Quantitative understanding of these effects is likely to be complicated, since they undoubtedly involve not only the structure and parameters of the local impurity levels (in the energy gap between the valence and conduction bands) but also the state of the surface, contact properties, and the like.

In summary, the research work on ferroelectric semiconductors in general and on SbSI in particular is basically of two types; firstly the determination of the influence of carriers on ferroelectric properties and secondly the study of the influence of the onset of ferroelectricity on typically semiconductive properties (such as electrical conductivity, energy gap, photoelectric effect, etc.). The first type of research can again be divided into two basic areas, one of the study of the ferroelectricity in a single-domain crystal and the other the description of polydomain structures. Naturally the single-domain problem must be understood before any realistic effort to describe the polydomain phenomena can be made, and as a result principal theoretical attention has to date been given to the single-domain problem. As for the effects of the onset of ferroelectricity on semiconducting properties the situation is fairly well defined in those cases where the question concerns static effects (equilibrium phenomena) but is much more complicated, both experimentally and theoretically, when effects possessing an essentially non-equilibrium character, such as those determined by carrier transport, are studied.

14.4. Germanium telluride and isomorphs

One of the more exciting recent discoveries in the field of ferroelectric semiconductors has been the existence of a series of IV–VI compounds which are not only narrow-band-gap semiconductors but also exhibit perhaps the simplest 'ferroelectric' lattice instability yet studied.† These materials, of which germanium telluride (GeTe) and tin telluride (SnTe) are the most studied, are examples of diatomic ferroelectrics, which are the simplest conceivable class of ferroelectrics, with the order parameter involving simply a relative displacement of the only two atoms in the unit cell. Although serious study of these ferroelectrics has only just begun, particularly as regards their semiconducting properties, the basic character of the lattice-dynamical instability is known and, by virtue of its extreme simplicity (coupled with the fact that band gaps involved are $\lesssim 0.5$ eV), the germanium telluride series of ferroelectric semiconductors is likely to become a prototype for study in this field.

Ferroelectric transitions in this series of materials have already been reported for GeTe (Goldak *et al.* 1966), SnTe (Brillson, Burstein, and Muldawer 1974, Iizumi *et al.* 1975), $Pb_{1-x}Sn_xTe$ (Bierly, Muldawer, and Beckman 1963), and $Pb_{1-x}Ge_xTe$ (Antcliffe, Bate, and Buss 1973). In these systems the high-temperature prototype phase is of cubic NaCl structure, and at the phase transition an essentially displacive trigonal distortion occurs to a polar phase (point group 3m) for which the direction of spontaneous polarization is along a trigonal [111] axis. The Curie temperature T_c for GeTe is about 670 K while that of SnTe is much smaller. In fact for the latter a positive value of T_c appears to require a relatively low concentration of carriers (T_c decreasing with increasing carrier concentration in qualitative accord with the theoretical arguments of § 14.2) and Iizumi *et al.* suggest that a $T_c \approx 100$ K may represent a fair approximation for the carrier-free limit. The system GeTe–SnTe also forms a complete solid solution with T_c falling smoothly and almost linearly with increasing tin content. The transition itself involves simply a small relative displacement of the Ge and Te sublattice arrays along a [111] direction and a small elongation of this axis relative to its cubic configuration. It also seems possible that the phase transitions are second order (in which case critical phenomena will be of

† The term ferroelectric is used here in a slightly more general sense than that defined in § 1.1.2, since although microscopically the transition is fully equivalent to that of a conventional ferroelectric the high electrical conductivity in all probability prevents a switching of orientational states by an external electric field for these systems.

great interest), but the issue is not definitely resolved and there are some indications from optic and X-ray studies that the phase change in at least some of these materials may be more complicated than a simple transition from a cubic to a trigonal structure. Whether these effects involve only domains and interface boundaries (as found for high-carrier SbSI samples) or concern a true intermediate structure is not yet known.

In the cubic high-temperature phase each of the two atoms in a primitive cell is located at a centre of symmetry if a displacive single-well potential is assumed. For such a configuration long-wavelength optic phonons are all Raman inactive. However, any relative displacement of the atoms which removes these centres of inversion renders the phonons Raman active. It might therefore be supposed that a careful study of Raman activity near the Curie temperature would provide information concerning any possible intermediate steps involved in the transition. Unfortunately there are several difficulties encountered for these narrow-band-gap materials (Shand, Burstein, and Brillson 1974). Most importantly, with band gaps of only a few tenths of an eV, the materials are opaque to laser radiation in the visible region. Skin depths are usually less than 100 nm so that scattering lengths (and therefore scattering intensities) are very small. In addition the large extinction coefficient means that there will be a breakdown of wavevector conservation and accordingly the normal wavevector-independent selection rules may not be applicable. Also, if the presence of carriers results in the creation of inhomogeneous structures and interphase boundaries near T_c (as outlined for SbSI) these also will obstruct a clear observation of any critical condensation. In spite of these difficulties clear first-order Raman scattering has been measured in the ordered phase in GeTe and in SnTe and a mode softening on approach to the phase transition observed at least qualitatively in the former.

The first use of Raman scattering in this context was by Steigmeier and Harbeke (1970) on GeTe at temperatures below 670 K. The equivalent work for SnTe has been published more recently by Brillson et al. (1974). The observation of Raman structure at low frequencies is difficult owing to background Rayleigh scattering, but clear observations of both A_1 and E symmetry LO phonon modes were found in both cases. In GeTe the LO modes of both symmetries appeared to be totally screened (by itinerant electrons) and therefore show a softening as the temperature is raised towards T_c. In SnTe, for which T_c varied from ≈ 4 K with a hole-carrier concentration $p = 2 \cdot 2 \times 10^{20}$ cm^{-3} to $T_c \approx 70$ K with $p = 1 \cdot 5 \times 10^{20}$ cm^{-3}, these LO modes were again observed but found to be temperature independent and essentially unscreened. No scattering arising from the soft TO phonon mode could be definitely established. In the SnTe experiments the Raman spectra for the higher-p sample disappeared quite quickly for temperatures above 4 K, but for the lower-p sample the Raman structure (though with decreasing intensity) remained to temperatures well in excess of what was deemed to be the transition temperature. In both SnTe samples the LO phonon energy (~ 130 cm^{-1}) was essentially equal to that observed by Pawley *et al.* (1966) who studied the inelastic neutron scattering spectra for a SnTe sample which was still just paraelectric at absolute zero. In the latter experiments, although a phase transition was not found, both the LO and TO phonon dispersion spectra were recorded and the softening of the TO modes as $T \to 0$ and screening of the LO mode measured (Fig. 14.3). The

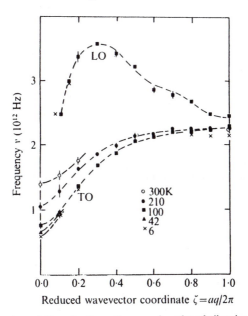

Fig. 14.3. The dispersion relations for the optic normal modes of vibration propagating along the [001] direction in SnTe at various temperatures: O 300 K; ● 210 K; ■ 100 K; ▲ 42 K; × 6 K. (After Pawley *et al.* 1966.)

square of the soft-mode frequency differs quite markedly from a linear relationship with temperature, much more so, for example, than the low-temperature soft mode in SrTiO$_3$. Iizumi *et al.* (1975) suggest that this may result from the large carrier concentration of Pawley's sample.

Screening by carriers is expected to affect primarily the longitudinal optic modes at long wavelengths since only they give rise to substantial macroscopic electric field distributions. Thus, whereas the TO phonon frequency $\Omega_T(\mathbf{q})$ is determined by the lattice force constants, the associated LO phonon frequency $\Omega_L(\mathbf{q})$ contains contributions from electric field as well. We write

$$\Omega_L^2(\mathbf{q}) = \Omega_T^2(\mathbf{q}) + \Omega^2(\mathbf{q}) \qquad (14.4.1)$$

where $\Omega^2(\mathbf{q})$ is the electric field contribution and can be shown to be inversely proportional to the 'background' dielectric constant $\varepsilon(\Omega_L, \mathbf{q})$ which contains a contribution from interband transitions and a contribution from free carriers. This background dielectric constant is large and negative at small $\mathbf{q}$ but becomes small and positive as $\mathbf{q}$ increases (Cochran, Dolling, and Elcombe 1966). Consequently at long wavelengths (or small $\mathbf{q}$) the LO phonon mode is close in frequency to the corresponding TO phonon (i.e. the macroscopic electric field is effectively totally screened by the itinerant electrons), while at shorter wavelengths the screening becomes much less effective. This in general results in a marked decrease in LO phonon frequency for a fixed temperature as $\mathbf{q} \to 0$ and the effect is quite clearly evident in Fig. 14.3.

The reasons why the expected screening of long-wavelength LO phonons was not seen by Brillson *et al.* (1974) is not altogether clear but some possible reasons were put forward by the authors. One possibility is that large-wavevector phonons are actually participating in the Raman scattering owing to the narrow penetration depth as referred to above. Alternatively the appearance of domains in the ordered phase may destroy translational invariance and also allow large-wavevector phonons to participate. Finally the high density of Sn lattice vacancies responsible for the p-type carrier densities of the samples may also result in a breakdown of $\mathbf{q}$-conservation (bulk SnTe crystals with $p < 10^{20}$ are difficult to fabricate). For a carrier concentration of $p = 2 \cdot 2 \times 10^{20}$ cm^{-3} the average Sn vacancy spacing is only about 2 nm. These effects may also help to explain the persistence of Raman scattering into the non-polar phase, although in the $T_c \sim 70$ K sample the LO phonon Raman peaks did eventually disappear above about 120 K.

From a lattice-dynamical point of view it would appear that the phase transitions in GeTe and SnTe have as much right to be classed as ferroelectric as the better known ones in the perovskite ferroelectrics for example. The primary difference, of course, relates to the very high electrical conductivity of the former which prevents the use of conventional dielectric methods for the measurement of susceptibility, and presumably also prevents the poling and switching of ordered phases by application of an externally applied electric field. Nevertheless, methods of measuring the essentially static dielectric susceptibility of these materials have been devised. For example, Bate, Carter, and Wrobel (1970) measured the capacitance–voltage relationship for graded junction diodes of PbTe and from its interpretation were able to deduce that PbTe is paraelectric with a reciprocal dielectric constant proportional to $T - T_0$ but with T_0 negative (~ -70 K). This interpretation has been qualitatively confirmed by a direct observation of the associated soft mode in PbTe using inelastic neutron scattering (Alperin *et al.* 1972). The complete TO and LO branches in the three principal cubic directions were measured down to 4 K and a soft $\mathbf{q} \to 0$

TO phonon frequency observed with frequency proportional to $(T - T_0)^{1/2}$, with T_0 negative but somewhat larger in magnitude than found from the susceptibility measurements. Via the Lyddane–Sachs–Teller relationship the observed soft-mode corresponds to a Curie–Weiss susceptibility and the quantitative discrepancies may only be due to neutron resolution corrections and broad linewidths in the scattering observations.

Although PbTe is not ferroelectric at real (positive) temperatures the negative Curie–Weiss temperature T_0 can be raised towards zero by forming solid solutions $Pb_{1-x}Sn_xTe$ or $Pb_{1-x}Ge_xTe$ with tin or germanium. In the latter case reasonably good crystals are available only with Ge content up to about 8 per cent but, not unexpectedly, in view of the high ferroelectric transition temperature of GeTe, the resulting crystals are ferroelectric above $T = 0$ K. Concentrations $x = 0.03$, 0.045, 0.06 showed dielectric peaks (measured as differential capacitance peaks of p–n junction diodes) which indicated Curie temperatures of order 90 K, 150 K, and 170 K respectively (Antcliffe et al. 1973). The phase transitions can also be observed by detailed lattice-parameter study (Hohnke, Holloway, and Kaiser 1972). The two methods are in essential agreement and indicate the critical value of x, at which $Pb_{1-x}Ge_xTe$ is just ferroelectric at absolute zero, to be less than 0.01. Evidence for an influence of the phase transition on band structure has also been obtained by Antcliffe et al. from a measurement of the band gap as a function of temperature (determined from the 50 per cent photovoltaic cutoff of a junction detector). Below T_c the band gap is small (a few tenths of an eV) and relatively independent of temperature while above T_c the gap increases rapidly. Attempts at Raman study have so far been unsuccessful, probably on account of the presence of compositional inhomogeneities in the sample volume.

Capacitance data on junction diodes have also suggested that there is an increasing tendency toward ferroelectricity when tin is added to PbTe. Since SnTe itself is only marginally ferroelectric for common carrier concentrations it comes as some surprise to find that $Pb_{1-x}Sn_xTe$ appears to be near ferroelectric for $x > 0.2$. PbTe mixes with SnTe as a solid solution in all concentration regions and is extremely interesting in that the band gap between valence and conduction bands decreases from about 0.2 eV at $x = 0$ to zero at about $x = 0.35$ (Butler 1969). With further increase of x the valence and conduction bands eventually change roles and the gap increases again. Recently Takano et al. (1974) have succeeded in measuring the dielectric constants of these extremely narrow-band-gap materials $(0 < x < 0.29)$ by interpreting magnetoplasma effects (a method first applied to PbTe by Sawada et al. (1964)). They found that in all the samples studied the dielectric constant at very low temperatures increased as a function of x, from ~1400 at $x = 0$ to nearly 11 000 at $x = 0.29$. For a given electron–lattice interaction one would expect from eqn (14.1.15) that the decrease in band gap alone, as x increased, would increasingly favour a potential ferroelectric instability. More accurately one might expect that the relevant parameter should be $E_G + 2E_F$, where E_F is the Fermi energy of the carriers measured from the band edge rather than the gap E_G itself (assuming that electron–hole pair excitations are playing an important role), and Takano *et al.* note that the reciprocal dielectric constant does vary fairly linearly with $E_G + 2E_F$ and extrapolates to zero when $E_G + 2E_F \to 0.07$ eV. In the absence of any carriers this would correspond to a tin concentration $x = 0.21$. The presence of extrinsic free carriers, as we noted in § 14.2, tends generally to make more difficult the production of a ferroelectric phase. Less qualitative comparisons of theory with experiment are not yet possible since extant theories are based at best on extremely simplified band models and contain several unknown parameters.

15

FILMS, CERAMICS, AND
METASTABLE POLARIZATION

15.1. Thin films

THE PARALLEL alignment of dipoles in a ferroelectric is due primarily to a relatively strong long-range interaction along the polar c-axis and a weaker shorter-range interaction normal to the axis. Physically this comes about because for electric dipolar forces the parallel alignment is energetically favourable for an isolated pair of c-axis dipoles but not for an isolated pair of dipoles normal to the polar axis. Theoretical and experimental estimates can be made of a correlation range, or distance over which near-neighbour-cell polar displacements ξ_i are strongly correlated in some sense. Although the quantitative manner in which the c-axis and (perpendicular) a-axis correlation ranges L_c and L_a are defined by different workers may vary a little, typical values for many ferroelectrics are $L_c \sim 10$–50 nm and $L_a \sim 1$–2 nm leading to correlation volumes (see eqn (12.5.7)) of 10–100 nm^3. We note that these correlation parameters concern near-neighbour motion and are to be distinguished from the critical correlation length ξ' of Chapter 11 which is defined with respect to correlations at infinity (e.g. eqn (11.3.2)). In particular the correlation volume $L_c L_a^2$ appears not to be divergent at T_c (Wemple 1970) and a sketch of the needle-shaped correlation volume is shown in Fig. 15.1. From the above estimates the transverse correlation range L_a is of the same order as the thickness of a 180° domain wall while the polar axis range L_c can amount to many tens of unit-cell lengths. Consequently as the physical dimensions of the crystal are reduced a change in the stability of the ferroelectric phase is to be expected.

A direct measure of correlation range is in principle possible by preparing thin ferroelectric films and studying the thickness dependence of the ferroelectric properties. This procedure, however, is usually complicated by additional surface effects which arise from incomplete neutralization of depolarizing fields (when the polar axis is perpendicular to the film) and space-charge fields arising from energy-band bending (see § 4.4) normal to the film surface. These effects greatly influence the ferroelectric behaviour in the surface region and no unambiguous results attributable to the correlation length in thin films have yet been obtained.

There have been many attempts to study the thickness dependence of ferroelectric properties of BaTiO$_3$ films prepared by vacuum-evaporation or sputtering techniques. Feuersanger (1969) reported that stoichiometric

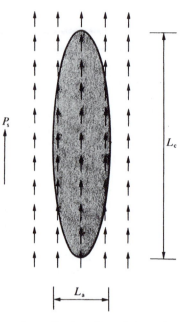

P_s

L_c

L_a

Fig. 15.1. The schematic form of a needle-shaped correlation volume oriented along the polar axis of a uniaxial ferroelectric.

BaTiO$_3$ films prepared by electron-beam evaporation were generally cubic at room temperature (for film thicknesses of about $0 \cdot 1 \, \mu$m) with small dielectric constant and showed no dielectric anomaly in the neighbourhood of the Curie temperature near 120°C. Thicker films ($\sim 1 \, \mu$m) showed essentially bulk dielectric permittivity at room temperature and an anomaly at T_c. This result suggested that the ferroelectric phase becomes unstable for film thicknesses below about $0 \cdot 1 \, \mu$m. On the other hand, similar measurements by Slack and Burfoot (1971) on flash-evaporated BaTiO$_3$ films showed that the effective permittivity remained approximately equal to the bulk value down to thicknesses of $0 \cdot 1 \, \mu$m below which it decreased, but even the thinnest films ($\sim 0 \cdot 04 \, \mu$m) showed ferroelectric switching behaviour. The decrease was attributed to saturation of the dielectric constant ε in the Schottky-barrier region below the film surfaces which creates a high impedance in series with the bulk capacitance. The temperature dependence of ε in this high-field region was found to be very weak even in the neighbourhood of T_c. This is consistent with the results obtained by Kahng and Wemple (1965) on Schottky-barrier diodes in semiconducting BaTiO$_3$ single crystals.

In contrast to these studies, recent measurements (Tomashpolski *et al.* 1974) on vacuum-deposited BaTiO$_3$ films showed sharp, well-defined, dielectric anomalies near 120°C down to film thicknesses of 0·023 μm. For a film thickness of 0·01 μm this anomaly disappeared but the anomaly at the transition near 0°C was still evident.

The discrepancies between the various studies must presumably be attributed to variation of the defect structure in films prepared by different techniques. Evaporation of ternary compounds like $BaTiO_3$ are complicated by the different volatilities of the constituents, so that the detailed defect structures of the films depend on many parameters such as evaporation rate, substrate temperature, and many other factors. In all the experiments outlined above X-ray determination of the structure of the polycrystalline films verified the $BaTiO_3$ phase but a comparison of the crystalline perfection was not possible.

The defects not only affect the ferroelectric behaviour directly (see §§ 4.3 and 4.4) but also the electronic conductivity. The thickness of the Schottky-barrier layers and the magnitude of the space charge varies significantly with free-carrier concentration. In the paraelectric phase both surfaces are equivalent and the expected band bending is shown in Fig. 15.2 (a). As the

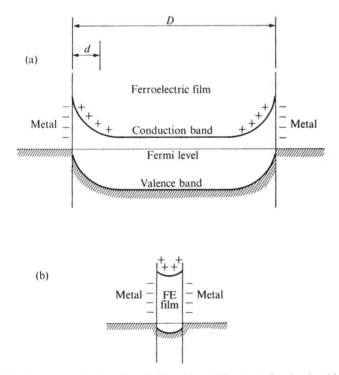

Fig. 15.2. Band structure of a ferroelectric film with metallic electrodes showing (a) Schottky barriers at both interfaces when the film thickness D is greater than the depletion layer of width d and (b) small band bending when $D < d$.

film thickness D is reduced below the free-carrier screening length d the depletion layer extends throughout the entire film as shown in Fig. 15.2(b). It might be anticipated that the effects of the surfaces on the phase transition would be less important in very thin films than in films of thickness comparable to the screening length. The disappearance of the dielectric anomaly at 120°C for film thicknesses below about 0.023 μm may thus be evidence of the ferroelectric phase becoming unstable owing to the finite size effect discussed earlier.

The preparation of thin ferroelectric films has to a great extent been motivated by the need for thin layers in several applications. The required degree of perfection varies according to the specific application. For instance for small volume capacitors a high dielectric constant and low dielectric loss are usually desirable and the variation of these parameters with temperature, frequency, and applied field are usually important considerations. While low-frequency (1 kHz) dielectric constants greater than 1000 have been obtained with stoichiometric polycrystalline films of thickness in the neighbourhood of 1 μm, dielectric absorption at higher frequencies gives rise to strong variation of both the dielectric constant and loss as a function of frequency. Small departures from stoichiometry usually result in decreased dielectric constant and increased loss.

For high-frequency piezoelectric filters and transducers and pyroelectric thermal detectors thin films can be used to advantage. In these cases a high piezoelectric coupling coefficient or pyroelectric coefficient are usually required together with a low dielectric loss. Polycrystalline films are adequate providing that the crystal axes can be suitably oriented during deposition or in subsequent poling treatments. Nevertheless, improved performance is expected with single-crystal films with well-defined pyroelectric and piezoelectric axes, not only because of increased coupling coefficients but also because of interfacial polarization effects in polycrystalline films. Most thin-film transducers have been fabricated from the simpler wurtzite compounds such as ZnO and CdS deposited in fibre-oriented polycrystalline form on sapphire or quartz substrates. By suitable choice of substrate and deposition conditions the polar hexagonal axis can be suitably oriented—usually normal to the film surface. Sputtered ZnO films have also been applied to pyroelectric detection (Roundy and Byer 1976) but the performance is greatly degraded by thermal conductivity and the heat capacity of the substrate. A large number of ferroelectric oxides possess much larger electromagnetic coupling coefficients and pyroelectric coefficients than the wurtzite compounds, but the few attempts (Foster 1970) to sputter oxides for these applications have not proved too successful.

The application of thin films to optical devices poses much more stringent requirements on the film properties and perfection, yet, because of the large effort devoted to optical memories and optical waveguides, films of high optical quality have been prepared with remarkable success. Early attempts at guiding optical beams in single-crystal films used ZnO grown epitaxially on sapphire substrates (Hammer *et al.* 1972) and mixed crystals of KH_2PO_4 (KDP) and $NH_4H_2PO_4$ (ADP) grown on KDP substrates (Ramaswamy 1972). The crystalline quality of these films was sufficiently good for reasonably low loss waveguides and represented a considerable improvement over the first attempts with polycrystalline films (Tien 1971). However, subsequent studies have been devoted primarily to materials with larger optical non-linearities which enable efficient modulation and processing within the film. High-optical-quality (low optical loss) single-crystal films of $LiNbO_3$ and mixed $LiNbO_3$–$LiTaO_3$ have been prepared for this purpose by a variety of techniques. $LiNbO_3$ and $LiTaO_3$ are isostructural with very similar lattice parameters and form a complete solid solution between the two end members. Consequently a number of schemes to obtain films of higher refractive index on a substrate of lower index, as required for guiding, are possible by varying the stoichiometry (i.e. the LiO/NbO_2 ratio—see § 8.3.2), the composition (i.e. TaO_2/NbO_2 ratio), or by the addition of other impurities. High-quality single-crystal films have been prepared by epitaxial growth from the melt (EGM) of $LiNbO_3$ on $LiTaO_3$ substrates (Miyazawa 1973, Ballman, Tien, and Riva-Sanseverino 1975 a), a liquid phase epitaxy (LPE) by dipping $LiTaO_3$ into a $LiNbO_3$ flux (Ballman, Tien, and Riva-Sanseverino 1975 b), out-diffusion of LiO from $LiNbO_3$ (Kaminow *et al.* 1973), in-diffusion of niobium metal or $LiNbO_3$ into $LiTaO_3$ (Hammer and Phillips 1974, Minakata, Noda, and Uchida 1975), and in-diffusion of transition metal impurities into $LiNbO_3$ (Schmidt and Kaminow 1974). In comparison, sputtered films have generally much higher optical loss (Takada *et al.* 1974), presumably owing to the poorer crystalline quality. The practical advantages arising from the simpler procedure for preparing sputtered films are to some extent offset by the need to pole them following deposition for most applications so that the high-temperature cycling and polishing procedures are not eliminated.

In each of the experiments involving $LiNbO_3/LiTaO_3$ structures the resulting changes of refractive index are consistent with the changes of film stoichiometry or composition and no effects due to the surface or due to the polarization divergence at the film–substrate interface have been detected. This is probably not surprising in view of the fact that the thinnest films have dimensions comparable to the wavelength of light, which we have seen previously is sufficient to show bulk characteristics. In all cases the films are prepared at high temperatures where the materials are electrically conducting and all depolarizing fields can be rapidly compensated by free charge. Nevertheless, large electric fields could be expected at the substrate–film interface during cooling of the structure due to pyroelectric effects— particularly on cooling through the ferroelectric transition of the $LiTaO_3$

substrate. Ballman (1975) deduced from the electro-optic behaviour of $LiNbO_3$ films grown by LPE that they are largely single domain following preparation, without any poling treatment, which may be attributable to effects of this kind.

The preceding discussion has focused almost entirely on the static properties of ferroelectric or pyroelectric films where the reversibility of the polarization was not an important factor—except possibly for a one-time poling of the film following preparation. For optical memories and display, however, polarization reversal plays an important role. Two materials which have been studied extensively in thin-film form for these applications are $Bi_4Ti_3O_{12}$ and a complex composition of lanthanum-doped $BiFeO_3–PbZrO_3–PbFe_{1/2}Nb_{1/2}O_3$, namely $Pb_{0.92}Bi_{0.07}La_{0.01}$ $(Fe_{0.405}Nb_{0.325}Zr_{0.27})O_3$. This latter composition (PBLFNZ) was developed as a result of an empirical and intuitive effort aimed at developing a low-coercive-field non-fatiguing film of moderate spontaneous polarization which showed the square hysteresis loops required for memories and display (Chapman 1969). Thin ceramic foil samples of the material have been prepared on conducting substrates with a perovskite structure and a Curie temperature of 150°C. The spontaneous polarization P_s and remanent polarization P_r derived from saturated loops are seen in Fig. 15.3 to increase

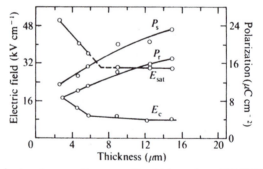

Fig. 15.3. 20 kHz hysteresis properties *versus* film thickness for a PBLFNZ composition (see text) on platinum. The electrode area is 2×10^{-3} cm^2. (After Chapman 1969.)

with increasing film thickness. Corresponding variation of the coercive field E_c and the field required to reach saturation E_{sat} are not so marked. Sputtered films of this same composition (Atkin 1972) have shown square loops with $P_s \sim P_r \sim 20$ μC cm^{-2} at 20 kHz even for film thicknesses down to 1 μm. The combined characteristics of microsecond switching times, low-fatigue square loops, and coercive voltages of 3 V with 5 μm thicknesses make these films attractive for a variety of applications.

To exploit the unusual switching behaviour of $Bi_4Ti_3O_{12}$ crystals (see § 16.3) it is necessary to produce large-area single-domain layers of $\langle 010 \rangle$

orientation displaying bulk single-crystal properties. Bulk crystals display two independently reversible components of polarization with magnitudes of $4 \, \mu C \, cm^{-2}$ and $50 \, \mu C \, cm^{-2}$ along the c- and a-axes respectively (Cummins and Cross 1968). Oriented polycrystalline epitaxial films reproducing these characteristics have been prepared by sputtering films onto MgO and $MgAl_2O_4$ substrates with thicknesses from 4 to $20 \, \mu m$ (Wu, Takei, and Francombe 1973, Francombe 1972). The coercive fields of $12 \, kV \, cm^{-1}$ and $90 \, kV \, cm^{-1}$ along the c- and a-axes were two to four times the single-crystal values and depended critically on the growth process. An interesting observation in these polycrystalline films is that individual domains may encompass several crystallites and that domain boundaries can move relatively unimpeded across crystallite boundaries.

15.2. Ceramics

For studies of the fundamental properties of a material large homogeneous single crystals are usually desirable to minimize the effects of surfaces and imperfections. Ferroelectric ceramics, on the other hand, have the advantage of being a great deal easier to prepare than their single-crystal counterparts. In many cases they show ferroelectric properties approaching quite closely those of the single crystal so that useful preliminary information on the bulk ferroelectric behaviour of the material may be obtained. In addition it is possible to prepare a wide range of ceramic compositions and to adjust the characteristics of the material for different applications. As we shall see the presence of grain boundaries gives rise to additional effects which are not present in single crystals and which have important practical implications.

Ceramics are generally prepared by first reacting well-mixed constituent powders at a relatively low temperature. The reacted powder is usually ball milled to increase the homogeneity and particle reactivity. It is then formed into some shape and fired at a higher temperature where sintering occurs. The various procedures for mixing, forming, and firing ceramics are reviewed by Jaffe, Cook, and Jaffe (1971). For most purposes a ceramic density in excess of 95 per cent of the single-crystal value is required. The resulting ceramic is basically isotropic and shows no directional behaviour such as piezoelectricity or pyroelectricity unless the material is poled in an external field. It is clear that ferroelectricity is thus a basic requirement for all these applications. Even following poling it is obvious that the anisotropy of the physical properties will not be as great as those of the single crystal.

15.2.1. Grain boundaries

The ferroelectric behaviour of the ceramic depends greatly on the crystallite (grain) size. The effects of surfaces and imperfections, which we have discussed in Chapter 4, play important roles in single crystals and frequently

dominate the dielectric behaviour of small-grain ceramics where a significant fraction of the material volume may be influenced by grain boundaries.
In some respects the grain boundaries are similar to ferroelectric surfaces. If
the polarization in neighbouring grains is not parallel, then the non-zero
polarization divergence at the interface gives rise to depolarizing fields
which in equilibrium may be compensated by free charge at the interface.
An important difference between grain boundaries and free surfaces, however, is that in high-density ceramics the material close to the grain boundaries is elastically clamped by the neighbouring grains at temperatures well
below the firing temperature. On cooling into the ferroelectric phase large
mechanical stresses may be generated by the anisotropic spontaneous strain,
and these stresses in turn affect the domain configurations and domain
dynamics within the grains. In the smallest grains the polarization may be
completely clamped by effects of this kind preventing domain reversal in
external fields. At intermediate grain sizes domain reversal becomes possible, but the phase transition is broadened over a large temperature range
and the peak dielectric constant is suppressed because of the inhomogeneous distribution of stresses and electric fields. Eventually, at the largest
grain sizes, bulk ferroelectric behaviour becomes dominant. The grain size
of the ceramic can be controlled during preparation by varying the firing
temperature and atmosphere, the sintering time, and the composition of the
material.

Poling a ceramic in an external field involves much more complicated
microscopic processes than those discussed for single crystals in Chapter 4.
Because of the random orientation of crystallographic axes of individual
grains complete poling involves orientation of the polarization of all the
individual grains to that allowed axis closest to the applied field. In the ideal
tetragonal perovskite ceramic with six permissible orientations of the polarization with respect to the crystallographic axes of the cubic paraelectric
phase one-sixth of all domains are already favourably oriented, one-sixth
reverse through 180°, and two-thirds rotate by 90°. The resulting maximum
macroscopic polarization is 83 per cent of the value within individual grains.
Domain reversal through 180° involves no change of spontaneous strain of a
grain with respect to its surroundings (except possibly during reversal, which
has the effect of increasing the coercive field), but 90° rotation is inhibited by
the elastic constraints at the grain boundaries. From geometrical considerations alone the poling strain across a completely poled tetragonal ceramic is
37 per cent of the single-crystal spontaneous strain. Corresponding measurements of the remanent strain and remanent polarization (Berlincourt and
Krueger 1959) demonstrate the difficulty of domain reorientation through
90° compared with 180°. Even if the applied field is sufficiently great to
orient 90° domains, when the field is removed the polarization frequently
reverses back to its initial state via ferroelastic interaction with the internal
stresses, either immediately or over a long period of time (ageing).

The remanent polarization is also affected by the degree of local charge compensation at the grain boundaries. In insulating materials local compensation only takes place very slowly so that at short times only the net field across the entire ceramic layer is compensated by free carriers at the electrodes. As a result the local fields do not vanish when the external field is removed as shown schematically in Fig. 15.4. In some grains the local field stabilizes the polarization while in others the field tends to reverse the polarization, resulting in a lower remanent polarization upon removal of the field. For maximum remanence the poling field must be applied for a sufficiently long time and at a sufficiently high temperature to allow both local charge compensation by conductivity and relaxation of the internal

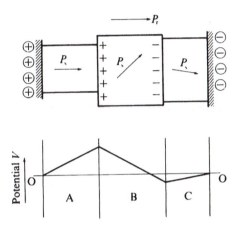

Fig. 15.4. Potential distribution in a poled ceramic when the electrodes are short-circuited and when charge compensation occurs only at the electrodes. In grains A and C the internal field is depolarizing while in B the internal field stabilizes the polarization.

stresses. Even under these circumstances some polarization relaxation may occur, especially in very-fine-grained ceramics, owing to an insufficient density of deep trapping states at the interface and the resulting incomplete compensation of fields (see § 4.4).

For alternating fields applied to insulating ceramics it is clear that neither dielectric nor elastic relaxation within the bulk is possible. Consequently hysteresis loops are generally less square than the single-crystal counterpart, saturation and remanent polarizations are lower, and the coercive fields are higher. Nevertheless significant progress has been made in preparing ceramics from materials with sufficiently small spontaneous strain and appropriate dielectric properties that remanent polarization approaching the saturation value can be obtained. An example of this is the PBLFNZ ceramic discussed in the preceding section.

15.2.2. Semiconducting ceramics

In semiconducting ceramics free carriers interact with the charged grain boundaries giving rise to anomalous behaviour of the electrical resistance and capacitance in the neighbourhood of the Curie temperature. For instance, in Fig. 15.5 the temperature dependence of the d.c. resistivity of a semiconducting $BaTiO_3$ ceramic doped with $0 \cdot 05$ per cent Sm^{3+} is compared with a single crystal of the same composition. The trivalent 'donor' Sm^{3+} impurities substitute for Ba^{2+} ions and the semiconductivity is attributed to the reduction of the $BaTiO_3$ lattice by the higher-valence impurity, i.e. the extra positive charge on the impurity is compensated by donor electrons in the Ti–3d conduction band. In the single crystal only a small change of the conductivity at the Curie temperature is observed in Fig. 15.5 which is

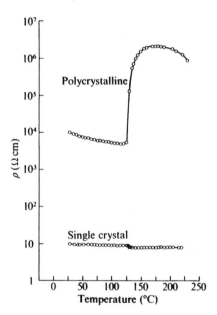

Fig. 15.5. Temperature dependence of the d.c. electrical resistivity of single-crystal and ceramic $Ba_{0 \cdot 9995}Sm_{0 \cdot 0005}TiO_3$. (After Goodman 1963.)

consistent with the change of electron mobility at the Curie temperature (Berglund and Baer 1967). The ceramic, on the other hand, shows a fairly normal negative temperature coefficient of resistance up to the Curie temperature where the resistivity suddenly increases by several orders of magnitude. Similar anomalous behaviour is observed in several other $BaTiO_3$ ceramics doped with trivalent impurities which substitute for Ba^{2+} ions or pentavalent impurities which substitute for Ti^{4+} ions. The first interpretation (Saburi 1961) attributed the effect to an abrupt change of

electron exchange between neighbouring Ti ions as the local field at the Ti
ion site changes at the Curie temperature. This is inconsistent with the
mobility measurements and the absence of the effect in single crystals.
Heywang (1961, 1971) proposed that the resistivity anomaly was due to
Schottky barriers at the grain boundaries and there is now considerable
experimental evidence supporting this interpretation.

In the paraelectric phase the band bending at the interface between grains
is determined by the number of deep interface states N_s which create a
depletion layer extending into the ceramic grains. The barrier height φ is
related to the width of the depletion layer d by Poisson's equation

$$\varphi = \frac{e^2 N_D d^2}{2\varepsilon\varepsilon_0} \tag{15.2.1}$$

where it has been assumed that all donors of concentration N_D are ionized
and that the boundary region is electrically neutral, i.e. $N_D d = N_s$. The
interface barrier is shown schematically in Fig. 15.6. The resistance per

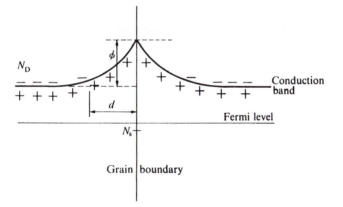

Fig. 15.6. Energy-level diagram near a grain boundary. (After Heywang 1971.)

square centimetre of the barrier is proportional to $\exp(\varphi/kT)$. The model
which now emerges is one in which the central core of the grains of relatively
high conductivity is surrounded by an insulating depletion layer. The total
conductivity of the ceramic is determined by the magnitude of φ. In the
region near the Curie temperature, where the dielectric constant diverges,
barrier height φ decreases according to eqn (15.2.1) corresponding to a drop
in resistivity. One difficulty of Heywang's model is that below the Curie
temperature the resistivity does not rise anomalously as the dielectric
constant decreases again. An explanation for this (Jonker 1964) is that, as
we have seen in Chapter 4, below the Curie temperature the polarization at
the interface strongly affects the barrier-layer region. It is clear that if the

density of trapped carriers in the interface states is just that required to compensate for the polarization divergence, then there will be no depletion layer. In general, however, the barrier height ϕ will be increased at some locations on the grain boundaries, decreased at others, and even become negative, corresponding to increased conductivity in the barrier region, depending on the exact details of the domain structure, the magnitude of the polarization, and the density of surface states. Since there is virtually no information on the latter it is not possible to make detailed predictions of the electrical behaviour in specific situations. There is now a large amount of experimental data on this anomalous positive temperature coefficient of resistance (PTC) in several ceramic compositions. The PTC anomaly is generally greater in smaller-grain ceramics (Hirose and Sasaki 1971) than in large-grain material as would be expected. The effect can be varied significantly by controlled diffusion of impurities into the surface region of the grains or by partial oxidation of the boundary layer. Treatments of this kind affect the conductivity of the boundary region by varying N_D, N_s, ε, or even P_s. Further complications which thwart quantitative analysis arise from the non-linear behaviour of ε in the barrier-layer region, the field and stress dependence of the band structure and the polarization, and the broadening of the ferroelectric transition over a wide temperature range. In fact in the simpler experimental situation of Schottky barriers at single-crystal semiconducting $BaTiO_3$ surfaces the dielectric constant in the barrier region was found to vary only slightly near the Curie temperature (Kahng and Wemple 1965). Nevertheless, there have been reports of PTC behaviour due to Schottky barriers in reduced $BaTiO_3$ single crystals (MacChesney and Potter 1965). In view of all these variables it is not surprising that the observed PTC anomalies may be abrupt near the Curie temperature, or more slowly varying over a wide temperature range, or even a combination of both.

From a practical point of view ceramics showing PTC anomalies are useful as temperature-control devices and current-limiting devices; an increase of the ceramic temperature increases its resistance thereby automatically decreasing power dissipation in a series circuit.

The electrical properties of a ceramic consisting of semiconducting grains of resistivity ρ_g and dielectric constant ε_g surrounded by insulating layers of resistivity $\rho_b \gg \rho_g$ and dielectric constant ε_b may be described by equations of Debye form (Billig and Plessner 1951)

$$\rho = \rho_g + \frac{X\rho_b}{1 + \omega^2 \tau_1^2} \qquad (15.2.2)$$

and

$$\varepsilon = \varepsilon_g + \frac{\varepsilon_b/X}{1 + \omega^2 \tau_2^2} \qquad (15.2.3)$$

where $\tau_1 = \varepsilon_0(\rho_b\rho_g\varepsilon_b^2/X)^{1/2}$, $\tau_2 = \varepsilon_0\rho_g\varepsilon_b/X$ are relaxation times, X is the ratio of the barrier thickness to the grain thickness, and it has been assumed that the boundary-layer resistance and capacitance are much greater than those within the grain bulk. It is clear that at very high frequencies the electrical properties are determined by the bulk properties of the grains, while at low frequencies boundary-layer effects are dominant. If the factor X is small the effective low-frequency dielectric constant becomes extremely large. As a result boundary-layer ceramics are finding increasing application as small-volume high-capacitance devices. However, as the low-frequency capacitance increases the dispersion frequency $f = 1/\tau_2$ decreases. For applications which require a high capacitance at the same time as high-frequency operation it is necessary to maximize the boundary-layer capacitance while maximizing the electrical conductivity within the bulk of the grain. To some extent these two requirements are compatible, since it is seen in § 4.4 that for Schottky barriers the barrier width decreases with increasing bulk conductivity.

With high-conductivity (Ba/Sr)TiO$_3$ mixtures values of ε in excess of 10^4, with dispersion frequencies in the GHz range, have been demonstrated (Masuno, Murakami, and Waku 1972). In this case grains were made conducting by Dy^{3+} doping and by reduction in an inert atmosphere, while insulating barrier layers were formed by subsequent diffusion of MnO$_2$ or CuO into the boundary region in an oxidizing atmosphere. In these materials the capacitance could be varied significantly with relatively small applied fields, which also has practical implications. Presumably this is attributable to the field dependence of the depletion layer width.

15.2.3. Effect of impurities

Ceramic research and technology has almost exclusively been devoted to the perovskite oxides—particularly compositions based on BaTiO$_3$, SrTiO$_3$, PbTiO$_3$, and PbZrO$_3$. These materials readily form solid solutions with each other and with a large number of other oxides and provide an enormous variety of ferroelectric properties. For instance the effect of forming isovalent solid solutions with BaTiO$_3$ on the ferroelectric phase transition is conveniently summarized in Fig. 15.7. The ability to vary the phase transition together with the magnitudes of the spontaneous polarization, the pyroelectric and piezoelectric coupling constants, thermal expansion coefficient, and so on provides for great versatility in the design of ferroelectric devices. For piezoelectric devices it is normally useful to have temperature-independent dielectric properties, so that it is useful to shift the phase transitions away from room temperature. On the other hand, for PTC devices it is usually desirable to have the phase transition close to the ambient temperature. Isovalent impurities in BaTiO$_3$ ceramics do not significantly affect the bulk electrical conductivity, while off-valent

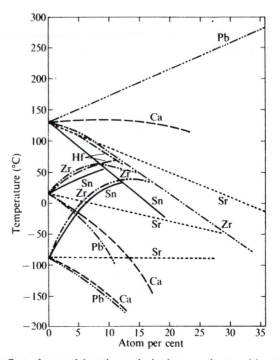

Fig. 15.7. The effect of several isovalent substitutions on the transition temperatures of BaTiO$_3$. All curves have been adjusted to transition temperatures for pure BaTiO$_3$ of 130, 15 and $-88°C$. (After Jaffe *et al.* 1971.)

impurities affect both the ferroelectric behaviour and the conductivity. In many ceramics the addition of impurities can significantly affect the crystallization kinetics of ceramics, since the diffusion kinetics during the sintering process can be greatly affected by the presence of impurities or charge-compensating defects. At the present time the approaches used to obtain the desired characteristics for specific applications have been largely intuitive and empirical.

To demonstrate the vast number of variable parameters available in a ceramic system we shall now consider in more detail the solid-solution series of PbTiO$_3$ and PbZrO$_3$ and particularly the effect of La^{3+} addition. These ceramics have found practical application because of their high piezoelectric coupling constants as well as their useful combination of electrical and optical characteristics. For information on other systems the reader is referred to the book by Jaffe *et al.* (1971).

Solid solutions of PbTiO$_3$ and PbZrO$_3$ (PZT) are particularly difficult to grow as single crystals (Ikeda and Fushimi 1962), and as a result little is known about the bulk dielectric characteristics. The sub-solidus phase diagram for the PZT system is shown in Fig. 15.8. At the two ends we have

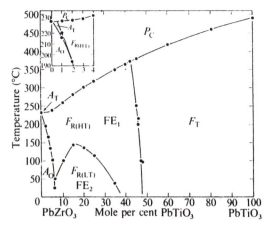

Fig. 15.8. The sub-solidus phase diagram for PbTiO₃–PbZrO₃. (After Jaffe *et al.* 1971.)

antiferroelectric $PbZrO_3$ and ferroelectric $PbTiO_3$ discussed in Chapter 8. The addition of $PbZrO_3$ to $PbTiO_3$ firstly has the effect of decreasing the room-temperature tetragonal c/a distortion of $PbTiO_3$ from 1·064 in the pure material to about 1·02 at a composition 53 per cent rich in $PbZrO_3$. At higher zirconate compositions the structure becomes a rhombohedral ferroelectric, finally becoming the orthorhombic antiferroelectric structure characteristic of pure lead zirconate at about 94 per cent $PbZrO_3$, the exact position of the phase boundary depending on the purity of the material. There are two distinct rhombohedral phases, both ferroelectric, which are structurally similar (see also § 8.2.2). The phases have been distinguished by neutron diffraction studies (Michel *et al.* 1969), electrical measurements (Berlincourt, Krueger, and Jaffe 1964), and thermal expansion studies (Shirane, Suzuki, and Takeda 1952). The great interest in PZT stemmed initially from their exceptional piezoelectric properties (Jaffe, Roth, and Marzullo 1955). Piezoelectric coupling coefficients as high as 0·6 have been measured (Weston, Webster, and McNamara 1967) near the rhombohedral–tetragonal phase boundary in undoped material and this may be still further increased by the addition of small amounts of impurities. A compositional change of 10 per cent from the phase boundary results in significant reduction of piezoelectric coupling. The remanent polarization also reaches a maximum value of about 47 $\mu C\,cm^{-2}$ (Berlincourt, Cmolik, and Jaffe 1960) for a rhombohedral composition near (but not at) the phase boundary. Since the position of the phase boundary is almost independent of temperature, the temperature dependence of the physical properties is not highly sensitive to small compositional variations.

The addition of off-valence impurities to PZT ceramics has quite a different effect from that in $BaTiO_3$. Undoped PZT ceramics usually have p-type conductivity which may possibly be attributed to an excess of Pb vacancies over oxygen vacancies due to loss of volatile PbO during firing. Electrons are then trapped at Pb vacancies leaving holes in the oxygen p-orbital valence band. It should be noted, however, that p-type conductivity of the same order is common in non-volatile oxides such as ZrO_2 (Gerson and Jaffe 1963). The addition of higher-valence impurities than the host cations tends to increase the resistivity of the ceramic by orders of magnitude. Presumably the stoichiometry is maintained by suitable volatilization of PbO rather than (as in $BaTiO_3$) reduction of the cations. PZT ceramics doped in this way with Bi^{3+} or La^{3+} at the Pb^{2+} ion sites or with Nb^{5+} or W^{6+} at Ti^{4+} or Zr^{4+} ion sites usually have sharp dielectric anomalies, lower coercive fields, square hysteresis loops, smaller ageing effects, and greater dielectric and acoustic loss than undoped material. Most of these effects can be explained as a result of vacancies facilitating domain-boundary motion (Gerson 1960), which in turn allows the relaxation of internal stresses and more efficient poling. On the other hand, impurities of lower valence than the host cations generally have the opposite effects and limited solid solubility.

15.2.4. Optical ceramics

Several trivalent impurities affect the crystallization of PZT in such a way that hot-pressed ceramics become reasonably transparent and exhibit interesting, and useful, electro-optic effects. The first studies of electro-optic properties of doped PZT ceramics were carried out with dopants such as bismuth and lanthanum (Land and Thacher 1969, Thacher and Land 1969). Lanthanum doping is particularly effective in producing a high-density (>99 per cent) ceramic body with good optical homogeneity and low insertion loss (low optical absorption and scattering) as well as a wide range of useful electro-optic characteristics (Haertling and Land 1971). For simplicity the compositions are designated by the proportional parts $x/y/z$ of the constituent La/Zr/Ti ions. The room-temperature phase diagram of La-doped PZT (PLZT) as reported by Haertling and Land (1971) is shown in Fig. 15.9 and the temperature dependence of the phases for constant La levels is shown in Fig. 15.10. Lanthanum addition is seen to extend the stability of the antiferroelectric phase to much higher titanate concentrations, increase the stability of the tetragonal phase over the rhombohedral phase, and lower the Curie temperatures of the ferroelectric phases. The cross-hatched region between the phase boundaries has been the subject of several investigations. It is tempting to suppose that this region consists of a mixture of the two phases on each side owing to small compositional variations. However, the situation is more complicated than that. It is now evident that the physical properties of this region depend on the thermal and electrical history of the crystal. Most of the studies of this region of the phase diagram have been

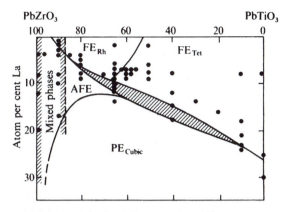

Fig. 15.9. The room-temperature phase diagram of PLZT ceramics. Points are actually prepared compositions. A mixed phase, rather than a solid solution, is observed at Zr/Ti ratios ≳90, and a moderately broad transition region (hatched) is found between the ferroelectric and antiferroelectric or paraelectric phases. (After Haertling and Land 1971.)

directed towards compositions $x/65/35$ with a lanthanum concentration x in the range $6 < x < 12$ per cent. Studies of mechanical (O'Bryan and Meitzler 1973), dielectric (Meitzler and O'Bryan 1973, Carl and Giesen 1973), ferroelectric, calorimetric, and optical properties (Keve and Annis 1973), strain measurements (Smith 1973), and X-ray diffraction studies (Keve and Bye 1975, O'Bryan and Meitzler 1973) have indicated that both

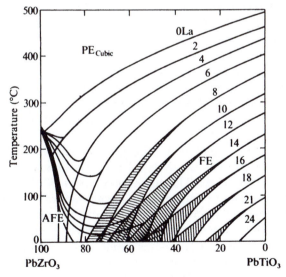

Fig. 15.10. Phase diagram of the PLZT system showing constant lanthanum concentration levels. (After Haertling and Land 1971.)

electric fields and mechanical stresses readily induce a phase transition for some compositions close to and within the hatched region of the phase diagram, but as yet there seems to be no universal agreement on the exact nature of the phases. Direct microscopic observations of thin (2 μm) ceramic plates of composition near 7/65/35 poled normally to the viewing direction have shown that the virgin material, and material which has been thermally depoled by heating into the cubic phase, are optically isotropic—or have a birefringence less than 0·003 (Keve and Bye 1975). The electrically poled ceramics, or material which has been electrically depoled by applying an alternating field of decreasing magnitude, is strongly birefringent. Keve and Bye's structural and optical studies suggest that the virgin and thermally depoled ceramic (α-phase) is a pure cubic phase in unstrained material, while the poled and electrically depoled states (β-phase) are pure orthorhombic for $x \sim 8$ and pure rhombohedral for $x \sim 7$. Because of residual strains in the ceramic it is difficult to eliminate completely the β-phase in thermally depoled material. O'Bryan and Meitzler interpreted their structural data in terms of a rhombohedral α-phase and a polymorphic mixture of tetragonal and rhombohedral phases, depending on the orientation of the grains with respect to the applied field in the poled state for $6 < x < 9$. Both sets of X-ray data indicate that the unit cell in the poled phase is greater than that in the thermally depoled phase, which is in conflict with the strain measurement of Smith (1973) which show that the ceramic volume decreases upon poling. These observations can be reconciled if the defect structure, not observed in the X-ray studies, changes significantly upon poling. More recent work (Keve 1975), however, has demonstrated that the structure of the β-phase depends on the magnitude of the applied field and the poling temperature. At high fields and poling at room temperature a tetragonal β-phase may be field induced which as a smaller unit-cell volume than the α-phase, and considerably smaller than the volume of the orthorhombic cell induced at low fields or higher temperatures. Comparison of the results of different workers must be made cautiously because of the great sensitivity of the ceramic properties to composition variations.

The dependence of the room-temperature behaviour of an 8/65/35 PLZT ceramic on thermal and electrical history is shown quite nicely by the dielectric data of Fig. 15.11. The phase transition at 55°C to the α-phase on heating does not occur upon cooling in the absence of an applied field. In addition the transition at 55°C can be prevented altogether by heating the ceramic in the presence of an applied field. However, at higher temperatures or higher lanthanum concentrations (well within the hatched region) application of a field only temporarily induces a ferroelectric phase. Upon removal of the field the remanent polarization is zero (Carl and Giesen 1973).

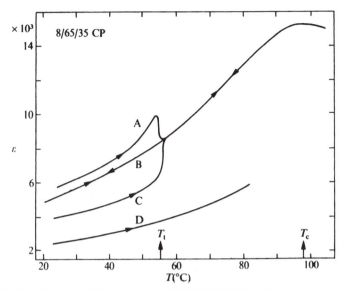

Fig. 15.11. Variation of the dielectric constant of 8/65/35 PLZT as a function of temperature in its (curve A) poled, (curve B) thermally depoled, (curve C) a.c. depoled states, and (curve D) with a 10 kV/cm^{-1} d.c. bias field applied. (After Keve and Annis 1973.)

The field- (or stress-) induced phase transition gives rise to an interesting electro-optical effect in coarse-grained ($\gtrsim 3$ μm) ceramics. The scattering of light by grain boundaries in a transparent ceramic body depends primarily on the magnitude of the refractive-index discontinuities at the domain walls and grain boundaries and on the total number of such boundaries (Dalisa and Seymour 1973). Accordingly in thermally depoled ceramics which have no, or at most low, birefringence, the index discontinuity and hence the optical scattering is weak. Upon electrically poling the material the field-induced β or polymorphic phase has a large birefringence and the scattering increases greatly since neighbouring grains do not have the same crystallographic orientation. In the electrically depoled state domains are more randomly oriented and probably smaller in size than in the poled state, resulting in even greater scattering. The transmission of a typical ceramic into a fixed angular aperture (the insertion loss) is shown in Fig. 15.12 as a function of the normalized remanent polarization, starting from the thermally depoled state A. The poled states at B, B' have opposite polarity, and D is the electrically depoled state. The ability to vary the insertion loss continuously between these two extremes has been used as the basis of optical memories (see Chapter 16).

As well as this electro-optic effect based on electrically controlled scattering, other PLZT ceramic compositions (particularly small-grain ceramics) show more conventional electro-optic behaviour based on field-controlled

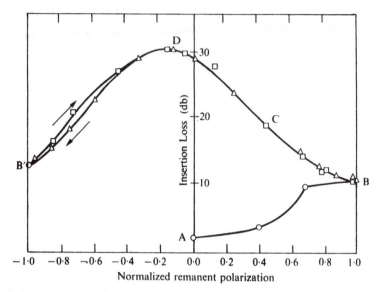

Fig. 15.12. Insertion loss *versus* normalized remanent polarization ($P_R = 1$ for a completely poled ceramic) for the longitudinal electro-optic scattering effect in a 277 μm thick 7/65/35 PLZT plate. The average grain size is 4·5 μm. (After Land 1974.)

birefringence. These effects have been thoroughly reviewed by Land (1974). To summarize briefly, PLZT ceramics may be generally categorized in the following way. Ferroelectric rhombohedral phase ceramics are suitable for electro-optic memories because of their relatively low coercive fields. Changes of the remanent polarization give rise to changes of birefringence due either to a field-induced transition or to 90° domain reversals which remain when the field is removed. Tetragonal PLZT compositions generally have higher coercive fields and exhibit conventional linear electro-optic behaviour (Pockels effect) at fields below the coercive field. In the cubic phase, close to the ferroelectric phase boundaries, PLZT compositions are highly polarizable, exhibiting slim ferroelectric hysteresis loops and large quadratic electro-optic effects. For certain experimental situations the ferroelastic nature of the ceramic can be used to advantage. Domains can be preferentially oriented by an external stress (strain bias). For instance domains can be oriented preferentially at right angles to the electric-field direction in order to maximize the birefringence change upon application of the field.

15.2.5. Glassy ceramics

In concluding this section we should mention another area of ceramics technology which adds yet another dimension to the versatility of these materials. These are the glassy ceramics in which crystallization is performed within a glassy matrix. The materials are initially formed as glasses that contain the constituents of a ferroelectric crystalline phase and usually a small amount of network-forming material such as SiO_2 is added. The glasses are then subjected to a heat treatment which produces a crystalline phase formed in a predominantly silica matrix. Several niobate compositions have shown ferroelectric behaviour (Borrelli and Layton 1971) such as dielectric anomalies, polarization reversal, and weak electro-optic behaviour. Devitrified $BaTiO_3$ also shows ferroelectric behaviour (Herczog 1964, Ulrich and Smoke 1966). Strontium titanate glass ceramics show a marked dielectric anomaly near 70 K (Lawless 1972) which has been proposed for several low-temperature applications. It has recently been shown that a wide variety of glasses (including $Pb_5Ge_3O_{11}$, $LiNbO_3$, and $LiTaO_3$) may be formed directly by rapidly cooling a melt containing only the ferroelectric constituents with no network-forming agents (Nassau and Glass 1976) and these materials can be completely crystallized to the ferroelectric phase (with the concomitant pyroelectric and piezoelectric behaviour) by suitable heat treatments.

In glassy ceramics it seems possible to achieve much smaller grain sizes than with conventional ceramic technology by suitably controlling the crystallization of the ferroelectric phase. Indeed, the available experimental data indicate that with decreasing grain size the phase transition becomes broadened and the polarization becomes unstable in the phase transition region. This behaviour may be indicative of the size effects discussed in § 15.1.

15.3. Metastable polarization

In this book we have only described polar crystalline solids which may be considered to be a matrix of dipoles oriented with the same sense owing to co-operative interaction between them. Below the Curie temperature a virgin crystal of this sort will always belong to a pyroelectric crystal class, even through a net polarization may not be observed owing to the presence of electrical twins. Traditionally ferroelectrics have been considered to be crystalline solids of this kind in which a net macroscopic polarization may be obtained and reversed in some temperature range by the application of an external field. The observation of domains, and a Curie–Weiss behaviour of the dielectric susceptibility at a paraelectric–ferroelectric phase transition, has been considered as supporting evidence for ferroelectricity. Nevertheless, materials such as GASH in which no ferroelectric phase transition can be observed since the crystal melts before such a temperature is reached, and materials with diffuse phase transitions, or ones in which domains have not been clearly identified are still normally considered to be ferroelectric because they have the characteristic of reversible polarization.

There has been increasing usage of the term ferroelectric to cover materials in which a persistent polarization may be induced and reversed by the application of an external field regardless of the crystal symmetry before the electric field was applied and of the microscopic processes which give rise to the polarization. This broader category of materials has traditionally been referred to as electrets, and includes both crystalline and amorphous materials in which the polarization may arise from field-induced alignment of dipoles or from ionic or even electronic charge trapped in the material during the poling procedure. While a complete discussion of electrets is outside the scope of this book, a brief summary of electret behaviour seems necessary, both because of several of the characteristics of ferroelectrics are also apparent in this broader group of materials and because some recent studies of polymers suggest that they may show both ferroelectric and electret behaviour.

15.3.1. Electrets

We are bound to encounter some difficulty in terminology in this section because there is no universally accepted definition of electrets or of ferroelectrics. Generally speaking electrets include all materials in which persistent polarization is observed following the application of an electric field. No microscopic mechanism is specified. With this description ferroelectrics are a subgroup of electrets with the more specific characteristics outlined in this book.

The report of a permanent electric moment in wax and rosin mixtures (Eguchi 1925) which were solidified in the presence of a strong electric field came only shortly after the discovery of ferroelectricity in Rochelle salt (Valasek 1920). It was soon recognized that there were two types of polarization generally present in electrets (Mikola 1925)—one due to homocharge, having the same polarity as the adjacent polarizing electrode, and the other due to heterocharge, having the opposite polarity to the adjacent polarizing electrode. The homocharge is attributed to real charge near the surface of the dielectric as shown in Fig. 15.13 (a) owing to conduction between the dielectric and the electrode, to field emission to or from the electrodes, or to electrical breakdown at the dielectric surface. Heterocharge, on the other hand, is attributed to bulk polarization of the dielectric owing to orientation of dipoles or by charge separation and trapping in the presence of the applied field as shown in Fig. 15.13 (b). The volume nature of the polarization was demonstrated (Jaeger 1934) by removing thin layers of an electret normal to the direction of the polarization and measuring the charge on the separate layers. Photoelectrets are heteropolar electrets in which the charge separation is due to photoexcitation of free carriers followed by their drift in the applied electric field. This group of electrets is the subject of a book by Fridkin and Zheludev (1961).

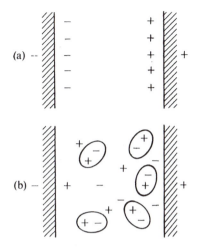

Fig. 15.13. (a) Model of a homopolar electret. Charges are injected either by conduction or breakdown at the electrodes during application of a field. (b) Model of a heteropolar electret. Dipoles are oriented in an external field or free charges are separated by the field and trapped.

Optically induced polarization in materials such as $LiNbO_3$ (see § 12.6) is clearly this kind of photoelectret behaviour.

The orientation of dipoles in amorphous heteropolar electrets resembles ferroelectricity but the origin of the polarization is quite different. Since the polarization is field induced and not an equilibrium property of the solid it will relax with the dipolar relaxation time of the solid. Although plastic materials have been found in which the room-temperature polarization is stable for many years, at temperatures close to the poling temperature the relaxation is very rapid. In this respect they differ from ferroelectrics which are thermodynamically stable when poled below the Curie temperature (or Curie range) because of the co-operative interaction between dipoles.

The procedure for obtaining a persistent polarization by dipole orientation in electrets is generally to cool the material in an applied electric field E from a temperature T_1, where the dipoles are free to rotate, to a temperature T_2, where the dipoles are frozen in position—clamped by the surrounding medium. An estimate of the magnitude of the polarization may be obtained directly from the Debye equation for dipole orientation

$$P = \frac{Np^2 E}{3kT_3} \tag{15.3.1}$$

provided that $Ep \ll kT$. Here, N is the concentration of polar molecules, each of dipole moment p, and T_3 is the temperature between T_1 and T_2 at which the dipoles become frozen. Since this actually occurs over a temperature range eqn (15.3.1) is only approximate. In fact this equation is only valid for a system of free dipoles, randomly oriented. Any potential which hinders dipole rotation will modify this expression in a straightforward way. An equivalent estimate of P which may be more useful in practice, especially if there is a variety of molecular species, is

$$P = \varepsilon_0 (\varepsilon_{T_1} - \varepsilon_{T_2}) E \qquad (15.3.2)$$

where ε_{T_1} and ε_{T_2} are the dielectric constants at T_1 and T_2. The value of the polarization P is considerably smaller than the saturation polarization obtained by complete alignment of all the dipoles (for practical values of applied field E), yet a polarization of $\sim 1 \, \mu C \, cm^{-2}$ can be obtained in several electrets.

Dielectric hysteresis is evident in electrets owing to viscous interaction of the rotating dipoles and the surrounding medium. The observation of dielectric hysteresis in CaF_2 and BeO crystals (Sawada *et al.* 1974), which have the centrosymmetric wurtzite structure, is presumably this kind of viscous hysteresis associated with defects and not ferroelectric hysteresis. Electric dipole complexes in these materials, and the more extensively studied alkali halides, result from the association of a charged defect (for instance an impurity cation of different valence from the host cation) with the charge-compensating defect.

For the experimentalist further complications arise when characterizing ferroelectrics in that electrets may show both piezoelectric and pyroelectric effects. Consider the simplest case of an electret with electroded surfaces as shown in Fig. 15.13 (b) in which N dipoles of moment $p = ql$ are oriented at an angle θ to the axis of polarization. Such a situation could arise for instance if the dipole orientation is constrained by a crystalline or viscous environment. Then the surface charge density σ_s at the electrodes is

$$\sigma_s = \frac{Q}{A} = \frac{Nql}{V} \cos\theta \qquad (15.3.3)$$

where A is the electroded area and V is the volume of electret between the electrodes.

The piezoelectric and pyroelectric effects can be identified as the pressure and temperature derivatives of the surface charge Q under short-circuit $(E = 0)$ conditions. The electrode charge Q depends only on the sample thickness $L = V/A$, the total number of dipoles between the electrodes, which is constant, and their dipole moment. A change of the dimension L due to applied pressure or thermal expansion gives rise to a change of surface charge

$$\Delta Q = N\Delta(qlL^{-1} \cos\theta). \qquad (15.3.4)$$

Three cases will be considered separately.

Case 1. Rigid dipoles of constant moment.

For this case

$$\Delta Q = Nql\Delta(L^{-1}\cos\theta) = |Q|\left[\frac{\Delta L}{L} - \tan\theta\ \Delta\theta\right]. \qquad (15.3.5)$$

A change of pressure or temperature is accompanied by a change of surface charge due both to a change of electret dimensions and to a rotation of dipoles. Thus heteropolar electrets in which the polarization is due to oriented molecular dipoles will in general show pyroelectric and piezoelectric effects.

Case 2. Trapped charges, q constant.

If the dipoles are not rigid, but change their length l by the same fraction as the change of the material dimension L while the effective charge remains constant and $\Delta\theta = 0$, then

$$\Delta Q = \{Nq\cos\theta\}\Delta\left(\frac{l}{L}\right) = 0. \qquad (15.3.6)$$

For homopolar or heteropolar electrets of uniform composition, where the polarization is due to charges of fixed magnitude trapped in the material either as a result of charge injection or charge separation, no piezoelectric or pyroelectric effects are observed. However, if the volume between the electrodes consists of an inhomogeneous mixture of materials or phases of different thermal expansion or elastic coefficients, then although l may change by the same fraction as the local change of material dimensions, this will in general not be the same fraction as the change of electrode separation L. In this case even trapped charge will give rise to piezoelectric and pyroelectric effects.

Case 3. Non-rigid dipoles, q not constant.

If the dipole length l changes in proportion to L as in Case 2, piezoelectric and pyroelectric effects can be observed if the effective charge of the dipole changes with its length. Then

$$\Delta Q = \frac{Nl}{L}\Delta q\cos\theta. \qquad (15.3.7)$$

The change of charge Δq can arise from a change of covalency of the dipoles with length and could give rise to pyroelectric and piezoelectric effects in covalent semiconductors of graded composition.

These results for the simple model of Fig. 15.13 also hold in essence for more general charge distributions and electrode configurations. In all cases the change of surface charge is completely reversible provided the electret polarization is stable. If a temperature increase results in the relaxation of a portion of the polarization, the resulting change of surface charge might appear to be pyroelectric, yet upon cooling the effect will not be reversible. Relaxation of the polarization will of course result in the loss of reversible pyroelectric and piezoelectric effects considered in Cases 1 and 3.

Two specific groups of materials will now be discussed which have received special attention because of their ferroelectric-like behaviour. These are fluorinated polymers and liquid crystals. At the present time it appears that most of these materials are best described as heteropolar electrets rather than as ferroelectrics, since the macroscopic polarization due to oriented molecular dipoles appears to be metastable and not an equilibrium property of the material.

15.3.2. Liquid crystals

The liquid-crystal, or mesomorphic, state is intermediate between a crystal-line solid and a normal isotropic liquid and has certain characteristics of each of these states. In contrast to isotropic liquids, in which there is a completely random arrangement of dipoles, liquid crystals have a considerable degree of molecular order. Friedel (1922) characterized liquid crystals into three main groups which he designated as smectic, nematic, and cholesteric phases. The simplest and most widely studied phase (particularly in the context of dielectric non-linearity) is the nematic phase, and it has been suggested that several nematic liquid crystals possess electrical properties similar to those of ferroelectrics (Williams and Heilmeier 1966, Kapustin and Vistin 1965).

Nematic liquid crystals consist of long molecules which maintain a parallel, or nearly parallel, arrangement to each other, yet they are mobile in three dimensions and free to rotate about the long molecular axis. Consequently these materials have the optical properties of a uniaxial crystal. If the molecules have a strong dipole moment along the long axis then the low-frequency dielectric constant parallel to this axis $\varepsilon_{\parallel}$ is greater than the perpendicular value $\varepsilon_{\perp}$. Such liquid crystals are said to have positive dielectric anisotropy. Conversely, if the molecules have their dipole moment

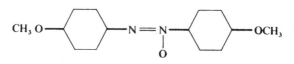

Fig. 15.14. Structure of p-azoxyanisole, a nematic liquid crystal of negative dielectric anisotropy.

normal to the axis, the $\varepsilon_{\parallel} < \varepsilon_{\perp}$ and the liquid crystals have negative dielectric anisotropy (see Fig. 15.14). Application of an electric field would be expected to align the molecules with the axis of greatest polarizability in the direction of the applied field. With alternating fields of sufficiently high frequency alignment of the molecules in this way has been confirmed, but in general the situation is considerably more complicated than this. With d.c. fields of sufficient magnitude applied to thin layers of nematic liquid crystals (Williams 1963) regular stationary domain patterns have been observed which resemble the domain structure of ferroelectrics. It has been suggested (Kapustin and Vistin 1965) that the molecular dipoles are aligned with the same sense within each domain but with an antiparallel alignment of dipoles in neighbouring domains. With electric fields of order 10^5 V cm^{-1} some co-operative interaction between the dipoles, which not only ordered the molecular axes but also the electrical sense of the dipoles, would have to exist for this interpretation. With higher electric fields the stationary domain pattern usually gives way to turbulence, thus producing dynamic scattering of light transmitted through the liquid crystal.

The observation of dielectric hysteresis in p-azoxyphenetole (Kapustin and Vistin 1965) and p-azoxyanisole (Williams and Heilmeier 1966) and the observation of transient current pulses following the reversal of the electric field was taken to be further evidence of the ferroelectric behaviour of these nematic liquid crystals. The saturation polarization of about $0.03 \mu C$ cm^{-2}

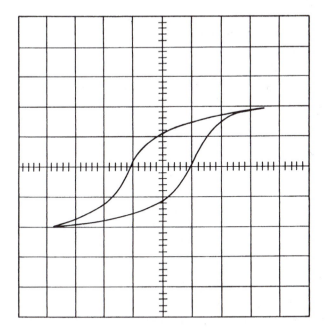

Fig. 15.15. Typical hysteresis loop for nematic p-azoxyanisole. Vertical scale $0.025 \mu C$ cm^{-2} per division, horizontal scale 10^4 V cm^{-1} per division. (After Williams and Heilmeier 1966.)

with fields of $\sim 2 \times 10^4$ V cm^{-1} measured in p-azoxyanisole (see Fig. 15.15), which has a dipole moment of $\sim 2 \cdot 5$ Debye, is about an order of magnitude greater than that expected from an isotropic liquid from eqn (15.3.1) yet is still some two orders of magnitude smaller than the polarization estimated for complete alignment of all the dipoles. These experiments are, however, inconclusive. Hysteresis loops have also been observed at temperatures well above the nematic–isotropic phase transition, and are observed even in azobenzene which does not have a liquid-crystal phase. In § 4.2 we have already shown how hysteresis loops can be obtained from a non-linear lossy material. A more satisfactory account of the domains and turbulence in liquid crystals in terms of complicated electrohydrodynamic effects has been given by Helfrich (1969). The anisotropy of the electrical conductivity as well as the dielectric properties of the liquid crystals gives rise to complex space-charge fields in the material when a uniform external field is applied.

Fairly conclusive evidence against the spontaneous parallel ordering of the electrical sense of the dipoles is afforded by seond-harmonic generation in liquid crystals. In the absence of an applied electric field no second-harmonic generation is observed in uniaxial liquid crystals, indicating that the dipole sense is randomly oriented at least over dimensions comparable to the wavelength of light (Durand and Lee 1968). Even in the presence of an applied electric field the efficiency for second-harmonic generation in several liquid crystals (as well as the isotropic azobenzene) is only consistent with the ordering expected for isotropic liquids.

It is interesting to note that this field-induced polarization can be frozen into the solid upon cooling the liquid crystal to the solid phase in the presence of the applied field. Consequently the solid is pyroelectric (Bini and Capoletti 1973) and the field-induced second-harmonic generation remains in the solid phase when the field is removed (Levine 1973 c). This behaviour is directly related to that of electrets rather than ferroelectrics.

Although the weight of evidence is therefore against the existence of ferroelectricity proper in nematic liquid crystals, very recent study of the smectic phase (Meyer *et al.* 1975) has established the probable existence of ferroelectricity in this more complex liquid-crystal phase. From a structural point of view all smectics are layered structures with a well-defined inter-layer spacing. They are therefore two-dimensional liquids and are correspondingly more ordered than nematics. There are several different types of smectic (e.g. optically uniaxial or biaxial) and we shall only briefly touch upon one particular type—so called smectic C—which is the simplest for which ferroelectricity has been claimed.

In the smectic C phase rod-like molecules are arranged in layers with the long molecular axes parallel to one another but tilted at an angle θ from the layer normal. If the phase is composed of molecules not superposable on their mirror image then only a single two-fold rotational axis (parallel to the layers and normal to the long molecular axis) remains as a symmetry element. This allows the existence of a permanent dipole parallel to this axis. Experimentally, Meyer *et al.* synthesized a material p-decloxybenzylidene p'-amino 2-methyl butyl cinnamate which was thought to optimize the conditions necessary for large polarization. Although experimentally the molecular tilt direction is found to precess around the layer normal, the helix is unwound by high applied fields to produce a typical mono-domain biaxial optical interference figure which reverses upon reversing the applied field. The helix appears to serve as an ideal ferroelectric domain structure, although it arises from local, not long-range, interactions.

The structure therefore has the characteristics of a two-dimensional ferroelectric (McMillan 1973). Blinc (1975) suggests that the phase can be thought of as a special class of extrinsic ferroelectric, driven by the inter-molecular forces producing the smectic tilting angle. It is possible that a mixture of different smectic C materials could be prepared for which the complications introduced by the existence of the helical 'domain' structure might be avoided by adjusting its pitch to approach infinity. In this manner it may eventually be possible to obtain a more conventional liquid-crystal ferroelectric with three-dimensional domains.

15.3.3. *Pyroelectric polymers*

Piezoelectricity (Kawai 1969) and pyroelectricity (Bergman, McFee, and Crane 1971, Nakamura and Wada 1971) have been induced in several polymers following the application of high electric fields at elevated temperatures. In some materials the polarization produced in this way is stable at room temperature for periods of at least a year. The polarization can be reversed at elevated temperatures. At the present time it is not clear whether these materials are properly classified as ferroelectric, since it has not been established whether there is any long-range co-operative interaction or whether the behaviour is just that characteristic of thermoelectrets in the more general sense. Even before pyroelectric behaviour was discovered in these materials it was proposed (Reddish 1966) that long-range interactions could account for the dielectric relaxation measurements in polyvinyl chloride, but there has been no corroborating evidence from other measurements.

Of the pyroelectric polymers polyvinylidene fluoride (PVF_2) has attracted particular attention because of the large piezoelectric and pyroelectric coefficients which can be induced in this material. Otherwise PVF_2 has the same general characteristics of the other polymers. X-ray diffraction and nuclear magnetic resonance determination of the structure of PVF_2 shows that drawn films of this polymer crystallize with two structures (Lando, Olf, and Peterlin 1966). Form I consists of a planar zigzag conformation of the polymer with two monomer CF_2-CH_2 units per unit cell belonging to the space group Cm2m, while in form II the molecules take a planar cis (2_1 helix) conformation, the tentative space-group determination being Pna2_1. Rolling and stretching the films at elevated temperatures during their preparation appears to enhance the crystallization of form I with polymer chains oriented preferentially along the stress axis.

The structural determination leaves little doubt that at least on a microscopic scale PVF$_2$ is pyroelectric with a net dipole moment of 2·1 Debye per monomer unit oriented perpendicularly to the polymer chain even in unpoled material. A model of the structure is shown in Fig. 15.16. Films

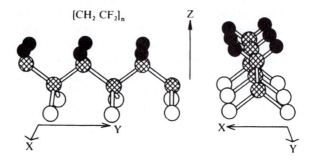

Fig. 15.16. A molecular model of polyvinylidene fluoride illustrating the planar zigzag conformation in form I. Fluorine atoms are the open circles below the cross-hatched carbon chain, and hydrogen atoms are the filled circles above the chain. The view on the right along Y is parallel to the chain; Z is the polar axis. (After McFee, Bergman, and Crane 1972.)

which have not been subjected to a poling field usually show no, or very slight, pyroelectricity or piezoelectricity. Using the pyroelectric scanning technique (see § 4.1.1 (e)) no domains have been observed to dimensions of about 10 μm so the dipolar ordering characteristic of the structure is on a scale smaller than this (Glass, McFee, and Bergman 1971). The absence of a net moment in unpoled material can be due to the rotation of the C—H and C—F dipoles about the chain axis as shown in Fig. 15.17. After an electric field of sufficient magnitude (usually greater than 10^5 V cm^{-1}) has been applied normal to the film at elevated temperatures (usually greater than 80°C), pyroelectric coefficients as large as $2 \cdot 4 \times 10^{-9}$ C cm^{-2} K^{-1} normal to the plane of the film and piezoelectric d_{31} coefficients as large as 20×10^{-8} c.g.s. units have been measured at room temperature, where the axes are indicated in Fig. 15.16. These results differ widely from one film to another poled in different ways or obtained from different sources, possibly owing to the varying degree of crystallinity of the films. In all material, however, the coefficients remain unchanged for periods up to at least a year. At lower temperatures (below ~50°C) the pyroelectric effect is completely reversible; that is, the surface which was poled with the positive electrode becomes positive on heating and negative upon cooling with the same amount of

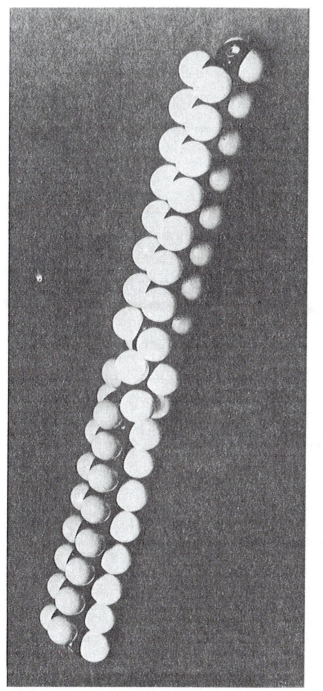

Fig. 15.17. A molecular model illustrating the rotation through 180° of molecular dipoles of polyvinylidene fluoride about the polymer chain axis. (After McFee *et al.* 1972.)

charge developed, but of opposite sign, for the same temperature difference. The pyroelectric current is proportional to the rate of change of temperature. The effect is thus not due to thermal liberation of charge trapped in the polymer during poling, but to a change of dipole moment per unit volume. The polarity of the piezoelectric effect also agrees with the dipolar origin of the polarization. Corresponding measurements of the piezoelectric and pyroelectric coefficients on the same film strongly suggest that the underlying mechanisms of the two effects are similar. Furthermore, corresponding measurements of the pyroelectric coefficient and the birefringence of the films (Bergman *et al.* 1971) indicate that the greater the degree of orientation of the polymer chains in the plane of the film produced during the film preparation, the greater the pyroelectric coefficient which can be induced during poling. This provides further evidence of the important role of the dipoles normal to the polymer chains. Typical birefringence measurements of uniaxial or biaxial films are $|n_{X,Y} - n_Z| \sim 0 \cdot 01 - 0 \cdot 02$.

The poling process which makes PVF_2 pyroelectric also makes the films active for second-harmonic generation. The three independent second-order non-linear coefficients of the point group mm2 of crystalline form I measured relative to crystal quartz are

$$d_{33} \sim 2d_{31} \sim d_{11}(SiO_2) \quad \text{and} \quad d_{32} \sim 0.$$

For both d_{33} and d_{31} the fundamental optical frequency is polarized with a component along the polarizable C—F and C—H bonds. The large values of these coefficients can be attributed to the high degree of orientation of these dipoles. Antiparallel alignment of the dipoles over dimensions less than the coherence length of the interaction of the fundamental and second-harmonic beams ($l_{33} \sim 30 \, \mu m$) would lead to cancellation of the second-harmonic intensity.

All these characteristics of PVF_2 closely resemble the characteristics of ferroelectrics, yet it would be premature to classify this and other polymers as ferroelectrics. As discussed earlier, these characteristics are also expected from certain heteropolar electrets. Even though the polarization is stable at room temperature in PVF_2, if the temperature is raised above about 80°C (in the temperature range within which the films can be poled), without an external field applied, the polarization relaxes. The rate of relaxation and the fraction of the polarization which relaxes depends on the annealing temperature and the poling conditions (Glass *et al.* 1971, Pfister, Abkowitz, and Crystal 1973). The variation of the stability of the polarization with poling treatment may be related to incomplete local compensation of the depolarizing fields at crystallite boundaries in the material in the same way as in ceramics. Internal fields are relaxed only for poling times long compared with the dielectric relaxation time. In any case the macroscopic polarization appears to be metastable even though the dipoles are ordered on a microscopic scale. There is no discrete Curie temperature. Dielectric measurements of PVF_2 show no evidence of Curie–Weiss behaviour or a marked dielectric anomaly in the temperature range where polarization relaxation occurs. There is a broad dielectric absorption peaking in the range 70–100°C at 300 Hz which has been associated with molecular motion within the crystalline regions of PVF_2 (Sasabe *et al.* 1969) as shown in Fig. 15.18. The higher the degree of crystallinity of the films the greater the

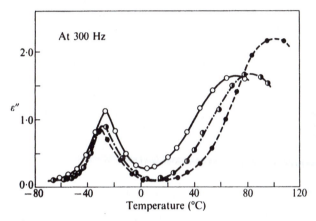

Fig. 15.18. Temperature dependence of the dielectric loss ε'' at 300 Hz for various samples of PVF_2 with different crystallinities: $\bigcirc$ 46 per cent; $\mathbb{O}$ 55 per cent; $\bullet$ 64 per cent. (After Sasabe *et al.* 1969.)

strength of the absorption and the higher the temperature of the peak absorption. The other absorption bands below room temperature have been associated with the amorphous region of the films and may account for the polarization effects observed in PVF_2 below room temperature (Pfister and Abkowitz 1974).

It is interesting to estimate the pyroelectric coefficients which would be expected in PVF_2 for a set of rigid C—F and C—H dipoles of constant moment in the two cases of 100 per cent alignment of all dipoles and the freezing in of randomly oriented dipoles according to the Debye eqn (15.3.1). For 100 per cent alignment, assuming 100 per cent crystallinity of the films and a dipole density of $1 \cdot 8 \times 10^{22}$ dipoles/cm^3 of moment $2 \cdot 1$ Debye, we find

$$P = 13 \ \mu\text{C cm}^{-2}$$

and from eqn (15.3.5) for dipoles of constant moment and no rotation

$$\frac{\text{d}P}{\text{d}T} = \alpha P = 1 \cdot 6 \ \text{nC cm}^{-2} \ \text{K}^{-1}$$

where $\alpha = 1\cdot2 \times 10^{-4}$ K^{-1} is the coefficient of linear expansion of PVF$_2$.†
This is remarkably close to the maximum experimental value of
$2\cdot4$ nC cm^{-2} K^{-1}.

On the other hand, using the Debye equation we find

$$P = 0\cdot6E \, \mu C \, cm^{-2}$$

where E is the applied field in MV cm^{-1}. This implies that saturating the
polarization by orienting all the dipoles in practical fields of $0\cdot1$–1 MV cm^{-1}
is not possible without large local-field corrections. This estimate neglects
the restrictions imposed on the motion of the dipoles by the polymer chain,
and assumes they are completely free. We now estimate for the pyroelectric
coefficient

$$\frac{\mathrm{d}P}{\mathrm{d}T} = 0\cdot07E \text{ nC cm}^{-2} \text{ K}^{-1}.$$

Even for the highest poling fields of 1 MV cm^{-1} this estimate is considerably
lower than the experimental value. While this might suggest that there is
some co-operative dipole interaction which leads to a greater degree of
dipole alignment, there is also the possibility that an additional mechanism
contributes to the pyroelectric effect. For instance the change of dipole
moment or rotation of the C—F and C—H bonds with temperature may be
the dominant effect and the assumption of fixed dipoles invalid. Alternatively, since PVF$_2$ is generally an inhomogeneous mixture of type-I and
type-II phases, trapped charge could play an important role.

Attempts to measure the polarization of PVF$_2$ have not been conclusive.
Polarization switching times are too long and the electrical conductivity at
poling temperatures is too high for most conventional measurements of the
polarization. Integration of the charge during depolarization of the films has
yielded values of order $0\cdot5$ μC cm^{-2} when poled with $0\cdot5$ MV cm^{-1} fields,
but these films also had small room-temperature pyroelectric coefficients
Pfister et al. 1973). These measurements also suggested that more than one
mechanism may be responsible for the polarization.

It will be clear from this section that a considerable amount of information
is still necessary before the behaviour of materials such as PVF$_2$ is understood. In the meantime, the ease of fabrication of the pyroelectric films may
make materials of this kind attractive candidates for pyroelectric applications.

† Kureha Chemical Co. data sheet.

16

APPLICATIONS OF FERROELECTRICS

16.1. Introduction

THROUGHOUT its history the study of ferroelectricity has been closely linked with device applications. The discovery of the first ferroelectric was made during studies of Rochelle salt which had excited a great deal of interest since it showed the largest piezoelectric effect known at the time. For several years Rochelle salt was widely used for phonograph pick-ups and microphones. Although Rochelle salt is now obsolete as *the* transducer material, other ferroelectrics—particularly ceramics—in which the electrical and mechanical impedances can be effectively matched for efficient power conversion, serve several useful functions in this field. $LiTaO_3$ also shows promise for certain piezoelectric-filter applications. However, piezoelectric applications will not be covered in this chapter since they are adequately documented in the literature and do not directly involve the ferroelectric or pyroelectric nature of the material.

Many other ferroelectrics were discovered during bursts of activity as concepts for new devices evolved for which ferroelectrics seemed suited. $BaTiO_3$ was accidentally discovered during a search for high-dielectric-constant materials for capacitors. Since TiO_2 had the largest known dielectric constant at the time it was natural to look for modifications of this material. Later came the search for computer memory elements. The bistable polarization of ferroelectrics made ferroelectrics attractive candidates for binary memories. The large amount of stored electrical energy would in principle allow enormous storage densities since a polarization of $10 \mu C\,cm^{-2}$ corresponds to about 10^{14} electrons/cm^2 so that the charge on $1 \mu m^2$ could be read out with high signal-to-noise ratio. The potential of this device was not realized for a variety of reasons. The ideal binary memory would have a square hysteresis loop with a well-defined coercive field allowing read-out of one element of an array without partial switching of other elements. No ferroelectric completely satisfied this criterion. Furthermore the crystals tended to fatigue after repeated polarization reversal which led to a higher, even less well-defined coercive field and slower switching times. The growth of crystals of sufficiently high quality and uniform properties was a problem with materials such as $BaTiO_3$ which were most promising for these applications. Eventually other technologies advanced faster and interest in ferroelectric memories based on polarization reversal declined.

The next burst of activity came in the mid-1960's when it was realized that the large nonlinear polarizability of ferroelectrics makes them one of the most promising classes of materials for electro-optical and optical parametric devices. However, the development of these devices had many ups and downs. For optical applications the quality of the crystals had to meet stringent requirements. The larger the polarizability of the material the more difficult it is to grow optically perfect crystals. This is because if the refractive index is to be sensitive to external fields it is also sensitive to internal stresses, non-uniform charge distribution, and compositional variations. Both $K(TaNb)O_3$ and $Sr_{1-x}Ba_xNb_2O_6$ excited interest because of their extremely high electro-optic coefficients. These crystals were difficult to grow with uniform optical properties, but even when this was achieved the optical quality degraded upon application of a field because of the electrical conductivity of the crystals and the development of space-charge fields. To avoid effects of this kind extremely-high-resistivity crystals are required ($\gg 10^{16}\ \Omega$ cm). In the case of $LiNbO_3$ the dark resistivity was high enough, but applications involving light at higher energies in the visible spectrum (wavelength shorter than about 500 nm) was limited by the so-called 'optical-damage' effect due to the excitation of photocarriers. Subsequently this effect has proved to be a promising mechanism for optical memories as discussed in § 16.3.

The tungsten-bronze ferroelectric $Ba_2NaNb_5O_{15}$ has one of the highest second-order nonlinear optical coefficients of any material in the visible spectrum and near infrared and does not show serious effects of optical damage in undoped crystals, but once again the growth of large, optically perfect crystals has proved to be an extremely difficult problem despite a great deal of effort. It is primarily because of this problem of crystal quality that materials such as KDP which can be grown from solution with high optical quality are still leaders in the field of non-linear optics despite their relatively low non-linear coefficients compared with the oxides.

The future of all these optical applications and further intensive effort to improve the materials now depends on the development of the entire field of 'photonics' or 'optoelectronics'.

During the ups and downs of the development of ferroelectrics for memories and optical devices the development of pyroelectrics as thermal detectors of infrared radiation has been steadily increasing. These devices have a useful combination of attributes which are not found in other infrared detectors. As thermal detectors they have a broad spectral response and room-temperature operation, but unlike other thermal detectors they have high frequency response. The more recent developments of infrared imaging devices are particularly promising applications since pyroelectrics have some obvious advantages over other technologies (see §16.2). Only uniform electrical properties of the pyroelectric target are required for these applications and the optical quality is unimportant, so the demands made on the crystal grower are much less severe.

It is evident that most of the applications of single-crystal ferroelectrics which have been proposed in recent years have not involved polarization reversal and thereby avoid the problem of fatigue. The majority of these are optical applications. However, with high-density ceramics fatigue may not be such a serious problem, for reasons which are not yet clear. Ceramics have for a long time been of interest for dielectric and piezoelectric applications, but only recently have optically transparent ceramics been developed which offer a wide range of possible electro-optic applications including several which involve polarization reversal. The advantages of ceramics in terms of ease of preparation and cost and of the variety of compositions and properties are obvious.

In this chapter we shall review three main areas of application of ferroelectrics which have received a great deal of attention, namely (a) pyroelectric detectors and imaging devices, (b) optical memories, and (c) modulators and deflectors. The basic ideas are emphasized rather than providing an extensive review of the literature. Of the remaining applications most can be inferred from the discussions in previous chapters, such as § 15.2 which includes discussion of PTC and barrier-layer devices, § 12.5.4 on ferroelectric lasers, and Chapter 13 on non-linear optics. Some important applications of non-linear optical materials which do not specifically involve ferroelectrics or pyroelectrics are not discussed in Chapter 13 (e.g. tunable parametric oscillators and infrared up-conversion), but these devices have been discussed extensively in other texts on non-linear optics.

16.2. Pyroelectric detection

We have already seen in Chapter 5 how the pyroelectric properties of a polar material are studied by measuring the current or voltage response of a crystal to a temperature change, either by continuous heating or by the absorption of sinusoidally modulated radiation. In the same way pyroelectric materials of known characteristics may be used for calorimetry and as thermal detectors of radiation. From eqns (5.4.1) and (5.4.2) we see that under short-circuit conditions ($E = 0$) the current response is given by

$$J_i = p_i \left(\frac{\mathrm{d}T}{\mathrm{d}t} \right). \tag{16.2.1}$$

In the presence of a constant external bias field E the temperature dependence of the dielectric constant gives an addition current

$$\{E(\partial \varepsilon / \partial T)\}(\mathrm{d}T/\mathrm{d}t)$$

but as we discussed in § 5.2 this non-linear dielectric term can be combined with the linear pyroelectric coefficient so that only one field-dependent pyroelectric coefficient p_i need be used in the analysis of pyroelectric detectors regardless of which contribution is dominant.

As with other thermal detectors pyroelectrics may be used to detect any radiation which results in a temperature change of the crystal, i.e. X-rays to microwaves and even particles. They also have the useful features of room-temperature operation (or any other convenient temperature), simplicity of construction and operation, and do not require an external bias field. However, in contrast with other thermal detectors, the pyroelectric current response depends on the rate of change of temperature rather than temperature itself. For this reason the maximum response is achieved at times shorter than the thermal relaxation time τ_T of the element so that pyroelectrics are basically much higher-frequency devices than other thermal detectors.

In this section we shall briefly describe the principles of operation of pyroelectric detectors, the noise limitations of the detectivity, and the materials requirements to optimize the detector performance. The extension of these principles to infrared imaging devices will also be described.

16.2.1. Responsivity

Typical detector geometries are shown in Fig. 16.1. A thin pyroelectric wafer is electroded normal to the polar axis which may be parallel (face electroded) or perpendicular (edge electroded) to the incident radiation.

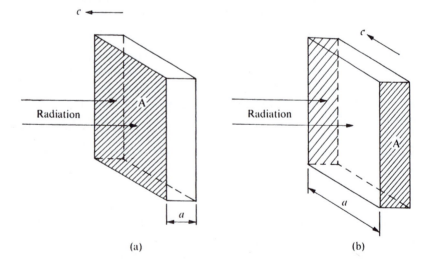

Fig. 16.1. Typical geometric configurations for pyroelectric detectors: (a) face electrodes; (b) edge electrodes.

With face electrodes the incident radiation may be absorbed at the electrode, which may be blackened, or within the wafer if the electrodes are transparent.

The electrical responsivity of a pyroelectric detector depends firstly on the thermal response of the detector element to the incident radiation and secondly on the pyroelectric response to the temperature change. The temperature change ΔT of a detector to incident radiation of power $W(t)$ is given by

$$\mathscr{C}\frac{dT}{dt}+G\Delta T = eW(t) \tag{16.2.2}$$

where e is the fraction of incident power which thermalizes in the crystals, $\mathscr{C}$ is the thermal capacity of the crystal, and G is the thermal conductance to the surroundings i.e. $\mathscr{C}/G = \tau_T$ is the thermal relaxation time of the crystal. If the incident radiation is 100 per cent amplitude modulated at frequency ω, i.e. $W(t) = W_0\{1+\exp(i\omega t)\}$ the solution of eqn (16.2.2) gives an average temperature change

$$\Delta T_{dc} = \frac{eW_0}{G} = \frac{eW_0\tau_T}{\mathscr{C}} \tag{16.2.3}$$

and a component at frequency ω

$$\Delta T(\omega) = \frac{eW_0}{(\omega^2\mathscr{C}^2+G^2)^{1/2}}\exp\{i(\omega t+\phi)\}. \tag{16.2.4}$$

where the phase angle $\phi = \tan^{-1}(\omega\tau_T)$.

The pyroelectric response to the temperature change given by eqns. (16.2.3) and (16.2.4) is an initial transient component due to ΔT_{dc} when the light is turned on, followed by an alternating current density $J(\omega)$ given by

$$J(\omega) = \frac{ep_iW_0}{\mathscr{C}(1+\omega^{-2}\tau_T^{-2})^{1/2}}\exp\{i(\omega t+\phi)\}. \tag{16.2.5}$$

It is clear that this is a maximum when $\omega\tau_T \gg 1$; then $|J(\omega)|$ is independent of ω, except for the variation of p_i with frequency. We can then define a current *responsivity* for a detector area A as

$$r_i = \frac{A|J|}{W_0} = Aep_i/\mathscr{C}. \tag{16.2.6}$$

If the detector is connected to a load consisting of a parallel RC circuit, as shown in Fig. 16.2, then a voltage responsivity can be defined in a similar manner as the ratio of the voltage output to the power input, i.e.

$$r_v = \frac{AeRp_i}{\mathscr{C}(1+\omega^2R^2C^2)^{1/2}(1+\omega^{-2}\tau_T^{-2})^{1/2}} \tag{16.2.7}$$

where R, C are the parallel resistance and capacitance of the detector and load circuit at frequency ω, as in Fig. 16.2, and A is the electroded area of the crystal normal to the pyroelectric axis. This expression was first

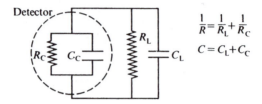

$$\frac{1}{R} = \frac{1}{R_L} + \frac{1}{R_C}$$

$$C = C_L + C_C$$

Fig. 16.2. Equivalent circuit of a pyroelectric detector of resistance R_C and capacitance C_C connected to a load of resistance R_L and capacitance C_L.

developed by Cooper (1962). The temperature change $\Delta T(\omega)$, the current responsivity r_i, and the voltage responsivity are plotted as a function of frequency in Fig. 16.3. The low-frequency roll-off due to thermal relaxation,

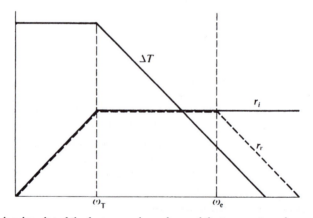

Fig. 16.3. A log–log plot of the frequency dependence of the temperature change ΔT, current responsivity r_i, and voltage responsivity r_v of a pyroelectric detector. The frequencies $\omega_T = \tau_T^{-1}$ and $\omega_e = \tau_e^{-1}$ are the reciprocal thermal and electrical time constants of the detector element.

characteristic of all thermal detectors, usually occurs at frequencies near 1–100 Hz while the high-frequency roll-off is determined by the electrical time constant τ_e of the detector–load circuit. The bandwidth of the detector can be increased by decreasing the load resistance R_L with a corresponding loss of voltage responsivity. For a maximum r_v the impedance of the detector–load circuit should be a maximum, and this is limited by the capacitance and dielectric loss of the crystal itself.

16.2.2. Noise limitations

The minimum detectable signal from a pyroelectric detector is limited by various noise sources in the detector element, the load, and the measuring circuit. If the pyroelectric-crystal impedance were purely capacitive, then the only source of noise from the crystal itself would be temperature noise due to power fluctuations of thermal radiation emitted and received by the detector elements (Smith, Jones, and Chasmar 1968)

$$\langle v_{\mathrm{T}}^2 \rangle = 2kr_v^2 \{T_{\mathrm{b}}^2 G(T_{\mathrm{b}}) - T^2 G(T)\} \frac{\Delta f}{e^2} \qquad (16.2.8)$$

where T_{b} is the temperature of the surroundings seen by the detector element and Δf is the bandwidth. The smallest value of this noise contribution for a given detector responsivity and temperature is when G is limited by radiative conductance. This is the fundamental noise limit of any thermal detector. If the detector element is mounted on a substrate, or suspended in a gas, conduction and convection will increase temperature noise.

In practice the pyroelectric crystal is not a perfect capacitor and dielectric loss in the crystal is a source of Johnson noise.

$$\langle v_{\mathrm{J}}^2 \rangle = \frac{4kTR\Delta f}{1 + \omega^2 R^2 C^2}. \qquad (16.2.9)$$

By including the parallel resistance and capacitance of the load, eqn (16.2.9) includes the Johnson noise from this source. Additional sources of noise are associated with the signal amplifier and are usually represented as a current generator $\langle i_{\mathrm{A}}^2 \rangle$ in parallel with the input circuit and a voltage generator $\langle v_{\mathrm{A}}^2 \rangle$ in series with the input, and are independent of the input circuit. For our purpose it is convenient to lump both terms together as a single noise source $\langle v_{\mathrm{A}}^2 \rangle$ which now depends on the detector circuit, so the amplifier must be specifically chosen for the particular detector element and application.

The minimum detectable power is thus

$$W_{\mathrm{m}}^{\mathrm{d}} = \frac{(\langle v_{\mathrm{T}}^2 \rangle + \langle v_{\mathrm{J}}^2 \rangle + \langle v_{\mathrm{A}}^2 \rangle)^{1/2}}{r_v} \qquad (16.2.10)$$

where the superscript d denotes direct detection in contrast to heterodyne detection. In most practical cases Johnson noise of the crystal element is more important than temperature noise, although in recent years the development of doped triglycine sulphate crystals with low dielectric loss brings these detectors close to the ideal temperature-noise limit.

When Johnson noise is dominant and amplifier noise can be neglected, from eqns (16.2.7), (16.2.9), and (16.2.10) we find that, for frequencies

$\omega \gg 1/\tau_{\mathrm{T}}$, the minimum detectable power

$$W_{\mathrm{m}}^{\mathrm{d}} = \frac{c_{\mathrm{p}}(Aa)^{1/2}\sigma^{1/2}}{ep_i}(4kT\Delta f)^{1/2} \qquad (16.2.11)$$

where we have written $\mathscr{C}$ in terms of the specific heat c_{p}, the crystal thickness a, and area A, and the resistance R is assumed to be limited by the crystal conductivity $\sigma = a/AR$.

Alternatively, a detectivity D^* is usually defined as

$$D^* = \frac{\Delta f^{1/2} A^{1/2}}{W_{\mathrm{m}}^{\mathrm{d}}} = \frac{ep_i(4kT)^{-1/2}}{\sigma^{1/2}c_{\mathrm{p}}a^{1/2}}. \qquad (16.2.12)$$

The conductivity σ includes both the d.c. conductivity and the dielectric loss $\omega\varepsilon''$ of the material. The origin of the loss may be defects or domain boundaries in imperfect crystals, but even in perfect crystals the polarization fluctuations contribute to the dielectric loss and of course become very large in the neighbourhood of a ferroelectric phase transition.

16.2.3. Pulse detection

With the advent of infrared lasers pyroelectric detectors have become increasingly important for fast pulse detection. As we have seen in eqn (16.2.5) the current response of the detector is independent of frequency above $1/\tau_{\mathrm{T}}$ so that the current pulse reproduces the incident pulse intensity provided p_i is constant over the frequency bandwidth of the input pulse. If the optical pulse has frequency components above the mechanical resonance of the detector element, then a slowly decaying oscillatory piezoelectric response may be superimposed on the pyroelectric response as shown in Fig. 5.2 if the detector element is not acoustically clamped. This effect can be avoided either by suitable choice of the detector material with a high acoustic loss or by detector construction. For instance the detector orientation can be chosen so that piezoelectric coupling is small, or the element can be elastically clamped by suitable mounting as discussed in § 5.3.

The bandwidth of the detector–amplifier combination can be increased either by lowering the load resistor to a value where $\omega RC \ll 1$ at the highest frequency component of the input pulse, or by using a frequency-compensating amplifier (Ludlow et al. 1967). The latter gives the optimum signal-to-noise performance, but at present such amplifiers are limited to bandwidths of a few megahertz. Pyroelectrics have useful capabilities at frequencies well above this range. For instance Fig. 16.4 shows the response of a $Ba_xSr_{1-x}Nb_2O_6$ detector to a train of pulses from a mode-locked CO_2 laser (Wood, Abrams, and Bridges 1970). The measured pulse width of about 1 ns was limited by the rise time of the oscilloscope. The ultimate frequency limitation of a pyroelectric detector is determined by the rate at

(a)

(b)

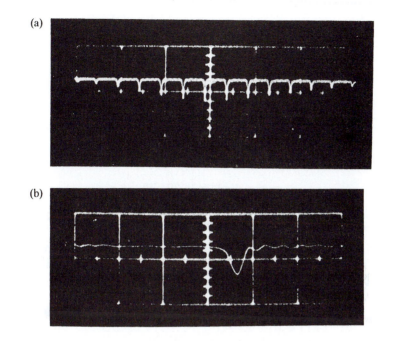

Fig. 16.4. The response of a $Ba_{0.4}Sr_{0.6}Nb_2O_6$ pyroelectric detector to a train of pulses from a mode-locked CO_2 laser: (a) 50 ns cm^{-1} showing the entire pulse train, and (b) 5 ns cm^{-1} showing a single pulse. Increasing intensity is downward and the vertical scale is of order 1 MW cm^{-1}. (After Wood *et al.* 1970.)

which the absorbed radiation thermalizes in the crystal. If the radiation is absorbed directly into an infrared absorption band of the crystal itself, then the upper frequency limit is determined by the coupling of these phonons with the anharmonic lattice modes. This is expected to be in the 10^{-12}–10^{13} s range. If the absorption is due to electronic transitions then the pyroelectric response is limited by a non-radiative electronic relaxation time (Auston, Glass, and LeFur 1973). In this case the electronic polarization may also contribute to the response (see § 16.2.7). If the radiation is absorbed at an electrode then the response is limited by a thermal diffusion time. For absorbing blacks this is typically 10^{-5}–10^{-7} s, but, by using thin metallic electrodes to absorb the radiation, subnanosecond response times have been achieved with a total reflection loss of only 50 per cent (Roundy and Byer 1972).

16.2.4. Heterodyne detection

Since the electric field generated by a pyroelectric detector is proportional to the square of the input field, considerable improvement of the performance for the detection of coherent radiation may be obtained by heterodyne detection, i.e. mixing the weak optical signal $E_s = A_s \cos \omega_s t$ with an intense local oscillator $E_L = A_L \cos \omega_L t$. Then the power absorbed by the detector for spatially coherent beams is

$$W \propto (E_s + E_L)^2 = A_L^2 \cos^2 \omega_L t + A_s^2 \cos^2 \omega_s t$$
$$+ A_L A_s [\cos\{(\omega_L + \omega_s)t\} + \cos\{\omega_L - \omega_s)t\}]. \quad (16.2.13)$$

Since the detector does not respond to optical frequencies the only output signal is due to the last term at the intermediate frequency $\omega_L - \omega_s$, while the first two terms thermally bias the detector. The peak output signal is then

$$v_h = r_v A_L A_s = 2r_v (W_s W_L)^{1/2} \quad (16.2.14)$$

where W_L, W_s are the average local-oscillator and signal powers. The gain in signal by heterodyning is thus $2(W_L W_s)^{1/2}/W_s = 2(W_L/W_s)^{1/2}$.

If the local oscillator were to introduce no additional noise then this ratio would be the improvement in signal-to-noise ratio. However, as W_L increases the detector temperature increases, thereby increasing both temperature and Johnson noise. In addition photon fluctuations in the local-oscillator beam give signal fluctuations (Abrams and Glass 1969)

$$\langle v_p^2 \rangle = \frac{2r_v^2 h\nu W_L \Delta f}{e} \quad (16.2.15)$$

where $h\nu$ is the photon energy. The ultimate limit of detectivity for a thermal heterodyne detector is when $\langle v_p^2 \rangle$ is the dominant source of noise, for which the minimum detectable power is

$$W_m^h = \frac{h\nu \Delta f}{e}. \quad (16.2.16)$$

It is possible in principle to approach this limit with practical pyroelectric detectors only at low intermediate frequencies ($\gtrsim 100$ Hz) when the detector responsivity r_v is a maximum, provided $h\nu \gg kT$ (Glass and Abrams 1970 b). There has been no attempt to achieve this limiting detectivity at low intermediate frequencies experimentally. However, considerable decrease of the minimum detectable power of both TGS and $Sr_{0.6}Ba_{0.4}Nb_2O_6$ detectors has been demonstrated by Leiba (1969) and Abrams and Glass (1969) by heterodyning at $10.6~\mu$m in the megahertz frequency range, and by Gebbie et al. (1967) with TGS at $337~\mu$m.

16.2.5. Optimizing detector performance

The important detector parameters which have been derived in the preceding sections are the responsivities r_i, r_v, and the minimum detectable power W_m^d. If we assume the detector is operated in the optimum manner, with the detector impedance much smaller than that of the load and $\omega \tau_T \gg 1$ and $\omega \tau_e \gg 1$, then from eqns (16.2.6), (16.2.7), and (16.2.11)

$$r_v = \frac{ep_i}{\omega c_p A \varepsilon'} \qquad r_i = \frac{ep_i}{c_p a} \qquad (W_m^d)^{-1} = \frac{ep_i \, \Delta f^{-1/2}}{c_p (4kT\sigma aA)^{1/2}}$$

where the detector impedance has been written in terms of the dielectric permittivity ε' and conductivity σ. In practice the a.c. conductivity is usually dominated by the dielectric loss and we can write $\sigma = \omega \varepsilon''$. From these figures of merit it is clear that optimizing the detector performance involves optimizing both the detector geometry and material characteristics.

The current responsivity is seen to vary inversely with electrode separation, the voltage responsivity varies inversely with electrode area, and the minimum detectable power in Johnson-noise-limited operation is proportional to the detector volume. Thus decreasing the detector dimensions generally improves the performance. However, no improvement is obtained if decreasing the dimensions increases the detector impedance above that of the load. Of the two configurations shown in Fig. 16.1 the face-electrode configuration is generally more useful, because this represents the lowest-impedance configuration and it allows the absorption to occur at the electrode. For high-power or very-high-frequency operation the edge-electrode geometry is more useful since the radiation is absorbed directly in the crystal volume. For this geometry only material of high dielectric constant can be used, otherwise fringing fields degrade the performance and the detector impedance becomes unacceptably high. Furthermore, with decreasing thickness it becomes increasingly difficult to absorb the radiation efficiently.

In optimizing the material characteristics the dielectric constant, dielectric loss, and pyroelectric coefficient must all be considered together. The materials figures of merit are $p_i/c_p \varepsilon'$ for high voltage responsivity, p_i/c_p for high current responsivity, and $p_i/c_p(\varepsilon'')^{1/2}$ for high detectivity (small W_m^d). Unfortunately an increase of p_i is usually accompanied by an increase of ε' and ε'', so that the figures of merit do not vary as widely from one material to another or with temperature as might otherwise be expected from the variation of the individual parameters. Some examples are listed in Table 16.1.

The usefulness of materials for specific applications does not necessarily depend on these figures of merit alone. Other considerations such as mechanical strength, chemical stability, matching to amplifiers, power-handling ability, and ease of fabrication may be important considerations. Triglycine sulphate has been the most extensively studied material for detector applications. The figures of merit of this material shown in Table 16.1 make it particularly suitable for use in low-power applications where

TABLE 16.1

Room temperature properties of various pyroelectric detector materials and some 'figures of merit' for their detector operation.

Material	p_i (nC cm^{-2} K^{-1})	$\varepsilon'/\varepsilon_0$	c_p (J cm^{-3} K^{-1})	$\varepsilon''/\varepsilon_0$	p_i/c_p (nA cm W^{-1})	$p_i/(c_p\varepsilon')$ (V cm^{-2} J^{-1})	$p_i/(c_p\varepsilon'')^{1/2}$ (cm^3 J^{-1})$^{1/2}$
TGS + alanine (a)	36	27	2·1	0·08	17·1	7100	0·203
TGS	30	50	1·7	0·16	17·8	4000	0·149
LiTaO$_3$	19	46	3·19	0·16	6·0	1470	0·050
Sr$_{1/2}$Ba$_{1/2}$Nb$_2$O$_6$	60	400	2·34	8·0	25·6	720	0·030
PLZT (6/80/20)	76	1000	2·57	8·8	29·9	340	0·034
PVF$_2$	3	11	2·4	0·25	1·3	1290	0·009

(a) Bye *et al.* (1974).
(b) The dielectric loss varies according to crystal preparation. The values listed here are typical values for comparison purposes only. After Liu (1976) except where otherwise stated. See text for further details.

high detectivities are required. Crystals are relatively easy to prepare with uniform properties. The addition of alanine impurities to TGS during the crystal growth (Lock 1971) was found greatly to improve the detector performance by lowering the dielectric loss and dielectric constant of the material (Keve, Bye, Whipps, and Annis 1971). Detectivities within a factor of 3 of the thermal limit have been achieved. Another benefit of alanine doping is that the dipolar impurities provide an internal bias of the polarization (see § 4.3) so that crystals heated accidentally above the Curie temperature (49°C) do not require repoling. A disadvantage of TGS crystals is that they are relatively fragile and have a fairly low thermal conductivity so that they are unable to withstand high incident power densities. Furthermore, because of their water solubility, some protective housing and optical window is required. These drawbacks are overcome with ferroelectric oxides such as $LiTaO_3$ and $Sr_{1-x}Ba_xNb_2O_6$ (SBN) which are mechanically strong and stable under atmospheric conditions. The current responsivity of SBN is particularly high because of the high pyroelectric coefficient and this material is convenient for the detection of short laser pulses requiring wide bandwidth and the use of low-impedance transmission lines. Alternatively, high-current responsivity may be advantageous at very low frequencies where amplifier-current noise is dominant. Many materials have ferroelectric phase transitions close to (but above) room temperature and have the high pyroelectric coefficients necessary for high current responsivity. Like the SBN system the properties of $K(TaNb)O_3$ can be 'tuned' over quite a large range by suitable choice of composition, but at the present time none of these compositions have yielded the detectivity of TGS at frequencies in the 10–1000 Hz range.

There has been a considerable amount of research into pyroelectric polymers (see § 15.3.3) and ceramics for detector application because of their ease of preparation in any shape or size. In the case of polyvinylidene fluoride (PVF_2) detectors (Glass, McFee, and Bergman 1971) extremely thin ($\sim 2\ \mu$m), freely suspended films are readily prepared by stretching and this may to some extent compensate for the lower figures of merit of this material. Improvements in the performance of PVF_2 have been due to improved poling procedures (Phelan *et al.* 1974), which improve the uniformity and pyroelectric properties, and to increasing the degree of crystallinity in the orthorhombic phase.

The interest in polymers for these applications is relatively recent and it seems likely that other polymers with improved characteristics will be forthcoming.

16.2.6. Pyroelectric energy conversion

It is interesting to consider the materials requirements for efficient power conversion from optical to electrical energy using pyroelectric detectors. We can write the conversion efficiency for $e = 1$

$$\eta = W_0 r_i r_v \tag{16.2.17}$$

which we see from eqns (16.2.5) to (16.2.7) may be written

$$\eta = \frac{W_0 p_i^2 R}{c_p^2 a^2 (1 + \omega^{-2} \tau_T^{-2})(1 + \omega^2 R^2 C^2)^{1/2}}. \tag{16.2.18}$$

For the internal efficiency of the detector element (high load impedance) we have

$$\eta = \frac{p_i^2}{c_p^2 (1 + \omega^{-2} \tau_T^{-2})(\varepsilon''^2 + \varepsilon'^2)^{1/2} \omega a} \left(\frac{W_0}{A}\right) \tag{16.2.19}$$

which is a maximum for $\omega \tau_T = 1$. Since the average temperature rise of the crystal is given by (16.2.3) we find that

$$\eta_{max} = \frac{p_i^2 \Delta T}{2(\varepsilon''^2 + \varepsilon'^2)^{1/2} c_p}. \tag{16.2.20}$$

In practice $\varepsilon''/\varepsilon' \ll 1$ so we see that the material figure of merit for power conversion is $p_i^2/c_p \varepsilon'$. For the materials listed in Table 16.1, and for a thermal relaxation time of 1 s (which is typical of freely suspended detectors) and an incident power density of 100 mW cm^{-2} (which is the intensity of solar radiation at the earth's surface), we find that $\eta \sim 10^{-3}$. The temperature rise of the detector element, which is typically 10^{-3} cm thick, is about 50°C under these circumstances, which obviously eliminates the materials with T_c close to room temperature for this application.

It has been shown (Zook and Liu 1976) that the ratio p_i^2/ε' is relatively insensitive to the shape of interionic potentials, Lorentz parameters, and the effective charges in pyroelectric crystals, and as a result this ratio is relatively constant for a variety of materials. It is not likely, therefore, that power conversion efficiencies greatly in excess of 10^{-3} will be realized under practical conditions by the use of the pyroelectric effect.

16.2.7. Excited-state polarization

The change of polarization upon electronic excitation which was discussed in § 12.6 may be used for detection in the same way as the pyroelectric effect. The equations for pyroelectric detection are applicable provided the ratio p_i/C is replaced by $\Delta \mu/h\nu$, where $\Delta \mu$ is the change of dipole moment per absorbed photon and $h\nu$ is the photon energy. The materials requirements are, however, different. By suitable choice of absorbing defect it is possible for a large $\Delta \mu$ to be obtained in a host of high impedance, which is not likely with the pyroelectric effect. Since electronic excitation usually occurs at wavelengths in the visible and ultraviolet spectral regions, where other detectors have high detectivity, this effect is only likely to find application at frequencies higher than the frequency response of other detectors. The ultimate frequency response in this case is not limited by phonon–phonon relaxation times, as with the pyroelectric effect, but by an electronic transition rate. Picosecond response times have been demonstrated using $LiTaO_3$ crystals doped with Cu^{2+} impurities to detect picosecond pulses from a mode-locked Nd : glass laser (Auston and Glass 1972). The resulting current pulses were 10 ps in duration (limited by the input pulse duration and the electro-optic measurement scheme) and 250 V in amplitude. Heterodyne detection (optical mixing) with two dye lasers in $LiNbO_3$, with differences tunable from 0 (d.c.) to 10^{12} Hz, was used to generate tunable far-infrared and microwave radiation (Auston *et al.* 1973). The $LiNbO_3$ crystals were reduced in hydrogen to make them absorbing in the visible spectrum. Since the absorption depth was less than the coherence length of the interaction, phase matching could be avoided without loss of conversion efficiency. These experiments demonstrate the feasibility of using the very fast response time of the excited-state polarization for certain limited applications.

16.2.8. Pyroelectric imaging

There are a variety of ways in which pyroelectric detectors can be used to record infrared images. One approach is to use a single-element detector and scan the image across the detector in a two-dimensional manner with appropriate infrared optics. A visible display can then be composed from the sequence of detected 'bits' by conventional means. An infrared camera based on these principles (Astheimer and Schwarz 1968) was capable of resolving 10^4 picture elements with a temperature resolution of 0.1 degC in the thermal field of view. However, the frame rate was only one frame per 30 s. To achieve a fast frame rate requires a large frequency bandwidth, typically several hundred kHz, and, as we have seen in the previous sections, at these frequencies and large bandwidths the minimum detectable power increases greatly above the low-frequency values. Reduction of the bandwidth can be obtained with a linear array of detectors and a one-dimensional image scan. The detector response may be scanned with solid-state integrated circuits.

To make use of the maximum detectivity of pyroelectrics at low frequencies the bandwidth can be further reduced by using a two-dimensional detector array, each detector corresponding to a single picture element. The rows and columns of the array are then scanned electronically.

Since most pyroelectric detector materials are reasonably good electrical insulators the construction of such an array is greatly simplified by using a single large wafer of the material onto which the infrared image is projected. If the front surface of the wafer is electroded then the pyroelectric charge pattern on the rear surface reproduces the infrared image. Instead of an electrode array the charge pattern may be read by any technique which is sensitive to the charge distribution or the electric field across the wafer. Read-out techniques based on electro-optic or electroluminescent displays coupled directly to the wafer have not proved successful, except in special situations involving high incident power densities, because of the low-power-conversion efficiency of the pyroelectric and the display device. One procedure for producing hard copies of the infrared image made use of the fact that electrostatically charged ink is directly attracted to regions of the pyroelectric wafer which are oppositely charged (Bergman *et al*, 1972), but this device also required intense illumination and could not be used for thermal imaging.

The most successful approach to thermal imaging has been the pyroelectric vidicon tube, in which the rear surface of the pyroelectric wafer is scanned with an electron beam as in a conventional television camera tube. This procedure was first proposed by Hadni *et al.* (1965), and general descriptions of the device and its application to thermal imaging have been published in articles by Holeman and Wreathall (1971), Tompsett (1971), and Putley, Watton, and Ludlow (1972). A typical device is shown in Fig. 16.5(a). Incoming radiation is focused by an infrared lens through a chopper onto the pyroelectric target which has its pyroelectric axis normal to the

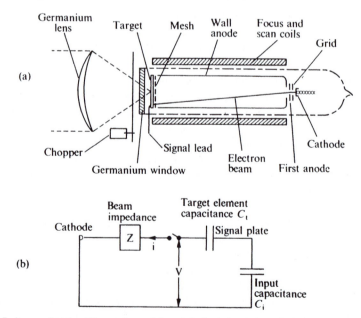

Fig. 16.5. A pyroelectric vidicon tube and the equivalent circuit for the target element. (After Taylor and Boot 1973, Logan and Watton 1972.)

wafer. The front electrode surface is covered with an absorbing film (or else a transparent electrode can be used if the target material absorbs in the infrared). In the dark the electron beam maintains the rear surface of the target at a reference potential. When the infrared image falls on the target the potential changes in proportion to the temperature change, the scanning electron beam then deposits sufficient charge to restore the original potential, and a video signal is generated in the circuit connected to the front electrode by capacitive coupling. The video signal is amplified and displayed with a conventional television display. The equivalent circuit is shown in Fig. 16.5 (b), where C_i is the input capacitance to the video amplifier, C_t is the target capacitance, and Z is the beam impedance which is the ratio of the target voltage to the beam current.

When the target is blocked by the chopper during the next half-cycle the target cools down and pyroelectric charge of the opposite sign appears on the rear surface. It is clear that if the target is not biased to a positive potential the electron beam can only address the target during one half-cycle, and operation is unstable. Two methods of electron beam read-out have been demonstrated. The first is cathode-potential-stabilized (CPS) operation in which the target is biased to a small positive value (pedestal) either by using a target of sufficiently low resistivity that the current through the target raises the potential so that the rear surface remains positive during both half cycles of operation, or else by introducing a gas into the tube which is ionized by the electron beam and deposits positively charged ions onto the pyroelectric surface. In either case the electron beam lands with near-zero velocity. Details of CPS operation are discussed by Logan and Watton (1972).

An alternative mode of operation is to use anode potential stabilization (APS) in which the target is held at a relatively high potential (~ 200 V), close to that of the grid. In this case the electrons land with high velocity and secondary electrons are emitted. In equilibrium as many secondaries return to the target as are emitted. If the target potential is changed there is a corresponding change in the secondary electron current.

APS operation has the problem that the random motion of the secondaries between the grid and target reduces the contrast and the high electron energies probably reduce the target lifetime. On the other hand, since most pyroelectrics have high resistivity, CPS operation requires a gassy tube which leads to a reduced tube lifetime and the shot noise in the ion current limits the signal-to-noise ratio. A scheme for producing the positive pedestal for CPS operation without introducing the gas was recently devised (Conklin et al. 1974) and uses an energetic electron beam to produce a secondary-electron coefficient at the target of greater than unity during flyback— between read-out scans. In this way the target is positively charged before the next read-out scan with a low-energy beam.

There are several requirements for the ideal material for vidicon targets. The material must be prepared as thin, large-area (~ 2 cm diameter) wafers with uniform and stable pyroelectric and dielectric characteristics. Variation of target properties appears as spatial noise. A high Curie temperature is helpful because the vidicon tube can then be baked out without repoling the material. In order to achieve high spatial resolution a low thermal conductivity is advantageous since lateral thermal diffusion in the target between frames reduces the resolution (Logan and McLean 1973). This can be overcome to some extent by reticulating the target (Singer and Lalak 1976) by photolithographic processes or laser machining, but some mechanical support of the target is necessary and any substrate material will contribute additional thermal capacitance.

It is not possible to specify a single figure of merit for the target material since the dominant source of noise depends on the mode of operation. In APS operation either shot noise in the electron-beam current or amplifier noise may be dominant. For CPS operation shot noise in the ion current (Logan and Moore 1973) or amplifier noise may be dominant and the efficiency of read-out of the pyroelectric charge depends on the target bias. The figures of merit may be the current responsivity proportional to p_i/c_p, voltage responsivity proportional to $p_i/c_p\varepsilon'$, or the ratio $p_i/\{c_p(\varepsilon'')^{1/2}\}$ depending on the details of operation. Details of these noise considerations are discussed by Taylor and Boot (1973).

At the present time the best performance has been achieved with tri-glycine sulphate. Resolvable temperatures in the field of view of under 1 degC have been reported for spatial frequencies of less than 4 line pairs per mm (Singer and Lalak 1976) with f/1 optics, but at higher spatial frequencies this performance deteriorates owing to thermal diffusion. A comparison of the minimum resolvable temperature of a single-crystal TGS target with a reticulated target is shown in Fig. 16.6. The figure shows clearly the improvement at higher spatial frequencies when thermal diffusion is inhibited by reticulation. At low spatial frequencies the reticulated target shows poorer performance due largely to reflection losses at the As_2S_3 substrate on which the target was mounted.

In Fig. 16.5 the necessary modulation of the infrared scene was obtained by chopping the incoming radiation. Alternative modulation schemes which eliminate the chopper involve moving the image over the target by panning or orbiting the tube. Somewhat better sensitivity is obtained in the panning mode (Watton *et al.* 1974) than in the straightforward chopping mode. Furthermore, flicker effects due to chopping are avoided. However, panning and orbiting have the disadvantage that for steady image display the motion must be compensated and, more seriously, a thermal blur trails behind image points having high contrast with the background due to the target motion. The pyroelectric charge on the retina requires several scans by the

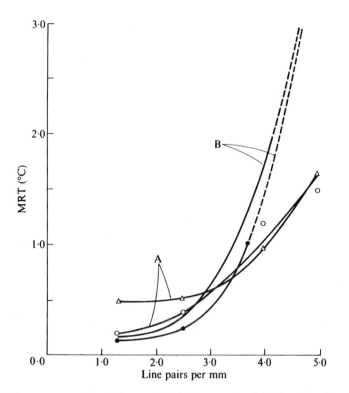

Fig. 16.6. Measurements of the minimum resolvable temperature (MRT) as a function of spatial frequency (line pairs per mm) for two reticulated TGS vidicons (curves A) compared with two typical TGS single-crystal vidicons (curves B). (After Singer and Lalak 1976.)

electron beam to be read out completely for a scan rate of typically 15 Hz. Chopping, on the other hand, produces a stable image and constant field of view. A particular advantage of the chopping mode is that spatial noise due to target non-uniformities and thermal blur of a moving target can be greatly diminished by suitable signal-processing techniques (Helmick and Wood-worth 1976).

A recent preliminary development in pyroelectric imaging techniques which avoids the use of an electron gun is worth noting. In this technique (Alting-Mees and Koda 1976) the rear surface of the pyroelectric target is coated with a grid of photoemissive material which is illuminated with an incandescent lamp. The photoelectrons emitted from the grid are imaged onto a visible emitting phosphor for direct viewing. An infrared image absorbed at the front electroded surface of the target produces a field at the rear surface which modulates the photoemission efficiency of the grid and hence the brightness of the phosphor. The estimate of the temperature resolution ($\sim 0 \cdot 1$ degC) looks promising but it is too early to determine the limitations of this device.

An alternative proposal for reading the charge developed on the pyroelectric target would use semiconducting charge-coupled devices directly on the target rear surface to read out the charge. This device would have the advantage of being all solid state (no electron gun) but its development at the time of writing is in its infancy and the future of the device cannot yet be determined.

16.3. Memories and display

The bistable polarization of ferroelectrics makes them candidates for binary memory systems in the same way as the bistable magnetization of ferromagnets. The memory is non-volatile and does not require a holding voltage. Yet for a variety of reasons which will become evident below ferroelectrics have been slow in realizing commercial application for memories and display.

To record information the polarization may be reversed or reoriented by application of an electric field greater than the coercive field. For erasure the polarization can be returned to its original state with an applied field of opposite polarity. To read the information there is a wide range of possibilities which may be divided into two broad classifications: those in which the stored information is retrieved by electrical means, and those in which it is retrieved by optical means via an electro-optic effect. In this latter category somewhat more versatility is afforded in the recording mechanism. For instance instead of polarization reversal the memory can be based on linear or quadratic electro-optic effects which arise from light-induced polarization changes.

16.3.1. Electrically read memories

A simple method of electrically reading the stored information is to apply an electric field greater than the coercive field to each memory element. A large displacement current accompanies polarization reversal in elements having a polarization opposing the applied field, while only a small displacement current corresponding to capacitative charging occurs in elements of opposite polarization. Read-out destroys the memory so the information must be immediately rewritten. More attractive are non-destructive read-out schemes in which another property of the crystal, which depends on the sense of the polarization, is measured. For instance if the entire memory is uniformly heated or stressed the pyroelectric or piezoelectric charge which develops across each element reproduces the memory. This charge can be read by one of the techniques described in the preceding section.

Another technique for reading the sign of the remanent polarization is the so-called ferroelectric field effect at a ferroelectric–semiconductor junction

(Godefroy 1956). As we discussed in § 4.4 the free charge required to neutralize the depolarization fields in the ferroelectric must be provided within the semiconductor if the ferroelectric is insulating. The resulting band bending in the semiconductor can greatly affect its conductivity. For instance, polarizing the ferroelectric with $-P_s$ toward a p-type semiconductor results in an enhanced hole concentration and increased conductivity, while polarization in the reverse direction results in a depletion of majority carriers and decreased conductivity. The conductivity can be measured with simple 'source' and 'drain' electrodes or else by more complicated field-effect-transistor structures (Zuleeg and Wieder 1966). For a stable polarization and a perfectly insulating ferroelectric the conductivity should remain unaltered with time. In practice slow changes of the conductivity are observed in the TGS prototype devices (Heyman and Heilmeier 1966, Teather and Young 1968) owing to changes of the remanent polarization (because of the non-vanishing field in the ferroelectric, see Section 4.4), changes of the ferroelectric—semiconductor interface, or because of finite conductivity of the ferroelectric.

At the present time there does not appear to be any advantage of this kind of ferroelectric memory over the well-established magnetic and semiconductor technologies. The capacity is no greater, the writing process is relatively slow (microseconds or longer), and many, if not most, ferroelectrics show fatigue with repeated cycling. As a result attention has steered toward optical memories where ferroelectrics compete more favourably with other materials. An excellent review of optical applications has been given by Anderson (1974).

16.3.2. Optically read memories

The polarization of a pyroelectric affects the optical properties in a variety of ways. For instance 180° reversal of $Gd_2(MoO_4)_3$ rotates the optical indicatrix through 90° in the plane perpendicular to the polarization vector, thereby producing a large change of birefringence for light propagating along the polar axis. In $Pb_5Ge_3O_{11}$ the sense of the spontaneous polarization is directly related to the sense of the optical rotary power. In $Bi_4Ti_3O_{12}$ the a-axis and c-axis components of polarization may be reversed independently through 180° so that the optical indicatrix may be rotated in the a–c plane in a variety of ways. Also, in Chapter 13 it was seen how the refractive index may be changed via the Pockels (linear) or Kerr (quadratic) effects, and in Chapter 15 it was shown how the optical transparency of some PLZT ceramics depends on the remanent polarization.

For memory applications the polarization change may be written into the ferroelectric by rotation or reversal of the spontaneous polarization in an external field or by the photorefractive effect and the resulting change of optical parameters may be read out by conventional optical means. Many

ferroelectrics are well suited for applications in opto-electronics because the
large coupling between the optical and electrical properties allows efficient
non-destructive read-out of the stored information.

16.3.3. Light valves

Conceptually the simplest kind of optical memory is the electrically addressed light valve. The example shown in Fig. 16.7 (a) utilizes $Bi_4Ti_3O_{12}$
with the incident light propagating along the b-axis initially polarized

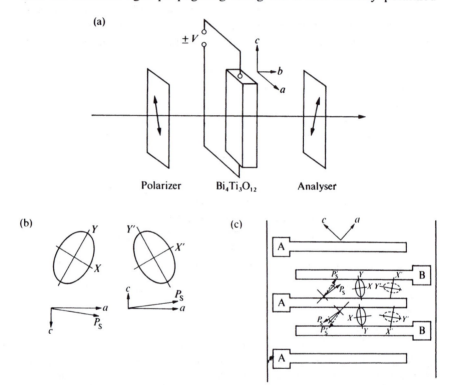

Fig. 16.7. (a) A simple light-valve arrangement utilizing $Bi_4Ti_3O_{12}$. The polarizer and analyser
are parallel and perpendicular to the major axis of the optical indicatrix. (b) Reversal of the
c-axis component of P_s rotates the optical indicatrix of $Bi_4Ti_3O_{12}$ through 50°. (After Cummins
and Luke 1972). (c) Set of interdigital electrodes on an epitaxial $Bi_4Ti_3O_{12}$ film. The film is
initially made optically 'homogeneous' by applying a large field across the A and B sets of
electrodes sufficient to orient both the a and c-axes. For switching only the c component of P_s is
reversed with a small field, resulting in the rotation of the optical indicatrix as shown by the
broken lines. (After Wu *et al.* 1973.)

parallel to the major or minor axis of the optical indicatrix and the analyser is
set for extinction. Reversal of the c-axis polarization by application of a
transverse electric field results in the rotation of the optic axis through 50° in

the a–c plane (Fig. 16.7 (b)). This is almost the optimum condition (45°) for a maximum transmission of light through the analyser. While $Bi_4Ti_3O_{12}$ single crystals grow readily from the flux as large-area c-axis plates it is difficult to grow crystals with the orientation necessary for large-area light valves. Progress in overcoming this difficulty has been made (Wu *et al.* 1972, 1973) by growing b-growing films epitaxially on $MgAl_2O_4$ substrates. The electric field is applied with a set of interdigital electrodes as shown in Fig. 16.7 (c).

Gadolinium molybdate single crystals can be used as light valves (Kumada 1972) with longitudinally applied fields and transparent electrodes as shown in Fig. 16.8. Reversal of the c-axis polarization interchanges the

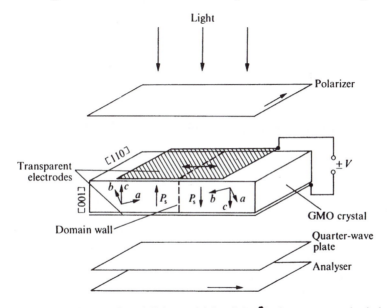

Fig. 16.8. Switching element of a gadolinium molybdate light valve between crossed polarizers. The quarter-wave plate compensates for the natural birefringence of the GMO plate in the single-domain OFF state. The a and b-axes are interchanged when c is reversed across a domain wall. The domain wall can be moved by changing the applied voltage. Outside the electroded region the domains have fixed orientation and act as nucleation sites for domain reversal.

a- and b-axes with a resulting birefringence change along the c-axis of $2\Delta n \sim 8 \times 10^{-4}$. An electrically switchable quarter-wave retardation is obtained with a crystal ~ 0.015 cm thick at 500 nm. The crystal is placed between crossed polarizers and the birefringence in the initial polarization state is compensated with a second crystal of identical thickness which is not switched by the field or by a quarter-wave plate. The transmission can then be switched on and off as the total retardation is switched from $\pi/2$ to zero.

A third candidate for light-valve applications which has the advantage of easy fabrication of large areas is PLZT ceramics. The large number of experimental configurations have been reviewed by Maldonado, Fraser, and Meitzler (1975) and Land (1974). Both transverse and longitudinal configurations similar to those of Figs. 16.7 and 16.8 have been demonstrated. However, in PLZT, unlike $Bi_4Ti_3O_{12}$, $Gd_2(MoO_4)_3$, and $Pb_5Ge_3O_{11}$, 180° domain reversal leaves the optical properties of the material unchanged, so that light-valve applications rely either on 90° switching or differences between the optical parameters of the poled and unpoled states. For instance transversely electroded light valves can make use of the fact that unpoled PLZT has a smaller birefringence than poled PLZT. Alternatively a longitudinal electro-optic effect may be obtained by applying a strain in the plane of the ceramic wafer (see § 15.2). The strain may be introduced either by deforming a transparent substrate on which the PLZT is mounted, or else by bonding a poled PLZT ceramic to a transparent substrate and utilizing the spontaneous strain; when the ceramic is depoled the planar dimensions of the ceramic change creating large strains in the wafer.

PLZT light valves based on the polarization dependence of the light scattering of ceramics have the advantage that no polarizers are required. The transmittance of a ceramic wafer into a fixed angular aperture depends on the value of the remanent polarization in the direction of the light path. Electrically depoled regions which are greatly scattering appear dark against the poled background which scatters less (or *vice versa*).

An advantage of PLZT devices over the single-crystal light valves is that it is relatively easy to obtain a continuous variation of light-valve transmission from a relatively low value (~ 1 per cent of the incident light) to almost 100 per cent (Haertling and Land 1971, Thacher and Land 1969). Single-element, large-area light valves can therefore be used as variable neutral-density applications such as welder's goggles. Although one would expect the resolution of single-crystal devices to be better than that of ceramics, this is not always the case. In devices involving polarization reversal the domains can be pinned at grain boundaries thereby reducing cross-talk between the elements which can occur in single crystals.

16.3.4. Light-valve arrays

For applications in information storage or display a two-dimensional array of light valves is required. A comprehensive review of various schemes has been presented by Taylor and Kosonocky (1972). For ease of fabrication it is usually desirable to have the entire $X-Y$ array on a single wafer of ferroelectric crystal or ceramic, or else to have two wafers with the valves arranged in rows and columns. Electrode configurations may be deposited onto the wafers as shown in Fig. 16.9 so that any element in the matrix may be individually switched by applying a positive voltage to the appropriate

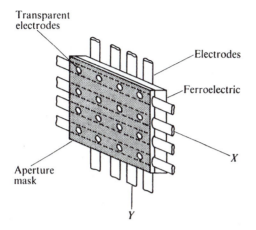

Fig. 16.9. Matrix-addressed light-valve array. The front row of electrodes (X) is transparent while the rear electrodes (Y) may be transparent or reflecting depending on the mode of operation. (After Anderson 1972.)

row on one side and a negative voltage to the appropriate column on the other so that the field across the element just exceeds the coercive field. All the other elements in the row and column only experience half the field so the switching process must be sufficiently non-linear to prevent partial switching of these other elements. The memory elements must be well enough separated to avoid interaction due to fringing fields. In PLZT, switching areas as small as 25 μm square have been reported. Mechanical cross-talk between elements in $Gd_2(MoO_4)_3$ due to the large strain associated with polarization reversal is a problem requiring the use of separate crystal plates for rows and columns.

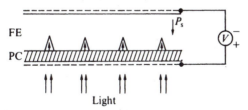

Fig. 16.10. Diagram of a ferroelectric (FE)–photoconductor (PC) sandwich structure for optical memories, showing wedge-shaped domain formation under illuminated regions. (After Keneman *et al.* 1972.)

The ferroelectric–photoconductor sandwich structure shown in Fig. 16.10 provides an alternative means for addressing individual elements in a light-valve array while eliminating the complications associated with the electrode matrix. With this kind of structure matrix memories based on a

longitudinal electro-optic effect are conveniently adapted for optical recording, with the information to be recorded carried by an optical beam rather
than by an electric field. A uniform electric field is applied to the entire
ferroelectric–photoconductor wafer with two transparent electrodes in the
direction necessary to turn the light valve on. However, domain reversal can
only take place under the illuminated regions of the photoconductor. This
layer must be sufficiently conducting in the illuminated regions so that (a) the
majority of the applied field falls across the ferroelectric layer and (b)
sufficient charge flows to allow compensation of the depolarizing fields of the
reversed domains during the switching time. In the dark regions the photoconductor must be a good insulator so that compensation is not possible
within this time, the polarization cannot reverse, and the magnitude of the
field across the ferroelectric is reduced. Photoconductors which have been
used for this application are ZnSe (Keneman, Miller, and Taylor 1972),
CdS/ZnS mixtures (Fraser 1973), or doped poly-n-vinyl carbazole (Meitzler, Maldonado, and Fraser 1970, Maldonado *et al.* 1975). Several designs
based on electro-optic effects in $Bi_4Ti_3O_{12}$ and PLZT, and light scattering in
PLZT have been reported. To obtain a longitudinal electro-optic effect in
$Bi_4Ti_3O_{12}$, c-axis platelets are used with the light propagating at a small
angle to the c-axis so that 180° reversal of the c-axis polarization results in a
change of birefringence. Repeated switching of the c-axis polarization
eventually results in a change of the c-axis component of polarization
resulting in slow degradation of the optical properties of the memory. A
difficulty of the $Bi_4Ti_3O_{12}$ device for applications involving high bit densities
is that polarization reversal only takes place near the crystal surface to a
depth somewhat greater than the diameter of the incident light spot
(~ 1/spatial frequency) as shown in Fig. 16.10. This is due to the fringing
fields and the nature of domain growth during polarization reversal. Consequently the domains are extremely small providing only small retardation of
the read-out beam. (For a birefringence change of $\sim 10^{-3}$ and a domain
length of $\sim 1\ \mu m$ the retardation is only $1/500$ of a wavelength at $0\cdot 5\ \mu m$.)
Furthermore, the retardation depends on the size of the light spot and not
only on the incident light intensity. As a result of these limitations
ferroelectric–photoconductor memories are best suited to applications
where low bit densities are required. They have the advantage that fairly low
optical energies are required with efficient photoconductors ($0\cdot 1$–
$1\ mJ\ cm^{-2}$) to record a page of information of about 10^4 to 10^5 bits.

   A device which is closely related to the ferroelectric–photoconductor
sandwich utilizes the Pockels effect rather than domain reversal (Feinleib
and Oliver 1972). The device is again a sandwich structure, but the two films
below the electrodes are insulating rather than photoconducting while the
electro-optic material itself is a photoconductor. Prototype materials are
ZnS and $Bi_{12}SiO_{20}$. Photoexcited carriers are swept to the insulating inter-

face by an externally applied field, thereby creating a variation of the field
and hence an index variation within the electro-optic material. Electro-optic
photoconductors with high dark resistance are used so that the space-charge
field and index variation remains after the external field is removed. The
structure is placed between crossed polarizers in such a manner that
transmission of the viewing light through the entire device is only possible
within the regions of induced retardation. Neither ZnS nor $Bi_{12}SiO_{20}$ have
large electro-optic effects so the induced retardation is small, but there is no
fundamental reason why other materials should not give improved performance.

16.3.5. High-capacity memories

The use of optical techniques for both recording and read-out offers the
capability of high-capacity, random-access memories without serious inter-
connection problems. In principle the dimensions of a memory bit are
limited only by the diffraction of light, thereby allowing storage densities in
excess of 10^8 bits cm^{-2}. This must be compared with $\sim 10^5$ bits cm^{-2} for
magnetic storage systems. Even higher storage densities can be achieved by
superimposing images in the same volume of the storage medium by
holographic means (see below).

Selective read-out of the desired bits could be obtained by directing a laser
beam to the appropriate address, but with such a scheme it is difficult to
provide rapid random access to a suitably large number of addresses. For
instance, for a random-access time of $1 \mu s$ currently available light deflectors are limited to less than 10^5 resolvable spots (Zook 1974), which is not
sufficient to utilize the high storage density of the recording medium if there
is only one bit at each address. The number of resolvable spots can only be
increased at the expense of access time. To overcome this difficulty the most
promising approach is the page-organized optical memory; an entire page of
about 10^5 bits is stored at each address of the deflected laser beam. The
storage plane is then a matrix of pages each of dimensions ~ 1 mm^2. Each
page can be the micro-image or Fourier transform of an $X-Y$ array of light
valves, as discussed above, or an analogue transparency. Recording the
Fourier transform by holography has the advantage of spatial redundancy
(each bit is recorded over the entire hologram), minimizing the effects of
dust spots or errors in beam positioning. One possible read–write holographic memory system is shown in Fig. 16.11. The deflector and optical
arrangement displaces the laser beam parallel to itself in the $X-Y$ plane. For
writing, the beam is split into an object beam and a reference beam, the
latter being directed to a particular element of the storage plane determined
by the deflector. The object beam is defocused by the fly's-eye lens to
illuminate the page composer or transparency. The object lens is placed
close to the page composer so that the Fourier transform is produced at the

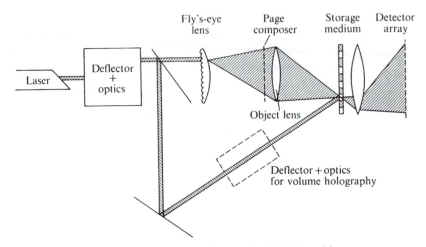

Fig. 16.11. A possible arrangement for a read–write holographic memory.

storage plane at the same point as the reference beam. The resulting interference pattern of the two beams is recorded in the storage medium. For reading, the reference beam only is diffracted from the hologram and the image of the page is projected onto the photodetector array. The signal from each detector corresponds to one bit of the page composer. Deflection of the laser to a new position allows recording or reading of a different page of the memory.

Diffraction from holograms recorded in thick storage media is only possible within a limited range of angles and a limited wavelength range. A storage medium is 'thick' if the interference-fringe spacing of the hologram is much less than the thickness of the medium. For faithful reconstruction of a thick hologram containing several spatial frequencies read-out with the same wavelength and angle is essential. To superimpose individual holograms either the wavelength or angle of the writing beams can be changed and each hologram can be reconstructed independently.

The materials requirements for the storage medium are particularly stringent for this type of application. The ideal storage medium would allow *in situ* erasure of the information (for up-dating the memory), a resolution of about 1000 lines/mm, long storage time, high recording sensitivity, high signal-to-noise ratio, and linear exposure characteristics.

Holographic storage with a resolution of 500 lines/mm has been accomplished with $Bi_4Ti_3O_{12}$ ferroelectric–photoconductor sandwich structures (Keneman *et al.* 1972), but the diffraction efficiency is low because of the domain-reversal problem discussed above. On the other hand, the signal-to-noise ratio can be quite good since the polarization of the diffracted light is perpendicular to the incident light and isolation can be obtained with a suitable analyser. Layer structures of this kind are restricted to the recording of thin holograms so that superposition of holograms is not possible.

The photorefractive effect (see § 12.8) can be used for storage of thick phase holograms with diffraction-limited resolution (Chen, LaMacchia, and Fraser 1968). The crystal is oriented so that the optically generated space-charge fields and the polarization of the reading light make use of the largest electro-optic coefficient of the material. For instance in $LiNbO_3$ the largest coefficient is r_{33}, so the holograms are recorded with the spontaneous polarization P_s of the crystal in the plane of the object and reference beam so that the interference fringes are normal to P_s as shown in Fig. 16.12 (a). The

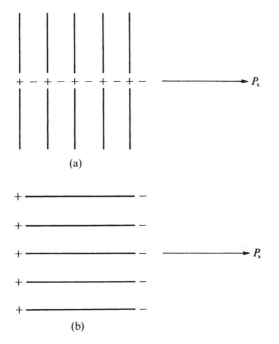

Fig. 16.12. Formation of space-charge fields during holographic recording using the photo-refractive effect: (a) optical interference grating perpendicular to P_s and (b) parallel to P_s.

reading light is polarized parallel to P_s. The optically generated carriers are displaced toward regions of lower light intensity creating space-charge fields as shown in the figure. If the interference fringes are parallel to P_s as in Fig. 16.12 (b) the photocarriers drift along the direction of the fringes and do not reproduce the optical fringe pattern (except for small diffusion effects). For transient holography using excited-state dipoles (see § 12.6) in which there is no macroscopic motion of free charge, diffraction is obtained as a consequence of the depolarization fields. However, the holograms in this case only persist for the lifetime of the excited state so are only useful for short-term memory applications in optical processing.

From eqn (12.8.7) the diffraction efficiency is seen to depend on both the absorption coefficient and the induced index change Δn. The index change itself depends on the absorption of light by charge transfer so that the recording sensitivity S, given by eqn (12.8.8), is constant for a specific material and dopant—at least for small Δn. For linear absorption (as opposed to multi-photon absorption) the maximum recording sensitivity is obtained for a crystal transmission of 33 per cent. Recording characteristics are sufficiently linear up to a few per cent for most applications. Greater efficiency can only be obtained at the expense of recording sensitivity. We saw in § 12.8 that the photorefractive efficiency depends on the magnitude of the dipole moment created per excited photoelectron. For holographic storage the maximum sensitivity is obtained when the drift length of the photoelectron is comparable with the interference fringe separation Λ (Young *et al.* 1974). For a quantum efficiency of unity the maximum photoinduced polarization is

$$\Delta P_3 = \frac{\alpha W_0 e \Lambda}{2\pi h \nu}. \tag{16.3.1}$$

Using eqn (12.8.9) and the observation (Wemple, DiDomenico, and Camlibel 1968) that the quantity

$$\frac{r_{33}}{\varepsilon_0(\varepsilon_3 - 1)} \sim \frac{1}{4} P_3$$

for a wide variety of oxide ferroelectrics within a factor of 2, where P_3 in units of $C\,m^{-2}$ is the total polarization of the crystal along the polar axis, we can estimate the maximum sensitivity

$$S_{\max} \sim \frac{1}{8} n_3^3 P_3 \frac{e \Lambda}{2\pi h \nu}.$$

Taking $n_3^3 = 10$, $P_3 = 10^{-4}\,C\,cm^{-2}$ which is greater than, but of the same order of magnitude as, the polarization of many ferroelectrics, then for a resolution of 1000 lines/mm at a wavelength of $0{\cdot}5\,\mu m$ we find $S_{\max} \sim 0{\cdot}08\,cm^3\,J^{-1}$. For a crystal transmission of 33 per cent and a diffraction efficiency of 1 per cent the required optical energy is about $60\,\mu J\,cm^{-2}$ which is comparable with holographic silver halide emulsions. In $LiNbO_3$ the drift length of excited carriers is too small to reach this ultimate sensitivity, even with externally applied fields, but in $K(Ta, Nb)O_3$ sensitivities approaching $S_{\max}$ can be obtained (Glass, Von der Linde, and Negran 1976).

A problem with thick holographic storage media of this kind which require no development is that, since the same wavelength light must be used for reading as for writing, the memory is degraded during read-out or during superposition of holograms. Several approaches have been taken to reduce this problem. Firstly, the hologram may be made insensitive to optical erasure by compensating for the electronic space-charge pattern with an ionic polarization pattern. For instance in $LiNbO_3$ the 'fixing' of holograms at elevated temperatures near 100°C was attributed to ionic conductivity (Staebler and Amodei 1972). In $BaTiO_3$ and other materials of low coercivity fixing has been accomplished (Micheron *et al.* 1974) by reversing the spontaneous polarization in regions of high-space-charge field. Following the fixing procedure the electronic-space-charge pattern can be partially removed by uniform illumination of the crystal leaving the polarization variation due primarily to ionic charge. Holograms fixed in this way can no longer be optically erased, but thermal or in some cases electrical erasure is still possible. Following fixing the material is still sensitive for recording so that many holograms may be superimposed. Some 500 holograms were successfully superimposed in $LiNbO_3$, each with more than 2·5 per cent diffraction efficiency (Staebler and Phillips 1974). The signal-to-noise ratio decreases with exposure during read-out because of interference of scattered light with the direct beam within the crystal, but this can be satisfactorily erased by illuminating the entire crystal with incoherent light, since the noise is not fixed.

An alternative approach to avoiding degradation during read-out of holograms stored by the photorefractive effect is by the use of two-photon absorption, as shown schematically in Fig. 12.15. Holograms are recorded with light of frequency ω_1 which by itself gives no photorefractive effect. Only in the presence of a second frequency ω_2 is the combined energy $\hbar(\omega_1 + \omega_2)$ sufficient to excite photocurrents (Von der Linde and Glass 1976). Frequency ω_1 alone can now be used for reconstruction as required by the Bragg conditions, but since the material is insensitive to this frequency alone unwanted erasure does not occur. The recording sensitivity S of $LiNbO_3$ has been found to be similar for single-photon absorption with iron doping, two-step absorption via real intermediate states of trivalent chromium impurities, and with intrinsic two-photon absorption. Although the energy requirements for two-photon absorption are the same, the peak intensity required is much greater, necessitating the use of short intense pulses.

Long-term storage based on electronic space charge is inevitably subject to decay by extrinsic electronic conductivity. In high-purity $LiNbO_3$, for instance, this can be as long as several months, yet in materials such as KTN and SBN decay times of about a day are more usual. In principle storage times can be increased by cooling the crystal below room temperature, by improving the crystalline perfection, or by suitable doping. On the other hand, holograms fixed by ionic motion appear to be quite stable.

16.3.6. Display

The ferroelectric light valves which have already been discussed in terms of their application to memories may equally well be used as the basis for digital and storage display devices. The simplest procedure conceptually (as shown in Fig. 16.13) is to illuminate the appropriately electroded ferroelectric element from the rear with an incandescent lamp using polarizers and a

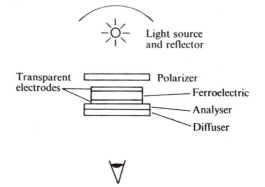

Fig. 16.13. A simple direct-view ferroelectric display element. (After Anderson 1972.)

diffuser as appropriate. Alternatively reflective structures can make use of ambient light. The requirements of light valves used for memories and display are not necessarily the same. For instance, for most display applications the storage time need not be very long, nor need the switching time be very fast although for serial recording of pictorial information switching times $\sim 10^{-8}$ s per bit would be helpful. For flicker-free animated display a suitable time scale for both switching and memory is about 1/30 s, since the eye cannot detect changes of information faster than this. Furthermore, the eye can tolerate a certain amount of misregistration and fairly large variations in light-valve characteristics, provided the eye can spatially smooth the variations.

For small-size (< 1 cm^2) direct-view data and alphanumeric displays with relatively few elements the cost is an important consideration which makes ceramic ferroelectrics, rather than single crystals, more promising candidates. They have the advantage over light-emitting devices in that the light is supplied by an external source and less power is needed for addressing. In addition the light-valve memory needs no holding voltage, further reducing power requirements.

Ferroelectric light valves have also been applied to large-screen projection displays where adequate brightness can only be achieved with incandescent lamps. An example of an electron-beam-addressed display using a deuterated KDP single crystal is shown in Fig. 16.14. The front surface of

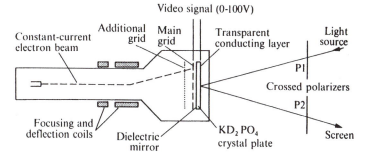

Fig. 16.14. Light-valve device using KD_2PO_4 operated just above T_c and addressed by means of a constant-current electron beam. (After Marie and Donjon 1973.)

the ferroelectric plate has a transparent electrode and the rear surface has a dielectric reflecting coating. The KD_2PO_4 crystal is operated just above its Curie temperature where it is optically isotropic in the absence of an applied field. The front surface is illuminated with polarized light from the projection lamp and the light-reflected from the dielectric coating passes through a second polarizer to the screen. The crystal is addressed from the rear with an electron beam of constant intensity which charges each point on the surface of the crystal to a potential determined by the instantaneous potential on the grid placed just before the crystal. The electron beam essentially short-circuits the crystal surface to the grid, so that for successive scans the previous information is automatically erased. The spatially modulated charge pattern on the crystal accordingly modulates the birefringence via the quadratic electro-optic (Kerr) effect. If the polarizers are crossed, then the birefringent regions of the crystal appear bright against a dark background. High-quality displays with up to 20 lines/mm (a total of ~750 horizontal lines) and a storage time of about 15 min (limited by crystal conductivity) have been achieved (Marie and Donjon 1973).

16.4. Electro-optic modulators

The feature of ferroelectrics which makes them useful for optical memories is the fact that the polarization, and hence the optical properties of the material, may be altered in some permanent or semi-permanent manner in such a way as to allow efficient read-out of the stored information. Even for most display applications a short-term memory is usually desirable since this eliminates the need for a holding voltage on the individual light-valve elements of an array.

Another important application of ferroelectrics is for the modulation of light, for which no memory is required and in fact would even be detrimental. Such modulators are required for most optical communications systems where the high-frequency laser carrier allows large-information bandwidths to be transmitted with complete isolation from electrical interference. There are several applications where some kind of rapid optical switching scheme is required for which the ferroelectric switches outlined in the preceding sections are too slow. Most of the practical modulators which make use of ferroelectrics are based on the linear electro-optic effect, although attempts have been made to exploit the quadratic electro-optic effect.

This section summarizes the principles of the various modulation techniques. For more detailed information the reviews of Kaminow and Turner (1966), Chen (1970), Spencer, Lenzo, and Ballman (1967), and Denton (1972) are particularly helpful.

16.4.1. The linear electro-optic modulator

It was shown in Chapter 13 how the non-linear interaction between a low-frequency (below the restrahl) electric field E and an optical frequency field in a non-centrosymmetric crystal may be expressed in terms of the distortion of the refractive-index ellipsoid, i.e.

$$\Delta\left(\frac{1}{n^2}\right)_{ij} = r_{ij,k}E_k \qquad (16.4.1)$$

from which we find in a principal-axis system that

$$\Delta n_{ij}^2 = -n_i^2 n_j^2 r_{ij,k}E_k. \qquad (16.4.2)$$

A plane wave propagating through the crystal polarized linearly along one of the principal axes of the index ellipsoid emerges from the crystal as a linearly polarized wave with a field-dependent phase

$$\Gamma_i = \frac{2\pi L}{\lambda}(n_i + \Delta n_i) = \frac{2\pi L}{\lambda}(n_i - \tfrac{1}{2}n_i^3 r_{ij,k}E_k) \qquad (16.4.3)$$

where L is the optical path length in the crystal in the region of the electric field and λ is the vacuum wavelength. In this section we assume that all the crystal dimensions are small compared with the wavelength associated with the modulating electric field (travelling-wave modulators are briefly considered in 16.4.4). For most applications it is necessary to obtain amplitude modulation of the wave unless a heterodyne detection scheme is used. Intensity modulation can be obtained by interfering two waves linearly polarized along two orthogonal principal axes of the index ellipsoid, propagating in the same direction through the crystal. Consider, for example, a uniaxial crystal such as $LiTaO_3$ with 3m symmetry with an electric field along the polar Z-axis and an incident plane wave polarized 45° to the X- and Z-axes as shown schematically in Fig. 16.15. The change of the ordinary and extraordinary indices of the crystal due to the field are (using reduced

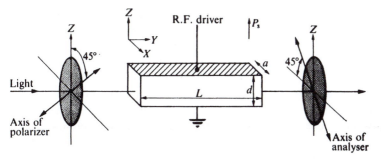

Fig. 16.15. Schematic of a transverse bulk electro-optic modulator.

notation for $r_{ij,k}$)

$$\Delta n_1 = -\tfrac{1}{2} n_1^3 r_{13} E_3$$

and

$$\Delta n_3 = -\tfrac{1}{2} n_3^3 r_{33} E_3. \qquad (16.4.4)$$

The relative phase change of the ordinary and extraordinary waves after travelling the distance L in the crystal, called the retardation, is thus

$$\Delta \Gamma = \Gamma_1 - \Gamma_3 = \frac{2\pi L}{\lambda}(n_1 + \Delta n_1 - n_3 - \Delta n_3) \qquad (16.4.5)$$

while their amplitudes are equal. The intensities of light emerging from the crystal polarized parallel and perpendicularly to the incident polarization are

$$I^{\parallel} = I_0 \cos^2\!\left(\frac{\Delta \Gamma}{2}\right) = \tfrac{1}{2} I_0 (1 + \cos \Delta \Gamma) \qquad (16.4.6)$$

and

$$I^{\perp} = I_0 \sin^2\!\left(\frac{\Delta \Gamma}{2}\right) = \tfrac{1}{2} I_0 (1 - \cos \Delta \Gamma) \qquad (16.4.7)$$

where I_0 is the incident intensity. The modulation can now be detected with an analyser placed after the crystal. In Fig. 16.15 the analyser is set to pass only the perpendicular polarization. Thus if the zero-field retardation $2\pi L(n_1 - n_3)/\lambda$ is an even multiple of π, then $I^{\perp} = 0$ in the absence of an applied field and for $E_3 \neq 0$

$$I^{\perp} = I_0 \sin^2\!\left\{\frac{\pi L}{2\lambda}(n_3^3 r_{33} - n_1^3 r_{13}) E_3\right\}. \qquad (16.4.8)$$

The voltage required to increase I from zero to unity is called the half-wave voltage V_π, i.e.

$$V_\pi = \frac{\lambda}{n_3^3 r_{33} - n_1^3 r_{13}} \left(\frac{d}{L}\right) \tag{16.4.9}$$

where d is the crystal thickness separating the electrodes. From eqn (16.4.8) it is seen that for small E_3 the transmitted intensity is a quadratic function of applied field. That is, for a sinusoidal modulating field $V_m = V_0 \sin \omega_m t$ the transmission is modulated at frequency 2ω. However, if the zero-field retardation is an odd multiple of $\pi/2$, i.e.

$$\frac{2\pi L}{\lambda}(n_1 - n_3) = (2m + 1)\frac{\pi}{2} \tag{16.4.10}$$

then

$$I^\perp = \tfrac{1}{2} I_0 \left\{ 1 + (-1)^{m+1} \sin\left(\frac{\pi V_m}{V_\pi}\right) \right\} \tag{16.4.11}$$

from which it is seen that $I^\perp$ is linear in V_m and hence E_3 for small retardation.

For a general modulating voltage the fractional modulation at the fundamental frequency ω_m can be written (Denton 1972)

$$\eta = \frac{I^\perp(\omega_m)}{\tfrac{1}{2} I_0} = 2J_1\left(\frac{\pi V_m}{V_\pi}\right) \tag{16.4.12}$$

where J_1 is a first-order Bessel function of the first kind. The nonlinearity of eqn (16.4.11) gives rise to harmonic distortion of the linear modulation, which is described by equations similar to eqn (16.4.12) but with higher-order Bessel functions. For $\eta = 0.3$ the third-harmonic intensity is 24 db below that of the fundamental and can be neglected for most practical purposes. The second-harmonic intensity is zero provided the zero-field retardation is exactly an odd multiple of $\pi/2$.

Setting the zero-field retardation to be an odd multiple of $\pi/2$ (or π) can be done by preparing the crystal to the appropriate length or else by means of a compensator which introduces additional retardation between the ordinary and extraordinary rays before the analyser. However, to keep $n_1 - n_3$ constant the crystal temperature must be precisely controlled. Although this has obvious disadvantages simple and practical modulators have been built with this configuration. One approach to overcome this problem is to compensate for the zero-field birefringence with a second identical crystal oriented as shown in Fig. 16.16 in such a way that when $E_3 = 0$, $\Delta\Gamma = 0$, yet the field-induced retardation in the two crystals is additive. A second approach which avoids temperature compensation is to

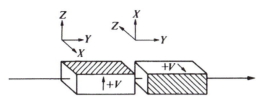

Fig. 16.16. Scheme for compensating the temperature dependence of the spontaneous birefringence.

use a crystal orientation which is optically isotropic to the light beam in the absence of an applied field.

Consider LiTaO$_3$ again, oriented so that light propagates along the polar Z-axis and with the field applied along the Y-axis. Now

$$\Delta n_2 = -\tfrac{1}{2}n_2^3 r_{22} E_2 \tag{16.4.13}$$

and if the light is polarized at 45° to the X- and Y-axes then

$$\Delta\Gamma = \frac{\pi L}{\lambda} n_2^3 r_{22} E_2. \tag{16.4.14}$$

LiTaO$_3$, however, is not a good choice for this geometry since the r_{22} coefficient is small, but the isostructural material LiNbO$_3$ has a usefully large value of r_{22} for this orientation.

In the geometry of Fig. 16.15 the applied field is transverse to the direction of light propagation. The symmetry of some materials, such as KDP at room temperature ($\bar{4}$2m), allows a longitudinal electro-optic configuration as shown in Figure 16.17. KDP has two non-vanishing coefficients

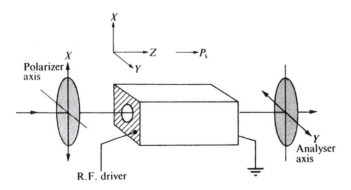

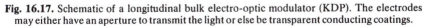

Fig. 16.17. Schematic of a longitudinal bulk electro-optic modulator (KDP). The electrodes may either have an aperture to transmit the light or else be transparent conducting coatings.

r_{41} and r_{63}. With a field along the Z-axis, light polarized along X- or Y-axes experiences a retardation

$$\Delta\Gamma = \frac{2\pi L}{\lambda} n_1^3 r_{63} E_3 \qquad (16.4.15)$$

and the half-wave voltage

$$V_\pi = \frac{\lambda}{2n_1^3 r_{63}} \qquad (16.4.16)$$

which is independent of the geometry of the crystal.

In most modulator geometries, whether longitudinal or transverse, the direction of light propagation is along a principal axis of the index ellipsoid to avoid walk-off due to double refraction, which limits the interaction length of the ordinary and extraordinary waves. The angular aperture of the modulator is restricted by the natural birefringence of the electro-optic crystal to within a few degrees so that these devices are best suited for the modulation of collimated laser beams. Nevertheless longitudinal KDP modulators were used long before the laser was invented and can be designed with quite respectable angular apertures and spectral bandwidths.

The minimum power required to drive a transverse modulator is determined by the half-wave voltage, the fractional modulation required, and the parallel impedance of the crystal and drive circuit. The minimum value of the circuit capacitance is the crystal capacitance $C = \varepsilon\varepsilon_0 La/d$, where a is the width of the electrodes, provided all parasitic capacitance can be neglected. If the peak modulating voltage V_m is to be provided by a voltage generator V_g of resistance R_g, then V_m/V_g is a maximum when R_g is matched to the termination resistance R at the modulator crystal by an ideal transformer. The modulation bandwidth, where V_m/V_g is reduced to half of its value at the matched frequency, is then

$$\Delta\omega = 2/RC \qquad (16.4.17)$$

so that R must be adjusted for the required bandwidth. Now the drive power required is

$$P = \frac{V_m^2}{2R} = \frac{CV_m^2 \Delta\omega}{4}. \qquad (16.4.18)$$

For small modulation index η we have $V_m = \eta V_\pi / \pi$ so that

$$P = \frac{\varepsilon_0\varepsilon}{4\pi^2} \frac{La}{d} (\eta V_\pi)^2 \Delta\omega. \qquad (16.4.19)$$

Alternatively, if we write $V_\pi = \lambda d/n^3 rL$ where the refractive indices and electro-optic coefficients determined by crystal symmetry and orientation have been abbreviated to n^3r, then

$$\frac{P}{\eta^2 \Delta\omega} = \frac{\varepsilon_0}{4\pi^2} \lambda^2 \left(\frac{\varepsilon}{n^6 r^2}\right)\left(\frac{ad}{L}\right). \tag{16.4.20}$$

Various terms have been combined in eqn (16.4.20) to demonstrate the dependence of drive power on the wavelength, the material properties, and the modulator geometry. The drive power is seen to increase linearly with bandwidth and quadratically with the modulation index so that the ratio $P/\eta^2 \Delta\omega$ can be used as a figure of merit for the electro-optic modulator material and construction.

16.4.2. Geometric considerations

From the last factors in eqns (16.4.9) and (16.4.20) it is seen that both the half-wave voltage and the drive power are reduced as the modulator length is increased and as the width and thickness are decreased. Increasing L/d also serves to increase the crystal capacitance, thereby reducing the effect of the parasitic capacitance of the circuit. The optimum geometry therefore is to have the aperture of the modulator just large enough so that the beam passes through the crystal. For a Gaussian beam the smallest cross-section over the crystal length L corresponds to a focused beam with a near-field distance equal to L as shown in Fig. 16.18. Then the beam diameter at each

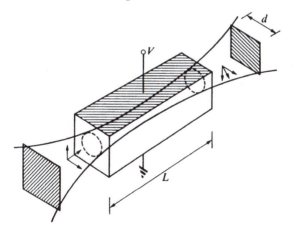

Fig. 16.18. Focused Gaussian beam passing through a bulk transverse electro-optic modulator. (After Kaminow 1975.)

end of the crystal, measured between the points where the amplitude is $1/e$ times the value on the beam axis, is (Kaminow and Turner 1966)

$$w = \left(\frac{2\lambda L}{\pi n}\right)^{1/2}. \tag{16.4.21}$$

The optimum crystal geometry for minimum power dissipation is given by the minimum value of the ratio ad/L which is

$$\left(\frac{ad}{L}\right)_{min} = \frac{4S^2\lambda}{n\pi}$$ (16.4.22)

where S is a safety factor. A value of S close to unity means that the alignment of the modulator is very critical if the insertion loss is to be low. Greater flexibility in alignment obtained by increasing S results in greater drive-power requirements. In practice a useful compromise is $3 < S < 10$. For the best practical bulk modulators the minimum value of the ratio $P/\eta^2\Delta\omega$ is about $10 \text{ mW MHz}^{-1} \text{ rad}^{-2}$ at $1\cdot06 \mu\text{m}$. Considerable improvement of this value has, however, been obtained with waveguide modulators where the diffraction limitation does not apply. These are discussed in § 16.4.6.

16.4.3. Material considerations

The improvement of the performance of electro-optic modulators depends to a large extent on optimizing the properties of the electro-optic crystals. In eqns (16.4.9) and (16.4.19) the half-wave voltage is decreased by maximizing the product of the appropriate electro-optic coefficient r and n^3, while reducing the modulation power P involves reducing the product εV_π^2 which is the stored energy of the modulator. For most applications there is no advantage in reducing V_π if this is accompanied by a large increase of the dielectric constant such that the product εV_π^2 increases. The linear electro-optic properties of a wide range of materials have been listed by Kaminow and Turner (1971), and a few of these are reproduced in Table 16.2 as representative examples of useful or potentially useful materials. Although the electro-optic r-coefficient and the dielectric constant vary considerably from one material to another, and even as a function of temperature in a given material, the ratio r/ε remains relatively constant. This empirical result has some theoretical justification within the anharmonic-oscillator model of § 13.2.2. Furthermore, the polarization potential description of electro-optic behaviour (§ 13.2.4) has shown that for the oxygen-octahedra ferroelectrics, which represent one of the most important groups of electro-optic materials, the polarization-optic f-coefficients are proportional to the spontaneous polarization (see eqn (13.2.69)), i.e.

$$f \sim gP_s \sim \frac{r}{\varepsilon_0(\varepsilon - 1)}$$ (16.4.23)

where the quadratic electro-optic g-coefficients are relatively invariant from one material to another except for the factor ζ close to unity related to the packing density of the oxygen octahedra. For $\varepsilon \gg 1$, eqn (16.4.23) shows that $r \propto \varepsilon P_s$.

TABLE 16.2

The room-temperature electro-optic properties of some important ferroelectrics†

Material T_c (K)		Linear electro-optic coefficients r_{lk} ($\times 10^{-12}$ m V^{-1})	Refractive indices	Dielectric constants
KDP ($\bar{4}$2m) 123	X†	$r_{63} = -10.5, r_{41} = +8.6$	$n_1 = 1.51$ $n_3 = 1.47$	$\varepsilon_3 = 21, \varepsilon_1 = 42$ $\varepsilon_3 = 21, \varepsilon_1 = 44$
	x	$r_{63} = 8.8$		
DKDP ($\bar{4}$2m) 222	X	$r_{63} = 26.4, r_{41} = 8.8$	$n_1 = 1.51$ $n_3 = 1.47$	$\varepsilon_3 = 50$ $\varepsilon_3 = 48, \varepsilon_1 = 58$
	x	$r_{63} = 24$		
LiNbO$_3$ (3m) ~1470	X	$r_{33} = +32.2, r_{13} = +10, r_{22} = 6.7, r_{51} = 32$	$n_1 = 2.27$ $n_3 = 2.19$	$\varepsilon_3 = 32, \varepsilon_1 = 78$ $\varepsilon_3 = 28, \varepsilon_1 = 43$
	x	$r_{33} = +30.8, r_{13} = +8.6, r_{22} = 3.4, r_{51} = +28$		
LiTaO$_3$ (3m) ~890	X		$n_1 = 2.183$ $n_3 = 2.188$	$\varepsilon_3 = 45, \varepsilon_1 = 51$ $\varepsilon_3 = 43, \varepsilon_1 = 41$
	x	$r_{33} = 30.3, r_{13} = 7, r_{22} \sim 1, r_{51} = 20$		
Sr$_{0.75}$Ba$_{0.25}$Nb$_2$O$_6$ (4mm) ~330	X	$r_{33} = 1340, r_{51} = 42, r_{13} = 67$	$n_1 = 2.312$ $n_3 = 2.299$	$\varepsilon_3 = 3400$ (15 MHz)

† Values are given at constant stress (X) and constant strain (x) and signs are given when known. All constants are given at or as close to 0.63 μm as quoted by Kaminow and Turner (1971).

Within this model the half-wave voltage and modulator power should vary as

$$V_\pi \propto 1/\varepsilon P_s \quad \text{and} \quad P \propto 1/\varepsilon P_s^2 \qquad (16.4.24)$$

since the refractive index does not vary greatly within this group of materials. The inverse relationship between V_π and ε has been verified experimentally (Van Uitert et al. 1969) for a variety of Ba, Sr, Na niobate tungsten bronzes. The room-temperature values of P_s were relatively constant for this group of materials although the Curie temperature ranged from 200°C to 560°C and ε varied by about an order of magnitude.

It would thus appear that a large spontaneous polarization and large dielectric constant are desirable for electro-optic modulators. Although the spontaneous polarization decreases as $T \to T_c$, the dielectric constant increases and the products εP_s and εP_s^2 generally increase as T_c is approached (Wemple and DiDomenico 1969). Thermodynamic theory actually predicts that the product εP_s^2 is independent of temperature as a second-order phase transition is approached, but this is only strictly true quite close to T_c.

Operation close to the Curie temperature is normally undesirable because of other material characteristics which do not enter into the figure of merit. For instance as ε increases close to T_c so does the dielectric loss, and heating of the crystal by the r.f. driver becomes a problem. Since only the surfaces can be heat sunk there are large temperature gradients in the crystal which degrade the optical properties. Also as T_c is approached the birefringence becomes increasingly sensitive to any spatial or temporal fluctuations of temperature and to any strains or compositional inhomogeneities.

Optimizing the properties of a practical modulator does not, therefore, just involve minimizing V_π and P. It is more important that crystals of sufficiently large dimensions should have high optical quality, small dielectric loss, and good thermal conductivity. It is also important that the optical quality does not degrade in the presence of a light beam and the r.f. field. Crystals of $Sr_{1-x}Ba_xNb_2O_6$ with low Curie temperatures and broad phase transitions have extremely large electro-optic coefficients (Spencer et al. 1967), but they tend to depole in the presence of a modulating field. Application of a d.c. bias field prevents depoling, but space-charge fields build up owing to the finite conductivity of the crystals and non-ohmic contacts, or non-uniform composition. These fields degrade the optical quality. In the same way light-induced space-charge fields can be a serious problem if the wavelength is sufficiently short to excite free carriers from impurities in the crystal (see § 12.8). This effect is particularly serious in $LiNbO_3$ which has otherwise useful electro-optic properties.

Finally, for wideband modulators the piezoelectric coupling coefficient must be sufficiently small to avoid mechanical resonances of the crystal. This same problem has already been discussed in connection with pyroelectric detectors, and can to some extent be avoided by proper mounting of the modulator crystal (Denton, Chen, and Ballman 1967).

At the present time KDP and DKDP are widely used as electro-optic modulator materials with both longitudinal and transverse configurations because of their high optical quality provided that the crystals are protected in a water-free environment. These crystals have the disadvantage of a relatively high half-wave voltage and the need for encapsulation. Lower half-wave voltages are readily obtainable with the chemically and mechanically stable $LiNbO_3$ and $LiTaO_3$ (which can be grown with sufficiently high optical quality for most applications) but at room temperature these crystals are limited to wavelengths longer than about $0 \cdot 6$ μm because of optically induced index changes (operation above $\sim 200°C$ avoids this problem). Although a variety of materials have been used specifically to make bulk modulators with low half-wave voltages (less than 100 V) most of these have not realized wide application because of difficulties with the optical quality.

16.4.4. Travelling-wave modulators

At very high modulation frequencies, usually in the microwave region, when the wavelength of the modulating field becomes comparable to or less than the dimensions of the modulator crystal, the modulator can no longer be considered as a lumped capacitor. In order to have efficient interaction between the modulating wave and the light beam, the group velocities of the two fields in the crystal must be matched. If dispersion over the modulation bandwidth is small, as required for undistorted modulation, then the condition for matching the group velocities is met if the phase velocities are matched. Phase matching has already been discussed in § 13.1.1. If the modulating wave and light wave are not phase matched then the maximum useful interaction length is the coherence length given by eqn (13.1.14).

For most electro-optic materials the microwave refractive index is much greater than the optical index because of the lattice contribution to the polarizability, so that the coherence length is quite short in a simple electroded geometry such as Fig. 16.15. However, phase matching can be achieved with the parallel-plate TEM waveguide shown in Fig. 16.19 which is partially filled with the crystal, the remainder being a material of lower dielectric constant. This has the effect of increasing the phase velocity of the modulating wave in the structure by approximately the dimensional ratio $(W/a)^{1/2}$, compared with that of the electroded crystal of Figure 16.15, if air is the material of low dielectric constant. (W is the width of the waveguide as shown in Fig. 16.19). Thus the phase velocities are matched by appropriate geometrical adjustments.

The structure of Fig. 16.19 is essentially dispersionless up to modulating frequencies for which the crystal dimension a becomes comparable with the

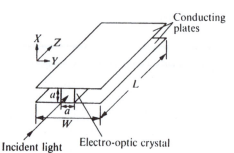

Fig. 16.19. A partially filed parallel-plate transmission line for travelling-wave modulation. (After Kaminow and Turner 1966.)

wavelength in the medium. The useful bandwidth may be written

$$\Delta\omega \sim \frac{c}{5a\sqrt{\varepsilon}}. \tag{16.4.25}$$

The crystal dimensions are then limited by diffraction and the required bandwidth if phase matching is perfect.

Travelling-wave modulators of this kind have been constructed using ADP and KDP (Peters 1963, 1965), KDA (Bicknell, Yap, and Peters 1967), and LiTaO$_3$ (White 1971) crystals with modulation bandwidths up to about 3 GHz.

Because of their large bandwidths two-plate waveguides are useful for very high-speed optical-switching applications. An example of such an application using a LiTaO$_3$ modulator (Auston and Glass 1972) is shown in Fig. 16.20 (a) where the optical beam is perpendicular to the waveguide and the direction of propagation of the modulating signal. A short electrical pulse of about 10 ps duration is generated at one end of the waveguide by absorbing a 10 ps optical pulse in a pyroelectric crystal bonded to the end of the transparent guide (see § 16.2). As the electrical pulse propagates down the waveguide the optical retardation changes instantaneously. If the waveguide is placed between crossed polarizers and illuminated transversely with a suitably synchronized light beam, the retardation can be observed as a short optical gate propagating along the guide as shown in Fig. 16.20 (b).

A few other schemes for travelling-wave modulation are reviewed by Kaminow and Turner (1966) and Chen (1970).

16.4.5. The quadratic electro-optic effect

Modulators have also been constructed using quadratic electro-optic materials in which the induced retardation is a quadratic function of applied field. To obtain a linear response the crystal must be biased with a d.c. electric field. This also has the effect of reducing the required modulation voltage.

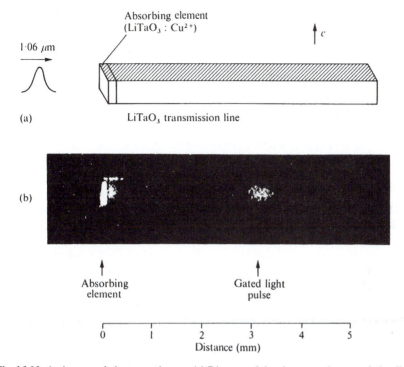

Fig. 16.20. A picosecond electro-optic gate. (a) Diagram of the electro-optic transmission line showing the absorbing pyroelectric element which generates the picosecond electrical pulse in response to a picosecond optical pulse. (b) Instantaneous photograph of the electro-optic gate viewed between crossed polarizers as the gate propagates along the transmission line. (After Auston and Glass 1972.)

The only materials which have sufficiently large quadratic electro-optic effects to be practical are ferroelectrics such as KTN (Chen *et al.* 1966) which are operated above, but close to, the Curie temperature. The Nb/Ta ratio of KTN crystals can be adjusted so that T_c lies just below room temperature.

For a transverse modulator with light polarized at 45° to the X- and Y-axes, the retardation induced by a field applied along the Z-axis is

$$\Delta\Gamma = \frac{\pi L n^3}{\lambda}(g_{11} - g_{12})P_3^2 \qquad (16.4.26)$$

where P_3 is the induced polarization including that part P_b due to the bias field. The half-wave voltage is then

$$V_\pi = \frac{\lambda}{2n^3(g_{11} - g_{12})P_b\varepsilon_b}\frac{d}{L} \qquad (16.4.27)$$

where ε_b is the dielectric constant in the presence of the bias field. Comparing eqn (16.4.27) with eqn (16.4.9) it is evident that the product $2(g_{11} - g_{12})P_b\varepsilon_b$ plays the same role as the r-coefficient in the linear modulator. Indeed we have already seen (see § 13.2) that in ferroelectrics with a centrosymmetric high-temperature phase the linear electro-optic effect is equivalent to the quadratic effect biased by the spontaneous polarization.

The requirement that the material be operated close to the Curie temperature in the presence of a d.c. bias field has all the disadvantages outlined in § 16.4.3, so that the material requirements for a modulator based on the quadratic electro-optic effect are particularly severe.

16.4.6. Waveguide modulators

With guided optical waves the diffraction limitations on the modulator geometry no longer apply and in principle the power P can be reduced without limit. Optical waveguides have been constructed both as planar guides which confine the light in one dimension only, allowing the beam to propagate in a plane, and with strip-guide configurations which confine the beam in both transverse dimensions. For a strip guide there are no diffraction limitations, but for the planar guide the unguided dimension is still limited by diffraction. The reduction of modulator drive power according to eqn (16.4.20) as the dimensions are reduced is only realized if the modulating field is confined to the same volume as the light beam. The construction of practical waveguide modulators therefore depends firstly on the production of thin films of electro-optic material with high optical quality, suitably large dimensions in the direction of light propagation within the film (0·1 to 1 cm), and extremely small dimensions in one or two transverse dimensions (~ 1 to 10 μm). Secondly an electrode configuration must be designed which provides efficient interaction between the modulating field and the guided optical wave.

For guiding light beams it is necessary to have a thin region of high refractive index bounded by regions of lower refractive index as shown in Fig. 16.21. The light can then propagate within the film as shown in the figure by total internal reflection at the boundaries. Because of interference between the rays within the guide only a discrete number of modes may propagate as determined by the boundary conditions and film thickness. For instance it is clear that the high-order modes become lossy when the angle ϕ which the rays make with the film normal becomes less than the critical angle for total internal reflection. The solution for the various modes of propagation can be obtained in much the same way as solving Schrödinger's equation, but instead of the electronic potential we have the refractive-index profile. Typical solutions for the lowest order TE_0 and TE_1 modes of a slab waveguide and a graded-index waveguide are shown in Fig. 16.21. The number of modes decreases as the slab thickness decreases, and the phase

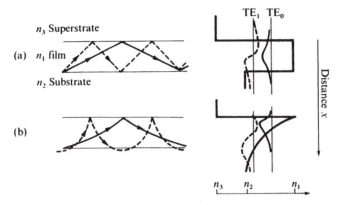

Fig. 16.21. Ray paths for (a) slab and (b) graded-index waveguides. The solid and broken rays represent low- and high-order modes. The wavefunctions for TE$_0$ and TE$_1$ modes in these waveguides are shown superposed on diagrams of the refractive-index profile. (After Kaminow 1975.)

velocity of the light along the guide axis decreases with increasing mode order (decreasing ϕ). A practical lower limit of the waveguide thickness is about 1 μm, since for thicknesses less than this coupling of the light beam into the guide becomes too lossy and the requirements on the uniformity of the refractive-index profile for low-loss guiding become particularly stringent.

Waveguides with thicknesses down to about 10 μm can be produced from single crystals by conventional polishing techniques. For thicknesses less than this the most successful procedures involve either epitaxial growth of a film on a substrate of lower refractive index or changing the composition of an electro-optic crystal in a small region below the surface by solid-state diffusion. Various procedures have been outlined in § 15.1. Waveguides produced by diffusion have a graded-index profile, the shape depending on the nature of the diffusion process. For instance if the index change is a linear function of concentration of the diffusant $c(x, t)$ given by the familiar error-function complement (Crank 1970), then

$$\Delta n(x, t) \propto c(x, t) = c(0, 0)\, \mathrm{erf}\!\left[\frac{x}{2(Dt)^{1/2}}\right] \qquad (16.4.28)$$

where $c(0, 0)$ is the (constant) surface concentration and D is the diffusion coefficient.

Low-loss waveguides have been produced in LiNbO$_3$ and LiTaO$_3$ single crystals by a variety of diffusion procedures which produce adequate index changes ($\Delta n \gtrsim 10^{-3}$). These have been summarized by Kaminow (1975) (see also § 15.1).

All these techniques are suitable for producing planar waveguides on a substrate. To produce stripguides part of the surface may be etched away leaving the desired structure, or else the guide may be formed by suitable masking of the surface during a diffusion process. Examples of planar and strip waveguide modulators are shown in Fig. 16.22. In the planar structure

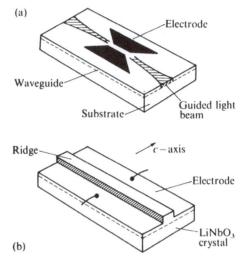

Fig. 16.22. Diagram of (a) a planar electro-optic waveguide modulator and (b) a ridge electro-optic waveguide. (After Kaminow 1975.)

the electrodes are deposited on the surface of the guide and the modulation relies on fringing fields penetrating below the crystal surface. For optimum interaction of the modulating field and the light beam the electrode separation should be of the order of the waveguide thickness. The length of the modulator is then determined by diffraction in the plane. With an out-diffused $LiNbO_3$ structure of this kind a modulating power of $0.4 \text{ mW MHz}^{-1} \text{ rad}^{-2}$ was obtained with a bandwidth of 1.6 GHz, which is a considerable improvement over the value of $10 \text{ mW MHz}^{-1} \text{ rad}^{-2}$ typical of good bulk modulators. With the ridge-guide configuration shown in Fig. 16.22 (b) the interaction of the modulating field and the guided wave is made considerably more efficient since the modulator length is not limited by diffraction and the interaction does not depend on fringing fields. By etching an out-diffused $LiNbO_3$ film to produce a modulator 19 mm wide and 11 mm long, a modulating power of only.20 $\mu\text{W MHz}^{-1}$ and a voltage of 1.2 V was sufficient to produce phase modulation of 1 rad with a potential bandwidth of 640 MHz.

To achieve amplitude modulation with a waveguide modulator a polarizer may be placed externally to the waveguide as in the bulk modulator. Alternatively there are a variety of schemes for producing amplitude

modulation within the waveguide, thereby eliminating the need for polarizing elements. These may involve coupling two waveguides so that the two
guided waves interfere. Any relative phase change of one beam compared
with the other then gives rise to amplitude modulation. There are also
schemes in which the waveguide modes are coupled by the electric field so
that lower-order modes may be converted to higher-order modes with
different propagation characteristics. These procedures are reviewed by
Kaminow (1975).

A very simple amplitude-modulation scheme suitable for planar
waveguides, which has been demonstrated experimentally in several systems, makes use of diffraction from an electro-optic phase grating. An
electrode array is deposited onto the waveguide surface and the guided
beam is diffracted as shown in Fig. 16.23. These schemes are not as efficient

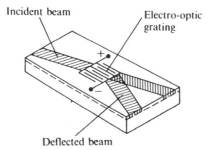

Fig. 16.23. Diagram of an electro-optic waveguide deflector.

as those for phase modulation discussed earlier but they have advantages for
optical switching. An interesting extension of such a device for wide-angle
switching makes use of an electro-optic deflector of this kind followed by a
thick-phase holographic grating stored in the waveguide. The hologram
diffracts the beam through a large angle only for a specific angle of incidence
determined by the deflector. The hologram is transparent for other angles of
incidence (Verber *et al.* 1976).

In this section we have only been concerned with ferroelectric waveguide
materials. At the present time no practical thin-film laser has been
developed using a ferroelectric material which can be monolithically integrated with thin-film modulators, which is desirable for several applications
(such as optical communications). A versatile competitor with ferroelectrics
for waveguides is the GaAs/AlAs system which has the advantage that
monolithic integration of both waveguide lasers and modulators is possible
for wavelengths in the red and near infrared, although this system also has
practical difficulties which are not yet resolved. However, there is no doubt
that as the field of integrated optics is advanced there will be considerable
versatility afforded by electro-optic waveguides which cannot be achieved
by their bulk-crystal counterparts.

APPENDIX A

The 32 crystallographic point groups arranged by crystal systems

Crystal system	Symbol		Pyro-electric	Piezo-electric	Centro-symmetric
	Inter-national	Schoen-flies			
Triclinic	1	C_1	✓	✓	
	$\bar{1}$	C_i			✓
Tetragonal	4	C_4	✓	✓	
	$\bar{4}$	S_4		✓	
	$4/m$	C_{4h}			✓
	422	D_4		✓	
	$4mm$	C_{4v}	✓	✓	
	$\bar{4}2m$	D_{2d}		✓	
	$4/mmm$	D_{4h}			✓
Hexagonal	6	C_6	✓	✓	
	$\bar{6}$	C_{3h}		✓	
	$6/m$	C_{6h}			✓
	622	D_6		✓	
	$6mm$	C_{6v}	✓	✓	
	$\bar{6}m2$	D_{3h}		✓	
	$6/mmm$	D_{6h}			✓
Monoclinic	2	C_2	✓	✓	
	m	C_s	✓	✓	
	$2/m$	C_{2h}			✓
Orthorhombic	222	D_2		✓	
	$mm2$	C_{2v}	✓	✓	
	mmm	D_{2h}			✓
Trigonal	3	C_3	✓	✓	
	$\bar{3}$	S_6			✓
	32	D_3		✓	
	$3m$	C_{3v}	✓	✓	
	$\bar{3}m$	D_{3d}			✓
Cubic	23	T		✓	
	$m3$	T_h			✓
	432	O			
	$\bar{4}3m$	T_d		✓	
	$m3m$	O_h			✓

A tick (√) indicates that the point group is pyroelectric, piezoelectric, or centrosymmetric, as the case may be.

APPENDIX B

Raman scattering tensors C_{mn}^j and irreducible representations of Raman active modes for the various crystal classes

The different elements of the 3×3 matrices C_{mn}^j for each active mode j are the nine components obtained by allowing m and n to take on the values x, y, z designating the three principal axes as in eqn (7.3.4). The component of phonon polarization for those modes which are also infrared active is given in brackets after the symbol of the irreducible representation. For crystals of triclinic symmetry, all elements of the scattering tensor are non-zero.

System	Point group	Raman scattering tensors and irreducible representations of Raman active modes
Monoclinic		$\begin{bmatrix} a & 0 & d \\ 0 & b & 0 \\ d & 0 & c \end{bmatrix}$ $\quad$ $\begin{bmatrix} 0 & e & 0 \\ e & 0 & f \\ 0 & f & 0 \end{bmatrix}$
	2(C_2)	$A(y)$ $\qquad\qquad$ $B(x,z)$
	m(C_s)	$A'(x,z)$ $\qquad\quad$ $A''(y)$
	2/m(C_{2h})	A_g $\qquad\qquad\quad$ B_g
Orthorhombic		$\begin{bmatrix} a & 0 & 0 \\ 0 & b & 0 \\ 0 & 0 & c \end{bmatrix}$ $\begin{bmatrix} 0 & d & 0 \\ d & 0 & 0 \\ 0 & 0 & 0 \end{bmatrix}$ $\begin{bmatrix} 0 & 0 & e \\ 0 & 0 & 0 \\ e & 0 & 0 \end{bmatrix}$ $\begin{bmatrix} 0 & 0 & 0 \\ 0 & 0 & f \\ 0 & f & 0 \end{bmatrix}$
	222(D_2)	A $\quad$ $B_1(z)$ $\quad$ $B_2(y)$ $\quad$ $B_3(x)$
	mm2(C_{2v})	$A_1(z)$ $\quad$ A_2 $\qquad$ $B_1(x)$ $\quad$ $B_2(y)$
	mmm(D_{2h})	A_g $\quad$ B_{1g} $\quad$ B_{2g} $\quad$ B_{3g}
Trigonal		$\begin{bmatrix} a & 0 & 0 \\ 0 & a & 0 \\ 0 & 0 & b \end{bmatrix}$ $\begin{bmatrix} c & d & e \\ d & -c & f \\ e & f & 0 \end{bmatrix}$ $\begin{bmatrix} d & -c & -f \\ -c & -d & e \\ -f & e & 0 \end{bmatrix}$
	3(C_3)	$A(z)$ $\qquad\quad$ $E(x)$ $\qquad\qquad$ $E(y)$
	$\bar{3}$(S_6)	A_g $\qquad\quad$ E_g $\qquad\qquad$ E_g

Raman scattering tensors and irreducible representations of Raman active modes

Trigonal

32(D_3), 3m(C_{3v}), $\bar{3}$m(D_{3d})

$$\begin{bmatrix} a & 0 & 0 \\ 0 & a & 0 \\ 0 & 0 & b \end{bmatrix} \quad \begin{bmatrix} c & 0 & 0 \\ 0 & -c & d \\ 0 & d & 0 \end{bmatrix} \quad \begin{bmatrix} 0 & -c & -d \\ -c & 0 & 0 \\ -d & 0 & 0 \end{bmatrix}$$

A_1	$E(x)$	$E(y)$
$A_1(z)$	$E(y)$	$E(-x)$
A_{1g}	E_g	E_g

Tetragonal

4(C_4), $\bar{4}$(S_4), 4/m(C_{4h})

$$\begin{bmatrix} a & 0 & 0 \\ 0 & a & 0 \\ 0 & 0 & b \end{bmatrix} \quad \begin{bmatrix} c & d & 0 \\ d & -c & 0 \\ 0 & 0 & 0 \end{bmatrix} \quad \begin{bmatrix} 0 & 0 & e \\ 0 & 0 & f \\ e & f & 0 \end{bmatrix} \quad \begin{bmatrix} 0 & 0 & -f \\ 0 & 0 & e \\ -f & e & 0 \end{bmatrix}$$

$A(z)$	B	$E(x)$	$E(y)$
A	$B(z)$	$E(x)$	$E(-y)$
A_g	B_g	E_g	E_g

4mm(C_{4v}), 422(D_4), $\bar{4}$2m(D_{2d}), 4/mmm(D_{4h})

$$\begin{bmatrix} a & 0 & 0 \\ 0 & a & 0 \\ 0 & 0 & b \end{bmatrix} \quad \begin{bmatrix} c & 0 & 0 \\ 0 & -c & 0 \\ 0 & 0 & 0 \end{bmatrix} \quad \begin{bmatrix} 0 & d & 0 \\ d & 0 & 0 \\ 0 & 0 & 0 \end{bmatrix} \quad \begin{bmatrix} 0 & 0 & e \\ 0 & 0 & 0 \\ e & 0 & 0 \end{bmatrix} \quad \begin{bmatrix} 0 & 0 & 0 \\ 0 & 0 & e \\ 0 & e & 0 \end{bmatrix}$$

$A_1(z)$	B_1	B_2	$E(x)$	$E(y)$
A_1	B_1	B_2	$E(-y)$	$E(x)$
A_1	B_1	$B_2(z)$	$E(y)$	$E(x)$
A_{1g}	B_{1g}	B_{2g}	E_g	E_g

Hexagonal

$6(C_6),\ \bar{6}(C_{3h}),\ 6/m(C_{6h})$

$$\begin{bmatrix} a & 0 & 0 \\ 0 & a & 0 \\ 0 & 0 & b \end{bmatrix} \quad \begin{bmatrix} 0 & 0 & c \\ 0 & 0 & d \\ c & d & 0 \end{bmatrix} \quad \begin{bmatrix} 0 & 0 & -d \\ 0 & 0 & c \\ -d & c & 0 \end{bmatrix} \quad \begin{bmatrix} e & f & 0 \\ f & -e & 0 \\ 0 & 0 & 0 \end{bmatrix} \quad \begin{bmatrix} f & -e & 0 \\ -e & -f & 0 \\ 0 & 0 & 0 \end{bmatrix}$$

Class					
$6(C_6)$	$A(z)$	$E_1(x)$	$E_1(y)$	E_2	E_2
$\bar{6}(C_{3h})$	A'	E''	E''	$E'(x)$	$E'(y)$
$6/m(C_{6h})$	A_g	E_{1g}	E_{1g}	E_{2g}	E_{2g}

$622(D_6),\ 6mm(C_{6v}),\ \bar{6}m2(D_{3h}),\ 6/mmm(D_{6h})$

$$\begin{bmatrix} a & 0 & 0 \\ 0 & a & 0 \\ 0 & 0 & b \end{bmatrix} \quad \begin{bmatrix} 0 & 0 & 0 \\ 0 & 0 & c \\ 0 & c & 0 \end{bmatrix} \quad \begin{bmatrix} 0 & 0 & -c \\ 0 & 0 & 0 \\ -c & 0 & 0 \end{bmatrix} \quad \begin{bmatrix} d & 0 & 0 \\ 0 & -d & 0 \\ 0 & 0 & 0 \end{bmatrix} \quad \begin{bmatrix} 0 & -d & 0 \\ -d & 0 & 0 \\ 0 & 0 & 0 \end{bmatrix}$$

Class					
$622(D_6)$	A_1	$E_1(x)$	$E_1(y)$	E_2	E_2
$6mm(C_{6v})$	$A_1(z)$	$E_1(y)$	$E_1(-x)$	E_2	E_2
$\bar{6}m2(D_{3h})$	A_1'	E''	E''	$E'(x)$	$E'(y)$
$6/mmm(D_{6h})$	A_{1g}	E_{1g}	E_{1g}	E_{2g}	E_{2g}

Cubic

$$\begin{bmatrix} a & 0 & 0 \\ 0 & a & 0 \\ 0 & 0 & a \end{bmatrix} \quad \begin{bmatrix} b & 0 & 0 \\ 0 & b & 0 \\ 0 & 0 & -2b \end{bmatrix} \quad \begin{bmatrix} -b\sqrt{3} & 0 & 0 \\ 0 & b\sqrt{3} & 0 \\ 0 & 0 & 0 \end{bmatrix} \quad \begin{bmatrix} 0 & 0 & 0 \\ 0 & 0 & d \\ 0 & d & 0 \end{bmatrix} \quad \begin{bmatrix} 0 & 0 & d \\ 0 & 0 & 0 \\ d & 0 & 0 \end{bmatrix} \quad \begin{bmatrix} 0 & d & 0 \\ d & 0 & 0 \\ 0 & 0 & 0 \end{bmatrix}$$

Class						
$23(T)$	A	E	E	$F(x)$	$F(y)$	$F(z)$
$m3(T_h)$	A_g	E_g	E_g	F_g	F_g	F_g
$432(O)$	A_1	E	E	$F_2(x)$	$F_2(y)$	$F_2(z)$
$\bar{4}3m(T_d)$	A_1	E	E	$F_2(x)$	$F_2(y)$	$F_2(z)$
$m3m(O_h)$	A_{1g}	E_g	E_g	F_{2g}	F_{2g}	F_{2g}

APPENDIX C

The possible ferroelectric states for continuous or quasi-continuous transitions from each of the 32 point groups of Appendix A.

Initial state	Possible ferroelectric states
C_1	—
C_i	C_1
C_4	C_1
S_4	C_2, C_1
C_{4h}	C_4, C_s, C_1
D_4	C_4, C_2, C_1
C_{4v}	C_s, C_1
D_{2d}	C_{2v}, C_2, C_s, C_1
D_{4h}	C_{4v}, C_{2v}, C_s, C_1
C_6	C_1
C_{3h}	C_3, C_s, C_1
C_{6h}	C_6, C_s, C_1
D_6	C_6, C_2, C_1
C_{6v}	C_s, C_1
D_{3h}	C_{3v}, C_s, C_1
D_{6h}	C_{6v}, C_{2v}, C_s, C_1
C_2	C_1
C_s	C_1
C_{2h}	C_2, C_s, C_1
D_2	C_2, C_1
C_{2v}	C_s, C_1
D_{2h}	C_{2v}, C_s, C_1
C_3	C_1
S_6	C_3, C_1
D_3	C_3, C_2, C_1
C_{3v}	C_s, C_1
D_{3d}	C_{3v}, C_2, C_s, C_1
T	C_3, C_2, C_1
T_h	C_3, C_{2v}, C_s, C_1
O	C_4, C_3, C_2, C_1
T_d	C_{3v}, C_{2v}, C_s, C_1
O_h	$C_{4v}, C_{3v}, C_{2v}, C_s, C_1$

APPENDIX D

The possible collinear antiferroelectric states for continuous or quasi-continuous transitions from each of the 32 point groups of Appendix A. Note that the existence of a centre of symmetry is not a necessary requirement (see for example Fig. 1.2).

Initial state	Possible antiferroelectric states
C_1	C_1
C_i	C_i
C_4	C_4, C_2, C_1
S_4	C_4, C_2, C_1
C_{4h}	C_{4h}, C_{2h}, C_i
D_4	D_4, D_2, C_2, C_1
C_{4v}	$C_{4v}, C_{2v}, C_2, C_s, C_1$
D_{2d}	$D_{2d}, D_2, C_{2v}, C_2, C_s, C_1$
D_{4h}	$D_{4h}, D_{2h}, C_{2h}, C_i$
C_6	C_6, C_2, C_1
C_{3h}	C_{3h}, C_s, C_1
C_{6h}	C_{6h}, C_{2h}, C_i
D_6	D_6, D_2, C_2, C_1
C_{6v}	$C_{6v}, C_{2v}, C_2, C_s, C_1$
D_{3h}	D_{3h}, C_2, C_s, C_1
D_{6h}	$D_{6h}, D_{2h}, C_{2h}, C_i$
C_2	C_2, C_1
C_s	C_s, C_1
C_{2h}	C_{2h}, C_i
D_2	D_2, C_2, C_1
C_{2v}	C_{2v}, C_2, C_s, C_1
D_{2h}	D_{2h}, C_{2h}, C_i
C_3	C_3, C_1
S_6	S_6, C_i
D_3	D_3, C_2, C_1
C_{3v}	C_{3v}, C_s, C_1
D_{3d}	D_{3d}, C_{2h}, C_2, C_i
T	D_2, C_3, C_2, C_1
T_h	S_6, D_{2h}, C_{2h}, C_i
O	D_4, D_3, D_2, C_2, C_1
T_d	$D_{2d}, C_{3v}, C_{2v}, C_2, C_1$
O_h	$D_4, D_{3d}, D_{2h}, C_{2h}, C_i$

APPENDIX E

Matrices for equilibrium properties in the 32 crystal classes

Key to notation
- · zero component   ● non-zero component   ●—● equal components
- ●—O components numerically equal, but opposite in sign
- ⊙ a component equal to twice the heavy dot component to which it is joined
- ⊚ a component equal to minus twice the heavy dot component to which it is joined
- × $2(s_{11} - s_{12})$

Each complete 10×10 matrix is symmetrical about the leading diagonal.
Taken from Nye 1964.

Triclinic system

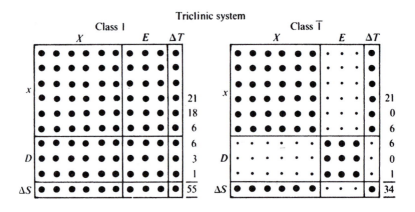

Monoclinic system

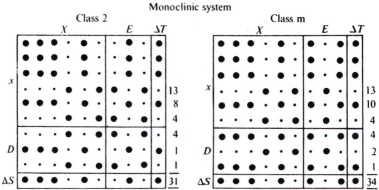

Monoclinic system—*continued*

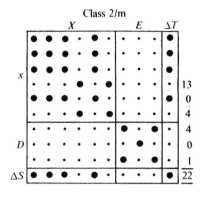

Class 2/m

Orthorhombic system

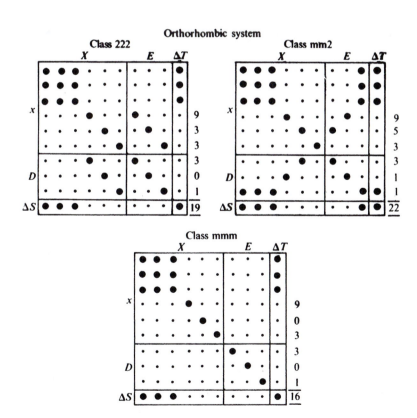

Class 222

Class mm2

Class mmm

Tetragonal system

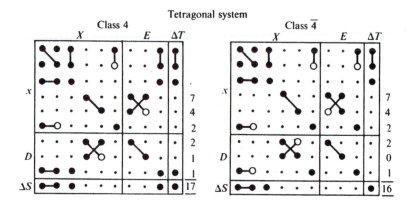

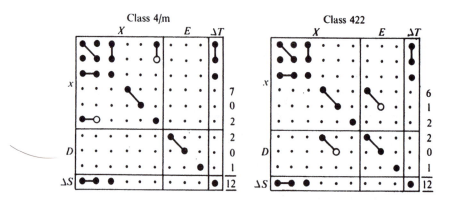

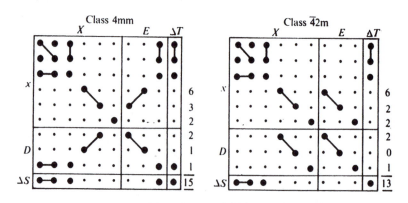

Tetragonal system—*continued*

Class 4/mmm

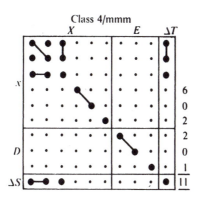

Trigonal system

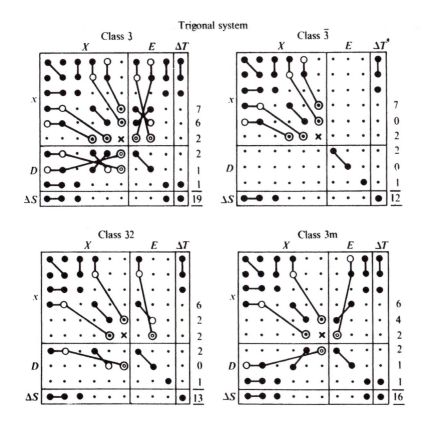

Trigonal system—*continued*

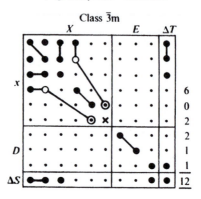

Class $\bar{3}$m

Hexagonal system

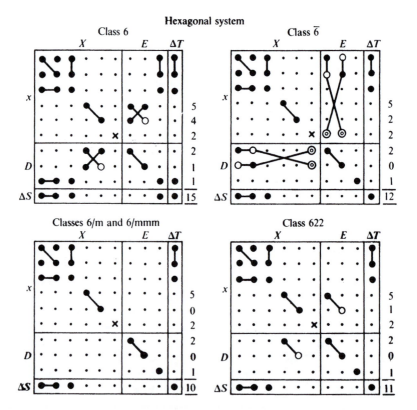

Class 6

Class $\bar{6}$

Classes 6/m and 6/mmm

Class 622

Hexagonal system—*continued*

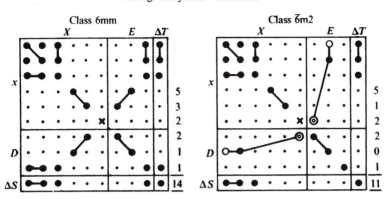

Class 6mm

Class 6̄m2

Cubic system

Classes 23 and 4̄3m

Classes m3, 432 and m3m

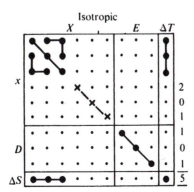

Isotropic

APPENDIX F

The major part of the following list of ferroelectric and antiferroelectric compounds was compiled by Subbarao (1973). A small additional list updating the table to 1975 has been added. References for this final section are quoted here; otherwise the original table of Subbarao should be consulted. Most solid solutions have been excluded. The table lists the highest Curie temperature $T_c(°C)$ and spontaneous polarization P_s (at temperature $T°C$) when these are known. In many cases P_s was obtained from hysteresis loops and represents a minimum value (see text). A question mark following a material indicates that ferroelectricity or antiferroelectricity is suspected, but has not been well established, for this material. The list of liquid crystals has been omitted from the original table of Subbarao for reasons outlined in the text.

More complete tables are available in the Landolt–Bornstein New Series Group III, Vol. 9, *Ferro- and Antiferroelectric Substances*, Springer-Verlag, Berlin (1975).

Ferroelectric compounds

Name	Formula	T_c (°C)	P_s (μC cm^{-2}) (at T°C)
A. Perovskite-type compounds			
Barium titanate	$BaTiO_3$	135	26·0 (23)
Lead titanate	$PbTiO_3$	490	>50 (23)
Strontium titanate?	$SrTiO_3$	−163	
Cadmium titanate?	$CdTiO_3$	110	
Sodium niobate	$NaNbO_3$	−200	12·0 (−200)
Potassium niobate	$KNbO_3$	435	30·0 (250)
Sodium tantalate?	$NaTaO_3$	480?	
Potassium tantalate?	$KTaO_3$	<−271·4	
Bismuth ferrite?	$BiFeO_3$	850	6·1
Potassium iodate	KIO_3	212	2·9 (−100)
Caesium germanium chloride	$CsGeCl_3$	155	15 (25)
Silver niobate?	$AgNbO_3$	325	
Silver tantalate?	$AgTaO_3$	370	
Dysprosium chromite?	$DyCrO_3$	575	
Holmium chromite?	$HoCrO_3$	495	
Ytterbium chromite?	$YbCrO_3$	515	
Lutetium chromite?	$LuCrO_3$	440	
Praseodymium chromite?	$PrCrO_3$	630	
Yttrium chromite?	$YCrO_3$	550	
Potassium bismuth titanate?	$(K_{1/2}Bi_{1/2})TiO_3$	270	
Sodium bismuth titanate?	$(Na_{1/2}Bi_{1/2})TiO_3$	200	$\approx$8·0 (116)
Lead cobalt tungstate	$Pb(Co_{1/2}W_{1/2})O_3$	−190 ... −170,	24 (−195)
Lead scandium niobate	$Pb(Sc_{1/2}Nb_{1/2})O_3$	90	3·6 (18)
Lead iron niobate	$Pb(Fe_{1/2}Nb_{1/2})O_3$	114	
Lead indium niobate?	$Pb(In_{1/2}Nb_{1/2})O_3$	90	
Lead scandium tantalate	$Pb(Sc_{1/2}Ta_{1/2})O_3$	26	
Lead iron tantalate	$Pb(Fe_{1/2}Ta_{1/2})O_3$	−40, −30?	28 (−170)

Name	Formula	T_c (°C)	P_s (μC cm^{-2}) (at T°C)
Lead magnesium niobate	$Pb(Mg_{1/3}Nb_{2/3})O_3$	−8?	24 (−170)
Lead zinc niobate	$Pb(Zn_{1/3}Nb_{2/3})O_3$	140	24 (25)
Lead cobalt niobate	$Pb(Co_{1/3}Nb_{2/3})O_3$	−98	
Lead nickel niobate	$Pb(Ni_{1/3}Nb_{2/3})O_3$	−120	
Lead cadmium niobate	$Pb(Cd_{1/3}Nb_{2/3})O_3$	270	0·65 (25)
Lead magnesium tantalate	$Pb(Mg_{1/3}Ta_{2/3})O_3$	−90	
Lead cobalt tantalate	$Pb(Co_{1/3}Ta_{2/3})O_3$	−140	
Lead nickel tantalate	$Pb(Ni_{1/3}Ta_{2/3})O_3$	−180	
Lead iron tungsten	$Pb(Fe_{2/3}W_{1/3})O_3$	−95	
Lead scandium tungstate?	$Pb(Sc_{2/3}W_{1/3})O_3$	−12	
Barium copper tungstate?	$Ba(Cu_{1/2}W_{1/2})O_3$	1200	
Strontium copper tungstate?	$Sr(Cu_{1/2}W_{1/2})O_3$	920	
Barium copper tantalate?	$Ba(Cu_{1/3}Ta_{2/3})O_3$	470	
Strontium copper tantalate?	$Sr(Cu_{1/3}Ta_{2/3})O_3$	1250	
Barium copper niobate?	$Ba(Cu_{1/3}Nb_{2/3})O_3$	380	
Strontium copper niobate?	$Sr(Cu_{1/3}Nb_{2/3})O_3$	390	
Barium bismuth oxide?	$BaBiO_{2·8}$	370	
Barium bismuth oxide?	$BaBiO_3$	340	
Barium bismuth niobate?	$Ba(Bi_{1/2}Nb_{1/2})O_3$	360	
Barium bismuth tantalate?	$Ba(Bi_{1/2}Ta_{1/2})O_3$	400	
Barium bismuth vanadate?	$Ba(Bi_{1/2}V_{1/2})O_3$	320	
Barium bismuth tungstate?	$Ba(Bi_{2/3}W_{1/3})O_3$	450	
Barium bismuth molybdate?	$Ba(Bi_{2/3}Mo_{1/3})O_3$	500	
Thallium zirconium tungstate?	$Tl(Zr_{1/2}W_{1/2})O_3$	310	
Cadmium hafnate?	$CdHfO_3$	605	
Cadmium iron niobate?	$Cd(Fe_{1/2}Nb_{1/2})O_3$	450	
Cadmium scandium niobate?	$Cd(Sc_{1/2}Nb_{1/2})O_3$	430	
Cadmium chromium niobate?	$Cd(Cr_{1/2}Nb_{1/2})O_3$	325	
Cadmium magnesium niobate?	$Cd(Mg_{1/3}Nb_{2/3})O_3$	295	
Lead chromium niobate?	$Pb(Cr_{1/2}Nb_{1/2})O_3$	5	
Lead lithium scandium tungstate?	$Pb(Li_{1/4}Sc_{1/4}W_{1/2})O_3$	−90	
Lead lithium iron tungstate?	$Pb(Li_{1/4}Fe_{1/4}W_{1/2})O_3$	−70	
Lead lithium cobalt tungstate?	$Pb(Li_{1/4}Co_{1/4}W_{1/2})O_3$	30	
Lead lithium indium tungstate?	$Pb(Li_{1/4}In_{1/4}W_{1/2})O_3$	−5	
Lead lithium yttrium tungstate?	$Pb(Li_{1/4}Y_{1/4}W_{1/2})O_3$	5	
Lead lithium terbium tungstate?	$Pb(Li_{1/4}Tb_{1/4}W_{1/2})O_3$	20	
Lead lithium ytterbium tungstate?	$Pb(Li_{1/4}Yb_{1/4}W_{1/2})O_3$	−40	
Lead lithium holmium tungstate?	$Pb(Li_{1/4}Ho_{1/4}W_{1/2})O_3$	0	
Lead lithium gadolinium tungstate?	$Pb(Li_{1/4}Gd_{1/4}W_{1/2})O_3$	0	
Lead lithium prasaedimium tungstate?	$Pb(Li_{1/4}Pr_{1/4}W_{1/2})O_3$	−18	
Lead lithium lanthanum tungstate?	$Pb(Li_{1/4}La_{1/4}W_{1/2})O_3$	−20	
Lead lithium samarium tungstate?	$Pb(Li_{1/4}Sm_{1/4}W_{1/2})O_3$	−15	
Lead sodium yttrium tungstate?	$Pb(Na_{1/4}Y_{1/4}W_{1/2})O_3$	120	
Lead sodium holmium tungstate?	$Pb(Na_{1/4}Ho_{1/4}W_{1/2})O_3$	150	

Name	Formula	T_c (°C)	P_s (μC cm^{-2}) (at T°C)
Lead lithium zirconium tungstate?	$Pb(Li_{1/3}Zr_{1/6}W_{1/2})O_3$	-30	
Lead cadmium niobium tungstate?	$Pb(Cd_{4/9}Nb_{2/9}W_{1/3})O_3$	495	
Lead scandium niobium tungstate?	$Pb(Sc_{5/9}Nb_{1/3}W_{1/9})O_3$	<-170	
Lead scandium chromium niobate?	$Pb(Sc_{1/4}Cr_{1/4}Nb_{1/2})O_3$	57	
Lead cadmium manganese niobate?	$Pb(Cd_{1/4}Mn_{1/4}Nb_{1/2})O_3$	20	
Lead magnesium manganese tungstate?	$Pb(Mg_{1/4}Mn_{1/4}W_{1/2})O_3$	>200	
Lead cadmium manganese tungstate?	$Pb(Cd_{1/4}Mn_{1/4}W_{1/2})O_3$	>250	
Lead cobalt manganese tungstate?	$Pb(Co_{1/4}Mn_{1/4}W_{1/2})O_3$	≈110	
Lead nickel manganese tungstate?	$Pb(Ni_{1/4}Mn_{1/4}W_{1/2})O_3$	≈100	
Lead nickel manganese niobate?	$Pb(Ni_{1/4}Mn_{1/4}Nb_{1/2})O_3$	-25	
Lead cobalt manganese niobate?	$Pb(Co_{1/4}Mn_{1/4}Nb_{1/2})O_3$	-20	
Lead magnesium manganese niobate?	$Pb(Mg_{1/4}Mn_{1/4}Nb_{1/2})O_3$	0	
Lead zinc manganese niobate?	$Pb(Zn_{1/4}Mn_{1/4}Nb_{1/2})O_3$	30	
lead lithium niobium tungstate?	$Pb(Li_{1/3}Nb_{1/3}W_{1/3})O_3$	-45	
Lead magnesium manganese tantalate?	$Pb(Mg_{1/4}Mn_{1/4}Ta_{1/2})O_3$	-135	
Lead nickel manganese tantalate?	$Pb(Ni_{1/4}Mn_{1/4}Ta_{1/2})O_3$	-180	
Bismuth manganate?	$BiMnO_3$	>500	
Tungsten trioxide	WO_3	-40	

B. Lithium-niobate-type compounds

Name	Formula	T_c (°C)	P_s (μC cm^{-2}) (at T°C)
Lithium niobate	$LiNbO_3$	1210	71(23)
Lithium tantalate	$LiTaO_3$	665	50
Lithium iron tantalum oxyfluoride	$Li(Fe_{1/2}Ta_{1/2})O_2F$	580	

C. Manganites

Name	Formula	T_c (°C)	P_s (μC cm^{-2}) (at T°C)
Yttrium manganite	$YMnO_3$	≈640	5·5
Erbium manganite	$ErMnO_3$	≈560	5·6
Holmium manganite	$HoMnO_3$	≈600	5·6
Thulium manganite	$TmMnO_3$	>300	
Ytterbium manganite	$YbMnO_3$	≈720	5·6
Lutetium manganite	$LuMnO_3$	>300	

D. Tungsten-bronze-type oxides

Name	Formula	T_c (°C)	P_s (μC cm^{-2}) (at T°C)
Lead (meta) niobate	$PbNb_2O_6$	570	

Name	Formula	T_c (°C)	P_s (μC cm^{-2}) (at T°C)
Lead (meta) tantalate	$PbTa_2O_6$	260	10·0 (25)
Potassium bismuth niobate	$K_2BiNb_5O_{15}$	350	
Potassium lanthanum niobate	$K_2LaNb_5O_{15}$	−120	
Rubidium strontium niobate	$RbSr_2Nb_5O_{15}$	139	
Potassium strontium niobate	$KSr_2Nb_5O_{15}$	156	8 (136)
Potassium strontium niobate	$KSr_{4.5}Nb_{10}O_{38}$	120	24
Potassium strontium niobium oxyfluoride	$K_2Sr_2Nb_5O_{14}F$	−178	
Sodium strontium niobate	$NaSr_2Nb_5O_{15}$	270	29
Lithium potassium strontium niobate	$LiKSr_4Nb_{10}O_{30}$	145	28
Lithium sodium strontium niobate	$LiNaSr_4Nb_{10}O_{30}$	260	18
Potassium barium niobate	$KBa_2Nb_5O_{15}$	373	
Sodium barium niobate (BaNaNa)	$NaBa_2Nb_5O_{15}$	560	40
Lithium barium niobate	$LiBa_2Nb_5O_{15}$	586	
Potassium lead niobate	$KPb_2Nb_5O_{15}$	374	
Barium magnesium niobate	$Ba_9MgNb_{14}O_{45}$	−25	4·0
Strontium magnesium niobate	$Sr_9MgNb_{14}O_{45}$	10	
Barium strontium niobate	$Ba_{1.3}Sr_{3.7}Nb_{10}O_{30}$	75	32
Barium strontium niobate	$Ba_2Sr_3Nb_{10}O_{30}$	78	34
Barium strontium niobate	$Ba_{2.7}Sr_{2.3}Nb_{10}O_{30}$	132	17 (100)
Potassium lithium niobate?	$K_3Li_2Nb_5O_{15}$	430	25
Potassium lithium tantalate	$K_3Li_2Ta_5O_{15}$	−266	
Barium sodium lanthanum niobate?	$Ba_2Na_3LaNb_{10}O_{30}$	−25	
Barium sodium lanthanum niobate?	$Ba_3NaLaNb_{10}O_{30}$	−50	
Barium sodium europium niobate?	$Ba_2Na_3EuNb_{10}O_{30}$	155	
Barium sodium gadolinium niobate?	$Ba_2Na_3GdNb_{10}O_{30}$	170	
Barium sodium gadolinium niobate?	$Ba_3NaGdNb_{10}O_{30}$	20	
Barium sodium dysprosium niobate?	$Ba_2Na_3DyNb_{10}O_{30}$	220	
Barium sodium yttrium niobate?	$Ba_2Na_3YNb_{10}O_{30}$	220	
Barium sodium yttrium niobate?	$Ba_3NaYNb_{10}O_{30}$	145	
Barium titanium niobate	$Ba_6Ti_2Nb_8O_{30}$	245	
Barium zirconium niobate	$BaZr_{0.25}Nb_{1.5}O_{5.25}$	none	18
Barium iron niobate	$Ba_6FeNb_9O_{30}$	−140	>1·6
Strontium titanium niobate	$Sr_6Ti_2Nb_3O_{30}$	130	
Barium neodymium iron niobate	$Ba_4Nd_2Fe_2Nb_8C_{30}$	≈55	>3
Barium gadolinium iron niobate	$Ba_4Gd_2Fe_2Nb_8O_{30}$	≈130	
Barium samarium iron niobate	$Ba_4Sm_2Fe_2Nb_8O_{30}$	≈130	>2
Strontium ytterbium iron niobate	$Sr_4Yb_2Fe_2Nb_8O_{30}$	≈10	

Name	Formula	$T_c(°C)$	$P_s\ (\mu C\,cm^{-2})$ (at $T°C$)
Barium bismuth titanium niobate?	$Ba_4Bi_2Ti_4Nb_6O_{30}$	−31	
Barium bismuth titanium niobate?	$Ba_5BiTi_3Nb_7O_{30}$	15	
Strontium bismuth titanium niobate?	$Sr_5BiTi_3Nb_7O_{30}$	10	
Barium lanthanum titanium niobate?	$Ba_3La_3Ti_5Nb_5O_{30}$	−130	
Barium lanthanum titanium niobate?	$Ba_4La_2Ti_4Nb_6O_{30}$	−80	
Barium lanthanum titanium niobate?	$Ba_5LaTi_3Nb_7O_{30}$	−55	
Strontium lanthanum titanium niobate?	$Sr_4La_2Ti_4Nb_6O_{30}$	−33	
Strontium lanthanum titanium niobate?	$Sr_5LaTi_3Nb_7O_{30}$	−7	
Potassium barium titanium niobate?	$KBa_5TiNb_9O_{30}$	290	
Sodium barium titanium niobate?	$NaBa_5TiNb_9O_{30}$	414	
Potassium strontium titanium niobate?	$KSr_5TiNb_9O_{30}$	118	
Sodium strontium titanium niobate?	$NaSr_5TiNb_9O_{30}$	157	
Sodium barium tungsten niobate?	$Na_3Ba_3WNb_9O_{30}$	460	
Sodium barium tungsten niobate?	$Na_4Ba_2WW_2Nb_8O_{30}$	365	
Potassium strontium tungsten niobate?	$K_3Sr_3WNb_9O_{30}$	70	
Potassium strontium tungsten niobate?	$K_4Sr_2W_2Nb_8O_{30}$	65	

E. Pyrochlore-type compounds

Name	Formula	$T_c(°C)$	$P_s\ (\mu C\,cm^{-2})$ (at $T°C$)
Cadmium (pyro)niobate	$Cd_2Nb_2O_7$	−88	≈6·0 (−185)
Cadmium niobium oxysulphide	$Cd_2Nb_2O_6S$	282	≈2·0
Cadmium chromium niobate	Cd_2CrNbO_6	−203	
Cadmium iron niobate	Cd_2FeNbO_6	−153	
Barium zinc tantalate?	$Ba_2Zn_{4/3}Ta_{2/3}O_6$	300	
Lead barium lithium niobate?	$Pb_{1.9}Ba_{0.1}Li_{0.5}Nb_{1.5}O_6$	≈100	
Lead bismuth niobate?	Pb_2BiNbO_6	475	
Lead bismuth tantalate?	Pb_2BiTaO_6	420	
Lead bismuth tungstate?	$Pb_2Bi_{1/3}W_{2/3}O_6$	400	
Lead bismuth molybdate?	$Pb_2Bi_{1/3}Mo_{2/3}O_6$	500	

F. Layer-structure oxides

Name	Formula	$T_c(°C)$	$P_s\ (\mu C\,cm^{-2})$ (at $T°C$)
Bismuth tungstate?	Bi_2WO_6	950	
Bismuth niobium oxyfluoride?	Bi_2NbO_5F	30	
Bismuth tantalum oxyfluoride?	Bi_2TaO_5F	10	
Bismuth titanium niobate?	Bi_3TiNbO_9	900...950	
Bismuth titanium tantalate?	Bi_3TiTaO_9	870	
Strontium bismuth niobate?	$SrBi_2Nb_2O_9$	440	

Name	Formula	$T_c(°C)$	$P_s (\mu C\ cm^{-2})$ (at T°C)
Strontium bismuth tantalate	$SrBi_2Ta_2O_9$	335	5·8 (25)
Barium bismuth niobate?	$BaBi_2Nb_2O_9$	200	
Barium bismuth tantalate?	$BaBi_2Ta_2O_9$	110	
Lead bismuth niobate?	$PbBi_2Nb_2O_9$	550	
Lead bismuth tantalate?	$PbBi_2Ta_2O_9$	430	
Bismuth titanate	$Bi_4Ti_3O_{12}$	675	>30
Praseodymium bismuth titanate?	$PrBi_3Ti_3O_{12}$.	400	
Holmium bismuth titanate?	$HoBi_3Ti_3O_{12}$	440	
Lanthanum bismuth titanate?	$LaBi_3Ti_3O_{12}$	315	
Calcium bismuth titanate?	$CaBi_4Ti_4O_{15}$	790	
Barium bismuth titanate?	$BaBi_4Ti_4O_{15}$	395	
Lead bismuth titanate?	$PbBi_4Ti_4O_{15}$	570	
Strontium bismuth titanate?	$SrBi_4Ti_4O_{15}$	330	
Lanthanum bismuth iron titanate?	$LaBi_4Ti_3FeO_{15}$	600	
Praseodymium bismuth iron titanate?	$PrBi_4Ti_3FeO_{15}$	740	
Barium bismuth titanate	$Ba_2Bi_4Ti_5O_{13}$	325	>2
Lead bismuth titanate	$Pb_2Bi_4Ti_5O_{18}$	310	6
Strontium bismuth titanate	$Sr_2Bi_4Ti_5O_{18}$	285	3·5
Bismuth iron titanate	$Bi_5Ti_3FeO_{15}$	750	
Bismuth iron titanate?	$Bi_6Ti_3FeO_{18}$	>750	
Bismuth iron titanate?	$Bi_5Bi_4Ti_3Fe_5O_{27}$	830	
Calcium bismuth iron titanate?	$CaBi_5Ti_4FeO_{18}$	≈770	
Strontium bismuth iron titanate?	$SrBi_5Ti_4FeO_{18}$	≈580	
Lead bismuth iron titanate?	$PbBi_5Ti_4FeO_{18}$	≈550	
Barium bismuth iron titanate?	$BaBi_5Ti_4FeO_{18}$	≈570	

G. Barium-fluoride-type compounds

Name	Formula	$T_c(°C)$	$P_s (\mu C\ cm^{-2})$ (at T°C)
Barium lithium aluminium oxyfluoride	$BaLi_{2x}Al_{2-2x}F_{4x}O_{4-4x}$		
	$x = 0·3$	148	0·15
	$x = 0·2$	127	0·12
Barium magnesium fluoride	$BaMgF_4$	none	7·7
Barium manganese fluoride	$BaMnF_4$	none	
Barium iron fluoride	$BaFeF_4$	none	
Barium cobalt fluoride	$BaCoF_4$	none	8·0
Barium nickel fluoride	$BaNiF_4$	none	6·7
Barium zinc fluoride	$BaZnF_4$	none	9·7

H. Molybdates

Name	Formula	$T_c(°C)$	$P_s (\mu C\ cm^{-2})$ (at T°C)
Samarium molybdate	$Sm_2(MoO_4)_3$	197	0·24 (50)
Europium molybdate	$Eu_2(MoO_4)_3$	180	0·14 (25)
Praseodymium molybdate	$Pr_2(MoO_4)_3$	235	
Neodymium molybdate	$Nd_2(MoO_4)_3$	225	
Dysprosium molybdate	$Dy_2(MoO_4)_3$	145	
Holmium molybdate	$Ho_2(MoO_4)_3$	134	
Gadolinium molybdate	$\beta\text{-}Gd_2(MoO_4)_3$	163	0·185 (25)
Terbium molybdate	$\beta\text{-}Tb_2(MoO_4)_3$	157	0·19 (25)

Name	Formula	$T_c(°C)$	$P_s\ (\mu\ cm^{-2})$ (at $T°C$)
I. Boracites			
Magnesium $Mg_3B_7O_{13}Cl$	$Mg_3B_7O_{13}Cl$	265	<2
Nickel boracite?	$Ni_3B_7O_{13}Cl$	337	
Nickel iodine boracite	$Ni_3B_7O_{13}I$	−289	0·08 (−269)
Cobalt iodine boracite	$Co_3B_7O_{13}I$	−81	1·5
Iron iodine boracite	$Fe_3B_7O_{13}I$	72	
J. Colemanite			
Colemanite	$Ca_2B_6O_{11}\cdot 5H_2O$	−7	0·65 (−70)
K. Miscellaneous oxides			
Rubidium tantalate?	$RbTaO_3$	247	
Sodium vanadate?	$NaVO_3$	380?	
Silver vanadate?	$AgVO_3$	170 . . . 180	
Lead germanate	$Pb_5Ge_3O_{11}$	178	4·6 (25)
Stibiotantalite	$Sb(TaNb)O_4$	400	16·8 (25)
Strontium tellurate	$SrTeO_3$	485	3·7 (312)
Strontium niobate	$Sr_2Nb_2O_7$		9 (25)
Lead niobate?	$PbNb_4O_{11}$		
Ammonium metaphosphate		none	
Sodium dithorium triphosphate	$NaTh_2(PO_4)_3$		
Silver dithorium triphosphate	$AgTh_2(PO_4)_3$		4·8 (25)
Sodium diuranium triphosphate	$NaU_2(PO_4)_3$		
L. Halides			
Hydrogen chloride	HCl	−175	1·2 (−190)
Deuterium chloride	DCl	−168	
Hydrogen bromide	HBr	−183	0·4 (−190)
Deuterium bromide	DBr	−179	
Caesium germanium chloride?	$CsGeCl_3$	155	
M. Antimony sulphide iodide type compounds			
Antimony sulphide bromide	SbSBr	−180	
Antimony sulphide iodide	SbSI	≈20	25 (0)
Antimony selenide iodide?	SbSeI	23	
Bismuth sulphide bromide	BiSBr	−170	
Bismuth sulphide iodide	BiSI	−160	
Iron sulphide	FeS	13	≈0·7
Bismuth trisulphide?	Bi_2S_3	≈50	
Antimony trisulphide?	Sb_2S_3	26	
Antimony penta iodide?	SbI_5		
N. Nitrite			
Sodium nitrite	$NaNO_2$	163	8 (100)
Potassium nitrite?	KNO_2	47	2·8 (25)
Silver sodium nitrite	$AgNa(NO_2)_2$	38	1·5 (37)

Name	Formula	$T_c(°C)$	$P_s\ (\mu C\ cm^{-2})$ (at $T°C$)
O. Nitrate			
Potassium nitrate	KNO$_3$ upper	124	6·3 (121)
	lower	110	
P. Potassium dihydrogen phosphate type compounds			
Potassium dihydrogen phosphate (KDP)	KH$_2$PO$_4$	−150	4·75 (−177)
Potassium dideuterium phosphate	KD$_2$PO$_4$	−60	4·83 (−93)
Rubidium dihydrogen phosphate	RbH$_2$PO$_4$	−126	5·6 (−183)
Rubidium dideuterium phosphate	RbD$_2$PO$_4$	−55	
Caesium dihydrogen phosphate	CsH$_2$PO$_4$	−114	
Potassium dihydrogen arsenate	KH$_2$AsO$_4$	−176	5·0 (−195)
Potassium dideuterium arsenate	KD$_2$AsO$_4$	−112 ... −114	
Rubidium dihydrogen arsenate	RbH$_2$AsO$_4$	−163	
Rubidium dideuterium arsenate	RbD$_2$AsO$_4$	−100	
Caesium dihydrogen arsenate	CsH$_2$AsO$_4$	−130	
Caesium dideuterium arsenate	CsD$_2$AsO$_4$	−61	
Q. Sulphates and related compounds			
Ammonium sulphate	(NH$_4$)$_2$SO$_4$	−49	0·62 (−52)
Ammonium fluorberyllate	(NH$_4$)$_2$BeF$_4$	−97	0·2 (−110)
Deuterated ammonium fluorberyllate	(ND$_4$)$_2$BeF$_4$	−94	
Potassium selenate	K$_2$SeO$_4$	−180	0·065 (−193)
Ammonium bisulphate	(NH$_4$)HSO$_4$ upper	−3	0·8 (−118)
	lower	−119	
Deuterated ammonium bisulphate	(ND$_4$)DSO$_4$ upper	−11	
	lower	−115	
Rubidium bisulphate	RbHSO$_4$	−15	0·65 (−170)
Diammonium dicadmium sulphate	(NH$_4$)$_2$Cd$_2$(SO$_4$)$_3$	−178	0·5 (−180)
Sodium ammonium sulphate dihydrate (Lecontite)	NaNH$_4$SO$_4$. 2H$_2$O	−172	
Sodium ammonium selenate dihydrate	NaNH$_4$SeO$_4$. 2H$_2$O	−93	0·48 (−100)
Lithium hydrazinium sulphate?	Li(N$_2$H$_5$)SO$_4$	164	0·3 (23)
R. Alums			
Ammonium iron alum	NH$_4$Fe(SO$_4$)$_2$. 12H$_2$O	−185	0·40 (−187)
Deuterated ammonium iron alum	ND$_4$Fe(SO$_4$)$_2$. 12D$_2$O	−185	0·40 (−187)
Ammonium vanadium alum	NH$_4$V(SO$_4$)$_2$. 12H$_2$O	−157	1·0 (−159)
Ammonium indium alum	NH$_4$In(SO$_4$)$_2$. 12H$_2$O	−146	1·2 (−148)

Name	Formula	$T_c(°C)$	P_s (μC cm^{-2}) (at $T°C$)
Methyl ammonium aluminium alum (MASD)	$CH_3NH_3Al(SO_4)_2 \cdot 12H_2O$	-96	$1\cdot0\,(-98)$
Deuterated methyl ammonium aluminium alum	$CD_3ND_3Al(SO_4)_2 \cdot 12D_2O$	-96	$1\cdot0\,(-98)$
Methyl ammonium gallium alum	$CH_3NH_3Ga(SO_4)_2 \cdot 12H_2O$	-102	
Methyl ammonium chrome alum	$CH_3NH_3Cr(SO_4)_2 \cdot 12H_2O$	-109	$1\cdot0\,(-111)$
Methyl ammonium iron alum	$CH_3NH_3Fe(SO_4)_2 \cdot 12H_2O$	-104	$1\cdot3\,(-106)$
Methyl ammonium vanadium alum	$CH_3NH_3V(SO_4)_2 \cdot 12H_2O$	-116	$0\cdot9\,(-118)$
Methyl ammonium indium alum	$CH_3NH_3In(SO_4)_2 \cdot 12H_2O$	-109	$1\cdot2\,(-111)$
Methyl ammonium aluminium selenate alum	$CH_3NH_3Al(SeO_4)_2 \cdot 12H_2O$	-57	$1\cdot2\,(-59)$
Urea chrome alum	$CO(NH_2)_2HCr(SO_4)_2 \cdot 12H_2O$	-113	$0\cdot2\,(-115)$

S. Guanidinium compounds

Name	Formula	$T_c(°C)$	P_s (μC cm^{-2}) (at $T°C$)
Guanidinium aluminium sulphate hexahydrate (GASH)	$C(NH_2)_3Al(SO_4)_2 \cdot 6H_2O$	none	$0\cdot35\,(23)$
Guanidinium chromium sulphate hexahydrate (GCrSH)	$C(NH_2)_3Cr(SO_4)_2 \cdot 6H_2O$	none	$0\cdot37\,(23)$
Guanidinium gallium sulphate hexahydrate (GGaSH)	$C(NH_2)_3Ga(SO_4)_2 \cdot 6H_2O$	none	$0\cdot36\,(23)$
Guanidinium vanadium sulphate hexahydrate (GVSH)	$C(NH_2)_3V(SO_4)_2 \cdot 6H_2O$	none	$0\cdot38\,(23)$
Guanidinium aluminium selenate hexahydrate (GAlSeH)	$C(NH_2)_3Al(SeO_4)_2 \cdot 6H_2O$	none	$0\cdot45\,(23)$
Guanidinium chromium selenate hexahydrate (GCrSeH)	$C(NH_2)_3Cr(SeO_4)_2 \cdot 6H_2O$	none	$0\cdot47\,(23)$
Guanidinium gallium selenate hexahydrate (GGaSeH)	$C(NH_2)_3Ga(SeO_4)_2 \cdot 6H_2O$	none	$0\cdot47\,(23)$
Deuterated guanidinium aluminium sulphate	$C(ND_2)_3Al(SO_4)_2 \cdot 6D_2O$	none	$0\cdot35\,(23)$

T. Selenites

Name	Formula	$T_c(°C)$	P_s (μC cm^{-2}) (at $T°C$)
Lithium trihydrogen selenite	$LiH_3(SeO_3)_2$		$15\,(23)$
Sodium trihydrogen selenite	$NaH_3(SeO_3)_2$	-79	
Sodium trideuterium selenite	$NaD_3(SeO_3)_2$	-4	
Potassium trihydrogen selenite?	$KH_3(SeO_3)_2$	-62	
Potassium trideuterium selenite?	$KD_3(SeO_3)_2$	12	
Rubidium trihydrogen selenite	$RbH_3(SeO_3)_2$	-115	$0\cdot013\,(-200)$

U. Potassium cyanides

Name	Formula	$T_c(°C)$	P_s (μC cm^{-2}) (at $T°C$)
Potassium manganese cyanide trihydrate	$K_4Mn(CN)_6 \cdot 3H_2O$	-40	
Potassium ferrocyanide trihydrate	$K_4Fe(CN)_6 \cdot 3H_2O$	$-24\cdot5$	$1\cdot45\,(250)$
Deuterated potassium ferrocyanide	$K_4Fe(CN)_6 \cdot 3D_2O$	-18	$1\cdot50(-40)$

Name	Formula	$T_c(°C)$	$P_s\ (\mu C\ cm^{-2})$ (at $T°C$)
Potassium ruthenium cyanide trihydrate	$K_4Ru(CN)_6 \cdot 3H_2O$	$-14\cdot5$	$1\cdot4\ (-65)$
Deuterated potassium ruthenium cyanide	$K_4Ru(CN)_6 \cdot 3D_2O$	$-7\cdot3$	$1\cdot5\ (-35)$
Potassium osmium cyanide trihydrate	$K_4Os(CN)_6 \cdot 3H_2O$	$-2\cdot4$	$3\cdot5\ (-45)$
Deuterated potassium osmium cyanide	$K_4Os(CN)_6 \cdot 3D_2O$	$1\cdot8$	$1\cdot3\ (-20)$

V. Triglycine sulphate and related compounds

Name	Formula	$T_c(°C)$	P_s
Triglycine sulphate (TGS)	$(NH_2CH_2COOH)_3 \cdot H_2SO_4$	49	$2\cdot8\ (20)$
Deuterated triglycine sulphate	$(ND_2CD_2COOD)_3 \cdot D_2SO_4$	60	$3\cdot2\ (-60)$
Triglycine selenate (TGSe)	$(NH_2CH_2COOH)_3 \cdot H_2SeO_4$	22	$3\cdot2\ (10)$
Triglycine fluoberyllate (TGFBe)	$(NH_2CH_2COOH)_3 \cdot H_2BeF_4$	75	$3\cdot2\ (20)$
Deuterated triglycine fluoberyllate	$(ND_2CD_2COOD)_3 \cdot D_2BeF_4$	77	
Diglycine nitrate (DGN)	$(NH_2CH_2COOH)_2 \cdot HNO_3$	-67	$1\cdot5(-190)$
Glycine silver nitrate	$NH_2CH_2COOH \cdot AgNO_3$	-55	$0\cdot6\ (-170)$
Diglycine manganous chloride trihydrate	$(NH_2CH_2COOH)_2 \cdot MnCl_2 \cdot 2H_2O$	none	$1\cdot3\ (25)$
Trissarcosine calcium chloride	$(CH_3NHCH_2COOH)_3 \cdot CaCl_2$	-146	$0\cdot27\ (-195)$

W. Rochelle salt and related compounds

Name	Formula		$T_c(°C)$	P_s
Sodium potassium tartrate tetrahydrate (Rochelle salt, RS)	$NaKC_4H_4O_6 \cdot 4H_2O$	upper lower	24 -18	$0\cdot25\ (5)$
Deuterated Rochelle salt	$NaKC_4D_4O_6 \cdot 4D_2O$	upper lower	35 -22	$0\cdot35\ (6)$
Sodium ammonium tartrate tetrahydrate	$NaNH_4C_4H_4O_6 \cdot 4H_2O$		-164	$0\cdot21\ (-181)$
Lithium ammonium tartrate monohydrate (LAT)	$LiNH_4C_4H_4O_6 \cdot H_2O$		-167	$0\cdot22\ (-178)$
Lithium thallium tartrate monohydrate (LTT)	$LiTlO_4H_4O_6 \cdot H_2O$		-263	$0\cdot14\ (-272)$

X. Metallic solids

Name	Formula	$T_c(°C)$
Tin telluride	SnTe	$-200 \ldots -273$
Germanium telluride?	GeTe	354
Vanadium silicide?	V_3Si	$-243 \ldots -253$

Y. Complex organic compounds

Name	Formula		$T_c(°C)$	P_s
Thiourea	$SC(NH_2)_2$		-205	$3\cdot5\ (-133)$
		lower upper	-97 -90	
Deuterated thiourea	$SC(ND_2)_2$		-88	
		lower upper	-81 -77	

Name	Formula	$T_c(°C)$	P_s (μC cm^{-2}) (at $T°$C)
Tetramethyl ammonium trichloro mercurate	$N(CH_3)_4 \cdot HgCl_3$	none	1·2 (23)
Tetra methyl ammonium tribromo mercurate	$N(CH_3)_4 \cdot HgBr_3$	none	1 (23)
Tetra methyl phosphonium tribromo mercurate	$P(CH_3)_4 \cdot HgBr_3$	none	3 (23)
Tetra methyl ammonium triiodo mercurate	$N(CH_3)_4HgI_3$	none	1·2 (−40)
Dicalcium strontium propionate	$Ca_2Sr(CH_3CH_2COO)_6$	10	0·3 (−15)
Ammonium monochloroacetate	$CH_2ClCOONH_4$	−150	0·1 (−170)
Ammonium dichloroacetate	$(CH_2ClCOO)_2H \cdot NH_4$	−145	0·18 (−195)
Phthalocyanine? Copper phthalocyanine?			0·6 (27)
Semicarbazide hydrochloride	$H_2NCONHNH_2 \cdot HCl$	21	

Name	Formula	T_c (°C)	P_s (μC cm^{-2}) (at $T°$C)	Ref.
Strontium tellurite	$SrTeO_3$	485	3·8	a
Lead germanium silicate	$Pb_5Ge_2SiO_{11}$	60	1·7	b
Lead niobate	$PbNb_4O_{11}$	660		c
Barium silver niobate	$Ba_2AgNb_5O_{15}$	420		d
Tanane	$C_9H_{18}NO$	14·5	~1	e
Lead monohydrogen phosphate	$PbHPO_4$	37	1·37 (0)	f
Lead monodeuterium phosphate	$PbDPO_4$	179	2·0 (0)	f
Dicadmium diammonium sulphate	$Cd_2(NH_4)_2(SO_4)_3$	−181		g
Dicadmium dithallium sulphate	$Cd_2Tl_2(SO_4)_3$	−143		h
Neodymium titanate	$Nd_2Ti_2O_7$	>1500	9	i
Lanthanum titanate	$La_2Ti_2O_7$	1500	5	j
Strontium niobate	$Sr_2Nb_2O_7$	1342	9	k

a Yamada, T. and Iwasaki, H. (1972). *Acta cryst.* **28A**, S181.
b Iwasaki, H., Miyazawa, S., Koizumi, H., Sugii, K., and Niizeki, N. (1972). *J. appl. Phys.* **43**, 4907.
c Kondo, Y. (1973). *J. phys. Soc. Japan* **35**, 1266.
d Sugai, T. and Wada, M. (1974). *Jap. J. appl. Phys.* **13**, 1291.
e Bordeaux, D., Bornarel, J., Capiomont, A., Lajzerowicz-Bonneteau, J., Lajzerowicz, J., and Legrand, J. F. (1973). *Phys. Rev. Lett.* **31**, 314.
f Negran, T. J., Glass, A. M., Brickenkamp, C. S., Rosenstein, R. D., Osterheld, R. K., and Susott, R. (1974). *Ferroel.* **6**, 179.
g Konak, C., Fousek, J., and Ivanov, N. R. (1974). *Ferroel.* **6**, 235.
h Brezina, B. and Glogarova, M. (1972). *Phys. Stat. Sol.* **11a**, K39.
i Kimura, M., Nanamatsu, S., Kawamura, T., and Matsushita, S. (1974). *Jap. J. appl. Phys.* **13**, 1473.
j Nanamatsu, S., Kimura, M., Doi, K., Matsushita, S., and Yamada, N. (1974). *Ferroel.* **8**, 511.
k Nanamatsu, S., Kimura, M., Doi, K., and Takahashi, M. (1971). *J. phys. Soc. Japan* **30**, 300.

Antiferroelectric compounds

Name (abbreviation)	Formula	T_c (°C)

A. Perovskite-type compounds

Sodium niobate	$NaNbO_3$	354
Lead zirconate	$PbZrO_3$	230
Lead hafnate?	$PbHfO_3$	163
Bismuth ferrite?	$BiFeO_3$	850
Cadmium titanate	$CdTiO_3$	960
Cadmium hafnate?	$CdHfO_3$	605
Caesium lead chloride?	$CsPbCl_3$	47
Potassium bismuth titanate?	$(K_{1/2}Bi_{1/2})TiO_3$	380, 410
Sodium bismuth titanate?	$(Na_{1/2}Bi_{1/2})TiO_3$	220
Lead magnesium tungstate	$Pb(Mg_{1/2}W_{1/2})O_3$	38
Lead cadmium tungstate?	$Pb(Cd_{1/2}W_{1/2})O_3$	400
Lead manganese tungstate?	$Pb(Mn_{1/2}W_{1/2})O_3$	150
Lead manganese tungstate?	$Pb(Mn_{2/3}W_{1/3})O_3$	200
Lead cobalt tungstate	$Pb(Co_{1/2}W_{1/2})O_3$	20, 32, 400
Lead manganese rhenate?	$Pb(Mn_{1/2}Re_{1/2})O_3$	120
Lead indium niobate?	$Pb(In_{1/2}Nb_{1/2})O_3$	90
Lead ytterbium niobate	$Pb(Yb_{1/2}Nb_{1/2})O_3$	302
Lead holmium niobate?	$Pb(Ho_{1/2}Nb_{1/2})O_3$	240
Lead lutetium niobate?	$Pb(Lu_{1/2}Nb_{1/2})O_3$	290
Lead ytterbium tantalate?	$Pb(Yb_{1/2}Ta_{1/2})O_3$	297
Lead lutetium tantalate?	$Pb(Lu_{1/2}Ta_{1/2})O_3$	300
Lead manganese tungstate?	$Pb(Mn_{2/3}W_{1/3})O_3$	200
Lead gallium niobate?	Pb_2GaNbO_6	100
Lead bismuth niobate?	Pb_2BiNbO_6	−235
Lead cadmium manganese niobate?	$Pb_2Cd_{1/2}Mn_{1/2}NbO_6$	−237
Lead cadmium titanium tantalate?	$Pb_2Cd_{1/2}Ti_{1/2}TaO_6$	−253
Lead lithium niobium tungstate	$Pb(Li_{1/3}Nb_{1/3}W_{1/3})O_3$	110
Lead cadmium manganese tungstate	$Pb(Cd_{1/3}Mn_{1/3}W_{1/3})O_3$	300
Cadmium scandium niobate	$CdSc_{1/2}Nb_{1/2}O_3$	−203

B. Pyrochlore-type oxides

Lead (pyro)niobate?	$Pb_2Nb_2O_7$	−258

C. Nitrates

Rubidium nitrate?	$RbNO_3$	219

D. Potassium-dihydrogen-phosphate-type compounds

Ammonium dihydrogen phosphate (ADP)?	$NH_4H_2PO_4$	−125
Deuterated ADP	$ND_4D_2PO_4$	−31
Ammonium dihydrogen arsenate?	$NH_4H_2AsO_4$	−57
Deuterated ammonium arsenate	$ND_4D_2AsO_4$	31

E. Selenites

Caesium trihydrogen selenite?	$CsH_3(SeO_3)_3$	−128

Name (abbreviation)	Formula	T_c (°C)
F. Formates		
Cupric formate tetrahydrate	$Cu(HCOO)_2 \cdot 4H_2O$	-37
Cupric formate tetradeuterate	$Cu(DCOO)_2 \cdot 4D_2O$	-27
G. Miscellaneous materials		
Silver trihydrogen periodate?	$Ag_2H_3IO_6$	$-68 \ldots -28$
Silver trideuterium periodate?	$Ag_2D_3IO_6$	$-28 \ldots 12$
Ammonium trihydrogen periodate?	$(NH_4)_2H_3IO_6$	-20
Deutero-ammonium trideuterium periodate	$(ND_4)_2D_3IO_6$	-7
Hexafluor phosphate salt?	$NH_4PF_6 \cdot NH_4F$	-45
Lead orthovanadate?	$Pb_3V_2O_8$	≈ 100
Lead silicate?	Pb_4SiO_6	≈ 155

REFERENCES

ABE, R. (1956). *J. phys. Soc. Japan* **11**, 104.
—— (1972). *Prog. theor. Phys. Kyoto* **47**, 1200.
ABEL, W. R. (1971). *Phys. Rev.* **B4**, 2696.
ABRAGAM, A. (1961). *The principles of nuclear magnetism*. Clarendon Press, Oxford.
ABRAGAM, A., and BLEANEY, B. (1970). *Electron paramagnetic resonance of transition metal ions*. Clarendon Press, Oxford.
ABRAHAMS, S. C. (1971). *Mater. Res. Bull.* **6**, 881.
ABRAHAMS, S. C., and BERNSTEIN, J. L. (1967). *J. Phys. Chem. Solids* **28**, 1685.
—— (1974). *Ferroelectrics.* **6**, 247.
ABRAHAMS, S. C., BUEHLER, E., HAMILTON, W. C., and LAPLACA, S. J. (1973). *J. Phys. Chem Solids* **34**, 521.
ABRAHAMS, S. C., HAMILTON, W. C., and REDDY, J. M. (1966). *J. Phys. Chem. Solids* **27**, 1013.
ABRAHAMS, S. C., HAMILTON, W. C., and SEQUEIRA, A. (1967). *J. Phys. Chem. Solids* **28**, 1693.
ABRAHAMS, S. C., JAMIESON, P. B., and BERNSTEIN, J. L. (1971). *J. chem. Phys.* **54**, 2355.
ABRAHAMS, S. C., and KEVE, E. T. (1971). *Ferroelectrics.* **2**, 129
ABRAHAMS, S. C., KURTZ, S. K., and JAMIESON, P. B. (1968). *Phys. Rev.* **172**, 551.
ABRAHAMS, S. C., LEVINSTEIN, H. J., and REDDY, J. M. (1966). *J. Phys. Chem. Solids* **27**, 1019.
ABRAHAMS, S. C., REDDY J. M., and BERNSTEIN, J. L. (1966). *J. Phys. Chem. Solids* **27**, 997.
ABRAMS, R. L., and GLASS, A. M. (1969). *Appl. Phys. Lett.* **15**, 251.
ACKERMANN, W. (1915). *Annln. Phys.* **46**, 197.
ADRIAENSSENS, G. J. (1975). *Phys. Rev.* **B12**, 5116.
AGRAWAL, D. K., and PERRY, C. H. (1971). *Phys. Rev.* **B4**, 1893.
AHARONY, A. (1973 a). *Phys. Rev.* **B8**, 4270.
—— (1973 b). *Phys. Rev.* **B8**, 3342.
—— (1973 c). *Phys. Rev.* **B8**, 3363.
—— (1973 d). *Phys. Rev.* **B8**, 3349.
—— (1973 e). *Phys. Rev.* **B8**, 4314.
—— (1976). *Phys. Rev.* **B13**, 2092.
AHARONY, A., and BRUCE, A. D. (1974). *Phys. Rev. Lett.* **33**, 427.
AHARONY, A., and FISHER, M. E. (1973). *Phys. Rev.* **B8**, 3323.
AHLERS, G., KORNBLIT, A., and GUGGENHEIM, H. J. (1975). *Phys. Rev. Lett.* **34**, 1227.
AHTEE, M., GLAZER, A. M., and MEGAW, H. D. (1972). *Phil. Mag.* **26**, 995.
AIZU, K. (1965). *J. phys. Soc. Japan* **20**, 959.
—— (1966 a). *J. phys. Soc. Japan* **21**, 1240.
—— (1966 b). *Phys. Rev.* **146**, 423.
—— (1969). *J. phys. Soc. Japan* **27**, 387, 1171.
—— (1971). *J. phys. Soc. Japan* **31**, 802.
—— (1972 a). *J. phys. Soc. Japan* **33**, 629.
—— (1972 b). *J. phys. Soc. Japan* **32**, 135.

ALEMANY, C., MENDIOLA, J., JIMENEZ, B., and MAURER, E. (1973). *Ferroelectrics.* **5**, 11.

ALPERIN, H. A., PICKART, S. J., RHYNE, J. J., and MINKIEWICZ, V. J. (1972). *Phys. Lett.* **40A**, 295.

ALTING-MEES, H. R., and KODA, N. J. (1976). *Ferroelectrics* **11**, 323.

AMIN, M., and STRUKOV, B. A. (1968). *Fiz. tverd. Tela* **10**, 3158 (*Sov. Phys.–Solid St.* **10**, 2498 (1969)).

—— (1970). *Fiz. tverd. Tela* **12**, 2035 (*Sov. Phys.–Solid State* **12**, 1616 (1971)).

AMIT, D. J., BERGMAN, D. J. and IMRY, Y. (1973). *J. Phys. C Solid St.* **6**, 2685.

ANDERSON, J. R., BRADY, G. W., MERZ, W. J., and REMEIKA, J. P. (1955). *J. appl. Phys.* **26**, 1387.

ANDERSON, L. K. (1972). *Ferroelectrics* **3**, 69.

—— (1974). *Ferroelectrics* **7**, 55.

ANDERSON, P. W. (1960). In *Fizika Dielektrikov* (ed. G. I. Skanavi). Akad. Nauk. SSSR, Moscow.

ANDRADE, P. DA R., KATIYAR, R. S., and PORTO, S. P. S. (1974). *Ferroelectrics* **8**, 637.

ANLIKER, M., BRUGGER, H. R., and KÄNZIG, W. (1954). *Helv. Phys. Acta* **27**, 99.

ANTCLIFFE, G. A., BATE, R. T., and BUSS, D. D. (1973). *Solid State Commun,* **13**, 1003.

ASCHER, E., RIEDER, H., SCHMID, H., and STÖSSEL, H. (1966). *J. appl. Phys.* **37**, 1404.

ASHKIN, A., BOYD, G. D., DZIEDZIC, J. M., SMITH, R. G., BALLMAN, A. A., LEVINSTEIN, H. J., and NASSAU, K. (1966). *Appl. Phys. Lett.* **9**, 72.

ASTHEIMER, R. W., and SCHWARZ, F. (1968) *Appl. Opt.* **7**, 1687.

ATKIN, R. B. (1972). *Ferroelectrics* **3**, 213.

AUSTON, D. H., and GLASS, A. M. (1972) *Appl. Phys. Lett.* **20**, 398.

AUSTON, D. H., GLASS, A. M., and BALLMAN, A. A. (1972). *Phys. Rev. Lett.* **28**, 897.

AUSTON, D. H., GLASS, A. M., and LEFUR, P. (1973). *Appl. Phys. Lett.* **23**, 47.

AUTIER, A., and PETROFF, J. E. (1964) *C. R. Acad. Sci. Paris* **258**, 4238.

AVIRAM, I., GOSHEN, S., MUKAMEL, D., and SHTRIKMAN, S. (1975). *Phys. Rev.* **B12**, 438.

AVOGADRO, A., BONERA, G., BORSA, F., and RIGAMONTI, A. (1973). *Phys. Rev.* **B9**, 3905.

AXE, J. D. (1967). *Phys. Rev.* **157**, 429.

—— (1968). *Phys. Rev.* **167**, 573.

AXE, J. D., DORNER, B., and SHIRANE, G. (1971). *Phys. Rev. Lett.* **26**, 519.

—— (1972). *Phys. Rev.* **B6**, 1950.

AXE, J. D., HARADA, J., and SHIRANE, G. (1970). *Phys. Rev.* **B1**, 1227.

AXE, J. D., SHIRANE, G., and MÜLLER, K. A. (1969). *Phys. Rev.* **183**, 820.

BACON, G. E., and PEASE, R. S. (1953). *Proc. R. Soc. Lond.* **A220**, 397.

—— (1955). *Proc. R. Soc. Lond.* **A230**, 359.

BAER, W. S. (1966). *Phys. Rev.* **144**, 734.

BAGDASAROV, K. S., BOGOMOLOVA, G. A., KAMINSKII, A. A., MELESHINA, V. A., PROKHORTSEVA, T. M., and SHUVALOV, L. A. (1971). *Bull. Acad. Sci. USSR Phys. Ser.* **35**, 1681.

BALAGUROV, B. Y., and VAKS, V. G. (1973). *Zh. eksp. teor. Fiz.* **65**, 1600 (*Sov. Phys. JETP* **38**, 799 (1974)).

BALDWIN, G. C. (1969). *An introduction to nonlinear optics*. Plenum Press, New York.

BALLANTYNE, J. M. (1964). *Phys. Rev.* **136**, 429.

BALLHAUSEN, C. J. (1962). *Introduction to ligand field theory*. McGraw-Hill, New York.

BALLMAN, A. A. (1975). Private communication.

BALLMAN, A. A., and BROWN, H. (1967). *J. cryst. Growth* **1**, 311.

—— (1972). *Ferroelectrics* **4**, 189.

BALLMAN, A. A., BROWN, H., TIEN, P. K., and RIVA-SANSEVERINO, S. (1975 a). *J. cryst. Growth* to be published.

—— (1975 b). *J. cryst. Growth* **29**, 289.

BALLMAN, A. A., KURTZ, S. K., and BROWN, H. (1971). *J. cryst. Growth* **10**, 185.

BALLMAN, A. A., LEVINSTEIN, H. J., CAPIO, C. D., and BROWN, H. (1967). *J. Am. ceram. Soc.* **50**, 657.

BANTLE, W. (1942). *Helv. Phys. Acta* **15**, 373.

BARANSKII, K. N., SHUSTIN, O. A., VELICHKINA, T. S., and YAKOVLEV, I. A. (1962). *Zh. eskp. teor. Fiz.* **43**, 730 (*Sov. Phys.–JETP* **16**, 518 (1963)).

BARKER, A. S. (1964). *Phys. Rev.* **136**, A1290.

—— (1966 a). *Phys. Rev.* **145**, 391.

—— (1966 b). In *Colloquium on the optical properties and electronic structure of metals and alloys* (ed. F. Abeles), p. 452. North-Holland, Amsterdam.

—— (1967). In *Ferroelectricity* (ed. E. F. Weller). Elsevier, New York.

—— (1976). To be published.

BARKER, A. S., BALLMAN, A. A., and DITZENBERGER, J. A. (1970). *Phys. Rev.* **B2**, 4233.

BARKER, A. S., and HOPFIELD, J. J. (1964). *Phys. Rev.* **135**, A1732.

BARKER, A. S., and LOUDON, R. (1967). *Phys. Rev.* **158**, 433.

—— (1972). *Rev. mod. Phys.* **44**, 18.

BARKER, A. A., and TINKHAM, M. (1963). *J. chem. Phys.* **38**, 2257.

BARKHAUSEN, H. (1919). *Phys. Szk.* **20**, 401.

BARNOSKI, M. K., and BALLANTYNE, J. M. (1968). *Phys. Rev.* **174**, 946.

BARNS, R. L., and CARRUTHERS, J. R. (1970). *J. appl. Crystallog.* **3**, 395.

BARRETT, J. H. (1952). *Phys. Rev.* **86**, 118.

BARTIS, F. J. (1973). *J. Phys. C. Solid St.* **6**, L295.

BATE, R. T., CARTER, D. L., and WROBEL, J. S. (1970). *Phys. Rev. Lett.* **25**, 159.

BATRA, I. P., WURFEL, P., and SILVERMAN, B. D. (1973). *Phys. Rev. Lett.* **30**, 384.

BAYER, H. (1951). *Z. Phys.* **130**, 227.

BENEPE, J. W., and REESE, W. (1971). *Phys. Rev.* **B3**, 3032.

BENGUIGUI, L. (1975 a). *Phys. Rev.* **B11**, 4547.

—— (1975 b). *J. Phys. C Solid St.* **8**, 17.

BERCHA, D. M., SLIVKA, V. Y., SYRBU, N. N., TURYANITSA, I. D., and CHEPUR, D. V. (1971). *Sov. Phys–Solid St.* **13**, 217.

BERGLUND, C. N., and BAER, W. S. (1967). *Phys. Rev.* **157**, 358.

BERGLUND, C. M., and BRAUN, H. J. (1967). *Phys. Rev.* **164**, 790.

BERGMAN, J. G. (1976). To be published.

BERGMAN, J. G., ASHKIN, A., BALLMAN, A. A., DZIEDZIC, J. M., LEVINSTEIN, H. J., and SMITH, R. G. (1968). *Appl. Phys. Lett.* **12**, 92.

BERGMAN, J. G., and CRANE, G. R. (1975). *J. solid st. Chem.* **12**, 172.

BERGMAN, J. G., CRANE, G. R., BALLMAN, A. A., and O'BRYAN, H. M. O. (1972). *Appl. Phys. Lett.* **21**, 497.

BERGMAN, J. G., McFEE, J. H., and CRANE, G. R. (1971). *Appl. Phys. Lett.* **18**, 203.

BERLINCOURT, D. (1956). *IRE trans. ultrasonic Engng* **UE-4**, 53.

BERLINCOURT, D., CMOLIK, C., and JAFFE, H. (1960). *Proc. IRE* **48**, 220.

BERLINCOURT, D., and KRUEGER, H. H. A. (1959). *J. appl. Phys.* **30**, 1804.

BERLINCOURT, D., KRUEGER, H. H. A., and JAFFE, B. (1964). *J. Phys. Chem. Solids* **25**, 659.

BERSUKER, I. B., and VEKHTER, B. G. (1967). *Fiz. tverd. Tela*, **9**, 2652.

—— (1969). *Izv. Akad. Nauk. SSSR Ser Fiz.* **33**, 199.

BERSUKER, I. B., VEKHTER, B. G., and MUZALEVSKII, A. A. (1974). *Ferroelectrics* **6**, 197.

BHIDE, V. G., and HEGDE, M. S. (1972). *Phys. Rev.* **B5,** 3488.

BHIDE, V. G., and MULTANI, M. S. (1965). *Phys. Rev.* **139**, A1983.

BICKNELL, W. E., YAP, B. K., and PETERS, C. J. (1967). *Proc. IEEE* **55**, 225.

BIERLY, J. N., MULDAWER, L., and BECKMAN, O. (1963). *Acta. Metall.* **11**, 447.

BILLIG, E., and PLESSNER, K. W. (1951). *Proc. phys. Soc.* **64B**, 361.

BINDER, K., MEISSNER, G. and MAIS, H. (1976). *Phys. Rev.* **B13**, 4890.

BINI, S., and CAPOLETTI, R. (1973). In *Electrets* (ed. M. M. Perlman), p. 66. Electrochemical Society, New York.

BIRGENEAU, R. L., KJEMS, J. K., SHIRANE, G., and VAN UITERT, L. G. (1974). *Phys. Rev.* **B10**, 2512.

BIRMAN, J. L. (1973). *Phys. Lett.* **45A**, 196.

BJORKSTAM, J. L. (1967). *Phys. Rev.* **153**, 599.

BJORKSTAM, J. L., and OETTEL, R. E. (1966). *Proc. 1st Int. Conf. on Ferroelectricity* (ed. V. Dvorak), vol. 2, p. 91. publ. Inst of Phys. of Czech. Acad. Sci.

BLANK, H., and AMELINCKX, S. (1963). *Appl. Phys. Lett.* **2**, 140.

BLINC, R. (1960). *J. Phys. Chem. Solids* **13**, 204.

—— (1968). In *Magnetic resonance* (ed. J. S. Waugh). Academic Press, New York.

—— (1975). *Phys. Stus. Solidi.* **70b**, K29.

BLINC, R., BRENMAN, M., and WAUGH, J. S. (1961). *J. chem. Phys.* **35**, 1770.

BLINC, R., BURGAR, M., and LEVSTIK, A. (1970). *Commun. solid state Phys.* **8**, 317.

—— (1973). *Commun. solid state Phys.* **12**, 573.

BLINC, R., CEVC, P., and SCHARA, M. (1967). *Phys. Rev.* **159**, 411.

BLINC, R., LAVRENCIC, B., LEVSTIK, I., SMOLEJ, V., and ZEKS, B. (1973). *Phys. Stus. Solidi* **60b**, 255.

BLINC, R., MALI, M., OSREDKAR, R., PRELENSNIK, A., ZUPANCIC, I., and EHRENBERG, L. (1971). *J. chem. Phys.* **55**, 4843.

BLINC, R., PINTAR, M., and ZUPANCIC, I. (1967). *J. Phys. Chem. Solids* **28**, 405.

BLINC, R., STEPISNIK, J., JAMSEK-VILFAN, M., and ZUMER, S. (1971). *J. chem. Phys.* **54**, 187.

BLINC, R., and SVETINA, S. (1966). *Phys. Rev.* **147**, 423, 430.

BLINC, R., SVETINA, S., and ZEKS, B. (1972). *Commun. solid state Phys.* **10**, 387.

BLINC, R., and ZEKS, B. (1972). *Adv. Phys,* **21**, 693.

—— (1974). *Soft modes in ferroelectrics and antiferroelectrics.* Elsevier, New York.

BLINC, R., ZUMER, S., and LAHAJNAR, G. (1970). *Phys. Rev.* **B1**, 4456.

BLOEMBERGEN, N. (1965). *Nonlinear optics.* Benjamin, New York.

BLOOMFIELD, P. E., LEFKOWITZ, I., and ARONOFF, A. D. (1971). *Phys. Rev.* **B4**, 974.

BLUME, M., and HUBBARD, J. (1970). *Phys. Rev.* **B1**, 3815.

BOCCARA, N., and SARMA, G. (1965 a). *Physics* **1**, 219.

—— (1965 b) In *Inelastic scattering of neutrons*, vol. 1, p. 313. IAEA, Vienna.

BOGOMOLOV, A. A., IVANOV, V. V., and RUDYAK, V. M. (1969). *Kristallog-rafiya* **14**, 1033 (*Sov. Phys.–Crystallography* **14**, 894 (1970)).
BOND, W. L. (1960). *Acta Crystallog.* **13**, 814.
BONERA, G., BORSA, F., and RIGAMONTI, A. (1970). *Phys. Rev.* **B2**, 2784.
BONILLA, I. R., HOLZ, A., and RUTT, H. N. (1974). *Phys. St. Solidi* **63b**, 297.
BORCHARDT, H. J., and BIERSTEDT, P. E. (1966). *Appl. Phys. Lett.* **8**, 50.
——(1967). *J. appl. Phys.* **38**, 2057.
BORDEAUX, D., BORNAREL, J., CAPIOMONT, A., LAJZEROWICZ-BONNETEAU, J., LAJZEROWICZ, J., and LEGRAND, J. F. (1973). *Phys. Rev. Lett.* **31**, 314.
BORN, M. (1945). *Rev. mod. Phys.* **17**, 245.
BORN, M., and HUANG, K. (1954). *The dynamical theory of crystal lattices.* Clarendon Press, Oxford.
BORN, M., and KARMAN, T. VON (1912). *Phys. Z.* **13**, 297.
BORRELLI, N. F., and LAYTON, M. M. (1971). *J. non-cryst. Solids* **6**, 197.
BORSA, F., CRIPPA, M. L., and DERIGHETTI, B. (1971). *Phys. Lett.* **34A**, 5.
BOUTIN, H., FRAZER, B. C., and JONA, F. (1963). *J. Phys. Chem. Solids* **24**, 1341.
BOYD, G. D., MILLER, R. C., NASSAU, K., BOND, W. L., and SAVAGE, A. (1964). *Appl. Phys. Lett.* **5**, 234.
BOYER, L. L., and HARDY, J. R. (1973). *Phys. Rev.* **B8**, 2205.
BRENIG, W. (1963). *Z. Phys.* **171**, 60.
BREWS, J. R. (1967). *Phys. Rev. Lett.* **18**, 662.
BREZINA, B., and GLOGAROVA, M. (1972). *Phys. St. Solidi*, **11a**, K.39.
BRIDENBAUGH, P. M., CARRUTHERS, J. R., DZIEDZIC, J. M., and NASH, F. R. (1970). *Appl. Phys. Lett.* **17**, 104.
BRILLSON, L. J., BURSTEIN, E., and MULDAWER, L. (1974). *Phys. Rev.* **B9**, 1547.
BROBERG, T. W., SHE, C. Y., WALL, L. S., and EDWARDS, D. F. (1972). *Phys. Rev.* **B6**, 3332.
BRODY, E. M., and CUMMINS, H. Z. (1968). *Phys. Rev. Lett.* **21**, 1263.
——(1969). *Phys. Rev. Lett.* **23**, 1039.
——(1974). *Phys. Rev.* **B9**, 179.
BROPHY, J. J. (1965). In *Fluctuation phenomena in solids* (ed. R. E. Burgess). Academic Press, New York.
BROPHY, J. J., and WEBB, S. L. (1962). *Phys. Rev.* **128**, 584.
BROSOWSKI, G., BUCHHEIT, W., MÜLLER, D., and PETERSSON, J. (1974). *Phys. Stus. Solidi* **62b**, 93.
BROUT, R. (1965). *Phase transitions.* Benjamin, New York.
BROUT, R., MÜLLER, K. A., and THOMAS, H. (1966). *Commun. solid state Phys.* **4**, 507.
BRUCE, A. D., and AHARONY, A. (1974). *Phys. Rev.* **B10**, 2078.
BRUCE, A. D., and COWLEY, R. A. (1973). *J. Phys. C Solid St.* **6**, 2422.
BRUNSTEIN, M., GRINBERG, J., PELAH, I., and WIENER, E. (1970). *Commun. solid state Phys.* **8**, 1211.
BUCKINGHAM, A. D., and ORR, B. J. (1967). *Q. Rev. chem. Soc. Lond.* **21**, 195.
BUERGER, M. J. (1963). *Elementary crystallography.* John Wiley, New York.
BURGESS, R. E. (1958). *Can. J. Phys.* **36**, 1569.
BURKE, W. J., and PRESSLEY, R. J. (1971). *Commun. solid state Phys.* **9**, 191.
BURNS, G. (1972), *Appl. Phys. Lett.* **20**, 230.
——(1973). *Phys. Lett.* **43A**, 271.
——(1974). *Phys. Rev.* **B10**, 1951.
BURNS, G., and BURSTEIN, E. (1974). *Ferroelectrics* **7**, 297.
BURNS, G., O'KANE, D. F. (1969). *Phys. Lett.* **28A**, 776.

BURNS, G., and SCOTT, B. A. (1970). *Phys. Rev. Lett.* **25**, 167.
—— (1971). *Commun. solid state Phys.* **9**, 813.
—— (1973 a). *Phys. Rev.* **B7**, 3088.
—— (1973 b). *Commun. solid state Phys.* **13**, 417, 423.
BURSIAN, E. V., GIRSHBERG, Y. G., and STAROV, E. N. (1972). *Fiz. tverd. Tela* **14**, 1019 (*Sov. Phys.–Solid St.* **14**, 872 (1972)).
BUSCH, G. (1938). *Helv. Phys. Acta* **11**, 269.
BUSCH, G., and SCHERRER, P. (1935). *Naturwissenschaft.* **23**, 737.
BUTLER, J. F. (1969). *Commun. solid state Phys.* **7**, 909.
BYE, K. L., and KEVE, E. T. (1972). *Ferroelectrics* **4**, 87.
BYE, K. L., WHIPPS, P. W., and KEVE, E. T. (1972). *Ferroelectrics* **4**, 253.
BYE, K. L., WHIPPS, P. W., KEVE, E. T., and JOSEY, M. R. (1974). *Ferroelectrics* **7**, 179.
BYER, R. L. and ROUNDY, C. B., (1972). *Ferroelectrics* **3**, 333.
CALLABY, D. R. (1966). *J. appl. Phys.* **37**, 2295.
CAMLIBEL, I. (1969). *J. appl. Phys.* **40**, 1690.
CANNER, J. P., YAGNIK, C. M., GERSON, R., and JAMES, W. J. (1971). *J. appl. Phys.* **42**, 4708.
CAPIZZI, M., and FROVA, A. (1971). *Nuovo Cim.* **5B**, 181.
CARDONA, M., (1965). *Phys. Rev.* **140** A651.
CARL, K., and GIESEN, K. (1973). *Proc. IEEE* **61**, 967.
CARRUTHERS, J. R., and GRASSO, M. (1970), *J. electrochem. Soc.* **117**, 1426.
CARRUTHERS, J. R., PETERSON, G. E., GRASSO, M., and BRIDENBAUGH, P. M. (1971). *J. appl. Phys.* **42**, 1846.
CARSLAW, H. S., and JAEGER, J. C. (1959). *Conduction of heat in solids.* Clarendon Press, Oxford.
CHANG, I. F., and MITRA, S. S. (1968). *Phys. Rev.* **172**, 924.
CHANG, T. S. (1972). *J. appl. Phys.* **43**, 3591.
CHAPMAN, D. W. (1969). *J. appl. Phys.* **40**, 2381.
CHAVES, A. S., and PORTO, S. P. S. (1973). *Commun. solid state Phys.* **13**, 865.
CHEN, A., and CHERNOW, F. (1967). *Phys. Rev.* **154**, 493.
CHEN, F. S. (1969) *J. appl. Phys.* **40**, 3389.
CHEN, F. S., GEUSIC, J. E., KURTZ, S. K., SKINNER, J. G., and WEMPLE, S. H. (1966). *J. appl. Phys.* **37**, 388.
CHEN, E. S., LAMACCHIA, J. T., and FRASER, D. B. (1968) *Appl. Phys. Lett.* **13**, 223.
CHENSKII, E. V. (1972). *Fiv. tverd. Tela* **14**, 2241. (*Sov. Phys.–Solid St.* **14**, 1940 (1973)).
CHIBA, T. (1964). *J. chem. Phys.* **41**, 1352.
CHOWDHURY, M. R., PECKHAM, G. E., ROSS, R. T., and SAUNDERSON, D. H. (1974). *J. Phys. C Solid State* **7**, L99.
CHRPA, E., IHRINGER, H., JAGODZINSKI, H., and KNEIFEL, A. (1972). *Z. Naturf.* **27a**, 469.
CHYNOWETH, A. G. (1956 a). *J. appl. Phys.* **27**, 78.
—— (1956 b). *Phys. Rev.* **102**, 705.
—— (1958). *Phys. Rev.* **110**, 1316.
—— (1959). *Phys. Rev.* **113**, 159.
—— (1960). *Phys. Rev.* **117**, 1235.
CHYNOWETH, A. G., and FELDMANN, W. L. (1960). *J. Phys. Chem. Solids* **15**, 225.
CLARK, M. G., DISALVO, F. J., GLASS, A. M., and PETERSON, G. E. (1973). *J. chem. Phys.* **59**, 6209.

CLARKE, R., and GLAZER, A. M. (1974). *J. Phys. C Solid State* **7**, 2147.
CLAUS, R., BORSTEL, G., WIESENDANGER, E., and STEFFAN, L. (1972). *Z. Naturf.* **27a**, 1187.
CLAUSER, M. J. (1970). *Phys. Rev.* **B1**, 357.
COCHRAN, W. (1960). *Adv. Phys.* **9**, 387.
—— (1961). *Adv. Phys.* **10**, 401.
—— (1963). *Rep. Prog. Phys.* **26**, 1.
—— (1968). *Phys. Stus. Solidi* **30**, K157.
—— (1969). *Adv. Phys.* **18**, 157.
—— (1971). In *Structural phase transitions and soft modes* (ed. E. J. Samuelsen). Universitetsvorleiget, Oslo.
COCHRAN, W., and COWLEY, R. A. (1962). *J. Phys. Chem. Solids* **23**, 447.
COCHRAN, W., COWLEY, R. A., DOLLING, G., and ELCOMBE, M. M. (1966). *Proc. R. Soc. Lond.* **293**, 433.
COCHRAN, W., and ZIA, A. (1968). *Phys. Stus. Solidi* **25**, 273.
COHEN, M., and EINSTEIN, T. L. (1973). *Phys. Rev.* **B7**, 1932.
COHEN, M. H., and KEFFER, F. (1955). *Phys. Rev.* **99**, 1128.
COLE, K. S., and COLE, R. H. (1941). *J. chem. Phys.* **9**, 341.
COLLINS, M. R., and TEH, H. C. (1973). *Phys. Rev. Lett.* **30**, 781.
COMES, R., DENOYER, F., DESCHAMPS, L., and LAMBERT, M. (1971). *Phys. Lett.* **34A**, 65.
COMES, R., LAMBERT, M., and GUINIER, A. (1968). *Commun. solid state Phys.* **6**, 715.
—— (1970). *Acta Crystallog.* **A26**, 244.
COMES, R., and SHIRANE, G., (1972). *Phys. Rev.* **B5**, 1886.
CONKLIN, T., SINGER, B., CROWELL, M. H., and KURCZEWSKI, P. (1974). *Tech. Digest IEDM*, p. 451.
CONNOLLY, T. F., TURNER, E., and HAWKINS, D. T. (1970, 1974). *Solid state physics literature guides*, vols. 1, 6. Plenum Press, New York.
COOMBS, G. J., and COWLEY, R. A. (1973). *J. Phys. C Solid State* **6**, 121, 143.
COOPER, J. (1962). *Rev. sci. Instrum.* **33**, 92.
COURDILLE, J. M., and DUMAS, J. (1971). *Commun. solid state Phys.* **9**, 609.
COUFOVA, P., and AREND, H. (1961). *Czech. J. Phys.* **B11**, 416.
—— (1962). *Czech. J. Phys.* **B12**, 308.
COWLEY, R. A. (1964). *Phys. Rev.* **134**, A981.
—— (1965). *Phil. Mag.* **11**, 673.
—— (1970). *J. phys. Soc. Japan* **28S**, 239.
—— (1976). *Phys. Rev. Lett.* **36**, 744.
COWLEY, R. A., and BRUCE, A. D. (1973). *J. Phys. C Solid State* **6**, L191.
COWLEY, R. A., BUYERS, W. J. L., and DOLLING, G. (1969). *Commun. solid state Phys.* **7**, 181.
COWLEY, R. A., COOMBS, G. J., KATIYAR, R. S., RYAN, J. F., and SCOTT, J. F. (1971). *J. Phys. C Solid St.* **4**, L203.
CRANK, J. (1970). *The mathematics of diffusion.* Clarendon Press, Oxford.
CROSS, L. E. (1956). *Phil. Mag.* **1**, 76.
—— (1967). *J. phys. Soc. Japan* **23**, 77.
CROSS, L. E., FOUSKOVA, A., and CUMMINS, S. E. (1968). *Phys. Rev. Lett.* **21**, 812.
CUMMINS, H. Z. (1969). In *Quantum optics* (ed. R. J. Glauber). Academic Press, New York.
CUMMINS, S. E. (1970). *Ferroelectrics*, **1**, 11.
CUMMINS, S. E., and CROSS, L. E. (1968). *J. appl. Phys.* **39**, 2268.

CUMMINS, S. E., and LUKE, T. E. (1972). *Ferroelectrics* **3**, 125.

CURIE, J., and CURIE, P. (1880). *C. R. Acad. Sci. Paris* **91**, 294. For a historical review see S. B. Lange, *Sourcebook of pyroelectricity*. Gordon and Breach. London (1974).

CURRAT, R., COMES, R., DORNER, B., and WIESENDANGER, E. (1974). *J. Phys. C. Solid State* **7**, 2521.

DALAL, N. S., and MCDOWELL, C. A. (1972). *Phys. Rev.* **B5**, 1074.

DALISA, A. L., and SEYMOUR, R. J. (1973). *Proc. IEEE* **61**, 981.

DALTON, N. W., and DOMB, C. (1966). *Proc. phys. Soc.* **89**, 859, 873.

DARLINGTON, C. N. W. (1971). *Ferroelectrics* **3**, 9.

DARLINGTON, C. N. W., FITZGERALD, W. J., and O'CONNOR, D. A. (1975). *Phys. Lett.* **54A**, 35.

DEGUCHI, K., and NAKAMURA, E. (1972 a). *Phys. Rev.* **B5**, 1072.

—— (1972 b). *Acta Crystallogr.* **A28**, S176.

DEMUROV, D. G., and VENEVTSEV, Y. N. (1971). *Fiz. tverd. Tela* **13**, 669. (*Sov. Phys.–Solid St.* **13**, 553 (1971)).

DENOYER, F., COMES, R., and LAMBERT, M. (1970). *Commun. solid state Phys.* **8**, 1979

—— (1971). *Acta Crystallog.* **A27**, 414.

DENTON, R. T. (1972). In *Laser Handbook* (eds. F. T. Arecchi and E. O. Shulz-Dubois), Elsevier, New York.

DENTON, R. T., CHEN, F. S., and BALLMAN, A. A. (1967). *J. appl. Phys.* **38**; 1611.

DEVINE, S., and PECKHAM, G. (1971). *J. Phys. C Solid State* **4**, 1091.

DEVONSHIRE, A. F. (1949). *Phil. Mag.* **40**, 1040.

—— (1951). *Phil. Mag.* **42**, 1065.

—— (1954). *Adv. Phys.* **3**, 85.

DICK, B. G., and OVERHAUSER, A. W. (1958). *Phys. Rev.* **112**, 90.

DICKENS, P. G., and WHITTINGHAM, M. S. (1968). *Q. Rev. Lond.* **22**, 30.

DIDOMENICO, M., EIBSCHÜTZ, M., GUGGENHEIM, H. J., and CAMLIBEL, I. (1969). *Commun. solid state Phys.* **7**, 1119.

DIDOMENICO, M., and WEMPLE, S. H. (1967). *Phys. Rev.* **155**, 539.

—— (1968). *Phys. Rev.* **166**, 565.

—— (1969). *J. appl. Phys.* **40**, 720.

DIDOMENICO, M., WEMPLE, S. H., PORTO, S. P. S., and BAUMAN, R. P. (1968). *Phys. Rev.* **174**, 522.

DIMIC, V., OSREDKAR, M., SLAK, J., and KANDUSAR, A. (1973). *Phys. Stus. Solidi* **59b**, 471.

DOLINO, G. (1973). *Appl. Phys. Lett.* **22**, 123.

DOMB, C., and GREEN, M. S. (1972). *Phase transitions and critical phenomena.* Academic Press, New York.

DOMB, E. R., MIHALISIN, T., and SKALYO, J. (1973). *Phys. Rev.* **B8**, 5837.

DÖNGES, E. (1950). *Z. anorg. Chem.* **263**, 112.

—— (1951). *Z. anorg. Chem.* **265**, 56.

DORNER, B., AXE, J., and SHIRANE, G. (1972). *Phys. Rev.* **B6**, 1950.

DOUGHERTY, J. P., SAWAGUCHI, E., and CROSS, L. E. (1972). *Appl. Phys. Lett.* **20**, 364.

DRAEGERT, D. A., and SINGH, S. (1971). *Commun. solid state Phys.* **9**, 595.

DROUGARD, M. E., LANDAUER, R., and YOUNG, D. R. (1955). *Phys. Rev.* **98**, 1010.

DURAND, G., and LEE, C. H. (1968). *Mol. Crystallog.* **5**, 171.

DVORAK, V. (1966). *Phys. Stus. Solidi* **14**, K161.
—— (1967). *Can. J. Phys.* **45**, 3903.
—— (1968). *Phys. Rev.* **167**, 525.
—— (1970). *Czech. J. Phys.* **B20**, 1.
—— (1971 a). *Phys. Stus. Solidi* **45b**, 147.
—— (1971 b). *Phys. Stus. Solidi* **46b**, 763.
—— (1971 c). *Czech. J. Phys.* **B21**, 1250.
—— (1971 d). *Czech. J. Phys.* **B21**, 836.
—— (1972 a). *Phys. Stus Solidi* **52b**, 93.
—— (1972 b). *Phys. Stus. Solidi* **51b**, K.129.
—— (1974). *Ferroelectrics* **7**, 1.
DVORAK, V., and PETZELT, J. (1971 a). *Phys. Lett.* **35A**, 209.
—— (1971 b). *Czech. J. Phys.* **B21**, 1141.
EGUCHI, M. (1925). *Phil. Mag*, **49**, 178.
EISENRIEGLER, E. (1974). *Phys. Rev.* **B9**, 1029.
ELLIOTT, R. J. (1971). In *Structural phase transitions and soft modes* (ed. E. J. Samuelsen), p. 235. Universitetsvorleiget, Oslo.
ELLIOTT, R. J., HARLEY, R. T., HAYES, W., and SMITH, S. R. P. (1972). *Proc. R. Soc.* **A328**, 217.
ELLIOTT, R. J., and WOOD, C. (1971). *J. Phys. C Solid State* **4**, 2359.
ENGLISH, F. L. (1968). *J. appl. Phys.* **39**, 3231.
ENNS, R. H., and HAERING, R. R. (1966). *Phys. Lett.* **21**, 534.
ENZ, C. P. (1974). *Commun. solid state Phys.* **15**, 459.
EVLANOVA, N. F., KOVALEV, A. S., KOPTSIK, V. A., KORNIENKO, L. S., PROKHOROV, A. M., and RASHKOVICH, L. N. (1967). *Zh. eksp. teor. Fiz. Pis.* **5**, 351 (*Sov. Phys.-JETP Lett.* **5**, 291 (1967)).
FAIRALL, C. W., and REESE, W. (1972). *Phys. Rev.* **B6**, 193.
—— (1974). *Phys. Rev.* **B10**, 882.
—— (1975). *Phys. Rev.* **B11**, 2066.
FATUZZO, E. (1960). *Proc. phys. Soc.* **76**, 797.
FATUZZO, E., HARBEKE, G., MERZ, W. J., NITSCHE, R., ROETSCHI, H., and RUPPEL, W. (1962). *Phys. Rev.* **127**, 2036.
FATUZZO, E., and MERZ, W. J. (1967). *Ferroelectricity*. North-Holland, Amsterdam.
FAY, H., ALFORD, W. J., and DESS, H. M. (1968). *Appl. Phys. Lett.* **12**, 89.
FEDER, J. (1971). *Commun. solid state Phys.* **9**, 2021.
—— (1973). *Commun. solid state Phys.* **13**, 1039.
FEDER, J., and PYTTE, E. (1970). *Phys. Rev.* **B1**, 4803.
FEINLEIB, J., and OLIVER, D. S. (1972). *Appl. Opt.* **11**, 2752.
FERER, M., MOORE, M. A., and WORTIS, M. (1973). *Phys. Rev.* **B8**, 5205.
FERRELL, R. A., MENYHARD, N., SCHMIDT, H., SCHWABL, F., and SZEPFALUSY, P. (1967). *Phys. Rev. Lett.* **18**, 891.
—— (1968). *Ann. Phys.* **47**, 565.
FEUERSANGER, A. E. (1969). In *Thin film dielectrics* (ed. F. Vratny), p. 209. Electrochemical, New York.
FISCHER, B., BAUERLE, D., and BUCKEL, W. J. (1974). *Commun. solid state Phys.* **14**, 291.
FISHER, M. E. (1964). *J. math. Phys.* **5**, 944.
—— (1968). *Phys. Rev.* **176**, 257.
FISHER, M. E., MA, S., and NICKEL, B. G. (1972). *Phys. Rev. Lett.* **29**, 917.
FLEURY, P. A. (1970). *Commun solid state Phys.* **8**, 601.

FLEURY, P. A., and LAZAY, P. D. (1971). *Phys. Rev. Lett.* **26**, 1331.

FLEURY, P. A. and LYONS, K. B. (1976). *Phys. Rev. Lett.* (in press).

FLEURY, P. A., SCOTT, J. F., and WORLOCK, J. M. (1968). *Phys. Rev. Lett.* **21**, 16.

FLEURY, P. A., and WORLOCK, J. M. (1967). *Phys. Rev. Lett.* **18**, 665.

—— (1968). *Phys. Rev.* **174**, 613.

FLUGGE, S., and MEYENN, K. V. (1972). *Z. Phys.* **253**, 369.

FLYTZANIS, C., and DUCUING, J. (1969). *Phys. Rev.* **178**, 1218.

FONTANA, M. P., and LAMBERT, M. (1972). *Commun. solid state. Phys.* **10**, 1.

FORSBERGH, P. W. (1956). *Hand. Phys.* **17**. Springer Verlag, Berlin.

FOSTER, N. F. (1970). In *Handbook of thin film technology* (eds. L. I. Maissel and R. Glang), p. 15. McGraw-Hill, New York.

FOUSEK, J. CROSS, L. E., and SEELY, K. (1970), *Ferroelectrics* **1**, 63.

FOUSEK, J., and JANOVEK, V. (1969). *J. appl. Phys.* **40**, 135.

FOUSEK, J. SAFRANKOVA, M., and KACZER, J. (1966). *Appl. Phys. Lett.* **8**, 192.

FRANCOMBE, M. H. (1960). *Acta Crystallog.* **13**, 131.

—— (1972). *Ferroelectrics.* **3**, 199.

FRANCOMBE, M. H., and LEWIS, B. (1958). *Acta Crystallog.* **11**, 696.

FRANZ, W. (1958). *Z. Naturf.* **13A**, 484.

FRASER, D. B. (1973). *Proc. IEEE* **61**, 1013.

FRAZER, B. C. (1971). In *Structural phase transitions and soft modes* (ed. E. J. Samuelsen), p. 43. Universitetsvorleiget, Oslo.

FRAZER, B. C., DANNER, H. R., and PEPINSKY, R. (1955). *Phys. Rev.* **100**, 745.

FRAZER, B. C., and PEPINSKY, R. (1953). *Acta Crystallog.* **6**, 273.

FREDERIKSE, H. P. R. (1969). In *Electronic structures in solids* (ed. E. D. Haidemenakis). Plenum Press, New York.

FREDERIKSE, H. P. R., HOSLER, W. R., THURBER, W. R., BABISKIN, J., and SIEBENMANN, P. G. (1967). *Phys. Rev.* **158**, 775.

FREDKIN, D. R., and WERTHAMER, N. R. (1965). *Phys. Rev.* **138**, A1527.

FRENZEL, C., PIETRASS, B., and HEGENBARTH, E. (1970). *Phys. Stus. Solidi* **2a**, 273.

FREUND, I. (1967). *Phys. Rev. Lett.* **19**, 1288.

FRIDKIN, V. M., GREKOV, A. A., KOSONOGOV, N. A., and VOLK, T. R. (1972) *Ferroelectrics* **4**, 169.

FRIDKIN, V. M., GREKOV, A. A., RODIN, A. I., SAVCHENKO, E. A., and VOLK T. R. (1973). *Ferroelectrics* **6**, 71.

FRIDKIN, V. M., and ZHELUDEV, I. S. (1961). *Photoelectrets and the electrophoto- graphic process.* Van Nostrand, New York.

FRIEDEL, G. (1922). *Annln. Phys.* **18**, 273.

FRIEDMAN, Z., and FELSTEINER, J. (1974). *Phys. Rev.* **B9**, 337.

FRITZ, I. J., and CUMMINS, H. Z. (1972). *Phys. Rev. Lett.* **28**, 96.

FROVA, A., and BODDY, P. J. (1967). *Phys. Rev.* **153**, 606.

FUJII, Y., HOSHINO, S., YAMADA, Y., and SHIRANE, G. (1974) *Phys. Rev.* **B9**, 4549.

FUJII, Y., and YAMADA, Y. (1971). *J. phys. Soc. Japan* **30**, 1676.

FUJIWARA, T. (1970). *J. phys. Soc. Japan.* **29**, 1282.

FURMAN, E., BRAFMAN, O., and MAKOVSKY, J. (1973). *Phys. Rev.* **B8**, 2341.

FURUHATA, Y., and TORIYAMA, K. (1973). *Appl. Phys. Lett.* **23**, 361.

FURUKAWA, M., FUJIMORI, Y., and HIRAKAWA, K. (1970). *J. phys. Soc. Japar* **29**, 1528.

GÄHWILLER, C. (1965). *Helv. Phys. Acta* **38**, 361.

—— (1967). *Commun. solid state Phys.* **5**, 65.

3

GALANOV, E. K., and KISLOVSKII, L. D. (1965). *Kristallografiya* **10**, 209 (*Sov. Phys.-Crystallography*, **10**, 156 (1965)).

GAMMON, R. W., and CUMMINS, H. Z. (1966). *Phys. Rev. Lett.* **17**, 193.

GARBER, S. R., and SMOLENKO, L. A. (1973). *Zh. eksp. teor. Fiz.* **64**, 181. (*Sov. Phys.-JETP* **37**, 94 (1973)).

GARLAND, C. W., and NOVOTNY, D. B. (1969). *Phys. Rev.* **177**, 971.

GARRETT, G. C. B. (1968). *IEEE J. Quant. Electron.* **4**, 70.

GARRETT, G. C. B., and ROBINSON, F. N. H. (1986). *IEEE J. Quantum Electron.* **2**, 328.

GAUGAIN, J. M. (1856). *C. R. Acad. Sci Paris* **42**, 1264; **43**, 916.

GAVRILYACHENKO, V. G., SPINKO, V. G., MARTYNENKO, R. I., and FESENKO, E. G. (1970). *Fiz. tverd. Tela* **12**, 1532 (*Sov. Phys.-Solid State.* **12**, 1203 (1970)).

GEBBIE, H. A., STONE, N. W. B., PUTLEY, E. H., and SHAW, N. (1967). *Nature* (*Lond.*) **214**, 165.

GENIN, D. J., O'REILLY, D. E., and TSANG, T. (1968). *Phys. Rev.* **167**, 445.

GENNES, P. G. DE, (1963). *Commun. solid st. Phys.* **1**, 132.

GERSON, R. (1960). *J. appl. Phys.* **31**, 1615.

GERSON, R., and JAFFE, H. (1963). *J. Phys. Chem. Solids* **24**, 979.

GERTHSEN, P., GROTH, R., HARDTL, K. H., HEESE, D., and REIK, H. G. (1965). *Commun. solid st. Phys.* **3**, 165.

GESI, K. (1965). *J. phys. Soc. Japan* **20**, 1764.

—— (1969). *J. phys. Soc. Japan* **26**, 953.

—— (1970). *J. phys. Soc. Japan.* **28**, 1365.

GESI, K., AXE, J. D., SHIRANE, G., and LINZ, A. (1972). *Phys. Rev.* **B5**, 1933.

GEUSIC, J. E., LEVINSTEIN, H. J., RUBIN, J. J., SINGH, S., and VAN UITERT, L. G. (1968). *Appl. Phys. Lett.* **11**, 269.

GILLIS, N. S., and KOEHLER, T. R. (1971). *Phys. Rev.* **B4**, 3971.

—— (1972 a). *Phys. Rev.* **B5**, 1925.

—— (1972 b). *Phys. Rev. Lett.* **29**, 369.

—— (1974). *Phys. Rev.* **B9**, 3806.

GINZBURG, V. L. (1945). *Zh. eksp. teor. Fiz.* **15**, 739.

—— (1949). *Zh. eksp. teor. Fiz.* **19**, 36.

—— (1955). *Dokl. Akad. Nauk. SSSR* **105**, 240

—— (1960). *Fiz. tverd. Tela* **2**, 2031 (*Sov. Phys.-Solid State* **2**, 1824 (1960)).

—— (1962). *Usp. Fiz. Nauk.* **77**, 621 (*Sov. Phys.-Uspekhi* **5**, 649 (1963)).

GINZBURG, V. L., and LEVANYUK, A. P. (1958). *J. Phys. Chem. Solids* **6**, 51.

—— (1974), *Phys. Lett.* **47A**, 345.

GIORDMAINE, J. A. (1962). *Phys. Rev. Lett.* **8**, 19.

GLADKII, V. V., and SIDNENKO, E. V. (1971). *Fiz. tverd. Tela* **13**, 3092 (*Sov. Phys.-Solid St.* **13**, 2592 (1972)).

GLASS, A. M. (1968). *Phys. Rev.* **172**, 564.

—— (1969 a). *J. appl. Phys.* **40**, 4699.

—— (1969 b). *J. chem. Phys.* **50**, 1501.

GLASS, A. M., and ABRAMS, R. L. (1970 a). *J. appl. Phys.* **41**, 4455.

—— (1970 b). *Proc. Symp. Submillimeter Waves*. Polytechnic Press, New York.

GLASS, A. M., and AUSTON, D. H. (1972). *Opt. Commun.* **5**, 45.

GLASS, A. M., and LINES, M. E. (1976), *Phys. Rev.* **B13**, 180.

GLASS, A. M., MCFEE, J. H., and BERGMAN, J. G. (1971). *J. appl. Phys.* **42**, 5219.

GLASS, A. M., LINDE, D. VON DER, AUSTON, D. H., and NEGRAN, T. (1976). *J. Electron. Mater.* in press.

GLASS, A. M., and LINDE, D. VON DER, (1976). *Ferroelectrics* **10**, 163.

—— (1974). *Appl. Phys. Lett.* **25**, 233.

GLAUBER, R. J. (1955). *Phys. Rev.* **98**, 1692.
—— (1963). *J. math. Phys.* **4**, 294.
GLAZER, A. M., and ISHIDA, K. (1974). *Ferroelectrics* **6**, 219.
GLAZER, A. M., and MEGAW, H. D. (1972). *Phil. Mag.* **25**, 1119.
GLEASON, T. G., and WALKER, J. C. (1969). *Phys. Rev.* **188**, 893.
GODEFROY, L. R. (1956). *Progress in Semiconductors*, Vol. **1**, p. 217. London-Heywood.
GOLDAK, J., BARRETT, C. S., INNES, D., and YOUDELIS, W. (1966). *J. chem. Phys.* **44**, 3323.
GOL'DMAN, I. I., KRIVCHENKOV, V. D., KOGAN, V. I., and GALITSKII, V. M. (1960). *Selected problems in quantum mechanics* (ed. D. ter Haar), Academic Press, New York.
GONZALO, J. A. (1970). *Phys. Rev.* **B1**, 3125.
—— (1974). *Phys. Rev.* **B9**, 3149.
GOODMAN, G. (1953). *J. Am. ceram. Soc.* **36**, 368.
—— (1957), *US Pat. No. 2805165*. United States Patent Office, Washington, DC.
—— (1963). *J. Am. ceram. Soc.* **46**, 48.
GOULPEAU, L. (1966). *Fiz. tverd. Tela* **8**, 2469. *Sov. Phys. Solid St.* **8**, 1970 (1967)).
GOYAL, S. G., AGARWAL, L. D., and VERMA, M. P. (1974). *Phys. Rev.* **B10**, 779.
GRIFFITHS, R. B. (1967). *Phys. Rev.* **158**, 176.
—— (1970). *Phys. Rev. Lett.* **24**, 1479.
GRIGAS, V. P., GRIGAS, I. P., and BELYATSKAS, R. P. (1967). *Fiz. tverd. Tela* **9**, 1532 (*Sov. Phys.–Solid St.* **9**, 1203 (1967)).
GRINDLAY, J. (1965). *Phys. Lett.* **18**, 239.
—— (1970). *An introduction to the phenomenological theory of ferroelectricity.* Pergamon Press, New York.
GURO, G. M., IVANCHIK, I. I., and KOVTONYUK, N. F. (1968). *Fiz. tverd. Tela* **10**, 135 (*Sov. Phys.–Solid St.* **10**, 100 (1968)).
—— (1969). *Fiz. tverd. Tela* **11**, 1956. (*Sov. Phys.–Solid St.* **11**, 1574 (1970)).
HAAS, C. (1965). *Phys. Rev.* **140**, A863.
HADNI, A. (1971) *Proc. Symp. Submillimeter waves, Microwave Res. Inst. Ser.*, Vol. 20, p. 251. Polytechnic Press, New York.
HADNI, A., HENNINGER, Y., THOMAS, R., VERGNAT, P., and WYNCKE, B. (1965). *J. Phys. Paris* **26**, 345.
HADNI, A., and THOMAS, R. (1972). *Ferroelectrics* **4**, 39.
HAERTLING, G. H., and LAND, C. E. (1971). *J. Am. ceram. Soc.* **54**, 1.
HALLERS, J. J., and CASPERS, W. J. (1969). *Phys. Stus. Solidi* **36**, 587.
HALLERS, J. J., and VERTOGEN, G. (1973). *Physica* **66**, 315.
HALPERIN, B. I. (1973). *Phys. Rev.* **B8**, 4437.
HALPERIN, B. I., and HOHENBERG, P. C. (1967). *Phys. Rev. Lett.* **19**, 700.
—— (1969). *Phys. Rev.* **177**, 952.
HALPERIN, B. I., HOHENBERG, P. C., and MA, S. (1974). *Phys. Rev.* **B10**, 139.
HAMANO, K. (1973). *J. phys. Soc. Japan* **35**, 157.
HAMMER. J. M., CHANNIN, D. J., DUFFY, M. T., and WITTKE, J. P. (1972). *Appl. Phys. Lett.* **21**, 358.
HAMMER, J. M., and PHILLIPS, W. (1974). *Appl. Phys. Lett.* **24**, 545.
HANDLER, P., MAPOTHER, D. E., and RAYL, M. (1967). *Phys. Rev. Lett.* **19**, 356.
HARADA, J., AXE, J. D., and SHIRANE, G. (1971). *Phys. Rev.* **B4**, 155.
HARADA, J., PEDERSEN, T., and BARNEA, Z. (1970). *Acta Crystallog.* **A26**, 336.
HARBEKE, G. (1963). *J. Phys. Chem. Solids* **24**, 957.
HARBEKE, G., STEIGMEIER, E. F., and WEHNER, R. K. (1970). *Commun. solid state Phys.* **8**, 1765

HARLEY, R. T., HAYES, W., PERRY, A. M., and SMITH, S. R. P. (1973). *J. Phys. C Solid St.* **6**. 2382.

HARRIS, S. (1969). *Proc. IEEE* **57**, 2096.

HARTWIG, C. M., WIENER-AVNEAR, E., and PORTO, S. P. S. (1972). *Phys. Rev.* **B5**, 79.

HATTA, I. (1968). *J. phys. Soc. Japan* **24**, 1043.

—— (1970). *J. phys. Soc. Japan* **28**, 1266.

HATTA, I., and IKUSHIMA, A. (1973). *J. Phys. Chem. Solids* **34**, 57.

HATTA, I., MATSUDA, M., and SAWADA, S. (1974) *J. Phys. C. Solid St.* **7**, L299.

HATTA, I., and SAWADA, S. (1965). *Jap. J. appl. Phys.* **4**, 389.

HAVLIN, S., LITOV, E., and SOMPOLINSKY, H. (1975). *Phys. Lett.* **51A**, 33.

—— (1976). *Phys. Rev.* **B13**, 4999.

HAVLIN, S., LITOV, E., and UEHLING, E. A. (1974). *Phys. Rev.* **B9**, 1024.

HAYASHI, M. (1972 a). *J. phys. Soc. Japan* **33**, 739.

—— (1972 b). *J. phys. Soc. Japan* **33**, 616.

HAZONY, Y., EARLS, D. E., and LEFKOWITZ, I. (1968). *Phys. Rev.* **166**, 507.

HEIMAN, D., and USHIODA, S. (1974). *Phys. Rev.* **B9**, 2122.

HEITLER, W. (1954). *Quantum theory of radiation*, p. 182 ff. Clarendon Press, Oxford.

HELFRICH, W. (1969). *J. chem. Phys.* **51**, 4092.

HELMICK, C. N., and WOODWORTH, W. H. (1976). *Ferroelectrics* **11**, 309.

HERCZOG, A. (1964). *J. Am. ceram. Soc.* **47**, 107.

HERZBERG, G. (1945). *Molecular spectra and molecular structure. II. Infrared and Raman spectra of polyatomic molecules*. Van Nostrand, New York.

HEWAT, A. W. (1973 a). *J. Phys. C Solid St.* **6**, 1074.

—— (1973 b). *J. Phys. C. Solid St.* **6**, 2559.

—— (1973 c). *Nature (Lond.)* **246**, 90.

—— (1974). *Ferroelectrics* **6**, 215.

HEWAT, A. W., ROUSE, K. D., and ZACCAI, G. (1972). *Ferroelectrics* **4**, 153.

HEYMAN, P. M., and HEILMEIER, G. H. (1966). *Proc. IEEE* **54**, 842.

HEYWANG, W. (1961). *Solid State Electron.* **3**, 51.

—— (1971). *J. Mater. Sci.* **6**, 1214.

HIKITA, T., SAKATA, K., and TATSUZAKI, I. (1973). *J. phys. Soc. Japan* **34**, 1248.

HILL, J. C., and MOHAN, P. V. (1971). *Ferroelectrics* **2**, 201.

HILL, R. M., and ICHIKI, S. K. (1963). *Phys. Rev.* **132**, 1963.

—— (1962). *Phys. Rev.* **128**, 1140.

HIPPEL, A. R. VON (1954). *Dielectrics and waves*. John Wiley, New York.

HIRAHARA, E., and MURAKAMI, M. (1958). *J. Phys. Chem. Solids* **7**, 281.

HIROSE, N., and SASAKI, H. (1971). *J. Am. ceram. Soc.* **54**, 320.

HÖCHLI, U. T. (1972). *Phys. Rev.* **B6**, 1814.

HOHNKE, D. K., HOLLOWAY, H., and KAISER, S. (1972). *J. Phys. Chem. Solids* **33**, 2053.

HOLAKOVSKY, J. (1973). *Phys. St. Solidi* **56b**, 615.

HOLAKOVSKY, J., BREZINA, B., and PACHEROVA, O. (1972). *Phys. St. Solidi* **53b**, K69.

HOLDEN, A. N., MATTHIAS, B. T., MERZ, W. J., and REMEIKA, J. P. (1955). *Phys. Rev.* **98**, 546.

HOLEMAN, B. R., and WREATHALL, W. M. (1971). *J. Phys. D appl. Phys.* **4**, 1898.

HOOTON, D. J. (1955 a). *Phil. Mag.* **46**, 7, 422, 433, 485.

—— (1955 b). *Z. Phys.* **142**, 42.

—— (1958). *Mag.* **3**, 49.

HORNIG, A. W., REMPEL, R. G., and WEAVER, H. E. (1959). *J. Phys. Chem. Solids* **10**, 1.

HOSHINO, S., and MOTEGI, H. (1967). *Jap. J. appl. Phys.* **6**, 708.
HOSHINO, S., OKAYA, Y., and PEPINSKY, R. (1959). *Phys. Rev.* **115**, 323.
HOUSTON, G. D., and BOLTON, H. C. (1971). *J. Phys. C Solid St.* **4**, 2894.
HUANG, C. C., and GRINDLAY, J. (1970). *Can. J. Phys.* **48**, 847.
HUBBARD, J. (1971). *J. Phys. C Solid St.* **4**, 53.
HÜLLER, A. (1969). *Z. Phys.* **220**, 145.
HULM, J. K., MATTHIAS, B. T., and LONG, E. A. (1950). *Phys. Rev.* **79**, 885.
HUSIMI, K., and KATAOKA, K. (1960). *Rev. sci. Instrum.* **31**, 418.
IIZUMI, M., HAMAGUCHI, Y., KOMATSUBARA, K. F., and KATO, Y. (1975). *J. phys. Soc. Japan* **38**, 443.
IKEDA, T., FUJIBAYASHI, K., NAGAI, T., and KOBAYASHI, J. (1973). *Phys. St. Solidi* **16a**, 279.
IKEDA, T., and FUSHIMI, S. (1962). *J. phys. Soc. Japan* **17**, 1202.
INDENBOM, V. L. (1960). *Isv. Akad. Nauk. SSSR ser. fiz.* **24**, 1180; *Kristallografiya* **5**, 115 (*Sov. Phys.–Crystallography* **5**, 106 (1960)).
INOUE, K. (1974). *Ferroelectrics* **7**, 107.
IRE standards on piezoelectric crystals. *Proc. IRE* **37**, 1378 (1949); **45**, 353 (1957); **46**, 764 (1958).
ISHIBASHI, Y., OHYA, S., and TAKAGI, Y. (1972). *J. phys. Soc. Japan* **33**, 1545.
—— (1974). *J. phys. Soc. Japan* **37**, 1035.
ISHIDA, K., and HONJO, G. (1973). *J. phys. Soc. Japan* **34**, 1279.
ITOH, K., and MITSUI, T. (1973). *Ferroelectrics* **5**, 235.
ITOH, S., and NAKAMURA, T. (1973). *Phys. Lett.* **44A**, 461.
—— (1974). *Commun. solid st. Phys.* **15**, 195.
IWASAKI, H., MIYAZAWA, S., KOIZUMI, H., SUGII, K., and NIIZEKI, N. (1972). *J. appl. Phys.* **43**, 4907.
IWASAKI, H., YAMADA, T., NIIZEKI, N., and TOYODA, H. (1970). *J. phys. Soc. Japan* **28S**, 306.
JACCARD, C., KÄNZIG, W., and PETER, M. (1953). *Helv. Phys. Acta.* **26**, 521.
JAEGER, P. (1934). *Annln. Phys.* **21**, 481.
JAFFE, B., COOK, W. R., and JAFFE, H. (1971). *Piezoelectric ceramics.* Academic Press, London.
JAFFE, B., ROTH, R. S., and MARZULLO, S. (1955) *J. Res. Nat. Bur. Stand.* **55**, 239.
JAIN, A. P., SHRINGI, S. N., and SHARMA, M. L. (1970). *Phys. Rev.* **B2**, 2756.
JAIN, Y. S., and BIST, H. D. (1974). *Commun. solid st. Phys.* **15**. 1229.
JAMIESON, P. B., ABRAHAMS, S. C., and BERNSTEIN, J. L. (1968). *J. chem. Phys.* **48**, 5048.
—— (1969). *J. chem. Phys.* **50**, 4352.
JANOVEC, V. (1966). *J. chem. Phys.* **45**, 1874.
JANTA, J. (1971). *Ferroelectrics* **2**, 299.
JEGGO, C. R., and BOYD, G. D. (1970). *J. appl. Phys.* **41**, 2741.
JEITSCHKO, W. (1972). *Acta Crystallog.* **B28**, 60.
JERPHAGNON, J. (1970). *Phys. Rev.* **B2**, 1091.
JERPHAGNON, J., and KURTZ, S. K. (1970). *J. appl. Phys.* **41**, 1667.
JOHNSON, C. J. (1965). *Appl. Phys. Lett.* **7**, 221.
JOHNSON, L. F., and BALLMAN, A. A. (1969). *J. appl. Phys.* **40**, 297.
JOHNSON, W. D. (1970). *J. appl. Phys.* **41**, 3279.
JOHNSTON, W. D., and KAMINOW, I. P. (1968). *Phys. Rev.* **168**, 1045.
JONA, F., and SHIRANE, G. (1962). *Ferroelectric crystals.* Macmillan, New York.
JONA, F., SHIRANE, G., MAZZI, F., and PEPINSKY, R. (1957). *Phys. Rev.* **105**, 849.

JONKER, G. H. (1964). *Solid. St. Electron.* **7**, 895.

KADANOFF, L. P., GOTZE, W., HAMBLEN, D., HECHT, R., LEWIS, E. A. S., PALCIAUSKAS, V. V., RAYL, M., SWIFT, J., ASPNES, D., and KANE, J. (1967). *Rev. mod. Phys.* **39**, 395.

KAHN, A. H., and LEYENDECKER, A. J. (1964). *Phys. Rev.* **135**, A1321.

KAHNG, D., and WEMPLE, S. H. (1965). *J. appl. Phys.* **36**, 2925.

KAMAEV, V. E. 1967 Candidates Dissertation, Voronezh Ped. Inst.

KAMINOW, I. P. (1965). *Phys. Rev.* **138**, A1539.

—— (1975) *Trans. IEEE. M.T.T.* **23**, 57.

KAMINOW, I. P., CARRUTHERS, J. R., TURNER, E. H., and STULZ, L. W. (1973). *Appl. Phys. Lett.* **22**, 540.

KAMINOW, I. P., and DAMEN, T. C. (1968). *Phys. Rev. Lett.* **20**, 1105.

KAMINOW, I. P., and JOHNSTON, W. D. (1967). *Phys. Rev.* **160**, 519.

KAMINOW, I. P., and TURNER, E. H. (1966). *Proc. IEEE* **54**, 1374.

—— (1971). In *Handbook of lasers* (ed. R. J. Pressley). Chemical Rubber Co., Cleveland, Ohio.

KAMINSKII, A. A. (1972). *Kristalografiya* **17**, 231 (*Sov. Phys.–Crystallography* **17**, 194 (1972)).

KÄNZIG, W. (1957). *Ferroelectrics and antiferroelectrics.* Academic Press, New York.

KAPUSTIN, A. P., and VISTIN, L. K. (1965). *Kristallografiya* **10**, 118 (*Sov. Phys.–Crystallography* **10**, 95 (1965)).

KASHIDA, S., HATTA, I., IKUSHIMA, A., and YAMADA, Y. (1973). *J. phys. Soc. Japan.* **34**, 997.

KATIYAR, R. S., RYAN, J. F., and SCOTT, J. F. (1971). *Phys. Rev.* **B4**, 2635.

KATO, T., and ABE, R. (1972). *J. phys. Soc. Japan* **32**, 717.

KAWADA, S., and IDA, M. (1965). *J. phys. Soc. Japan* **20**, 1287.

KAWAI, H. (1969). *Jap. J. appl. Phys.* **8**, 975.

KAWASAKI, K. (1967). *J. Phys. Chem. Solids* **28**, 1277.

—— (1968). *Prog. theor. Phys. Kyoto* **39**, 285.

KAY, M. I. (1972). *Ferroelectrics* **4**, 235.

KAY, M. I., and KLEINBERG, R. (1973). *Ferroelectrics* **5**, 45.

KELLERMANN, E. W. (1940). *Phil. Trans R. Soc.* **A238**, 513.

KENEMAN, S. A., MILLER, A., and TAYLOR, G. W. (1972). *Ferroelectrics* **3**, 131.

KEVE, E. T. (1975). *Appl. Phys. Lett.* in press.

KEVE, E. T., and ABRAHAMS, S. C. (1970). *Ferroelectrics* **1**, 243.

KEVE, E. T., ABRAHAMS, S. C., and BERNSTEIN, J. L. (1970). *J. chem. Phys.* **53**, 3279.

—— (1971). *J. chem. Phys.* **54**, 3185.

KEVE, E. T., ABRAHAMS, S. C., NASSAU, K., and GLASS, A. M. (1970). *Commun. solid st. Phys.* **8**, 1517.

KEVE, E. T., and ANNIS, A. D. (1973). *Ferroelectrics* **5**, 77.

KEVE, E. T., and BYE, K. L. (1975). *J. appl. Phys.* **46**, 810.

KEVE, E. T., BYE, K. L., WHIPPS, P. W., and ANNIS, A. D. (1971). *Ferroelectrics* **3**, 39.

KHMEL'NITSKII, D. E., and SHNEERSON, V. L. (1973). *Fiz. tverd. Tela* **15**, 1838 (*Sov. Phys.–Solid St.* **15**, 1226 (1973)).

KIKUCHI, A., OKA, Y., and SAWAGUCHI, E. (1967). *J. phys. Soc. Japan* **23**, 337.

KIM, Q., POWELL, R. C., and WILSON, T. M. (1974). *Commun. solid st. Phys.* **14**, 541.

KIMURA, M., NANAMATSU, S., KAWAMURA, T., and MATSUSHITA, S. (1974). *Jap. J. appl. Phys.* **13**, 1473.

KINASE, W., and TAKAHASI, H. (1957). *J. phys. Soc. Japan* **12**, 464.

KITTEL, C. (1946). *Phys. Rev.* **70**, 965.

—— (1951). *Phys. Rev.* **82**, 729.

—— (1972). *Commun. solid st. Phys.* **10**, 119.

KITTEL, C., and SHORE, H. (1965). *Phys. Rev.* **138**, A.1165.

KJEMS, J. K., SHIRANE, G., MÜLLER, K. A., and SCHEEL, H. J. (1973). *Phys. Rev.* **B8**, 1119.

KLEINMAN, D. A. (1962). *Phys. Rev.* **128**, 1761.

KOBAYASHI, J., ENOMOTO, Y., and SATO, Y. (1972). *Phys. St. Solidi* **50b**, 335.

KOBAYASHI, J., and MIZUTANI, I. (1970). *Phys. St. Solidi* **2a**, K89.

KOBAYASHI, J., SATO, Y., and SCHMID, H. (1972). *Phys. St. Solidi* **10a**, 259.

KOBAYASHI, J., SCHMID, H., and ASCHER, E. (1968). *Phys. St. Solidi* **26**, 277.

KOBAYASHI, J., YAMADA, N., and NAKAMURA, T. (1963). *Phys. Rev. Lett.* **11**, 508.

KOBAYASHI, K. K. (1968). *J. phys. Soc. Japan* **24**, 497.

KOCINSKI, J., and WOJTCZAK, L. (1973). *Phys. Lett.* **43A**, 215.

KOEHLER, T. R., and GILLIS, N. S. (1973). *Phys. Rev.* **B7**, 4980.

KONAK, C., FOUSEK, J., and IVANOV, N. R. (1974). *Ferroelectrics* **6**, 235.

KONDO, Y. (1973). *J. phys. Soc. Japan* **35**, 1266.

KONSIN, P. I., and KRISTOFFEL, N. N. (1972). *Fiz. tverd. Tela* **14**, 2873 (*Sov. Phys.–Solid St.* **14**, 2484 (1973)).

KOPSKY, V. (1971). *Phys. St. Solidi* **5a**, K69.

KOTZLER, J., and SCHEITHE, W. (1973). *Commun. solid st. Phys.* **12**, 643.

KRÄTZIG, E., and KURZ, H. (1976). *Ferroelectrics* **10**, 159.

KRISTOFFEL, N. N., and KONSIN, P. I. (1967). *Izv. Akad. Nauk. Est. SSR Ser. fiz. math.* **16**, 429.

—— (1968). *Phys. St. Solidi* **28**, 731.

—— (1969). *Izv. Akad. Nauk. SSSR. Ser fiz. math.* **18**, 439.

—— (1971). *Fiz. tverd. Tela.* **13**, 3513 (*Sov. Phys.–Solid St.* **13**, 2969 (1972)).

—— (1973). *Ferroelectrics,* **6**, 1.

KRUMHANSL, J. A. and SCHRIEFFER, J. R. (1975). *Phys. Rev.* **B11**, 3535.

KUBO, R. (1957). *J. phys. Soc. Japan* **12**, 570.

—— (1966). *Rep. Prog. Phys.* **29**, 255.

KUDZIN, A. Y., and PANCHENKO, T. V. (1972). *Fiz. tverd. Tela* **14**, 1843 (*Sov. Phys.–Solid St.* **14**, 1599 (1972)).

KUMADA, A. (1969). *Phys. Lett.* **30A**, 186.

—— (1972). *Ferroelectrics* **3**, 115.

KUMADA, A., YUMOTO, H., and ASHIDA, S. (1970). *J. phys. Soc. Japan* **28S**, 351.

KURCHATOV, I. V. (1933). Segnetoelektriki Moscow: C–M GTTI, abbreviated French version publ. Hermann and Cie, Paris, 1936.

KURTZ, S. K. (1966). *Proc. 1st. Int. Conf. on Ferroelectricity* (ed. V. Dvorak), vol. 1, p. 413. publ. Inst. of Physics of the Czech. Acad. Sci.

—— (1972). In *Laser handbook* (eds. F. T. Arecchi and E. O. Schulz-Dubois), vol. 1, p. 924. Elsevier, New York.

—— (1976). *Trans. Am. crystallog. Ass.* in press.

KURTZ, S. K., and PERRY, T. T. (1968). *J. appl. Phys.* **39**, 3798.

KURTZ, S. K., and ROBINSON, F. N. H. (1967). *Appl. Phys. Lett.* **10**, 62.

KUSHIDA, T. (1955). *J. Sci. Hiroshima Univ.* **A19**, 327.

KWOK, P. C., and MILLER, P. B. (1966). *Phys. Rev.* **151**, 387.

LAGAKOS, N., and CUMMINS, H. Z. (1974). *Phys. Rev.* **B10**, 1063.
—— (1975). *Phys. Rev. Lett.* **34**, 883.
LAMBERT, M., and COMES, R. (1969). *Commun. solid st. Phys.* **7**, 305.
LAMOTTE, B., GAILLARD, J., and CONSTANTINESCU, O. (1972). *J. chem. Phys.* **57**, 3319.
LAND, C. E. (1974). *Ferroelectrics* **7**, 45.
LAND, C. E., and THACHER, P. D. (1969). *Proc. IEEE* **57**, 751.
LANDAU, L. D. (1937 a). *Phys. Z. Sowjun.* **11**, 26, 545.
—— (1937 b). *Zh. eksp. teor. Fiz.* **7**, 19, 627.
—— (1965). *Collected papers of L. D. Landau* (ed. D. ter Haar). Gordon and Breach, New York.
LANDAU, L. D., and KHALATNIKOV, I. M. (1954). *Dokl. Akad. Nauk. SSSR.* **96**, 469.
LANDAUER, R. (1957). *J. appl. Phys.* **28**, 227.
LANDO, J. B., OLF, H. G., and PETERLIN, A. (1966). *J. Polymer Sci. A-1* **4**, 941.
LANG, S. B. (1971). *Phys. Rev.* **B4**, 3603.
—— (1974). *Sourcebook of pyroelectricity.* Gordon and Breach, New York.
LANG, S. B., and STECKEL, F. (1965). *Rev. sci. Instrum.* **36**, 929.
LARKIN, A. I., and KHMEL'NITSKII, D. E (1969). *Zh. eksp. teor. Fiz.* **56**, 2087 (*Sov. Phys.–JETP* **29**, 1123 (1969)).
—— (1968). *Zh. eksp. teor. Fiz.* **55**, 2345 (*Sov. Phys.–JETP* **28**, 1245 (1969)).
LAUBEREAU, A., and ZUREK, R. (1970). *Z. Naturf.* **25a**, 391.
LAVAL, J. (1958). *Rev. mod. Phys.* **30**, 222.
LAWLESS, W. N. (1972). *Ferroelectrics* **3**, 287.
—— (1976). *Phys. Rev.* **B14**, 134.
LAX, M., and NELSON, D. F. (1974). In *Polaritons* (eds. E. Burstein and F. DeMartini). Pergamon Press, New York.
LAZAY, P. D., and FLEURY, P. A. (1971). *Proc. 2nd. Int. Conf. on Light Scattering in Solids* (ed. M. Balkanski). Flammarion Sciences, Paris.
LEFKOWITZ, I. (1959). *J. Phys. Chem. Solids* **10**, 169.
LEFKOWITZ, I., LUKASZEWICZ, K., and MEGAW, H. D. (1966). *Acta Crystallog.* **20**, 670.
LEIBA, E. (1969). *C. R. Acad. Sci. Paris* **268**, B31.
LERNER, P., LEGRAS, C., and DUMAS, J. P. (1968). *J. cryst. Growth* **3**, 231.
LEVANYUK, A. P. (1965). *Zh. eksp. teor. Fiz.* **49**, 1304 (*Sov. Phys.–JETP* **22**, 901 (1966)).
LEVANYUK, A. P., and SANNIKOV, D. G. (1968). *Zh. eksp. teor. Fiz.* **55**, 256 (*Sov. Phys.–JETP* **28**, 134 (1969)).
—— (1970). *Fiz. tverd. Tela* **12**, 2997 (*Sov. Phys.–Solid St,* **12**, 2418 (1971)).
LEVANYUK, A. P., and SHCHEDRINA, N. V. (1972). *Fiz. tverd. Tela* **14**, 3012 (*Sov. Phys.–Solid St.* **14**, 2581 (1973)).
LEVIN, S., PELAH, I., and WIENER-AVNEAR, E. (1973). *Phys. St. Solidi* **58b**, 61.
LEVINE, B. F. (1973 a). *Phys. Rev.* **B7**, 2600.
—— (1973 b). *J. chem. Phys.* **59**, 1463.
—— (1973 c). Private communication.
LIEB, E. H. (1969). In *Lectures in theoretical physics.* University of Colorado Press, Boulder, Colo.
LIEB, E. H., and WU, F. Y. (1972). *Two dimensional ferroelectric models in phase transitions and critical phenomena* (eds. C. Domb and M. S. Green). Academic Press, New York.
LINES, M. E. (1969 a). *Phys. Rev.* **177**, 797.

—— (1969 b). *Phys. Rev.* **177**, 812, 819.
—— (1970 a). *Phys. Rev.* **B2**, 690.
—— (1970 b). *Phys. Rev.* **B2**, 698.
—— (1972 a). *Phys. Rev.* **B5**, 3690.
—— (1972 b). *Commun. solid st. Phys.* **10**, 793.
—— (1974). *Phys. Rev.* **B9**, 950.
LIPKIN H. J. (1960). *Ann. Phys. N.Y.* **9**, 332.
—— (1964). *Ann. Phys. N.Y.* **26**, 115.
LIPPINCOTT, E. R., and SCHROEDER, R. (1955). *J. chem. Phys.* **23**, 1099.
LITOV, E., and HAVLIN, S. (1974). *Phys. Lett.* **47A**, 57.
LITOV, E., and UEHLING, E. A. (1968). *Phys. Rev. Lett.* **21**, 809.
—— (1970). *Phys. Rev.* **B1**, 3713.
LITTLE, E. A. (1955). *Phys. Rev.* **98**, 978.
LIU, S. T. (1976). *Ferroelectrics* **10**, 83.
LIU, S. T., and ZOOK, J. D. (1974). *Ferroelectrics.* **7**, 171.
LOCK, P. J. (1971). *Appl. Phys. Lett.* **19**, 390.
LOGAN, R. M., and MCLEAN, T. P. (1973). *Infrared Phys.* **13**, 15.
LOGAN, R. M., and MOORE, K. (1973). *Infrared Phys.* **13**, 37.
LOGAN, R. M., and WATTON, R. (1972). *Infrared Phys.* **12**, 17.
LOUDON, R. D. (1964). *Ad. Phys.* **13**, 423.
LOVELUCK, J. M., and SOKOLOFF, J. B. (1973). *J. Phys. Chem. Solids* **34**, 869.
LOWNDES, R. P., AND RASTOGI, A. (1973). *J. Phys. C. Solid St.* **6**, 932.
LOWNDES, R. P., TORNBERG, N. E., and LEUNG, R. C. (1974). *Phys. Rev.* **B10**, 911.
LUBENSKY, T. C. (1975). *Phys. Rev.* **B11**, 3573.
LUDLOW, J. H., MITCHELL, W. H., PUTLEY, E. H., and SHAW, N. (1967). *J. sci. Instrum.* **44**, 694.
LUTHER, G. (1972). *J. de Phys. Paris* **33**, C2–221.
LUTHER, G., and MUSER, H. E. (1969). *Z. Naturf.* **24a**,389.
—— (1970). *Z. angew. Phys.* **29**, 237.
LYONS, K. B., MOCKLER, R. C., and O'SULLIVAN, W. J. (1973). *J. Phys. C Solid St.* **6**, L420.
MA, S. (1973). *Rev. mod. Phys.* **45**, 589.
MACCHESNEY, J. B., and POTTER, J. F. (1965). *J. Am. ceram. Soc.* **48**, 81.
MCFEE, J. H., BERGMAN, J. G., and CRANE, G. R. (1972). *Ferroelectrics* **3**, 305.
MCMILLAN, W. L. (1973). *Phys. Rev.* **A8**, 1921.
MAGNELI, A., (1949). *Ark. Kem.* **1**, 213.
MAHAN, G. D. (1966). *Phys. Rev.* **145**, 602.
MAKER, P. D. (1970). *Phys. Rev.* **A1**, 923.
MAKER, P. D., TERHUNE, R. W., NISENOFF, M., and SAVAGE, C. M. (1962). *Phys. Rev. Lett.* **8**, 21.
MAKISHIMA, S., YAMAMOTO, H., TOMOTSU, T., and SHIONOYA, S. (1965). *J. phys. Soc. Japan* **20**, 2147.
MALDONADO, J. R., FRASER, D. B., and MEITZLER, A. H. (1975). *Advances in image pickup and display devices.* Academic Press, New York.
MANSINGH, A., and LIM, K. O. (1972). *J. phys. Soc. Japan* **33**, 747.
MANY, A., GOLDSTEIN, Y., and GROVER, N. B. (1965). *Semiconductor surfaces.* North-Holland, Amsterdam.
MARADUDIN, A. A. (1964). *Rev. mod. Phys.* **36**, 417.
—— (1966). *Solid St. Phys.* **18**, 273 (eds. F. Seitz and D. Turnbull). Academic Press, New York.

MARADUDIN, A. A., and FEIN, A. E. (1962). *Phys. Rev.* **128**, 2589.
MARIE, G., and DONJON, J. (1973). *Proc. IEEE* **61**, 942.
MARTIN, R. M. (1974). *Phys. Rev.* **B9**, 1998.
MARTIN, T. P. (1970). *Phys. Rev.* **B1**, 3480.
MARTIN, T. P., and GENZEL, L. (1974) *Phys. St. Solidi* **61b**, 493.
MASON, W. P. (1947). *Phys. Rev.* **72**, 854.
—— (1950). *Piezoelectric crystals and their application to ultrasonics.* Van Nostrand, Princeton, N.J.
MASON, W. P., and JAFFE, H. (1954). *Proc. IRE* **42**, 921.
MASUNO, K., MURAKAMI, T., and WAKU, S. (1972). *Ferroelectrics* **3**, 315.
MATTHEISS, L. F. (1972). *Phys. Rev.* **B6**, 4718, 4740.
MATTHIAS, B. T. (1949). *Phys. Rev.* **75**, 1771.
MATTHIAS, B. T., MILLER, C. E., and REMEIKA, J. P. (1956). *Phys. Rev.* **104**, 849.
MATTHIAS, B. T., and REMEIKA, J. P. (1956). *Phys. Rev.* **103**, 262.
—— (1949). *Phys. Rev.* **76**, 1886.
MAVRIN, B. N., ABRAMOVICH, T. E., and STERIN, K. E. (1972). *Fiz. tverd. Tela* **14**, 1810 (*Sov. Phys.–Solid St.* **14**, 1562 (1972)).
MEGAW, H. D. (1968). *Acta Crystallog.* **A24**, 589.
MEISTER, H., SKALYO, J., FRAZER, B. C., and SHIRANE, G. (1969). *Phys. Rev.* **184**, 550.
MEITZLER, A. H., and O'BRYAN, H. M. (1973). *Proc. IEEE* **61**, 959.
MEITZLER, A. H., MALDONADO, J. R., and FRASER, D. B. (1970). *Bell Syst. Tech. J* **49**, 953.
MERZ, W. J. (1954). *Phys. Rev.* **95**, 690.
—— (1956). *J. appl. Phys.* **27**, 938.
—— (1958). *Helv. Phys. Acta* **31**, 625.
MEYER, R. B., LIEBERT, L., STRZELECKI, L., and KELLER, P. (1975). *J. Phys. Lett.* **36**, L69.
MICHEL, C., MOREAU, J. M., ACHENBACH, G. D., GERSON, R., and JAMES, W. J. (1969). *Commun. solid st. Phys.* **7**, 865.
MICHERON, F., MAYEUX, C., and TROTIER, J. C. (1974). *Appl. Optics* **13**, 784.
MIKOLA, S. (1925). *Z. Phys.* **32**, 476.
MILLER, P. B., and KWOK, P. C. (1967). *Commun. solid st. Phys.* **5**, 57.
—— (1968). *Phys. Rev.* **175**, 1062.
MILLER, R. C. (1964 a). *Phys. Rev.* **134**, A1313.
—— (1964 b). *Appl. Phys. Lett.* **5**, 17.
—— (1958). *Phys. Rev.* **111**, 736.
MILLER, R. C., BOYD, G. D., and SAVAGE, A. (1965). *Appl. Phys. Lett.* **6**, 77.
MILLER, R. C., and WEINREICH, G. (1960). *Phys. Rev.* **117**, 1460.
MINAEVA, K. A., and LEVANYUK, A. P. (1965). *Bull. Acad. Sci. USSR* **29**, 978.
MINAEVA, K. A., LEVANYUK, A. P., STRUKOV, B. A., and KOPTSIK, V. A. (1967). *Fiz. tverd. Tela* **9**, 1220 (*Sov. Phys.–Solid St.* **9**, 950 (1967)).
MINAEVA, K. A., STRUKOV, B. A., and VARNSTORFF, K. (1968). *Fiz. tverd. Tela* **10**, 2125 (*Sov. Phys.–Solid St.* **10**, 1665 (1969)).
MINAKATA, M., NODA, J., and UCHIDA, N. (1975). *Appl. Phys. Lett* **26**, 395.
MITSUI, T., and FURUICHI, J. (1953). *Phys. Rev.* **90**, 193.
MITSUI, T., NAKAMURA, E., and TOKUNAGA, M. (1973). *Ferroelectrics* **5**, 185.
MIYAZAWA, S. (1973). *Appl. Phys. Lett.* **23**, 198.
MONTANO, P. A., SHECHTER, H., and SHIMONY, U. (1971). *Phys. Rev.* **B3**, 858.
MOORE, M. A., and WILLIAMS, H. C. W. L. (1972). *J. Phys. C Solid St.* **5**, 3168, 3185, 3222.

MORAVETS, F., and KONSTANTINOVA, V. P. (1968). *Kristallografiya* **13**, 284 (*Sov. Phys.–Crystallography* **13**, 221 (1968)).

MORI, T., and TAMURA, H. (1964). *J. phys. Soc. Japan* **19**, 1247.

MORI, T., TAMURA, H., and SAWAGUCHI, E. (1965). *J. phys. Soc. Japan* **20**, 281.

MORITA, M., ONO, K., and MURATA, K. (1974). *Ferroelectrics* **8**, 425.

MORLON, B., COQUET, E., and DEVIN, A. (1970). *C. R. Acad. Sci. Paris* **B270**, 283.

MOTT, N. F., and GURNEY, R. W. (1940). *Electronic processes in ionic crystals.* Clarendon Press, Oxford.

MUELLER, H. (1940 a). *Phys. Rev.* **57**, 829.

—— (1940 b). *Phys. Rev.* **58**, 565, 805.

MÜLLER, K. A. (1971). In *Structural phase transitions and soft modes* (ed. E. J. Samuelsen). Universitetsvorleiget, Oslo.

MÜLLER, K. A., and BERLINGER, W. (1971). *Phys. Rev. Lett.* **26**, 13.

—— (1972). *Phys. Rev. Lett.* **29**, 715.

MÜLLER, K. A., BERLINGER, W., CAPIZZI, M., and GRÄNICHER, H. (1970) *Commun. solid st. Phys.* **8**, 549.

MÜLLER, K. A., BERLINGER, W., and WALDNER, F. (1968). *Phys. Rev. Lett.* **21**, 814.

MÜLLER, K. A., DALAL, N. S., and BERLINGER, W. (1976). *Phys. Rev. Lett.* **36**, 1504.

MUZIKAR, C., JANOVEC, V., and DVORAK, V. (1963). *Phys. St. Solidi* **3**, K9.

NAGAMIYA, T. (1952). *Prog. theor. Phys. Kyoto* **7**, 275.

NAKAMURA, E., NAGAI, T., ISHIDA, K., ITOH, K., and MITSUI, T. (1970). *J. phys. Soc. Japan* **28S**, 271.

NAKAMURA, K., and WADA, Y. (1971). *J. Polymer Sci. A*-2 **9**, 161.

NAKO, K., and BALKANSKI, M. (1973). *Phys. Rev.* **B8**, 5759.

NANAMATSU, S., KIMURA, M., DOI, K., MATSUSHITA, S., and YAMADA, N. (1974). *Ferroelectrics* **8**, 511.

NANAMATSU, S., KIMURA, M., DOI, K., and TAKAHASHI, M. (1971). *J. phys. Soc. Japan* **30**, 300.

NASH, F. R., BOYD, G. D., SARGENT, M., and BRIDENBAUGH, P. M. (1970). *J. appl. Phys.* **41**, 2564.

NASSAU, K. (1967). In *Ferroelectricity* (ed. E. F. Weller). Elsevier, Amsterdam.

NASSAU, K., and GLASS, A. M. Unpublished Data.

NASSAU, K., and LINES, M. E. (1970). *J. appl. Phys.* **41**, 533.

NATTERMANN, T. (1975). *Ferroelectrics.* **9**, 229.

NEGRAN, T. J., GLASS, A. M., BRICKENCAMP, C. S., ROSENSTEIN, R. D., OSTERHELD, R. K., and SUSOTT, R. (1974). *Ferroelectrics* **6**, 235.

NELMES, R. J. (1974). *J. Phys. C Solid St.* **7**, 3840.

NELMES, R. J., EIRIKSSON, V. R., and ROUSE, K. D. (1972). *Commun. solid st. Phys.* **11**, 1261.

NELMES, R. J., and THORNLEY, F. R. (1975). *J. Phys. C Solid St.* **7**, 3855.

NELSON, D. F., and LAX, M. (1976). *Phys. Rev.* **B13**, in press.

NETTLETON, R. E. (1969). *Z. Phys.* **220**, 401.

—— (1971 a). *Ferroelectrics* **2**, 77.

—— (1971 b). *Z. Phys.* **248**, 101.

—— (1974). *J. Phys. C Solid St.* **7**, 3785.

—— (1975). *J. Phys. C Solid St.* **8**, 943.

NEUMARK, G. F. (1962). *Phys. Rev.* **125**, 838.

NIIZEKI, N., and HASEGAWA, M. (1964). *J. phys. Soc. Japan* **19**, 550.

NIKIFOROV, I. Y., and KHASABOV, A. G., (1971). *Fiz. tverd. Tela* **13**, 3589 (*Sov. Phys.–Solid St.* **13**, 3030).

NISHIMURA, K., and HASHIMOTO, T. (1973). *J. phys. Soc. Japan* **35**, 1699.

NITSCHE, R., and MERZ, W. J. (1960). *J. Phys. Chem. Solids.* **13**, 154.

NITSCHE, R., ROETSCHI, H., and WILD, P. (1964). *Appl. Phys. Lett.* **4**, 210.

NOVAKOVIC, L. (1966). *J. Phys. Chem. Solids* **27**, 1469.

NUNES, A. C., AXE, J. D., and SHIRANE G. (1971). *Ferroelectrics* **2**, 291.

NYE. J. F. (1964). *Physical properties of crystals.* Clarendon Press, Oxford.

O'BRIEN, E. J., and LITOVITZ, T. A. (1964). *J. appl. Phys.* **35**, 180.

O'BRYAN, H. M. (1973). *J. Am. ceram. Soc.* **56**, 385.

O'BRYAN, H. M., and MEITZLER, A. H. (1973). *Conf. on Phase Transitions and their Applications to Materials Science* (eds. H. K. Henish, R. Roy, and L. E. Cross). Pergamon Press, New York.

O'HARE, J. M., and HURST, R. P. (1967). *J. chem. Phys.* **46** 2356.

OKADA, K. (1961). *J. phys. Soc. Japan* **16**, 1647.

—— (1969). *J. phys. Soc. Japan* **27**, 420.

—— (1970). *J. phys. Soc. Japan* **28S**, 58.

—— (1974). *J. phys. Soc. Japan* **37**, 1226.

OKADA, K., and SUGIE, H. (1971). *Phys. Lett.* **37A**, 337.

ONODERA, Y. (1970). *Prog. theor. Phys. Kyoto* **44**, 1477.

ONSAGER, L. (1936). *J. Am. chem. Soc.* **58**, 1486.

ORMANCEY, G., and GODEFROY, G. (1974). *J. Phys.* **35**, 135.

OTTO, A. (1968). *Z. Phys.* **216**, 398.

PAK, K. N. (1973). *Phys. St. Solidi* **60b**, 233.

PASYNKOV, R. E. (1973). *Ferroelectrics* **6**, 19.

PATHRIA, R. K. (1972). *Statistical mechanics.* Pergamon Press, Oxford.

PATTERSON, A. L. (1934). *Phys. Rev.* **46**, 372.

PAUL, G. L., COCHRAN, W., BUYERS, W. J. L., and COWLEY, R. A. (1970). *Phys. Rev.* **B2**, 4603.

PAWLEY, G. S., COCHRAN, W., COWLEY, R. A., and DOLLING, G. (1966). *Phys. Rev. Lett.* **17**, 753.

PEARSON, G. L., and FELDMANN, W. L. (1958). *J. Phys. Chem. Solids* **9**, 28.

PEERCY, P. S. (1975 a). *Commun. solid. st. Phys.* **16**, 439.

—— (1975 b). *Phys. Rev. Lett.* **35**, 1581.

—— (1974). *Phys. Rev.* **B9**, 4868.

—— (1973). *Phys. Rev. Lett.* **31**, 379.

PEERCY, P. S., and SAMARA, G. A. (1973). *Phys. Rev.* **B8**, 2033.

PENNA, A. F., CHAVES, A., ANDRADE, P. DA R., and PORTO, S. P. S. (1976). *Phys. Rev.* **B13**, 4907.

PENNA, A. F., CHAVES, A., and PORTO, S. P. S. (1976). *Commun. solid st. Phys.* **19**, 491.

PENSAK, L. (1958). *Phys. Rev.* **109**, 601.

PEPINSKY, R., JONA, F., and SHIRANE, G. (1956). *Phys. Rev.* **102**, 1181.

PERRY, C. H., and MCNELLY, T. F. (1967). *Phys. Rev.* **154**, 456.

PESHIKOV, E. V. (1972). *Fiz. tverd. Tela* **14**, 1597 (*Sov. Phys.–Solid St.* **14**, 1377 (1972)).

PETERS, C. J. (1963). *Proc. IEEEE* **51**, 147.

—— (1965). *Proc. IEEE* **53**, 455.

PETERSON, G. E., BALLMAN, A. A., LENZO, P. V., and BRIDENBAUGH, P. M. (1964). *Appl. Phys. Lett.* **5**, 62.

PETERSON, G. E., and BRIDENBAUGH, P. M. (1968). *J. chem. Phys.* **48**, 3402.

PETERSON, G. E., BRIDENBAUGH, P. M., and GREEN, P. (1967). *J. chem. Phys.* **46**, 4009.

PETERSON, G. E., and CARNIVALE, A. (1972). *J. chem. Phys.* **56**, 4848.

PETERSON, G. E., and CARRUTHERS, J. R. (1969). *J. solid st. Chem.* **1**, 98.

PETERSON, G. E., GLASS, A. M., and NEGRAN, T. J. (1971). *Appl. Phys. Lett.* **19**, 130.

PETZELT, J. (1969). *Phys. St. Solidi* **36**, 321.

—— (1971). *Commun. solid st. Phys.* **9**, 1485.

PETZELT, J., and DVORAK, V. (1971). *Phys. St. Solidi* **46b**, 413.

——, —— (1976). *J. Phys. C. Solid St.* **9**, 1571.

PETZELT, J., GRIGAS, J., and MAYEROVA, I. (1974). *Ferroelectrics* **6**, 225.

PFEUTY, P., and ELLIOTT, R. J. (1971). *J. Phys. C Solid St.* **4**, 2370.

PFISTER, G., and ABKOWITZ, M. A. (1974). *J. appl. Phys.* **45**, 1001.

PFISTER, G., ABKOWITZ, M. A., and CRYSTAL, R. G. (1973). *J. appl. Phys.* **44**, 2064.

PHELAN, R. J., PETERSON, R. L., HAMILTON, C. A., and DAY, G. W. (1974). *Ferroelectrics* **7**, 375.

PHILLIPS, J. C., and VAN VECHTEN, J. A. (1969). *Phys. Rev.* **183**, 709.

PHILLIPS, W., AMODEI, J. J., and STAEBLER, D. L. (1972). *RCA Rev.* **33**, 94.

PIETRASS, B. (1972). *Phys. St. Solidi* **53b**, 279.

PIKKA, T. A., and FRIDKIN, V. M. (1968). *Fiz. tverd. Tela* **10**, 3378 (*Sov. Phys.–Solid. St.* **10**, 2668 (1969).

PINNOW, D. A. (1971). In *Handbook of lasers*, p. 478. (ed. R. J. Pressley). Chemical Rubber Co., Cleveland, Ohio.

PISAREV, R. V., DRUZHININ, V. V., PROCHOROVA, S. D., NESTEROVA, N. N., and ANDREEVA, G. T. (1969). *Phys. St. Solidi* **35**, 145.

PLAKHTY, V., and COCHRAN, W. (1968). *Phys. St. Solidi* **29**, K81.

PLESSER, T., and STILLER, H. (1969). *Commun. solid st. Phys.* **7**, 323.

POSLEDOVICH, M., WINTER, F. X., BORSTEL, G., and CLAUS, R. (1973). *Phys. St. Solidi* **55b**, 711.

PURA, B., and PRZEDMOJSKI, J. (1973). *Phys. Lett.* **43A**, 217.

PUTLEY, E. H., WATTON, R., and LUDLOW, J. H. (1972). *Ferroelectrics* **3**, 263.

PYTTE, E. (1970 a). *Phys. Rev.* **B1**, 924.

—— (1970 b). *Commun. solid. st. Phys.* **8**, 2101.

—— (1971 a). In *Structural transitions and soft modes* (ed. E. J. Samuelsen). Universitetsvorleiget, Oslo.

—— (1971 b). *Phys. Rev.* **B3**, 3503.

—— (1972 a). *Phys. Rev. Lett.* **28**, 895.

—— (1972 b). *Phys. Rev.* **B5**, 3758.

—— (1973). *Phys. Rev.* **B8**, 3954.

PYTTE, E., and FEDER, J. (1969). *Phys. Rev.* **187**, 1077.

PYTTE, E., and THOMAS, H. (1968). *Phys. Rev.* **175**, 610.

QUÈZEL, G., and SCHMID, H. (1968). *Commun. solid st. Phys.* **6**, 447.

QUITTET, A. M., and LAMBERT, M. (1973). *Commun. solid. st. Phys.* **12**, 1053.

RAMAN, C. V., and NEDUNGADI, T. M. K. (1940). *Nature (Lond.)* **145**, 147.

RAMASWAMY, V. (1972). *Appl. Phys. Lett.* **21**, 183.

RAPOPORT, E. (1966). *J. chem. Phys.* **45**, 2721.

REDDISH, W. (1966). *J. polymer Sci.* **C14**, 123.

REDFIELD, D. (1963). *Phys. Rev.* **130**, 916.

REESE, R. L., FRITZ, I. J., and CUMMINS, H. Z. (1973). *Phys. Rev.* **B7**, 4165.

REESE, W. (1969 a). *Phys. Rev.* **181**, 905.

—— (1969 b) *Commun. solid st. Phys.* **7**, 969.
REESE, W., and MAY, L. F. (1967). *Phys. Rev.* **162**, 510.
—— (1968). *Phys. Rev.* **167**, 504.
REIK, H. G. (1963). *Commun. solid st. Phys.* **1**, 67.
REMEIKA, J. P., and GLASS, A. M. (1970). *Mater. Res. Bull.* **5**, 37.
RESIBOIS, P. and DE LEENER, M. (1966). *Phys. Rev.* **152**, 305.
—— (1969). *Phys. Rev.* **178**, 819.
RIETVELD, H. M. (1969). *J. appl. Crystallog.* **2**, 65.
RIGAMONTI, A. (1967). *Phys. Rev. Lett.* **19**, 436.
RIMAI, L., and MARS, G. A. DE (1962). *Phys. Rev.* **127**, 702.
RISTE, T., SAMUELSEN, E. J., OTNES, K., and FEDER, J. (1971). *Commun. solid st. Phys.* **9**, 1455.
RITCHIE, R. H. (1973). *Surface Sci.* **34**, 1.
ROBINSON, F. N. H. (1967). *Bell Syst. Tech. J.* **46**, 913.
—— (1968). *Phys. Lett.* **26A**, 435.
ROUNDY, C. B., and BYER, R. L. (1972). *Appl. Phys. Lett.* **21**, 512.
—— (1976). *Ferroelectrics* **10**, in press.
ROZHDESTVENSKAYA, M. Y., SHEFTEL', I. T., STOGOVA, V. A., KOZYREVA, M. S., and KRAYUKHINA, E. K. (1970). *Fiz. tverd. Tela* **12**, 873 (*Sov. Phys.-Solid. St.* **12**, 674 (1970)).
RUBIN, J. J., VAN UITERT, L. G., and LEVINSTEIN, H. J. (1967). *J. cryst. Growth* **1**, 315.
RUDYAK, V. M. (1970). *Usp. Fiz. Nauk.* **101**, 429 (*Sov. Phys.-Uspekhi* **13**, 461 (1971)).
RUDYAK, V. M., and BOGOMOLOV, A. A. (1967). *Fiz. tverd. Tela* **9**, 3336 (*Sov. Phys.-Solid St.* **9**, 2624 (1968)).
RUDYAK, V. M., KUDZIN, A. Y., and PANCHENKO, T. V. (1972). *Fiz. tverd. Tela* **14**, 2441 (*Sov. Phys.-Solid. St.* **14**, 2112 (1973)).
RUPPIN, R. (1970). *Commun. solid st. Phys.* **8**, 1129.
RUPPIN, R., and ENGLMAN, R. (1970). *Rep. Prog. Phys.* **33**, 149.
RYAN, J. F., KATIYAR, R. S., and TAYLOR, W. (1972). *J. Phys. Paris* **33**, C2-49.
SABURI, O. (1961). *J. Am. ceram. Soc.* **44**, 54.
SAKOWSKI-COWLEY, A. C., LUKASZEWICZ, K., and MEGAW, H. D. (1969). *Acta Crystallog.* **B25**, 851.
SAKUDO, T. (1963). *J. phys. Soc. Japan* **18**, 1626.
SAKUDO, T., and UNOKI, H. (1971). *Phys. Rev. Lett.* **26**, 851.
SAKURAI, J., COWLEY, R. A., and DOLLING, G., (1970). *J. phys. Soc. Japan* **28**, 1426.
SALAMON, M. B. (1970). *Phys. Rev.* **B2**, 214.
—— (1973). *Commun. solid st. Phys.* **13**, 1741.
SAMARA, G. A. (1966), *Phys. Rev.* **151**, 378.
—— (1968). *Phys. Lett,* **27A**, 232.
—— (1969). In *Advances in high pressure research* (ed. R. S. Bradley), vol. 3, chap. 3. Academic Press, New York.
—— (1970 a). *J. phys. Soc. Japan.* **28S**, 399.
—— (1970 b). *Phys. Rev.* **B1**, 3777.
—— (1971). *Phys. Rev. Lett.* **27**, 103.
—— (1973). *Ferroelectrics*, **5**, 25.
—— (1974). *Ferroelectrics.* **7**, 221.
—— (1975). *Ferroelectrics*, **9**, 209.
SAMARA, G. A., and MOROSIN, B. (1973). *Phys. Rev.* **B8**, 1256.
SAMUEL, E. A., and SUNDARAM, V. S. (1974). *Phys. Lett.* **47A**, 421.

SAMUELSEN, E. J. (1971). In *Structural phase transitions and soft modes* (ed. E. J. Samuelsen), p. 189. Universitetsvorleiget, Oslo.

SANNIKOV, D. G. (1962). *Fiz. tverd. Tela* **4**, 1619. (*Sov. Phys.–Solid St.* **4**, 1187 (1962)).

SASABE, H., SAITO, S., ASAHINA, M., and KAKUTANI, H. (1969). *J. polymer Sci A-2*, **7**, 1405.

SASTRY, M. D. (1972). *Commun. solid st. Phys* **11**, 1671.

SAVAGE, C. M., and MAKER, P. D. (1970). *Appl. Opt.* **10**, 965.

SAWADA, S., and TAKAGI, Y. (1972). *J. phys. Soc. Japan* **33**, 1071.

SAWADA, A., TAKAGI, Y., and ISHIBASHI, Y. (1973). *J. phys. Soc. Japan* **34**, 748.

SAWADA, S., HIROTSU, S., TAKASHIGE, M., SHIROISHI, Y., and IWAMURA, H. (1974). *J. phys. Soc. Japan* **36**, 1211.

SAWADA, S., NOMURA, S., FUJII, S., and YOSHIDA, I. (1958). *Phys. Rev. Lett.* **1**, 320.

SAWADA, Y., BURSTEIN, E., CARTER, D. L., and TESTARDI, L. (1964). In *Plasma effects in solids*, p. 71. Dunod, Paris.

SAWAGUCHI, E., and CROSS, L. E. (1971). *Appl. Phys. Lett.* **18**, 1.

SAWAGUCHI, E., MANIWA, H., and HOSHINO, S. (1951). *Phys. Rev.* **83**, 1078.

SAWYER, C. B., and TOWER, C. H. (1930). *Phys. Rev.* **35**, 269.

SCHEMPP, E., PETERSON, G. E., and CARRUTHERS, J. R. (1970). *J. chem. Phys.* **53**, 306.

SCHLOSSER, H., and DROUGARD, M. E. (1961). *J. appl. Phys.* **32**, 1227.

SCHMID, H. (1970). *J. phys. Soc. Japan* **28S**, 354.

SCHMID, H., RIEDER, H., and ASCHER, E. (1965). *Commun. solid st. Phys.* **3**, 327.

SCHMIDT, R. V., and KAMINOW, I. P. (1974). *Appl. Phys. Lett.* **25**, 458.

SCHMIDT, V. H., and UEHLING, E. A. (1962). *Phys. Rev.* **126**, 447.

SCHNEIDER, T., and STOLL, E. (1973). *Phys. Rev. Lett.* **31**, 1254.

——, —— (1976). *Phys. Rev. Lett.* **36**, 1501.

SCHRÖDER, U. (1966). *Commun. solid st. Phys.* **4**, 347.

SCHURMANN, H. K., GILLESPIE, S., GUNTON, J. D., and MIHALISIN, T. (1973). *Phys. Lett.* **45A**, 417.

SCHWABL, F. (1973). *Phys. Rev.* **B7**, 2038.

SCOTT, B. A., and BURNS, G. (1972 a). *J. Am. ceram. Soc.* **55**, 331.

—— (1972 b). *J. Am. ceram. Soc.* **55**, 225.

SCOTT, B. A., GIESS, E. A., OLSON, B. L., BURNS, G., SMITH, A. W., and O'KANE, D. F. (1970). *Mater. Res. Bull.* **5**, 47.

SCOTT, J. F., and WILSON, C. M. (1972). *Commun. solid st. Phys.* **10**, 597.

SCOTT, J. F., and WORLOCK, J. M. (1973). *Commun. solid st. Phys.* **12**, 67.

SEKIDO, T., and MITSUI, T. (1967). *J. Phys. Chem. Solids* **28**, 967.

SELYUK, B. V. (1971). *Kristallografiya* **16**, 356 (*Sov. Phys.–Crystallography*, **16**, 292 (1971)).

—— (1973). *Ferroelectrics* **6**, 37.

SERDOBOL'SKAYA, O. Y., and KUAK TKHI TAM (1972). *Fiz. tverd. Tela* **14**, 2443 (*Sov. Phys.–Solid St.* **14**, 2114 (1973)).

SHAKMANOV, V. V., SPIVAK, G. V., and YAKUNIN, S. I. (1970) *Fiz. tverd. Tela* **12**, 2286 (*Sov. Phys.–Solid St.* **12**, 1827 (1971)).

SHAND, M. L., BURSTEIN, E., and BRILLSON, L. J. (1974). *Ferroelectrics* **7**, 283.

SHAPKIN, V. V., GROMOV, B. A., and PETROV, G. T. (1973). *Fiz. tverd. Tela* **15**, 1401 (*Sov. Phys.–Solid St.* **15**, 947 (1973)).

SHAPIRO, S. M., AXE, J. D., SHIRANE, G., and RISTE, T. (1972). *Phys. Rev.* **B6**, 4332.

SHE, C. Y., BROBERG, T. W., WALL, L. S., and EDWARDS, D. F. (1972). *Phys. Rev.* **B6**, 1847.

SHEPHERD, I. W., and BARKLEY, J. R. (1972). *Commun. solid st. Phys.* **10**, 123.

SHIBUYA, I., and MITSUI, T. (1961). *J. phys. Soc. Japan* **16**, 479.

SHIGENARI, T., and TAKAGI, Y. (1971). *J. phys. Soc. Japan* **31**, 312.

SHIOZAKI, Y. (1971). *Ferroelectrics* **2**, 245.

SHIRANE, G., and AXE, J. D. (1971). *Phys. Rev. Lett.* **27**, 1803.

SHIRANE, G., AXE, J. D., HARADA, J., and LINZ, A. (1970). *Phys. Rev.* **B2**, 3651.

SHIRANE, G., AXE, J. D., HARADA, J., and REMEIKA, J. (1970). *Phys. Rev.* **B2**, 155.

SHIRANE, G., DANNER, H., and PEPINSKY, R. (1957). *Phys. Rev.* **105**, 856.

SHIRANE, G., HOSHINO, S., and SUZUKI, K. (1950). *Phys. Rev.* **80**, 1105.

SHIRANE, G., NATHANS, R., and MINKIEWICZ, V. J. (1967). *Phys. Rev.* **157**, 396.

SHIRANE, G., and PEPINSKY, R. (1953). *Phys. Rev.* **91**, 812.

SHIRANE, G., PEPINSKY, R., and FRAZER, B. C. (1955). *Phys. Rev.* **97**, 1179.

—— (1956). *Acta Crystallog.* **9**, 131.

SHIRANE, G., SUZUKI, K., and TAKEDA, A. (1952). *J. phys. Soc. Japan* **7**, 12.

SHIRANE, G., and YAMADA, Y. (1969). *Phys. Rev.* **177**, 858.

SHUKLA, G. C., and SINHA, K. P. (1966). *J. Phys. Chem. Solids*, **27**, 1837.

SHUVALOV, L. A. (1970). *J. phys. Soc. Japan* **28S**, 38.

SHUVALOV, L. A., RUDYAK, V. M., and KAMAEV, V. E. (1965). *Dokl. Akad. Nauk. SSSR* **163**, 347 (*Sov. Phys.–Doklady* **10**, 639 (1966)).

SILSBEE, H. B., UEHLING, E. A., and SCHMIDT, V. H. (1964). *Phys. Rev.* **133**, A165.

SILVERMAN, B. D. (1964). *Phys. Rev.* **135**, A1596.

—— (1970). *Phys. Rev. Lett.* **25**, 107.

SILVERMAN, B. D., and JOSEPH, R. I. (1963). *Phys. Rev.* **129**, 2062.

—— (1964). *Phys. Rev.* **133**, A207.

SINGER, B., and LALAK, J. (1976). *Ferroelectrics* **10**, 103.

SINGH, S., DRAEGERT, D. A., and GEUSIC, J. E. (1970). *Phys. Rev.* **B2**, 2709.

SINGH, S., and SINGH, K. (1974). *J. phys. Soc. Japan* **36**, 1588.

SINGWI, K. S., and SJÖLANDER, A. (1960). *Phys. Rev.* **120**, 1093.

SINHA, J. K. (1965). *J. sci. Instrum.* **42**, 696.

SKALYO, J., FRAZER, B. C., and SHIRANE, G. (1970). *Phys. Rev.* **B1**, 278.

SLACK, J. R., and BURFOOT, J. C. (1971). *J. Phys. C Solid St.* **4**, 898.

SLATER, J. C. (1941). *J. chem. Phys.* **9**, 16.

—— (1950). *Phys. Rev.* **78**, 748.

SLOTFELDT-ELLINGSEN, D., and PEDERSEN, B. (1974). *Phys. St. Solidi* **24a**, 191.

SMITH, A. W., and BURNS, G. (1969). *Phys. Lett.* **28A**, 501.

SMITH, R. A., JONES, F. E., and CHASMAR, R. P. (1968). *The detection and measurement of infrared radiation*. Clarendon Press, Oxford.

SMITH, W. D. (1973). *In Phase transitions* (ed. H. K. Henish, R. Roy, and L. E. Cross), p. 71. Pergamon Press, New York.

SMOLENSKII, G. A. (1970). *J. phys. Soc. Japan* **28S**, 26.

SMOLENSKII, G. A. and AGRANOVSKAYA, A. I. (1954). *Dokl. Akad. Nauk. SSSR* **97**, 237.

SMOLENSKII, G. A., and ISUPOV, V. A. (1954). *Dokl. Akad. Nauk. SSSR* **97**, 653.

SMOLENSKII, G. A., ISUPOV, V. A., AGRANOVSKAYA, A. A., and POPOV, S. N. (1960). *Fiz. tverd. Tela.* **2**, 2906.

SMOLENSKII, G. A., ISUPOV, V. A., KRANIK, N. N., and AGRANOVSKAYA, A. I. (1961). *Izv. Akad. Nauk. SSSR Ser. Fiz.* **25**, 1333.

SMOLENSKII, G. A., and JOFFE, V. A. (1958). *Commun. Colloq. Int. de Magnetism, Grenoble, Commun. No.* 71.

SMOLENSKII, G. A., and KOZLOVSKII, V. (1954). *Zh. eksp. teor. Fiz.* **26**, 684.

SMUTNY, F., and FOUSEK, J. (1970). *Phys. St. Solidi* **40**, K13.

SOKOLOV, A. I. (1974). *Fiz. tverd. Tela* **16**, 733 (*Sov. Phys.–Solid St.* **16**, 478 (1974)).

SOKOLOV, A. I., and VENDIK, O. G. (1974). *Ferroelectrics* **7**, 391.

SOREF, R. A. (1969). *IEEE J. quant. Electron* **5**, 126.

SPECTOR, H. N. (1974). *Commun. solid st. Phys.* **14**, 537.

SPENCER, E. G., LENKO, P. V., and BALLMAN, A. A. (1967). *Proc. IEEE* **55**, 2074.

SPITZER, W. G., MILLER, R. C., KLEINMAN, D. A., and HOWARTH, L. E. (1962). *Phys. Rev.* **126**, 7110.

SPIVAK, G. V., IGRAS, E., PRYAMKOVA, I. A., and ZHELUDEV, I. S. (1959). *Kristallografiya* **4**, 115 (*Sov. Phys.–Crystallography* **4**, 123 (1959)).

STADLER, H. L. (1958). *J. appl. Phys.* **29**, 1485.

STADLER, H. L. and ZACHMANIDIS, P. J. (1963). *J. appl. Phys.* **34**, 3255.

STAEBLER, D. L., and AMODEI, J. J. (1972). *Ferroelectrics* **3**, 107.

STAEBLER, D. L., and PHILLIPS, W. (1974). *Appl. Optics* **13**, 793.

STANLEY, H. E. (1971 a). *Introduction to phase transitions and critical phenomena.* Clarendon Press, Oxford.

—— (1971 b). In *Structural phase transitions and soft modes* (ed. E. J. Samuelsen), pp. 271, 289. Universitetsvorleiget, Oslo.

STEIGMEIER, E. F., and AUDERSET, H. (1973). *Commun. solid st. Phys.* **12**, 565.

STEIGMEIER, E. F., AUDERSET, H., and HARBEKE, G. (1973). *Commun. solid st. Phys.* **12**, 1077.

—— (1975). *Phys. St. Solidi* **70b**, 705.

STEIGMEIER, E. G., and HARBEKE, G. (1970). *Commun. solid st. Phys.* **8**, 1275.

—— (1972). *J. Phys. Paris* **33**, C2–55.

STEIGMEIER, E. F., HARBEKE, G., and WEHNER, R. K. (1971). *Proc. 2nd. Int. Conf. on Light Scattering in Solids* (ed. M. Balkanski). Paris: Flammarion.

STINCHCOMBE, R. B. (1973). *J. Phys. C Solid St.* **6**, 2459.

STIRLING, W. G. (1972). *J. Phys. C Solid St.* **5**, 2711.

STOKOWSKI, S. E., and SCHAWLOW, A. L. (1969). *Phys. Rev.* **178**, 457, 464.

STRATTON, J. A. (1941). *Electromagnetic theory.* McGraw-Hill, New York.

STRUKOV, B. A. (1964). *Fiz. tverd. Tela* **6**, 2862 (*Sov. Phys.–Solid St.* **6**, 2278 (1965)).

—— (1966 a). *Kristallografiya* **11**, 892 (*Sov. Phys.–Crystallography* **11**, 757 (1967)).

—— (1966 b). *Phys. St. Solidi.* **14**, K135.

STRUKOV, B. A., BADDUR, A., ZINENKO, V. I., MISCHENKO, A. V., and KOPTSIK, V. A. (1973b). *Fiz. tverd. Tela* **15**, 1388 (*Sov. Phys.–Solid St.* **15**, 939 (1973)).

STRUKOV, B. A., BADDUR, A., ZINENKO, V. I., MIKHAILOV, V. K., and KOPTSIK, V. A. (1973 a). *Fiz. tverd. Tela* **15**, 2018 (*Sov. Phys.–Solid St.* **15**, 1347 (1974)).

STRUKOV, B. A., KORZHUEV, M. A., BADDUR, A., and KOPTSIK, V. A. (1972). *Sov. Phys.–Solid St.* **13**, 1569.

STRUKOV, B. A., TARASKIN, S. A., and KOPTSIK, V. A. (1966). *Zh. eksp. teor. Fiz.* **51**, 1037 (*Sov. Phys.–JETP* **24**, 692 (1967)).

SUBBARAO, E. C. (1973). *Ferroelectrics.* **5**, 267.

SUGAI, T., and WADA, M. (1974). *Jap. J. appl. Phys.* **13**, 1291.

SUGAWARA, F., and NAKAMURA, T. (1972). *J. Phys. Chem. Solids* **33**, 1665.
SUGII, K., IWASAKI, H., ITOH, Y., and NIIZEKI, N. (1972). *J. cryst. Growth* **16**, 291.
SUZUKI, M., and KUBO, R. (1968). *J. phys. Soc. Japan* **24**, 51.
SZIGETI, B., (1975). *Phys. Rev. Lett.* **35**, 1532.
TAKADA, S., OHNISHI, M., HAYAKAWA, H., and MIKOSHIBA, N. (1974). *Appl. Phys. Lett.* **24**, 490.
TAKAGI, M., SUZUKI, S., and WATANABE, H. (1970). *J. phys. Soc. Japan.* **28S**, 369.
TAKAGI, Y. (1948). *J. phys. Soc. Japan* **3**, 271.
—— (1952). *Phys. Rev.* **85**, 315.
TAKANO, S., HOTTA, S., KAWAMURA, H., KATO, Y., KOBAYASHI, K. L. I., and KOMATSUBARA, K. F. (1974). *J. phys. Soc. Japan* **37**, 1007.
TANAKA, M., and HONJO, G. (1964). *J. phys. Soc. Japan* **19**, 954.
TANI, K., and TSUDA, N. (1969). *J. phys. Soc. Japan* **26**, 113.
TAYLOR, G. W., and KOSONOCKY, W. F. (1972). *Ferroelectrics* **3**, 81.
TAYLOR, R. G. F., and BOOT, H. A. H. (1973). *Contemporary Phys.* **14**, 55.
TEATHER, G. G., and YOUNG, L. (1968). *Solid St. Electron.* **11**, 527.
TELLO, M. J., and HERNANDEZ, E. (1973). *J phys. Soc. Japan* **35**, 1289.
TENG, M. K., BALKANSKI, M., and MASSOT, M. (1972). *Phys. Rev.* **B5**, 1031.
TENNERY, V. J. (1966). *J. Am. ceram. Soc.* **49**, 483.
TERAUCHI, H., and YAMADA, Y. (1972). *J. phys. Soc. Japan.* **33**, 446.
TERHUNE, R. W., MAKER, P. D., and SAVAGE, C. M. (1965). *Phys. Rev. Lett.* **14**, 681.
THACHER, P. D., and LAND, C. E. (1969). *IEEE Trans. Electron Dev.* **16**, 515.
THAXTER, J. B., and KESTIGIAN, M. (1974). *Appl. Opt.* **13**, 913.
THOMAS, H. (1969). *IEEE Trans. Magn.* **5**, 874.
—— (1971). In *Structural phase transitions and soft modes* (ed. E. J. Samuelsen), p. 15. Universitetsvorleiget, Oslo.
TIEN, P. K. (1971). *Appl. Optics* **10**, 2395.
TOKUNAGA, M. (1966). *Progr. theor. Phys. Kyoto* **36**, 857.
—— (1970). *Ferroelectrics* **1**, 195.
TOKUNAGA, M., and MATSUBARA, T. (1966). *Prog. theor. Phys. Kyoto* **35**, 581.
TOLEDANO, J. C. (1975). *Phys. Rev.* **B12**, 943.
TOMASHPOLSKI, Y. Y., SEVOSTIANOV, M. A., PENTAGOVA, M. V., SOROKINA, L. A., and VENEVSTSEV, Y. N. (1974). *Ferroelectrics* **7**, 257.
TOMINAGA, Y., and NAKAMURA, T. (1974). *Commun. solid st. Phys.* **15**, 1193.
TOMPSETT, M. F. (1971). *IEEE Trans. Electron Dev.* **18**, 1070.
TOWNSEND, R. L., and LAMACCHIA, J. T. (1970). *J. appl. Phys.* **41**, 5188.
TRIEBWASSER, S. (1959). *Phys. Rev.* **114**, 63.
—— (1960). *Phys. Rev.* **118**, 100.
ULLMAN, F. G., HOLDEN, B. J., GANGULY, B. N., and HARDY, J. R. (1973). *Phys. Rev.* **B8**, 2991.
ULRICH, D. R., and SMOKE, E. J. (1966). *J. Am. ceram. Soc.* **49**, 210.
UNOKI, H., and SAKUDO, T. (1973). *J. phys. Soc. Japan* **35**, 1128.
—— (1974). *J. phys. Soc. Japan.* **37**, 145.
—— (1967). *J. phys. Soc. Japan* **23**, 546.
UNRUH, H. G., and WAHL, H. J. (1972). *Phys. St. Solidi* **9a**, 119.
URBACH, F. (1953). *Phys. Rev.* **92**, 1324.
UWE, H., UNOKI, H., FUJII, Y., and SAKUDO, T. (1973). *Commun. solid st. Phys.* **13**, 737.

VAKS, V. G. (1968). *Zh. eksp. teor. Fiz.* **54**, 910 (1968). (*Sov. Phys.–JETP* **27**, 486 (1968)).

VAKS, V. G., GALITSKII, V. M., and LARKIN, A. I. (1966). *Zh. eksp. teor. Fiz* **51**, 1592 (*Sov. Phys.–JETP* **24**, 1071 (1967)).

—— (1968). *Zh. eksp. teor. Fiz.* **54**, 1172 (*Sov. Phys.–JETP* **27**, 627 (1968)).

VAKS, V. G., LARKIN, A. I., and PIKIN, S. A. (1966). *Zh. eksp. teor. Fiz.* **51**, 361 (*Sov. Phys.–JETP* **24**, 240 (1967)).

—— (1967) *Zh. eksp. teor. Fiz.* **53**, 281 (*Sov. Phys.–JETP* **26**, 188 (1968)).

VAKS, V. G., and ZINENKO, V. I. (1973). *Zh. eksp. teor. Fiz.* **64**, 650 (*Sov. Phys.–JETP* **37**, 330 (1973)).

VALASEK, J. (1920). *Phys. Rev.* **15**, 537.

—— (1921). *Phys. Rev.* **17**, 475.

VALLADE, M. (1975). *Phys. Rev.* **B12**, 3755.

VAN DEN BERG, C. B. (1972). *Ferroelectrics* **4**, 103, 195.

VAN HOVE, L. (1954). *Phys. Rev.* **95**, 249.

VAN UITERT, L. G., LEVINSTEIN, H. J., RUBIN, J. J., CAPIO, C. D., DEARBORN, E. F., and BONNER, W. A. (1968). *Mater. Res. Bull.* **3**, 47.

VAN UITERT, L. G., RUBIN, J. J., GRODKIEWICZ, W. H., and BONNER, W. A. (1969). *Mater. Res. Bull.* **4**, 63.

VAN UITERT, L. G., SINGH, S., LEVINSTEIN, H. J., GEUSIC, J. E., and BONNER, W. A. (1967). *Appl. Phys. Lett.* **11**, 161.

VAN VECHTEN, J. A. (1969). *Phys. Rev.* **182**, 891; **187**, 1009.

VARMA, C. M. (1976). *Phys. Rev.* **B14**, 244.

VENTURINI, E. L., SPENCER, E. G., LENZO, P. V., and BALLMAN, A. A. (1968). *J. appl. Phys.* **39**, 343.

VERBER, C. M., WOOD, V. E., KENAN, R. P., and HARTMAN, N. F. (1976). *Ferroelectrics* **10**, 253.

VERBITSKAYA, T. N., LAVERKO, E. N., POLYAKOV, S. M., ROZORENOVA, L. A., and RAEVSKAYA, E. B. (1971). *Bull. Acad. Sci. USSR. Phys. Ser.* **35**, 1792.

VILLAIN, J., and STAMENOVIC, S. (1966). *Phys. St. Solidi* **15**. 585.

VISSCHER, W. M. (1960). *Annln. Phys.* **9**, 194.

VOGT, H. (1974). *Appl. Phys.* **5**, 85.

VOLKOFF, G. M., PETCH, H. E., and SMELLIE, D. W. L. (1952). *Can. J. Phys.* **30**, 270.

VON DER LINDE, D., and GLASS, A. M. (1976). *Ferroelectrics* **10**, 5.

VON DER LINDE, D., GLASS, A. M., and RODGERS, K. F. (1974). *Appl. Phys. Lett.* **25**, 155; **26**, 22.

VOUSDEN, P. (1951). *Acta Crystallog.* **4**, 545.

VUL, B. M., GURO, G. M., and IVANCHIK, I. I. (1973). *Ferroelectrics* **6**, 29.

WALDKIRCH, T. VON, MULLER, K. A., and BERLINGER, W. (1973). *Phys. Rev.* **B7**, 1052.

WALDKIRCH, T. VON, MULLER, K. A., BERLINGER, W., and THOMAS, H. (1972). *Phys. Rev. Lett.* **28**, 503.

WALLACE, C. A. (1970). *J. appl. Cryst.* **3**, 546.

WALLACE, E. A., COCHRAN, W., and STRINGFELLOW, M. (1972). *J. Phys. Paris* **33S**, C2–59.

WANG, Y. L., and COOPER, B. R. (1968). *Phys. Rev.* **172**, 539.

WARTBURG, W. VON (1974). *Phys. St. Solidi* **21a**, 557.

WATTON, R., SMITH, C., HARPER, B., and WREATHALL, W. M. (1974). *IEEE Trans. Electron Dev.* **21**, 462.

WEBER, M. J., and SCHAUFELE, R. F. (1965). *Phys. Rev.* **138**, A1544.

WEMPLE, S. H. (1970). *Phys. Rev.* **B2**, 2679.

WEMPLE, S. H., and DIDOMENICO, M. (1972). *Appl. solid. st. Sci.* (ed. R. Wolfe) **3**, 263.

—— (1969). *J. appl. Phys.* **40**, 735.

WEMPLE, S. H., DIDOMENICO, M., and CAMLIBEL, I. (1968). *Appl. Phys. Lett.* **12**, 209.

WEMPLE, S. H., DIDOMENICO, M., and JAYARAMAN, A. (1969). *Phys. Rev.* **180**, 547.

WERTHAMER, N. R. (1969). *Am. J. Phys.* **37**, 763.

—— (1970). *Phys. Rev,* **B1**, 572.

WERTHEIM, G. K. (1964). *Mössbauer effect: principles and applications.* Academic Press, New York.

WESTON, T. B., WEBSTER, A. H., and MCNAMARA, V. M., (1967). *J. Can. ceram. Soc.* **36**, 15.

WHITE, G. (1971). *Bell Syst. Tech. J.* **40**, 2607.

WIENER-AVNEAR, E., LEVIN, S., and PELAH, I. (1966). *Phys. Lett.* **23**, 533.

—— (1970). *J. chem. Phys.* **52**, 2891.

WILLIAMS, R. (1965). *J. Phys. Chem. Solids* **26**, 399.

—— (1963). *J. chem. Phys.* **39**, 384.

WILLIAMS, R., and HEILMEIER, G. (1966). *J. chem. Phys.* **44**, 638.

WILSON, C. M. (1971). *PhD thesis.* Johns Hopkins University, Baltimore, Md.

WILSON, K. G. (1971). *Phys. Rev.* **B4**, 3174, 3184.

WILSON, K. G., and FISHER, M. E. (1972). *Phys. Rev. Lett.* **28**, 240.

WINTER, F. X., and CLAUS, R. (1972). *Opt. Commun.* **6**, 22.

WITTELS, M. C., and SHERRILL, F. A. (1957). *J. appl. Phys.* **28**, 606.

WOOD, E. A., MILLER, R. C., and REMEIKA. J. P. (1962). *Acta. Crystallog.* **15**, 1273.

WOOD, O. R., ABRAMS, R. L., and BRIDGES, T. J. (1970). *Appl. Phys. Lett.* **17**, 376.

WOODS, A. D. B., COCHRAN, W., and BROCKHOUSE, B. N. (1960). *Phys. Rev.* **119**, 980.

WORLOCK, J. M., and FLEURY, P. A. (1967). *Phys. Rev. Lett* **19**, 1176.

WORLOCK, J. M., and OLSEN, D. H. (1971). *Proc. 2nd. Int. Conf. on Light Scattering* (ed. M. Balkanski), p. 410. Flammarion, Paris.

WU, S. Y., TAKEI, W. J., and FRANCOMBE, M. H. (1973). *Appl. Phys. Lett.* **22**, 26.

WU, S. Y., TAKEI, W. J., FRANCOMBE, M. H., and CUMMINS, S. E. (1972). *Ferroelectrics* **3**, 217.

WUL, B., and GOLDMAN, I. M. (1945). *C. R. Acad. Sci. URSS.* **46**, 139; **49**, 177.

—— (1946). *C. R. Acad. Sci. URSS* **51**, 21.

WURFEL, P., and BATRA, I. P. (1973). *Phys. Rev.* **B8**, 5126.

WYNNE, J. J., and BLOEMBERGEN, N. (1969). *Phys. Rev.* **188**, 1211.

YAKOVLEV, L. A., VELICHKINA, T. S., and MIKHEEVA, L. F. (1956). *Sov. Phys.–Crystallography.* **1**, 91.

YAKUNIN, S. I., SHAKMANOV, V. V., SPIVAK, G. V., and VASIL'EVA, N. V. (1972). *Fiz. tverd. Tela* **14**, 373 (*Sov. Phys.–Solid St.* **14**, 310 (1972)).

YAMADA, T., and IWASAKI, H. (1972). *Acta Crystallog.* **28A**, 5181.

YAMADA, Y., FUJII, Y., and HATTA, I. (1968). *J. phys. Soc. Japan* **24**, 1053.

YAMADA, Y., SHIBUYA, I., and HOSHINO, S. (1963). *J. phys. Soc. Japan* **18**, 1594.

YAMADA, Y., and SHIRANE, G. (1969). *J. phys. Soc. Japan* **26**, 396.

YAMADA, Y., SHIRANE, G., and LINZ, A. (1969). *Phys. Rev.* **177**, 848.

YAMADA, Y., and YAMADA, T. (1966). *J. phys. Soc. Japan* **21**, 2167.

YELON, W. B., COCHRAN, W., SHIRANE, G., and LINZ, A. (1971). *Ferroelectrics* **2**, 261.

YMRY Y., PELAH, J., WIENER-AVNEAR, E., and ZAFRIR, H. (1967). *Comm. solid st. Phys.* **5**, 41.

YOUNG, A. P., and ELLIOTT, R. J. (1974). *J. Phys. C. Solid St.* **7**, 2721.

YOUNG, L., WONG, W. K. Y., THEWALT, M. L. W., and CORNISH, W. D. (1974). *Appl. phys. Lett.* **24**, 254.

YURKEVICH, V. E., and ROLOV, B. N. (1975). *Czech J. Phys.* **B25**, 701.

ZACCAI, G., and HEWAT, A. W. (1974). *J. Phys. C Solid St.* **7**, 15.

ZHELUDEV, I. S., PROSKURNIN, M. A., IURNIN, V. A., and BABERKIN, A. S. (1955). *Dokl. Akad. Nauk. SSSR* **103**, 207.

ZHIRNOV, V. A. (1958). *Zh. eksp. teor. Fiz.* **35**, 1175 (*Sov. Phys.–JETP* **8**, 822 (1959)).

ZIMAN, J. M. (1960). *Electrons and phonons*. Clarendon Press, Oxford.

ZOOK, J. D. (1974). *Appl. Opt.* **13**, 875.

ZOOK, J. D., and CASSELMAN, T. N. (1966). *Phys. Rev. Lett.* **17**, 960.

ZOOK, J. D., and LIU, S. T. (1976). *Ferroelectrics* **11**, 371.

ZUBAREV, D. N. (1960). *Usp. Fiz. Nauk. SSSR*. **71**, 71 (*Sov. Phys.–Uspekhi* **3**, 320 (1960)).

ZULEEG, R., and WIEDER, H. H. (1966). *Solid St. Electron.* **9**, 657.

ZVEREV, G. M., LEVCHUK, E. A., PASHKOV, V. A., and PORYADIN, Y. D. (1972). *Zh. eskp. teor. Fiz.* **62**, 307 (*Sov. Phys.–JETP* **35**, 165).

AUTHOR INDEX

SUBJECT INDEX

lithium niobate—*contd.*
  electro-optic properties of, 460 ff., 473 ff., 599
  excited state polarization in, 450 ff., 573
  low temperature polarization of, 271 ff.
  n.q.r. in, 419 f.
  soft modes in, 221, 236 f.
  stoichiometry, 275 ff.
  structure and basic properties of, 185, 266 ff.
  use in electro-optic modulators, 595, 600 f., 605
  use in memories, 587, 589
lithium sulphate monohydrate, 150
lithium tantalate
  dielectric constant near T_c, 134 ff.
  electro-optic properties of, 460, 481, 496, 599
  excited state polarization in, 450 ff., 573
  low temperature polarization of, 271 ff.
  n.q.r. in, 420
  piezoelectric coefficients and spontaneous strain, 156 ff.
  pyroelectric properties, 131 ff., 148
  specific heat, 151 f.
  structure and basic properties of, 266 ff.
  use as electro-optic modulator material, 592, 595, 601 f., 605
  X-ray structure refinement near T_c, 187 f.
lithium terbium tetrafluoride, 388
lithium thallium tartrate, 173
lithium trihydrogen selenate, 161
local mode approximation, 17
Lorentz field, 24, 243, 290
Lyddane–Sachs–Teller relation,
  antimony sulphoiodide, 516
  disordered materials, 291
  frequency dependent damping, 216 ff.
  high field studies, 172
  high pressure studies, 162
  lead telluride, 523
  perovskites, 231, 253, 256
  sum rules, 212

macroscopic theory and phenomenology, 59 ff.
magnetic impurities in ferroelectrics, 422 ff., 443 ff.
magnetic resonance,
  electron paramagnetic resonance (e.p.r.), 320 f., 326, 391, 422 ff.
  nuclear magnetic resonance (n.m.r.), 271, 277 f., 409 ff., 417 f.
  nuclear quadrupole resonance (n.q.r.), 326 f, 331, 409 ff., 418
  proton resonance, 420 f.
magnetoplasma effects, 523
maker fringes, 128, 471
Maxwell relations, 64
mean field theory of soft modes and phase transitions, 24 ff., 45 ff., 52 ff.
mechanical work, 61 ff.
mechanical resonance and relaxation, 129 f.
memories, 7, 460, 528, 530, 543, 559 f.,
  electrically read, 578 f.
  optically read, 579 f.
  high capacity, 585 f.
  see also light valves
metastable polarization, 545 ff.
methylammonium chrome alum, 445

Miller indices, 176, 185
Miller's delta, 478 ff., 486, 490, 496
mobility, 534
model ferroelectric Hamiltonian, 15 ff.
mode strength, 210 f., 221, 246, 270, 363
modulators
  see electro-optic modulators
Mössbauer fraction, 429 ff.
Mössbauer spectroscopy, 265, 320, 428 ff.

narrow band gap ferroelectrics, 503 ff.
neutron irradiation, 116
neutron scattering,
  elastic, 183 ff., 245, 265 ff., 293, 321 f., 329, 331, 369, 392, 414
  incoherent, 184, 208, 305, 320
  inelastic, 199 ff., 245, 247, 254, 269, 305, 311, 330, 356, 398 ff., 521 f., 539
  profile refinement technique, 189, 246, 316
  quasi-elastic or diffuse, 203, 206 f., 307, 311, 320, 329, 392, 399
nickel iodine boracite, 353, 369, 445
niobium tin alloy, 400 f.
n.m.r.
  see magnetic resonance
noise, 134, 562, 565 f., 575 f.
non-linear acoustics, 347
non-linear optics, 128, 467 ff., 560, 587 ff.
Nyquist's theorem, 134

optical absorption edge, 286, 441 f.
optical ceramics, 540 ff., 579 ff.
optical damage
  see laser damage, photorefractive effect
optical indicatrix, 479, 579
optical mixing, 469
optical rotation, 90, 579
order–disorder ferroelectrics, 10, 45 ff., 52 ff., 293 ff., 403 ff.

packing density, 499 f.
page composers
  see light valves
paraelectric phase, definition, 10
parametric oscillators, 266, 448
Patterson function, 177
perovskite structure, 241
  see also individual members
phase boundary, 65
phase matching, 91, 280, 448, 470 ff., 601
phase transitions,
  first order, 76 ff.
  second order, 24 ff., 45 ff., 52 ff., 72 ff., 371 ff.
phenomenological theories,
  of antiferroelectrics, 81 ff.
  of ferroelectrics, first order, 76 ff.
  of ferroelectrics, second order, 71 ff.
Phillips–Van Vechten dielectric theory, 489, 493 ff.
photochromic effect, 446 f.
photoconductivity, 456, 462
  devices using, 583 f.
photodeformation effect, 518
photodomain effect, 518
photoelastic effect, 469